Probability Models in Engineering and Science

MECHANICAL ENGINEERING
A Series of Textbooks and Reference Books

Founding Editor

L. L. Faulkner

Columbus Division, Battelle Memorial Institute
and Department of Mechanical Engineering
The Ohio State University
Columbus, Ohio

1. *Spring Designer's Handbook*, Harold Carlson
2. *Computer-Aided Graphics and Design*, Daniel L. Ryan
3. *Lubrication Fundamentals*, J. George Wills
4. *Solar Engineering for Domestic Buildings*, William A. Himmelman
5. *Applied Engineering Mechanics: Statics and Dynamics*, G. Boothroyd and C. Poli
6. *Centrifugal Pump Clinic*, Igor J. Karassik
7. *Computer-Aided Kinetics for Machine Design*, Daniel L. Ryan
8. *Plastics Products Design Handbook, Part A: Materials and Components; Part B: Processes and Design for Processes*, edited by Edward Miller
9. *Turbomachinery: Basic Theory and Applications*, Earl Logan, Jr.
10. *Vibrations of Shells and Plates*, Werner Soedel
11. *Flat and Corrugated Diaphragm Design Handbook*, Mario Di Giovanni
12. *Practical Stress Analysis in Engineering Design*, Alexander Blake
13. *An Introduction to the Design and Behavior of Bolted Joints*, John H. Bickford
14. *Optimal Engineering Design: Principles and Applications*, James N. Siddall
15. *Spring Manufacturing Handbook*, Harold Carlson
16. *Industrial Noise Control: Fundamentals and Applications*, edited by Lewis H. Bell
17. *Gears and Their Vibration: A Basic Approach to Understanding Gear Noise*, J. Derek Smith
18. *Chains for Power Transmission and Material Handling: Design and Applications Handbook*, American Chain Association
19. *Corrosion and Corrosion Protection Handbook*, edited by Philip A. Schweitzer
20. *Gear Drive Systems: Design and Application*, Peter Lynwander
21. *Controlling In-Plant Airborne Contaminants: Systems Design and Calculations*, John D. Constance
22. *CAD/CAM Systems Planning and Implementation*, Charles S. Knox
23. *Probabilistic Engineering Design: Principles and Applications*, James N. Siddall
24. *Traction Drives: Selection and Application*, Frederick W. Heilich III and Eugene E. Shube

25. *Finite Element Methods: An Introduction*, Ronald L. Huston and Chris E. Passerello
26. *Mechanical Fastening of Plastics: An Engineering Handbook*, Brayton Lincoln, Kenneth J. Gomes, and James F. Braden
27. *Lubrication in Practice: Second Edition*, edited by W. S. Robertson
28. *Principles of Automated Drafting*, Daniel L. Ryan
29. *Practical Seal Design*, edited by Leonard J. Martini
30. *Engineering Documentation for CAD/CAM Applications*, Charles S. Knox
31. *Design Dimensioning with Computer Graphics Applications*, Jerome C. Lange
32. *Mechanism Analysis: Simplified Graphical and Analytical Techniques*, Lyndon O. Barton
33. *CAD/CAM Systems: Justification, Implementation, Productivity Measurement*, Edward J. Preston, George W. Crawford, and Mark E. Coticchia
34. *Steam Plant Calculations Manual*, V. Ganapathy
35. *Design Assurance for Engineers and Managers*, John A. Burgess
36. *Heat Transfer Fluids and Systems for Process and Energy Applications*, Jasbir Singh
37. *Potential Flows: Computer Graphic Solutions*, Robert H. Kirchhoff
38. *Computer-Aided Graphics and Design: Second Edition*, Daniel L. Ryan
39. *Electronically Controlled Proportional Valves: Selection and Application*, Michael J. Tonyan, edited by Tobi Goldoftas
40. *Pressure Gauge Handbook*, AMETEK, U.S. Gauge Division, edited by Philip W. Harland
41. *Fabric Filtration for Combustion Sources: Fundamentals and Basic Technology*, R. P. Donovan
42. *Design of Mechanical Joints*, Alexander Blake
43. *CAD/CAM Dictionary*, Edward J. Preston, George W. Crawford, and Mark E. Coticchia
44. *Machinery Adhesives for Locking, Retaining, and Sealing*, Girard S. Haviland
45. *Couplings and Joints: Design, Selection, and Application*, Jon R. Mancuso
46. *Shaft Alignment Handbook*, John Piotrowski
47. *BASIC Programs for Steam Plant Engineers: Boilers, Combustion, Fluid Flow, and Heat Transfer*, V. Ganapathy
48. *Solving Mechanical Design Problems with Computer Graphics*, Jerome C. Lange
49. *Plastics Gearing: Selection and Application*, Clifford E. Adams
50. *Clutches and Brakes: Design and Selection*, William C. Orthwein
51. *Transducers in Mechanical and Electronic Design*, Harry L. Trietley
52. *Metallurgical Applications of Shock-Wave and High-Strain-Rate Phenomena*, edited by Lawrence E. Murr, Karl P. Staudhammer, and Marc A. Meyers
53. *Magnesium Products Design*, Robert S. Busk
54. *How to Integrate CAD/CAM Systems: Management and Technology*, William D. Engelke
55. *Cam Design and Manufacture: Second Edition; with cam design software for the IBM PC and compatibles*, disk included, Preben W. Jensen
56. *Solid-State AC Motor Controls: Selection and Application*, Sylvester Campbell

57. *Fundamentals of Robotics*, David D. Ardayfio
58. *Belt Selection and Application for Engineers*, edited by Wallace D. Erickson
59. *Developing Three-Dimensional CAD Software with the IBM PC*, C. Stan Wei
60. *Organizing Data for CIM Applications*, Charles S. Knox, with contributions by Thomas C. Boos, Ross S. Culverhouse, and Paul F. Muchnicki
61. *Computer-Aided Simulation in Railway Dynamics*, by Rao V. Dukkipati and Joseph R. Amyot
62. *Fiber-Reinforced Composites: Materials, Manufacturing, and Design*, P. K. Mallick
63. *Photoelectric Sensors and Controls: Selection and Application*, Scott M. Juds
64. *Finite Element Analysis with Personal Computers*, Edward R. Champion, Jr. and J. Michael Ensminger
65. *Ultrasonics: Fundamentals, Technology, Applications: Second Edition, Revised and Expanded*, Dale Ensminger
66. *Applied Finite Element Modeling: Practical Problem Solving for Engineers*, Jeffrey M. Steele
67. *Measurement and Instrumentation in Engineering: Principles and Basic Laboratory Experiments*, Francis S. Tse and Ivan E. Morse
68. *Centrifugal Pump Clinic: Second Edition, Revised and Expanded*, Igor J. Karassik
69. *Practical Stress Analysis in Engineering Design: Second Edition, Revised and Expanded*, Alexander Blake
70. *An Introduction to the Design and Behavior of Bolted Joints: Second Edition, Revised and Expanded*, John H. Bickford
71. *High Vacuum Technology: A Practical Guide*, Marsbed H. Hablanian
72. *Pressure Sensors: Selection and Application*, Duane Tandeske
73. *Zinc Handbook: Properties, Processing, and Use in Design*, Frank Porter
74. *Thermal Fatigue of Metals*, Andrzej Weronski and Tadeusz Hejwowski
75. *Classical and Modern Mechanisms for Engineers and Inventors*, Preben W. Jensen
76. *Handbook of Electronic Package Design*, edited by Michael Pecht
77. *Shock-Wave and High-Strain-Rate Phenomena in Materials*, edited by Marc A. Meyers, Lawrence E. Murr, and Karl P. Staudhammer
78. *Industrial Refrigeration: Principles, Design and Applications*, P. C. Koelet
79. *Applied Combustion*, Eugene L. Keating
80. *Engine Oils and Automotive Lubrication*, edited by Wilfried J. Bartz
81. *Mechanism Analysis: Simplified and Graphical Techniques, Second Edition, Revised and Expanded*, Lyndon O. Barton
82. *Fundamental Fluid Mechanics for the Practicing Engineer*, James W. Murdock
83. *Fiber-Reinforced Composites: Materials, Manufacturing, and Design, Second Edition, Revised and Expanded*, P. K. Mallick
84. *Numerical Methods for Engineering Applications*, Edward R. Champion, Jr.
85. *Turbomachinery: Basic Theory and Applications, Second Edition, Revised and Expanded*, Earl Logan, Jr.
86. *Vibrations of Shells and Plates: Second Edition, Revised and Expanded*, Werner Soedel
87. *Steam Plant Calculations Manual: Second Edition, Revised and Expanded*, V. Ganapathy

88. *Industrial Noise Control: Fundamentals and Applications, Second Edition, Revised and Expanded*, Lewis H. Bell and Douglas H. Bell
89. *Finite Elements: Their Design and Performance*, Richard H. MacNeal
90. *Mechanical Properties of Polymers and Composites: Second Edition, Revised and Expanded*, Lawrence E. Nielsen and Robert F. Landel
91. *Mechanical Wear Prediction and Prevention*, Raymond G. Bayer
92. *Mechanical Power Transmission Components*, edited by David W. South and Jon R. Mancuso
93. *Handbook of Turbomachinery*, edited by Earl Logan, Jr.
94. *Engineering Documentation Control Practices and Procedures*, Ray E. Monahan
95. *Refractory Linings Thermomechanical Design and Applications*, Charles A. Schacht
96. *Geometric Dimensioning and Tolerancing: Applications and Techniques for Use in Design, Manufacturing, and Inspection*, James D. Meadows
97. *An Introduction to the Design and Behavior of Bolted Joints: Third Edition, Revised and Expanded*, John H. Bickford
98. *Shaft Alignment Handbook: Second Edition, Revised and Expanded*, John Piotrowski
99. *Computer-Aided Design of Polymer-Matrix Composite Structures*, edited by Suong Van Hoa
100. *Friction Science and Technology*, Peter J. Blau
101. *Introduction to Plastics and Composites: Mechanical Properties and Engineering Applications*, Edward Miller
102. *Practical Fracture Mechanics in Design*, Alexander Blake
103. *Pump Characteristics and Applications*, Michael W. Volk
104. *Optical Principles and Technology for Engineers*, James E. Stewart
105. *Optimizing the Shape of Mechanical Elements and Structures*, A. A. Seireg and Jorge Rodriguez
106. *Kinematics and Dynamics of Machinery*, Vladimír Stejskal and Michael Valásek
107. *Shaft Seals for Dynamic Applications*, Les Horve
108. *Reliability-Based Mechanical Design*, edited by Thomas A. Cruse
109. *Mechanical Fastening, Joining, and Assembly*, James A. Speck
110. *Turbomachinery Fluid Dynamics and Heat Transfer*, edited by Chunill Hah
111. *High-Vacuum Technology: A Practical Guide, Second Edition, Revised and Expanded*, Marsbed H. Hablanian
112. *Geometric Dimensioning and Tolerancing: Workbook and Answerbook*, James D. Meadows
113. *Handbook of Materials Selection for Engineering Applications*, edited by G. T. Murray
114. *Handbook of Thermoplastic Piping System Design*, Thomas Sixsmith and Reinhard Hanselka
115. *Practical Guide to Finite Elements: A Solid Mechanics* Approach, Steven M. Lepi
116. *Applied Computational Fluid Dynamics*, edited by Vijay K. Garg
117. *Fluid Sealing Technology*, Heinz K. Muller and Bernard S. Nau
118. *Friction and Lubrication in Mechanical Design*, A. A. Seireg
119. *Influence Functions and Matrices*, Yuri A. Melnikov

120. *Mechanical Analysis of Electronic Packaging Systems*, Stephen A. McKeown
121. *Couplings and Joints: Design, Selection, and Application, Second Edition, Revised and Expanded*, Jon R. Mancuso
122. *Thermodynamics: Processes and Applications*, Earl Logan, Jr.
123. *Gear Noise and Vibration*, J. Derek Smith
124. *Practical Fluid Mechanics for Engineering Applications*, John J. Bloomer
125. *Handbook of Hydraulic Fluid Technology*, edited by George E. Totten
126. *Heat Exchanger Design Handbook*, T. Kuppan
127. *Designing for Product Sound Quality*, Richard H. Lyon
128. *Probability Applications in Mechanical Design*, Franklin E. Fisher and Joy R. Fisher
129. *Nickel Alloys*, edited by Ulrich Heubner
130. *Rotating Machinery Vibration: Problem Analysis and Troubleshooting*, Maurice L. Adams, Jr.
131. *Formulas for Dynamic Analysis*, Ronald L. Huston and C. Q. Liu
132. *Handbook of Machinery Dynamics*, Lynn L. Faulkner and Earl Logan, Jr.
133. *Rapid Prototyping Technology: Selection and Application*, Kenneth G. Cooper
134. *Reciprocating Machinery Dynamics: Design and Analysis*, Abdulla S. Rangwala
135. *Maintenance Excellence: Optimizing Equipment Life-Cycle Decisions*, edited by John D. Campbell and Andrew K. S. Jardine
136. *Practical Guide to Industrial Boiler Systems*, Ralph L. Vandagriff
137. *Lubrication Fundamentals: Second Edition, Revised and Expanded*, D. M. Pirro and A. A. Wessol
138. *Mechanical Life Cycle Handbook: Good Environmental Design and Manufacturing*, edited by Mahendra S. Hundal
139. *Micromachining of Engineering Materials*, edited by Joseph McGeough
140. *Control Strategies for Dynamic Systems: Design and Implementation*, John H. Lumkes, Jr.
141. *Practical Guide to Pressure Vessel Manufacturing*, Sunil Pullarcot
142. *Nondestructive Evaluation: Theory, Techniques, and Applications*, edited by Peter J. Shull
143. *Diesel Engine Engineering: Thermodynamics, Dynamics, Design, and Control*, Andrei Makartchouk
144. *Handbook of Machine Tool Analysis*, Ioan D. Marinescu, Constantin Ispas, and Dan Boboc
145. *Implementing Concurrent Engineering in Small Companies*, Susan Carlson Skalak
146. *Practical Guide to the Packaging of Electronics: Thermal and Mechanical Design and Analysis*, Ali Jamnia
147. *Bearing Design in Machinery: Engineering Tribology and Lubrication*, Avraham Harnoy
148. *Mechanical Reliability Improvement: Probability and Statistics for Experimental Testing*, R. E. Little
149. *Industrial Boilers and Heat Recovery Steam Generators: Design, Applications, and Calculations*, V. Ganapathy
150. *The CAD Guidebook: A Basic Manual for Understanding and Improving Computer-Aided Design*, Stephen J. Schoonmaker

151. *Industrial Noise Control and Acoustics*, Randall F. Barron
152. *Mechanical Properties of Engineered Materials*, Wolé Soboyejo
153. *Reliability Verification, Testing, and Analysis in Engineering Design*, Gary S. Wasserman
154. *Fundamental Mechanics of Fluids: Third Edition*, I. G. Currie
155. *Intermediate Heat Transfer*, Kau-Fui Vincent Wong
156. *HVAC Water Chillers and Cooling Towers: Fundamentals, Application, and Operation*, Herbert W. Stanford III
157. *Gear Noise and Vibration: Second Edition, Revised and Expanded*, J. Derek Smith
158. *Handbook of Turbomachinery: Second Edition*, Revised and Expanded, edited by Earl Logan, Jr. and Ramendra Roy
159. *Piping and Pipeline Engineering: Design, Construction, Maintenance, Integrity, and Repair*, George A. Antaki
160. *Turbomachinery: Design and Theory*, Rama S. R. Gorla and Aijaz Ahmed Khan
161. *Target Costing: Market-Driven Product Design*, M. Bradford Clifton, Henry M. B. Bird, Robert E. Albano, and Wesley P. Townsend
162. *Fluidized Bed Combustion*, Simeon N. Oka
163. *Theory of Dimensioning: An Introduction to Parameterizing Geometric Models*, Vijay Srinivasan
164. *Handbook of Mechanical Alloy Design*, edited by George E. Totten, Lin Xie, and Kiyoshi Funatani
165. *Structural Analysis of Polymeric Composite Materials*, Mark E. Tuttle
166. *Modeling and Simulation for Material Selection and Mechanical Design*, edited by George E. Totten, Lin Xie, and Kiyoshi Funatani
167. *Handbook of Pneumatic Conveying Engineering*, David Mills, Mark G. Jones, and Vijay K. Agarwal
168. *Clutches and Brakes: Design and Selection, Second Edition*, William C. Orthwein
169. *Fundamentals of Fluid Film Lubrication: Second Edition*, Bernard J. Hamrock, Steven R. Schmid, and Bo O. Jacobson
170. *Handbook of Lead-Free Solder Technology for Microelectronic Assemblies*, edited by Karl J. Puttlitz and Kathleen A. Stalter
171. *Vehicle Stability*, Dean Karnopp
172. *Mechanical Wear Fundamentals and Testing: Second Edition, Revised and Expanded*, Raymond G. Bayer
173. *Liquid Pipeline Hydraulics*, E. Shashi Menon
174. *Solid Fuels Combustion and Gasification*, Marcio L. de Souza-Santos
175. *Mechanical Tolerance Stackup and Analysis*, Bryan R. Fischer
176. *Engineering Design for Wear*, Raymond G. Bayer
177. *Vibrations of Shells and Plates: Third Edition, Revised and Expanded*, Werner Soedel
178. *Refractories Handbook*, edited by Charles A. Schacht
179. *Practical Engineering Failure Analysis*, Hani M. Tawancy, Anwar Ul-Hamid, and Nureddin M. Abbas
180. *Mechanical Alloying and Milling*, C. Suryanarayana
181. *Mechanical Vibration: Analysis, Uncertainties, and Control, Second Edition, Revised and Expanded*, Haym Benaroya

182. *Design of Automatic Machinery*, Stephen J. Derby
183. *Practical Fracture Mechanics in Design: Second Edition, Revised and Expanded*, Arun Shukla
184. *Practical Guide to Designed Experiments*, Paul D. Funkenbusch
185. *Gigacycle Fatigue in Mechanical Practive*, Claude Bathias and Paul C. Paris
186. *Selection of Engineering Materials and Adhesives*, Lawrence W. Fisher
187. *Boundary Methods: Elements, Contours, and Nodes*, Subrata Mukherjee and Yu Xie Mukherjee
188. *Rotordynamics*, Agnieszka (Agnes) Muszńyska
189. *Pump Characteristics and Applications: Second Edition*, Michael W. Volk
190. *Reliability Engineering: Probability Models and Maintenance Methods*, Joel A. Nachlas
191. *Industrial Heating: Principles, Techniques, Materials, Applications, and Design*, Yeshvant V. Deshmukh
192. *Micro Electro Mechanical System Design*, James J. Allen
193. *Probability Models in Engineering and Science*, Haym Benaroya and Seon Mi Han
194. *Damage Mechanics*, George Z. Voyiadjis and Peter I. Kattan

Probability Models in Engineering and Science

Haym Benaroya
and Seon Mi Han

Taylor & Francis
Taylor & Francis Group

Boca Raton   London   New York   Singapore

A CRC title, part of the Taylor & Francis imprint, a member of the
Taylor & Francis Group, the academic division of T&F Informa plc.

Cover: The autocorrelation function when

$$S_0(w) = \frac{1}{w_2 - w_1} \text{ for } w_2 < |w| < w_2$$

and zero elsewhere. When $w_2 \simeq w_1$, the process $X(t)$ is a narrow band process, and when w_2 is far from w_1, the process $X(t)$ is a broad band process. See also Figures 5.28 and 5.29.

Published in 2005 by
CRC Press
Taylor & Francis Group
6000 Broken Sound Parkway NW, Suite 300
Boca Raton, FL 33487-2742

© 2005 by Taylor & Francis Group, LLC
CRC Press is an imprint of Taylor & Francis Group

No claim to original U.S. Government works
Printed in the United States of America on acid-free paper
10 9 8 7 6 5 4 3 2 1

International Standard Book Number-10: 0-8247-2315-5 (Hardcover)
International Standard Book Number-13: 978-0-8247-2315-6 (Hardcover)
Library of Congress Card Number 2005043711

This book contains information obtained from authentic and highly regarded sources. Reprinted material is quoted with permission, and sources are indicated. A wide variety of references are listed. Reasonable efforts have been made to publish reliable data and information, but the author and the publisher cannot assume responsibility for the validity of all materials or for the consequences of their use.

No part of this book may be reprinted, reproduced, transmitted, or utilized in any form by any electronic, mechanical, or other means, now known or hereafter invented, including photocopying, microfilming, and recording, or in any information storage or retrieval system, without written permission from the publishers.

For permission to photocopy or use material electronically from this work, please access www.copyright.com (http://www.copyright.com/) or contact the Copyright Clearance Center, Inc. (CCC) 222 Rosewood Drive, Danvers, MA 01923, 978-750-8400. CCC is a not-for-profit organization that provides licenses and registration for a variety of users. For organizations that have been granted a photocopy license by the CCC, a separate system of payment has been arranged.

Trademark Notice: Product or corporate names may be trademarks or registered trademarks, and are used only for identification and explanation without intent to infringe.

Library of Congress Cataloging-in-Publication Data

Benaroya, Haym, 1954-
  Probability models in engineering and science / Haym Benaroya, Seon Han.
    p. cm. -- (Mechanical engineering ; v. 192)
  Includes bibliographical references and index.
  ISBN 0-8247-2315-5
  1. Reliability (Engineering)--Mathematical models. I. Han, Seon Mi. II. Title. III. Mechanical engineering series (Boca Raton, Fla.) ; v. 192.

TA169.B464 2005
620'.000452'015118--dc22   2005043711

Taylor & Francis Group
is the Academic Division of T&F Informa plc.

Visit the Taylor & Francis Web site at
http://www.taylorandfrancis.com

and the CRC Press Web site at
http://www.crcpress.com

The first author dedicates this book to his parents Esther and Alfred Benaroya, to whom he is grateful for much. The second author dedicates this book to her son James, who brought so much joy to her life.

Preface

There are not many texts that develop applied probability for engineering and scientific applications. Our intention in writing this text is to provide engineers and scientists a self-contained introduction to probabilistic modeling.

This book can be studied in two semesters, one undergraduate and one graduate. Distinguishing features worth noting include: numerous example problems from simple to challenging, short biographies and portraits of some of the key "names" mentioned, and the completely self-contained nature of the presentation, thus making this book very suitable for self-study. Also included are end-of-chapter problems, plus footnotes to the literature. These footnotes have two purposes. The first is to provide proper citation and expanded discussion. There is no separate list of references in this text, and the footnotes serve as such attribution. The second is to introduce the reader to the relevant journal literature and to some of the very useful texts.

The biographical summaries provided cannot do justice to the richness of the history of the field or its applications. Rather, the intent is to add, for the reader, the essential human connection to this subject. The history of the subject provides valuable insights to the subject, even though we rarely spend much time in its consideration when we are introduced to the field. These biographies, and portraits, downloaded from the World Wide Web, are included here by courtesy and permission of Professors E.F. Robertson and J.J. O'Connor, School of Mathematical and Computer Sciences, University of St. Andrews, St. Andrews, Scotland. The Web site is at HTTP://WWW-GROUPS.DCS.ST-ANDREWS.AC.UK/HISTORY/INDEX.HTML.

Finally, that this text is self-contained means that the student may start at the beginning and continue to the end with rare need to refer to other works, except to find additional perspectives on the subject. But then, no one text can cover all aspects of a subject as broad as probability. Where more details become necessary, other works are cited where the reader will find additional information.

The instructor may choose a variety of options regarding the use of this text. Generally, an undergraduate introduction will include Chapters 1-4 on probability, plus parts of Chapter 5 on random processes, Chapter 6 on single-degree-of-freedom random vibration, and Chapter 9 on reliability. A graduate, or second, course will include much of the second part of the text, Chapters 5-13.

We have decided to create a Web page for this text where corrections and additional notes will be placed. On occasion interesting items will also be added. That site is found listed in HTTP://CSXE.RUTGERS.EDU/LINKS.HTML.

Readers' comments are most welcome, as are any suggestions and corrections. All of these would be most appreciated. The authors may be reached at BENAROYA@RCI.RUTGERS.EDU, and SEONHAN@COE.TTU.EDU. All messages will be acknowledged.

Authors

Dr. **Haym Benaroya** received the B.E. from The Cooper Union for the Advancement of Science and Art, in 1976, and his M.S. and Ph.D. from the University of Pennsylvania, in 1977 and 1981, respectively. He worked for Weidlinger Associates, Consulting Engineers, New York, between 1981 and 1989, at which time he joined Rutgers University. He is currently a Professor of Mechanical and Aerospace Engineering at Rutgers, the State University of New Jersey. His research interests include structures and vibration, offshore structural dynamics, fluid-structure interaction, aircraft structures, and the development of concepts for lunar structures. Related interests include science, space and defense policy, and educational methods and policy. He is the director of the newly-formed Center for Structures in Extreme Environments, about which more information can be obtained through HTTP://CSXE.RUTGERS.EDU/INDEX.HTML.

Professor Benaroya is the author of numerous publications and other books, and is a Fellow of the American Society of Mechanical Engineers, a Fellow of the British Interplanetary Society, an Associate Fellow of the American Institute of Aeronautics and Astronautics, and a Corresponding Member of the International Academy of Astronautics.

Dr. **Seon Mi Han** received the B.E. from The Cooper Union for the Advancement of Science and Art in 1996, and her M.S. and Ph.D. from Rutgers, the State University of New Jersey, in 1998 and 2001, respectively. She was a National Science Foundation summer intern at Korea Institute of Technology and Tokyo Institute of Technology. She received the Woods Hole Oceanographic Institution Postdoctoral Scholarship between 2001 and 2003, and she is currently an Assistant Professor of Mechanical Engineering at Texas Tech University. Her research interests include vibration and dynamics of offshore and marine structures.

Acknowledgments

No project of this magnitude is completed without the explicit and implicit assistance of others.

We acknowledge the input and efforts made by Professors M. Nagurka, M. Noori, and A. Zerva, who spent their valuable time reviewing an early manuscript. This is very much appreciated. We thank Professors E.F. Robertson and J.J. O'Connor who graciously allowed us to use the fruits of their labor, the biographical sketches included here. We are also grateful to John Corrigan, who originally signed this book for Marcel Dekker.

We thank our students who have read sections of the text and provided useful feedback: Rene Gabbai, Mangala Gadagi, Subramanian Ramakrishnan, and Seyul Son.

The first author wishes to acknowledge the Department of Mechanical and Aerospace Engineering, at the School of Engineering at Rutgers University, that provides an excellent scholarly environment, and especially Professor A. Zebib for many years of encouragement and support. The second author would like to acknowledge the Mechanical Engineering Department at Texas Tech University, and the support and encouragement of its former chairman, Dr. Thomas Burton.

The first author is also pleased to thank Dr. T. Swean, of the Office of Naval Research, for the continued support of his research activity. Such research permits the refinement and development of concepts necessary for the advancement of engineering, in particular, and society, in general. It also is needed for the creation of texts such as this.

Contents

1 Introduction   **1**
  1.1 Applications 1
      1.1.1 Random Vibration 2
      1.1.2 Fatigue Life 3
      1.1.3 Ocean-Wave Forces 5
      1.1.4 Wind Forces 8
      1.1.5 Material Properties 8
      1.1.6 Statistics and Probability 11
  1.2 Units 12
  1.3 Organization of the Text 13
  1.4 Problems 13

2 Events and Probability   **15**
  2.1 Sets 15
      2.1.1 Basic Events 16
      2.1.2 Operational Rules 21
  2.2 Probability 26
      2.2.1 Axioms of Probability 29
      2.2.2 Extensions from the Axioms 30
      2.2.3 Conditional Probability 31
      2.2.4 Statistical Independence 33
      2.2.5 Total Probability 35
      2.2.6 Bayes' Theorem 37
  2.3 Concluding Summary 45
  2.4 Problems 45

3 Random Variable Models   **49**
  3.1 Probability Distribution Function 50
  3.2 Probability Density Function 52
  3.3 Mathematical Expectation 57
      3.3.1 Variance 60

3.4 Useful Probability Densities 66
    3.4.1 The Uniform Density 66
    3.4.2 The Exponential Density 69
    3.4.3 The Normal (Gaussian) Density 71
    3.4.4 The Lognormal Density 84
    3.4.5 The Rayleigh Density 87
    3.4.6 Probability Density Functions of a Discrete Random Variable 90
    3.4.7 Moment-Generating Functions 99
3.5 Two Random Variables 101
    3.5.1 Covariance and Correlation 110
3.6 Concluding Summary 117
3.7 Problems 118

4 Functions of Random Variables   125
4.1 Exact Functions of One Variable 125
4.2 Functions of Two or More RVs 132
    4.2.1 General Case 149
4.3 Approximate Analysis 161
    4.3.1 Direct Methods 161
    4.3.2 Mean and Variance of a General Function of X to Order σ_X^2 165
    4.3.3 Mean and Variance of a General Function of n RVs . . 168
4.4 Monte Carlo Methods 179
    4.4.1 Independent Uniform Random Numbers 179
    4.4.2 Independent Normal Random Numbers 183
    4.4.3 A Discretization Procedure 185
4.5 Concluding Summary 188
4.6 Problems 188
4.7 The Standard Normal Table 192

5 Random Processes   195
5.1 Basic Random Process Descriptors 195
5.2 Ensemble Averaging 196
5.3 Stationarity 201
5.4 Derivatives of Stationary Processes 208
5.5 Fourier Series and Fourier Transforms 210
5.6 Harmonic Processes 229
5.7 Power Spectra 230
    5.7.1 Narrow- and Broad-Band Processes 254
    5.7.2 White Noise Processes 257

       5.7.3 Spectral Densities of Derivatives of Stationary Random Processes 259
  5.8 Fourier Representation of a Random Process 261
       5.8.1 Borgman's Method of Frequency Discretization 267
  5.9 Concluding Summary 271
  5.10 Problems 271

6 Single-Degree-of-Freedom Dynamics 275
  6.1 Motivating Examples 277
       6.1.1 Transport of a Satellite 277
       6.1.2 Rocket Ship 277
  6.2 Deterministic SDoF Vibration 278
       6.2.1 Free Vibration With No Damping 288
       6.2.2 Harmonic Forced Vibration With No Damping 289
       6.2.3 Free Vibration With Damping 291
       6.2.4 Forced Vibration With Damping 292
       6.2.5 Impulse Excitation 296
       6.2.6 Arbitrary Loading: Convolution 299
       6.2.7 Frequency Response Function 303
  6.3 SDoF: The Response to Random Loads 307
       6.3.1 Formulation 307
       6.3.2 Derivation of Equations 308
       6.3.3 Response Correlations 309
       6.3.4 Response Spectral Density 312
  6.4 Response to Two Random Loads 325
  6.5 Concluding Summary 333
  6.6 Problems 333

7 Multidegree-of-Freedom Vibration 339
  7.1 Deterministic Vibration 339
       7.1.1 Solution by Frequency Response Function 341
       7.1.2 Modal Analysis 343
       7.1.3 Advantages of Modal Analysis 349
  7.2 Response to Random Loads 351
       7.2.1 Response due to a Single Random Force 353
       7.2.2 Response to Multiple Random Forces 356
  7.3 Periodic Structures 373
       7.3.1 Perfect Lattice Models 375
       7.3.2 Effects of Imperfection 378
  7.4 Inverse Vibration 378
       7.4.1 Deterministic Inverse Vibration Problem 381
       7.4.2 Effect of Uncertain Data 384

7.5	Random Eigenvalues	389
	7.5.1 A Two-Degree-of-Freedom Model	393
7.6	Concluding Summary	395
7.7	Problems	395

8 Continuous System Vibration — 403

8.1	Deterministic Continuous Systems	404
	8.1.1 Strings	404
	8.1.2 Axial Vibration of Beams	406
	8.1.3 Transversely Vibrating Beams	408
8.2	Sturm-Liouville Eigenvalue Problem	411
	8.2.1 Orthogonality	413
	8.2.2 Natural Frequencies and Mode Shapes	414
8.3	Deterministic Vibration	420
	8.3.1 Free Response	420
	8.3.2 Forced Response via Eigenfunction Expansion	422
8.4	Random Vibration of Continuous Systems	428
	8.4.1 Derivation of Response Spectral Density	429
8.5	Beams with Complex Loading	439
	8.5.1 Transverse Vibration of Beam with Axial Force	439
	8.5.2 Transverse Vibration of Beam on Elastic Foundation	442
	8.5.3 Response of a Beam to a Traveling Force	446
8.6	Concluding Summary	451
8.7	Problems	452

9 Reliability — 455

9.1	Introduction	455
9.2	First Excursion Failure	457
	9.2.1 Exponential Failure Law	462
	9.2.2 Modified Exponential Failure Law	465
	9.2.3 Calculation of Up-Crossing Rate	466
	9.2.4 Narrow-Band Process – Envelope Function	472
	9.2.5 Rice's Envelope Function for Gaussian Narrow-Band Process $X(t)$	474
	9.2.6 Other Failure Laws	488
9.3	Fatigue Life Prediction	493
	9.3.1 Failure Curves	495
	9.3.2 Peak Distribution for Stationary Random Process	497
	9.3.3 Peak Distribution of a Gaussian Process	501
9.4	Concluding Summary	512
9.5	Problems	512

10 Nonlinear Dynamic Models — 515
- 10.1 Examples of Nonlinear Vibration ... 516
- 10.2 Fundamental Nonlinear Equations ... 519
- 10.3 Statistical Equivalent Linearization ... 521
  - 10.3.1 Equivalent Nonlinearization ... 530
- 10.4 Perturbation Methods ... 532
  - 10.4.1 Lindstedt-Poincaré Method ... 534
  - 10.4.2 Forced Oscillations of Quasiharmonic Systems ... 539
  - 10.4.3 Jump Phenomenon ... 543
  - 10.4.4 Periodic Solutions of Nonautonomous Systems ... 544
  - 10.4.5 Random Duffing Oscillator ... 552
- 10.5 The van der Pol Equation ... 555
  - 10.5.1 Limit Cycles ... 556
  - 10.5.2 The Forced van der Pol Equation ... 557
- 10.6 Markov Process-Based Models ... 561
  - 10.6.1 Probability Background ... 561
  - 10.6.2 The Fokker-Planck Equation ... 565
- 10.7 Concluding Summary ... 587
- 10.8 Problems ... 587

11 Nonstationary Models — 591
- 11.1 Some Applications ... 592
- 11.2 Envelope Function Model ... 598
  - 11.2.1 Transient Response ... 599
  - 11.2.2 MS Nonstationary Response ... 605
- 11.3 Nonstationary Generalizations ... 607
  - 11.3.1 Discrete Model ... 608
  - 11.3.2 Complex-Valued Stochastic Processes ... 610
  - 11.3.3 Continuous Model ... 610
- 11.4 Priestley's Model ... 612
  - 11.4.1 The Stieltjes Integral: An Aside ... 612
  - 11.4.2 Priestley's Model ... 614
- 11.5 SDoF Oscillator Response ... 615
  - 11.5.1 Stationary Case ... 615
  - 11.5.2 Nonstationary Case ... 616
  - 11.5.3 Undamped Oscillator ... 618
  - 11.5.4 Underdamped Oscillator ... 619
- 11.6 Multi DoF Oscillator Response ... 622
  - 11.6.1 Input Characterization ... 622
  - 11.6.2 Response Characterization ... 624
- 11.7 Nonstationary and Nonlinear Oscillator ... 625
  - 11.7.1 The Nonstationary and Nonlinear Duffing ... 627

    11.8 Concluding Summary 629
    11.9 Problems 629

12 The Monte Carlo Method   631
    12.1 Introduction 631
    12.2 Random-Number Generation 635
        12.2.1 Standard Uniform Random Numbers 635
        12.2.2 Generation of Nonuniform Random Variates 637
        12.2.3 Composition Method 648
        12.2.4 Von Neumann's Rejection-Acceptance Method ... 651
    12.3 Joint Random Numbers 659
        12.3.1 Inverse Transform Method 660
        12.3.2 Linear Transform Method 661
    12.4 Error Estimates 664
    12.5 Applications 670
        12.5.1 Evaluation of Finite-Dimensional Integrals 670
        12.5.2 Generating a Time History for a Stationary Random Process Defined by a Power Spectral Density 673
    12.6 Concluding Summary 676
    12.7 Problems 676

13 Fluid-Induced Vibration   679
    13.1 Ocean Currents and Waves 679
        13.1.1 Spectral Density 680
        13.1.2 Ocean Wave Spectral Densities 684
        13.1.3 Approximation of Spectral Density from Time Series . 689
        13.1.4 Generation of Time Series from a Spectral Density .. 691
        13.1.5 Short-Term Statistics 692
        13.1.6 Long-Term Statistics 698
        13.1.7 Wave Velocities via Linear Wave Theory 700
    13.2 Fluid Forces in General 702
        13.2.1 Wave Force Regime 703
        13.2.2 Wave Forces on Small Structures – Morison Equation 705
        13.2.3 Vortex-Induced Vibration 711
    13.3 Examples 713
        13.3.1 Static Configuration of a Towing Cable 713
        13.3.2 Fluid Forces on an Articulated Tower 717
        13.3.3 Weibull and Gumbel Wave Height Distributions ... 720
        13.3.4 Reconstructing Time Series for a Given Significant Wave Height 722
    13.4 Available Numerical Codes 723

Index   **727**

Chapter 1

Introduction

Probability and random processes are the quantifications of uncertainties. There is a steep learning curve to be covered before it becomes possible to consider any significant problem that includes a probabilistic component. We must learn a new way of thinking with uncertainty as our *paradigm*.[1] The probabilistic paradigm is not the most comfortable for engineers since we are raised to believe that, given enough experimental data and theoretical development, any problem is solvable exactly, or at least to within measurement tolerances. Reality tells us that certainty exists only in idealized models, not in the actual physical systems that must be understood and designed. While an exact quantity exists only in our imagination, sometimes uncertainties can be ignored for particular applications. In our studies here, we begin to take account of uncertainty and learn some basic concepts in probability.

First, let us consider some motivating examples.

1.1 Applications

To demonstrate the importance of uncertainty modeling in mechanical systems, a number of examples are chosen for brief and qualitative discussion.

---

[1] A paradigm, pronounced "para–dime," is a way of thinking. It may be viewed as the beliefs, values, and techniques shared by a particular group of people. Therefore, a new paradigm in a technical area implies a completely new way of thinking about that area. A recent example of a *paradigm shift* is the development of the field of *chaos* in nonlinear dynamics.

1.1.1 Random Vibration

The discipline of random vibration of structures was borne of the need to understand how structures respond to dynamic loads that are too complex to model deterministically. Examples include aerodynamic loading on aircraft and earthquake loading of structures. Essentially, the question that must be answered is: *Given the statistics (read: uncertainties) of the loading, what are the statistics (read: most likely values with bounds) of the response?* Generally, for engineering applications the statistics of greatest concern are the mean, or average value, and the variance, or scatter. These concepts are discussed in detail subsequently. There are a number of texts on random vibration[2] where the subject is explored fully. A significant portion of this book also deals with this important subject. But it is important to note that the methodologies introduced and developed in this text are applicable to a wide variety of problems in engineering and the physical sciences. Random vibration is a wonderful way to study linear systems, that is linear equations of all sorts.

Suppose that we are aircraft designers currently working on the analysis and design of a wing for a new airplane. As engineers, we are very familiar with the mechanics of solids, and can size the wing for static loads. Also, we have vibration experience and can evaluate the response of the wing to a harmonic or impulsive forcing. But this wing needs to provide lift to an airplane flying through a turbulent atmosphere. The fluid dynamicists in the design group know that turbulence is a very complicated physical process. In fact, the fluid (air) motion is so complicated that probabilistic models are required to model the behavior. *Here, a plausibly deterministic but very complicated dynamic process is taken to be random for purposes of modeling.* Wing design requires force data resulting from the interaction between fluid and structure. Such data can be shown as the time history in Figure 1.1.

The challenge is to make sense of such intricate fluctuations. The analyst and designer must run scale model tests. A wing section is set up in the wind tunnel and representative aerodynamic forces are generated. Data on wind forces and structural response are gathered and analyzed. With additional data analysis, it is possible to estimate the force magnitudes. Estimates of mean values of these forces can be calculated, as well as of the range of

---

[2]A very useful one that includes a broad spectrum of theory and application is by I. Elishakoff, *Probabilistic Methods in the Theory of Structures*, Wiley-Interscience, 1983, now in a Dover edition. Two exceptionally clear early books on random vibration are worth reading. The first is *Random Vibration in Mechanical Systems* by S.H. Crandall and W.D. Mark, Academic Press, 1963. The other is *An Introduction to Random Vibration* by J.D. Robson, Elsevier, 1964. Another introduction to the subject is *An Introduction to Random Vibrations and Spectral Analysis* by D.E. Newland, Longman, 1975, now in its third edition.

1.1. APPLICATIONS

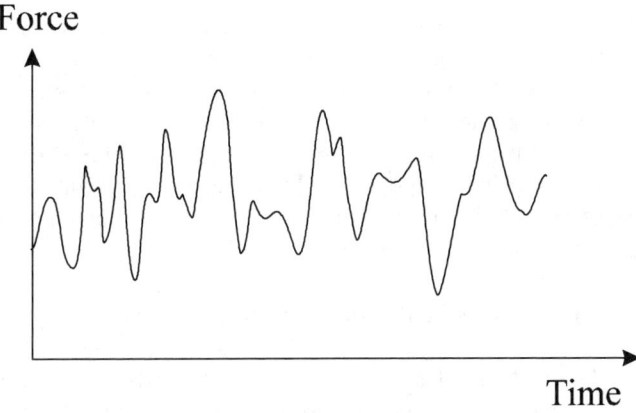

Figure 1.1: Time history of turbulent force.

possible forces. With these estimates, it is possible to study the complex physical problem of the behavior of the wing under a variety of realistic loading scenarios using probabilistic and statistical methods. Probability provides the vehicle for quantifying uncertainties.

This text introduces the use of probabilistic information in mechanical systems analysis and design, primarily structural and dynamic systems. However, we emphasize that these tools are applicable to all the sciences and engineering, even though the focus in this text is the mechanical sciences and engineering.

1.1.2 Fatigue Life

The fatigue life of mechanical components and structures[3] depends on many factors such as material properties, temperature, corrosion environment, and also on vibration history. A first step in estimating fatigue life involves the characterization of the static and dynamic loading cycles the structure has experienced. Are there many cycles, what are the amplitude ranges, and is the loading harmonic or of broad frequency band? The estimates of fatigue life estimates are extremely important for the proper operation of a modern industrial society. Such estimates are intimately linked to the reliability of machines and structures. They determine the frequency with which components need to be replaced, the economics of the operation of the machine, and hence the insurance costs.

---

[3] A very useful book with which to begin the study of fatigue is by V.V. Bolotin, *Prediction of Service Life for Machines and Structures*, ASME Press, 1989.

Anyone studying fatigue life data will be immediately struck by the significant scatter. Components normally considered to be identical can have a wide range of lives. As engineers, we are concerned about having a rigorous basis for estimating the fatigue lives of nominally identical manufactured components. Eventually, it is necessary to relate the life estimate of the structure to that of its components. This is generally a difficult task, one that requires the ability to evaluate structural and machine response to random forces. Chapter 9 is devoted to reliability.

Example 1.1 Miner's Rule for Fatigue Damage

One of the most important early works on the estimation of fatigue life is by Miner,[4] who was a strength test engineer with the Douglas Aircraft Company. Miner's rule is a deterministic way to deal with the uncertainties of structural damage and fatigue. The phenomenon of cumulative damage under repeated loads is assumed to be related to the total work absorbed by a test specimen. The number of loading cycles applied, expressed as a fraction of the number of cycles to failure at a given stress level, is taken to be proportional to the useful structural life expended. When the total damage reaches 1, the fatigue test specimen is assumed to fail. Miner presented validation of his theory with data from experiments on Aluminum sheets.

At a certain stress level for a specific material and geometry,[5] this rule estimates the number of cycles to failure. Mathematically, this can be written as

$$\frac{n}{N} = 1, \qquad (1.1)$$

where n equals the number of cycles undergone by the structure at a specific stress level, and N is the experimentally known number of cycles to failure at that stress level. Since most structures undergo a mixture of loading cycles at different stress levels, Equation 1.1 must be written for each stress level i as follows,

$$\sum_i \frac{n_i}{N_i} = 1, \qquad (1.2)$$

where each fraction represents the percentage of life used up at each stress level. Suppose the structure is loaded at two stress levels, $i = 1, 2$, with corresponding cycles to failure $N_1 = 100$ and $N_2 = 50$. According to Equation 1.2, the following relation,

$$\frac{n_1}{100} + \frac{n_2}{50} = 1,$$

---

[4] M.A. Miner, "Cumulative Damage in Fatigue," pp. A159-A164, *Journal of Applied Mechanics*, September 1945.

[5] Corners and discontinuities result in high stress concentrations and lower fatigue life.

1.1. APPLICATIONS

holds between the number of possible cycles n_1 and n_2 for each stress level. There are numerous combinations that lead to failure. For example: $(n_1, n_2) = (50, 25)$, $(n_1, n_2) = (100, 0)$, $(n_1, n_2) = (0, 50)$, with others not hard to find. Miner realized that these summations were only approximations. His experiments showed that sometimes a component failed before the sum totaled 1, and, at other times, did not fail until the sum was greater than 1. Furthermore, failure by this rule is independent of the ordering of the stress cycles. This means that fatigue life is the same whether high stress cycles precede or follow lower stress cycles. We know, however, that stress history affects fatigue life. In the last 50 years since Miner's paper, despite the vast amount of work that has been done to build on his and other work to better understand fatigue, Miner's rule and its variants remain widely utilized practical methods.

1.1.3 Ocean-Wave Forces

There are similarities between approaches used to model ocean-wave forces on structures and those for wind forces. The differences are primarily due to the added mass[6] of the water, and the different structural types designed for use in the ocean. The calculation of wave-induced forces is a very important aspect of ocean engineering.[7] As might be expected, many engineering disciplines are utilized in ocean engineering. The estimation of wave forces on offshore oil-drilling platforms, ships, and other ocean and hydraulic structures, such as water channel spillways and dams, is the basis for the design of such structures. Without these estimates, there is no way to design or analyze the structure. The estimation of loads is always first on the list of tasks for an engineer. Figure 1.2 is a schematic showing a structural cylinder in the ocean that is subjected to random and harmonic waves, superimposed on the still water line.

---

[6] The force required to accelerate a rigid body with mass m in a fluid medium that is normally at rest is

$$(m + \text{added mass of the surrounding fluid}) \times \ddot{y}.$$

The additional force is required to accelerate the surrounding fluid to $\ddot{y}$. If the fluid medium is air, the force required to accelerate the surrounding air is often neglected.

[7] Some useful books, among hundreds of available volumes, are the following:

J.F. Wilson, *Dynamics of Offshore Structures*, John Wiley & Sons, 1984. This book is out of print, but there is a second edition.

O.M. Faltinsen, *Sea Loads on Ships and Offshore Structures*, Cambridge University Press, 1990.

S. Gran, *A Course in Ocean Engineering*, Elsevier Science, 1992. This is a very thorough book.

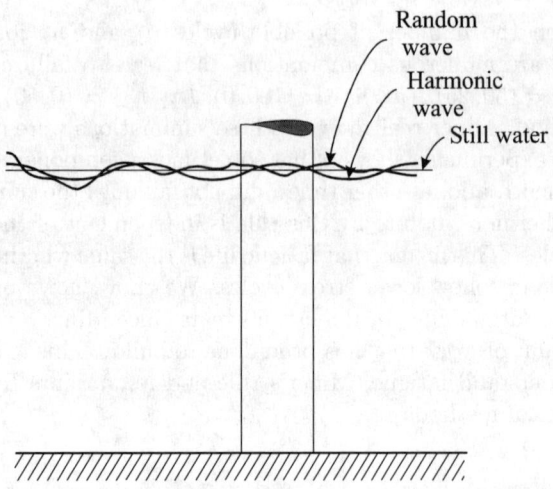

Figure 1.2: Schematic of structural cylinder in the ocean subjected to random and harmonic waves.

Example 1.2 Wave Forces on an Oil-Drilling Platform

The need to drill for oil in the oceans has driven our ability to design ocean structures for sites of ever-increasing depths. Today's fixed-bottom ocean structures, when taken with their foundations, are taller than our tallest skyscrapers. As might be expected, the dynamic response of these towers to ocean waves and currents is significant and must be understood and analyzed. Consider the ocean-wave force on an ocean platform such as the one shown in Figure 1.3.

The single most important paper[8] on the force exerted by ocean waves on fixed structures, even though it was written half a century ago, derived what became to be universally known as the Morison equation. In this paper, after much experimental work, Morison and his colleagues came to the conclusion that the force exerted by unbroken surface waves on a circular cylindrical column that extends from the bottom of the ocean upward above the wave crest is made up of two components:

- a drag force proportional to the square of the water particle velocity, with proportionality represented by a drag coefficient having substantially the same value as for steady flow, and

---

[8] J.R. Morison, M.P. O'Brien, J.W. Johnson, and S.A. Schaff, "The Force Exerted by Surface Waves on Piles," pp. 149-154, *Petroleum Transactions*, Vol. 189, 1950.

1.1. APPLICATIONS

Figure 1.3: Offshore platform.

- an inertia force proportional to the horizontal component of the inertia force exerted on the mass of water displaced by the column, with proportionality represented by an inertia coefficient.

The drag force on an element of length dx is given by

$$dF_D = C_D \rho D \frac{u|u|}{2} dx, \qquad (1.3)$$

where C_D is the experimentally determined drag coefficient, ρ is the density of water, D is the diameter of the cylinder, and u is the instantaneous horizontal water particle velocity. The term $u|u|$ ensures that the direction of the force is in the direction of the flow, where $|u|$ is the absolute value of the particle velocity. The inertia force on an element of length dx is given by

$$dF_I = C_I \rho \frac{\pi D^2}{4} \dot{u}\, dx, \qquad (1.4)$$

where $\dot{u} \equiv \partial u/\partial t$ is the instantaneous horizontal water particle acceleration and C_I is the inertia coefficient. The dimensionless drag and inertia coefficients are functions of flow characteristics, cylinder diameter, and fluid density. Depending on the application, the analyst may assume them to be effectively constant, or may need to account for the scatter in their values.

The Morison force equation is the sum of the above drag and inertia components. Classical deterministic fluid mechanics is used to derive wave particle velocities and accelerations.

Since many tall ocean structures oscillate appreciably, the relative velocity and acceleration between fluid and structure is commonly used in the Morison equation, where u is replaced by $u - \dot{x}$ and $\dot{u}$ is replaced by $(\dot{u} - \ddot{x})$, where $\dot{x}$ and $\ddot{x}$ are the structural velocity and acceleration, respectively.

Also, in order to better characterize the complexity of the wave motion, the fluid velocity, acceleration, and resulting force are modeled as random functions of time. We begin to explore the concept of random functions in detail in Chapter 5. More details on the vibration of structures in fluids can be found in numerous books.[9] Chapter 13 provides an introduction to fluid induced vibration.

❋

1.1.4 Wind Forces

Engineering structures such as cooling towers, aircraft, skyscrapers, rockets, and bridges are all exposed to wind and aerodynamic loads. Examples of these engineering structures are shown in Figures 1.4-1.7. Wind is the natural movement of the atmosphere due to temperature and pressure gradients. Aerodynamic loads are the atmospheric forces resulting from the interaction of wind and structure.

While we know how to write an equation for a harmonic force, what does an equation for wind force look like? Due to the complexity of the fluid mechanics of wind, it is generally necessary to approximate the force due to wind. There are various levels of approximate relations, depending on the application. In all instances, the force relation includes at least one experimentally determined parameter or coefficient. Such *semiempirical* force equations are very valuable in wind engineering practice. These look very much like the Morison equation discussed in the last section.

An excellent monograph is by Simiu and Scanlan.[10]

1.1.5 Material Properties

While the modeling of randomness in material properties is beyond our scope in this book, it is worthwhile to briefly mention this type of modeling

---

[9] These two books are worth having a look at: R.D. Blevins, *Flow-Induced Vibration*, van Nostrand Reinhold, 1977. (There is a second edition.) A.T. Ippen, Editor, *Estuary and Coastline Hydrodynamics*, McGraw-Hill, 1966.

[10] *Wind Effects on Structures: Fundamentals and Applications to Design*, E. Simiu, R.H. Scanlan, Third Edition, 1996, Wiley-Interscience.

1.1. APPLICATIONS

Figure 1.4: Cooling tower.

Figure 1.5: Vortex study in aircraft (courtesy of NASA, Dryer Flight Research Center).

Figure 1.6: Saturn V Apollo 11.

Figure 1.7: Charleston Grace Memorial Bridge.

1.1. APPLICATIONS

because of the importance of many new materials that have *effective* properties, that is, properties that are an average over a cross-section. These include various *composites* and *tailored materials*, modern materials designed for particular structural applications, especially where high strength and durability are needed concomitantly with light weight. Such designs require a complicated mix of fibers and substrates configured to obtain particular properties. Defining stress-strain relations and Young's modulus for such components is not straightforward. It is sometimes necessary that properties be averaged or effective properties be defined.

The soil is an example of a naturally occurring material that is extremely complex and cannot be modeled in a traditional manner. It is common that two nearby volumes of soil have very different mechanical properties. Therefore, in structural dynamics applications, such as earthquake engineering, the loading is effectively random because, in part, by the time it reaches the structure, the force has traversed a complex topology of the Earth.[11]

Data on the variability of material properties are tabulated in numerous references.[12] From Haugen, for example, hot rolled 1035 steel round bars of diameters in the range 1-9 in have yield strengths of between 40,000-60,000 psi, with an average yield of just under 50,000 psi. In addition, the variability can change appreciably depending on temperature. A titanium-aluminum-lead alloy has an ultimate shear strength of between 88,000-114,000 psi at 90°F but at 1000°F the strength drops to between 42,000-60,000 psi.

The obvious conclusion is that variability can be significant and is a function of different factors. Analysis and design require knowledge of the environment within which the structure will operate. While temperature and thermal effects are topics for specialized texts, these can be critical factors in many advanced aerospace and machine designs, and therefore the reader needs to be aware of the importance of these aspects.

1.1.6 Statistics and Probability

The previous examples of natural forces all have one factor in common. It is that they depend on experimentally determined parameters. Just as linearity of vibration depends on small oscillations, these semiempirical equations are valid only for a particular range in the data. While deterministic models also depend greatly on experimental data for their formulation and

---

[11] The area of research known as *earthquake engineering* and the specific study of how energy propagates through complex materials such as soils is known as the study of *waves in random media*.

[12] One can begin with the text by E.B. Haugen, *Probabilistic Mechanical Design*, Wiley-Interscience, 1980. There are interesting applications of probability to mechanical engineering, primarily based on the Gaussian distribution.

ultimately their validity, random models are an attempt to *explicitly* deal with observed scatter in the data and with very intricate dynamic behavior. Random models also allow an estimate of how data scatter affects response scatter.

Data are always our link to valid probabilistic models, their derivation and validation. While this is not our focus in this text, it is important that the reader is at least aware that this step precedes any valid probabilistic models.

Example 1.3 From Data to Model and Back to Data

Modeling can be as much an art as a science. Engineers are generally handed a problem that needs to be solved, not an equation, not even a well-thought-out description of the problem. For example: *We need to go to the Moon in ten years*!

Engineering is predicated on understanding how structures, machines, and materials behave under various operating conditions. This understanding is based on theory and data. Many experiments have been performed to get us to our current level of understanding and intuition. The experiments suggest cause and effect between variables. They provide us with parameter values. Finally, they are the basis for the equations we derive.

We know that data has scatter, and the significance of the scatter to a particular problem determines whether it can be ignored. If it cannot be ignored, then the data is used to estimate the statistics of the randomness. The resulting probabilistic model is used to study the particular problem at hand, and the model's validity is established by comparing its predictions with available data. Such comparisons help define the limits of model validity.

In this way, a full circle is achieved. Data gives birth to understanding and parameter values, which lead to governing equations and their predictions, and finally validity is established by comparing model predictions with new data that is not part of the original set.

⊛

1.2 Units

All physical parameters have units that tie them to a particular system. There are primarily two systems of units, the *English System* and the *SI System*, where *SI* stands for *Système International*. The *SI* units are considered modern and more appropriate. In this book, both systems are used since both are used in practice in the United States. In Table 1.1, the English and SI system of units are shown for certain key physical parameters

Table 1.1: *English* and *SI* Units for Key Physical Parameters

Parameter	English	SI
Force	1 lb	4.448 N (kg m/s^2)
Mass	1 slug (lb s^2/ft)	14.59 kg (kilogram)
	1 lbm	0.455 kg
Length	1 ft	0.3048 m (meter)
Acceleration	1 ft/s^2	0.3048 m/s^2
Spring constant	1 lb/in	175.12 N/m
Torsional spring constant	1 lb in/rad	0.1130 N m/rad
Damping constant	1 lb s/in	175.12 N s/m
Mass moment of inertia	1 lb in s^2	0.1130 kg m^2
Angle	1 deg	0.0175 rad
Pressure	1/6894.757 psi	1 N/m^2 = 1 Pascal

that we will encounter in this text. The SI system has taken hold in the scientific and engineering communities in English-speaking countries over the past several decades, but the English system continues to be popular.

1.3 Organization of the Text

There are many chapters in this book and many topics that, at first sight, may be difficult to organize in the reader's mind. The major groupings are methods for "static" problems, the first four chapters, where time does not play a role, and "dynamic" problems where time is a critical factor, the remaining chapters. While this text contains many standard items and topics, it also includes sections on important material that is not exactly mainstream, as well as topics that break into more advanced disciplines.

The first four chapters introduce basic concepts in probability and variables that are best described as random. The remaining chapters introduce time as a critical parameter in the randomness. An undergraduate course can be based on Chapters 1-5, and a graduate course based on Chapters 5 and subsequent.

1.4 Problems

1. Identify engineering and scientific applications where uncertainties can be ignored. Explain.

2. Identify engineering and scientific applications where uncertainties cannot be ignored. Explain.

3. Discuss how an engineer may ascertain whether uncertainties are important or can be ignored in an analysis and design. Use examples in your discussion.

4. If Miner's rule for fatigue damage has to be extended to cases where the order of loading cycles is important, how can this be accomplished? Explain with or without using equations.

5. If Miner's rule for fatigue damage has to be extended to cases where a cycle at stress $n\sigma$ causes n times as much damage as a cycle at stress σ, how can this be accomplished? Explain with or without using equations.

6. Which variables or parameters in Equations 1.3 and 1.4 are better assumed to be random variables? Explain your choices.

Chapter 2

Events and Probability

2.1 Sets

Set theory is a useful concept for understanding probability, and its building block is the *event*. Events are collections of outcomes on which probabilities have been assigned.

Why is probability necessary? Suppose a set of experiments to measure the diameters of a group of rods are performed. Even though the rods are supposed to have the same dimensions, they do not because of manufacturing imperfections and measurement errors. Therefore, if the diameter for a design is to be specified, there is no single value, but many possible values. These possible values are events.

As a way to organize this information, a *new language* that can accommodate all the possibilities is needed. The language of probability is able to account for all the possibilities in a mathematical way so that computation is possible.

A *set* is a collection of things that share certain characteristics. Included are objects, numbers, colors, essentially any of the items and concepts we use to define our environment.

A *sample space* is the set of all possibilities. The group of possible diameters is such a sample space. Or it can be the values a variable may take. Each of these possibilities is called a *sample point*. If one of the sample points occur, then a *realization* of that sample point has occurred. In this case, the sample points are discrete-valued, but in other applications they may be continuous-valued. A subset of the sample space is an *event* within the sample space.

We will examine the connection between events and probability, but first some examples of events are considered. In engineering reliability, the goal is to estimate the probability that the system will perform as designed. This means that it performs at or better than design specifications for a certain design life with a certain reliability level. The technical term for this characteristic of a structure or a component is its *safety*. The probability that the system does not perform as designed is given by its *probability of failure*. Generally, it is very difficult to estimate such a quantity due to the complexity of engineered systems and the uncertainties associated with operating environments.

Example 2.1 Probabilities of Interest

We have just mentioned safety and failure as two very important probabilities that define the robustness of a particular design. Other probabilities of interest might be: *(i)* the live load[1] is less than 10 N/m^2, *(ii)* the temperature is greater than 50°C, *(iii)* the frequency of vibration is in the range $\omega_1 < \omega < \omega_2$, and *(iv)* the flow rate is less than 3 ft^3/s. The safety of a complex machine is a function of the reliability of all its components and how they are connected to each other. First, these individual safeties must be determined.

The range $\omega_1 < \omega < \omega_2$ is a continuous sample space. A particular value, $\omega_1 < 33.01 < \omega_2$, is a sample point whose occurrence is a realization.
⊛

2.1.1 Basic Events

Several events are needed for a complete modeling framework. They are:

1. The *impossible event,* denoted by $\{\}$ or ϕ, is the event with no sample point. In the sample space, the impossible or *null* event is an empty set.

2. The *certain* or *universal event*, denoted by S or Ω, is the event that contains all the sample points. The certain event is in fact the sample space itself.

3. The *complementary event* is denoted by the event symbol with an overbar. For event E, the complementary event is $\overline{E}$, meaning that it contains all sample points in S that are not in E. Therefore, $\overline{E}$ is *not E*.

---

[1] A live load is one that is due to transient effects, for example, a person walking across a floor. The counterpart is - you guessed it - the dead load. For example, the weight of the floor, which does not change.

2.1. SETS

A useful visual representation of sets, spaces, and their manipulations is provided by the *Venn diagram.* The sample space or the universe is represented by a rectangle. An event is represented by a closed region within the rectangle. The complement of this event will be the remaining region. See Figure 2.1.

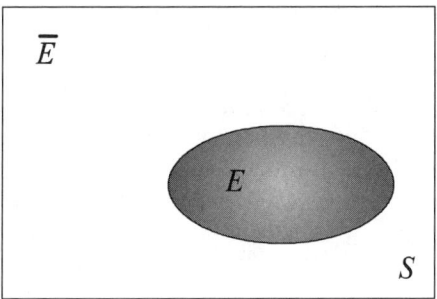

Figure 2.1: Venn diagram with event E.

The concepts of impossible, universal, and complementary events are demonstrated by an example.

Example 2.2 Impossible, Universal, and Complementary Events

Consider an experiment where a coin is tossed three times consecutively and the total number of heads counted. The certain or universal event is the set of all possible outcomes, the total number of heads counted,

$$S = \{0, 1, 2, 3\}.$$

If A is the event that the total number of heads is greater than 3, then A is an impossible event, written as

$$A = \{\} \text{ or } A = \phi.$$

Keep in mind that $\{0\}$ is not an impossible event, but the event that no heads appeared.

Let B be the event that the total number of heads is 2. That is,

$$B = \{2\}.$$

The complementary event of B is then

$$\bar{B} = \{0, 1, 3\}.$$

It should be noted that the impossible event is a complementary event of the universal event and vice versa,

$$\bar{S} = \phi, \;\; \bar{\phi} = S.$$

In many instances, it is necessary to consider more than one event and how the different events are coupled or affect each other. For example, a high-strength alloy has a range of yield stresses that depend on temperature. The two events are yield stresses and temperature. It is important to be able to make simultaneous or joint statements on both events. To do this, we need to operate with sets using the following rules on the combination of events. These rules can be easily seen utilizing the Venn diagrams in Figure 2.2.

1. The *union* of two events E_1 and E_2, denoted as $E_1 \cup E_2$, is defined as E_1 *or* E_2, or *both*. This means that either or both events may occur.

2. The *intersection* of events E_1 and E_2, denoted as $E_1 \cap E_2$ or $E_1 E_2$, is defined as E_1 *and* E_2. This means that both events occur.

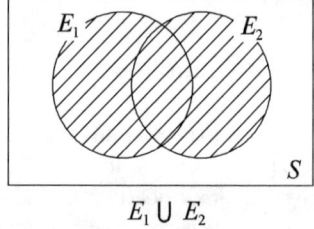 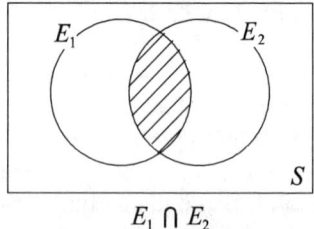

Figure 2.2: The union and intersection of E_1 and E_2.

The concept of the event can be used to signify certain regions of interest within the sample space S. Note the following: $E_i \cup S = S$ and $E_i \cap \phi = \phi$. Also, if two events E_1, E_2 are *mutually exclusive* or *disjoint* then $E_1 \cap E_2 = \phi$. All these operations can be easily visualized using the Venn diagram.

Example 2.3 Union and Intersection I

Conduct an experiment in which two dice are tossed at the same time. The experiment is repeated many times, and the numbers that show are then added. S is the universal event,

$$S = \{2, 3, 4, 5, 6, 7, 8, 9, 10, 11, 12\}.$$

2.1. SETS

Let A be the event that the sum is greater than 7, and B the event that the sum is less than 11. Then,

$$A = \{8, 9, 10, 11, 12\} \text{ and } B = \{2, 3, 4, 5, 6, 7, 8, 9, 10\}.$$

Obtain the unions and intersections of A and B, $\bar{A}$ and B, A and $\bar{B}$, and $\bar{A}$ and $\bar{B}$.

Solution Using the definitions of union and intersection,

$$A \cup B = \{2, 3, 4, 5, 6, 7, 8, 9, 10, 11, 12\}$$
$$\bar{A} \cup B = \{2, 3, 4, 5, 6, 7, 8, 9, 10\}$$
$$A \cup \bar{B} = \{8, 9, 10, 11, 12\}$$
$$\bar{A} \cup \bar{B} = \{2, 3, 4, 5, 6, 7, 11, 12\},$$

and

$$A \cap B = \{8, 9, 10\}$$
$$\bar{A} \cap B = \{2, 3, 4, 5, 6, 7\}$$
$$A \cap \bar{B} = \{11, 12\}$$
$$\bar{A} \cap \bar{B} = \phi.$$

✱

Example 2.4 Union and Intersection II

Let E_1, E_2, and E_3 be three events. Express the following statements in set notation, *(i)* at least one event occurs, *(ii)* all three events occur, *(iii)* exactly one of the three events occurs.

Solution

(i) At least one event occurring is same as either E_1, E_2, or E_3 occurring. That is, $E_1 \cup E_2 \cup E_3$.

(ii) All three events occurring is same as all E_1, E_2, and E_3 occurring. That is, $E_1 \cap E_2 \cap E_3$.

(iii) There are three ways that exactly one of three events occurs. The first is when E_1 occurs, and E_2 and E_3 do not. The second is when E_2 occurs, and the others do not. The third is when E_3 occurs, and the others do not. Call these events A_1, A_2, and A_3, respectively,

$$A_1 = E_1 \cap \bar{E_2} \cap \bar{E_3}$$
$$A_2 = \bar{E_1} \cap E_2 \cap \bar{E_3}$$
$$A_3 = \bar{E_1} \cap \bar{E_2} \cap E_3.$$

The event that exactly one of three mutually exclusive events occurs is the union of A_1, A_2, and A_3, that is, $A_1 \cup A_2 \cup A_3$, or

$$\left(E_1 \cap \overline{E_2} \cap \overline{E_3}\right) \cup \left(\overline{E_1} \cap E_2 \cap \overline{E_3}\right) \cup \left(\overline{E_1} \cap \overline{E_2} \cap E_3\right).$$

⊛

Figure 2.3: John Venn (1834-1923).

Venn John Venn came from a Low Church Evangelical background and when he entered Gonville and Caius College Cambridge in 1853 he had so slight an acquaintance with books of any kind that he may be said to have begun there his knowledge of literature.

He graduated in 1857, was elected a Fellow in that year and two years later was ordained a priest. For a year he was curate at Mortlake.

In 1862 he returned to Cambridge University as a lecturer in Moral Science, studying and teaching logic and probability theory. He developed Boole's mathematical logic and is best known for his diagrammatic way of representing sets and their unions and intersections.

Venn wrote *Logic of Chance* in 1866, which Keynes described as *strikingly original and considerably influenced the development of the theory of statistics.*

Venn published *Symbolic Logic* in 1881 and *The Principles of Empirical Logic* in 1889. The second of these is rather less original but the first was described by Keynes as *probably his most enduring work on logic.*

2.1. SETS

In 1883 Venn was elected a Fellow of the Royal Society. About this time his career changed direction. He had already left the Church in 1870 but his interest now turned to history. He wrote a history of his college, publishing *The Biographical History of Gonville and Caius College 1349-1897* in 1897.

He then undertook the immense task of compiling a history of Cambridge University, the first volume of which was published in 1922. He was assisted by his son in this task, which was described by another historian in these terms: *It is difficult for anyone who has not seen the work in its making to realize the immense amount of research involved in this great undertaking.*

Venn had other skills and interests, too, including a rare skill in building machines. He used his skill to build a machine for bowling cricket balls, which was so good that when the Australian Cricket team visited Cambridge in 1909, Venn's machine clean bowled one of its top stars four times.

2.1.2 Operational Rules

As with the mathematical operations of addition and multiplication, set operations have similar rules. We have started operating with sets in the last section. By doing so, it is possible to derive additional information. For example, sets A and B are known. By performing the union $A \cup B$, information about a new set is found, $C = A \cup B$.

In engineering a similar expansion of knowledge is possible. The design life and reliability of an aircraft wing depends on our knowledge of the states of stress in the wing. We need to know how the stress event varies with time and location on the wing. Stress, however, depends on many other events. Two such events are the wind velocity and the surrounding temperature distribution. While it is possible to estimate the velocity and temperature, these must be related to the stress state in the structure.

This is what is meant by operating with events. Such operations allow us to make statements on other related and more important events. Some of the more important properties of unions and intersections follow. It is worthwhile to use the Venn diagram symbolism to verify the properties developed next.

The *commutative* property is a statement of whether the *ordering* of a union or intersection changes the outcome. It does not, as shown for the two events E_1 and E_2,

$$E_1 \cup E_2 = E_2 \cup E_1$$
$$E_1 \cap E_2 = E_2 \cap E_1.$$

The *associative* property is a statement of whether operations *grouping* affects the result. It does not, as shown for events E_1, E_2, and E_3,

$$E_1 \cup (E_2 \cup E_3) = (E_1 \cup E_2) \cup E_3$$
$$E_1 \cap (E_2 \cap E_3) = (E_1 \cap E_2) \cap E_3.$$

The *distributive* property is a statement about mixed operations, for example, intersections of sums and sums of intersections. For events E_1, E_2, and E_3,

$$E_1 \cap (E_2 \cup E_3) = (E_1 \cap E_2) \cup (E_1 \cap E_3) \qquad (2.1)$$
$$E_1 \cup (E_2 \cap E_3) = (E_1 \cup E_2) \cap (E_1 \cup E_3). \qquad (2.2)$$

Using the shorthand notation for intersection, Equation 2.1 can be written as $E_1(E_2 \cup E_3) = E_1 E_2 \cup E_1 E_3$. New rules can be obtained by interchanging $\cup$ and $\cap$. *De Morgan's rules* formalize this:

$$\overline{E_1 \cup E_2} = \overline{E_1} \cap \overline{E_2} \qquad (2.3)$$
$$\overline{E_1 \cap E_2} = \overline{E_1} \cup \overline{E_2}. \qquad (2.4)$$

These rules can be generalized to n events. The validity of Equation 2.3 is shown in the Venn diagram of Figure 2.4. De Morgan's rules demonstrate a duality relation: the complement of unions and intersections is equal to the intersections and unions of complements. This property can simplify a calculation since in numerous instances is it simpler to evaluate the complement of an event than the event itself.

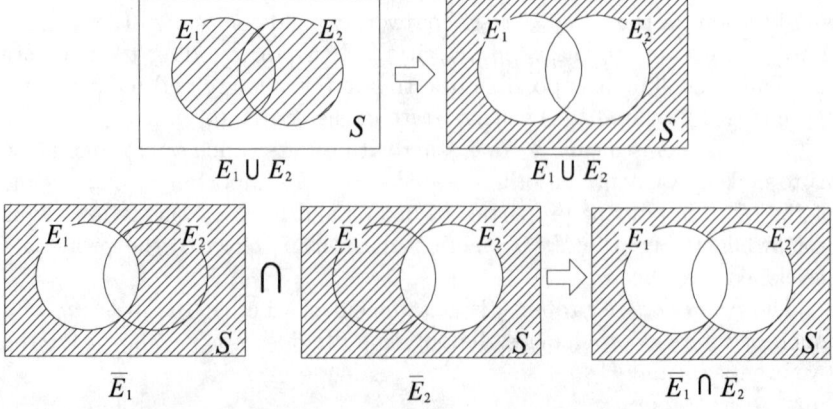

Figure 2.4: De Morgan's rule visualized using Venn diagrams.

2.1. SETS

Example 2.5 Failure of Drive Train

In a drive train, two drive shafts are connected via a joint that permits the transmission of power even if the shafts are naturally misaligned, as shown in Figure 2.5. The drive shafts transmit a torque T. If one of the shafts fails, then the drive train is no longer operable. Formulate this problem and find the failure event for the drive train in terms of the events that define the breakage of each shaft individually.

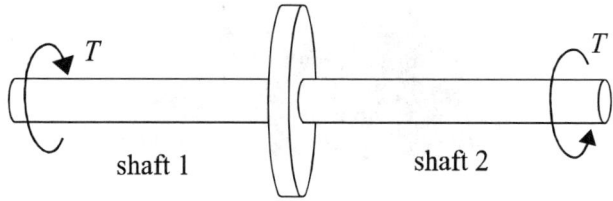

Figure 2.5: Drive train under torque.

Solution   Define the following events:

$$E_1 = \text{breakage of shaft 1}$$
$$E_2 = \text{breakage of shaft 2}.$$

The failure of the drive train is defined as the failure of either of the two shafts, or of both shafts. Using set notation,

$$\text{failure of drive train } = E_1 \cup E_2.$$

Therefore, the event of no failure is given by

$$\text{no failure of drive train } = \overline{E_1 \cup E_2}.$$

Also, in another way of looking at the event of no failure, if the drive train is operational, this implies that both shafts are operational, or,

$$\text{no failure of drive train } = \overline{E_1} \cap \overline{E_2}.$$

Since these two events must be equal to each other, we have an illustration of De Morgan's rule,

$$\overline{E_1 \cup E_2} = \overline{E_1} \cap \overline{E_2}.$$

✲

Figure 2.6: Augustus De Morgan (1806-1871).

De Morgan   Augustus De Morgan, the fifth child of Lieutenant-Colonel John De Morgan, was born while his father was stationed in India. While he was stationed there his fifth child Augustus was born. Augustus lost the sight of his right eye shortly after birth and, when seven months old, returned to England with the family. His father died when Augustus was 10 years old.

De Morgan did not excel at school and, because of his physical disability he did not join in the sports of other boys. He was even made the victim of cruel practical jokes by some school fellows.

In 1823 at the age of 16, De Morgan entered Trinity College Cambridge where he was taught by Peacock and Whewell; the three became lifelong friends. He received his B.A. but, because an M.A. required a theological test, something to which De Morgan strongly objected despite being a member of the Church of England, he could go no further at Cambridge, being not eligible for a Fellowship without his M.A.

In 1826 he returned to his home in London and entered Lincoln's Inn to study for the Bar. In 1827 (at the age of 21) he applied for the chair of mathematics in the newly founded University College London and, despite having no mathematical publications, he was appointed.

In 1828 De Morgan became the first professor of mathematics at University College. He gave his inaugural lecture *On the study of mathematics*. De Morgan was to resign his chair, on a matter of principle, in 1831. He was appointed to the chair again in 1836 and held it until 1866 when he was to resign for a second time, again on a matter of principle.

2.1. SETS

His book, *Elements of Arithmetic* (1830), was his second publication and was to see many editions.

In 1838 he defined and introduced the term "mathematical induction," putting a process that had been used without clarity on a rigorous basis. The term first appears in De Morgan's article, *Induction (Mathematics)*, in the *Penny Cyclopedia*. (Over the years he was to write 712 articles for the *Penny Cyclopedia*.) The *Penny Cyclopedia* was published by the Society for the Diffusion of Useful Knowledge, set up by the same reformers who founded London University, and that Society also published the famous work by De Morgan, *The Differential and Integral Calculus*.

De Morgan was always interested in odd numerical facts and writing in 1864 he noted that he had the distinction of being x years old in the year x^2.

In 1849 he published *Trigonometry and Double Algebra* in which he gave a geometric interpretation of complex numbers. He recognized the purely symbolic nature of algebra and he was aware of the existence of algebras other than ordinary algebra. He introduced De Morgan's laws and his greatest contribution is as a reformer of mathematical logic.

De Morgan corresponded with Charles Babbage and gave private tutoring to Lady Lovelace who, it is claimed, wrote the first computer program for Babbage. De Morgan also corresponded with Hamilton and, like Hamilton, attempted to extend double algebra to three dimensions.

In 1866 he was a cofounder of the London Mathematical Society and became its first president. De Morgan's son, George, a very able mathematician, became its first secretary. In the same year De Morgan was elected a Fellow of the Royal Astronomical Society.

De Morgan was never a Fellow of the Royal Society as he refused to let his name be put forward. He also refused an honorary degree from the University of Edinburgh. He was described by Thomas Hirst thus: *A dry dogmatic pedant I fear is Mr. De Morgan, notwithstanding his unquestioned ability.*

Macfarlane remarks that ... *De Morgan considered himself a Briton unattached, neither English, Scottish, Welsh or Irish.* He also says: *He disliked the country and while his family enjoyed the seaside, and men of science were having a good time at a meeting of the British Association in the country, he remained in the hot and dusty libraries of the metropolis. He had no ideas or sympathies in common with the physical philosopher. His attitude was doubtless due to his physical infirmity, which prevented him from being either an observer or an experimenter. He never voted in an election, and he never visited the House of Commons, or the Tower, or Westminster Abbey.*

2.2 Probability

Let us define probability. Believe it or not, much dispute has centered on the definition of probability, primarily pitting those who view it as a subjective quantity against others who believe that only with experimentation can a rigorously derived probability be possible. The former will counter that it is often not possible to perform enough experiments to arrive at that rigorous probability and judgment must be used. Fortunately, our purposes here do not require us to resolve this debate. We can assume that in some manner it is possible to obtain the probabilities necessary for our computations, usually based on data analysis.[2]

Think of a randomly vibrating oscillator where random behavior implies unpredictable periods, amplitudes, and frequencies. These all appear to vary from one instant of time to the next, as shown in Figure 2.7. As another example, what do you suppose is the temperature of the flame in a fire? Clearly, the temperature at a particular point within the flame varies in a very complex way, fluctuating in much the same way as the amplitude of Figure 2.7.

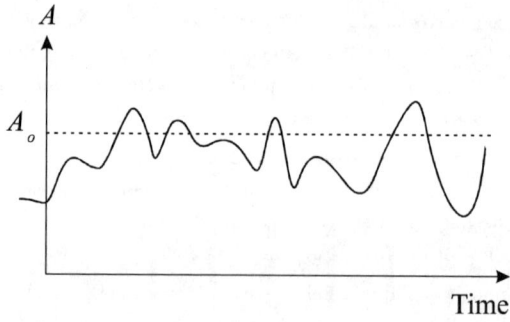

Figure 2.7: Amplitude variation in time of a random oscillator.

How can we answer a question such as: *What is the probability that the value of amplitude A is greater than a specific number A_0?* Begin by expressing the question using probability notation, $\Pr(A > A_0)$. The question can be restated as: for how much time is the oscillator at amplitudes greater than A_0? This implies a *fraction* or *frequency* interpretation for probability. Looking at a long-time history of the oscillation, it is possible to estimate the amount of time the amplitude is greater than A_0. That

---

[2] J.S. Bendat and A.G. Piersol, *Random Data: Analysis and Measurement Procedures*, Second Edition, John Wiley & Sons, 1986. This book provides an excellent development of the theory and techniques of data analysis.

2.2. PROBABILITY

excursion frequency is the probability estimate,

$$\Pr(A \geq A_0) \simeq \frac{\text{amount of time } A > A_0}{\text{total time}}.$$

For example, if the oscillation time history is 350 hr long and for 37 of those hours $A > A_0$, then $\Pr(A > A_0) \simeq 37/350 = 0.106$; the probability that $A > A_0$ is estimated to be 10.6%. This is only an estimate since, if the test lasted 3500 hr instead of 350 hr, we expect some change in the probability, although hopefully slight so that there is confidence in the estimate. The key to a good estimate is that the test is long enough so the probability estimate has approximately converged.

Figure 2.8 depicts this procedure of estimating probabilities by frequencies of occurrence for the case of a discrete process that can have one of 7 outcomes. The frequency of each outcome is counted and the ratio with respect to the total number of all outcomes provides an estimate of the occurrence probability. Figure 2.8 is known as a *histogram*. The frequency interpretation for estimating probabilities is the most useful, and is the approach used in this text.

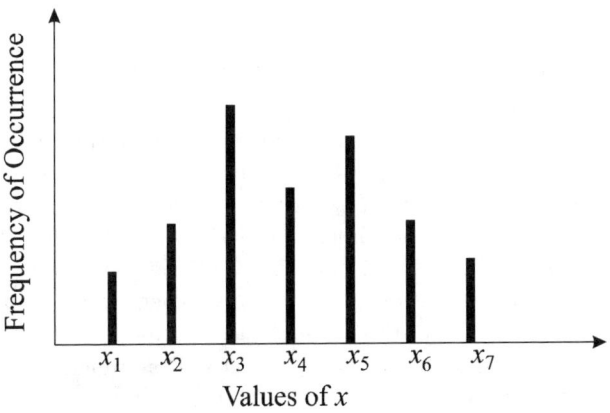

Figure 2.8: Relative frequency of occurrence.

Probabilities may or may not involve time. Probabilistic models of mechanical systems are a natural result of the observation that most physical variables may take on a range of possible values. For example, if 100 machine shafts are manufactured, there will be 100 different diameters if enough significant figures are kept in the measurements. Figure 2.9 depicts a histogram of diameter data where three significant figures are kept. As expected, the diameters are very close in value, but not exactly the same.

Note that the sum of all frequencies must add up to 1 since all of the possible outcomes have been included.

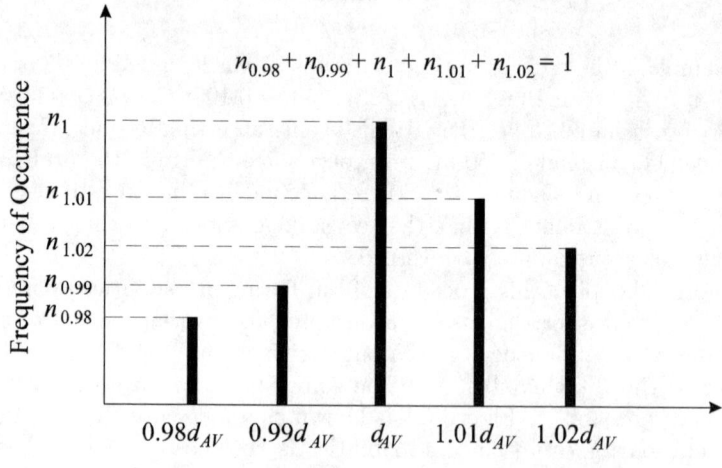

Figure 2.9: Histogram of machine shaft diameters.

How is such a spread of values considered in measuring the strength of the shaft in torsion? What numbers should be substituted into the stress-strain relation? Similarly, running ultimate tensile-strength tests a number of times on "identical specimens" will show no two identical results. Small differences in dimensions, material properties, and boundary conditions make it impossible to exactly duplicate experimental results. *There will always be some scatter.* How should this information be utilized?

Randomness is possible for constants as well as functions (of time or space). A constant with a scatter of possible values is called a *random variable*. A function of time with a scatter is generally called a *random*[3] *process*. Random variables are those that can only be prescribed to a certain level of certainty. Important examples are material yield characteristics that define the transition from elastic to plastic behavior. *Random processes* are time-dependent (or space-dependent) phenomena that, with repeated observation under essentially identical conditions, do not show the same time histories.

For increasingly complex engineering demands, it is important to understand, and be able to model, uncertainties and qualitative information for

---

[3] Or the Greek *stochastic*: στοκος.

2.2. PROBABILITY

analysis and design. An example of qualitative information is a verbal description of size or strength. Developing the ability to analyze uncertainties allows the engineer to decide for which applications they are insignificant and may be ignored. For many applications discussed in this text, scatter cannot be ignored because of the resulting variability of response.

2.2.1 Axioms of Probability

Mathematical probability is a major branch of mathematics. Yet all of the vast understanding can be traced to the three *axioms of probability*. An axiom is a stipulation of a property that is not proven. It is postulated. It is used as a foundation upon which a deductive framework is built. The veracity and usefulness of an axiom is linked to the veracity and usefulness of the framework that is created. All of mathematics is based on various axioms.

The axioms of probability are the following:

1. For every event E in a sample space S, the probability of the event is governed by $\Pr(E) \geq 0$.

2. The probability of the certain event S is given by $\Pr(S) = 1$.

3. For two events E_1, E_2, that are *mutually exclusive*, that is, $E_1 \cap E_2 = \emptyset$, the probability of the occurrence of either or both events is given by

$$\Pr(E_1 \cup E_2) = \Pr(E_1) + \Pr(E_2).$$

Using these axioms along with the previous combination rules allows us to derive all the rules of probability.

Example 2.6 The Third Axiom of Probability

Machine shafts are being manufactured. The shaft can be rejected if the diameter is less than 98% or greater than 102% of its nominal value. The "nominal" value is the desired or design value. The probability that a shaft is being rejected because the diameter is less than 98% of its nominal value is given as 0.02 and the probability that a shaft is being rejected because the diameter is greater than 102% of its nominal value is given as 0.015. What is the probability that a shaft will be rejected?

Solution Let E_1 be the event that the diameter of the shaft is less than 98% of its nominal value, and E_2 be the event that the diameter of the shaft is greater than 102% of its nominal value. The probability of each event is

$$\Pr(E_1) = 0.02, \text{ and } \Pr(E_2) = 0.015.$$

A shaft will be rejected if either event E_1 or E_2 occurs. That is, the probability that a shaft will be rejected equals the probability of the union E_1 and E_2, that is, $\Pr(E_1 \cup E_2)$. Since the shaft diameter cannot be too small and too large at the same time, the events E_1 and E_2 are mutually exclusive, and using the third axiom of probability,

$$\Pr(E_1 \cup E_2) = \Pr(E_1) + \Pr(E_2) = 0.035.$$

※

2.2.2 Extensions from the Axioms

A probability is numerically in the range $0 \leq \Pr(E) \leq 1$. To show this, write the following equivalent relations:

$$\Pr(E \cup \overline{E}) = \Pr(E) + \Pr(\overline{E}) \quad \text{by axiom } 3$$
$$\Pr(E \cup \overline{E}) = \Pr(S) = 1 \quad \text{by axiom } 2.$$

Equate the two probability statements to find

$$\Pr(E) + \Pr(\overline{E}) = 1 \quad \text{and therefore} \quad 0 \leq \Pr(E) \leq 1.$$

These two probability rules are universal and used often. An extension of axiom 3 for the case where $E_1 \cap E_2 \neq \emptyset$ is very useful in applications since in most instances events that are part of the same system are not disjoint. For overlapping events, as shown in the Venn diagram of Figure 2.10,

$$\Pr(E_1 \cup E_2) = \Pr(E_1) + \Pr(E_2) - \Pr(E_1 \cap E_2). \tag{2.5}$$

It is clear from the Venn diagram that the event $E_1 \cap E_2$ must be subtracted since it has been counted twice by adding E_1 and E_2. Equation 2.5 can be extended to any number of variables as shown here,

$$\begin{aligned}
\Pr(E_1 \cup E_2 \cup E_3) &= \Pr([E_1 \cup E_2] \cup E_3) \\
&= \Pr(E_1 \cup E_2) + \Pr(E_3) - \Pr([E_1 \cup E_2] \cap E_3) \\
&= \Pr(E_1) + \Pr(E_2) - \Pr(E_1 \cap E_2) + \Pr(E_3) \\
&\quad - \Pr([E_1 \cap E_3] \cup [E_2 \cap E_3]) \\
&= \Pr(E_1) + \Pr(E_2) + \Pr(E_3) - \Pr(E_1 \cap E_2) \\
&\quad - \Pr(E_1 \cap E_3) - \Pr(E_2 \cap E_3) + \Pr(E_1 \cap E_2 \cap E_3).
\end{aligned}$$

Estimating probabilities such as $\Pr(E_i \cap E_j)$ and $\Pr(E_i \cap E_j \cap E_k)$ requires information, usually experimentally derived, on whether these events "depend" on each other. If they are disjoint, $E_i \cap E_j = \phi$ and $E_i \cap E_j \cap E_k = \phi$,

2.2. PROBABILITY

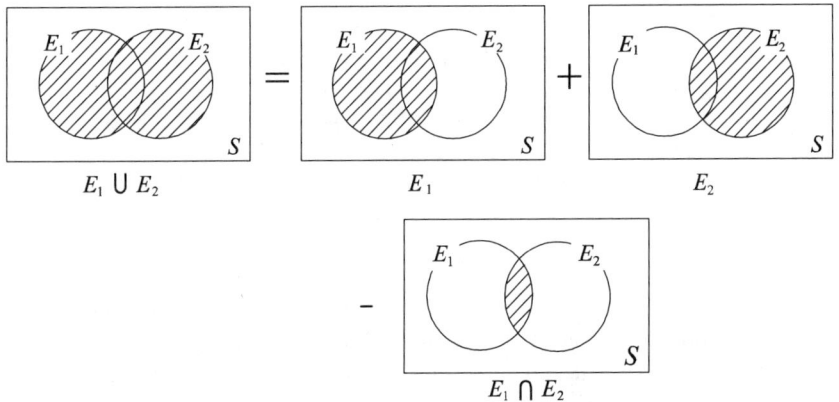

Figure 2.10: Union of E_1 and E_2 visualized using Venn diagrams.

then their probabilities equal zero. Using De Morgan's rule, Equation 2.3, we also find that

$$\Pr(E_1 \cup E_2 \cup E_3) = 1 - \Pr(\overline{E_1 \cup E_2 \cup E_3})$$
$$= 1 - \Pr(\overline{E_1} \cap \overline{E_2} \cap \overline{E_3}).$$

Example 2.7 Extension of Probability Axioms

Show that $\Pr(E_1 \cup E_2) \leq \Pr(E_1) + \Pr(E_2)$.
Solution From the first axiom, we know that $\Pr(E_1 \cap E_2) \geq 0$. Therefore,

$$\Pr(E_1) + \Pr(E_2) - \Pr(E_1 \cap E_2) \leq \Pr(E_1) + \Pr(E_2). \quad (2.6)$$

Using Equation 2.5, the left-hand side of Equation 2.6 is identical to the probability $\Pr(E_1 \cup E_2)$. Thus,

$$\Pr(E_1 \cup E_2) \leq \Pr(E_1) + \Pr(E_2).$$

⊛

2.2.3 Conditional Probability

Thinking about the implications of Equation 2.5 leads us to consider the modeling of *causality* probabilistically. It is important to be able to estimate the probability that one event occurs *given* that another event *has already occurred.* Formally, the *conditional probability* is defined as the probability of E_1 occurring given that E_2 has occurred, and is given by

$$\Pr(E_1|E_2) = \frac{\Pr(E_1 \cap E_2)}{\Pr(E_2)}. \tag{2.7}$$

This equation will sometimes be useful in its equivalent form:

$$\Pr(E_1 \cap E_2) = \Pr(E_1|E_2)\Pr(E_2).$$

Note that $\Pr(E_1 \cap E_2)$ also equals $\Pr(E_2|E_1)\Pr(E_1)$ since $E_1 \cap E_2 = E_2 \cap E_1$. Interpret Equation 2.7 in terms of the sample space S and the relative frequency interpretation of probability. Recall that in this interpretation, the ratio of desired outcomes to the complete space S is a measure of the probability of realization of that outcome. Since the conditional probability states that event E_2 has occurred, the complete space is reduced from S to E_2. The event that E_1 occurs given that E_2 has occurred is given by the intersection $E_1 \cap E_2$ and the probability equals the ratio of the probability of the intersection to the probability of E_2.

Given what we have just said about the conditional probability makes it clear that we have been actually using conditional probabilities when describing probabilities, except that the condition was on the whole space S. That is, with $E_2 = S$,

$$\begin{aligned}\Pr(E_1|S) &= \frac{\Pr(E_1 \cap S)}{\Pr(S)} \\ &= \frac{\Pr(E_1)}{1} \\ &= \Pr(E_1),\end{aligned}$$

so that the probability of an event E_1, as given by $\Pr(E_1)$, is with respect to the space S.

Example 2.8 Conditional Probability

Light bulbs are tested to establish lifetimes. Suppose a light bulb can fail prematurely for a variety of reasons, one reason being a defective filament. The probability that the part will fail for any reason is 0.01. If the bulb happens to have a defective filament, then the probability that it will fail prematurely is 0.1. If the bulb fails, then the probability that the cause of failure is the defective filament is 0.05. What is the probability that the filament is defective?

Solution Let A be the event that the bulb fails prematurely and B the event that the filament is defective. The given probabilities are

$$\Pr(A) = 0.01, \ \Pr(A|B) = 0.1, \ \text{and} \ \Pr(B|A) = 0.05.$$

2.2. PROBABILITY

Find $\Pr(B)$.

Using the definition of conditional probability in Equation 2.7, we find that
$$\Pr(B \cap A) = \Pr(B|A)\Pr(A)$$
$$= 0.05 \times 0.01$$
$$= 0.0005.$$

The probability that the bulb fails and also has a defective filament is 0.0005. Again using the definition of conditional probability,
$$\Pr(B) = \frac{\Pr(A \cap B)}{\Pr(A|B)}$$
$$= \frac{0.0005}{0.1}$$
$$= 0.005.$$

The probability that any given bulb has a defective filament is 0.005.
⊛

2.2.4 Statistical Independence

Two events are said to be *statistically independent* if the occurrence of one event is *not dependent on,* and *does not affect* the occurrence of, the other event. From the definition of conditional probability, for statistically independent events E_1 and E_2,
$$\Pr(E_1|E_2) = \Pr(E_1)$$
$$\Pr(E_2|E_1) = \Pr(E_2),$$
and therefore,
$$\Pr(E_1 \cap E_2) = \Pr(E_1|E_2)\Pr(E_2) = \Pr(E_1)\Pr(E_2)$$
$$\Pr(E_2 \cap E_1) = \Pr(E_2|E_1)\Pr(E_1) = \Pr(E_2)\Pr(E_1).$$

Statistical independence can then be defined as
$$\Pr(E_1 \cap E_2) = \Pr(E_1)\Pr(E_2). \qquad (2.9)$$

Statistical independence relates how the probability of the joint event $E_1 \cap E_2$ is related to the probabilities of the individual events E_1 and E_2. On the other hand, mutual exclusiveness is a statement of whether the joint event $E_1 \cap E_2$ exists. Mutually exclusive events cannot both happen at the same time. That is, if E_1 and E_2 are mutually exclusive, $\Pr(E_1 \cap E_2) = 0$. For nontrivial events, $\Pr(E_1) \neq 0$ and $\Pr(E_2) \neq 0$, the right-hand side of Equation 2.9 cannot be zero, and in this case mutually exclusive events cannot be statistically independent.

Example 2.9 Statistically Independent versus Mutually Exclusive Events

Toss a coin once. E_1 is the event that the result is heads, and E_2 is the event that the result is tails

$$\Pr(E_1) = 0.5$$
$$\Pr(E_2) = 0.5.$$

These two events are mutually exclusive since both cannot happen at the same time, $\Pr(E_1 \cap E_2) = 0$. These events are not statistically independent since if E_1 occurs, E_2 cannot occur.

Now toss two fair coins. Let E_1 be the event that the result of the first coin toss is heads, and E_2 be the event that the result of the second coin toss is heads. These two events are statistically independent since the outcome of one does not affect the outcome of the other. These events can also happen at the same time. That is, these events are not mutually exclusive, but are statistically independent.

Example 2.10 Statistical Independence

A die and a coin are tossed. Assume that they are thrown such that they do not interfere with each other. What is the probability of rolling a 2 provided that the coin comes up heads. Also, what is the probability that the die shows an even number and the coin comes up heads?

Solution Let A be the event that the die will show 2 and B the event that the coin shows heads. Assuming a fair coin and die, the probability of each event is

$$\Pr(A) = 1/6 \text{ and } \Pr(B) = 1/2.$$

The two events can be assumed to be independent. That is, the outcome of one event does not affect the outcome of the other. Therefore, the probability that the die will show 2 is not affected by the outcome of the coin. The answer is 1/6. We just showed

$$\Pr(A|B) = \Pr(A).$$

The probability that the die shows 2 and the coin shows heads can be written as $\Pr(A \cap B)$. The probability is $1/6 \times 1/2 = 1/12$.

2.2.5 Total Probability

Most applications that engineers encounter, and must design for, contain many components and processes. For each of these many events we can define their operating characteristics and reliability. Suppose there is an event E_0 that affects the design, but it cannot be measured or determined directly. Its occurrence, however, is always accompanied by the occurrence of one or several other events. For example, perhaps the pressure is needed but cannot be measured, whereas the temperature that can be related to the pressure is measurable.

Let these other possible events be denoted by the mutually exclusive events $E_1, E_2, \cdots, E_n$, as shown in Figure 2.11.

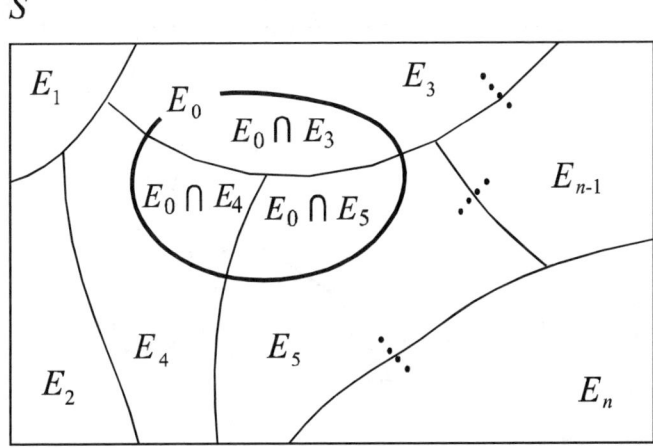

Figure 2.11: Constructing the theorem of total probability.

E_0 can overlap any of the other events in the Venn diagram. Conditional probabilities can be used to construct the equivalent statements,

$$\begin{aligned}
\Pr(E_0) &= \Pr(E_0 E_1) + \Pr(E_0 E_2) + \cdots + \Pr(E_0 E_n) \\
&= \Pr(E_0|E_1)\Pr(E_1) + \Pr(E_0|E_2)\Pr(E_2) + \\
&\quad \cdots + \Pr(E_0|E_n)\Pr(E_n).
\end{aligned} \qquad (2.10)$$

This equation is called the *theorem of total probability*, which is very useful in deducing probabilities of an event in terms of its dependence on other events.

Example 2.11 Inspection of a Manufactured Product via Total Probability

The process of manufacturing an item requires that two of the components be welded, after which a visual inspection is performed. The item fails the inspection if either the weld has gaps or the components are misaligned. The manufacturing history allows us to estimate the following probabilities. It is estimated that 1% of the welds have gaps and that 4% of the components are misaligned. It is also known that if a weld is misaligned that it is 30% more likely to have gaps than if it were not misaligned. We need to estimate the probability that a given weld will pass inspection.

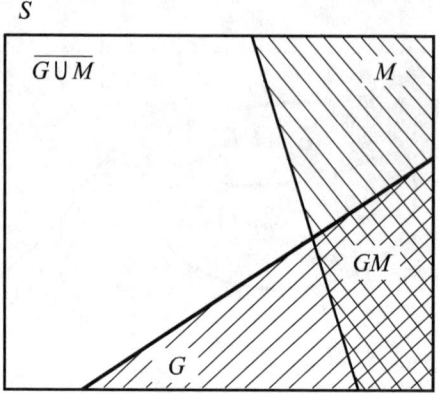

Figure 2.12: Schematic of regions: G, M, GM, and $\overline{G \cup M}$.

Solution First find the probability of failure, and then use this to answer the question. Define the following events: G = gap in weld, M = misaligned component. We are given the probabilities $\Pr(G) = 0.01$ and $\Pr(M) = 0.04$ as shown schematically on the Venn diagram of Figure 2.12. We are also given a relation between misalignment and gaps, that is, $\Pr(G|M) = 1.3\Pr(G|\overline{M})$. Since failure occurs when either or both events G and M occur, we need to evaluate

$$\Pr(G \cup M) = \Pr(G) + \Pr(M) - \Pr(GM),$$

where $\Pr(GM) = \Pr(G|M)\Pr(M)$.

But $\Pr(G|M)$ is unknown. Therefore, another equation is needed since there are two unknowns. The theorem of total probability is that equation,

2.2. PROBABILITY

stating that a gap may occur with or without a mismatch,

$$\Pr(G) = \Pr(G|M)\Pr(M) + \Pr(G|\overline{M})\Pr(\overline{M})$$
$$0.01 = \Pr(G|M)0.04 + \frac{1}{1.3}\Pr(G|M)(1-0.04)$$
$$0.01 = 0.778\Pr(G|M)$$
$$0.0128 = \Pr(G|M).$$

This is the value of the needed probability, and

$$\Pr(G \cup M) = \Pr(G) + \Pr(M) - \Pr(G|M)\Pr(M)$$
$$= 0.01 + 0.04 - 0.0128(0.04)$$
$$= 0.0495.$$

This is the probability that the weld will fail inspection. The probability that it will pass inspection is

$$\Pr(\overline{G \cup M}) = 1 - 0.0495 = 0.9505,$$

or 95.05%. If the designer or the end user believes this to be too small a reliability, then the previous set of calculations provide guidance where manufacturing improvements need to be made.

In applications, such precise numbers may not be obtainable. Then one needs to make additional assumptions. For example, if G and M are assumed to be statistically independent, then $\Pr(GM) = \Pr(G)\Pr(M)$. Performing the same calculations as above leads to the probability $\Pr(G \cup M) = 0.0496$. If on the other hand the unrealistic assumption is made that G and M are mutually exclusive (they cannot both occur), then $\Pr(GM) = 0$, and $\Pr(G \cup M) = 0.0500$. For this example these assumptions do not significantly alter the results. Other problems may not be so forgiving.

※

2.2.6 Bayes' Theorem

It is possible to combine the expressions for conditional probability and the theorem of total probability into one that provides a mechanism for updating probabilities as new information becomes available. Since $\Pr(E_1 \cap E_2) = \Pr(E_2 \cap E_1)$, then $\Pr(E_1 \cap E_2) = \Pr(E_1|E_2)\Pr(E_2) = \Pr(E_2|E_1)\Pr(E_1)$. For an arbitrary event E_a,

$$\Pr(E_a|E_i)\Pr(E_i) = \Pr(E_i|E_a)\Pr(E_a)$$
$$\implies \Pr(E_i|E_a) = \frac{\Pr(E_a|E_i)\Pr(E_i)}{\Pr(E_a)}, \quad (2.11)$$

where $i = 1, ..., n$, and $n =$ *number of possible events*. This equation is known as *Bayes' theorem*. The theorem of total probability can replace $\Pr(E_a)$ in the denominator, resulting in the generalized Bayes' theorem,

$$\Pr(E_i|E_a) = \frac{\Pr(E_a|E_i)\Pr(E_i)}{\sum_{j=1}^n \Pr(E_a|E_j)\Pr(E_j)}.$$

The meaning of this equation is not obvious. We first explain it qualitatively, and then fix these ideas with examples.

Rewrite Equation 2.11 for the case $E_i = E_0$:

$$\Pr(E_0|E_a) = \frac{\Pr(E_a|E_0)\Pr(E_0)}{\Pr(E_a)}. \tag{2.12}$$

The initial probability estimate that E_0 occurs is $\Pr(E_0)$. To improve this estimate, a test is performed on the system resulting in event E_a. Figure 2.11 provides a hint at where we are heading. In the theorem of total probability, $\Pr(E_0)$ is found by evaluating the intersections of E_0 with all the other events in the space. The occurrence, or nonoccurrence, of any of these other events provides additional information about $\Pr(E_0)$. In a similar way here, knowing that E_a occurred with probability $\Pr(E_a)$ provides additional information. The estimate $\Pr(E_0)$ can be *updated* with the additional knowledge that E_a has taken place with $\Pr(E_a)$, that is, $\Pr(E_0|E_a)$. To signify this updating, rewrite Equation 2.12 in the following way:

$$\Pr(E_0|E_a) = \left[\frac{\Pr(E_a|E_0)}{\Pr(E_a)}\right]\Pr(E_0),$$

where the term in the square brackets is sometimes called a *likelihood function* or *transfer function* and represents the effect of the new information on the initial estimate $\Pr(E_0)$. The following example helps to demonstrate this procedure.

Example 2.12 Bayes' Rule in Manufacturing

Cameras are being manufactured by two factories, A and B. It is known that the probability that a defective unit will still pass the final inspection is 0.01. If the unit is defective, the probability that it came from factory A is 0.3. If the unit is not defective, the probability that it came from factory B is 0.8. Suppose that a consumer can find out where the camera came from. This additional information can help a customer to decide which camera he or she will buy. In order to decide which camera to buy, we need to find the probability that the camera is defective if it came from factory A, and the probability that the camera is defective if it came from factory B.

2.2. PROBABILITY

Solution Let F be the event that a camera is defective, and A and B the events that a camera came from factory A and B, respectively. We are given

$$\Pr(F) = 0.01, \ \Pr(A|F) = 0.3, \text{ and } \Pr(B|\bar{F}) = 0.8.$$

Also, if the camera is not manufactured by factory A, then it must be manufactured by factory B,

$$\bar{A} = B \text{ and } A = \bar{B}.$$

We are asked to find $\Pr(F|A)$ and $\Pr(F|B)$.

It is given that 30% of the defective items are from factory A, which leaves 70% for factory B. Based on this figure, the consumer may be tempted to choose a camera from factory A. However, if factory B manufactures substantially more products than factory A, it may safer to choose the product from factory B. Using Bayes' rule, the following conditional probabilities are known,

$$\Pr(F|A) = \frac{\Pr(A|F)\Pr(F)}{\Pr(A|F)\Pr(F) + \Pr(A|\bar{F})\Pr(\bar{F})}$$

$$\Pr(F|B) = \frac{\Pr(B|F)\Pr(F)}{\Pr(B|F)\Pr(F) + \Pr(B|\bar{F})\Pr(\bar{F})}.$$

Since A and B are complementary,

$$\Pr(A|\bar{F}) = 1 - \Pr(B|\bar{F}) = 0.2,$$

the probability $\Pr(\bar{F}) = 0.99$, and the probability,

$$\Pr(B|F) = 1 - \Pr(A|F) = 0.7.$$

The probability that the camera from factory A is defective is

$$\Pr(F|A) = \frac{0.3 \times 0.01}{0.3 \times 0.01 + 0.2 \times 0.99}$$
$$= 0.0149,$$

and the probability that the camera from factory B is defective is

$$\Pr(F|B) = \frac{0.7 \times 0.01}{0.7 \times 0.01 + 0.8 \times 0.99}$$
$$= 0.00876.$$

Therefore, a consumer is better off with a camera from factory B.

Table 2.1: Probabilities

	1	2	3	4	5	6
Die 1	1/12	1/6	1/12	1/3	1/6	1/6
Die 2	1/6	1/6	1/6	1/12	1/12	1/3
Die 3	1/3	1/6	1/6	1/6	1/12	1/12

When the origin of the product is not known, the consumer is faced with a probability of 0.01 that the product is defective. However, with the new piece of information, buying the camera from factory B reduced the probability that the product is defective to 0.00876. In this way, additional information can be used to refine and make more precise the probability estimates, thus updating them.

⊛

Example 2.13 Bayes' Rule and the Loaded Dice

Table 2.1 shows the probabilities of all the realizations for three dice. The dice are all loaded, meaning that they are unbalanced in some way leading to an unequal probability of any number coming up. Note that summing each row still must yield 1, since all the numbers can appear.

(i) Find the probability of rolling a 6 if one die is selected at random. Use the theorem of total probability. Let B equal the event of rolling a 6; A_1, A_2, A_3 equal the events of selecting die 1, 2, or 3, respectively. Since the die is selected at random, $\Pr(A_1) = \Pr(A_2) = \Pr(A_3) = 1/3$. From the table, $\Pr(B|A_1) = 1/6$, $\Pr(B|A_2) = 1/3$, $\Pr(B|A_3) = 1/12$. Substitute these into the total probability theorem to obtain the probability of rolling a 6,

$$\Pr(B) = \Pr(B|A_1)\Pr(A_1) + \Pr(B|A_2)\Pr(A_2) + \Pr(B|A_3)\Pr(A_3)$$
$$= \left(\frac{1}{6} \cdot \frac{1}{3}\right) + \left(\frac{1}{3} \cdot \frac{1}{3}\right) + \left(\frac{1}{12} \cdot \frac{1}{3}\right)$$
$$= \frac{7}{36} = 0.19444.$$

If all the die were fair, we would have found $\Pr(B) = 3 \times (1/18) = 1/6 = 0.16667$.

(ii) Determine the probability that die 2 was chosen if a 6 was rolled with the randomly selected die. This question looks at the problem in a reverse way than did the last one. The required probability is $\Pr(A_2|B)$, given by

$$\Pr(A_2|B) = \frac{\Pr(B|A_2)\Pr(A_2)}{\Pr(B)},$$

2.2. PROBABILITY

using Bayes' rule. Note that $\Pr(B)$ is required in order to evaluate this expression, and, therefore, even if the previous question was not asked, we would have had to apply the total probability theorem. We have then,

$$\Pr(A_2|B) = \frac{\frac{1}{3} \cdot \frac{1}{3}}{\frac{7}{36}} = \frac{4}{7} = 0.57143.$$

This result can also be obtained by considering the reduced space of probabilities for a roll of 6. From the table, the probabilities under the 6 are the reduced space, thus,

$$\Pr(A_2|B) = \frac{\frac{1}{3}}{\frac{1}{6} + \frac{1}{3} + \frac{1}{12}} = 0.57143.$$

Applying Bayes' rule, we can similarly find the following,

$$\Pr(A_1|B) = \frac{\frac{1}{6} \cdot \frac{1}{3}}{\frac{7}{36}} = 0.28571$$

$$\Pr(A_3|B) = \frac{\frac{1}{12} \cdot \frac{1}{3}}{\frac{7}{36}} = 0.14286.$$

These results can be verified by adding $\Pr(A_1|B)+\Pr(A_2|B)+\Pr(A_3|B) = 1$, as they should, signifying that the realizations have been tracked accurately.
⊛

Example 2.14 Liquid Sloshing in a Container[4]

The sloshing of a liquid in a container is a very important problem in engineering design. A number of key engineering systems contain an enclosed fluid, for example, the water tower on top of a building, fuel in aircraft wings and fuselage, fuel in spacecraft flying in low or microgravity, liquid flow in pipes. Each of these is an example of liquid sloshing in a container. A common characteristic is that the amount of liquid varies with time, the sloshing (moving) liquid has inertia, and the amount of liquid affects the vibration characteristics of the enclosing structure.

The purpose of this problem is to consider the simplified problem of a water tower in a seismic zone. Studies have been performed on the strength of the water tower to the shock waves of various earthquakes. There are two primary uncertainties in this problem, the strength and arrival time of the earthquake, and the amount of water in the tower at the seismic event. The arrival time uncertainty and the quantity of water in the tower are really

---
[4]Problem based on A.H-S. Ang and W.H. Tang, *Probability Concepts in Engineering and Planning, Vol.1 Basic Principles*, John Wiley & Sons, 1975.

the same uncertainty, because if we knew the arrival time of the quake, we could extrapolate the amount of water in the tower.

From test data on the tower and historical data on the seismicity of the region, the following information is available. When an earthquake occurs, the probability that the tower fails depends on the magnitude of the earthquake and on the amount of water in the tower at that time. Since the earthquake magnitude and the fullness of the tower can be any of a broad range of numbers, we simplify these by assuming that the tower is either full or half-full with a relative likelihood of 1:3. The lowest magnitude earthquake that is of significance is called a weak earthquake, and the one of highest magnitude is called a strong earthquake. The relative likelihood is, respectively, 8:2. It is determined that if a strong earthquake hits the tower, it collapses regardless of the quantity of water in the tank. We also know that if a weak earthquake hits the tower and it is at most half-full, then it will definitely survive the event.[5] However, if during a weak seismic event the tower is full, then it has only a 1 in 2 chance of survival. (This 50% chance of survival implies that within the group of weak earthquakes there are characteristic differences that can be substantial and can sometimes lead to tower failure.)

The two questions of interest about this system and its response to its environment are: *(i)* what is the probability of tower collapse, and *(ii)* if the tower collapsed during a seismic event, what is the probability that the tank was full at the time of the quake?

Solution The first step is to define all possible events along with their respective probabilities. Then the theorem of total probability is used along with Bayes' rule. Define the following events:

$$F = \text{full tower},\ H = \text{half-full tower},\ C = \text{tower collapse},$$
$$S = \text{strong seismic event},\ W = \text{weak seismic event}.$$

Using the relative likelihood information given in the problem statement, the following probabilities can be deduced:

$$\Pr(F) = 0.25, \quad \Pr(H) = 0.75$$
$$\Pr(S) = 0.20, \quad \Pr(W) = 0.80$$
$$\Pr(C|SF) = \Pr(C|SH) = 1$$
$$\Pr(C|WF) = \Pr(\overline{C}|WF) = 0.5$$
$$\Pr(C|WH) = 0.$$

---

[5] This bit of information suggests to a designer that the tower be sensored so that in the event of an earthquake, enough water be released into the sewers so that the tower is less than half full.

2.2. PROBABILITY

To calculate the probability of collapse, use the total probability theorem, summing all the possible ways collapse can occur,

$$\begin{aligned}\Pr(C) &= \Pr(CSH) + \Pr(CSF) + \Pr(CWH) + \Pr(CWF) \\ &= \Pr(C|SH)\Pr(SH) + \Pr(C|SF)\Pr(SF) \\ &\quad + \Pr(C|WH)\Pr(WH) + \Pr(C|WF)\Pr(WF).\end{aligned}$$

At this point a physical judgement is required about the joint event (seismic event, quantity of water in tank). It is reasonable to assume that these two events are statistically independent. Thus,

$$\begin{aligned}\Pr(C) &= 1 \times 0.20 \times 0.75 + 1 \times 0.20 \times 0.25 + 0 + 0.5 \times 0.80 \times 0.25 \\ &= 0.3.\end{aligned}$$

The probability that the tower is full, assuming a failure event, is given by

$$\begin{aligned}\Pr(F|C) &= \Pr(FS|C) + \Pr(FW|C) \\ &= \frac{\Pr(C|FS)\Pr(FS)}{\Pr(C)} + \frac{\Pr(C|FW)\Pr(FW)}{\Pr(C)} \\ &= \frac{1 \times 0.25 \times 0.20}{0.3} + \frac{0.5 \times 0.25 \times 0.80}{0.3} \\ &= 0.5.\end{aligned}$$

While this is the answer to the problem, a designer would look at the results and think about how the probability of collapse can be reduced. The general equation for $\Pr(C)$ shows how each component probability on the right hand side adds up to the total probability. It suggests weaknesses in the structure. Therefore, the above procedure can be used iteratively as a design tool until $\Pr(C)$ is sufficiently small.

※

Bayes Thomas Bayes was ordained a Nonconformist minister like his father, who was one of the six Nonconformist ministers to be ordained in England. Thomas was educated privately, something that appears to have been necessary for the son of a Nonconformist minister at that time. Nothing is known of his tutors, but Barnard points out the intriguing possibility that he could have been tutored by de Moivre, who was certainly giving private tuition in London at this time.

At first Bayes assisted his father in Holborn. In the late 1720s he became minister of the Presbyterian Chapel in Tunbridge Wells, 35 miles southeast of London. On August 24, 1746 William Wiston describes having breakfast with Bayes, who he says is ... *a dissenting Minister at Tunbridge Wells, and a Successor, though not immediate, to Mr. Humphrey Ditton, and like him a very good mathematician.*

Figure 2.13: Thomas Bayes (1702-1761).

Bayes apparently tried to retire from the ministry in 1749 but remained minister at Tunbridge Wells until retirement in 1752, but continued to live in Tunbridge Wells.

Bayes set out his theory of probability in *Essay towards solving a problem in the doctrine of chances* published in the *Philosophical Transactions of the Royal Society of London* in 1764. The paper was sent to the Royal Society by Richard Price, a friend of Bayes, who wrote: *I now send you an essay which I have found among the papers of our deceased friend Mr. Bayes, and which, in my opinion, has great merit. In an introduction which he has writ to this Essay, he says, that his design at first in thinking on the subject of it was, to find out a method by which we might judge concerning the probability that an event has to happen, in given circumstances, upon supposition that we know nothing concerning it but that, under the same circumstances, it has happened a certain number of times, and failed a certain other number of times.*

Bayes' conclusions were accepted by Laplace in a 1781 memoir, rediscovered by Condorcet (as Laplace mentions), and remained unchallenged until Boole questioned them in the Laws of Thought. Since then Bayes' techniques have been subject to controversy.

Bayes also wrote an article *An Introduction to the Doctrine of Fluxions, and a Defence of the Mathematicians Against the Objections of the Author of The Analyst* (1736) attacking Berkeley for his attack on the logical foundations of the calculus. Bayes writes that Berkeley ...*represents the disputes and controversies among mathematicians as disparaging the*

2.3. CONCLUDING SUMMARY 45

evidence of their methods and ... he represents Logics and Metaphysics as proper to open their eyes, and extricate them from their difficulties. If the disputes of the professors of any science disparage the science itself, Logics and Metaphysics are much more disparaged than Mathematics, why, therefore, if I am half blind, must I take for my guide one that cannot see at all?

Bayes was elected a Fellow of the Royal Society in 1742 despite the fact that at that time he had no published works on mathematics, indeed none were published in his lifetime under his own name. The article on fluxions referred to above was published anonymously. Another mathematical publication on asymptotic series appeared after his death.

2.3 Concluding Summary

This chapter has introduced the underlying basis for probability theory, sets and operations with sets. These and the axioms of probability permit the development of the subsequent rules for working with uncertain parameters and variables. The interpretation of probability as a frequency of occurrence is provided as a useful practical definition. In addition, some basic probabilistic concepts have been introduced, in particular, conditional probability and independence, very important concepts that are used in the following chapters.

2.4 Problems

Section 2.1: Sets

1. Give 3 examples of impossible events from everyday life.

2. Give 3 examples of impossible events from fluids engineering.

3. Give 3 examples of impossible events from materials engineering.

4. Give 3 examples of impossible events from strength of materials.

5. Give 3 examples of impossible events from mechanical vibration.

6. Give 3 examples of impossible events from thermal engineering.

7. Give 3 examples of certain events from everyday life.

8. Give 3 examples of certain events from fluids engineering.

9. Give 3 examples of certain events from materials engineering.

10. Give 3 examples of certain events from strength of materials.

11. Give 3 examples of certain events from mechanical vibration.

12. Give 3 examples of certain events from thermal engineering.

13. Give an example of complementary events from everyday life.

14. Give an example of complementary events from fluids engineering.

15. Give an example of complementary events from materials engineering.

16. Give an example of complementary events from strength of materials.

17. Give an example of complementary events from mechanical vibration.

18. Give an example of complementary events from thermal engineering.

19. Given the three events:

$$X = \{\text{odd numbers}\}$$
$$Y = \{\text{even numbers}\}$$
$$Z = \{\text{negative numbers}\},$$

obtain the following:

(i) $X \cup Y$ (ii) $X \cap Y$
(iii) $\overline{X}$ (iv) $\overline{Y}$
(v) $\overline{Z}$ (vi) $Y \cap Z$.

20. Extend Example 2.5 to the case where there are three shafts connecting two rotors (instead of two shafts connected to one rotor). The shafts are numbered from left to right as 1, 2, 3. The failure of the drive train is defined as the failure of either of the three shafts, with events defined by E_1, E_2, and E_3, respectively. Find the following events:

(i) failure of the drive train,

(ii) no failure of the drive trains, and

(iii) show an illustration of de Morgan's rule.

Section 2.2: Probability

21. In your own words, explain the essential ideas of the Theorem of Total Probability, and briefly discuss its importance.

2.4. PROBLEMS

22. Consider Figure 2.7 where $d_{av} = 50$ mm, and suppose 50 shafts are manufactured. From the measurements we observe that

   25 have the diameter d_{av}
   10 have the diameter $1.01\, d_{av}$
   6 have the diameter $1.02\, d_{av}$
   5 have the diameter $0.99\, d_{av}$
   4 have the diameter $0.98\, d_{av}$.

   Sketch the frequency diagram showing appropriate numbers along the axes. Using the frequency interpretation for probability, calculate the probability of occurrence for each shaft size and verify that the sum of these probabilities equals 1.

23. Suppose $\Pr(E_1) = 0.20$, and $\Pr(E_2) = 0.30$.

   (i) If E_1 and E_2 relate to a particular process, are any events not accounted for here? Why?

   (ii) If $\Pr(E_1 \cup E_2) = 0.90$ are these processes mutually exclusive? Why?

   (iii) If $\Pr(E_1 \cup E_2) = 0.50$, then what is the value of $\Pr(E_1 E_2)$?

24. Suppose $\Pr(A) = 0.5$, $\Pr(A|B) = 0.3$, and $\Pr(B|A) = 0.1$, calculate $\Pr(B)$.

25. Suppose $\Pr(A) = 0.5$, $\Pr(A|B) = 0.5$, and $\Pr(B|A) = 0.1$, calculate $\Pr(B)$. What can be concluded about the statistical relationship, if any, between A and B.

26. Continuing Example 2.13, let G equal the event of rolling a 3; A_1, A_2, A_3 equal the events of selecting die 1, 2, or 3, respectively.

   (i) Find the probability of rolling a 3 if one die is selected at random. Use the theorem of total probability.

   (ii) Determine the probability that die 2 was chosen if a 3 was rolled with the randomly selected die.

27. Consider the drive train Example 2.5 from a different perspective. The drive train consists of a rotor R and turbine blades B. The components are manufactured in high precision. How well the system operates depends on the precision of the manufactured components. Testing of the individual components by the manufacturer yields the following information:

   0.1% of R have imperfections
   0.01% of B fail.

Also, it is determined that if R has imperfections, then the blades B are 50% more likely to fail due to the additional vibration forces that result. Determine the probability that the system will pass inspection.

Chapter 3

Random Variable Models

In our previous discussions, a language based on set theory was introduced to work with probability and uncertainties. Several key equations were used to relate probabilities with each other. But these probabilities are about "extreme" events. For example, a component fails or not, with some probability. What we would like to be able to do is to provide a complete probabilistic description of a parameter, not just the probability of a particular realization. This complete probabilistic description is in the form of a function over the whole range of realizations.

Begin by exploring the properties of *random variables* and the functions that define them. Probability affords a framework for defining and utilizing such variables in the models developed for engineering analysis and design. Mathematical models of physical phenomena are essentially relationships between variables. Where some of these variables have associated uncertainties, there are a multiplicity of possible values for each random variable. An example is the set of possible values of Young's modulus determined from a series of experiments on "identical" test specimens. This multiplicity is represented by the *probability functions* introduced next. A random variable may be *discrete*, *continuous*, or *mixed*. Once a parameter is taken to be random, its probability functions provide a complete description of its variability.

The importance of these probability functions is that they can be used in mathematical operations involving randomness.

3.1 Probability Distribution Function

The likelihood of a random variable taking on a particular range of values is defined by its *probability distribution function*,[1] defined as

$$F_X(x) = \Pr(X \le x), \tag{3.1}$$

where $\Pr(X \le x)$ is the probability that random variable X is less than or equal to the number x. This probability is, of course, a function of the particular value x. Based on the axioms of probability,[2] it can be shown that $F_X(x)$ is an increasing function of x, and is bound by 0 and 1. The *impossible* event has a zero probability, and the *certain* event has a probability of one. In particular,

$$\lim_{x \to -\infty} F_X(x) = 0$$

since $\Pr(X < -\infty) = 0$; all *realizations* of the random variable must be greater than negative infinity. A *realization* is one of the many possible values of a random variable. Similarly,

$$\lim_{x \to +\infty} F_X(x) = 1$$

since $\Pr(X < +\infty) = 1$; all realizations of the random variable must be less than positive infinity. Thus, bounds on $F_X(x)$ are $0 \le F_X(x) \le 1$, and for $x_1 \le x_2$,

$$F_X(x_1) \le F_X(x_2),$$

since $\Pr(X \le x_1) \le \Pr(X \le x_2)$. See Figure 3.1. Note that the probability distribution function is *nondecreasing*. Where the context is clear, the subscript X is omitted.

While the above definitions apply to all types of random variables, if the variable is discrete, Equation 3.1 becomes

$$\begin{aligned} F_X(x) &= \sum_{\text{all } x_i \le x} \Pr(X = x_i) \\ &= \sum_{\text{all } x_i \le x} p_X(x_i), \end{aligned} \tag{3.2}$$

---

[1] $F(x)$ is sometimes called the *cumulative distribution function*, since probability is accumulated as x becomes larger.

[2] An excellent book on the basics of probabilistic modeling is by A. Papoulis, *Probability, Random Variables, and Stochastic Processes*, McGraw–Hill. There are several editions. We find the first edition most readable. We would also encourage the reader to look up some of the other fine texts by Papoulis on probability and stochastic processes; they are among the best. A different approach to explaining probability is offered by C. Ash in *The Probability Tutoring Book: An Intuitive Course for Engineers and Scientists (and everyone else!)*, IEEE Press, 1993. It offers an introduction through problem solving.

3.1. PROBABILITY DISTRIBUTION FUNCTION

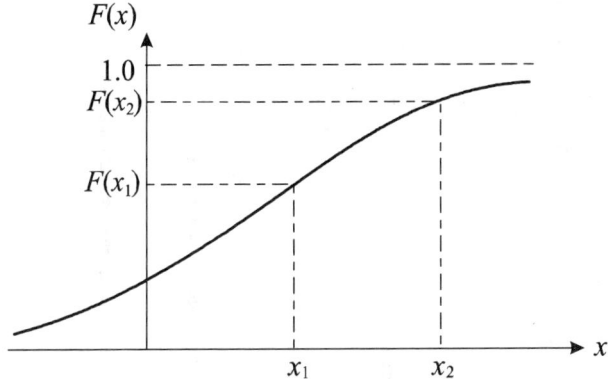

Figure 3.1: A continuous probability distribution function.

where $p_X(x_i)$ are the individual probabilities of each realization. A hint of the cumulative distribution function can be found in the histogram of Figure 2.9. Each frequency is a probability, $p_X(x_i)$, and the sum from left to right is the cumulative distribution function. The histogram is a practical way to build a distribution function.

Example 3.1 From Histogram to Cumulative Distribution Function

The cumulative distribution function corresponding to the histogram of Figure 2.9 can be drawn using Equation 3.2 . For example, $F(0.98 d_{AV}) = n_{0.98}$, $F(0.99 d_{AV}) = n_{0.98} + n_{0.99}$, ..., and $F(1.02 d_{AV}) = n_{0.98} + ... + n_{1.02} = 1$. See Figure 3.2.

⊛

Example 3.2 Discrete and Continuous Random Variables

A packaging machine fills containers with bolts. If the machine does not count the number of bolts it places in the containers, then that number is approximate. The number of bolts in each package is a discrete random variable because only integer values are allowed. On the other hand, the diameter of each bolt is only approximately the same. The diameter is a continuous random variable because the diameter can take on any value over a range.

⊛

CHAPTER 3. RANDOM VARIABLE MODELS

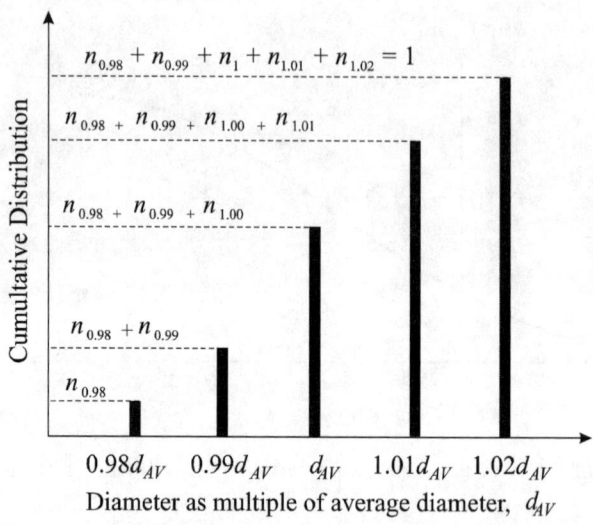

Figure 3.2: The discrete cumulative distribution function.

3.2 Probability Density Function

The *probability density function* presents the same information contained in the probability distribution function, but in a more useful form. Assuming continuity of the distribution,[3] the density function $f_X(x)$ is defined as

$$f_X(x) = \frac{dF_X(x)}{dx}.$$

Alternatively, by integrating both sides and rearranging, the distribution can be related to the density,

$$F_X(x) = \Pr(X \leq x) = \int_{-\infty}^{x} f_X(\xi) d\xi. \qquad (3.3)$$

The probability density function $f_X(x)$ is analogous to the individual probabilities $p_X(x_i)$ of a discrete random variable.

Equation 3.3 provides a useful interpretation of the density function: *the probability that a continuous random variable X has a value less than or equal to the number x is equal to the area under the density function for*

---

[3] The distribution function does not have to be a continuous function. In many instances it may have discrete jumps where a finite probability exists for a certain realization. It is just easier to work initially with a continuous function.

3.2. PROBABILITY DENSITY FUNCTION

values less than x. Similarly, for arbitrary x_1 and x_2, the probability that $x_1 < X \leq x_2$ is given by the integral,

$$\Pr(x_1 < X \leq x_2) = \int_{x_1}^{x_2} f(x)dx, \qquad (3.4)$$

as shown in Figure 3.3, and for small $dx = x_2 - x_1$, the probability can be

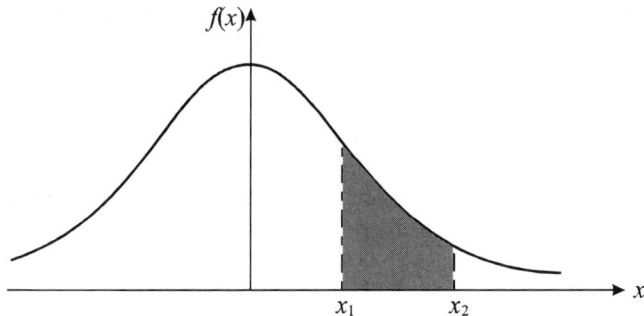

Figure 3.3: The probability that $x_1 < X \leq x_2$ is given by the shaded area under the probability density function.

estimated as $\Pr(x_1 < X \leq x_2) \simeq f_X(x)\,dx$. The probability of any specific realization equals zero, that is,

$$\Pr(X = \lambda) = \int_\lambda^\lambda f(x)dx = 0.$$

Note the important *normalization* property of the probability density,

$$\int_{-\infty}^{\infty} f(x)dx = 1, \qquad (3.5)$$

signifying that the density function is representative of all possible outcomes or realizations of the random variable. The area under the density function is *normalized* to 1. Since probability is numerically in the range 0 to 1, the density function must be a positive semidefinite[4] function: $f(x) \geq 0$. It is important to recall that the random variable has a static property. That is, the shape of the density function does not change with time. Where the density function is time-dependent, the variable is called a *random* or *stochastic process*. This more advanced topic will be discussed in the second part of this book, beginning with Chapter 5.

---

[4] A *positive definite* function is one that has all values greater than zero. If it is positive *semidefinite*, then it may also be equal to zero.

For two values a and b, where $b > a$,

$$\Pr(a < X \le b) = \int_{-\infty}^{b} f(x)dx - \int_{-\infty}^{a} f(x)dx$$
$$= F(b) - F(a).$$

In some special cases, the random variable X may assume discrete values as well as all values in some interval. X is then both discrete and continuous. The probability $\Pr(a < X \le b)$ can then be written as

$$\Pr(a < X \le b) = \int_{a}^{b} f(x)dx + \sum_{a < x_i \le b} p(x_i).$$

If the probability that $X = x_0$ is p, then the probability distribution function $F(x)$ has a jump at x_0 as shown in Figure 3.4, and

$$\begin{aligned}\Pr(X = x_0) &= p \\ \Pr(X \le x_0) &= p_0 + p \\ \Pr(X < x_0) &= p_0.\end{aligned}$$

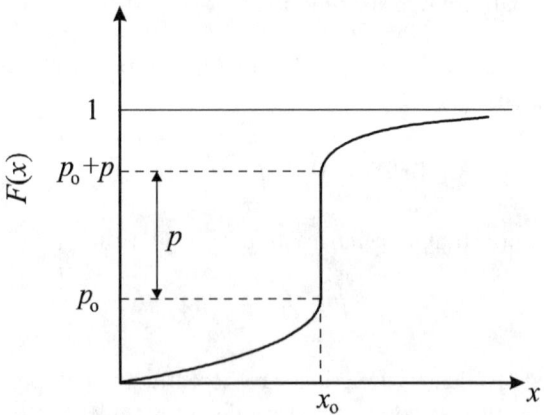

Figure 3.4: Random variable with mixed distribution.

Example 3.3 An Exponential Density Function

Suppose the probability density function of random variable X is $f(x) = c\exp(-|x|)$. Evaluate the constant c, find $\Pr(-2 < X \le 2)$, and derive the probability distribution function $F(x)$.

3.2. PROBABILITY DENSITY FUNCTION

Solution   The constant c must be first evaluated using the normalization property before the density function can be used to derive probabilities of events. Using Equation 3.5,

$$c \int_{-\infty}^{\infty} \exp(-|x|)dx = 1$$

$$2c \int_{0}^{\infty} \exp(-x)dx = 1$$

$$\Longrightarrow c = \frac{1}{2}.$$

Therefore, $f(x) = \frac{1}{2}\exp(-|x|)$, which is plotted in Figure 3.5. Now,

$$\Pr(-2 < x \leq 2) = \int_{-2}^{2} \frac{1}{2}\exp(-|x|)dx = 1 - e^{-2} = 0.86466.$$

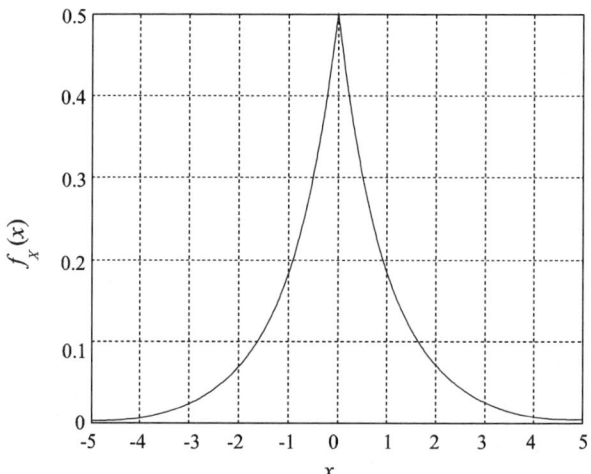

Figure 3.5: $f(x) = \frac{1}{2}\exp(-|x|)$.

The probability distribution function is given by

$$F(x) = \int_{-\infty}^{x} \frac{1}{2}\exp(-|x|)dx$$

$$= \begin{cases} \frac{1}{2}e^{x} & \text{for } x \leq 0 \\ 1 - \frac{1}{2}e^{-x} & \text{for } x \geq 0. \end{cases}$$

Note that $\Pr(-2 < x \leq 2) = F(2) - F(-2)$.

Such density functions are useful for numerous engineering applications. We will see how an exponential density is used for reliability analysis in a subsequent section.

Example 3.4 Another Density Function

Suppose the density function is given by $f(x) = 100/x^2$, $x \geq 100$. Find the distribution function, and plot both functions.

Solution The distribution function is found by integrating the density function as follows,

$$F(x) = \int_{-\infty}^{x} f(x)dx = \int_{100}^{x} \frac{100}{x^2} dx$$
$$= \frac{x - 100}{x}, \qquad x \geq 100.$$

The plots of these two functions are given in Figures 3.6 and 3.7.

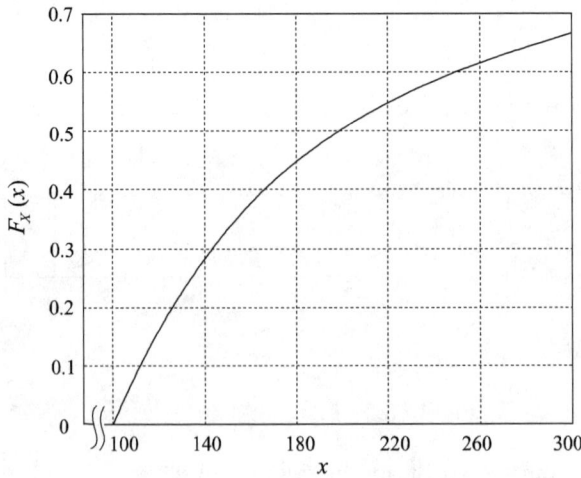

Figure 3.6: $F(x) = (x - 100)/x$, $x \geq 100$.

Before examining some important densities, we need to define an averaging procedure known as the mathematical expectation for probabilistic variables.

3.3. MATHEMATICAL EXPECTATION

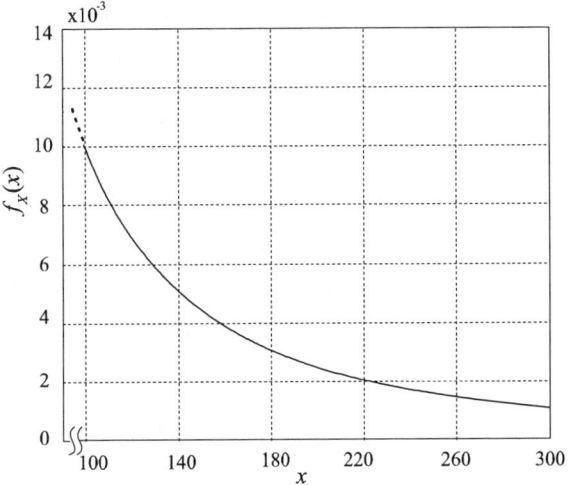

Figure 3.7: $f(x) = 100/x^2$, $x \geq 100$.

3.3 Mathematical Expectation

The single most important descriptor of a random variable is its *mean* or *expected value*. This defines the most likely value of the variable. However, numerous random variables may have the same mean, but the *spread of possible values* or their *variances* can be considerably different. The variance is the next most important statistical descriptor. The mean and variance of a random variable are statistical averages, and can be evaluated using the concept of the *mathematical expectation of a function of random variable* X, defined as

$$E\{g(X)\} = \int_{-\infty}^{\infty} g(x)f(x)dx. \quad (3.6)$$

The *expected* or *mean value* is defined using Equation 3.6,

$$\mu = E\{X\} = \int_{-\infty}^{\infty} xf(x)dx. \quad (3.7)$$

The expected value is a constant, first-order statistic, and is also known as the *first moment* because the variable x appears to the first power. The term moment is used by analogy to the center of mass in the mechanics of solids. The result $E\{X\}$ is the center of "probability mass." The density function acts as a probabilistic "weighting" function. It is a larger factor in the integral for more probable values of the random variable. Also, the expected value of a constant is itself, that is $E\{c\} = c$.

Example 3.5 Expected Value for a Continuous Variable

The random pressure fluctuations P in a fuel tank are governed by the density,

$$f_P(p) = \frac{1}{10}, \quad 100 \text{ psi} \leq p \leq 110 \text{ psi}.$$

The expected value is calculated as follows,

$$\begin{aligned} E\{P\} &= \int_{100}^{110} p f_P(p) dp \\ &= \int_{100}^{110} p \frac{1}{10} dp = 105. \end{aligned}$$

The constant density is known as a uniform density, and it will be introduced formally in a subsequent section.

Example 3.6 Expected Value for a Discrete Variable

Equation 3.7 is written for a continuous random variable. There are instances when the variable of interest is discrete. How is a discrete variable analyzed?

Solution For a discrete random variable, the integral in the mathematical expectation becomes a summation. Suppose an experiment for the yield strength of a material is run ten times with the following strength data:

10.0, 9.8, 11.1, 9.1, 9.9, 9.7, 10.3, 10.1, 9.9, 10.0.

To find the expected value use the discrete counterpart of Equation 3.7,

$$\mu = E\{X\} = \sum_{i=1}^{10} x_i p_X(x_i), \qquad (3.8)$$

where x_i are the test results and $p_X(x_i)$ are the probability weights, in this case the fraction of times that a particular value occurred. For the data listed, test results that occur once have a probability of 1/10. Results such as 9.9 and 10.0 that occur twice have a probability 2/10. Note that $\sum_{i=1}^{10} p_X(x_i) = 1$, signifying that all possible outcomes have been included.

3.3. MATHEMATICAL EXPECTATION

Then,

$$\mu = \left(10.0 \times \frac{2}{10}\right) + \left(9.8 \times \frac{1}{10}\right) + \left(11.1 \times \frac{1}{10}\right) + \left(9.1 \times \frac{1}{10}\right)$$
$$+ \left(9.9 \times \frac{2}{10}\right) + \left(9.7 \times \frac{1}{10}\right) + \left(10.3 \times \frac{1}{10}\right) + \left(10.1 \times \frac{1}{10}\right)$$
$$= \frac{99.9}{10} = 9.99.$$

On the other hand, if the yield strength is assumed to be continuous and uniform[5] between the values 9.1 and 11.1 with continuous density $f(x) = 1/(11.1 - 9.1) = 1/2$, then

$$\mu = \int_{9.1}^{11.1} x \cdot \frac{1}{2} dx = 10.1.$$

This result is slightly different than for the discrete case due to the uneven discrete distribution.

✽

Sometimes, for n discrete variables, $p_X(x_i)$ are written as $\Pr(x_i)$, and Equation 3.8 becomes

$$\mu = E\{X\} = \sum_{i=1}^{n} x_i \Pr(x_i).$$

As can be observed in the above example for the discrete variable, the expected value is not necessarily a possible realization.

Related definitions are the *mode*, which is the most probable value or the most frequently appearing value, and the *median* value, for which $F(x_m) = 0.5$, that is, half of the probability space is below x_m and half above. The median can be identified by either of the following equations,

$$F_X(x_m) = 0.5$$
$$\int_{-\infty}^{x_m} f_X(x) dx = \int_{x_m}^{\infty} f_X(x) dx = 0.5.$$

Example 3.7 Mean, Mode, and Median

Dried banana chips are being packaged. Ten packages are randomly selected, and their weights are recorded keeping one significant digit after the decimal. The weights are, in ounces,

$$14.1, 14.7, 14.7, 15.0, 15.1, 15.1, \ 15.2, \ 15.2, \ 15.2, \ 15.7.$$

---

[5] A detailed description of the uniform probability density is given in Section 3.4.1.

The mean weight is

$$\mu = 14.1\frac{1}{10} + 14.7\frac{2}{10} + 15.0\frac{1}{10} + 15.1\frac{2}{10} + 15.2\frac{3}{10} + 15.7\frac{1}{10}$$
$$= 15.0 \text{ oz},$$

the mode is 15.2 oz, and the median is 15.1 oz, by observation.

⊛

Now that the question of the most likely value of a random variable has been addressed using the expected value, we derive in the following section an equation that provides a measure of the scatter about the mean value.

3.3.1 Variance

The *variance* is a *second-order moment*. It is defined as

$$Var\{X\} = E\{(X - E\{X\})^2\} = \int_{-\infty}^{\infty} (x - \mu)^2 f(x) dx.$$

Expand the squared term in parentheses, integrate term by term, and find the variance to equal

$$Var\{X\} = E\{X^2\} - (E\{X\})^2. \tag{3.9}$$

The variance is the difference between the *mean-square value* and the *mean value squared*. Here, the mechanical analogy is with the *mass moment of inertia*.

So that the measure of dispersion has the same dimensions as the random variable, the *standard deviation* is defined as the positive square root of the variance,

$$\sigma = +\sqrt{Var\{X\}}. \tag{3.10}$$

An important dimensionless parameter is the *coefficient of variation*,

$$\delta = \frac{\sigma}{\mu}.$$

It is used as a nondimensional measure of the degree of uncertainty in a parameter, that is, the scatter of its data. Note that by this definition, two different random variables may have the same scatter σ, but their δ value can be very different by virtue of their mean values. Such cases are shown in the following table.

	σ	μ	δ
Random Variable A	1	4	0.25
Random Variable B	1	100	0.01
Random Variable C	2	100	0.02

3.3. MATHEMATICAL EXPECTATION

Random variables A and B have the same standard deviation. Random variable B is considered to be less random than A because its δ value is much smaller. Random variables B and C have the same mean, but different standard deviations and therefore different δ. Again, random variable B is considered to be less random than C because its δ value is smaller.

In engineering practice, one expects a δ value of between 0.05-0.15, or 5-15%. Values larger than this imply that a serious lack of knowledge exists about the system itself and its underlying physics. If this is the case, then a significant program of experiments is necessary before one can consider the analysis and design of such a system.

Example 3.8 Detection of fracture cracks in a material

As structures such as bridges and aircraft age, in many instances beyond their design life, it becomes increasingly important to be able to detect and repair damage before it debilitates the use of the structure. Repairs are always easier and less costly when performed earlier. In this example, we consider the effectiveness of a *nondestructive testing* (NDT) device. Cracks in a material have a variety of shapes and sizes, leading to the statistic that the NDT device has a probability of 0.9 of detecting a crack of a certain threshold size. The following questions are of interest.

(i) What is the probability of detecting two cracks in a material? Define the event D as the detection of a single crack. Then to detect two cracks we have $\Pr(\text{detect two cracks}) = \Pr(D_1 D_2)$. Since the joint event was not given, and we have no further data, we make the assumption that the NDT device detects each crack independently. Assuming statistical independence, $\Pr(\text{detect two cracks}) = \Pr(D_1)(D_2) = \Pr(D)\Pr(D) = 0.9 \times 0.9 = 0.81$, and the probability of not detecting two cracks is $1 - 0.81 = 0.19$.

(ii) Generally, the number of cracks is not known *a priori*. The number of cracks is a random number governed by a discrete probability density function. *If the number of cracks N can be either $0, 1, 2,$ or 3, and the probability mass function is as given in* Figure 3.8, *find the probability that the NDT device will fail to detect any crack.*

Let A be the detection of any crack. Then, use the theorem of total probability to evaluate $\Pr(\overline{A})$,

$$\Pr(\overline{A}) = \Pr(\overline{A}|N=0)\Pr(N=0) + \Pr(\overline{A}|N=1)\Pr(N=1)$$
$$+ \Pr(\overline{A}|N=2)\Pr(N=2) + \Pr(\overline{A}|N=3)\Pr(N=3)$$
$$= (1.0 \times 0.15) + (1 - 0.9) \times 0.25$$
$$+ \left(1 - 0.9^2\right) \times 0.50 + \left(1 - 0.9^3\right) \times 0.10$$
$$= 0.2971.$$

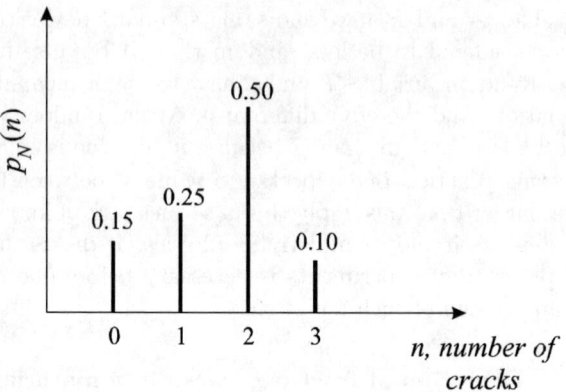

Figure 3.8: Probability mass function (density).

It is assumed that if $N = 0$, $\Pr(\overline{A}) = 1$; that is, $\Pr(\overline{A}|N = 0) = 1$ and there are no false-positive readings.

(iii) *Determine the mean, variance, and coefficient of variation of the random variable N.* From the respective definitions, the mean value is determined as follows,

$$\mu_N = \sum_{i=1}^{4} n_i p_i$$
$$= 0 \times 0.15 + 1 \times 0.25 + 2 \times 0.50 + 3 \times 0.1$$
$$= 1.55,$$

and the variance is given by

$$Var\{N\} = \sum_{i=1}^{4} (n_i - \mu_N)^2 p_i$$
$$= (0 - 1.55)^2 \times 0.15 + (1 - 1.55)^2 \times 0.25$$
$$+ (2 - 1.55)^2 \times 0.50 + (3 - 1.55)^2 \times 0.10$$
$$= 0.7475.$$

The standard deviation equals the positive square root of the variance,

$$\sigma_N = \sqrt{0.7475} = 0.86458$$

and the coefficient of variation equals the ratio of standard deviation to mean value,

$$\delta_N = \frac{0.86458}{1.55} = 0.55779.$$

3.3. MATHEMATICAL EXPECTATION

(iv) There is concern that if the device detects no cracks that, perhaps, it is not operating. *If the device fails to detect any crack, what is the probability that it is actually operating properly and there are no cracks?* Using Bayes' theorem,

$$\Pr(N = 0|\overline{A}) = \frac{\Pr(\overline{A}|N = 0)\Pr(N = 0)}{\Pr(\overline{A})}$$

$$= \frac{1.0 \times 0.15}{0.2971} = 0.50488,$$

or only a 50% probability that the device is accurate. Clearly these numbers indicate that the device is of poor quality and not a reliable instrument.
⊛

Example 3.9 Assembly Line in Two Segments

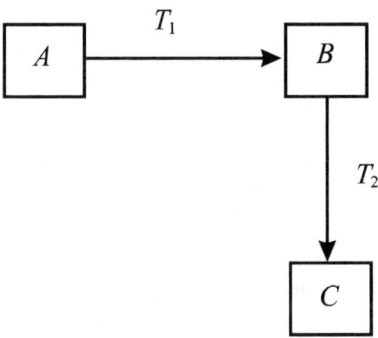

Figure 3.9: Two-stage assembly line.

Consider the two-stage assembly line shown in Figure 3.9. The assembly line operates in two distinct stages. In the first stage, raw materials are introduced at location A. During time span T_1 the materials undergo preliminary assembly, leaving completion to the second stage of duration T_2. Both time spans are approximate. At location B the unfinished product undergoes a preliminary inspection.

It is found that by undertaking a preliminary inspection of the product before it is completed, any shortcomings can be corrected at a much lower expense than if the inspection is left to the end. At C another inspection is performed. The inspection at B is well defined and takes exactly one hour. The probability mass functions for the two random variables T_1 and T_2 are given in Figure 3.10.

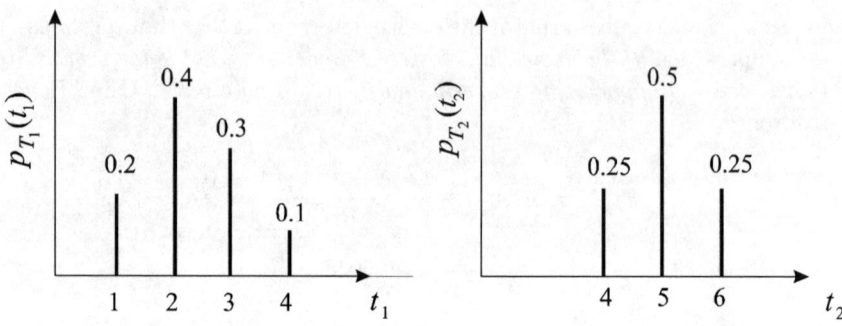

Figure 3.10: Probability mass functions for T_1 and T_2.

Table 3.1: Total Time and the Probability Mass

T_1	T_2	T	Probability Mass	T_1	T_2	T	Probability Mass
1	4	6	$0.2 \times 0.25 = 0.05$	3	4	8	$0.3 \times 0.25 = 0.075$
1	5	7	$0.2 \times 0.50 = 0.10$	3	5	9	$0.3 \times 0.50 = 0.150$
1	6	8	$0.2 \times 0.25 = 0.05$	3	6	10	$0.3 \times 0.25 = 0.075$
2	4	7	$0.4 \times 0.25 = 0.10$	4	4	9	$0.1 \times 0.25 = 0.025$
2	5	8	$0.4 \times 0.50 = 0.20$	4	5	10	$0.1 \times 0.50 = 0.050$
2	6	9	$0.4 \times 0.25 = 0.10$	4	6	11	$0.1 \times 0.25 = 0.025$

The probability mass functions indicate that the range of times for T_1 is between 1 and 4 hours, and for the second stage the range of times is between 4 and 6 hours. There are several questions that need to be addressed about the component processes as well as the complete process between positions A and C.

(i) Calculate the mean value, variance, standard deviation, and coefficient of variation for time spent on the first stage, T_1.

$$\mu_{T_1} = \sum_{i=1}^{4} t_i p_i = 1 \times 0.2 + 2 \times 0.4 + 3 \times 0.3 + 4 \times 0.1 = 2.3 \text{ hr}$$

$$\begin{aligned}
Var\{T_1\} &= \sum_{i=1}^{4} \left(t_i - \mu_{T_1}\right)^2 p_i \\
&= (1-2.3)^2 \, 0.2 + (2-2.3)^2 \, 0.4 + (3-2.3)^2 \, 0.3 + (4-2.3)^2 \, 0.1 \\
&= 0.81 \text{ hr}^2
\end{aligned}$$

3.3. MATHEMATICAL EXPECTATION

$$\sigma_{T_1} = \sqrt{0.81} = 0.9 \text{ hr}$$

$$\delta_{T_1} = \frac{\sigma_{T_1}}{\mu_{T_1}} = \frac{0.9}{2.3} = 0.3913.$$

(ii) Next, derive the probability mass function for the total time the product takes to reach C. Set up Table 3.1 of all the possible combinations of times for the two manufacturing segments. Let T be the total time event, $T = T_1 + 1 + T_2$, where the value 1 is the deterministic time spent for inspection at B. The two random variables are assumed to be independent.

Note that the sum of all these probability masses is 1, as required. Add the probabilities of all identical times T as follows,

$$\Pr(T = 6) = 0.05$$
$$\Pr(T = 7) = 2 \times 0.10 = 0.2$$
$$\Pr(T = 8) = 0.05 + 0.20 + 0.075 = 0.325$$
$$\Pr(T = 9) = 0.10 + 0.150 + 0.025 = 0.275$$
$$\Pr(T = 10) = 0.075 + 0.05 = 0.125$$
$$\Pr(T = 11) = 0.025,$$

and plot this as a probability mass function in Figure 3.11.

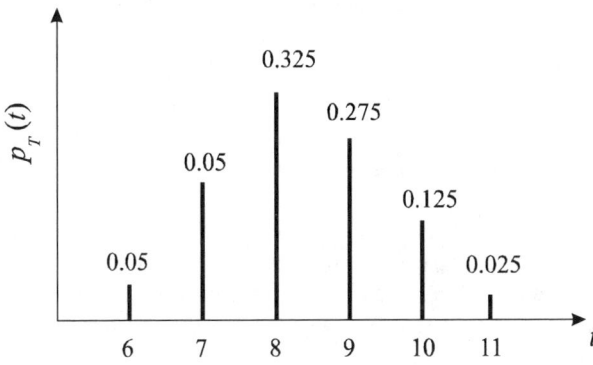

Figure 3.11: Probability mass function for T.

Using this probability mass function we can make probability statements on possible durations for the whole process. For example, the probability that the total manufacturing process will be at least 10 hr is given by $\Pr(T \geq 10) = 0.125 + 0.025 = 0.150$. Had we asked "greater than," the probability becomes $\Pr(T > 10) = 0.025$.

✳

Which density functions are of use in engineering applications? To be able to design a product such as a structure or a machine, one needs to be able to understand the behavior of materials, the characteristics of a vibrating system, and the external forces. Usually the largest uncertainties occur with load definition. Even so, in practice, we expect that the graphs of probability densities will have most of their area about the mean value, that is, with a small variance.

Sometimes in engineering, we only know the *high/low* values of a variable. In this instance all intermediate values are equally probable. This property leads to the uniform probability density. Other times our experience tells us that parameter values significantly different from the mean can happen, even if these are unlikely. This characteristic implies the Gaussian density, studied in Section 3.4.3. Data from testing and design experience helps considerably in the choice of the most physically realistic probability density function. Section 3.4 provides the details.

3.4 Useful Probability Densities

It turns out that a handful of density functions are sufficient for probabilistic modeling in most engineering applications. Here, five of these are discussed: the uniform, exponential, normal or Gaussian, lognormal, and the Rayleigh density functions.

3.4.1 The Uniform Density

The *uniform density* is a good model for a variable with known upper and lower bounds, and *equally likely* values within the range. From Figure 3.12 it can be seen that for any range Δx, the area under the density curve (*a horizontal line*) is the same. The left-hand side of Equation 3.4 must be a constant so that there is equal probability for the variable to be in any range.

Suppose then that X is a continuous random variable that can have any value in the interval $[a, b]$, where both a and b are finite. If the probability density function is given by

$$f(x) = \begin{cases} 1/(b-a), & a \leq x \leq b \\ 0, & \text{otherwise}, \end{cases}$$

then X is *uniformly distributed*. The probability distribution function for a

3.4. USEFUL PROBABILITY DENSITIES

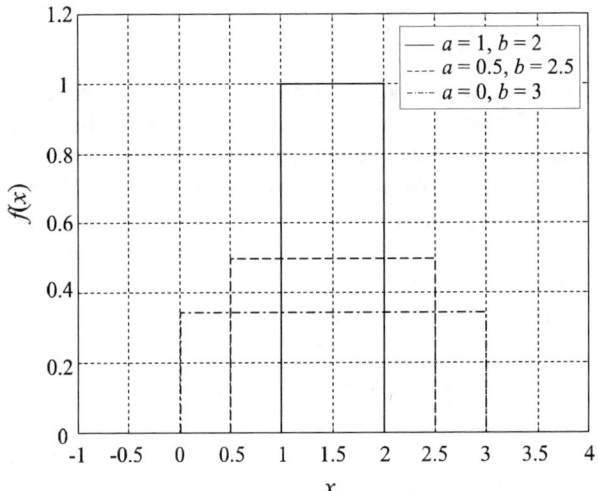

Figure 3.12: Three examples of a uniform density function.

uniformly distributed random variable is

$$F(x) = \Pr(X \leq x) = \int_{-\infty}^{x} f(s)ds$$

$$= \begin{cases} 0, & x < a \\ (x-a)/(b-a), & a \leq x < b \\ 1, & x \geq b. \end{cases}$$

The mean and mean-square values of uniformly distributed random variable X are given by

$$E\{X\} = \int_a^b x \frac{1}{b-a} dx = \frac{b+a}{2} \tag{3.11}$$

$$E\{X^2\} = \int_a^b x^2 \frac{1}{b-a} dx = \frac{b^2 + ab + a^2}{3}. \tag{3.12}$$

The standard deviation is then found to be

$$\sigma = \frac{b-a}{\sqrt{12}} = 0.289(b-a).$$

As expected, the statistics are functions of only the upper and the lower bounds.

Example 3.10 Uniform Density

A constant force P is known to have a value between 10 N and 25 N, but no additional information is available. All values in the range are of equal probability, given existing knowledge. Find the most likely value for P, and estimate the probability that the force $P > 20$ N. Also, calculate the variance and coefficient of variation.

Solution For such cases, where any value in a range is equally likely, a uniform density is chosen, $f_P(p) = c$, where c is a constant. Use the normalization property of the density to find the value of c,

$$\int_{10}^{25} f_P(p)\,dp = \int_{10}^{25} c\,dp = 1;$$

therefore $c = 1/15$. The most likely, or mean, value for a uniform density is the midpoint between the upper and lower bounds. By Equation 3.11, the mean is $\mu_P = (25 + 10)/2 = 17.5$ N. Or, just use the definition of mean value, $E\{P\} = \int_{10}^{25} p f_P(p)\,dp$, and come to the same result.

The probability that force $P > 20$ N is given by

$$\Pr(P > 20) = \int_{20}^{25} \frac{1}{15}\,dp = \frac{1}{3}.$$

The variance is evaluated as

$$\sigma_P^2 = E\{P^2\} - \mu_P^2 = 18.75\,\text{N}^2,$$

and the coefficient of variation is then

$$\delta = \frac{\sigma_P}{\mu_P} = \frac{4.3}{17.5} = 0.25,$$

or 25%. This is a relatively large scatter about the mean value. In engineering applications, coefficients of variation greater than 15% or, 0.15, imply a need for further data gathering.

⊛

Example 3.11 Quadratic Density

For comparison, suppose that instead of a uniform density, P is distributed according to a quadratic law, $f_P(p) = \alpha p^2$, also with $10 \leq P \leq 25$. Following the above procedure, it is straightforward to find $\alpha = 0.00021$, $\mu_P = 19.98$ N, and

$$\Pr(P > 20) = \int_{20}^{25} 0.00021 p^2\,dp = 0.53,$$

3.4. USEFUL PROBABILITY DENSITIES

which makes sense since much more of the area under the density function is located near the upper end of the range.

Here, the variance is $\sigma_P^2 = 6.76\,\text{N}^2$, a much smaller value than for the uniform density, and the coefficient of variation is $\delta = 0.13$ or 13%, again signifying that the spread of values is much smaller for the quadratic than the uniform. Figure 3.13 is a sketch of the quadratic superimposed on the uniform, providing confirmation that for the quadratic density a large fraction of the area is clustered at the extreme values of P.

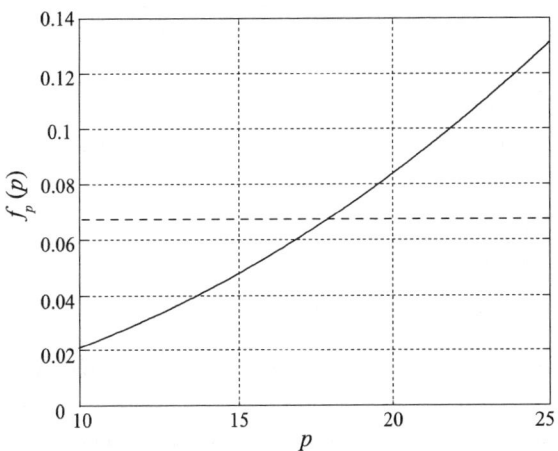

Figure 3.13: $f_P(p) = 0.00021p^2$ and $f_P(p) = \frac{1}{15}$ for $10 \leq p \leq 25$.

3.4.2 The Exponential Density

For mechanical reliability,[6] the exponential function is most commonly used to estimate failure times. The failure density is given by

$$f(t) = \lambda e^{-\lambda t}, \qquad \lambda > 0, \quad t \geq 0, \tag{3.13}$$

where λ is a constant (failure) rate per unit time, and $1/\lambda$ is the mean (time to failure). A sample exponential density is plotted for $\lambda = 1$ in Figure 3.14.

---

[6] A good starting point for studying reliability is the book by B.S. Dhillon, *Mechanical Reliability: Theory, Models and Applications*, AIAA, 1988.

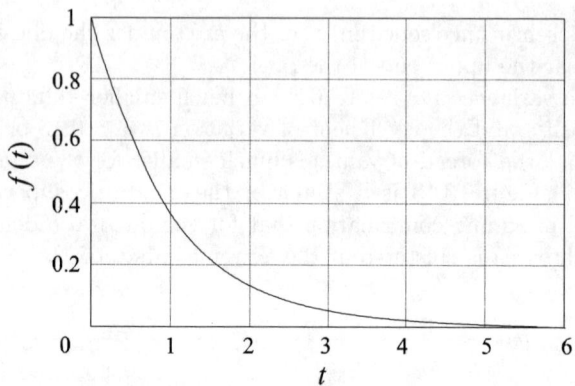

Figure 3.14: Exponential distribution with $\lambda = 1$.

Example 3.12 Time to Failure

A pump is known to fail according to the exponential density with a mean of 1000 hr. Then $\lambda = 1/1000$. Suppose that a critical mission requires the pump to operate 200 hr. Calculate the failure probability.

Solution The probability density function $f(t)$ tells us how the failure time is distributed. As long as the failure time is greater than 200 hr, the pump's performance is satisfactory. The failure probability is then the probability that the failure time is less than or equal to 200 hr,

$$\Pr(\text{failure}) = \Pr(t \leq 200) \text{ or } F(200),$$

where $F(t)$ is the cumulative distribution function corresponding to the density $f(t)$. For an exponential density, the probability distribution function is

$$F(t) = 1 - e^{-\lambda t}.$$

For this example,

$$F(200) = 1 - e^{-200/1000} = 0.1813.$$

The probability that the pump will fail during the first 200 hr is 0.1813 or 18.13%. Knowing this value will help in making the decision whether a backup pump needs to be on hand.

✵

3.4.3 The Normal (Gaussian) Density

Many physical variables are assumed to be governed by the normal or Gaussian density. See Figure 3.15. There are two reasons for this: the broad applicability of the central limit theorem,[7] and the Gaussian is mathematically tractable and tabulated.

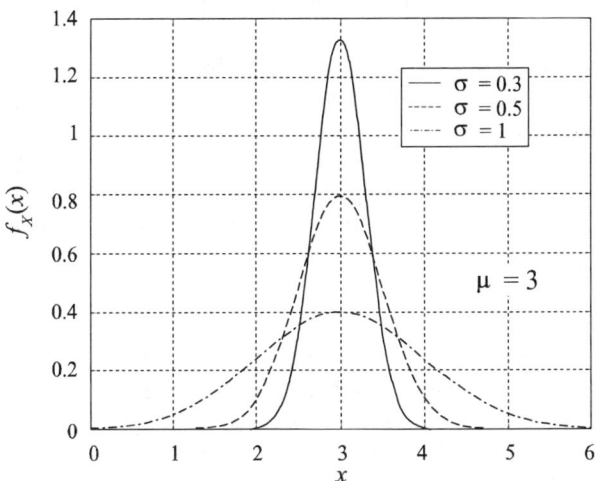

Figure 3.15: The normal or Gaussian density function.

A random variable governed by the Gaussian density has the probability density function,

$$f(x) = \frac{1}{\sigma\sqrt{2\pi}} \exp\left\{-\frac{1}{2}\left(\frac{x-\mu}{\sigma}\right)^2\right\}, \quad -\infty < x < \infty,$$

where the meaning of parameters μ and σ are found by taking the expected

---

[7] The central limit theorem states that under very general conditions, as the number of variables in a sum becomes large, the density of the sum of random variables will approach the Gaussian regardless of the individual densities. Examples of variables that arise as the sum of a number of random effects, where no one effect dominates, are noise generated by falling rain, the effects of a turbulent boundary layer, and the response of linear structures to a turbulent environment. Many naturally occurring physical processes approach a Gaussian density.

values[8] of X and X^2, respectively,

$$E\{X\} = \frac{1}{\sqrt{2\pi}} \int_{-\infty}^{\infty} (\sigma y + \mu) e^{-y^2/2} dy = \mu$$

$$E\{X^2\} = \frac{1}{\sqrt{2\pi}} \int_{-\infty}^{\infty} (\sigma y + \mu)^2 e^{-y^2/2} dy = \mu^2 + \sigma^2.$$

Thus, the mean value is μ and, using Equation 3.10, the standard deviation is σ.

Note that the Gaussian density extends from $-\infty$ to ∞, and therefore, cannot represent any physical variable except approximately. Since there are no physical parameters that can take on all possible values on the real number line, we may rightly wonder how good a model is the Gaussian for physical processes. But the approximation, in many instances, turns out to be very good. For example, consider a positive-definite random variable X that is modeled as a Gaussian with coefficient of variation $\delta = 0.20$, or $\mu = 5\sigma$. How significant is the area under the density function in the negative X region? Integrating numerically for $x < 0$, one finds an area of approximately 24×10^{-8}, a negligible probability for most purposes. The suitability of the Gaussian model depends on the application, and how much the tails extend into physically forbidden regions.

When it is not possible to accept any negative values,[9] the analyst sometimes resorts to the truncated Gaussian,

$$f(x) = \frac{A}{\sigma\sqrt{2\pi}} \exp\left[-\frac{(x-x_0)^2}{2\sigma^2}\right], \quad 0 \leq x_1 \leq x \leq x_2,$$

and zero elsewhere. (If $x_1 \to -\infty$ and $x_2 \to \infty$, then $A \to 1$, and X becomes a Gaussian random variable with $E\{X\} = x_0$ and $Var(X) = \sigma^2$. See Figure 3.16.)

For ease in applications, the Gaussian variable X is sometimes transformed so that the resulting variable S is *zero mean* with *unit variance*,

$$S = \frac{X - \mu_X}{\sigma_X},$$

resulting in the *standard normal density*,

$$f_S(s) = \frac{1}{\sqrt{2\pi}} e^{-s^2/2}.$$

---

[8] Make use of the transformation of variables: $y = (x-\mu)/\sigma$ and note that $dx = \sigma dy$.
[9] This is especially true in reliability calculations where the probabilities of failure may be very small, even on the order of 10^{-8}, and extra care must be taken to ensure that the density function is suitable.

3.4. USEFUL PROBABILITY DENSITIES

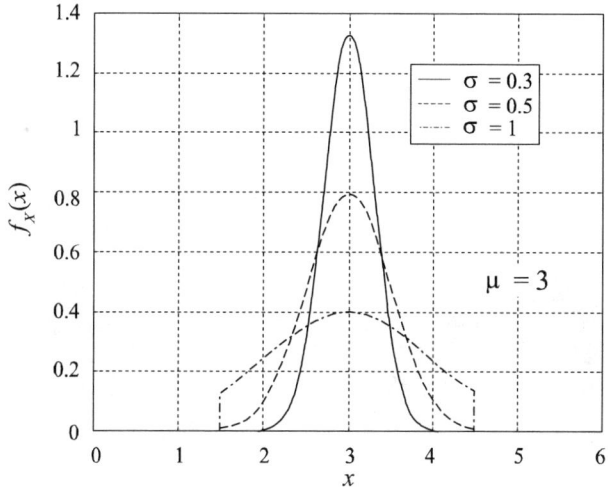

Figure 3.16: The truncated Gaussian for $1.5 \leq X \leq 4.5$.

The probability distribution is then

$$F_S(s) = \Pr(S \leq s) = \int_{-\infty}^{s} \frac{1}{\sqrt{2\pi}} e^{-s^2/2} ds,$$

where $F_S(s)$ is often denoted as $\Phi(s)$ and is tabulated. See Table 4.7.

Finally, compute the probability that X is within k standard deviations of its mean. This probability is expressed as

$$\Pr(\mu_X - k\sigma_X < X \leq \mu_X + k\sigma_X) = \Pr\left(-k < \frac{X - \mu_X}{\sigma_X} \leq k\right)$$
$$= \Pr(-k < S \leq k),$$

which is equal to the probability that the standard normal random variable S is between $-k$ and k. Note that the probability depends only on k and is independent of μ_X and σ_X. In Figure 3.17, the shaded areas under the curves are equal even though μ_X and σ_X are different. It will be shown in Example 3.14 that the probabilities that X is within $1\sigma_X$, $2\sigma_X$, and $3\sigma_X$ are 0.6827, 0.9545, and 0.9973, respectively, and that 95% of the area under the normal distribution lies within $1.96\sigma_X$ of the mean. See Figure 3.18.

Gaussian Table

When utilizing the Gaussian model, it has become the practice to utilize tables of values of the *Standard Normal Cumulative Distribution Function*.

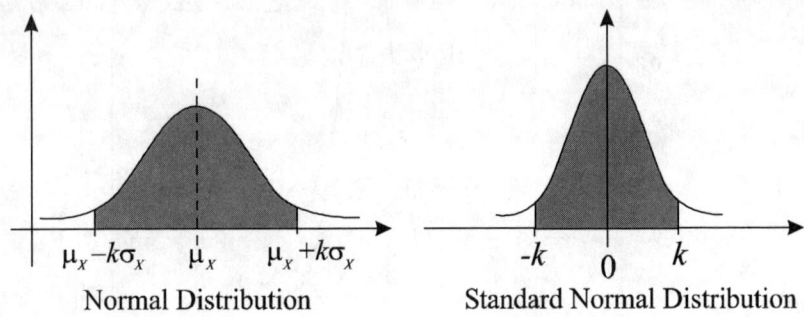

Figure 3.17: Equal shaded areas.

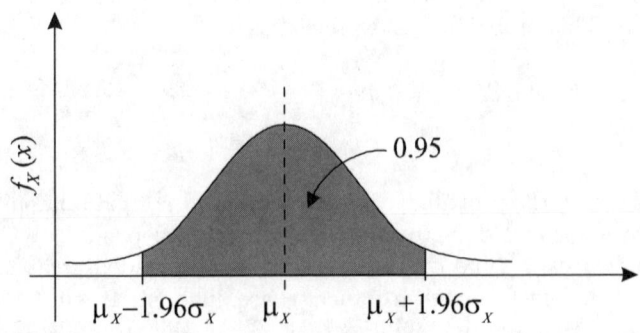

Figure 3.18: Area under the normal density.

3.4. USEFUL PROBABILITY DENSITIES

Utilizing the table is generally easier than integrating the Gaussian function over a range of values. The tabulated probability values are for the standard normal function. Suppose the normal random variable of interest X has a mean μ and a standard deviation σ, in shorthand $N(\mu, \sigma)$. This can be transformed into the standard normal variable S by using the relation,

$$S = \frac{X - \mu}{\sigma}.$$

Therefore, the event that $X \leq x$ is probabilistically equivalent to the event that $S \leq (x - \mu)/\sigma$, for which values are tabulated. A common notation used for this is

$$\Pr\left(S \leq \frac{x - \mu}{\sigma}\right) = \Phi\left(\frac{x - \mu}{\sigma}\right), \tag{3.14}$$

where $\Phi(-a) = 1 - \Phi(a)$. The argument of Φ is the number of standard deviations above or below the mean value. The values of $\Phi(s)$ are tabulated in Section 4.7. This is called the Gaussian table or the standard normal table. Note that the numbers 0 to 9 across the table are the second digits after the decimal.

Example 3.13 Reading from the Standard Normal Table

Using the Standard Normal Table in Section 4.7, obtain probabilities (*i*) $\Pr(S \leq -2.03)$, (*ii*) $\Pr(S \leq 1.76)$, and (*iii*) $\Pr(S \geq -1.58)$.
Solution
(*i*) From the standard normal table, $\Pr(S \leq -2.03) = 0.0212$.
(*ii*) $\Pr(S \leq 1.76)$ can be found using the fact that

$$\Pr(S \leq 1.76) = 1 - \Pr(S < -1.76).$$
$$= 1 - 0.0392$$
$$= 0.9608.$$

This result can be confirmed by looking up $\Phi(1.76)$ in the Standard Normal Table.
(*iii*) $\Pr(S \geq -1.58)$ can be written as

$$\Pr(S \geq -1.58) = 1 - \Pr(S < -1.58).$$

$\Pr(S \leq -1.58) = 0.0570$ from the Standard Normal Table. Therefore,

$$\Pr(S \geq -1.58) = 1 - 0.0570$$
$$= 0.9430.$$

⊛

CHAPTER 3. RANDOM VARIABLE MODELS

Example 3.14 Probabilities Within k Standard Deviations

Find the probabilities that X is within $1\sigma_X$, $2\sigma_X$, and $3\sigma_X$, respectively. Also, show that 95% of the area under the normal density lies within $1.96\sigma_X$ of the mean.

Solution The probability that X is within $1\sigma_X$ is calculated as follows,

$$\begin{aligned}
\Pr(\mu_X - \sigma_X < X \le \mu_X + \sigma_X) &= \Pr(-1 < S \le 1) \\
&= \Pr(S \le 1) - \Pr(S < -1) \\
&= \Phi(1) - \Phi(-1) \\
&= 2\Phi(1) - 1 \\
&= 0.6826.
\end{aligned}$$

Similarly, the probabilities that X is within $2\sigma_X$ and $3\sigma_X$ are

$$\begin{aligned}
\Pr(\mu_X - 2\sigma_X < X \le \mu_X + 2\sigma_X) &= 2\Phi(2) - 1 \\
&= 0.9544,
\end{aligned}$$

and

$$\begin{aligned}
\Pr(\mu_X - 2\sigma_X < X \le \mu_X + 2\sigma_X) &= 2\Phi(3) - 1 \\
&= 0.9974.
\end{aligned}$$

It can be shown that 95% of the area under the normal distribution lies within $1.96\sigma_X$ of the mean,

$$\begin{aligned}
\Pr(\mu_X - k\sigma_X < X \le \mu_X + k\sigma_X) &= 0.95 \\
2\Phi(k) - 1 &= 0.95 \\
\Phi(k) &= 0.975.
\end{aligned}$$

From the Gaussian table, $k = 1.96$, and thus about 96% of the area under the normal density lies within $1.96\,\sigma_X$ of the mean value.
⊛

Example 3.15 Cable Failure

The cable of a suspension bridge is taken from a lot delivered by the manufacturer who guarantees that, statistically, the lot has the following properties. The cables have a yield stress that is normal with $N(50\text{ ksi}, 5\text{ ksi})$. If a load of 40 ksi acts on one cable, what is the probability of failure?

Solution Let Y be the yield event. The transformation is given as

$$S = \frac{Y - 50}{5}.$$

3.4. USEFUL PROBABILITY DENSITIES

The probability of failure under the given load can be calculated in the following way,

$$\begin{aligned}\Pr(Y < 40) &= \Pr\left(S < \frac{40-50}{5}\right)\\ &= \Phi(-2.0)\\ &= 0.0228.\end{aligned}$$

This is the probability that the yield stress of the given cable is less than 40 ksi, that is, 2.28%.

⊛

Example 3.16 Rocket Ship Traffic Volume

It is the year 2150. Permanent habitats, really small cities, have been growing on the Moon after the first groundbreaking in 2019. On Mars there are now 3 cities, the oldest one of which dates back to 2028. Populations are growing and people are being born on both the Moon and Mars. Interplanetary commerce has become a significant fraction of total human commerce. There is even talk about sending settlers to permanent sites on one of the Jovian Moons, but really, that sounds incredible. We are being asked to perform a traffic study for Lunar Spaceport Alpha. As the oldest and largest spaceport on the Moon, it needs to be expanded. Here is what we know, and what we need to estimate.

On Lunar Spaceport Alpha, the current volume V (rocket ship landings and blastoffs) during the peak hour is $N(50, 10)$. The present capacity is for a total of 80 operations per hour. If there are more than 80 operations, there is a backlog of blastoffs and rockets entering the lunar space must orbit until clearance is granted.

(i) Find the current probability of congestion. Since the volume is a normal random variable, we have the option of utilizing the normal density function, or using the standard normal table. Let us do both. For the normal density function, the required probability is given by

$$\begin{aligned}\Pr(V > 80) &= \frac{1}{\sigma\sqrt{2\pi}} \int_{80}^{\infty} \exp\left[-\frac{1}{2}\left(\frac{V-\mu}{\sigma}\right)^2\right] dV\\ &= \frac{1}{10\sqrt{2\pi}} \int_{80}^{\infty} \exp\left[-\frac{1}{2}\left(\frac{V-50}{10}\right)^2\right] dV\\ &= 1.3499 \times 10^{-3}.\end{aligned}$$

This is an extremely small number. If we are to use the standard normal table, we need to transform as follows,

$$S = \frac{V - \mu}{\sigma} = \frac{V - 50}{10}.$$

Then,

$$\begin{aligned}
\Pr(V > 80) &= 1 - \Pr(V < 80) \\
&= 1 - \Pr\left(S < \frac{80 - 50}{10}\right) \\
&= 1 - \Pr(S < 3.0) \\
&= 1 - \Phi(3.0),
\end{aligned}$$

where from the standard normal tables $\Phi(3.0) = 0.9987$. This is approximately the same number we obtained above. Note that the argument of the function Φ is in reality the number of standard deviations above or below the mean value. Therefore, currently congestion is not a problem. But wait!

(ii) It is estimated that the mean volume will increase by 5% of the current volume per year while the coefficient of variation will stay the same. Perform the same computations as above to estimate the probability of congestion in 10 years time assuming that capacity is not increased.

The coefficient of variation is $\delta = \sigma/\mu = 10/50 = 0.2$. The mean volume in 10 years time is $50 + (0.05 \times 50) \times 10 = 75.0$. The standard deviation will be $\sigma_{10} = \delta \mu_{10} = 0.2 \times 75 = 15.0$, that is, $V_{10} = N(75, 15)$. Now we can repeat the calculations of part one. First, use the density function,

$$\Pr(V_{10} > 80) = \frac{1}{15\sqrt{2\pi}} \int_{80}^{\infty} \exp\left[-\frac{1}{2}\left(\frac{V - 75}{15}\right)^2\right] dV = 0.36944,$$

and then using the standard normal tables,

$$\begin{aligned}
\Pr(V_{10} > 80) &= 1 - \Pr(V_{10} < 80) \\
&= 1 - \Pr\left(S_{10} < \frac{80 - 75}{15}\right) \\
&= 1 - \Phi(0.33) \\
&= 1 - 0.6293 = 0.3707,
\end{aligned}$$

which is approximately the same, the difference being due to the roundoff in the table lookup.

(iii) Finally, since such a congestion probability is too large, find the required capacity in 10 years time such that the present service conditions are maintained. Let C_{10} be the necessary capacity after ten years. It is

3.4. USEFUL PROBABILITY DENSITIES

required that

$$\Pr(V_{10} > C_{10}) = 1.3499 \times 10^{-3}$$
$$1 - \Phi\left(\frac{C_{10} - 75}{15}\right) = 1 - \Phi(3.0)$$
$$\frac{C_{10} - 75}{15} = 3.0$$
$$C_{10} = 120.$$

Therefore, capacity must be increased to 120 landing and blast-off operations per hour in order to maintain the same low congestion rate.
⊛

Figure 3.19: Carl Friedrich Gauss (1777-1855).

Gauss At the age of seven, Carl Friedrich Gauss started elementary school, and his potential was noticed almost immediately. His teacher, Büttner, and his assistant, Martin Bartels, were amazed when Gauss summed the integers from 1 to 100 instantly, spotting that the sum was 50 pairs of numbers each pair summing to 101.

In 1788 Gauss began his education at the Gymnasium with the help of Büttner and Bartels, where he learned High German and Latin. After receiving a stipend from the Duke of Brunswick-Wolfenbüttel, Gauss entered Brunswick Collegium Carolinum in 1792. At the academy Gauss independently discovered Bode's law, the binomial theorem, and the arithmetic-geometric mean, as well as the law of quadratic reciprocity and the prime number theorem.

In 1795 Gauss left Brunswick to study at Göttingen University. Gauss' teacher there was Kaestner, whom Gauss often ridiculed. His only known friend among the students was Farkas Bolyai. They met in 1799 and corresponded with each other for many years.

Gauss left Göttingen in 1798 without a diploma, but by this time he had made one of his most important discoveries, the construction of a regular 17-gon by ruler and compasses. This was the most major advance in this field since the time of Greek mathematics and was published as Section VII of Gauss' famous work, *Disquisitiones Arithmeticae*.

Gauss returned to Brunswick where he received a degree in 1799. After the Duke of Brunswick had agreed to continue Gauss' stipend, he requested that Gauss submit a doctoral dissertation to the University of Helmstedt. He already knew Pfaff, who was chosen to be his advisor. Gauss' dissertation was a discussion of the fundamental theorem of algebra.

With his stipend to support him, Gauss did not need to find a job so devoted himself to research. He published the book *Disquisitiones Arithmeticae* in the summer of 1801. There were seven sections, all but the last section, referred to above, being devoted to number theory.

In June 1801, Zach, an astronomer whom Gauss had come to know two or three years previously, published the orbital positions of Ceres, a new "small planet," which was discovered by G. Piazzi, an Italian astronomer on January 1, 1801. Unfortunately, Piazzi had only been able to observe 9 degrees of its orbit before it disappeared behind the Sun. Zach published several predictions of its position, including one by Gauss, which differed greatly from the others. When Ceres was rediscovered by Zach on December 7, 1801 it was almost exactly where Gauss had predicted. Although he did not disclose his methods at the time, Gauss had used his least squares approximation method.

In June 1802 Gauss visited Olbers who had discovered Pallas in March of that year and Gauss investigated its orbit. Olbers requested that Gauss be made director of the proposed new observatory in Göttingen, but no action was taken. Gauss began corresponding with Bessel, whom he did not meet until 1825, and with Sophie Germain.

Gauss married Johanna Ostoff on October 9, 1805. Despite Gauss having a happy personal life for the first time, his benefactor, the Duke of Brunswick, was killed fighting for the Prussian army. In 1807 Gauss left Brunswick to take up the position of director of the Göttingen observatory.

Gauss arrived in Göttingen in late 1807. In 1808 his father died, and a year later Gauss' wife Johanna died after giving birth to their second son, who was to die soon after her. Gauss was shattered and wrote to Olbers asking him to give him a home for a few weeks, *to gather new strength in the arms of your friendship - strength for a life which is only valuable*

3.4. USEFUL PROBABILITY DENSITIES 81

because it belongs to my three small children.

Gauss was married for a second time the next year, to Minna, the best friend of Johanna, and although they had three children, this marriage seemed to be one of convenience for Gauss.

Gauss' work never seemed to suffer from his personal tragedy. He published his second book, *Theoria Motus Corporum Coelestium in Sectionibus Conicis Solem Ambientium*, in 1809, a major two-volume treatise on the motion of celestial bodies. In the first volume he discussed differential equations, conic sections, and elliptic orbits, while in the second volume, the main part of the work, he showed how to estimate and then to refine the estimation of a planet's orbit. Gauss' contributions to theoretical astronomy stopped after 1817, although he went on making observations until the age of 70.

Much of Gauss' time was spent on a new observatory, completed in 1816, but he still found the time to work on other subjects. His publications during this time include *Disquisitiones generales circa seriem infinitam*, a rigorous treatment of series and an introduction of the hypergeometric function, *Methodus nova integralium valores per approximationem inveniendi*, a practical essay on approximate integration, *Bestimmung der Genauigkeit der Beobachtungen*, a discussion of statistical estimators, and *Theoria attractionis corporum sphaeroidicorum ellipticorum homogeneorum methodus nova tractata*. The latter work was inspired by geodesic problems and was principally concerned with potential theory. In fact, Gauss found himself more and more interested in geodesy in the 1820s.

Gauss had been asked in 1818 to carry out a geodesic survey of the state of Hanover to link up with the existing Danish grid. Gauss was pleased to accept and took personal charge of the survey, making measurements during the day and reducing them at night, using his extraordinary mental capacity for calculations. He regularly wrote to Schumacher, Olbers, and Bessel, reporting on his progress and discussing problems.

Because of the survey, Gauss invented the heliotrope, which worked by reflecting the Sun's rays using a design of mirrors and a small telescope. However, inaccurate base lines were used for the survey and an unsatisfactory network of triangles. Gauss often wondered if he would have been better advised to have pursued some other occupation but he published over 70 papers between 1820 and 1830.

In 1822 Gauss won the Copenhagen University Prize with *Theoria Attractionis ...* together with the idea of mapping one surface onto another so that the two *are similar in their smallest parts*. This paper was published in 1825 and led to the much later publication of *Untersuchungen über Gegenstände der Höheren Geodäsie* (1843 and 1846). The paper *Theoria combinationis observationum erroribus minimis obnoxiae* (1823), with its supplement (1828), was devoted to mathematical statistics, in particular to the least-squares method.

From the early 1800s Gauss had an interest in the question of the possible existence of a non-Euclidean geometry. He discussed this topic at length with Farkas Bolyai and in his correspondence with Gerling and Schumacher. In a book review in 1816 he discussed proofs that deduced the axiom of parallels from the other Euclidean axioms, suggesting that he believed in the existence of non-Euclidean geometry, although he was rather vague. Gauss confided in Schumacher, telling him that he believed his reputation would suffer if he admitted in public that he believed in the existence of such a geometry.

In 1831 Farkas Bolyai sent to Gauss his son János Bolyai's work on the subject. Gauss replied *to praise it would mean to praise myself*.

Again, a decade later, when he was informed of Lobachevsky's work on the subject, he praised its "genuinely geometric" character, while in a letter to Schumacher in 1846, states that he *had the same convictions for 54 years* indicating that he had known of the existence of a non-Euclidean geometry since he was 15 years of age (this seems unlikely).

Gauss had a major interest in differential geometry, and published many papers on the subject. *Disquisitiones generales circa superficies curva* (1828) was his most renowned work in this field. In fact, this paper rose from his geodesic interests, but it contained such geometrical ideas as Gaussian curvature. The paper also includes Gauss' famous *theorema egregrium*:

If an area in E3 can be developed (i.e., mapped isometrically) into another area of E3, the values of the Gaussian curvatures are identical in corresponding points.

The period 1817-1832 was a particularly distressing time for Gauss. He took in his sick mother in 1817, who stayed until her death in 1839, while he was arguing with his wife and her family about whether they should go to Berlin. He had been offered a position at Berlin University and Minna and her family were keen to move there. Gauss, however, never liked change and decided to stay in Göttingen. In 1831 Gauss' second wife died after a long illness.

In 1831, Wilhelm Weber arrived in Göttingen as physics professor filling Tobias Mayer's chair. Gauss had known Weber since 1828 and supported his appointment. Gauss had worked on physics before 1831, publishing *Über ein neues allgemeines Grundgesetz der Mechanik*, which contained the principle of least constraint, and *Principia generalia theoriae figurae fluidorum in statu aequilibrii,* which discussed forces of attraction. These papers were based on Gauss' potential theory, which proved of great importance in his work on physics. He later came to believe his potential theory and his method of least squares provided vital links between science and

3.4. USEFUL PROBABILITY DENSITIES

nature.

In 1832, Gauss and Weber began investigating the theory of terrestrial magnetism after Alexander von Humboldt attempted to obtain Gauss' assistance in making a grid of magnetic observation points around the Earth. Gauss was excited by this prospect and by 1840 he had written three important papers on the subject: *Intensitas vis magneticae terrestris ad mensuram absolutam revocata* (1832), *Allgemeine Theorie des Erdmagnetismus* (1839), and *Allgemeine Lehrsätze in Beziehung auf die im verkehrten Verhältnisse des Quadrats der Entfernung wirkenden Anziehungs- und Abstossungskräfte* (1840). These papers all dealt with the current theories on terrestrial magnetism, including Poisson's ideas, absolute measure for magnetic force and an empirical definition of terrestrial magnetism. Dirichlet's principle was mentioned without proof.

Allgemeine Theorie showed that there can only be two poles in the globe and went on to prove an important theorem, which concerned the determination of the intensity of the horizontal component of the magnetic force along with the angle of inclination. Gauss used the Laplace equation to aid him with his calculations, and ended up specifying a location for the magnetic South Pole.

Humboldt had devised a calendar for observations of magnetic declination. However, once Gauss' new magnetic observatory (completed in 1833, free of all magnetic metals) had been built, he proceeded to alter many of Humboldt's procedures, not pleasing Humboldt greatly. However, Gauss' changes obtained more accurate results with less effort.

Gauss and Weber achieved much in their six years together. They discovered Kirchhoff's laws, as well as building a primitive telegraph device that could send messages over a distance of 5000 ft. However, this was just an enjoyable pastime for Gauss. He was more interested in the task of establishing a worldwide net of magnetic observation points. This occupation produced many concrete results. The Magnetischer Verein and its journal were founded, and the atlas of geomagnetism was published, while Gauss and Weber's own journal in which their results were published ran from 1836 to 1841.

In 1837, Weber was forced to leave Göttingen when he became involved in a political dispute and, from this time, Gauss' activity gradually decreased. He still produced letters in response to fellow scientists' discoveries, usually remarking that he had known the methods for years but had never felt the need to publish. Sometimes he seemed extremely pleased with advances made by other mathematicians, particularly that of Eisenstein and Lobachevsky.

Gauss spent the years from 1845 to 1851 updating the Göttingen University widow's fund. This work gave him practical experience in financial matters, and he went on to make his fortune through shrewd investments in bonds issued by private companies.

Two of Gauss' last doctoral students were Moritz Cantor and Dedekind. Dedekind wrote a fine description of his supervisor: ... *usually he sat in a comfortable attitude, looking down, slightly stooped, with hands folded above his lap. He spoke quite freely, very clearly, simply and plainly: but when he wanted to emphasize a new viewpoint ... then he lifted his head, turned to one of those sitting next to him, and gazed at him with his beautiful, penetrating blue eyes during the emphatic speech. ... If he proceeded from an explanation of principles to the development of mathematical formulas, then he got up, and in a stately very upright posture he wrote on a blackboard beside him in his peculiarly beautiful handwriting: he always succeeded through economy and deliberate arrangement in making do with a rather small space. For numerical examples, on whose careful completion he placed special value, he brought along the requisite data on little slips of paper.*

Gauss presented his golden jubilee lecture in 1849, fifty years after his diploma had been granted by Hemstedt University. It was appropriately a variation on his dissertation of 1799. From the mathematical community only Jacobi and Dirichlet were present, but Gauss received many messages and honors.

From 1850 onwards Gauss' work was again of nearly all of a practical nature although he did approve Riemann's doctoral thesis and heard his probationary lecture. His last known scientific exchange was with Gerling. He discussed a modified Foucalt pendulum in 1854. He was also able to attend the opening of the new railway link between Hanover and Göttingen, but this proved to be his last outing. His health deteriorated slowly, and Gauss died in his sleep early in the morning of February 23, 1855.

3.4.4 The Lognormal Density

Sometimes it is important to strictly limit possible values of a parameter to the positive range. In such instances, a likely choice is the lognormal density, depicted in Figure 3.20. Applications for the lognormal include material strength, fatigue life, loading intensity, time to the occurrence of an event, and volumes and areas. A random variable X has a *lognormal* probability density function if $\ln X$ is normally distributed,[10] that is,

$$f_X(x) = \frac{1}{x\zeta\sqrt{2\pi}} \exp\left\{-\frac{1}{2}\left(\frac{\ln x - \lambda}{\zeta}\right)^2\right\}, \quad 0 < x < \infty, \qquad (3.15)$$

---

[10] The probability density function of $\ln X$ is derived from the normal density in Example 4.1.

3.4. USEFUL PROBABILITY DENSITIES

where $\lambda = E\{\ln X\}$ is the mean value of $\ln X$ and $\zeta = \sqrt{Var\{\ln X\}}$ is the standard deviation of $\ln X$.

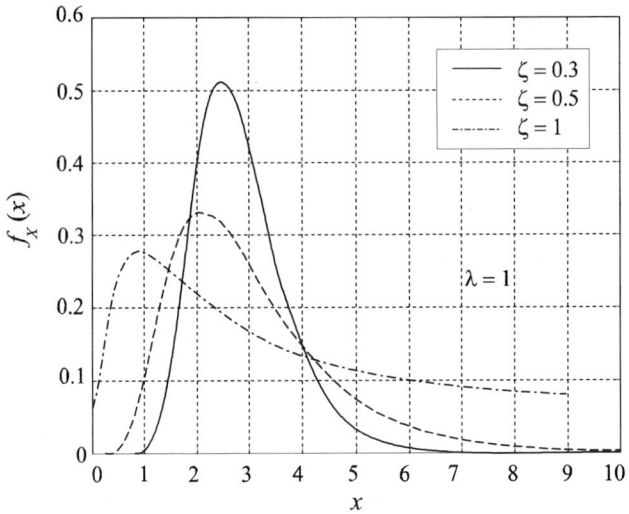

Figure 3.20: The lognormal density.

It is of interest to relate λ and ζ to the mean and standard deviation of X. Define $Y = \ln X$. From the definition of lognormal, Y is a normal random variable with mean λ and standard deviation ζ, that is, $Y = N(\lambda, \zeta)$. Solve for X by taking the exponential of both sides; $X = \exp(Y)$. Define the following,

$$\mu_X = E\{X\}$$
$$Var\{X\} = E\{X^2\} - E^2\{X\}.$$

Therefore, using the property that Y is normal,

$$\mu_X = E\{\exp(Y)\} = \int_{-\infty}^{\infty} \exp(y) \frac{1}{\zeta\sqrt{2\pi}} \exp\left[-\frac{1}{2}\left(\frac{y-\lambda}{\zeta}\right)^2\right] dy$$

$$= \int_{-\infty}^{\infty} \frac{1}{\zeta\sqrt{2\pi}} \exp\left[y - \frac{1}{2}\left(\frac{y-\lambda}{\zeta}\right)^2\right] dy$$

$$= \exp(\lambda + \frac{1}{2}\zeta^2),$$

and solving for λ,

$$\lambda = \ln \mu_X - \frac{1}{2}\zeta^2.$$

Similarly, $E\{X^2\} = \exp\left[2\left(\lambda + \zeta^2\right)\right]$, and the variance is given by $Var\{X\} = \mu_X^2 \left[\exp\left(\zeta^2\right) - 1\right]$. Then,

$$\sigma_X = +\sqrt{Var\{X\}}$$
$$= \sqrt{\mu_X^2 \left[\exp\left(\zeta^2\right) - 1\right]}$$
$$\zeta^2 = \ln\left(1 + \frac{\sigma_X^2}{\mu_X^2}\right),$$

and $\lambda = \ln\mu_X - \frac{1}{2}\ln\left(1 + \sigma_X^2/\mu_X^2\right)$. For most engineering applications, the ratio σ_X/μ_X is on the order of 0.1. Given this, $1 \gg \sigma_X^2/\mu_X^2$, and using the expansion[11] for $\ln(1 + x)$, we have

$$\ln\left(1 + \frac{\sigma_X^2}{\mu_X^2}\right) \simeq \frac{\sigma_X^2}{\mu_X^2};$$
$$\zeta \simeq \frac{\sigma_X}{\mu_X},$$

which is the coefficient of variation of X; the mean value of Y is approximately equal to the coefficient of variation of X.

Example 3.17 Cable Failure Revisited

The cable of a suspension bridge studied in Example 3.15, demonstrating the use of the Gaussian density, is now assumed to have a yield stress that is lognormal with mean 50 ksi and standard deviation 5 ksi. If a load of 40 ksi acts on one cable, what is the probability of failure?

Solution   Let X be the yield stress with $\mu_X = 50$ ksi and $\sigma_X = 5$ ksi.[12] Let $Y = \ln X$, where Y is normally distributed with a distribution $N(\lambda, \zeta)$. From the figure it is clear that the area under the lognormal and to the left of $x = 40$ ksi equals the probability that the yield stress is less than the applied stress. Therefore, to find $\Pr(X < 40)$ evaluate

$$\Pr(X < 40) = \int_{-\infty}^{40} \frac{1}{x\zeta\sqrt{2\pi}} \exp\left\{-\frac{1}{2}\left(\frac{\ln x - \lambda}{\zeta}\right)^2\right\} dx, \qquad (3.16)$$

where λ and ζ are mean and the standard variation of $\ln X$ or Y. Rather than integrating Equation 3.16, the Gaussian table can be used to evaluate the integral. The probability that X is less than 40 equals the probability that Y is less than $\ln 40$,

$$\Pr(X < 40) = \Pr(Y < \ln 40).$$

---
[11] The expansion is $\ln(1 + x) = x - x^2/2 + x^3/3 - ..., -1 < x \leq 1$.
[12] ksi is shorthand for kips per square inch, and kip is one thousand.

3.4. USEFUL PROBABILITY DENSITIES

To evaluate $\Pr(Y < \ln 40)$ using the Gaussian tables, use the transformation,

$$S = \frac{Y - \lambda}{\zeta},$$

where

$$\lambda = \ln \mu_X - \frac{1}{2}\zeta^2$$
$$\zeta = \sqrt{\ln(1 + \delta_X^2)}.$$

Substitute the values $\mu_X = 50$ ksi and $\sigma_X = 5$ ksi, and find that

$$\begin{aligned}\zeta &= \sqrt{\ln(1 + (5/50)^2)} \\ &= 9.9751 \times 10^{-2} \text{ ksi},\end{aligned}$$

and

$$\begin{aligned}\lambda &= \ln 50 - 0.5 \times (9.9751 \times 10^{-2})^2 \\ &= 3.907 \text{ ksi}.\end{aligned}$$

Then, the probability of failure under the given load can be calculated from the Gaussian tables in the following way:

$$\begin{aligned}\Pr(Y < \ln 40) &= \Pr\left(S < \frac{\ln 40 - 3.907}{9.9751 \times 10^{-2}}\right) \\ &= \Phi(-2.18) \\ &= 0.0146.\end{aligned}$$

Compare this with 0.0228 where the variable was assumed normal.

Note that the probability is also equivalent to the area under the probability density function. Figure 3.21 shows the probability density function, and the shaded area is the probability that X is less than 40.
⊛

3.4.5 The Rayleigh Density

The *Rayleigh* density, like the lognormal, is also limited to strictly positive-valued random variables,

$$f(x) = \frac{x}{\sigma^2} \exp\left\{-\frac{x^2}{2\sigma^2}\right\}, \quad x \geq 0.$$

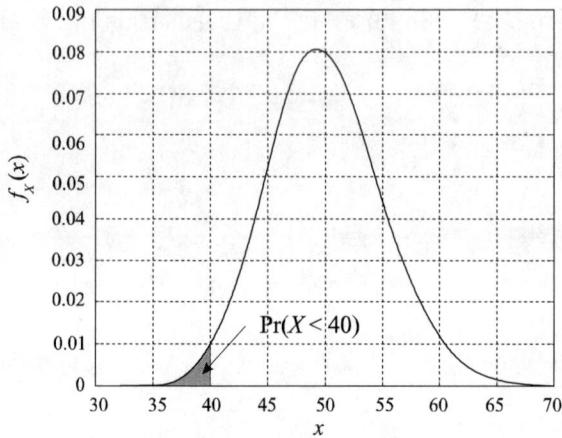

Figure 3.21: The lognormal density for $\zeta = 9.9751 \times 10^{-2}$ and $\lambda = 3.907$. The shaded area equals $\Pr(X < 40)$.

The first- and second-order statistics can be derived to be

$$E\{X\} = \sqrt{\frac{\pi}{2}}\sigma$$

$$E\{X^2\} = 2\sigma^2$$

$$\sigma_X = \sqrt{\frac{4-\pi}{2}}\sigma \simeq 0.655\sigma.$$

Figure 3.22 shows the Rayleigh density with three different values of σ. As an example of where the Rayleigh density is a good model, consider a random oscillation, or any time dependent process that is governed by the Gaussian. This oscillation has peaks that are distributed randomly as well. The Rayleigh density is a good model for the distribution of these peak values.

Lord Rayleigh John William Strutt, better known as Lord Rayleigh, suffered from poor health so his schooling at Eton and Harrow was disrupted and for four years he had a private tutor. He entered Trinity College, Cambridge, in 1861, graduating in 1864.

His first paper in 1865 was on Maxwell's electromagnetic theory. He worked on propagation of sound and, while on an excursion to Egypt taken for health reasons, Strutt wrote *Treatise on Sound* (1870-1871). In 1879 he wrote a paper on travelling waves. This theory has now developed into

3.4. USEFUL PROBABILITY DENSITIES

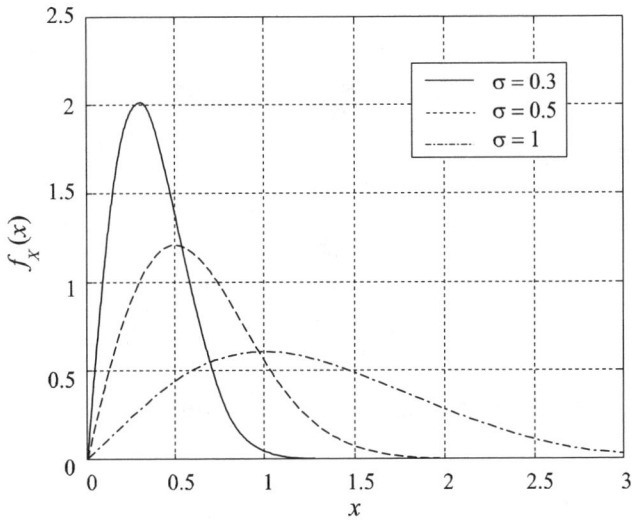

Figure 3.22: The Rayleigh probability density function.

Figure 3.23: John William Strutt, Lord Rayleigh (1842-1919).

the theory of solitons. His theory of scattering (1871) was the first correct explanation of why the sky is blue.

In 1873 he succeeded to the title of Baron Rayleigh. From 1879 to 1884 he was the second Cavendish professor of experimental physics at Cambridge, succeeding Maxwell. Then in 1884 he became secretary of the Royal Society. Rayleigh discovered the inert gas argon in 1895, work that earned him a Nobel Prize in 1904.

He was awarded the De Morgan Medal of the London Mathematical Society in 1890 and was president of the Royal Society between 1905 and 1908. He became chancellor of Cambridge University in 1908.

3.4.6 Probability Density Functions of a Discrete Random Variable

So far we have discussed the probability density functions of continuous random variables. Here, we consider such functions for discrete random variables.

The Binomial Distribution

Perform an experiment where the outcome can be either event A or $\bar{A}$. The probability that event A occurs is p_o. If this experiment is repeated n times, what will be the probability that event A occurs exactly k times? Note that each repetition is assumed statistically independent and that probability p_o is constant between experiments. If X is the binomial variable, based on n repetitions, the probability density function of X is given by[13]

$$p_X(x=k) = \binom{n}{k} p_o^k (1-p_o)^{n-k}, \qquad (3.17)$$

where

$$\binom{n}{k} = \frac{n!}{k!\,(n-k)!} = \frac{n(n-1)\cdots(n-k+1)}{k!}.$$

---

[13] The reason why this distribution is called binomial is that the kth term in the summation is the same as the kth term in the binomial expansion of $(p_o + (1-p_o))^n$.

3.4. USEFUL PROBABILITY DENSITIES

For $n = 5$ and $p_o = 0.3$, the probabilities are given by

$$p_X(x = 0) = \frac{5!}{0!5!} \times 0.3^0 \times (1 - 0.3)^5 = 0.16807$$

$$p_X(x = 1) = \frac{5!}{1!4!} \times 0.3^1 \times (1 - 0.3)^4 = 0.36015$$

$$p_X(x = 2) = \frac{5!}{2!3!} \times 0.3^2 \times (1 - 0.3)^3 = 0.30870$$

$$p_X(x = 3) = \frac{5!}{3!2!} \times 0.3^3 \times (1 - 0.3)^2 = 0.13230$$

$$p_X(x = 4) = \frac{5!}{4!1!} \times 0.3^4 \times (1 - 0.3)^1 = 0.02835$$

$$p_X(x = 5) = \frac{5!}{5!0!} \times 0.3^5 \times (1 - 0.3)^0 = 0.00243.$$

The reader can verify that $\sum_{k=0}^{5} p_X(x = k) = 1$. The expected value of the binomial distribution is

$$\mu_X = \sum_{k=0}^{n} k p_X(x = k)$$

$$= \sum_{k=0}^{n} k \frac{n!}{k!(n-k)!} p_o^k (1 - p_o)^{n-k}.$$

Since the first term equals zero, μ_X can be written as

$$\mu_X = \sum_{k=1}^{n} k \frac{n!}{k!(n-k)!} p_o^k (1 - p_o)^{n-k}.$$

Replace k with $l + 1$,

$$\mu_X = \sum_{l=0}^{n-1} (l+1) \frac{n!}{(l+1)!(n-1-l)!} p_o^{l+1} (1 - p_o)^{n-1-l}$$

$$= n p_o \sum_{l=0}^{n-1} \frac{(n-1)!}{l!(n-1-l)!} p_o^l (1 - p_o)^{n-1-l}.$$

Further replace $n - 1$ with m,

$$\mu_X = n p_o \sum_{l=0}^{m} \frac{m!}{l!(m-l)!} p_o^l (1 - p_o)^{m-l}$$

$$\mu_X = n p_o,$$

because the summation of all the individual probabilities add to 1.

Similarly, the variance is found to be $np_o(1-p_o)$. The reader is encouraged to derive this value.

Example 3.18 Binomial Distribution

Suppose that 300 cars are being assembled every day in an assembly line. After they are assembled, each vehicle is tested. On average, 2% of the tested vehicles are rejected. The rejected vehicles are then taken apart and reassembled. Find the expected value and standard deviation of the number of vehicles that will pass the test. Also, what is the probability that all 300 vehicles will pass the test?

Solution Let X be the number of vehicles that will pass the test. X has a binomial distribution. The expected value and the standard deviation are given by

$$\mu_X = np_o = 294$$
$$\sigma_X = \sqrt{np_o(1-p_o)} = 2.425,$$

and the probability that k vehicles will pass the test is

$$p_X(x=k) = \binom{300}{k} 0.98^k 0.02^{300-k}.$$

The probability that all 300 vehicles will pass the test is

$$p_X(x=300) = \binom{300}{300} 0.98^{300} 0.02^0$$
$$= 2.3325 \times 10^{-3},$$

a very small value. It is highly unlikely that all 300 vehicles will pass the test on any given day.
⊛

The Poisson Distribution

Consider experiments similar to those that led to the binomial distribution. The outcome of the experiment can be either event A or $\bar{A}$. The probability that the event A occurs is p_o. If this experiment is repeated an infinite number of times instead of n times, what will be the probability that the event A occurs exactly k times? Each repetition is assumed statistically independent and outcome probability p_o is constant between experiments. Further, assume that $p_o \to 0$ as $n \to \infty$ such that the expected value np_o stays constant and equal to λ.

3.4. USEFUL PROBABILITY DENSITIES

Replace p_o with λ/n in Equation 3.17,

$$p_{binomial\ X}(x=k)$$
$$= n(n-1)\cdots(n-k+1)\frac{1}{k!}\left(\frac{\lambda}{n}\right)^k\left(1-\frac{\lambda}{n}\right)^{n-k}$$
$$= 1\cdot\left(1-\frac{1}{n}\right)\cdots\left(1-\frac{k-1}{n}\right)\frac{1}{k!}(\lambda)^k\left(1-\frac{\lambda}{n}\right)^{n-k}$$
$$= 1\cdot\left(1-\frac{1}{n}\right)\cdots\left(1-\frac{k-1}{n}\right)\frac{1}{k!}\left(\frac{\lambda n}{n-\lambda}\right)^k\left(1-\frac{\lambda}{n}\right)^n.$$

Let $n \to \infty$,

$$\lim_{n\to\infty} p_{binomial\ X}(x=k) = \frac{e^{-\lambda}\lambda^k}{k!},$$

where $\lim_{n\to\infty}(1-\lambda/n)^n = e^{-\lambda}$. This is called the *Poisson distribution*,

$$p_{Poisson\ X}(x=k) = \frac{e^{-\lambda}\lambda^k}{k!}, \quad k=0,1,2,3,\cdots,$$

with mean and variance both given by

$$\mu_X = \sigma_X^2 = \lambda.$$

This shows that, in the limit, the Poisson distribution converges to the binomial distribution as $n \to \infty$, $p_o \to 0$, and $np_o = \lambda$. This interpretation of the Poisson distribution is useful when we approximate the value of the binomial distribution for a large value of n. The term $n!$ in the binomial distribution can be quite large for a moderate value of n, and approximations may be in order.

Sometimes, λ is written as νt, where ν equals the average number of occurrences per unit time or over a physical space, and t is a parameter such as time or physical space. The Poisson distribution is particularly useful in reliability theory. The subject of reliability is explored in Chapter 9.

Poisson   Siméon-Denis Poisson was born (June 21, 1781) in Pithiviers, France, and died (April 25, 1840) in Sceaux near Paris, France. Poisson's parents were not from the nobility and it was becoming increasingly difficult to distinguish between the nobility and the bourgeoisie in France in the years prior to the Revolution. Nevertheless the French class system still had a major influence on his early years. The main reason for this was that the army was one of the few occupations where the nobility enjoyed significant institutional privileges and Poisson's father had been a

Figure 3.24: Siméon-Denis Poisson (1781-1840).

soldier. Certainly Poisson's father was discriminated against by the nobility in the upper ranks of the army and this made a large impression on him. After retiring from active service he was appointed to a lowly administrative post that he held at the time that his son Siméon-Denis was born. There is no doubt that Siméon-Denis' family put a great deal of energy into helping him to a good start in life.

Now Siméon-Denis was not the first of his parents' children but several of his older brothers and sisters had failed to survive. Indeed his health was also very fragile as a child and he was fortunate to pull through. This may have been because his mother, fearing that her young child would die, entrusted him to the care of a nurse to bring him through the critical period. His father had a large influence on his young son, devoting time to teach him to read and write.

Siméon-Denis was eight years old when the Parisian insurrection of July 14, 1789, heralded the start of the French Revolution. As might be expected of someone who had suffered discrimination at the hands of the nobility, Poisson senior was enthusiastic about the political turn of events. One immediate consequence for his support of the Revolution was the fact that he became president of the district of Pithiviers, which is in central France, about 80 km south of Paris. From this position he was able to influence the future career of his son.

Poisson's father decided that the medical profession would provide a secure future for his son. An uncle of Poisson's was a surgeon in Fontainebleau

3.4. USEFUL PROBABILITY DENSITIES 95

and Poisson was sent there to become an apprentice surgeon. However, Poisson found that he was ill suited to be a surgeon. First, he lacked coordination to quite a large degree, which meant that he completely failed to master the delicate movements required. Second, it was quickly evident that, although he was a talented child, he had no interest in the medical profession. Poisson returned home from Fontainebleau having essentially failed to make the grade in his apprenticeship and his father had to think again to find a career for him.

Times were changing quite quickly in France, which was by this time a republic. No longer were certain professions controlled by the nobility as they had been and there had been moves toward making education available to everyone. In 1796 Poisson was sent back to Fontainebleau by his father, this time to enroll in the École Centrale. On the one hand he had shown a great lack of manual dexterity, but he now showed that he had great talents for learning, especially mathematics. His teachers at the École Centrale were extremely impressed and encouraged him to sit for the entrance examinations for the École Polytechnique in Paris. He proved his teachers right, for although he had far less formal education than most of the young men taking the examinations he achieved the top place.

Few people can have achieved academic success as quickly as Poisson did. When he began to study mathematics in 1798 at the École Polytechnique, he was therefore in a strong position to cope with the rigors of a hard course, yet overcome the deficiencies of his early education. There were certainly problems for him to overcome for he had little experience in the social or academic environment into which he was suddenly thrust. It was therefore to his credit that he was able to undertake his academic studies with great enthusiasm and diligence, yet find time to enjoy the theater and other social activities in Paris. His only weakness was the lack of coordination that had made a career as a surgeon impossible. This was still a drawback to him in some respects, for drawing mathematical diagrams was quite beyond him.

His teachers Laplace and Lagrange quickly saw his mathematical talents. They were to become friends for life with their extremely able young student and they gave him strong support in a variety of ways. A memoir on finite differences, written when Poisson was 18, attracted the attention of Legendre. However, Poisson found that descriptive geometry, an important topic at the École Polytechnique because of Monge, was impossible for him to succeed with because of his inability to draw diagrams. This would have been an insurmountable problem had he been going into public service, but those aiming at a career in pure science could be excused the drawing requirements, and Poisson was not held back.

In his final year of study he wrote a paper on the theory of equations and Bezout's theorem, and this was of such quality that he was allowed to graduate in 1800 without taking the final examination. He proceeded immediately to the position of répétiteur in the École Polytechnique, mainly on the strong recommendation of Laplace. It was quite unusual for anyone to gain a first appointment in Paris, most of the top mathematicians having to serve in the provinces before returning to Paris.

Poisson was named deputy professor at the École Polytechnique in 1802, a position he held until 1806 when he was appointed to the professorship at the École Polytechnique that Fourier had vacated when he had been sent by Napoleon to Grenoble. In fact Poisson had little time for politics; his energies were directed to support mathematics, science, education and the École Polytechnique. When the students at the École had been about to publish an attack on Napoleon's ideas for the Grand Empire in 1804, Poisson had managed to stop them, not because he supported Napoleon's views but rather because he saw that the students would damage the École Polytechnique by their actions. Poisson's motives were not understood by Napoleon's administration, however, and they saw Poisson as a supporter, which did his career no harm at all.

During this period Poisson studied problems relating to ordinary differential equations and partial differential equations. In particular he studied applications to a number of physical problems such as the pendulum in a resisting medium and the theory of sound. His studies were purely theoretical, however, for as we mentioned above, he was extremely clumsy with his hands: *Poisson was content to remain totally unfamiliar with the vicissitudes of experimental research. It is quite unlikely that he ever attempted an experimental measurement, nor did he try his hand at drafting experimental designs.*

His first attempt to be elected to the Institute was in 1806 when he was backed by Laplace, Lagrange, Lacroix, Legendre, and Biot for a place in the Mathematics Section. Bossut was 76 years old at the time and, had he died, Poisson would have gained a place. However Bossut lived for another seven years so there was no route into the mathematics section for Poisson. He did, however, gain further prestigious posts. In addition to his professorship at the École Polytechnique, in 1808 Poisson became an astronomer at Bureau des Longitudes. In 1809 he added another appointment, namely that of the chair of mechanics in the newly opened Faculté des Sciences.

In 1808 and 1809 Poisson published three important papers with the Academy of Sciences. In the first, *Sur les inégalités des moyens mouvement des planètes*, he looked at the mathematical problems that Laplace and Lagrange had raised about perturbations of the planets. His approach to these problems was to use series expansions to derive approximate solutions. This was typical of the type of problem that he found interesting. Libri

3.4. USEFUL PROBABILITY DENSITIES

wrote: ... *he especially liked unresolved questions that had been treated by others or areas in which there was still work to be done.*

In 1809 he published two papers, the first, *Sur le mouvement de rotation de la terre,* and the second, *Sur la variation des constantes arbitraires dans les questions de méchanique,* were a direct consequence of developments in Lagrange's method of variation of arbitrary constants, which had been inspired by Poisson's 1808 paper. In addition he published a new edition of *Clairaut's Théorie de la Figure de la Terre* in 1808. The work had been first published by Clairaut in 1743 and it confirmed the Newton-Huygens belief that the Earth was flattened at the poles. In 1811 Poisson published his two-volume treatise, *Traité de Mécanique,* which was an exceptionally clear treatment based on his course notes at the École Polytechnique.

Malus was known to have a terminal illness by 1811 and his death would leave a vacancy in the physics section of the Institute. The mathematicians, aiming to have Poisson fill that vacancy when it occurred, set the topic for the Grand Prix on electricity so as to maximize Poisson's chances. The topic for the prize was as follows: To determine by calculation and to confirm by experiment the manner in which electricity is distributed at the surface of electrical bodies considered either in isolation or in the presence of each other, for example, at the surface of two electrified spheres in the presence of each other. In order to simplify the problem, one needed to look only at an examination of cases where the electricity spread on each surface remains always of the same kind.

Poisson had made considerable progress with the problem before Malus died on February 24, 1812. Poisson submitted the first part of his solution to the Academy on 9 March entitled *Sur la distribution de l'électricité à la surface des corps conducteurs.* As the mathematicians had intended, this was the deciding factor in Poisson being elected to the physics section of the Institute to replace Malus. It also marked a move away from experimental research toward theoretical research in what was considered to constitute physics, and in this the Institute was following the lead given by Laplace.

Poisson continued to add various responsibilities to his already busy life. In 1815 he became examiner for the École Militaire and in the following year he became an examiner for the final examinations at the École Polytechnique.

It is remarkable how much work Poisson put into his research, his teaching, and into playing an ever increasingly important role in the organization of mathematics in France. When he married Nancy de Bardi in 1817 he found that family life put yet another pressure on him, yet somehow he survived the pressures, continuing to take on further duties. His research contributions covered a wide range of applied mathematics topics. Although he devised no innovative new theories, he made major contributions to further develop the theories of others, often being the first to exhibit their real significance. We mention now just a few of the topics he studied after his election to the Academy.

In 1813 Poisson studied the potential in the interior of attracting masses, producing results that would find application in electrostatics. He produced major work on electricity and magnetism, followed by work on elastic surfaces. Papers followed on the velocity of sound in gasses, on the propagation of heat, and on elastic vibrations. In 1815 he published a work on heat, which annoyed Fourier, who wrote: *Poisson has too much talent to apply it to the work of others. To use it to discover what is already known is to waste it ... .* Fourier went on to make valid objections to Poisson's arguments, which he corrected in later memoirs of 1820 and 1821.

In 1823 Poisson published on heat, producing results that influenced Sadi Carnot. Much of Poisson's work was motivated by results of Laplace, in particular his work on the relative velocity of sound and his work on attractive forces. This latter work was not only influenced by Laplace's work but also by the earlier contributions of Ivory. Poisson's work on attractive forces was itself a major influence on Green's major paper of 1828 although Poisson never seems to have discovered that Green was inspired by his formulations.

In *Recherchés sur la probabilité des jugements en matière criminelle et matière civile*, an important work on probability published in 1837, the Poisson distribution first appears. The Poisson distribution describes the probability that a random event will occur in a time or space interval under the conditions that the probability of the event occurring is very small, but the number of trials is very large so that the event actually occurs a few times. He also introduced the expression "law of large numbers." Although we now rate this work as of great importance, it found little favor at the time, the exception being in Russia where Chebyshev developed his ideas.

It is interesting that Poisson did not exhibit the chauvinistic attitude of many scientists of his day. Lagrange and Laplace recognized Fermat as the inventor of the differential and integral calculus; he was French after all while neither Leibniz nor Newton were! Poisson, however, wrote in 1831: *This [differential and integral] calculus consists in a collection of rules ... rather than in the use of infinitely small quantities ... and in this regard its creation does not predate Leibniz, the author of the algorithm and of the notation that has generally prevailed.*

He published between 300 and 400 mathematical works in all. Despite this exceptionally large output, he worked on one topic at a time. Libri writes: *Poisson never wished to occupy himself with two things at the same time; when, in the course of his labors, a research project crossed his mind that did not form any immediate connection with what he was doing at*

3.4. USEFUL PROBABILITY DENSITIES

the time, he contented himself with writing a few words in his little wallet. The persons to whom he used to communicate his scientific ideas know that as soon as he had finished one memoir, he passed without interruption to another subject, and that he customarily selected from his wallet the questions with which he should occupy himself. To foresee beforehand in this manner the problems that offer some chance of success, and to be able to wait before applying oneself to them, is to show proof of a mind both penetrating and methodical.

Poisson's name is attached to a wide variety of ideas, for example: Poisson's integral, Poisson's equation in potential theory, Poisson brackets in differential equations, Poisson's ratio in elasticity, and Poisson's constant in electricity. However, he was not highly regarded by other French mathematicians either during his lifetime or after his death. His reputation was guaranteed by the esteem that he was held in by foreign mathematicians who seemed more able than his own colleagues to recognize the importance of his ideas. Poisson himself was completely dedicated to mathematics. Arago reported that Poisson frequently said: *Life is good for only two things: to study mathematics and to teach it.*

3.4.7 Moment-Generating Functions

The *moment-generating function* $M_X(t)$ of random variable X is very useful because all the moments of X can be derived directly from this function. It is defined as

$$\begin{aligned} M_X(t) &= E\{\exp(tX)\} \\ &= \int_{-\infty}^{\infty} \exp(tx) f_X(x)\,dx, \end{aligned} \quad (3.18)$$

where $f_X(x)$ is the probability density function of X, and $M_X(t)$ is its Laplace transform. Write the exponential function in its power series form,

$$\exp(tx) = \sum_{k=0}^{\infty} \frac{1}{k!}(tx)^k,$$

and substitute into Equation 3.18 to find that

$$\begin{aligned} M_X(t) &= \int_{-\infty}^{\infty} \left(1 + tx + \frac{1}{2!}(tx)^2 + \frac{1}{3!}(tx)^3 + \cdots\right) f_X(x)\,dx \\ &= 1 + tE\{X\} + \frac{1}{2!}t^2 E\{X^2\} + \frac{1}{3!}t^3 E\{X^3\} + \frac{1}{4!}t^4 E\{X^4\} \cdots. \end{aligned}$$

All the moments can be evaluated using the appropriate derivative of the moment generating function,

$$E\{X^k\} = \left|\frac{d^k}{dt^k} M_X(t)\right|_{t=0}. \tag{3.19}$$

It is interesting to note that if experimental data can be used to estimate the moments of X, then Equation 3.19 can be used to find $M_X(t)$ and finally, *in principle*, by an inverse Laplace transform $f_X(x)$ can be recovered.

Example 3.19 Moment-Generating Function

For X governed by the standard normal density,

$$E\{X\} = 0$$
$$E\{X^2\} = 1,$$

with all higher-order moments related to the first two moments by

$$E\{X^n\} = 1 \cdot 3 \cdots (n-1)\sigma^n \quad \text{for } n \text{ even},$$

and zero for n odd, the moment generating function is given by

$$\begin{aligned}M_Z(t) &= 1 + \frac{1}{2!}t^2(\sigma^2) + \frac{1}{4!}t^4(3\sigma^4) \cdots \\ &= 1 + \frac{1}{2!}t^2 + \frac{3}{4!}t^4 \cdots \\ &= \exp\left(\frac{1}{2}t^2\right).\end{aligned}$$

❊

Characteristic Function

A related function is the *characteristic function*, defined as

$$\Phi_X(t) = E\{\exp(itX)\},$$

the Fourier transform of $f_X(x)$. The characteristic function exists for all probability density functions, whereas this is not true for the moment-generating function. The moments can be derived using this function, as shown in the example below.

3.5. TWO RANDOM VARIABLES

Example 3.20 Characteristic Function

Using the definition of characteristic function,

$$\Phi_X(t) = 1 + (it) E\{X\} + \frac{1}{2!}(it)^2 E\{X^2\} + \cdots.$$

Then,

$$E\{X^n\} = \frac{1}{i^n} \frac{d^n}{dt^n} \Phi_X(t)\bigg|_{t=0}.$$

For the Gaussian,

$$\begin{aligned}\Phi_X(t) &= 1 + \frac{1}{2!}(it)^2 (\sigma^2) + \frac{1}{4!}(it)^4 (3\sigma^4)\cdots \\ &= \exp\left(-\frac{1}{2}t^2\right).\end{aligned}$$

Therefore, for the (standard) Gaussian,

$$\begin{aligned}E\{X\} &= \frac{1}{i}\frac{d}{dt}\exp\left(-\frac{1}{2}t^2\right)\bigg|_{t=0} \\ &= \frac{1}{i}\frac{d}{dt}\exp\left(-\frac{1}{2}t^2\right)\bigg|_{t=0} \\ &= \frac{1}{i}\left(-t\exp\left(-\frac{1}{2}t^2\right)\right)\bigg|_{t=0} \\ &= 0.\end{aligned}$$

and

$$\begin{aligned}E\{X^2\} &= \frac{1}{i^2}\frac{d^2}{dt^2}\exp\left(-\frac{1}{2}t^2\right)\bigg|_{t=0} \\ &= -\frac{d^2}{dt^2}\exp\left(-\frac{1}{2}t^2\right)\bigg|_{t=0} \\ &= \exp\left(-\frac{1}{2}t^2\right) - t^2 \exp\left(-\frac{1}{2}t^2\right)\bigg|_{t=0} \\ &= 1.\end{aligned}$$

⊛

3.5 Two Random Variables

We have already introduced concepts that relate two random variables. These are the concepts of statistical independence and conditional probability. Earlier concepts and definitions need to be generalized so that it is possible to address questions such as: *How do the possible values of random variable X affect the possible values of related random variable Y?* Generalizations for two random variables are utilized to answer these questions since these permit exploration of the necessary generalizations without needlessly crowding the concepts with too much algebra.

Consider the two random variables X and Y. The *joint cumulative distribution function* completely defines their probabilistic properties, and is denoted by

$$F_{XY}(x,y) = \Pr(X \le x, Y \le y).$$

This function defines the probability that random variable X is less than or equal to x *and* random variable Y is less than or equal to y. The definition is valid for discrete and continuous random variables. Based on this definition, the following general statements can be made about $F_{XY}(x,y)$,

$$F_{XY}(-\infty,-\infty) = 0 \qquad F_{XY}(\infty,\infty) = 1$$
$$F_{XY}(-\infty,y) = 0 \qquad F_{XY}(x,-\infty) = 0$$
$$F_{XY}(\infty,y) = F_Y(y) \qquad F_{XY}(x,\infty) = F_X(x).$$

As for a single random variable, $F_{XY}(x,y)$ is *nondecreasing* and $F_{XY}(x,y) \ge 0$.

Generalizing from our study of one random variable, and of joint discrete variables, the *joint density function* for a continuous pair of random variables is for small dx and dy, approximately defined by

$$f_{XY}(x,y)dx\,dy \simeq \Pr(x < X \le x+dx, y < Y \le y+dy),$$

and exactly equal to

$$\Pr(a < X \le b, c < Y \le d) = \int_c^d \int_a^b f_{XY}(x,y)dx\,dy. \qquad (3.20)$$

The joint distribution function is then

$$F_{XY}(x,y) = \int_{-\infty}^y \int_{-\infty}^x f_{XY}(u,v)du\,dv.$$

Conversely,

$$f_{XY}(x,y) = \frac{\partial^2 F_{XY}(x,y)}{\partial x\,\partial y}.$$

Equation 3.20 is the probability that two random variables are *simultaneously* within a certain range, specifically, that $a < X \le b$ and $c < Y \le d$. Such information is useful when considering problems that have two or more variables where it is necessary to establish how the values of one variable affect those of the other variables. This topic is considered in Section 3.5.1.

3.5. TWO RANDOM VARIABLES

For a process with two random variables, the volume under the density function must equal 1.

Suppose X and Y are discrete random variables. There then exists a *joint probability mass function* $p_{XY}(x,y)$, defined as

$$p_{XY}(x,y) = \Pr(X = x, Y = y).$$

The joint cumulative distribution function for these discrete random variables is given by the summation,

$$F_{XY}(x,y) = \sum_{x_i \leq x,\, y_i \leq y} p_{XY}(x_i, y_i).$$

A conditional probability mass function $p_{X|Y}(x|y)$ is defined, referring to Equation 2.7, as

$$p_{X|Y}(x|y) = \Pr(X = x|Y = y)$$
$$= \frac{p_{XY}(x,y)}{p_Y(y)}.$$

A *marginal probability mass function* can be derived using the theorem of total probability, $\Pr(A) = \sum_{i=1}^{n} \Pr(A|B_i)\Pr(B_i)$,

$$p_X(x) = \sum_{y_i} \Pr(X = x|Y = y_i)\Pr(Y = y_i)$$
$$= \sum_{y_i} \Pr(X = x, Y = y_i)$$
$$= \sum_{y_i} p_{XY}(x, y_i).$$

If X and Y are statistically independent, then $p_{X|Y}(x|y) = p_X(x)$, $p_{Y|X}(y|x) = p_Y(y)$, and

$$p_{XY}(x,y) = p_X(x)p_Y(y) \tag{3.21}$$

for all possible combinations of x and y.

Example 3.21 Discrete Marginal and Total Probabilities from Probability Mass Data

Suppose data are obtained from a series of experiments on two parameters, X and Y. For example, in the flow of atmosphere over a wing, X could be the normalized flow velocity and Y the turbulence fluctuation length scale. The data is tabulated as follows:

$y \backslash x$	25	27	29	$p_Y(y)$
1.0	7/30	1/30	0	[8/30]
1.5	0	5/30	1/30	[6/30]
2.0	1/30	6/30	1/30	[8/30]
2.5	0	1/30	7/30	[8/30]
$p_X(x)$	[8/30]	[13/30]	[9/30]	$\sum = 1$

X takes on values $x = $ 25, 27, and 29, and Y takes on the values $y = 1.0$, 1.5, 2.0, and 2.5. Some sample joint probabilities are

$$\Pr(X = 25, Y = 2.0) = p_{XY}(25, 2.0) = \frac{1}{30}$$

$$\Pr(X = 29, Y = 2.5) = p_{XY}(29, 2.5) = \frac{7}{30}.$$

The marginal probability $p_X(27)$ is found by summing the joint probabilities,

$$\begin{aligned} p_X(27) &= \Pr(X = 27) \\ &= \Pr(X = 27, Y = 1.0) + \Pr(X = 27, Y = 1.5) \\ &+ \Pr(X = 27, Y = 2.0) + \Pr(X = 27, Y = 2.5) \\ &= \frac{13}{30}, \end{aligned}$$

which is obtained by summing down the column under $X = 27$. Similarly, $p_X(25) = 8/30$ and $p_X(29) = 9/30$. Note that the marginal probability masses sum to 1, as they should since the complete probability space is covered. One can check whether X and Y are statistically independent by applying Equation 3.21 to all the respective probability mass values, for example, from the table,

$$p_{XY}(25, 1.0) \neq p_X(25) p_Y(1.0),$$

and therefore the random variables are not statistically independent. Note that for statistical independence to hold, it must hold for

$$p_{XY}(x_i, y_j) = p_X(x_i) p_Y(y_j),$$
$$\text{for all } (i, j).$$

3.5. TWO RANDOM VARIABLES

Example 3.22 Wind Speed Amplification due to Large Structures

We have all experienced high wind speeds around large structures such as skyscrapers. This amplification depends on the size and shape of the structure as well as on the surrounding terrain, and atmospheric conditions such as temperature (gradients). The possible speeds are modeled by a probability density, in this case a joint density in the plane of ground level. Suppose the component speeds in two orthogonal directions are the random variables U and V, and the joint probability density function is a decaying exponential, $f_{UV}(u,v) = C\exp(-3[u+v])$, $u \geq 0, v \geq u$. This density and the range on the variables has been derived for a specific site. Find the value of C. Calculate the probability $\Pr(V < 3U)$, and the probability that the maximum speed in both directions is less than 5.

Solution The value of C is obtained by requiring the joint density to satisfy the relation $\int_u \int_v f_{UV}(u,v)\,dv\,du = 1$. For this problem,

$$\int_0^\infty \int_u^\infty C\exp(-3[u+v])\,dv\,du = 1$$

$$C = \frac{1}{\int_0^\infty \int_u^\infty \exp(-3[u+v])\,dv\,du} = 18.$$

To calculate the probability $\Pr(V < 3U)$, the limits of the integration are needed, and can be found by sketching the lines $u = v$ and $v = 3u$. See Figure 3.25.

Integrate between these lines,

$$\Pr(V < 3U) = \int_0^\infty \int_u^{3u} 18\exp(-3[u+v])\,dv\,du = \frac{1}{2}.$$

Similarly, the probability that both speeds are less than 5 is

$$\Pr(U, V < 5) = \int_0^5 \int_u^5 18\exp(-3[u+v])\,dv\,du$$
$$= \exp(-30) - 2\exp(-15) + 1$$
$$\simeq 1.0.$$

⊛

Continuing the study of joint random variables, define and understand conditional densities. Recall $\Pr(A|B) = \Pr(AB)/\Pr(B)$. The conditional probability is taken on a reduced sample space, here $\Pr(B)$. The conditional density function is needed to find probabilities such as

$$\Pr(a < X \leq b | Y = y).$$

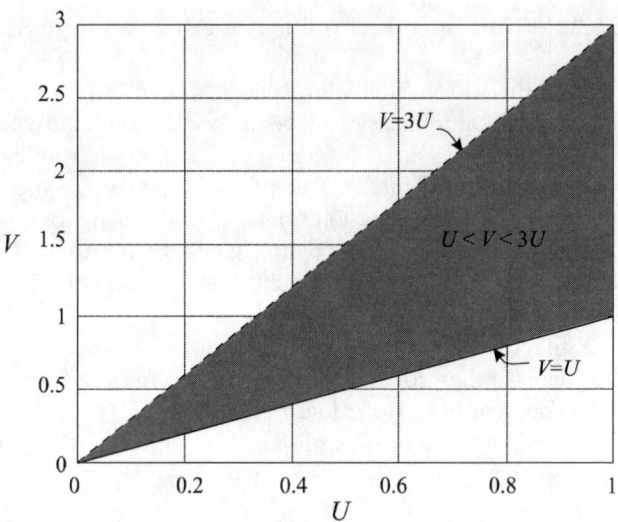

Figure 3.25: Domain of integration bounded by $V = 3U$ and $V = U$.

For continuous random variables $\Pr(Y = y) = 0$, therefore, define the conditional density function to be

$$\Pr(a < X \leq b | Y = y) = \int_a^b f(x|y) dx,$$

where $f(x|y)$ is sometimes written as $f_{X|Y}$ and is *the conditional density of X given Y*. Given that the area under a region of the density function is a measure of the respective probability, then

$$f(x|y) dx \simeq \Pr(x < X \leq dx | Y = y)$$
$$= \lim_{dy \to 0} \frac{\Pr(x < X \leq dx, y < Y \leq y + dy)}{\Pr(y < Y \leq y + dy)}$$
$$= \lim_{dy \to 0} \frac{f(x,y) dx\, dy}{f(y) dy} = \frac{f(x,y) dx}{f(y)}$$
$$f(x|y) = \frac{f(x,y)}{f(y)}. \tag{3.22}$$

Similarly, $f(y|x) = f(x,y)/f(x)$. These expressions lead to the relation that defines statistical independence in the following way. Rewrite Equation 3.22 as $f(x,y) = f(x|y)f(y)$. If X and Y are *statistically independent*, then $f(x|y) = f(x)$ and $f(x,y) = f(x)f(y)$.

3.5. TWO RANDOM VARIABLES

Another version of the total probability theorem can be obtained if the variable B is taken to be continuous. The summation becomes an integral, found by rewriting Equations 2.9 and 2.10, leading to

$$f_{XY}(x,y) = f_{X|Y}(x|y)f_Y(y)$$
$$f_X(x) = \int_{-\infty}^{\infty} f_{XY}(x,y)dy, \qquad (3.23)$$

where $f_X(x)$ is called the *marginal probability density function*. Similarly, $f_Y(y) = \int_{-\infty}^{\infty} f_{XY}(x,y)dx$.

Example 3.23 Marginal Densities

Given that the joint density of X and Y is uniform over the domain $Y \geq X^2 - 1$, $Y \leq 0$, as shown in Figure 3.26, find the density function and both marginal densities.

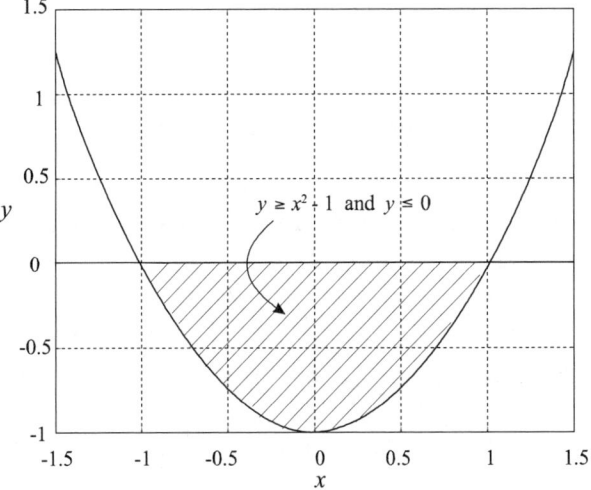

Figure 3.26: Domain of integration bounded by $Y \geq X^2 - 1$ and $Y \leq 0$.

Solution Since the density function is uniform, it equals the inverse of the area shown in the sketch. That is,

$$\text{Area} = \left| \int_{-1}^{1} (x^2 - 1) \, dx \right| = \frac{4}{3},$$

and the joint density $f_{XY}(x,y) = 1/\text{Area} = 3/4$. The marginal densities can now be evaluated,

$$f_X(x) = \int_{x^2-1}^{0} f_{XY}(x,y)\,dy$$

$$= \int_{x^2-1}^{0} \frac{3}{4}\,dy$$

$$= \frac{3}{4} - \frac{3}{4}x^2,$$

$$f_Y(y) = \int_{-\sqrt{y+1}}^{\sqrt{y+1}} f_{XY}(x,y)\,dx$$

$$= \int_{-\sqrt{y+1}}^{\sqrt{y+1}} \frac{3}{4}\,dx$$

$$= \frac{3}{2}\sqrt{(y+1)}.$$

The conclusion is that X and Y are not statistically independent, that is, $f_{XY}(x,y) \neq f_X(x)f_Y(y)$.

⊛

Example 3.24 Two Geometrical Parameters

Data taken from a manufacturing site concludes that two geometrical parameters primarily affect the design life. These two parameters, H and W, are jointly distributed according to

$$f_{WH}(w,h) = \begin{cases} 2w^2 e^{-w(1+2h)}, & w,h > 0 \\ 0, & \text{otherwise.} \end{cases}$$

Find the probability $\Pr(W \leq 0.1, H \leq 2.0)$. Derive the marginal densities. Evaluate the probability $\Pr(H \leq 2.0 \mid 0.1 < W \leq 0.2)$.

Solution The joint density is plotted in Figure 3.27. Then,

$$\Pr(W \leq 0.1, H \leq 2.0) = \int_0^{0.1} \int_0^2 2w^2 e^{-w(1+2h)}\,dh\,dw$$

$$= 1.0707 \times 10^{-3}.$$

The marginal densities are obtained using Equation 3.23 for each variable,

$$f_W(w) = \int_0^\infty f_{WH}(w,h)\,dh$$

$$= \int_0^\infty 2w^2 e^{-w(1+2h)}\,dh$$

$$= we^{-w}.$$

3.5. TWO RANDOM VARIABLES

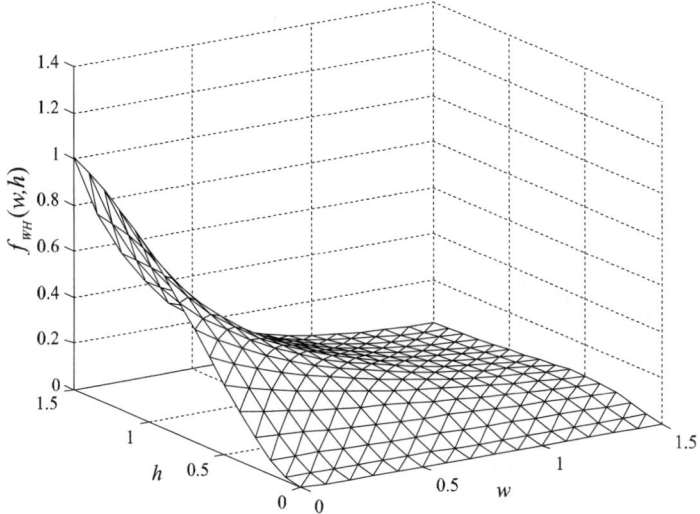

Figure 3.27: The joint density function for random variables w and h: $f_{WH}(w,h) = 2w^2 e^{-w(1+2h)}, \quad w, h > 0$.

Similarly,

$$f_H(h) = \int_0^\infty f_{WH}(w,h)\,dw$$
$$= \int_0^\infty 2w^2 e^{-w(1+2h)}\,dw$$
$$= \frac{4}{(1+2h)^3}.$$

Finally,

$$\Pr(H \leq 2.0 \,|\, 0.1 < W \leq 0.2) = \frac{\Pr(H \leq 2.0, 0.1 < W \leq 0.2)}{\Pr(0.1 < W \leq 0.2)}$$
$$= \frac{\int_0^2 \int_{0.1}^{0.2} f_{WH}(w,h)\,dw\,dh}{\int_{0.1}^{0.2} \left\{\int_0^\infty f_{WH}(w,h)\,dh\right\}\,dw}$$
$$= \frac{\int_0^2 \int_{0.1}^{0.2} 2w^2 e^{-w(1+2h)}\,dw\,dh}{\int_{0.1}^{0.2} \left\{\int_0^\infty 2w^2 e^{-w(1+2h)}\,dh\right\}\,dw}$$
$$= \frac{5.8828 \times 10^{-3}}{1.2844 \times 10^{-2}} = 0.45801.$$

where Equation 3.23 is used. Note that W and H are not statistically independent.

✱

3.5.1 Covariance and Correlation

Proceeding from models with one random variable to those with two requires the introduction of the concept of *covariance*, and the related parameter, the *correlation coefficient*. These are measures of how *linear* the relationship is between the two random variables. Consider the two random variables X and Y with second joint moment,

$$E\{XY\} = \int_{-\infty}^{\infty} \int_{-\infty}^{\infty} xy f_{XY}(x,y) dx\, dy. \qquad (3.24)$$

If X and Y are statistically independent, then the joint density function can be separated into the product of the respective marginal densities, $f_{XY}(x,y) = f_X(x)f_Y(y)$, and by Equation 3.24, the integrals can be separated, resulting in the relation $E\{XY\} = E\{X\}E\{Y\}$. This is a one-way relation; the product of the densities implies the product of the expectations but not the other way around. It is possible to have $E\{XY\} = E\{X\}E\{Y\}$ even when X and Y are not statistically independent.

Suppose $Z = X + Y$, where the mean values and standard deviations of X and Y are given. Find the mean value and standard deviation of the *derived variable* Z by taking the expectation of the equation,

$$E\{Z\} = E\{X + Y\} = E\{X\} + E\{Y\}$$
$$\mu_Z = \mu_X + \mu_Y.$$

This is true regardless of whether X and Y are statistically independent. To find σ_Z requires a bit more algebra,

$$\begin{aligned} Var\{Z\} &= E\{[Z - \mu_Z]^2\} \\ &= E\left\{[X + Y - \mu_X - \mu_Y]^2\right\} \\ &= E\{X^2\} - \mu_X^2 + E\{Y^2\} - \mu_Y^2 + 2E\{XY\} - 2\mu_X\mu_Y \\ &= Var\{X\} + Var\{Y\} + 2E\{XY\} - 2\mu_X\mu_Y. \end{aligned}$$

To evaluate the variance of the sum, Z, it is necessary to know the joint statistical properties embodied in $E\{XY\}$. If X and Y are statistically independent, then $E\{XY\} = E\{X\}E\{Y\}$ and

$$Var\{Z\} = Var\{X\} + Var\{Y\}$$
$$\sigma_Z^2 = \sigma_X^2 + \sigma_Y^2. \qquad (3.25)$$

3.5. TWO RANDOM VARIABLES

The *covariance* is defined as the second joint moment about mean values μ_X and μ_Y,

$$Cov\{X,Y\} = E\{(X-\mu_X)(Y-\mu_Y)\} = E\{XY\} - \mu_X\mu_Y. \qquad (3.26)$$

Note that if the variables are statistically independent, $Cov\{X,Y\} = 0$. The *correlation coefficient* is defined as the normalized (dimensionless) covariance,

$$\rho = \frac{Cov\{X,Y\}}{\sigma_X \sigma_Y}. \qquad (3.27)$$

To better understand the correlation coefficient, assume that X and Y are linearly related by the equation $X = aY$, where a is a constant. Then, the variance is calculated to be

$$\begin{aligned} Var\{X\} &= E\{X^2\} - E^2\{X\} \\ &= E\{a^2 Y^2\} - E^2\{aY\} \\ &= a^2 \left(E\{Y^2\} - E^2\{Y\}\right) \\ &= a^2 Var\{Y\}. \end{aligned}$$

Taking the square root of both sides, the standard deviations of X and Y are related by

$$\sigma_X = |a|\sigma_Y.$$

Similarly, the covariance is calculated,

$$\begin{aligned} Cov\{X,Y\} &= E\{XY\} - E\{X\}E\{Y\} \\ &= a\left(E\{Y^2\} - E^2\{Y\}\right) \\ &= a Var\{Y\} = a\sigma_Y^2. \end{aligned}$$

Therefore, Equation 3.27 becomes

$$\rho = \frac{a\sigma_Y^2}{\sigma_X \sigma_Y} = \frac{a\sigma_Y}{\sigma_X} = \frac{a}{|a|}.$$

If $a > 0$, the correlation coefficient $\rho = +1$. The random variables X and Y are in perfect positive correlation. If $a < 0$, the correlation coefficient $\rho = -1$, or X and Y are in perfect negative correlation.[14]

---

[14]It is important to realize that a high value for ρ may indicate strong correlation, but not direct *cause* and *effect* since X and Y may be correlated by virtue of being related to some third variable. Also, if X and Y are independent, $\rho = 0$, as we see in the following example. The converse is not necessarily true. $\rho = 0$ indicates the absence of a *linear* relationship; a random or a nonlinear functional relationship between X and Y is still a possibility.

Even though it is not shown here, it can be shown that these values, $\rho = +1$ and -1, are the extreme values of ρ. That is, $-1 \leq \rho \leq 1$. Figure 3.28 depicts representative correlations between data points for nonlinear and perfect linear relationships. Note that if the sample points for nonlinear relationships happen to be symmetrical around either $X = \mu_X$ or $Y = \mu_Y$, the correlation coefficient is zero.

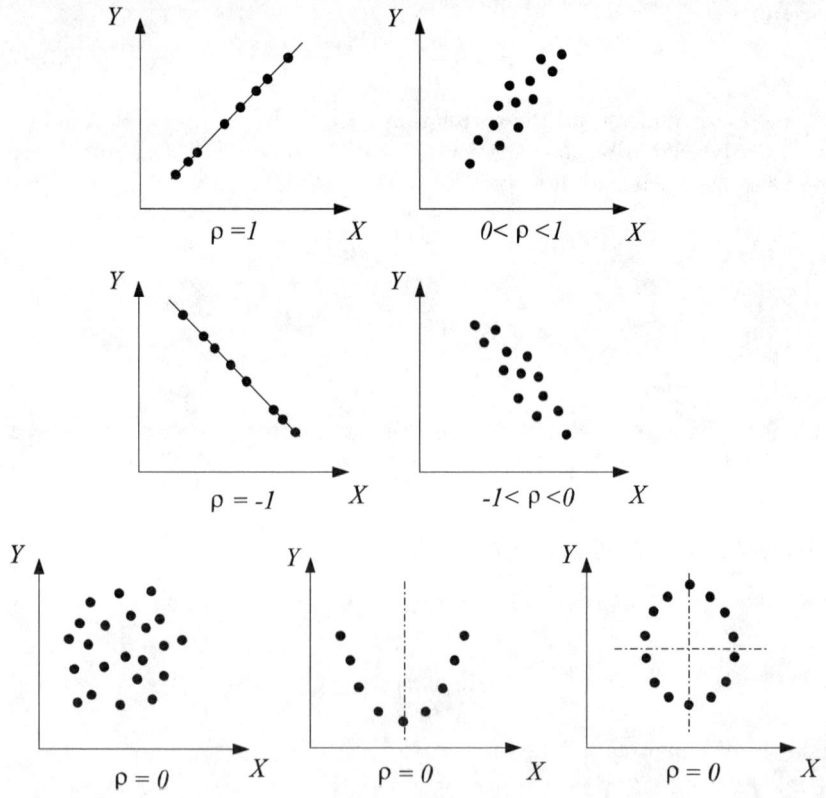

Figure 3.28: Discrete data distributions for various correlation coefficients.

Example 3.25 Jointly Distributed Variables

Two random variables X and Y are jointly distributed according to the joint density $f_{XY}(x, y) = \frac{1}{2}e^{-y}$, $y > |x|$, $-\infty < x < \infty$. See Figure 3.29. Compute the marginal densities and the covariance.

3.5. TWO RANDOM VARIABLES

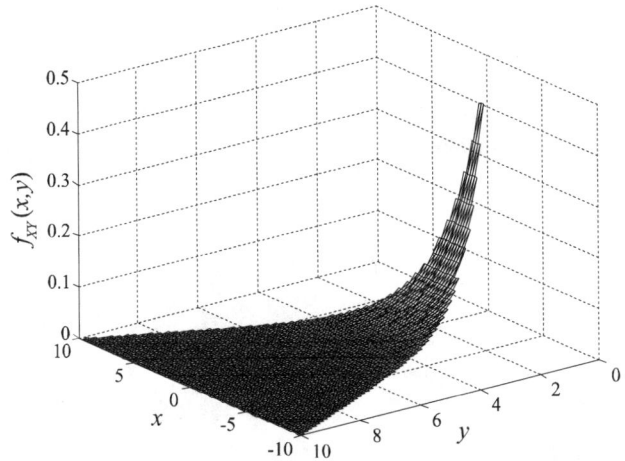

Figure 3.29: Joint probability density function, $f_{XY}(x,y) = \frac{1}{2}e^{-y}$ for $y > |x|$ and $-\infty < x < \infty$.

Solution The marginal densities are defined and evaluated:

$$f_X(x) = \int_{-\infty}^{\infty} f_{XY}(x,y)dy = \int_{|x|}^{\infty} \frac{1}{2}e^{-y}dy = \frac{1}{2}e^{-|x|}, \qquad -\infty < x < \infty$$

$$f_Y(y) = \int_{-\infty}^{\infty} f_{XY}(x,y)dx = \int_{-y}^{y} \frac{1}{2}e^{-y}dx = ye^{-y}, \qquad y > 0.$$

Since the joint density is not equal to the product of the two marginal densities, the variables are not statistically independent. Covariance Equation 3.26 requires an evaluation of the second joint moment,

$$E\{XY\} = \int_0^{\infty} \int_{-y}^{y} \frac{1}{2}xye^{-y}dx\,dy$$

$$= \frac{1}{2}\int_0^{\infty} ye^{-y}\int_{-y}^{y} x dx\, dy = 0,$$

and the respective mean values,

$$E\{X\} = \int_{-\infty}^{\infty} xf_X(x)dx = \int_{-\infty}^{0} \frac{1}{2}xe^{x}dx + \int_0^{\infty} \frac{1}{2}xe^{-x}dx = 0,$$

and similarly, $E\{Y\} = 2$. Therefore, $Cov\{X,Y\} = 0$, not because the two variables are independent, but because $E\{XY\} = E\{X\}E\{Y\}$ due to the particular numbers in the problem. $E\{XY\}$ can be equal to $E\{X\}E\{Y\}$ even when the two variables are not statistically independent.

⊛

Example 3.26 Statistics of Support Reaction of Simply Supported Beam

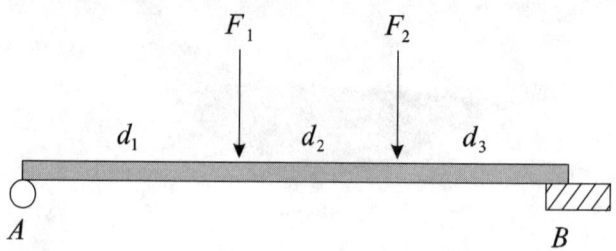

Figure 3.30: Simply supported beam loaded by random forces F_1 and F_2.

Consider the simply supported beam in Figure 3.30. There are two loads acting on the beam, F_1 at a distance of d_1 from the left end and F_2 at a distance of d_3 from the right end. The length of the beam is $L = d_1 + d_2 + d_3$. Given are the following loading statistics:

$$\begin{aligned} \mu_1 &= E\{F_1\} \\ \sigma_1 &= \sqrt{E\{F_1^2\} - \mu_1^2} \\ \mu_2 &= E\{F_2\} \\ \sigma_2 &= \sqrt{E\{F_2^2\} - \mu_2^2}. \end{aligned}$$

Derive the statistics of the reactions at A and B, R_A and R_B, assuming that the forces are statistically independent.

Solution From a static analysis of the free body of the beam assuming the forces to be exactly known, the functional relation between the forces, the dimensions and the reactions are

$$R_A = \frac{(F_1+F_2)\,d_3 + F_1 d_2}{L} = \frac{F_1\,(d_3+d_2) + F_2 d_3}{L}$$

$$R_B = \frac{(F_1+F_2)\,d_1 + F_2 d_2}{L} = \frac{F_2\,(d_1+d_2) + F_1 d_1}{L}.$$

While the forces are given as statistically independent, the reactions are dependent because they are both functions of the same parameters F_1

3.5. TWO RANDOM VARIABLES

and F_2. This can be shown by deriving the correlation coefficient. For this problem it is also assumed that the distances d_i are known exactly. Then, the mean values of the reactions are straightforwardly given by

$$\mu_A = \frac{(\mu_1 + \mu_2)\, d_3 + \mu_1 d_2}{L}$$
$$\mu_B = \frac{(\mu_1 + \mu_2)\, d_1 + \mu_2 d_2}{L}.$$

The variances can be calculated for these sums of independent random variables using Equation 3.25,

$$\sigma_A^2 = \frac{1}{L^2}\left[(d_3 + d_2)^2\, \sigma_1^2 + d_3^2 \sigma_2^2\right]$$
$$\sigma_B^2 = \frac{1}{L^2}\left[d_1^2 \sigma_1^2 + (d_1 + d_2)^2\, \sigma_2^2\right].$$

The standard deviations are the positive square roots of the respective variances. To find the correlation coefficient first find the expression for $E\{R_A R_B\}$,

$$E\{R_A R_B\} = E\left\{\frac{[F_1(d_3 + d_2) + F_2 d_3][F_2(d_1 + d_2) + F_1 d_1]}{L^2}\right\}$$
$$= \frac{1}{L^2} E\left\{\left(2d_1 d_3 + d_2 d_3 + d_1 d_2 + d_2^2\right) F_1 F_2\right.$$
$$\left. + d_1(d_3 + d_2) F_1^2 + d_3(d_1 + d_2) F_2^2\right\}$$
$$= \frac{1}{L^2}\left[\left(2d_1 d_3 + d_2 d_3 + d_1 d_2 + d_2^2\right) E\{F_1 F_2\}\right.$$
$$\left. + d_1(d_3 + d_2) E\{F_1^2\} + d_3(d_1 + d_2) E\{F_2^2\}\right]$$
$$= a_1 E\{F_1 F_2\} + a_2 E\{F_1^2\} + a_3 E\{F_2^2\},$$

where a_i are constant coefficients introduced to simplify the expression. We make use of independence, $E\{F_1 F_2\} = E\{F_1\} E\{F_2\}$. (If F_1 and F_2 are not statistically independent, their correlation coefficient $\rho_{F_1 F_2}$ is needed.) Also,

$$E\{F_1^2\} = \sigma_1^2 + \mu_1^2$$
$$E\{F_2^2\} = \sigma_2^2 + \mu_2^2.$$

Then,

$$E\{R_A R_B\} = a_1 \mu_1 \mu_2 + a_2\left(\sigma_1^2 + \mu_1^2\right) + a_3\left(\sigma_2^2 + \mu_2^2\right).$$

The covariance is given by

$$\text{Cov}\{R_A, R_B\} = E\{R_A R_B\} - \mu_A \mu_B,$$

and the correlation coefficient can now be expressed as

$$\rho_{R_A R_B} = \frac{Cov\{R_A, R_B\}}{\sigma_A \sigma_B},$$

which can be used for numerical evaluation. Consider the case where $\sigma_1 = \sigma_2 = \sigma$,

$$\sigma_A^2 = \frac{1}{L^2}\left[(d_3+d_2)^2 \sigma^2 + d_3^2\sigma^2\right] = \frac{\sigma^2\left(2d_3^2 + 2\,d_3d_2 + d_2^2\right)}{L^2}$$

$$\sigma_B^2 = \frac{1}{L^2}\left[d_1^2\sigma^2 + (d_1+d_2)^2\sigma^2\right] = \frac{\sigma^2\left(2d_1^2 + 2\,d_1d_2 + d_2^2\right)}{L^2},$$

and

$$\rho_{R_A R_B} = \frac{L^2\left[a_1\mu_1\mu_2 + a_2\left(\sigma^2 + \mu_1^2\right) + a_3\left(\sigma^2 + \mu_2^2\right) - \mu_A\mu_B\right]}{\sigma^2\sqrt{(2d_3^2 + 2\,d_3d_2 + d_2^2)(2d_1^2 + 2\,d_1d_2 + d_2^2)}}.$$

⊛

Example 3.27 Correlation Coefficient and Reliability

A number of examples have been drawn from the discipline of reliability, in part because it is very important in all aspects of engineering. Reliability is also a function of vibration characteristics, which depend on uncertainties in loading and material parameter values. Suppose that we have been tasked with the job of estimating the reliability of a component that is to be used in a particular stress environment. Assume that tests are performed on both component and loading to gather data. How do we proceed?

Solution   Define the strength of the component as X psi and the stress it experiences due to loading as Y psi. The strength may be a yield stress or an ultimate stress. Test a sufficient number of "identical" components in order to establish its strength probability density function, $f_X(x)$. Similarly, loading data leads us to a loading or stress density function $f_Y(y)$.

The strength is designed to exceed the stress for all but the most rare of cases, as one would expect. This is shown schematically in Figure 3.31.

The shaded region in the figure represents the realizations where the loading stress is greater than the component strength. This is a failure of the component. The probability of failure equals $\Pr([X-Y]\leq 0)$. Define $Z = X - Y$ with the goal of finding $\Pr(Z \leq 0)$. The reliability of the component is then defined as $R = 1 - \Pr(Z \leq 0)$, which is designed to be a very large number.

From Equations 3.26 and 3.27 we have

$$E\{XY\} = \rho\sigma_X\sigma_Y + \mu_X\mu_Y,$$

3.6. CONCLUDING SUMMARY

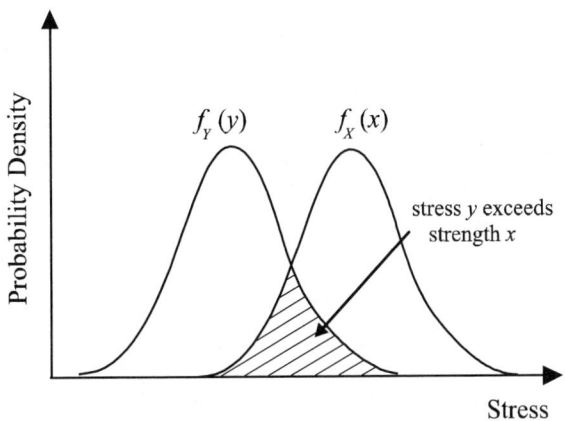

Figure 3.31: Reliability schematic where shaded region is a measure of the probability of failure.

and using Equation 3.9, the variance of the new variable Z is

$$Var\{Z\} = Var\{X\} + Var\{Y\} - 2\rho\sigma_X\sigma_Y.$$

Since strength and loading stress are uncorrelated, $\rho = 0$, and the variance of Z equals the variance of X plus the variance of Y, or

$$\sigma_Z = \sqrt{\sigma_X^2 + \sigma_Y^2}.$$

The mean value of Z is

$$\mu_Z = \mu_X - \mu_Y.$$

These are the mean and variance of the probability of failure. Given the density function $f_Z(z)$, the reliability of the component is

$$R = 1 - \int_{-\infty}^{0} f_Z(z)dz. \tag{3.28}$$

Suppose that both X and Y are Gaussian. Then it is known, and not proven here, that Z is also Gaussian. In this instance, $f_Z(z)$ is fully defined given μ_Z and σ_Z, and Equation 3.28 can then be evaluated.

※

3.6 Concluding Summary

In this chapter, the functions used to define random variables, their distribution functions, are defined. The probability distribution function or the probability density function fully describe the probabilistic properties of the random variable. The mathematical expectation is defined, a function that is utilized to derive the moments of the random variable. Various specific densities are defined and worked with, showing how the expectation function can be utilized. Additional descriptors are introduced in order to be able to work with models of more than one random variable. It becomes necessary to account for how the variables interact with each other. To address such questions, the concept of correlation is introduced.

3.7 Problems

Section 3.1: Probability Distribution Function

1. Which of the following Figures 3.32 *(a)*, *(b)*, *(c)* represent possible probability distribution functions?

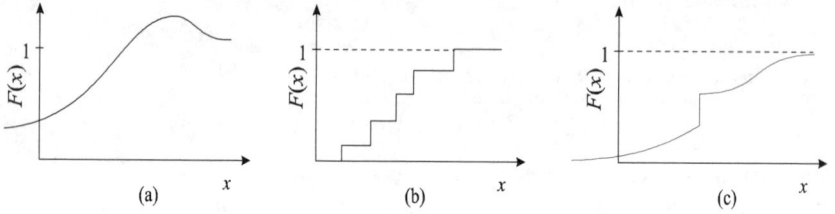

Figure 3.32: Possible distributions.

2. Show that the expected value of a constant equals that constant.

3. Suppose X is distributed as indicated in Figure 3.33. All lines are straight except for the exponential curve in the range $1 < x < 3$. Express $F_X(x)$ algebraically and then find the following probabilities:

   $\Pr(X = 1/3)$ $\Pr(X = 1)$ $\Pr(X < 1/3)$
   $\Pr(X \sim 1/3)$ $\Pr(X < 1)$ $\Pr(X \leq 1)$
   $\Pr(1 < X \leq 2)$ $\Pr(1 \leq X \leq 2)$ $\Pr(X = 1 \text{ or } 1/4 < X < 1/2)$.

Section 3.2: Probability Density Function

4. For each of the following functions, state why they are valid or invalid probability density functions. If arbitrary constants are present, evaluate them. Sketch the probability density functions. Then, find the cumulative distribution functions and sketch.

3.7. PROBLEMS

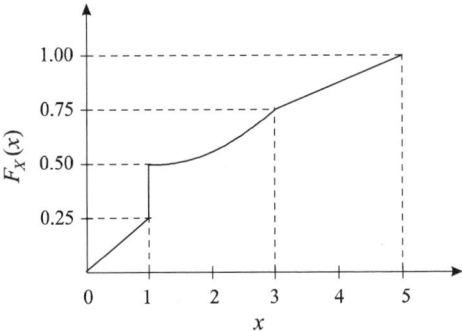

Figure 3.33: Distribution $F_X(x)$.

(i) $f_X(x) = 1/10, \quad 0 \le x \le 10,$
(ii) $f_X(x) = c\exp(-\lambda x), \quad x \ge 0,$
(iii) $f_X(x) = c + 1/20, \quad 0 \le x \le 10.$

Section 3.3: Mathematical Expectation

5. Find the expectation and variance of the following functions and data sets:

   (i) X, $f_X(x) = 1/3, \quad 1 \le x \le 4,$
   (ii) X^2, $f_X(x) = 1/3, \quad 1 \le x \le 4,$
   (iii) $2, 2.5, 2.5, 4, 4.3, 4.9, 7, 10, 10, 11$

6. Derive Equation 3.9.

7. For the random variable X, we are given the following probability density function,
$$f(x) = \frac{c}{x^2}, \quad x \ge 10.$$
Derive the cumulative distribution function, $F(x)$, and sketch both functions. Find the expected value of X and the standard deviation of X. What is the coefficient of variation? Is this a large number?

8. Two different probability density functions are recommended as models for a random variable that is representative of a particular material's strength, in nondimensional units:
$$\begin{aligned} f(x) &= c_1 \exp(-x), \quad 0 \le x \le 15, \\ f(x) &= c_2 x, \quad 0 \le x \le 15. \end{aligned}$$

Discuss how these models are different. Refer to their mean values, standard deviations, coefficients of variation. Plot both functions. Provide your recommendations as to which you think is a better physical model, and justify.

9. Materials and structures age with time. The rate of aging is variable and depends on many factors beyond the control of the designer. It is reasonable to say that this aging process is a random function of time. First, state a definition of aging: *How does an engineer define the aging of a material or a structure such as an aircraft?* Second, can one use any of the probability functions we have studied to date, such as the cumulative distribution function or the probability density function, to model such an aging process?

10. The time duration T of a force acting on a structure is a random variable having the probability density function,

$$f_T(t) = \begin{cases} at^2 & 0 \leq t \leq 12 \\ b & 12 \leq t \leq 16 \\ 0 & \text{elsewhere.} \end{cases}$$

(*i*) Determine the appropriate values for constants a and b.

(*ii*) Calculate the mean value and variance of T.

(*iii*) Calculate $\Pr(T \geq 6)$.

11. A strong wind at a particular site may come from any direction between due East at $\theta = 0°$ and due North at $\theta = 90°$. The wind speed V can vary between 0 and 150 mph.

(*i*) Sketch the sample space for wind speed and direction in one figure.

(*ii*) Let

$$\begin{aligned} A &= \{V \geq 20 \text{ mph }\} \\ B &= \{12 \text{ mph } \leq V \leq 30 \text{ mph }\} \\ C &= \{\theta \leq 30°\}. \end{aligned}$$

Identify the events $A, B, C, \overline{A}$ in the sample space sketched in part (*i*).

(*iii*) Use new sketches to identify the following events:

$$\begin{aligned} D &= A \cap C \\ E &= A \cup B \\ F &= A \cap B \cap C. \end{aligned}$$

(*iv*) Are the events D and E mutually exclusive? Are the events A and C mutually exclusive?

3.7. PROBLEMS

12. A fiber-optic cable must be manufactured to the highest tolerances. An important parameter is cable diameter D. Testing of the manufactured product finds the diameter to be normally distributed with a mean diameter of 0.1 m with a coefficient of variation of 0.02. A cable is considered unacceptable if its diameter is 3% off the mean value, that is, more than 3% above or below the mean value. What is the probability of a cable being unacceptable? Sketch the probability density function and the unacceptable region.

13. Two cables are used to lift load W. Normally, only cable A will be carrying the load; cable B is slightly longer than A, so it does not participate in carrying the load. But if cable A breaks, then B will have to carry the full load until A is replaced. We are given the following information: The probability that cable A will break is 0.02. The probability that B will fail if it has to carry the load by itself is 0.30.

$$
\begin{aligned}
A &= \text{cable } A \text{ breaks} \\
B &= \text{cable } B \text{ breaks} \\
\Pr(A) &= 0.02 \\
\Pr(B|A) &= 0.30 \qquad B \text{ would break only if } A \text{ already broke.}
\end{aligned}
$$

(i) What is the probability that both cables will fail?

(ii) If the load remains lifted, what is the probability that none of the cables have failed?

Section 3.4: Useful Probability Densities

14. Verify that the variance of a random variable governed by the binomial density equals $np_o(1 - p_o)$.

15. In Example 3.12, another design offers two pumps in parallel in the pipe. As in the example, each has a mean time to failure of 1000 hr, governed by an exponential density. State any assumptions made, and calculate the probability that both pumps will survive the first 200 hr of operation.

Section 3.5: Two Random Variables

16. Consider again Example 3.26 of the simply supported beam, with $\mu_1 = 30$, $\sigma_1 = 10$ and $\mu_2 = 50$, $\sigma_2 = 5$. Calculate the statistics of the two reactions and the correlation coefficient between the two reactions.

17. Consider again Example 3.26. Suppose the external forces are due to related causes, and therefore are statistically dependent. Explain what differences occur in the derivation.

18. For the statistically dependent version of Example 3.26, derived in the previous problem, calculate the statistics of the two reactions and their correlation coefficient for $\mu_1 = 30$, $\sigma_1 = 10$ and $\mu_2 = 50$, $\sigma_2 = 5$, and $E\{F_1 F_2\} = 75$.

19. Random variables X and Y have the following joint density function:

$$f(x,y) = K\exp[-(x+y)], \qquad x \geq 0, \quad y \geq x.$$

Sketch the domain of X, Y on the xy plane, and do the following:

(i) find the value of K that makes this function a valid probability density function,

(ii) find $\Pr(Y < 2X)$, and

(iii) find the probability that the maximum of X or Y is ≤ 4.

In parts (ii) and (iii) sketch the domains of integration in the xy plane.

20. The random variables X and Y have the joint uniform probability density function over the region bounded by the x axis and the curve $1 - x^2$. Sketch this domain in the xy plane. Find the joint probability density function $f(x, y)$, and the marginal densities: $f(x)$ and $f(y)$. Sketch the domains of integration.

21. Find the marginal density of X and the conditional density of Y given X, where

$$f_{XY}(x,y) = \begin{cases} 2 & x+y < 1,\ x \geq 0,\ y > 0 \\ 0 & \text{otherwise.} \end{cases}$$

22. The random variables X, Y are jointly distributed according to

$$f_{XY}(x,y) = a\cos x, \qquad \begin{cases} 0 \leq x \leq \pi/2 \\ 0 \leq y \leq 1. \end{cases}$$

(i) sketch this joint density function
(ii) find the marginal density functions f_X and f_Y
(iii) find $\Pr(X \leq \pi/4, Y \leq 1/2)$
(iv) find $\Pr(X > \pi/4, Y > 1/2)$
(v) are X, Y statistically independent?
(vi) what is the covariance $\text{Cov}(X, Y)$?

3.7. PROBLEMS

23. In Example 3.26, rederive the statistics assuming that the forces are correlated. Then reduce these equations for the cases $\rho_{12} = 1$ and $\rho_{12} = -1$. Compare results and discuss the importance of the correlation coefficient to the final results.

24. For Example 3.26, evaluate the statistics for the parameter values:

Parameter	Value
F_1	10
F_2	20
d_1	1
d_2	1
d_3	1

    Do this for the uncorrelated statistics. Then for the correlated statistics of the previous problem. Discuss the importance of correlation to the final results.

Chapter 4

Functions of Random Variables

Most applications require engineers to be able to relate the values of one variable to the values of another. Sometimes there are many variables involved in this relation called a function. When one or more of the variables are randomly distributed according to particular densities, there is the need to derive the densities of the derived functions in terms of the densities that are given. For the simplest case of $Y = g(X)$, given $f_X(x)$ and the functional relationship $g(X)$ between the variables X and Y, find the density $f_Y(y)$, or at least the mean value and standard deviation of Y. How much we are able to derive depends on the complexity of the function $g(X)$.

In some instances, it is not possible to find the density $f_X(x)$, but rather the statistical measures μ_X and σ_X. In such instances, the goal is to estimate μ_Y and σ_Y. In this chapter, general techniques for the exact and approximate evaluation of *derived densities* and *derived approximate statistical measures* are developed.

4.1 Exact Functions of One Variable

Given $f_X(x)$ and $g(X)$, where $Y = g(X)$, there is an interest in finding $f_Y(y)$. A general relation can be derived based on the assumption that $g(X)$ is simple enough to allow for a calculation of the inverse $X = g^{-1}(Y)$. The approach is best motivated using the graphical representation shown in Figure 4.1.

Using the figure as a reference, the strategy is to find the probability $\Pr(y < Y \leq y + dy)$ which, for sufficiently small dy, is approximately equal to $f_Y(y)\,dy$. In deriving $f_Y(y)$ it is useful to find an equivalent statement

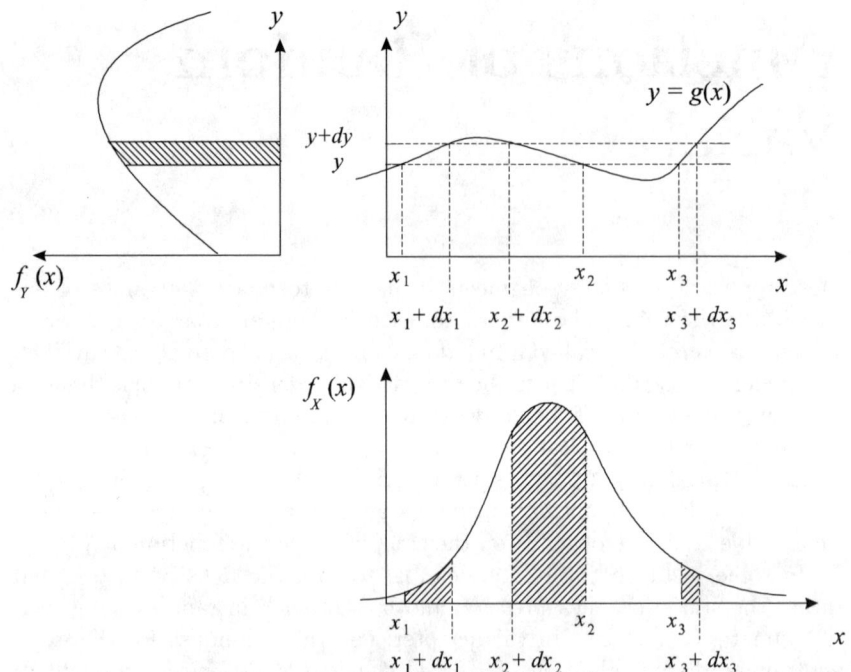

Figure 4.1: Transformation procedure schematic for a function of a random variable using the equation $y = g(x)$.

4.1. EXACT FUNCTIONS OF ONE VARIABLE

to $\Pr(y < Y \leq y + dy)$. From the figure it is observed that there is an equivalence between the following events,

$$\{y < Y \leq y + dy\} = \{x_1 < X \leq x_1 + dx_1\} + \{x_2 + dx_2 < X \leq x_2\} + \{x_3 < X \leq x_3 + dx_3\},$$

where $dx_1 > 0$, $dx_2 < 0$, and $dx_3 > 0$. Then,

$$\Pr(y < Y \leq y + dy) = \Pr(x_1 < X \leq x_1 + dx_1) + \Pr(x_2 + dx_2 < X \leq x_2) + \Pr(x_3 < X \leq x_3 + dx_3),$$

where the right-hand side equals the sum of the shaded areas in the figure, thus,

$$f_Y(y)\, dy \simeq f_X(x_1)\, dx_1 + f_X(x_2)\, |dx_2| + f_X(x_3)\, dx_3 \qquad (4.1)$$

is an approximation of the equality. To complete this derivation it is necessary to relate the increments dx_i to the functional relation between X and Y. From the graph,

$$g'(X) \equiv \frac{dg}{dX} \equiv \frac{dy}{dX},$$

and, therefore,

$$g'(x_i)\, dX|_{X=x_i} = dy,$$

since all the values of dy are identical. Let $dx_i \equiv dX|_{X=x_i}$ and replace each dx_i by $|dx_i|$ in Equation 4.1 for generality. Then,

$$f_Y(y)\, dy = \frac{f_X(x_1)}{|g'(x_1)|} dy + \frac{f_X(x_2)}{|g'(x_2)|} dy + \frac{f_X(x_3)}{|g'(x_3)|} dy.$$

In general, for any number n of roots x_i, the relation becomes

$$f_Y(y) = \sum_{i=1}^{n} \frac{f_X(x_i)}{|g'(x_i)|}.$$

Again, the use of this equation requires that $g^{-1}(X)$ exist. The roots x_i are implicitly real since for application to probability, only the real roots have any significance. In addition, the physical statement of the problem sometimes rules out certain roots as being not physically feasible.

The following are examples of this procedure for some special cases of $g(X)$ and $f_X(x)$.

Example 4.1 Lognormal Density

The lognormal density in Section 3.4.4 is revisited. Two random variables Y and X are related by $Y = e^X$. The random variable X is normally distributed with mean λ and standard deviation ζ. Find $f_Y(y)$.

Solution   The probability density function of X is given by

$$f_X(x) = \frac{1}{\zeta\sqrt{2\pi}} \exp\left\{-\frac{1}{2}\left(\frac{x-\lambda}{\zeta}\right)^2\right\}.$$

The transformation function is $g(X) = e^X$. Solving for x that corresponds to y, there is only one root, $x_1 = \ln y$. The derivative of $g(X)$ evaluated at x_1 is given by

$$\left.\frac{dg}{dX}\right|_{X=x_1} = e^{x_1} = y.$$

The probability density function of Y is then given by

$$f_Y(y) = \frac{f_X(x_1)}{|g'(x_1)|} = \frac{1}{y\zeta\sqrt{2\pi}} \exp\left\{-\frac{1}{2}\left(\frac{\ln y - \lambda}{\zeta}\right)^2\right\},$$

which is identical to Equation 3.15 with x and y reversed.
❋

Sums of Random Variables

Let the relation between X and Y be given by $Y = aX + b$, where a and b are constants. Here $g(X) = aX + b$ and $g'(X) = a$. Solving for X we find one root, $x_1 = (y-b)/a$ for every y. Therefore, for any density $f_X(x)$, the density of Y is then

$$f_Y(y) = \frac{f_X(x_1)}{|g'(x_1)|} = \frac{1}{|a|} f_X\left(\frac{y-b}{a}\right).$$

Now substitute the density function of X to find the density function of Y. As a simple example, suppose that $f_X(x) = 1/(m-n)$. Note that for the uniform density, the argument makes no difference since the density is a constant value. Then $f_X((y-b)/a) = 1/(m-n)$, and

$$f_Y(y) = \frac{1}{|a|} \frac{1}{m-n}, \qquad n < \frac{y-b}{a} < m,$$

or $an + b < y < am + b$.

Inverse Relation between Random Variables

Suppose the relation between X and Y is given by $Y = a/X$, where a is a constant and $X, Y > 0$. This relation also has only one root, $x_1 = a/y$ for

4.1. EXACT FUNCTIONS OF ONE VARIABLE

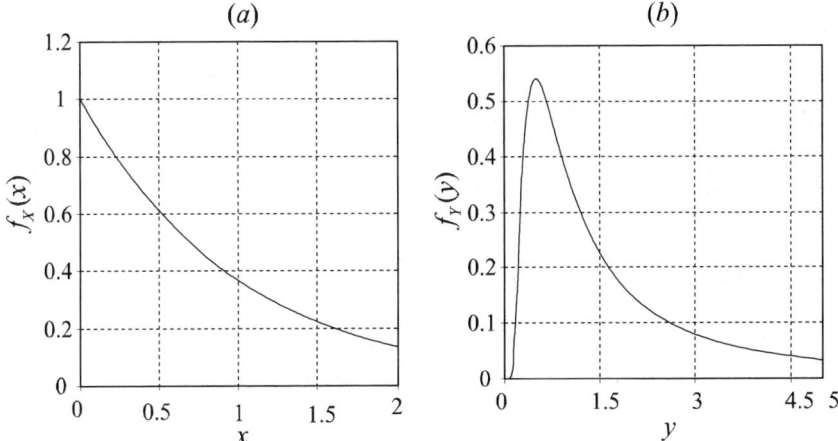

Figure 4.2: Density function for $Y = a/X$, (a) $f_X(x) = \exp(-x)$, (b) $f_Y(y) = (1/y^2)\exp(-1/y)$.

every y. Since $g(X) = a/X$, the derivative is $g'(X) = -a/X^2 = -Y^2/a$. Therefore,

$$f_Y(y) = \frac{f_X(x_1)}{|g'(x_1)|} = \frac{|a|}{y^2} f_X\left(\frac{a}{y}\right), \qquad y > 0.$$

If the density function of X is an exponential, $f_X(x) = \lambda \exp(-\lambda x)$, then $f_Y(y) = (|a|/y^2)\lambda \exp(-(a/y)\lambda)$. For numerical comparison, let $a = 1$ and $\lambda = 1$. The respective densities are shown in Figures 4.2(a) and (b).

Parabolic Transformation of Random Variables

In this case, the random variables X and Y are related by the parabolic equation $Y = aX^2$, $a > 0$. Since only the real roots are needed, and there are no real solutions if $Y < 0$, then $f_Y(y) = 0$ for this domain. If $Y \geq 0$, there are two solutions,

$$x_1 = +\sqrt{\frac{y}{a}} \qquad x_2 = -\sqrt{\frac{y}{a}}.$$

The functional relation is $g(X) = aX^2$, with its derivative $g'(X) = 2aX = 2a\sqrt{Y/a} = 2\sqrt{aY}$. Therefore, the general transformation is given by

$$f_Y(y) = \sum_{i=1}^{2} \frac{f_X(x_i)}{|g'(x_i)|} = \frac{1}{2\sqrt{ay}}\left\{f_X\left(\sqrt{\frac{y}{a}}\right) + f_X\left(-\sqrt{\frac{y}{a}}\right)\right\}, \qquad y \geq 0. \tag{4.2}$$

For X normally distributed, the probability density for X is given by

$$f_X(x) = \frac{1}{\sigma_X\sqrt{2\pi}} \exp\left\{-\frac{(x-\mu_X)^2}{2\sigma_X^2}\right\}, \qquad -\infty < x < \infty,$$

and the density for Y is

$$f_Y(y) = \frac{1}{\sigma_X\sqrt{2\pi ay}} \exp\left\{-\frac{\left(\sqrt{y/a}-\mu_X\right)^2}{2a\sigma_X^2}\right\}, \qquad y > 0. \tag{4.3}$$

For the case $\sigma_X = 1$, $\mu_X = 0$ and $a = 1$, the plots of these densities are given in Figures 4.3 (a) and (b).

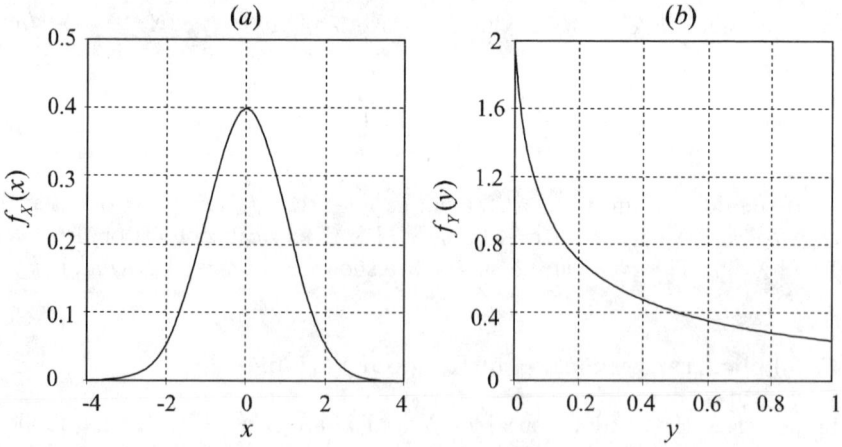

Figure 4.3: Density function for $Y = aX^2$, (a) $f_X(x) = \frac{1}{\sqrt{2\pi}}\exp\{-x^2/2\}$, (b) $f_Y(y) = \frac{1}{\sqrt{2\pi y}}\exp\{-y/2\}$.

The parabolic transformation has important applications, such as the drag force due to flow around a body. This drag force is proportional to the square of the velocity $F \sim V^2$. The proportionality constant is the drag coefficient C_D, that is, $F = C_D V^2$. Given the density of V, then using the

4.1. EXACT FUNCTIONS OF ONE VARIABLE

above discussion the density of F can be derived. With this force density function the analyst can calculate the probability that the force is in a certain range and base the design of the structure to a certain probability of failure. This subject of structural reliability is examined in more depth in Chapter 9.

Example 4.2 Parabolic Transformation

Consider a case where X is uniformly distributed in $(0, d)$. That is,

$$f_X(x) = 1/d, \qquad 0 < x < d.$$

The random variables Y and X are related by $Y = aX^2$. The probability density function for Y is then given by Equation 4.2,

$$f_Y(y) = \frac{1}{2\sqrt{ay}} \left\{ f_X\left(\sqrt{\frac{y}{a}}\right) + f_X\left(-\sqrt{\frac{y}{a}}\right) \right\}.$$

Note that $f_X\left(\sqrt{y/a}\right) = 1/d$. However, $f_X\left(-\sqrt{y/a}\right) = 0$ since $f_X(X)$ is defined only between 0 and d. Therefore,

$$f_Y(y) = \frac{1}{2\sqrt{ayd}}, \qquad 0 \leq y \leq ad^2. \tag{4.4}$$

⊛

Harmonic Transformation of Random Variables

In this case, the transformation leads to an infinity of roots due to the periodic nature of the harmonic function. Consider the relation,

$$Y = a \sin(X + \theta), \qquad a > 0. \tag{4.5}$$

There are roots only when $|Y| < a$. These roots are given by

$$x_i = \arcsin\left(\frac{y}{a}\right) - \theta, \qquad i = \cdots -1, 0, 1, \cdots,$$

with $g'(x_i) = a \cos(x_i + \theta)$. Then,

$$f_Y(y) = \frac{1}{\sqrt{a^2 - y^2}} \sum_{i=-\infty}^{\infty} f_X(x_i), \qquad |y| < a. \tag{4.6}$$

Use the fact that $g^2 + (g')^2 = a^2$. If $|y| > a$, there are no real solutions and $f_Y(y) = 0$. Note that $f_Y(\pm a) = \infty$. Since there are no probability masses at $\pm a$, this implies that $\Pr(y = \pm a) = 0$. Even though it appears that in Equation 4.5 Y can equal $\pm a$, we must be careful to distinguish between a deterministic and a random equation.

The example that follows shows how the physics of the problem helps limit the number of roots.

Example 4.3 Harmonic Transformation

An interesting application of this transformation is the problem of the path of a projectile. Suppose a particle is ejected at an initial velocity v and an angle of θ to the horizontal, from a point in space that we label as the origin. It is assumed that v is a constant, but θ is uniformly distributed between $(0, \pi/2)$. Find the density of the horizontal distance Y the projectile will fly before it returns to the horizontal, and find the probability that this distance is less than or equal to a particular value y. Basic physics yields the relation,

$$Y = \frac{v^2}{g} \sin 2\theta. \qquad (4.7)$$

This can be written more simply as $Y = a \sin \phi$, where $a = v^2/g$, and $\phi = 2\theta$, where ϕ is uniformly distributed between $(0, \pi)$. Note that the maximum value of Y is a. Using Equation 4.6, we find

$$f_Y(y) = \frac{1}{\sqrt{a^2 - y^2}} \left(\frac{1}{\pi} + \frac{1}{\pi} \right), \qquad 0 < y < a.$$

The probability $\Pr(Y \leq y)$ can now be found by integrating the area under the density function,

$$\Pr(Y \leq y) = F_Y(y) = \int_0^y f_Y(y_o) \, dy_o$$

$$= \int_0^y \frac{1}{\sqrt{a^2 - y_o^2}} \frac{2}{\pi} dy_o$$

$$= \frac{2}{\pi} \arcsin \frac{y}{a}, \qquad 0 < y < a.$$

For $a = 1$, $f_Y(y)$ and $F_Y(y)$ are plotted in Figures 4.4 (a) and (b).
⊛

4.2 Functions of Two or More RVs

The technique developed above can, in principle, be extended to two or more variables. However, as a practical matter, only simple functions are

4.2. FUNCTIONS OF TWO OR MORE RVS

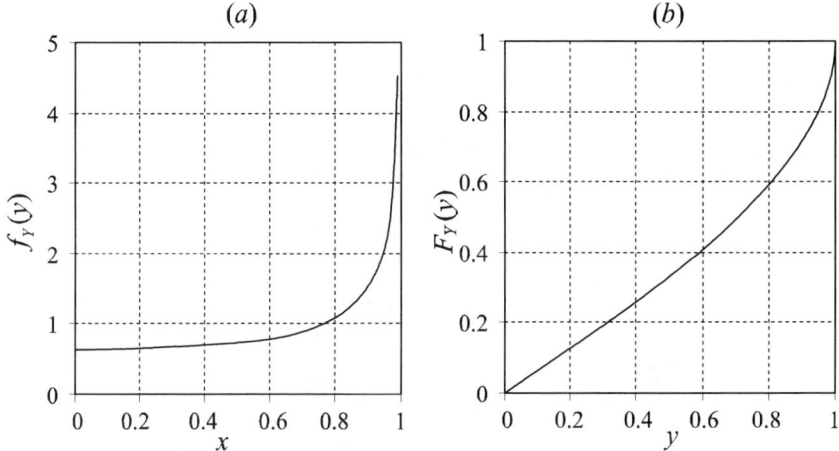

Figure 4.4: Density function for projectile Equation 4.7, (a) $f_Y(y) = 2/(\pi\sqrt{1-y^2})$, (b) $F_Y(y) = (2/\pi)\sin^{-1} y$.

Table 4.1: Table of Discrete Probabilities, $p_{XY}(x_i, y_i)$

Y\X	X = 0	X = 1	X = 2	X = 3	Sum
Y = 0	0.01	0.03	0.08	0.13	0.25
Y = 1	0.02	0.04	0.08	0.12	0.26
Y = 2	0.02	0.05	0.08	0.10	0.25
Y = 3	0.02	0.05	0.07	0.10	0.24
Sum	0.07	0.17	0.31	0.45	1

amenable to inversion. The procedure is developed for any functional relation $Z = g(X, Y)$. It is useful, however, to have some function in mind, for example,

$$Z = aX + bY, \qquad (4.8)$$

where X, Y, and Z are random variables, and a and b are given constants. Given the joint density function $f_{XY}(x, y)$, find $f_Z(z)$.

Start with an example with discrete variables. Given $p_{XY}(x_i, y_i)$, find $p_Z(z_i)$. Suppose that $a = b = 1$ and (X, Y) has the discrete distribution given in Table 4.1. $Z = X + Y$ can take numbers from 0 to 6. For instance, $z = 0$ when $(X, Y) = (0, 0)$, $z = 1$ when $(X, Y) = (0, 1)$ or $(1, 0)$, $z = 2$ when $(X, Y) = (0, 2)$, $(1, 1)$, or $(2, 1)$, and so on. In general, for $Z = Z_j$ these set of points are the points along the line $Y = Z_j - X$ or the diagonal lines in Figure 4.5.

CHAPTER 4. FUNCTIONS OF RANDOM VARIABLES

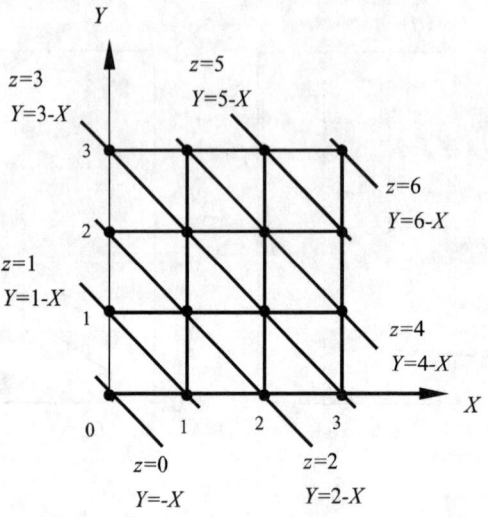

Figure 4.5: The line $Y = Z_j - X$.

The individual probabilities for Z_j can be obtained by adding the individual probabilities of each point along the line $Y = Z_j - X$. The individual probabilities are

$p_Z(z = 0) = 0.01$, $p_Z(z = 1) = 0.05$, $p_Z(z = 2) = 0.14$,
$p_Z(z = 3) = 0.28$, $p_Z(z = 4) = 0.25$, $p_Z(z = 5) = 0.17$, $p_Z(z = 6) = 0.10$.

The equation used to obtain these values is

$$p_Z(z_j) = \sum_{x_i=0}^{3} p_{XY}(x_i, y_i)|_{y_i = z_j - x_i}$$
$$= \sum_{x_i=0}^{3} p_{XY}(x_i, z_j - x_i).$$

For example,

$$p_Z(z=3) = \sum_{x_i=0}^{3} p_{XY}(x_i, 3 - x_i)$$
$$= p_{XY}(0,3) + p_{XY}(1,2) + p_{XY}(2,1) + p_{XY}(3,0).$$

4.2. FUNCTIONS OF TWO OR MORE RVS

Note that x_i can eliminated instead of y_i so that

$$p_Z(z_j) = \sum_{y_i=0}^{3} p_{XY}(x_i, y_i)\Big|_{x_i=z_j-y_i}$$

$$= \sum_{y_i=0}^{3} p_{XY}(z_j - y_i, y_i).$$

To demonstrate the procedure for continuous random variables, assume the transformation function $g(X,Y) = aX + bY$. Begin with the respective cumulative distribution,

$$F_Z(z) = \Pr(Z \leq z)$$
$$= \Pr(g(X,Y) \leq z)$$
$$= \iint_{g(x,y) \leq z} f_{XY}(x,y)\, dx\, dy.$$

The procedure developed for discrete variables was based on rewriting the required probability on Z, that is, $\Pr(Z \leq z)$, in terms of its equivalent, $\Pr(g(X,Y) \leq z)$, for which $f_{XY}(x,y)$ is known. Figure 4.6 shows the region of integration, $ax + by \leq z$.

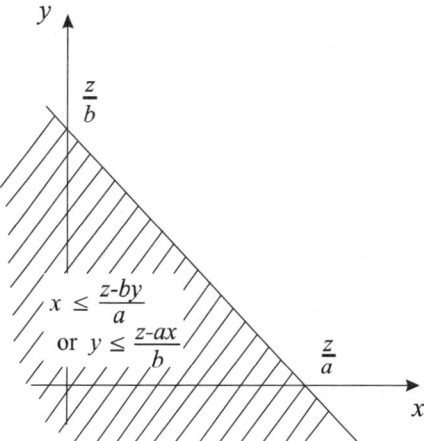

Figure 4.6: Region of integration $g(x,y) \leq z$. The equation of the boundary line is $z = ax + by$.

The limits on x can be found by solving for x in terms of y and z. Solving for x in Equation 4.8, we have $x = (z - by)/a$. Therefore, x is integrated from $-\infty$ to this value of x,

$$F_Z(z) = \int_{-\infty}^{\infty} \int_{-\infty}^{(z-b\tilde{y})/a} f_{XY}(\tilde{x}, \tilde{y}) \, d\tilde{x} \, d\tilde{y}, \qquad (4.9)$$

where $\tilde{x}$ and $\tilde{y}$ are dummy variables. The goal is to derive an expression for $f_Z(z)$. Therefore, replace $d\tilde{x}$ by its equivalent in $(\tilde{y}, \tilde{z})$. The dummy variables are also related by $\tilde{x} = (\tilde{z} - b\tilde{y})/a$. Therefore,

$$\begin{aligned} d\tilde{x} &= \frac{\partial \tilde{x}}{\partial \tilde{z}} d\tilde{z} \\ &= \frac{\partial}{\partial \tilde{z}} \left(\frac{\tilde{z} - b\tilde{y}}{a} \right) d\tilde{z} \\ &= \frac{1}{a} d\tilde{z}, \end{aligned}$$

which is interpreted as $|1/a|$ so that the probability functions are strictly positive regardless of the value of a. We do all this so that the result after integration is only a function of z. The limits of integration on $\tilde{x}$ are transformed to

$$\int_0^{(z-b\tilde{y})/a} d\tilde{x} \to \int_0^z d\tilde{z}.$$

Then,

$$F_Z(z) = \int_{-\infty}^{\infty} \int_{-\infty}^{z} \frac{1}{|a|} f_{XY}\left(\frac{\tilde{z} - b\tilde{y}}{a}, \tilde{y} \right) d\tilde{z} \, d\tilde{y},$$

and

$$\begin{aligned} f_Z(z) &= \frac{dF_Z(z)}{dz} \\ &= \int_{-\infty}^{\infty} \frac{1}{|a|} f_{XY}\left(\frac{z - by}{a}, y \right) dy. \qquad (4.10) \end{aligned}$$

The same result can be obtained by taking the derivative of Equation 4.9 with respect to z. Leibniz's rule is useful here. In general, it is given by the relation,

$$\frac{d}{dz} \int_{a(z)}^{b(z)} f(x, z) \, dx = \frac{db(z)}{dz} f(b(z), z) - \frac{da(z)}{dz} f(a(z), z) + \int_{a(z)}^{b(z)} \frac{\partial f(x, z)}{\partial z} dx.$$

4.2. FUNCTIONS OF TWO OR MORE RVS

Using Leibniz's rule, the probability density function $f_Z(z)$ is given by

$$f_Z(z) = \frac{dF_Z(z)}{dz}$$
$$= \frac{d}{dz}\int_{-\infty}^{\infty}\int_{-\infty}^{(z-by)/a} f_{XY}(x,y)\,dx\,dy$$
$$= \int_{-\infty}^{\infty} \frac{1}{|a|} f_{XY}\left(\frac{z-by}{a}, y\right) dy, \qquad (4.11)$$

where the absolute value of a is used to ensure that $f_Z(z)$ remains positive. This is identical to Equation 4.10.

In general terms, following the above procedure, the relation is $x = h_1(y, z)$, and

$$F_Z(z) = \int_{-\infty}^{\infty}\int_{-\infty}^{x=h_1(y,z)} f_{XY}(x,y)\,dx\,dy,$$

with

$$dx = \frac{\partial h_1(y,z)}{\partial z} dz.$$

The density function is then found to be

$$f_Z(z) = \frac{dF_Z(z)}{dz}$$
$$= \int_{-\infty}^{\infty} f_{XY}[h_1(y,z), y] \left|\frac{\partial h_1(y,z)}{\partial z}\right| dy. \qquad (4.12)$$

Solving for y instead of x above leads to the following equivalent relations,

$$f_Z(z) = \int_{-\infty}^{\infty} f_{XY}[x, h_2(x,z)] \left|\frac{\partial h_2(x,z)}{\partial z}\right| dx \qquad (4.13)$$
$$= \int_{-\infty}^{\infty} \frac{1}{|b|} f_{XY}\left(x, \frac{z-ax}{b}\right) dx, \qquad (4.14)$$

where Equation 4.14 is the counterpart to Equation 4.11.

Equations 4.12 and 4.13 are completely general, but they require that $g(X, Y)$ be invertible.

> **Leibniz**   Gottfried Leibniz was the son of Friedrich Leibniz, a professor of moral philosophy at Leipzig. Friedrich Leibniz ...*was evidently a competent though not original scholar, who devoted his time to his offices and to his family as a pious, Christian father.* Leibniz's mother was Catharina Schmuck, the daughter of a lawyer and Friedrich Leibniz's third

Figure 4.7: Gottfried Leibniz (1646-1716)

wife. However, Friedrich Leibniz died when Leibniz was only six years old, and he was brought up by his mother. Certainly, Leibniz learned his moral and religious values from her, which would play an important role in his life and philosophy.

At the age of seven, Leibniz entered the Nicolai School in Leipzig. Although he was taught Latin at school, Leibniz had taught himself far more advanced Latin and some Greek by the age of 12. He seems to have been motivated by wanting to read his father's books. In particular, he read metaphysics books and theology books from both Catholic and Protestant writers. As he progressed through school, he was taught Aristotle's logic and theory of categorizing knowledge. Leibniz was clearly not satisfied with Aristotle's system and began to develop his own ideas on how to improve on it. In later life, Leibniz recalled that, at this time, he was trying to find ordering on logical truths that were the ideas behind rigorous mathematical proofs.

In 1661, at the age of fourteen, Leibniz entered the University of Leipzig. It may appear that this was a truly exceptionally early age for anyone to enter university, but it is fair to say that by the standards of the time, although he was quite young, there would be others of a similar age. He studied philosophy, which was well taught at the University of Leipzig, and mathematics, which was very poorly taught. Among the other topics that were included in this two year general degree course were rhetoric, Latin,

4.2. FUNCTIONS OF TWO OR MORE RVS

Greek, and Hebrew. He graduated with a bachelors degree in 1663, with the thesis *De Principio Individui* (*On the Principle of the Individual*) which ... *emphasized the existential value of the individual, who is not to be explained either by matter alone or by form alone but rather by his whole being.* In this work one finds the beginning of his notion of "monad."

Leibniz then went to Jena to spend the summer term of 1663. At Jena, the professor of mathematics was Erhard Weigel, who was also a philosopher. Through him Leibniz began to understand the importance of the method of mathematical proof for subjects such as logic and philosophy. Weigel believed that the number was the fundamental concept of the universe, and his ideas were to have considerable influence on Leibniz.

After being awarded a bachelor's degree in law, Leibniz worked on his habilitation in philosophy. His work was to be published in 1666 as *Dissertatio de Arte Combinatoria* (*Dissertation on the Combinatorial Art*). In this work, Leibniz aimed to reduce all reasoning and discovery to a combination of basic elements such as numbers, letters, sounds, and colors.

He was awarded his Master's Degree in philosophy for a dissertation that combined aspects of philosophy and law, studying the relations between these subjects and mathematical ideas that he had learned from Weigel. A few days after Leibniz presented his dissertation, his mother died.

By October 1663, Leibniz was back in Leipzig starting his studies towards a doctorate in law. Despite his growing reputation and acknowledged scholarship, Leibniz was refused the doctorate in law at Leipzig. It is unclear why this happened. He served as secretary to the Nuremberg alchemical society for a while. Then, he met Baron Johann Christian von Boineburg. By November 1667, Leibniz was living in Frankfurt, employed by Boineburg. During the next few years Leibniz undertook a variety of different projects, scientific, literary, and political. He also continued his law career taking up residence at the courts of Mainz before 1670. One of his tasks there, undertaken for the Elector of Mainz, was to improve the Roman civil law code for Mainz but *Leibniz was also occupied by turns as Boineburg's secretary, assistant, librarian, lawyer and advisor, while at the same time a personal friend of the Baron and his family.*

Boineburg was a Catholic, while Leibniz was a Lutheran, but Leibniz had as one of his lifelong aims the reunification of the Christian Churches, and ... *with Boineburg's encouragement, he drafted a number of monographs on religious topics, mostly to do with points at issue between the churches....*

Another of Leibniz's lifelong aims was to collate all human knowledge. Certainly, he saw his work on Roman civil law as part of this effort. Leibniz also tried to bring the work of the learned societies together to coordinate research.

Leibniz began to study motion, and although he had in mind the problem of explaining the results of Wren and Huygens on elastic collisions, he began with abstract ideas of motion. In 1671, he published *Hypothesis Physica Nova* (*New Physical Hypothesis*). In this work he claimed, as had Kepler, that movement depends on the action of a spirit. He communicated with Oldenburg, the secretary of the Royal Society of London, and dedicated some of his scientific works to the Royal Society and the Paris Academy. Leibniz was also in contact with Carcavi, the Royal Librarian in Paris. As Ross explains ... *Although Leibniz's interests were clearly developing in a scientific direction, he still hankered after a literary career. All his life he prided himself on his poetry (mostly Latin), and boasted that he could recite the bulk of Virgil's "Aeneid" by heart. During this time with Boineburg, he would have passed for a typical late Renaissance humanist.*

Leibniz's work followed parallel paths of scientific and nonscientific inquiry. He wished to visit Paris to make more scientific contacts and had started the construction of a calculating machine that he hoped would be of interest. In 1672, Leibniz went to Paris on behalf of Boineburg to try to use his plan to divert Louis XIV from attacking German areas. He formed a political plan to try to persuade the French to attack Egypt and this provided the reason for visiting Paris. His first objective in Paris was to make contact with the French government but, while waiting for such an opportunity, Leibniz made contact with mathematicians and philosophers there, in particular Arnauld and Malebranche, discussing with Arnauld a variety of topics, but particularly church reunification.

In Paris, beginning in the autumn of 1672, Leibniz studied mathematics and physics under Christiaan Huygens. On Huygens' advice, Leibniz read Saint-Vincent's work on summing series and made some discoveries of his own in this area. Also in the autumn of 1672, Boineburg's son was sent to Paris to study under Leibniz, which meant that his financial support was secure. Accompanying Boineburg's son was Boineburg's nephew on a diplomatic mission to try to persuade Louis XIV to set up a peace congress. Boineburg died on December 15, but Leibniz continued to be supported by the Boineburg family.

In January, 1673, Leibniz and Boineburg's nephew went to England to try the same peace mission, the French one having failed. Leibniz visited the Royal Society, and demonstrated his incomplete calculating machine. He also talked with Hooke, Boyle, and Pell. While explaining his results on series to Pell, he was told that these were to be found in a book by Mouton. The next day he consulted Mouton's book and found that Pell was correct. At the meeting of the Royal Society on February 15, which Leibniz did not attend, Hooke made some unfavorable comments on Leibniz's calculating

4.2. FUNCTIONS OF TWO OR MORE RVS

machine.

Leibniz returned to Paris on hearing that the Elector of Mainz had died. Leibniz realized that his knowledge of mathematics was less than he would have liked so he redoubled his efforts on the subject.

The Royal Society of London elected Leibniz a fellow on April 19, 1673. Leibniz met Ozanam and solved one of his problems. He also met again with Huygens who gave him a reading list including works by Pascal, Fabri, Gregory, Saint-Vincent, Descartes, and Sluze. He began to study the geometry of infinitesimals and wrote to Oldenburg at the Royal Society in 1674. Oldenburg replied that Newton and Gregory had found general methods. Leibniz was, however, not on best terms with the Royal Society, since he had not kept his promise of finishing his mechanical calculating machine. Nor was Oldenburg to know that Leibniz had changed from the rather ordinary mathematician who visited London, into a creative mathematical genius. In August, 1675, Tschirnhaus arrived in Paris, forming a close friendship with Leibniz that proved mathematically beneficial to both.

It was during this period in Paris that Leibniz developed the basic features of his version of the calculus. In 1673, he was still struggling to develop a good notation for his calculus and his first calculations were clumsy. On November 21, 1675, he wrote a manuscript using the $\int f(x)dx$ notation for the first time. In the same manuscript, the product rule for differentiation is given. By autumn 1676, Leibniz discovered the familiar $d(x^n) = nx^{(n-1)}dx$ for both integral and fractional n.

Newton wrote a letter to Leibniz, through Oldenburg, which took some time to reach him. The letter listed many of Newton's results but it did not describe his methods. Leibniz replied immediately, but Newton, not realizing that his letter had taken a long time to reach Leibniz, thought he had six weeks to work on his reply. Certainly one of the consequences of Newton's letter was that Leibniz realized that he must quickly publish a fuller account of his own methods.

Newton wrote a second letter to Leibniz on October 24, 1676, that did not reach Leibniz until June, 1677, by which time Leibniz was in Hanover. This second letter, although polite in tone, was clearly written by Newton believing that Leibniz had stolen his methods. In his reply, Leibniz gave some details of the principles of his differential calculus, including the rule for differentiating a function of a function.

Newton was to claim, with justification, that ... *not a single previously unsolved problem was solved* ... by Leibniz's approach. But the formalism was to prove vital in the latter development of the calculus. Leibniz never thought of the derivative as a limit. This concept does not appear until the work of d'Alembert.

Leibniz would have liked to have remained in Paris in the Academy of Sciences, but it was believed locally that there were already enough foreigners there and so no invitation came. Reluctantly, Leibniz accepted the position of librarian and Court Councillor at Hanover from the Duke of Hanover, Johann Friedrich. He left Paris in October, 1676, making the journey to Hanover via London and Holland. The rest of Leibniz's life was spent at Hanover, except for the many travels that he made.

His duties at Hanover ... *as librarian were onerous, but fairly mundane: general administration, purchase of new books and second-hand libraries, and conventional cataloging.*

He undertook a whole collection of other projects, however. For example, one major project started in 1678-1679 involved draining water from the mines in the Harz mountains. His idea was to use wind and water power to operate pumps. He designed many different types of windmills, pumps, and gears, but ... *every one of these projects ended in failure. Leibniz himself believed that this was because of deliberate obstruction by administrators and technicians, and the worker's fear that technological progress would cost them their jobs.*

In 1680, Duke Johann Friedrich died and his brother Ernst August became the new Duke. The Harz project had always been difficult, and it failed by 1684. However, Leibniz had achieved important scientific results, becoming one of the first people to study geology through the observations he compiled for the Harz project. During this work he formed the hypothesis that the Earth was at first molten.

Another of Leibniz's great achievements in mathematics was his development of the binary system of arithmetic. He perfected his system by 1679, but he did not publish anything until 1701, when he sent the paper, *Essay d'une Nouvelle Science des Nombres,* to the Paris Academy to mark his election to the Academy. His major work on determinants arose from his methods to solve systems of linear equations. Although he never published this work in his lifetime, he developed many different approaches to the topic with many different notations being tried out to find the one that was most useful. An unpublished paper dated January 22, 1684, contains his notation and results.

Leibniz continued to perfect his metaphysical system in the 1680s, attempting to reduce reasoning to an algebra of thought. Leibniz published *Meditationes de Cognitione, Veritate et Ideis (Reflections on Knowledge, Truth, and Ideas),* which clarified his theory of knowledge. In February, 1686, Leibniz wrote his *Discours de Métaphysique.*

Another major project which Leibniz undertook, this time for Duke Ernst August, was writing the history of the Guelf family, of which the House of Brunswick was a part. He made a lengthy trip to search archives for material on which to base this history, visiting Bavaria, Austria, and

4.2. FUNCTIONS OF TWO OR MORE RVS

Italy between November, 1687 and June, 1690. As always, Leibniz took the opportunity to meet with scholars of many different subjects on these journeys. In Florence, for example, he discussed mathematics with Viviani, who had been Galileo's last pupil. Although Leibniz published nine large volumes of archival material on the history of the Guelf family, he never wrote the work that was commissioned.

In 1684, Leibniz published details of his differential calculus, *Nova Methodus pro Maximis et Minimis, Itemque Tangentibus*, in *Acta Eruditorum*, a journal established in Leipzig two years earlier. The paper contained the familiar d notation, and the rules for computing the derivatives of powers, products, and quotients. However, it contained no proofs, and Jacob Bernoulli called it an enigma rather than an explanation.

In 1686, Leibniz published, in *Acta Eruditorum*, a paper dealing with the integral calculus where the notation $\int$ first appeared.

Newton's *Principia* appeared the following year. Newton's "method of fluxions" was written in 1671, but Newton failed to get it published, and it did not appear in print until John Colson produced an English translation in 1736. This time delay in the publication of Newton's work resulted in a dispute with Leibniz.

Another important piece of mathematical work undertaken by Leibniz was his work on dynamics. He criticized Descartes' ideas of mechanics and examined what are effectively kinetic energy, potential energy, and momentum. This work was started in 1676, but he returned to it at various times, in particular, while he was in Rome in 1689. It is clear that while he was in Rome, in addition to working in the Vatican library, Leibniz worked with members of the Accademia, to which he was elected a member at this time. Also while in Rome he read Newton's *Principia*. His two part treatise, *Dynamica,* studied abstract dynamics and practical dynamics, and is written in a style somewhat similar to Newton's *Principia*.

Ross writes ... *although Leibniz was ahead of his time in aiming at a genuine dynamics, it was this very ambition that prevented him from matching the achievement of his rival Newton. ... It was only by simplifying the issues... that Newton succeeded in reducing them to manageable proportions.*

Leibniz put much energy into promoting scientific societies. He was involved in moves to set up academies in Berlin, Dresden, Vienna, and St. Petersburg. He began a campaign for an academy in Berlin in 1695, and visited Berlin in 1698 as part of his efforts. In 1700, he finally persuaded Friedrich to found the Brandenburg Society of Sciences, on July 11. Leibniz was appointed its first president, this being an appointment for life. However, the Academy was not particularly successful, and only one volume of the proceedings were ever published. It did lead to the creation of the

Berlin Academy some years later.

Other attempts by Leibniz to found academies were less successful. He was appointed as Director of a proposed Vienna Academy in 1712, but Leibniz died before the Academy was created. Similarly, he did much of the work to prompt the setting up of the St. Petersburg Academy, but again it did not come into existence until after his death.

It is no exaggeration to say that Leibniz corresponded with most of the scholars in Europe. He had over 600 correspondents. Among the mathematicians with whom he corresponded was Grandi. The correspondence started in 1703, and later concerned the results obtained by putting $x=1$ into $1/(1+x) = 1 - x + x^2 - x^3 + $ .... Leibniz also corresponded with Varignon on this paradox. Leibniz discussed logarithms of negative numbers with Johann Bernoulli.

In 1710, Leibniz published *Théodicée,* a philosophical work intended to tackle the problem of evil in a world created by a good God. Leibniz claims that the universe had to be imperfect, otherwise it would not be distinct from God. He then claims that the universe is the best possible without being perfect. Leibniz is aware that this argument looks unlikely – surely a universe in which nobody is killed by floods is better than the present one, but still not perfect. His argument here is that the elimination of natural disasters, for example, would involve such changes to the laws of science that the world would be worse. In 1714, Leibniz wrote *Monadologia,* which synthesized the philosophy of his earlier work, *Théodicée.*

Much of the mathematical activity of Leibniz's last years involved the priority dispute over the invention of the calculus. In 1711, he read the paper by Keill in the *Transactions of the Royal Society of London* that accused Leibniz of plagiarism. Leibniz demanded a retraction, saying that he had never heard of the calculus of fluxions until he had read the works of Wallis. Keill replied to Leibniz, saying that ... *the two letters from Newton, sent through Oldenburg, had given ... pretty plain indications... whence Leibniz derived the principles of that calculus, or at least could have derived them.*

Leibniz wrote again to the Royal Society asking them to correct the wrong done to him by Keill's claims. In response to this letter, the Royal Society set up a committee to pronounce on the priority dispute. It was totally biased, not asking Leibniz to give his version of the events. The report of the committee, finding in favor of Newton, was written by Newton himself, and published as *Commercium Epistolicum* near the beginning of 1713, but not seen by Leibniz until the autumn of 1714. He learned of its contents in 1713, in a letter from Johann Bernoulli, reporting on the copy of the work brought from Paris by his nephew Nicolaus(I) Bernoulli. Leibniz then published an anonymous pamphlet, *Charta Volans,* setting

4.2. FUNCTIONS OF TWO OR MORE RVS

out his side, in which a mistake by Newton in his understanding of second and higher derivatives, spotted by Johann Bernoulli, is used as evidence of Leibniz's case.

The argument continued with Keill, who published a reply to *Charta Volans*. Leibniz refused to carry on the argument with Keill, saying that he could not reply to an idiot. However, when Newton wrote to him directly, Leibniz did reply and gave a detailed description of his discovery of the differential calculus. From 1715 up until his death, Leibniz corresponded with Samuel Clarke, a supporter of Newton, on time, space, freewill, gravitational attraction across a void, and other topics.

Leibniz is described as follows: *Leibniz was a man of medium height with a stoop, broad-shouldered but bandy-legged, as capable of thinking for several days sitting in the same chair as of travelling the roads of Europe summer and winter. He was an indefatigable worker, a universal letter writer (he had more than 600 correspondents), a patriot and cosmopolitan, a great scientist, and one of the most powerful spirits of Western civilization.*

Ross points out that Leibniz's legacy may have not been quite what he had hoped for: *It is ironical that one so devoted to the cause of mutual understanding should have succeeded only in adding to intellectual chauvinism and dogmatism. There is a similar irony in the fact that he was one of the last great polymaths – not in the frivolous sense of having a wide general knowledge, but in the deeper sense of one who is a citizen of the whole world of intellectual inquiry. He deliberately ignored boundaries between disciplines, and lack of qualifications never deterred him from contributing fresh insights to established specialisms. Indeed, one of the reasons why he was so hostile to universities as institutions was because their faculty structure prevented the cross-fertilization of ideas which he saw as essential to the advance of knowledge and of wisdom. The irony is that he was himself instrumental in bringing about an era of far greater intellectual and scientific specialism, as technical advances pushed more and more disciplines out of the reach of the intelligent layman and amateur.*

Example 4.4 Kinetic Energy Density

A particle of mass m moving in the xy plane has a kinetic energy $T = mZ^2/2$ where Z is the resultant velocity that is related to the component velocities (speed) in each coordinate direction by $Z = \sqrt{\dot{X}^2 + \dot{Y}^2}$. Suppose that $\dot{X}$ and $\dot{Y}$ are statistically independent (for convenience) and each is distributed as a standard normal random variable, that is, zero mean and unitary standard deviation. Derive the density of the kinetic energy.

Solution Use the result from Section 4.1 where a parabolic transformation of densities was derived. To do this, write the kinetic energy as

$$T = \frac{1}{2}mZ^2 = \frac{1}{2}m\left(\dot{X}^2 + \dot{Y}^2\right)$$
$$= U + V.$$

The first step is to apply Equation 4.3 to transform between $\dot{X}^2$ and U, and $\dot{Y}^2$ and V. The resulting densities are

$$f_U(u) = \frac{1}{\sqrt{\pi m u}} \exp\left(-\frac{u}{m}\right), \qquad u \geq 0, \qquad (4.15)$$

$$f_V(v) = \frac{1}{\sqrt{\pi m v}} \exp\left(-\frac{v}{m}\right), \qquad v \geq 0. \qquad (4.16)$$

Now following the procedure of this section, solve for either u or v, say,

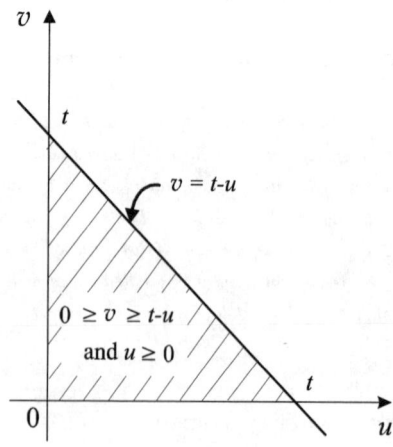

Figure 4.8: Domain of integration for Example 4.4.

$v = t - u$, where $t - u \geq 0$, or $t \geq u$. This relation is useful in setting the integration limits. The domain of integration is shown in Figure 4.8.

4.2. FUNCTIONS OF TWO OR MORE RVS

Applying Equation 4.12, the kinetic energy density is given by

$$f_T(t) = \int_0^t f_U(u) f_V(t-u) \, du$$

$$= \int_0^t \frac{1}{\sqrt{\pi m u}} \exp\left(-\frac{u}{m}\right) \frac{1}{\sqrt{\pi m (t-u)}} \exp\left(-\frac{t-u}{m}\right) du$$

$$= \frac{1}{m} \exp(-t/m), \qquad (4.17)$$

which was integrated directly using MAPLE, or could be integrated by transforming according to $r = u/t$ and $du = t\,dr$, resulting in a beta function.[1] Equations 4.15–4.17 are *chi-squared* densities with *one* and *two* degrees of freedom.

⊛

Example 4.5 Products and Quotients

Two other important transformations are for the functions $Z = XY$ and $Z = X/Y$. Products and quotients are everywhere in mathematics. For the product functional relations, the following procedure is developed,

$$x = z/y \equiv h_1(y, z)$$

$$\frac{\partial h_1}{\partial z} = \frac{1}{y}$$

$$f_Z(z) = \int_{-\infty}^{\infty} \frac{1}{|y|} f_{XY}\left(\frac{z}{y}, y\right) dy$$

$$\text{or} \ = \int_{-\infty}^{\infty} \frac{1}{|x|} f_{XY}\left(x, \frac{z}{x}\right) dx.$$

The last equality was obtained by solving for $y = z/x = h_2(x, z)$ and using $\partial h_2 / \partial z = 1/x$.

---

[1]

$$f_T(t) = \frac{1}{\pi m} \exp\left(-\frac{t}{m}\right) \int_0^t u^{-1/2} (t-u)^{-1/2} \, du$$

$$= \frac{1}{\pi m} \exp\left(-\frac{t}{m}\right) \int_0^1 r^{-1/2} (1-r)^{-1/2} \, dr$$

$$= \frac{1}{\pi m} \exp\left(-\frac{t}{m}\right) B\left(\frac{1}{2}, \frac{1}{2}\right)$$

$$= \frac{1}{m} \exp\left(-\frac{t}{m}\right).$$

Considering next the quotient $Z = X/Y$, follow the procedure,

$$x = zy \equiv h_1(y,z)$$

$$\frac{\partial h_1}{\partial z} = y$$

$$f_Z(z) = \int_{-\infty}^{\infty} |y| f_{XY}(zy, y)\, dy$$

$$= \int_{-\infty}^{\infty} |x| f_{XY}(x, zx)\, dx.$$

Suppose that $Z = XY$, X, and Y are statistically independent, and their probability density functions are given by

$$f_X(x) = \frac{x}{\sigma^2} \exp\left\{-\frac{x^2}{2\sigma^2}\right\}, \quad x \geq 0$$

$$f_Y(y) = \frac{1}{y} \ln\frac{b}{a}, \quad 0 < a \leq y \leq b.$$

The cumulative distribution $F_Z(z)$ is then

$$F_Z(z) = \iint_{xy \leq z} f_X(x) f_Y(y)\, dx\, dy.$$

The domain of integration is defined by

$$x \geq 0,\ a \leq y \leq b,\ \text{and}\ xy \leq z,$$

and is shown in Figure 4.9. Therefore, $F_Z(z)$ can be written as

$$F_Z(z) = \int_a^b \int_0^{z/y} f_X(x) f_Y(y)\, dx\, dy.$$

Differentiating $F_Z(z)$ with respect to z, obtain

$$f_Z(z) = \frac{d}{dz} F_Z(z)$$

$$= \int_a^b \frac{1}{|y|} f_X\left(\frac{z}{y}\right) f_Y(y)\, dy$$

$$= \int_a^b \frac{1}{|y|} \frac{z}{y\sigma^2} \exp\left\{-\frac{z^2}{2\sigma^2 y^2}\right\} \frac{1}{y} \ln\frac{b}{a}\, dy.$$

Since y is always positive, the absolute value sign can be omitted. The resulting probability density is

$$f_Z(z) = \frac{1}{z} \ln\frac{b}{a} \int_a^b \frac{z^2}{y^3 \sigma^2} \exp\left\{-\frac{z^2}{2\sigma^2 y^2}\right\} dy$$

$$= \frac{1}{z} \ln\frac{b}{a} \left(\exp\left\{-\frac{z^2}{2\sigma^2 b^2}\right\} - \exp\left\{-\frac{z^2}{2\sigma^2 a^2}\right\}\right).$$

4.2. FUNCTIONS OF TWO OR MORE RVS

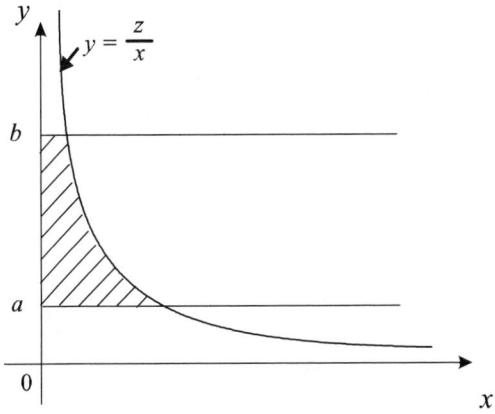

Figure 4.9: Domain of integration for Example 4.5.

With the density function for Z, moments of the variable can be calculated.

⊛

The examples above are only meant to demonstrate the basic procedure, which is generalizable to any number of random variables, and can by way of a sequence of transformations be applied to very complex functional relationships. For example, a complex relation can be built in the following way, where the arrows indicate the direction of the density transformation,

$$Y_1 \longleftarrow X_1^2$$
$$Y_2 \longleftarrow Y_1 + X_2$$
$$Y_3 \longleftarrow Y_2^2$$
$$Y_4 \longleftarrow X_3 X_4$$
$$Y_5 \longleftarrow Y_3 + Y_4$$
$$\implies Y_5 = \left(X_1^2 + X_2\right)^2 + X_3 X_4.$$

4.2.1 General Case

Consider the two-dimensional continuous random variable, (X, Y), with joint probability density function $f_{XY}(x, y)$. Z and W are two-dimensional continuous random variables that are functions of X and Y such that

$$Z = H_1(X, Y) \text{ and } W = H_2(X, Y).$$

Then, how do we find the probability density function $f_{ZW}(z, w)$? In order to answer this question, a similar procedure is followed as that for the function of one variable, demonstrated in Section 4.1.

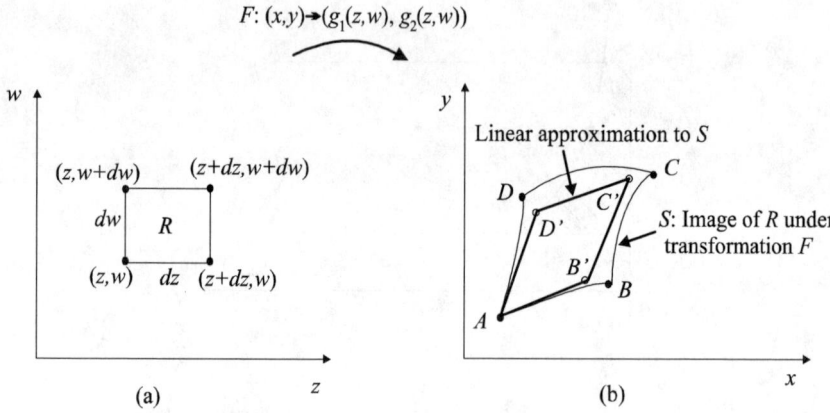

Figure 4.10: Transformation of variables.

Assume that there is a one-to-one relationship between (X,Y) and (Z,W) such that the inverse transform exists,

$$X = g_1(Z,W) \text{ and } Y = g_2(Z,W). \tag{4.19}$$

Consider the probability,

$$\Pr(x < X \leq x + dx \text{ and } y < Y \leq y + dy).$$

For sufficiently small dx and dy, the probability can be approximated by $f_{XY}(x,y)\,dx\,dy$. The same probability can be written as

$$\Pr(z < Z \leq z + dz \text{ and } w < W \leq w + dw)$$

and approximated by $f_{ZW}(z,w)\,dz\,dw$. That is,

$$f_{XY}(x,y)\,dx\,dy = f_{ZW}(z,w)\,|dz\,dw|, \tag{4.20}$$

which is equivalent to Equation 4.1 for probability density functions of one variable only. The next step is to find $dx\,dy$ in terms of $dz\,dw$. This is not as simple as in the previous case. The term $dz\,dw$ is the incremental area R shown in Figure 4.10 (a). Figure 4.10 (b) shows the same incremental area in xy plane after it is transformed using the inverse transformation defined in Equation 4.19. The actual transformation, denoted as F, results in the

4.2. FUNCTIONS OF TWO OR MORE RVS

image S, defined by the vertices $ABCD$. The corners of S are

$$
\begin{align}
A &: \quad [g_1(z,w), g_2(z,w)] \\
B &: \quad [g_1(z+dz,w), g_2(z+dz,w)] \\
C &: \quad [g_1(z+dz,w+dw), g_2(z+dz,w+dw)] \\
D &: \quad [g_1(z,w+dw), g_2(z,w+dw)].
\end{align}
$$

Using the Taylor series about (z,w), we can approximate

$$
\begin{align}
g_1(z+dz, w+dw) &\simeq g_1(z,w) + \frac{\partial g_1}{\partial z}dz + \frac{\partial g_1}{\partial w}dw \\
g_1(z, w+dw) &\simeq g_1(z,w) + \frac{\partial g_1}{\partial w}dw \\
g_1(z+dz, w) &\simeq g_1(z,w) + \frac{\partial g_1}{\partial z}dz.
\end{align}
$$

The same expansion can be performed for g_2. The area S is approximated by the parallelogram with vertices $A'B'C'D'$. The corners of the parallelogram are

$$
\begin{align}
A' &= A: (g_1(z,w), g_2(z,w)) \\
B' &: \left(g_1(z,w) + \frac{\partial g_1}{\partial z}dz,\, g_2(z,w) + \frac{\partial g_2}{\partial z}dz\right) \\
C' &: \left(g_1(z,w) + \frac{\partial g_1}{\partial z}dz + \frac{\partial g_1}{\partial w}dw,\, g_2(z,w) + \frac{\partial g_2}{\partial z}dz + \frac{\partial g_2}{\partial w}dw\right) \\
D' &: \left(g_1(z,w) + \frac{\partial g_1}{\partial w}dw,\, g_2(z,w) + \frac{\partial g_2}{\partial w}dw\right).
\end{align}
$$

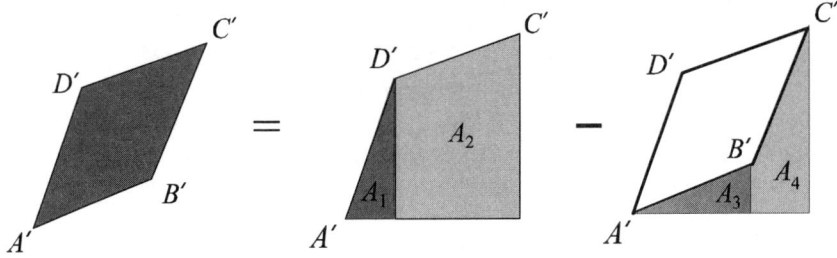

Figure 4.11: Area of the parallelogram.

The area of the parallelogram, $A'B'C'D'$ can be obtained by performing the operations $(A_1 + A_2) - (A_3 + A_4)$, as shown in Figure 4.11. The areas

A_1, A_2, A_3, and A_4 are given by

$$A_1 = \frac{1}{2}\frac{\partial g_1}{\partial w}dw\frac{\partial g_2}{\partial w}dw$$

$$A_2 = \frac{1}{2}\frac{\partial g_1}{\partial z}dz\left(\frac{\partial g_2}{\partial w}dw + \frac{\partial g_2}{\partial z}dz + \frac{\partial g_2}{\partial w}dw\right)$$

$$A_3 = \frac{1}{2}\frac{\partial g_1}{\partial z}dz\frac{\partial g_2}{\partial z}dz$$

$$A_4 = \frac{1}{2}\frac{\partial g_1}{\partial w}dw\left(\frac{\partial g_2}{\partial z}dz + \frac{\partial g_2}{\partial z}dz + \frac{\partial g_2}{\partial w}dw\right).$$

Therefore, the area of the parallelogram is given by

$$\begin{aligned}&\frac{1}{2}\frac{\partial g_1}{\partial w}\frac{\partial g_2}{\partial w}dw^2 + \frac{\partial g_1}{\partial z}\frac{\partial g_2}{\partial w}dz\,dw + \frac{1}{2}\frac{\partial g_1}{\partial z}\frac{\partial g_2}{\partial z}dz^2 \\ &-\frac{1}{2}\frac{\partial g_1}{\partial z}\frac{\partial g_2}{\partial z}dz^2 - \frac{\partial g_1}{\partial w}\frac{\partial g_2}{\partial z}dz\,dw - \frac{1}{2}\frac{\partial g_1}{\partial w}\frac{\partial g_2}{\partial w}dw^2 \\ &= \left(\frac{\partial g_1}{\partial z}\frac{\partial g_2}{\partial w} - \frac{\partial g_1}{\partial w}\frac{\partial g_2}{\partial z}\right)dz\,dw.\end{aligned}$$

That is, the area $dz\,dw$ has increased by a factor of

$$\left(\frac{\partial g_1}{\partial z}\frac{\partial g_2}{\partial w} - \frac{\partial g_1}{\partial w}\frac{\partial g_2}{\partial z}\right)$$

under the transformation. This factor is the Jacobian J, or the determinant of the matrix

$$\begin{bmatrix}\partial g_1/\partial z & \partial g_1/\partial w \\ \partial g_2/\partial z & \partial g_2/\partial w\end{bmatrix},$$

and the incremental area S in the xy plane is $dx\,dy$. Therefore,

$$\begin{aligned}dx\,dy &= \det\begin{bmatrix}\partial g_1/\partial z & \partial g_1/\partial w \\ \partial g_2/\partial z & \partial g_2/\partial w\end{bmatrix}dz\,dw \\ &= J\,dz\,dw.\end{aligned}$$

Rewriting Equation 4.20, we obtain

$$f_{ZW}(z,w) = f_{XY}(g_1(z,w), g_2(z,w))|J|, \qquad (4.21)$$

where $|J|$ is the absolute value of the Jacobian. The use of Equation 4.21 is demonstrated in the following example.

4.2. FUNCTIONS OF TWO OR MORE RVS

Example 4.6 Functions of a Two-Dimensional Random Variable

Suppose (X, Y) is a random variable with probability density function $f_{XY}(x, y)$ defined in a circle with radius 1 and center at $(0, 0)$. The joint probability density function is given by

$$f_{XY}(x, y) = \begin{cases} (1/2\pi\sigma^2) \exp\left\{-\frac{1}{2}\left(\frac{x^2+y^2}{\sigma^2}\right)\right\}, & (x, y) \text{ inside the circle} \\ 0, & (x, y) \text{ elsewhere.} \end{cases}$$

Suppose that it is of interest to express the probability density function in terms of the distance from the center, R, and the angle from the positive axis, Θ. The random variables R and Θ are related to X and Y by

$$R = \sqrt{X^2 + Y^2}$$
$$\Theta = \tan^{-1}\frac{Y}{X}.$$

Find the probability density function $f_{R\Theta}(r, \theta)$.
Solution Solving for X and Y, we obtain

$$X = R\cos\Theta$$
$$Y = R\sin\Theta.$$

The Jacobian is given by

$$J = \det \begin{bmatrix} \partial x/\partial r & \partial x/\partial \theta \\ \partial y/\partial r & \partial y/\partial \theta \end{bmatrix}$$
$$= \det \begin{bmatrix} \cos\theta & -r\sin\theta \\ \sin\theta & r\cos\theta \end{bmatrix}$$
$$= r.$$

Then the joint probability density function $f_{R\Theta}(r, \theta)$ is given by

$$f_{R\Theta}(r, \theta) = |r| f_{XY}(r\cos\theta, r\sin\theta)$$
$$= \frac{r}{2\pi\sigma^2} \exp\left\{-\frac{1}{2}\left(\frac{r^2}{\sigma^2}\right)\right\} \quad \text{for } 0 \le r < 1,\ 0 \le \theta < 2\pi.$$

⊛

Example 4.7 Distribution of a Sum of Random Variables

Suppose that (X, Y) is a random variable with density $f_{XY}(x, y)$. We seek to find the probability density $f_Z(z)$ where Z is defined as follows,

$$Z = aX + bY.$$

Let $f_{XY}(x,y)$ be defined on $0 < X < 1$ and $0 < Y < 1$.

Solution A similar problem was solved previously where $f_{XY}(x,y)$ was defined on the infinite domain, and the results shown in Equations 4.10 and 4.14. In order to use the general method, we define another variable W,

$$W = X.$$

The choice of W is arbitrary, and so a convenient function is chosen.

Solve for X and Y,

$$X = W$$

$$Y = \frac{Z - aW}{b}.$$

The Jacobian is evaluated as

$$J = \det \begin{bmatrix} 0 & 1 \\ 1/b & -a/b \end{bmatrix} = -\frac{1}{b},$$

and the joint probability density is then

$$f_{ZW}(z,w) = \left| -\frac{1}{b} \right| f_{XY}\left(w, \frac{z - aw}{b}\right).$$

If $f_{XY}(x,y)$ is defined on the infinite domain, $-\infty < X < \infty$ and $-\infty < Y < \infty$, then the corresponding domain in W and Z is also the infinite domain,

$$(-\infty < X < \infty \text{ and } -\infty < Y < \infty) \to (-\infty < W < \infty, -\infty < Z < \infty).$$

The marginal density $f_Z(z)$ is then

$$f_Z(z) = \int_{-\infty}^{\infty} f_{ZW}(z,w)\,dw = \int_{-\infty}^{\infty} \left| -\frac{1}{b} \right| f_{XY}\left(w, \frac{z - aw}{b}\right) dw.$$

This is an alternate method to obtain the result in Equation 4.14.

If $f_{XY}(x,y)$ is defined on the rectangle defined by $0 < X < 1$ and $0 < Y < 1$, obtaining the corresponding domain in W and Z and the marginal density $f_Z(z)$ requires more work. Start with the domain of W and Z. Since $X = W$, we know $0 < W < 1$. The range for Z can be derived from that of Y. It is given that $0 < Y < 1$. Writing Y in terms of Z and W, results in

$$0 < Y < 1 \to 0 < \frac{Z - aW}{b} < 1.$$

Solving for Z,

$$aW < Z < b + aW.$$

4.2. FUNCTIONS OF TWO OR MORE RVS

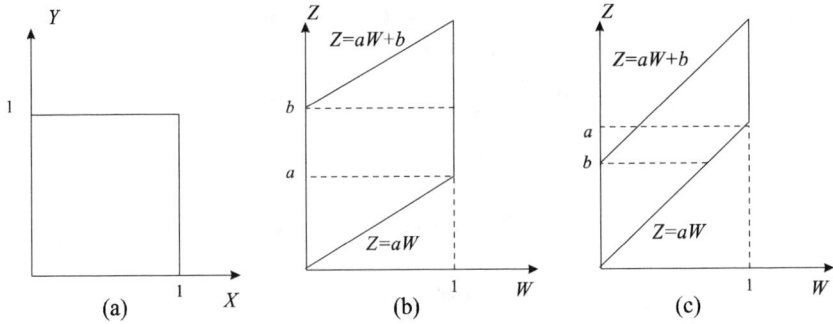

Figure 4.12: Transformation of domains.

Z ranges between two lines: $Z = aW$ and $Z = b + aW$. Therefore, $f_{ZW}(z,w)$ is defined in a domain that resembles a parallelogram, as shown in Figure 4.12(b) or (c).

In order to obtain the marginal density, the joint distribution $f_{ZW}(z,w)$ needs to be integrated over W. The range of W depends on Z. If $a < b$ as shown in Figure 4.12(b), W ranges on

$$0 < W < \frac{Z}{a}, \qquad 0 < Z < a$$
$$0 < W < 1, \qquad a < Z < b$$
$$\frac{Z-b}{a} < W < 1, \qquad b < Z < a+b.$$

Replace $f_{ZW}(z,w)$ by its equivalent joint density f_{XY} to obtain the marginal density,

$$f_Z(z) = \int_0^{z/a} \left|-\frac{1}{b}\right| f_{XY}\left(w, \frac{z-aw}{b}\right) dw, \qquad 0 < z < a$$
$$= \int_0^1 \left|-\frac{1}{b}\right| f_{XY}\left(w, \frac{z-aw}{b}\right) dw, \qquad a < z < b$$
$$= \int_{(z-a)/b}^1 \left|-\frac{1}{b}\right| f_{XY}\left(w, \frac{z-aw}{b}\right) dw, \qquad b < z < a+b.$$

If $a > b$ as shown in Figure 4.12(c), W ranges on

$$0 < W < \frac{Z}{a}, \qquad 0 < Z < b$$
$$\frac{Z-b}{a} < W < \frac{Z}{a}, \qquad b < Z < a$$
$$\frac{Z-b}{a} < W < 1, \qquad a < Z < 1,$$

and the marginal density is given by

$$f_Z(z) = \int_0^{z/a} \left|-\frac{1}{b}\right| f_{XY}\left(w, \frac{z-aw}{b}\right) dw, \qquad 0 < z < b$$
$$= \int_{(z-a)/b}^{z/a} \left|-\frac{1}{b}\right| f_{XY}\left(w, \frac{z-aw}{b}\right) dw, \qquad b < z < a$$
$$= \int_{(z-a)/b}^{1} \left|-\frac{1}{b}\right| f_{XY}\left(w, \frac{z-aw}{b}\right) dw, \qquad a < z < a+b.$$

If $a = b$, the marginal density is given by

$$f_Z(z) = \int_0^{z/a} \left|-\frac{1}{b}\right| f_{XY}\left(w, \frac{z-aw}{b}\right) dw, \qquad 0 < z < a$$
$$= \int_{(z-a)/a}^{1} \left|-\frac{1}{b}\right| f_{XY}\left(w, \frac{z-aw}{b}\right) dw, \qquad a < z < a+b.$$

For numerical purposes, assume that $a = b = 1$, and $f_{XY}(x, y)$ is distributed uniformly. Then

$$f_{XY}(x, y) = 1, \quad 0 < x < 1 \text{ and } 0 < y < 1.$$

The marginal density $f_Z(z)$ is

$$f_Z(z) = \begin{cases} \int_0^z f_{XY}(w, z-w) \, dw, & 0 < z < 1 \\ \int_{z-1}^1 f_{XY}(w, z-w) \, dw, & 1 < z < 2, \end{cases}$$

which is integrated to

$$f_Z(z) = \begin{cases} z, & 0 < z < 1 \\ 2 - z, & 1 < z < 2. \end{cases}$$

4.2. FUNCTIONS OF TWO OR MORE RVS

Figure 4.13: Brook Taylor (1685-1731).

Taylor   Brook Taylor was the son of John Taylor and Olivia Tempest. His father was the son of Nathaniel Taylor who was recorder of Colchester and a member representing Bedfordshire in Oliver Cromwell's Assembly, while his mother was the daughter of Sir John Tempest. Brook was, therefore, born into a family that was on the fringes of the nobility and certainly was fairly wealthy.

Taylor was brought up in a household where his father ruled as a strict disciplinarian, yet he was a man of culture with interests in painting and music. Although John Taylor had some negative influences on his son, he also had some positive ones, particularly giving his son a love of music and painting. Brook Taylor grew up not only to be an accomplished musician and painter, but he applied his mathematical skills to both these areas later in his life.

As Taylor's family were well off they could afford to have private tutors for their son and in fact this home education was all that Brook enjoyed before entering St. John's College Cambridge on April 3, 1703. By this time he had a good grounding in classics and mathematics. At Cambridge Taylor became highly involved with mathematics. He graduated with an LL.B. in 1709 but by this time (in 1708) he had already written his first important mathematics paper, although it would not be published until 1714. We know something of the details of Taylor's thoughts on various mathematical problems from letters he exchanged with Machin and Keill beginning in his undergraduate years.

In 1712 Taylor was elected to the Royal Society, and clearly it was an election based more on the expertise that Machin, Keill, and others knew that Taylor had, rather than on his published results. For example, Taylor wrote to Machin in 1712 providing a solution to a problem concerning Kepler's second law of planetary motion. Also in 1712 Taylor was appointed to the committee set up to adjudicate on whether the claim of Newton or of Leibniz to have invented the calculus was correct.

The paper we referred to above as being written in 1708 was published in the *Philosophical Transactions of the Royal Society* in 1714. The paper gives a solution to the problem of the center of oscillation of a body, and it resulted in a priority dispute with Johann Bernoulli. We shall say a little more below about disputes between Taylor and Johann Bernoulli. Returning to the paper, it is a mechanics paper that rests heavily on Newton's approach to the differential calculus.

The year 1714 also marks Taylor's election as Secretary to the Royal Society. It was a position that Taylor held from January 14 of that year until October 21, 1718, when he resigned, partly for health reasons, partly due to his lack of interest in the rather demanding position. The period during which Taylor was Secretary to the Royal Society does mark what must be considered his most mathematically productive time. Two books that appeared in 1715, *Methodus Incrementorum Directa et Inversa* and *Linear Perspective,* are extremely important in the history of mathematics. Second editions would appear in 1717 and 1719, respectively. We discuss the content of these works in some detail below.

Taylor made several visits to France. These were made partly for health reasons and partly to visit the friends he had made there. He met Pierre Rémond de Montmort and corresponded with him on various mathematical topics after his return. In particular they discussed infinite series and probability. Taylor also corresponded with de Moivre on probability and at times there was a three-way discussion going on between these mathematicians.

Between 1712 and 1724 Taylor published thirteen articles on topics as diverse as describing experiments in capillary action, magnetism, and thermometers. He gave an account of an experiment to discover the law of magnetic attraction (1715) and an improved method for approximating the roots of an equation by giving a new method for computing logarithms (1717). His life, however, suffered a series of personal tragedies beginning around 1721. In that year he married Miss Brydges from Wallington in Surrey. Although she was from a good family, it was not a family with money and Taylor's father strongly objected to the marriage. The result was that relations between Taylor and his father broke down and there was no contact between father and son until 1723. It was in that year that Taylor's wife died in childbirth. The child, who would have been their first, also died.

4.2. FUNCTIONS OF TWO OR MORE RVS

After the tragedy of losing his wife and child, Taylor returned to live with his father and relations between the two were repaired. Two years later, in 1725, Taylor married again to Sabetta Sawbridge from Olantigh in Kent. This marriage had the approval of Taylor's father, who died four years later on April 4, 1729. Taylor inherited his father's estate of Bifons but further tragedy was to strike when his second wife Sabetta died in childbirth in the following year. On this occasion the child, a daughter Elizabeth, did survive.

Taylor added to mathematics a new branch now called the "calculus of finite differences," invented integration by parts, and discovered the celebrated series known as Taylor's expansion. These ideas appear in his book *Methodus Incrementorum Directa et Inversa* of 1715 referred to above. In fact the first mention by Taylor of a version of what is today called Taylor's Theorem appears in a letter which he wrote to Machin on 26 July 1712. In this letter Taylor explains carefully where he got the idea from.

It was, wrote Taylor, due to a comment that Machin made in Child's Coffeehouse when he had commented on using "Sir Isaac Newton's series" to solve Kepler's problem, and also using "Dr. Halley's method of extracting roots" of polynomial equations. There are, in fact, two versions of Taylor's Theorem given in the 1715 paper, which to a modern reader look equivalent but which, Feigenbaum[2] argues convincingly, were differently motivated. Taylor initially derived the version which occurs as Proposition 11 as a generalization of Halley's method of approximating roots of the Kepler equation, but soon discovered that it was a consequence of the Bernoulli series. This is the version that was inspired by the Coffeehouse conversation described above. The second version occurs as Corollary 2 to Proposition 7 and was thought of as a method of expanding solutions of fluxional equations in infinite series.

We must not give the impression that this result was one which Taylor was the first to discover. James Gregory, Newton, Leibniz, Johann Bernoulli, and de Moivre had all discovered variants of Taylor's Theorem. Gregory, for example, knew that

$$\tan^{-1} x = x - x^3/3 + x^5/5 - x^7/7 + \cdots,$$

and his methods are discussed by Jones.[3] The differences in Newton's ideas of Taylor series and those of Gregory are discussed by Petrova and Romanovska.[4] All of these mathematicians had made their discoveries in-

---

[2] L Feigenbaum, "Brook Taylor and the Method of Increments," *Arch. Hist. Exact Sci.* 34 (1-2) (1985), 1-140.

[3] P.S. Jones, "Brook Taylor and the Mathematical Theory of Linear Perspective," *Amer. Math. Monthly* 58 (1951), 597-606.

[4] S.S. Petrova and D.A. Romanovska, "On the History of the Discovery of Taylor Series (Russian)," *Istor.-Mat. Issled.* No. 25 (1980), 10-24; 378.

dependently, and Taylor's work was also independent of that of the others. The importance of Taylor's Theorem remained unrecognized until 1772 when Lagrange proclaimed it the basic principle of the differential calculus. The term "Taylor's series" seems to have been used for the first time by Lhuilier in 1786.

There are other important ideas that are contained in the *Methodus Incrementorum Directa et Inversa* of 1715, which were not recognized as important at the time. These include singular solutions to differential equations, a change of variables formula, and a way of relating the derivative of a function to the derivative of the inverse function. Also contained is a discussion on vibrating strings, an interest that almost certainly came from Taylor's early love of music.

Taylor, in his studies of vibrating strings, was not attempting to establish equations of motion, but was considering the oscillation of a flexible string in terms of the isochrony of the pendulum. He tried to find the shape of the vibrating string and the length of the isochronous pendulum rather than to find its equations of motion.

Taylor also devised the basic principles of perspective in *Linear Perspective* (1715). The second edition has a different title, being called *New Principles of Linear Perspective*. The work gives the first general treatment of vanishing points. Taylor had a highly mathematical approach to the subject and made no concessions to artists who should have found the ideas of fundamental importance to them. At times it is very difficult for even a mathematician to understand Taylor's results. The phrase "linear perspective" was invented by Taylor in this work and he defined the vanishing point of a line, not parallel to the plane of the picture, as the point where a line through the eye parallel to the given line intersects the plane of the picture. He also defined the vanishing line to a given plane, not parallel to the plane of the picture, as the intersection of the plane through the eye parallel to the given plane. He did not invent the terms vanishing point and vanishing line, but he was one of the first to stress their importance. The main theorem in Taylor's theory of linear perspective is that the projection of a straight line not parallel to the plane of the picture passes through its intersection and its vanishing point.

There is also the interesting inverse problem, which is to find the position of the eye in order to see the picture from the viewpoint that the artist intended. Taylor was not the first to discuss this inverse problem but he did make innovative contributions to the theory of such perspective problems. One could certainly consider this work as laying the foundations for the theory of descriptive and projective geometry.

Taylor challenged the "non-English mathematicians" to integrate a certain differential. One has to see this challenge as part of the argument

4.3. APPROXIMATE ANALYSIS

between the Newtonians and the Leibnitzians. Conte discusses the answers given by Johann Bernoulli and Giulio Fagnano to Taylor's challenge. We mentioned above the arguments between Johann Bernoulli and Taylor. Taylor, although he did not win all the arguments, could certainly dispute with Johann Bernoulli on fairly equal terms. Jones says,[5] *Their debates in journals occasionally included rather heated phrases and, at one time, a wager of fifty guineas. When Bernoulli suggested in a private letter that they couch their debate in more gentlemanly terms, Taylor replied that he meant to sound sharp and to "show an indignation."*

Jones also explains that Taylor was a mathematician of far greater depth than many have given him credit for: *A study of Brook Taylor's life and work reveals that his contribution to the development of mathematics was substantially greater than the attachment of his name to one theorem would suggest. His work was concise and hard to follow. The surprising number of major concepts that he touched upon, initially developed, but failed to elaborate further leads one to regret that health, family concerns and sadness, or other unassessable factors, including wealth and parental dominance, restricted the mathematically productive portion of his relatively short life.*

4.3 Approximate Analysis

For many more complicated functions, it is necessary to resort to approximate approaches to the estimation of output statistics. It is also common, even where the functional relations are relatively simple, that the density functions are not known for any of the random variables. At best, one may know the mean value and the variance for each random variable. In such instances, other approaches need to be developed and applied.

4.3.1 Direct Methods

Direct methods imply the direct application of the expectation to the equation that relates the random variables. For instance, we already know how to derive the density function $f_Y(y)$ for the function $Y = aX + b$, where $f_X(x)$ is known as are constants a, b. Suppose, however, that only μ_X and σ_X^2 are known. In general it may not be possible to calculate the mean and

---

[5]Biography in *Dictionary of Scientific Biography* (New York 1970-1990).

variance of Y. For the simple function $Y = aX + b$ it is possible to do so,

$$E\{Y\} = E\{aX + b\}$$
$$= \int_{-\infty}^{\infty} (ax + b) f_X(x) dx$$
$$\mu_Y = a\mu_X + b. \tag{4.22}$$

Similarly for the variance,

$$Var\{Y\} = E\{(Y - E\{Y\})^2\} = \int_{-\infty}^{\infty} (y - \mu_Y)^2 f_Y(y) dy$$
$$= E\left\{(aX + b - a\mu_X - b)^2\right\}$$
$$= a^2 E\left\{X^2 - 2\mu_X X + \mu_X^2\right\}$$
$$= a^2 E\left\{(X - \mu_X)^2\right\}$$
$$= a^2 Var\{X\}$$
$$\sigma_Y^2 = a^2 \sigma_X^2.$$

Therefore,
$$\sigma_Y = a\sigma_X. \tag{4.23}$$

Note that the mean-square value can be written as

$$E\{X^2\} = \sigma_X^2 + \mu_X^2,$$

and then $\sigma_Y^2 = a^2 \left[E\{X^2\} - \mu_X^2 \right]$.

Example 4.8 Beam Supported by Cables under Random Load

Consider the beam supported by two cables, as shown in Figure 4.14. The force F on the beam has uncertainties associated with its magnitude. It is known that the mean value is $\mu_F = 100$ N and the standard deviation is $\sigma_F = 12$ N. Estimate the mean and standard deviation of the tension in each cable. Such knowledge is needed to design the cable for tensile strength and to properly design the supporting structure. Assume that the beam is rigid and that its weight is small enough when compared to the force magnitude that it can be neglected.

Solution   Assume that the given angles are exact. First solve the respective deterministic problem by using the free body diagram of the beam to relate the tensions in the cables to the external force,

$$T_1 = 0.732F \qquad T_2 = 0.517F. \tag{4.24}$$

4.3. APPROXIMATE ANALYSIS

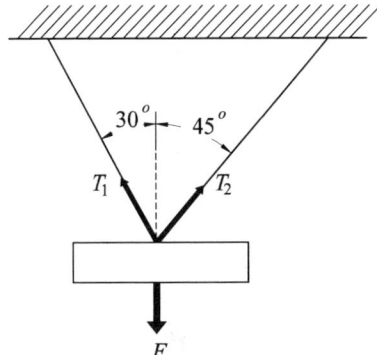

Figure 4.14: A beam supported by two cables under random load F. The free body tension forces are superimposed on the cables.

Next find the mean values and standard deviations of T_1 and T_2 in terms of those for F. Using Equations 4.22 and 4.23 yields

$$E\{T_1\} = 0.732\mu_F = 73.2 \text{ N}$$
$$E\{T_2\} = 0.517\mu_F = 51.7 \text{ N}$$

and

$$\sigma_{T_1} = 0.732\sigma_F = 8.784 \text{ N}$$
$$\sigma_{T_2} = 0.517\sigma_F = 6.204 \text{ N}.$$

Since Equations 4.24 are linear, T_1 and T_2 have density functions of the same shape as the density of F, multiplied by the respective constant factors.

⊛

Suppose the following sum is formed, $Y = a_1 X_1 + a_2 X_2$. The mean value of Y is easily found,

$$E\{Y\} = a_1 E\{X_1\} + a_2 E\{X_2\}$$
$$\mu_Y = a_1 \mu_{X_1} + a_2 \mu_{X_2}.$$

The variance requires a bit more work,

$$\begin{aligned} Var\{Y\} &\equiv E\left\{[Y - \mu_Y]^2\right\} \\ &= E\left\{\left[a_1 X_1 + a_2 X_2 - a_1 \mu_{X_1} - a_2 \mu_{X_2}\right]^2\right\} \\ &= E\left\{a_1^2 \left(X_1 - \mu_{X_1}\right)^2 + a_2^2 \left(X_2 - \mu_{X_2}\right)^2 \right. \\ &\quad \left. + 2a_1 a_2 \left(X_1 - \mu_{X_1}\right)\left(X_2 - \mu_{X_2}\right)\right\}. \end{aligned}$$

The first two terms are the respective variances of X_1 and X_2. Using the definitions of covariance, Equation 3.26, and of correlation coefficient, Equation 3.27,

$$E\{2a_1a_2(X_1 - \mu_{X_1})(X_2 - \mu_{X_2})\} = 2a_1a_2 Cov\{X_1, X_2\}$$
$$= 2a_1a_2 \rho_{X_1 X_2} \sigma_{X_1} \sigma_{X_2}.$$

Therefore,

$$Var\{Y\} = a_1^2 Var\{X_1\} + a_2^2 Var\{X_2\} + 2a_1a_2 \rho_{X_1 X_2} \sigma_{X_1} \sigma_{X_2}$$

or

$$\sigma_Y^2 = (a_1 \sigma_{X_1})^2 + (a_2 \sigma_{X_2})^2 + 2a_1 a_2 \rho_{X_1 X_2} \sigma_{X_1} \sigma_{X_2}.$$

If X_1 and X_2 are statistically independent, the covariance equals zero and $\rho_{X_1 X_2} = 0$. For the relation $Y = a_1 X_1 - a_2 X_2$,

$$\mu_Y = a_1 \mu_{X_1} - a_2 \mu_{X_2}$$
$$Var\{Y\} = a_1^2 Var\{X_1\} + a_2^2 Var\{X_2\} - 2a_1 a_2 \rho_{X_1 X_2} \sigma_{X_1} \sigma_{X_2}.$$

The mean and variance for the sum of two random variables can be generalized for the sum of n random variables,

$$Y = \sum_{i=1}^{n} a_i X_i$$

$$E\{Y\} = \sum_{i=1}^{n} a_i E\{X_i\}$$

$$Var\{Y\} = \sum_{i=1}^{n} a_i^2 Var\{X_i\} + \sum_{i=1}^{n}\sum_{\substack{j=1 \\ j \neq i}}^{n} a_i a_j Cov\{X_i, X_j\}$$

$$= \sum_{i=1}^{n} a_i^2 \sigma_{X_i}^2 + \sum_{i=1}^{n}\sum_{\substack{j=1 \\ j \neq i}}^{n} a_i a_j \rho_{ij} \sigma_{X_i} \sigma_{X_j},$$

where ρ_{ij} is the correlation coefficient between X_i and X_j.

Product functions are more difficult to work with because the expected value of a product requires joint densities. But if the random variables are statistically independent, products become workable. Suppose for the relation $Z = X_1 X_2 \cdots X_n$, the n variables X_i are statistically independent. Then,

$$E\{Z\} = E\{X_1\} E\{X_2\} \cdots E\{X_n\}$$
$$\mu_Z = \mu_{X_1} \mu_{X_2} \cdots \mu_{X_n}$$
$$\sigma_Z^2 = E\{X_1^2\} E\{X_2^2\} \cdots E\{X_n^2\} - (\mu_{X_1} \mu_{X_2} \cdots \mu_{X_n})^2.$$

Thus, mean values and variances are straightforward to calculate.

4.3. APPROXIMATE ANALYSIS

4.3.2 Mean and Variance of a General Function of X to Order σ_X^2

As a prelude to the derivation of expressions for the mean and variance of a general function, consider the more limited relation previously studied, $Y = g(X)$.

The mean value of Y is found by performing the integral $\int g(x) f_X(x)\, dx$ over the domain of x. Suppose the function $g(X)$ is "relatively smooth" in the region of the mean value μ_X (say, $\mu_X \pm \sigma_X$) as shown in Figure 4.15.

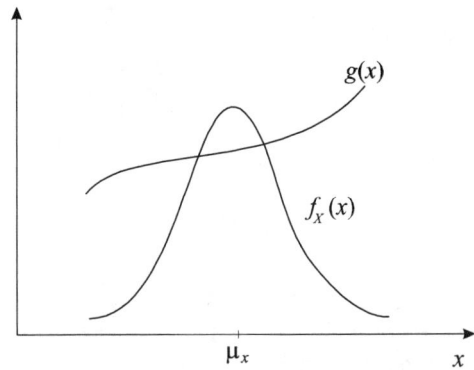

Figure 4.15: A general function, $g(x)$, and the density function $f_X(x)$ of X.

If this is the case, then the following *one term approximation* of the mean of Y can be made,

$$E\{Y\} \simeq g(\mu_X) \int_{-\infty}^{\infty} f_X(x)\, dx = g(\mu_X).$$

This is seen to be the first term of a Taylor series[6] expansion of Y about the mean value of X. A full expansion of $Y = g(X)$ about the mean value μ_X yields

$$Y = g(X)|_{X=\mu_X} + (X - \mu_X) \left.\frac{dg}{dx}\right|_{X=\mu_X} + \frac{1}{2!}(X - \mu_X)^2 \left.\frac{d^2g}{dx^2}\right|_{X=\mu_X} + \cdots.$$

---

[6] The Taylor series of function $g(x)$ expanded about the number a is given by $g(a) + g'(a)(x-a) + g''(a)(x-a)^2/2! + \cdots$, where prime denotes differentiation with respect to x.

Take the expected value of both sides of this equation to find

$$E\{Y\} = E\{g(X)|_{X=\mu_X}\} + E\left\{(X - \mu_X)\left.\frac{dg}{dx}\right|_{X=\mu_X}\right\}$$

$$+ E\left\{\frac{1}{2!}(X - \mu_X)^2 \left.\frac{d^2g}{dx^2}\right|_{X=\mu_X}\right\} + \cdots,$$

or

$$\mu_Y = g(\mu_X) + \frac{\sigma_X^2}{2}g''(\mu_X) + \cdots, \tag{4.25}$$

where the following is used,

$$\begin{aligned} E\{X - \mu_X\} &= E\{X\} - \mu_X \\ &= \mu_X - \mu_X \\ &= 0, \end{aligned}$$

and the derivatives of g are evaluated at $X = \mu_X$, thus making them constants.

Generally, two terms are sufficient for practical applications, but this will depend on the "smoothness" of $g(X)$. The linear approximation is $\mu_Y = g(\mu_X)$. To estimate the variance of Y, begin with the definition of variance, $Var\{Y\} = E\left\{(Y - \mu_Y)^2\right\} = E\{Y^2\} - \mu_Y^2$, where $Y = g(X)$. Defining the new variable $h(X) \equiv g^2(X)$ leads to the use of Equation 4.25 to evaluate the variance,

$$E\{Y^2\} = E\{h(X)\}$$

$$\simeq h(\mu_X) + \frac{\sigma_X^2}{2}h''(\mu_X)$$

$$= g^2(\mu_X) + \frac{\sigma_X^2}{2}\left\{2\left[g'(\mu_X)\right]^2 + 2g(\mu_X)g''(\mu_X)\right\}.$$

$Var\{Y\}$ is then

$$\begin{aligned} \sigma_Y^2 &= E\{Y^2\} - \mu_X^2 \\ &\simeq \left[g^2(\mu_X) + \sigma_X^2\left\{\left[g'(\mu_X)\right]^2 + g(\mu_X)g''(\mu_X)\right\}\right] \\ &\quad - \left[g(\mu_X) + \frac{\sigma_X^2}{2}g''(\mu_X)\right]^2. \end{aligned}$$

Expand the right-hand side of this equation and assume that $\sigma_X^4 \ll \sigma_X^2$ and can be ignored. The approximation to order σ_X^2 for the variance is

4.3. APPROXIMATE ANALYSIS

then given by the relation,

$$\sigma_Y^2 \simeq \sigma_X^2 \left[g'(\mu_X)\right]^2$$
$$\text{or} \quad \sigma_Y \simeq \sigma_X \left[g'(\mu_X)\right]. \quad (4.26)$$

Sometimes, one term approximations are used for both mean and variance.

Example 4.9 A Simple Example of a One-Term Approximation

The moment of inertia of a solid circular cylinder is $I = \pi D^4/64$, where D is the diameter. The first-order approximations for the mean and standard deviation are

$$E\{I\} \simeq \frac{\pi}{64}[E\{D\}]^4$$

$$\sigma_I \simeq \sigma_D \left.\frac{d\left(\pi D^4/64\right)}{dD}\right|_{D=E\{D\}} = \frac{\pi[E\{D\}]^3}{16}\sigma_D.$$

Suppose, $E\{D\} = 10$ in and $\sigma_D = 1$ in. Then

$$E\{I\} = \pi(10)^4/64 = 490.87 \text{ in}^4$$
$$\sigma_I = \pi(10)^3(1)/16 = 196.35 \text{ in}^4.$$

If, from Equation 4.25, we retain the σ^2 order term in the mean value, $\sigma_D^2 g''(\mu_D)/2$, the mean value $E\{I\}$ is increased by $(1)^2 3\pi\mu_D^2/(16\cdot 2) = 29.453$ in^4, which is a 6% increase in value.

If $f_D(d)$ is available, then it is possible to evaluate these expressions exactly. Suppose, $f_D(d) = 1/(d_2 - d_1)$, a uniform density. Then

$$E\{I\} \simeq \frac{\pi}{64}E\{D^4\}$$
$$= \frac{\pi}{64}\frac{1}{d_2 - d_1}\int_{d_1}^{d_2} d^4 dd$$
$$= \frac{\pi}{64}\frac{1}{5}\left(d_1^4 + d_1^3 d_2 + d_1^2 d_2^2 + d_1 d_2^3 + d_2^4\right).$$

⊛

Example 4.10 One-Term Estimates of the Statistics of an Area

From the equation for the area of a circle, $A = \pi R^2$, assuming μ_r and σ_r are known, the following first-order approximations can be found,

$$\mu_A \simeq \pi\mu_R^2$$
$$\sigma_A \simeq \sigma_R\left(2\pi\mu_R\right).$$

Note that if the area equation used is $A = \pi D^2/4$, the average area is the same as above, and the standard deviation becomes $\sigma_D(\pi\mu_D)$, where $\sigma_D = 2\sigma_R$. For a particular density function $f_R(r)$, we have learned how to find the density $f_A(a)$ exactly.

✱

4.3.3 Mean and Variance of a General Function of n RVs

It is useful to extend the discussion of the last section to approximations for completely general functions,

$$Y = g(X_1, X_2, \ldots, X_n),$$

where $g(\cdot)$ is an arbitrary nonlinear function of n random variables. Utilize the Taylor series expansion of a function of n variables, expanded about the mean value of each variable. As a prelude, the Taylor series of a function of two variables $g(X_1, X_2)$ about their respective means (μ_1, μ_2), is given by the series,

$$g(X_1, X_2) = g(\mu_1, \mu_2) + (X_1 - \mu_1)\left.\frac{\partial g}{\partial X_1}\right|_{(\mu_1,\mu_2)} + (X_2 - \mu_2)\left.\frac{\partial g}{\partial X_2}\right|_{(\mu_1,\mu_2)}$$

$$+ \frac{1}{2!}\left[(X_1 - \mu_1)^2 \left.\frac{\partial^2 g}{\partial X_1^2}\right|_{(\mu_1,\mu_2)} + (X_2 - \mu_2)^2 \left.\frac{\partial^2 g}{\partial X_2^2}\right|_{(\mu_1,\mu_2)}\right.$$

$$\left. + 2(X_1 - \mu_1)(X_2 - \mu_2)\left.\frac{\partial^2 g}{\partial X_1 \partial X_2}\right|_{(\mu_1,\mu_2)}\right] + \cdots. \quad (4.27)$$

The expected value and variance of Y are obtained using the expanded function on the right-hand side.

Generalizing the expansion to functions of n random variables, the mean value is

$$E\{Y\} = g(\mu_1, \mu_2, \ldots, \mu_n)$$
$$+ \frac{1}{2}\sum_{i=1}^{n}\sum_{j=1}^{n}\left(\frac{\partial^2 g(\mu_1, \mu_2, \ldots, \mu_n)}{\partial X_i \partial X_j}\right) Cov\{X_i, X_j\}$$
$$+ \cdots, \quad (4.28)$$

where $Cov\{X_i, X_j\} = E\{X_i X_j\} - \mu_i\mu_j$ and $\rho_{ij} = Cov\{X_i, X_j\}/\sigma_i\sigma_j$, and

$$\left(\frac{\partial^2 g(\mu_1, \mu_2, \ldots, \mu_n)}{\partial X_i \partial X_j}\right) \equiv \left.\frac{\partial^2 g}{\partial X_i \partial X_j}\right|_{(\mu_1,\ldots,\mu_n)}.$$

4.3. APPROXIMATE ANALYSIS

The notation is simplified using $\sigma_i \equiv \sigma_{X_i}$, $\rho_{ij} \equiv \rho_{X_i X_j}$. The variance of Y is found to be

$$\begin{aligned}Var\{Y\} &\equiv \sigma_Y^2 \\ &= \sum_{i=1}^{n} \left(\frac{\partial g(\mu_1, \mu_2, \ldots, \mu_n)}{\partial X_i}\right)^2 \sigma_i^2 \\ &+ \sum_{i=1}^{n} \sum_{j=1, j\neq i}^{n} \left(\frac{\partial g(\mu_1, \mu_2, \ldots, \mu_n)}{\partial X_i}\right) \left(\frac{\partial g(\mu_1, \mu_2, \ldots, \mu_n)}{\partial X_j}\right) \rho_{ij}\sigma_i\sigma_j \\ &+ \ldots\end{aligned} \quad (4.29)$$

If any two of the random variables X_i and X_j are statistically independent, then $\rho_{X_i X_j} = 0$. Also, if any of the variables X_i are deterministic, then $\mu_i = X_i$ and $\sigma_i = 0$. We retain only the terms shown above in our approximations. To have a better feel for how to work with Equations 4.28 and 4.29, expand them for the case $n = 2$:

$$\begin{aligned}E\{Y\} &\simeq g(\mu_1, \mu_2) \\ &+ \frac{1}{2}\left[\frac{\partial^2 g(\mu_1, \mu_2)}{\partial X_1^2} Cov\{X_1, X_1\} + \frac{\partial^2 g(\mu_1, \mu_2)}{\partial X_1 \partial X_2} Cov\{X_1, X_2\}\right. \\ &\left. + \frac{\partial^2 g(\mu_1, \mu_2)}{\partial X_2 \partial X_1} Cov\{X_2, X_1\} + \frac{\partial^2 g(\mu_1, \mu_2)}{\partial X_2^2} Cov\{X_2, X_2\}\right] \\ &= g(\mu_1, \mu_2) + \frac{1}{2}\left[\frac{\partial^2 g(\mu_1, \mu_2)}{\partial X_1^2}\sigma_1^2 + 2\frac{\partial^2 g(\mu_1, \mu_2)}{\partial X_1 \partial X_2}\rho_{12}\sigma_1\sigma_2\right. \\ &\left. + \frac{\partial^2 g(\mu_1, \mu_2)}{\partial X_2^2}\sigma_2^2\right]\end{aligned} \quad (4.30)$$

$$\sigma_Y^2 \simeq \left(\frac{\partial g(\mu_1, \mu_2)}{\partial X_1}\right)^2 \sigma_1^2 + 2\left(\frac{\partial g(\mu_1, \mu_2)}{\partial X_1}\right)\left(\frac{\partial g(\mu_1, \mu_2)}{\partial X_2}\right)\rho_{12}\sigma_1\sigma_2 + \left(\frac{\partial g(\mu_1, \mu_2)}{\partial X_2}\right)^2 \sigma_2^2. \quad (4.31)$$

Note that in the above expressions, only terms to order σ^2 have been retained.

Example 4.11 A Simple Function of Two Random Variables

Apply Equations 4.30 and 4.31 to the relation for the moment of inertia of a rectangular section $I = bh^3/12$, where the base, b, and the height, h, are random variables with known means and variances. The one term and order σ^2 approximations of the mean are given by

$$E\{I\}_1 \simeq \frac{1}{12}\mu_b\mu_h^3$$

$$E\{I\}_2 \simeq \frac{1}{12}\mu_b\mu_h^3 + \frac{1}{2}\left[0\cdot\sigma_b^2 + 2\frac{\mu_h^2}{4}\rho_{bh}\sigma_b\sigma_h + \frac{\mu_b\mu_h}{2}\sigma_h^2\right].$$

For the variance,

$$Var\{I\} \simeq \left(\frac{1}{12}\mu_h^3\right)^2\sigma_b^2 + 2\left(\frac{1}{12}\mu_h^3\right)\left(\frac{1}{4}\mu_b\mu_h^2\right)\rho_{bh}\sigma_b\sigma_h + \left(\frac{1}{4}\mu_b\mu_h^2\right)^2\sigma_h^2.$$

If b and h are uncorrelated, then $\rho_{bh} = 0$. Assume the following numerical values: $\mu_b = 2$, $\mu_h = 3$, $\rho_{bh} = 0.5$, $\sigma_b = 0.2$, and $\sigma_h = 0.3$. Then,

$$E\{I\}_1 \simeq \frac{1}{12}2\cdot 3^3 = 4.5$$

$$E\{I\}_2 \simeq 4.5 + \frac{1}{2}\left[2\frac{3^2}{4}0.5\cdot 0.2\cdot 0.3 + \frac{2\cdot 3}{2}0.3^2\right]$$

$$= 4.5 + 0.2025$$

$$= 4.7025;$$

$$Var\{I\} \simeq \left(\frac{1}{12}3^3\right)^2 0.2^2 + 2\left(\frac{1}{12}3^3\right)\left(\frac{1}{4}2\cdot 3^2\right)0.5\cdot 0.2\cdot 0.3$$

$$+ \left(\frac{1}{4}2\cdot 3^2\right)^2 0.3^2$$

$$= 0.2025 + 0.6075 + 1.8225$$

$$= 2.6325$$

$$\sigma_I = \sqrt{2.6325}$$

$$= 1.6225.$$

Generally, one finds that the second-order contributions are significantly smaller than the first-order term.
⊛

Example 4.12 Elongation of a Tension Element

Tension elements such as those found in trusses are often encountered. The extension Δ is found by combining the following relations: stress-strain $\varsigma = E\varepsilon$, stress-force $\varsigma = F/A$, and strain-displacement $\varepsilon = \Delta/L$, resulting in

$$\Delta = g(F, L, A, E) = \frac{FL}{AE}.$$

4.3. APPROXIMATE ANALYSIS

Using one-term approximations only, and assuming that all the random variables $(X_i,\ i=1,2,3,4=F,L,A,E)$ are uncorrelated, the mean value is approximated by

$$\mu_\Delta \simeq \frac{\mu_F \mu_L}{\mu_A \mu_E}. \tag{4.32}$$

To order σ^2, the variance is approximated by

$$\begin{aligned}
\sigma_\Delta^2 &= \sum_{i=1}^n \left(\frac{\partial g(\mu_1,\mu_2,\mu_3,\mu_4)}{\partial X_i}\right)^2 \sigma_i^2 \\
&= \left(\frac{\partial g}{\partial F}\right)^2 \sigma_F^2 + \left(\frac{\partial g}{\partial L}\right)^2 \sigma_L^2 + \left(\frac{\partial g}{\partial A}\right)^2 \sigma_A^2 + \left(\frac{\partial g}{\partial E}\right)^2 \sigma_E^2 \\
&= \left(\frac{\mu_L}{\mu_A \mu_E}\right)^2 \sigma_F^2 + \left(\frac{\mu_F}{\mu_A \mu_E}\right)^2 \sigma_L^2 + \left(-\frac{\mu_F \mu_L}{\mu_A^2 \mu_E}\right)^2 \sigma_A^2 \\
&\quad + \left(-\frac{\mu_F \mu_L}{\mu_A \mu_E^2}\right)^2 \sigma_E^2.
\end{aligned} \tag{4.33}$$

If any of the variables X_i is deterministic, then set $\sigma_i = 0$. Also note that in general the parameter with the largest variance contributes the most to the output variance σ_Δ^2. Assume the following parameter values:

$$\begin{aligned}
(\mu_F, \sigma_F) &= (15000, 750) \text{ lb} \\
(\mu_A, \sigma_A) &= (1, 0.05) \text{ in}^2 \\
(\mu_L, \sigma_L) &= (250, 2.5) \text{ in} \\
(\mu_E, \sigma_E) &= (30 \times 10^6, 0.45 \times 10^6) \text{ psi}.
\end{aligned}$$

Then,

$$\mu_\Delta \simeq \frac{15000 \times 250}{1 \times 30 \times 10^6} = 0.125 \text{ in}$$

$$\begin{aligned}
\sigma_\Delta^2 &= \left(\frac{250}{1 \times 30 \times 10^6}\right)^2 750^2 + \left(\frac{15000}{1 \times 30 \times 10^6}\right)^2 2.5^2 \\
&\quad + \left(-\frac{15000 \times 250}{1^2 \times 30 \times 10^6}\right)^2 0.05^2 + \left(\frac{15000 \times 250}{1 \times (30 \times 10^6)^2}\right)^2 (0.45 \times 10^6)^2 \\
&= 3.9063 \times 10^{-5} + 1.5625 \times 10^{-6} + 3.9063 \times 10^{-5} + 3.5156 \times 10^{-6} \\
&= 8.3203 \times 10^{-5} \text{ in}^2
\end{aligned}$$

$$\sigma_\Delta = 9.1216 \times 10^{-3} \text{ in}.$$

In this way it is possible to determine which data has a significant contribution to the overall variance. Here, the contributions of the force and area variances are an order of magnitude greater than those of the length and the modulus of elasticity. Furthermore, if it is assumed that most of the realizations of Δ are (Gaussian-like) within three standard deviations, then from $(\mu_\Delta, \sigma_\Delta) = (0.125 \text{ in}, 9.1216 \times 10^{-3} \text{ in})$ it can be stated that

$$\begin{aligned}\Delta &= \mu_\Delta \pm 3\sigma_\Delta \\ &= 0.125 \pm 0.027365 \text{ in.}\end{aligned}$$

⊛

Example 4.13   Function of Four Random Variables with Correlation

Suppose it is of interest to examine the effects of correlation between the four random variables in the previous example. First generalize Equations 4.32 and 4.33 to include correlations between all pairs of random variables. From Equations 4.28 and 4.29, the general relations for any function of four correlated random variables, $Y = g(X_1, X_2, X_3, X_4)$, to order σ^2 are

$$E\{Y\} \simeq g(\mu_1, \mu_2, \mu_3, \mu_4)$$
$$+ \frac{1}{2}\left\{\frac{\partial^2 g}{\partial X_1^2}\sigma_1^2 + \frac{\partial^2 g}{\partial X_2^2}\sigma_2^2 + \frac{\partial^2 g}{\partial X_3^2}\sigma_3^2 + \frac{\partial^2 g}{\partial X_4^2}\sigma_4^2\right.$$
$$+ 2\frac{\partial^2 g}{\partial X_1 \partial X_2}\rho_{12}\sigma_1\sigma_2 + 2\frac{\partial^2 g}{\partial X_1 \partial X_3}\rho_{13}\sigma_1\sigma_3 + 2\frac{\partial^2 g}{\partial X_1 \partial X_4}\rho_{14}\sigma_1\sigma_4$$
$$\left.+ 2\frac{\partial^2 g}{\partial X_2 \partial X_3}\rho_{23}\sigma_2\sigma_3 + 2\frac{\partial^2 g}{\partial X_2 \partial X_4}\rho_{24}\sigma_2\sigma_4 + 2\frac{\partial^2 g}{\partial X_3 \partial X_4}\rho_{34}\sigma_3\sigma_4\right\},$$

$$Var\{Y\} \simeq \left(\frac{\partial g}{\partial X_1}\right)^2\sigma_1^2 + \left(\frac{\partial g}{\partial X_2}\right)^2\sigma_2^2 + \left(\frac{\partial g}{\partial X_3}\right)^2\sigma_3^2 + \left(\frac{\partial g}{\partial X_4}\right)^2\sigma_4^2$$
$$+ 2\frac{\partial g}{\partial X_1}\frac{\partial g}{\partial X_2}\rho_{12}\sigma_1\sigma_2 + 2\frac{\partial g}{\partial X_1}\frac{\partial g}{\partial X_3}\rho_{13}\sigma_1\sigma_3$$
$$+ 2\frac{\partial g}{\partial X_1}\frac{\partial g}{\partial X_4}\rho_{14}\sigma_1\sigma_4 + 2\frac{\partial g}{\partial X_2}\frac{\partial g}{\partial X_3}\rho_{23}\sigma_2\sigma_3$$
$$+ 2\frac{\partial g}{\partial X_2}\frac{\partial g}{\partial X_4}\rho_{24}\sigma_2\sigma_4 + 2\frac{\partial g}{\partial X_3}\frac{\partial g}{\partial X_4}\rho_{34}\sigma_3\sigma_4.$$

Then, for the function of the previous example, $\Delta = g(F, L, A, E) = FL/AE$,

4.3. APPROXIMATE ANALYSIS

the approximate mean and variance are given by

$$\mu_\Delta \simeq g(\mu_F, \mu_L, \mu_A, \mu_E)$$
$$+ \frac{1}{2}\left\{ \frac{\partial^2 g}{\partial F^2}\sigma_F^2 + \frac{\partial^2 g}{\partial L^2}\sigma_L^2 + \frac{\partial^2 g}{\partial A^2}\sigma_A^2 + \frac{\partial^2 g}{\partial E^2}\sigma_E^2 \right.$$
$$+ 2\frac{\partial^2 g}{\partial F \partial L}\rho_{FL}\sigma_F\sigma_L + 2\frac{\partial^2 g}{\partial F \partial A}\rho_{FA}\sigma_F\sigma_A + 2\frac{\partial^2 g}{\partial F \partial E}\rho_{FE}\sigma_F\sigma_E$$
$$\left. + 2\frac{\partial^2 g}{\partial L \partial A}\rho_{LA}\sigma_L\sigma_A + 2\frac{\partial^2 g}{\partial L \partial E}\rho_{LE}\sigma_L\sigma_E + 2\frac{\partial^2 g}{\partial A \partial E}\rho_{AE}\sigma_A\sigma_E \right\},$$

and

$$Var\{Y\} \simeq \left(\frac{\partial g}{\partial F}\right)^2 \sigma_F^2 + \left(\frac{\partial g}{\partial L}\right)^2 \sigma_L^2 + \left(\frac{\partial g}{\partial E}\right)^2 \sigma_E^2 + \left(\frac{\partial g}{\partial A}\right)^2 \sigma_A^2$$
$$+ 2\frac{\partial g}{\partial F}\frac{\partial g}{\partial L}\rho_{FL}\sigma_L + 2\frac{\partial g}{\partial F}\frac{\partial g}{\partial E}\rho_{FE}\sigma_F\sigma_E$$
$$+ 2\frac{\partial g}{\partial F}\frac{\partial g}{\partial A}\rho_{FA}\sigma_F\sigma_A + 2\frac{\partial g}{\partial L}\frac{\partial g}{\partial E}\rho_{LE}\sigma_L\sigma_E$$
$$+ 2\frac{\partial g}{\partial L}\frac{\partial g}{\partial A}\rho_{LA}\sigma_L\sigma_A + 2\frac{\partial g}{\partial E}\frac{\partial g}{\partial A}\rho_{EA}\sigma_E\sigma_A.$$

Replace $g(F, L, A, E)$ by FL/AE; then the mean and the variance are given by

$$\mu_\Delta \simeq \frac{\mu_F \mu_L}{\mu_A \mu_E} + \frac{1}{2}\left\{ 0 + 0 + 2\frac{\mu_F \mu_L}{\mu_A^3 \mu_E}\sigma_A^2 + 2\frac{\mu_F \mu_L}{\mu_A \mu_E^3}\sigma_E^2 \right.$$
$$+ 2\frac{1}{\mu_A \mu_E}\rho_{FL}\sigma_F\sigma_L - 2\frac{\mu_L}{\mu_A^2 \mu_E}\rho_{FA}\sigma_F\sigma_A - 2\frac{\mu_L}{\mu_A \mu_E^2}\rho_{FE}\sigma_F\sigma_E$$
$$\left. - 2\frac{\mu_F}{\mu_A^2 \mu_E}\rho_{LA}\sigma_L\sigma_A - 2\frac{\mu_F}{\mu_A \mu_E^2}\rho_{LE}\sigma_L\sigma_E + 2\frac{\mu_F \mu_L}{\mu_A^2 \mu_E^2}\rho_{AE}\sigma_A\sigma_E \right\},$$

and

$$Var\{\Delta\} \simeq \left(\frac{\mu_L}{\mu_A \mu_E}\right)^2 \sigma_F^2 + \left(\frac{\mu_F}{\mu_A \mu_E}\right)^2 \sigma_L^2$$
$$+ \left(-\frac{\mu_L \mu_F}{\mu_A^2 \mu_E}\right)^2 \sigma_A^2 + \left(-\frac{\mu_L \mu_F}{\mu_A \mu_E^2}\right)^2 \sigma_E^2$$
$$+ 2\frac{\mu_L \mu_F}{\mu_A^2 \mu_E^2}\rho_{FL}\sigma_F\sigma_L - 2\frac{\mu_L^2 \mu_F}{\mu_A^3 \mu_E^2}\rho_{FA}\sigma_F\sigma_A - 2\frac{\mu_L^2 \mu_F}{\mu_A^2 \mu_E^3}\rho_{FE}\sigma_F\sigma_E$$
$$- 2\frac{\mu_L \mu_F^2}{\mu_A^3 \mu_E^2}\rho_{LA}\sigma_L\sigma_A - 2\frac{\mu_L \mu_F^2}{\mu_A^2 \mu_E^3}\rho_{LE}\sigma_L\sigma_E + 2\frac{\mu_L^2 \mu_F^2}{\mu_A^3 \mu_E^3}\rho_{AE}\sigma_A\sigma_E.$$

CHAPTER 4. FUNCTIONS OF RANDOM VARIABLES

The parameter values given above are repeated, now also including the respective correlation coefficients,

$$(\mu_F, \sigma_F) = (15000, 750) \text{ lb}$$
$$(\mu_A, \sigma_A) = (1, 0.05) \text{ in}^2$$
$$(\mu_L, \sigma_L) = (250, 2.5) \text{ in}$$
$$(\mu_E, \sigma_E) = (30 \times 10^6, 0.45 \times 10^6) \text{ psi}$$

$$\rho_{FL} = 0.1 \quad \rho_{FA} = 0.1 \quad \rho_{FE} = 0.1$$
$$\rho_{LA} = 0.5 \quad \rho_{LE} = 0.1 \quad \rho_{AE} = 0.1.$$

Substitute them into the approximation for the mean, to find

$$\mu_\Delta \simeq \frac{15000 \times 250}{1 \times 30 \times 10^6} + \frac{1}{2}\left\{ 0 + 0 + \frac{2 \times 15000 \times 250}{1^3 \times 30 \times 10^6} \times 0.05^2 \right.$$
$$+ \frac{2 \times 15000 \times 250}{1 \times (30 \times 10^6)^3}(0.45 \times 10^6)^2 + 2 \times \frac{1}{1 \times 30 \times 10^6} \times 0.1 \times 750 \times 2.5$$
$$- 2\frac{250}{1^2 \times 30 \times 10^6} \times 0.1 \times 750 \times 0.05$$
$$- 2 \times \frac{250}{1 \times (30 \times 10^6)^2} \times 0.1 \times 750 \times 0.45 \times 10^6$$
$$- 2 \times \frac{15000}{1^2 \times 30 \times 10^6} \times 0.05 \times 2.5 \times 0.05$$
$$- 2 \times \frac{15000}{1 \times (30 \times 10^6)^2} \times 0.1 \times 2.5 \times 0.45 \times 10^6$$
$$\left. + 2 \times \frac{15000 \times 250}{1^2 \times (30 \times 10^6)^2} \times 0.1 \times 0.5 \times 0.45 \times 10^6 \right\}$$
$$= 0.125 + \frac{1}{2}\left\{ 6.25 \times 10^{-4} + 5.625 \times 10^{-5} + 1.25 \times 10^{-5} \right.$$
$$- 6.25 \times 10^{-5} - 1.875 \times 10^{-5}$$
$$\left. -6.25 \times 10^{-6} - 3.75 \times 10^{-6} + 1.875 \times 10^{-4} \right\}$$
$$= 0.125 + \mathbf{3.95 \times 10^{-4}}$$
$$= 0.1254,$$

where the bold term is the contribution of the σ^2 order terms, and

$$Var\{\Delta\} \simeq 8.3203 \times 10^{-5} - \mathbf{1.4531 \times 10^{-5}}$$
$$= 6.8672 \times 10^{-5}$$
$$\sigma_\Delta \simeq 8.2869 \times 10^{-3},$$

4.3. APPROXIMATE ANALYSIS

where the correlation term is bold. Thus, for the mean, the correction was to the fourth significant digit of $3.95 \times 10^{-4}/0.125 = 0.00316$ or 0.316%, and $1.4531/8.3203 = 0.17465$ or -17.5% for the variance.

⊛

Example 4.14 Projectile Motion in Two Random Variables

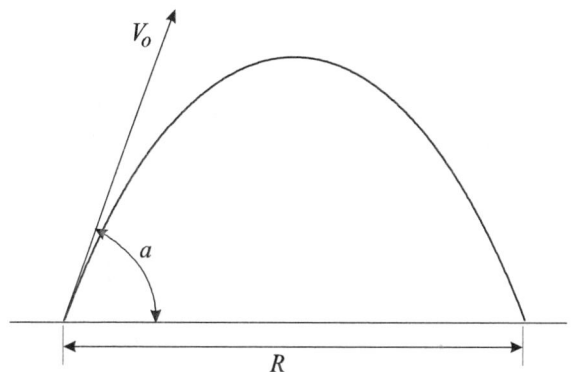

Figure 4.16: Projectile with inital velocity V_0 and angle a as random variables.

Consider the projectile with initial velocity vector and trajectory shown in Figure 4.16. The range R is a function of the initial velocity V_0 and the initiation angle a, $R = g(V_0, a)$. Estimate R, given the following information,

$$\mu_a = 30° = \pi/6 \text{ rad}$$
$$\delta_a = 0.10$$

and

$$\mu_{V_0} = 300 \text{ m/s}$$
$$\sigma_{V_0} = 35 \text{ m/s}.$$

Elementary physics provides the relation between these three variables,

$$R = \frac{V_0^2}{\mathfrak{g}} \sin 2a, \qquad (4.34)$$

where $\mathfrak{g}$ is the gravitational acceleration. The one-term approximation to the mean value is obtained by replacing each random variable by its mean value in Equation 4.34,

$$\mu_R = \frac{\mu_{V_0}^2}{g} \sin 2\mu_a$$
$$= \frac{300^2}{9.8} \sin \frac{\pi}{3}$$
$$= 7953.3 \text{ m}.$$

The σ^2 order estimate of the variance of this function (of two random variables) is given by

$$\sigma_R^2 \simeq \left(\frac{\partial g\left(\mu_{V_0}, \mu_a\right)}{\partial V_0}\right)^2 \sigma_{V_0}^2$$
$$+ 2\left(\frac{\partial g\left(\mu_{V_0}, \mu_a\right)}{\partial V_0}\right)\left(\frac{\partial g\left(\mu_{V_0}, \mu_a\right)}{\partial a}\right) \rho_{V_0 a} \sigma_{V_0} \left(\delta_a \mu_a\right)$$
$$+ \left(\frac{\partial g\left(\mu_{V_0}, \mu_a\right)}{\partial a}\right)^2 \left(\delta_a \mu_a\right)^2.$$

If V_0 and a are statistically independent, then $\rho_{V_0 a} = 0$, and the variance is estimated by

$$\sigma_R^2 \simeq \left(\frac{\partial g\left(\mu_{V_0}, \mu_a\right)}{\partial V_0}\right)^2 \sigma_{V_0}^2 + \left(\frac{\partial g\left(\mu_{V_0}, \mu_a\right)}{\partial a}\right)^2 \left(\delta_a \mu_a\right)^2$$
$$= \left(\frac{2\mu_{V_0}}{g} \sin 2\mu_a\right)^2 \sigma_{V_0}^2 + \left(\frac{2\mu_{V_0}^2}{g} \cos 2\mu_a\right)^2 \left(\delta_a \mu_a\right)^2$$
$$= \left(\frac{2 \cdot 300}{9.8} \sin \frac{\pi}{3}\right)^2 35^2 + \left(\frac{2 \cdot 300^2}{9.8} \cos \frac{\pi}{3}\right)^2 (0.1 \cdot \pi/6)^2$$
$$= 3.4439 \times 10^6 + 2.3122 \times 10^5$$
$$= 3.6751 \times 10^6$$
$$\sigma_R \simeq \sqrt{3.6751 \times 10^6}$$
$$= 1917.1 \text{ m}.$$

To gauge the effect of any correlation between V_0 and a, calculate σ_R

4.3. APPROXIMATE ANALYSIS

for the cases $\rho_{V_0 a} = \pm 1$, respectively,

$$\sigma_R^2 \simeq \left(\frac{2\mu_{V_0}}{\mathrm{g}} \sin 2\mu_a\right)^2 \sigma_{V_0}^2$$
$$+ 2\left(\frac{2\mu_{V_0}}{\mathrm{g}} \sin 2\mu_a\right)\left(\frac{2\mu_{V_0}^2}{\mathrm{g}} \cos 2\mu_a\right) \rho_{V_0 a} \sigma_{V_0} (\delta_a \mu_a)$$
$$+ \left(\frac{2\mu_{V_0}^2}{\mathrm{g}} \cos 2\mu_a\right)^2 (\delta_a \mu_a)^2$$
$$= \left(\frac{2 \cdot 300}{9.8} \sin \frac{\pi}{3}\right)^2 35^2$$
$$+ 2\left(\frac{2 \cdot 300}{9.8} \sin \frac{\pi}{3}\right)\left(\frac{2 \cdot 300^2}{9.8} \cos \frac{\pi}{3}\right)(+1)(35)(0.1 \cdot \pi/6)$$
$$+ \left(\frac{2 \cdot 300^2}{9.8} \cos \frac{\pi}{3}\right)^2 (0.1 \times \pi/6)^2$$
$$= 3.4439 \times 10^6 + 1.7847 \times 10^6 + 2.3122 \times 10^5$$
$$= 5.4598 \times 10^6$$
$$\sigma_R \simeq \sqrt{5.4598 \times 10^6}$$
$$= 2336.6 \text{ m};$$

$$\sigma_R^2 \simeq \left(\frac{2 \cdot 300}{9.8} \sin \frac{\pi}{3}\right)^2 35^2$$
$$+ 2\left(\frac{2 \cdot 300}{9.8} \sin \frac{\pi}{3}\right)\left(\frac{2 \cdot 300^2}{9.8} \cos \frac{\pi}{3}\right)(-1)(35)(0.1 \cdot \pi/6)$$
$$+ \left(\frac{2 \cdot 300^2}{9.8} \cos \frac{\pi}{6}\right)^2 (0.1 \times \pi/6)^2$$
$$= 3.4439 \times 10^6 - 1.7847 \times 10^6 + 2.3122 \times 10^5$$
$$= 1.8904 \times 10^6$$
$$\sigma_R \simeq \sqrt{1.8904 \times 10^6}$$
$$= 1374.9 \text{ m}.$$

The correlation between the two random variables has a significant effect on the variance of the range. This set of calculations provides the analyst with bounds on the variance for the full range of correlation coefficients, knowledge that is useful as part of a design process.

✳

If the σ^2 order term is retained in the expansion of $Y = g(X)$,

$$Y = g(\mu_X) + (X - \mu_X)\left.\frac{dg}{dx}\right|_{X=\mu_X} + \frac{1}{2!}(X - \mu_X)^2\left.\frac{d^2g}{dx^2}\right|_{X=\mu_X} + \cdots$$
$$\simeq g(\mu_X) + (X - \mu_X)g'(\mu_X) + \frac{1}{2!}(X - \mu_X)^2 g''(\mu_X), \qquad (4.35)$$

then the σ^2 order estimate of the mean is given by Equation 4.25, $\mu_Y = g(\mu_X) + \sigma_X^2 g''(\mu_X)/2$. The variance based on Equation 4.35 is found by definition,

$$Var\{Y\} \simeq \int [y - \mu_Y]^2 f_Y(y)\,dy$$

$$= \int [g(x) - \mu_Y]^2 f_X(x)\,dx$$

$$= \int \left[g(\mu_X) + (x - \mu_X)g'(\mu_X) + \frac{(x-\mu_X)^2}{2!}g''(\mu_X)\right.$$
$$\left. - g(\mu_X) - \frac{\sigma_X^2}{2}g''(\mu_X)\right]^2 f_X(x)\,dx$$

$$= \int \left[(x - \mu_X)g'(\mu_X) + \frac{1}{2!}(x - \mu_X)^2 g''(\mu_X) - \frac{\sigma_X^2}{2}g''(\mu_X)\right]^2 f_X(x)\,dx$$

$$= \int \left\{ (x - \mu_X)^2 [g'(\mu_X)]^2 + \frac{1}{4}(x - \mu_X)^4 [g''(\mu_X)]^2 + \frac{\sigma_X^4}{4}[g''(\mu_X)]^2 \right.$$
$$+ (x - \mu_X)^3 g'(\mu_X)g''(\mu_X) - (x - \mu_X)g'(\mu_X)\sigma_X^2 g''(\mu_X)$$
$$\left. - \frac{1}{2}(x - \mu_X)^2 \sigma_X^2 [g''(\mu_X)]^2 \right\} f_X(x)\,dx$$

$$= \sigma_X^2 [g'(\mu_X)]^2 + \frac{1}{4}E\{(X - \mu_X)^4\}[g''(\mu_X)]^2 - \frac{\sigma_X^4}{4}[g''(\mu_X)]^2$$
$$+ E\{(x - \mu_X)^3\}g'(\mu_X)g''(\mu_X).$$

Retention of the σ^2 order term in the expansion for $g(X)$ requires the third- and fourth-order expectations. These may be difficult to determine experimentally. In practice, if such techniques are utilized, it is rare for one to retain terms of order greater than σ^2.

4.4 Monte Carlo Methods

Monte Carlo methods are attractive numerical methods that are used to solve practical and complex problems involving randomness[7] where exact and approximate methods are not useful. Monte Carlo methods are based on an understanding of the underlying deterministic relations between the variables of interest. Monte Carlo analysis may be viewed as an experiment carried out on the computer.

The sequence of steps in the procedure follows:

- define the deterministic problem and the relevant equations

- identify the random parameters and their probability densities or their mean values and variances

- use a random number generator to generate a list of random numbers that are uniformly distributed between 0 and 1

- use these random numbers to select realizations of the random variables in the problem to be analyzed; this may require a transformation to the appropriate density

- substitute these realizations into the aforementioned deterministic equations and solve for the required outputs

- repeat this process many times to generate a set of output values

- perform a statistical analysis of this set of output values to find their means and variances

This procedure is used next for a number of simple problems for variables with uniform and Gaussian densities.

4.4.1 Independent Uniform Random Numbers

Consider Figure 4.17 of the cumulative distribution function for a random variable X with a uniform probability density in the range $[a, b]$. The function is a straight line through the coordinates $[a, 0]$ and $[b, 1]$.

When the random number is selected, it is interpreted as a probability p_i to be entered on the vertical axis. Following this value across to the distribution curve, and then down to the intercept results in a realization

---

[7]Applying statistical approaches to solving complex deterministic problems is a well-known process. Monte Carlo methods were originally used to estimate definite integrals.

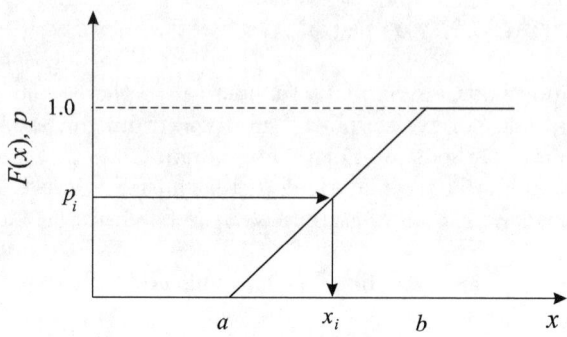

Figure 4.17: Cumulative uniform distribution function. Each p_i corresponds to an x_i.

Table 4.2: Sample Uniform Random Numbers between 0 and 1

0.42742	0.32111	0.34363	0.47426	0.55846	0.74675	0.03206
0.25981	0.31008	0.79718	0.03917	0.08843	0.96050	0.81292
0.95105	0.14649	0.15559	0.42939	0.52543	0.27260	0.21976
0.28134	0.79250	0.75121	0.62836	0.31375	0.00586	0.07481
0.13554	0.99164	0.45261	0.74287	0.50254	0.19905	0.82524
0.79539	0.62948	0.73645	0.40159	0.17831	0.91483	0.28131
0.60513	0.55292	0.89152	0.62032	0.91285	0.52572	0.44747
0.15763	0.88250	0.33906	0.06591	0.11952	0.57416	0.15065
0.68159	0.70974	0.23110	0.69380	0.18830	0.01082	0.05431
0.02581	0.12102	0.66529	0.09740	0.78042	0.98779	0.67420
0.72297	0.45375	0.67598	0.64384	0.99190	0.45410	0.98149

of the random variable, x_i. This can be accomplished graphically, but more easily using the equation for the distribution. In this case,

$$x_i = a + (b-a)\,p_i, \qquad (4.36)$$

where $p_i \equiv F_X(x_i) = \Pr(X \leq x_i)$. So it comes down to the generation of random numbers uniformly on the range $[0, 1]$. Table 4.2 provides a list of random numbers generated using MAPLE.

Example 4.15 Generate 5 Uniform Random Numbers

Use Table 4.2 to generate 5 random numbers that are uniformly distributed with a mean of 9.76 and a standard deviation of 1.15.

4.4. MONTE CARLO METHODS

Table 4.3: Sample Realizations

Sample Number i	Random Number p_i	Generated Realization x_i with $\mu = 9.76$ and $\sigma = 1.15$
1	0.42742	9.4708
2	0.25981	8.8031
3	0.95105	11.557
4	0.28134	8.8889
5	0.13554	8.308

Solution Using the properties of the uniform density, the mean and variance are given by

$$\mu = \frac{a+b}{2}$$

$$\sigma^2 = \frac{1}{12}(b-a)^2.$$

Solve for the upper and lower bounds of the uniform density to find $b = 11.752$ and $a = 7.768$. Select the first 5 random numbers from the table and generate the first 5 realizations using

$$x_i = 7.768 + (11.752 - 7.768)\, p_i.$$

The results are shown in Table 4.3. Any five random numbers from the table could have been used. Calculating x_i is the first step in a Monte Carlo procedure. Realizations x_i are substituted into the equation being studied.

Example 4.16 Monte Carlo Method for a Static Beam

Consider a cantilever beam of length $L = 25$ ft, fixed at the left end and loaded by a force R acting upward at the right end as shown in Figure 4.18. The beam has a dead weight of $w = 100$ lb/ft. R is a random variable with mean value $\mu_R = 9{,}760$ lb and a coefficient of variation $\delta_R = 0.11783$. Estimate the shear V and moment M at the fixed end.

Solution Do this using generated random numbers p_i from Table 4.2.

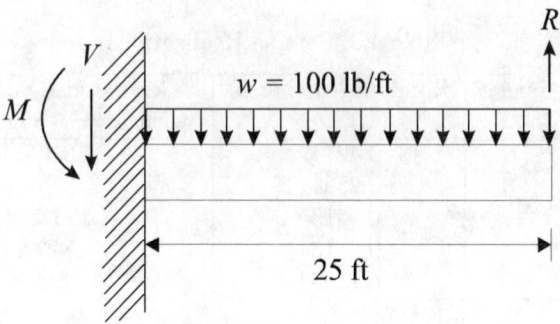

Figure 4.18: Cantilever beam under deterministic uniform load w and concentrated random force R.

The shear and moment at the fixed end are, respectively,

$$V = R - wL$$
$$= R - 2500 \text{ lb},$$
$$M = -RL + wL\left(\frac{L}{2}\right)$$
$$= -25R + 31250 \text{ lb-ft}.$$

The shear and moment are functions of R, which is a random variable. Using the given μ_R and δ we can find $\sigma_R = \mu_R \delta = 9760 \times 0.11783 = 1150$. As before, the upper and lower bounds of the uniform density are found to be $a = 7768$ and $b = 11752$. Thus, utilize Table 4.2 to first generate realizations of R. For each realization R_i, calculate V_i and M_i using the following transformations:

$$V_i = R_i - 2500$$
$$M_i = -25R_i + 31250,$$

where

$$R_i = a + (b-a)p_i.$$

Sample realizations of the shear and moment are shown in Table 4.4. These sample values of V and M can now be averaged to estimate their respective

4.4. MONTE CARLO METHODS

Table 4.4: Sample Realizations of V and M

p_i	R_i	$V_i = R_i - 2500$	$M_i = -25R_i + 31250$
0.42742	9470.8	6970.8	-2.0552×10^5
0.25981	8803.1	6303.1	-1.8883×10^5
0.95105	11557	9057.0	-2.5768×10^5
0.28134	8888.9	6388.9	-1.9097×10^5
0.13554	8308.0	5808.0	-1.7633×10^5
0.32111	9047.3	6547.3	-1.9493×10^5
0.31008	9003.4	6503.4	-1.9384×10^5
0.14649	8351.6	5851.6	-1.7754×10^5
0.79250	10925	8425.0	-2.4188×10^5
0.99164	11719	9219.0	-2.6173×10^5

means and standard deviations with results shown in Table 4.5.[8] We realize that 10 sample points are by no means sufficient in an applied problem. As part of the computation, the mean and variance at the end of each cycle can be calculated and in this way one can observe if the values converge to within a prescribed value. Once the change in mean and variance is below this prescribed value, the program can stop the simulation.

Note that the mean is a negative quantity, the standard deviation is (always) a positive quantity, and the coefficient of variation is also a positive quantity since it is a measure of the scatter normalized by the mean.

4.4.2 Independent Normal Random Numbers

Many applications have random variables that are not uniform, but rather are distributed in a variety of other ways. The normal density is a popular model of randomness. There are two straightforward ways to go about

---

[8] Note that the mean and the variance of discrete data are obtained using

$$\mu_X = \frac{1}{n} \sum_{i=1}^{n} x_i$$

$$\sigma_X = \sqrt{\sum_{i=1}^{n} \frac{(x_i - \mu_x)^2}{n-1}}.$$

Table 4.5: Sample Statistics for V and M

p_i	V (lb)	M (lb-ft)
μ	7107.4	-2.0892×10^5
σ	1296.2	0.3076×10^5
δ	0.18237	0.14721

Table 4.6: Standard Normal Random Numbers

0.58885	−0.48187	−0.50652	−0.13616	0.96148
−1.3286	1.5813	−0.23290	−0.42309	0.16112
4.1952×10^{-2}	0.77828	−0.19540	0.91004	0.85910
−0.46021	0.52954	0.19337	2.0304	−0.36170

generating random numbers according to a distribution. The first is to generate the standard uniform random numbers, and then enter them into the standard normal table of values to look up the respective standard normal random numbers. The other approach is to generate the normal random numbers using a random-number generator.

In the first method:

- select a random number p_i from a standard uniform list of numbers

- use this number as the entry point in a table of standard normal random numbers and work the table in an inverse way, entering the cumulative probability p_i and reading the standard normal number s_i off the scale

- use s_i to find the realization x of the normal variable X with mean μ_X and standard deviation σ_X using the transformation $x = \mu_X + s\sigma_X$ between normal and standard normal

For example, take the first number from the standard uniform table, 0.42742, use it as the entry point into a standard normal table,[9] and find the value of $s \simeq -0.18$. Therefore, $x = \mu_X - 0.18\sigma_X$.

By the second approach, generate 20 standard normal random numbers as shown in Table 4.6. (These standard normal random numbers can also be generated using the Box-Muller method demonstrated in Example 12.8 in Section 12.2.2.) Suppose 5 normal random numbers with mean $\mu_X = 15$ and standard deviation $\sigma_X = 3$ need to be generated. The way to proceed

---
[9] The standard normal table is different from a table of numbers that follow the standard normal density.

4.4. MONTE CARLO METHODS

is to pick 5 standard normal random numbers from Table 4.6 and use the transformation between normal x and standard normal variables s,

$$s = \frac{x - \mu_x}{\sigma_x} = \frac{x - 15}{3},$$

or $x = 3s + 15$. Using the first five values of s, the values of x we are seeking are shown in Table 4.7. These values of x would then be used as realizations of the variable X.

Table 4.7: Normal Random Numbers x with $\mu_X = 15$ and $\sigma_X = 3$

s	$x = \mu_X + s\sigma_X$
0.58885	16.767
-1.3286	11.014
4.1952×10^{-2}	15.126
-0.46021	13.619
-0.48187	13.554

4.4.3 A Discretization Procedure

Sometimes in applications the distribution is quite complex since it is created from experimental data. Such a complex distribution curve is difficult to work with. In this instance, the complicated curve can be discretized into a sequence of straight lines. This arbitrary cumulative probability distribution can be discretized into as many straight line segments as necessary to maintain any level of accuracy. After discretization, the arbitrary cumulative probability distribution function will look like the sketch in Figure 4.19.

Suppose the curve is discretized into m segments. This is equivalent to discretizing the vertical probability axis into m segments, $\Delta p = 1/m$. The subscript j in the figure is used for these points. Each p_j corresponds to an x_j. When an explicit expression for $F^{-1}(x)$ is not known, x_j are obtained numerically.

For any standard uniform random number p_i between p_j and p_{j+1}, the equation for the straight line segment is

$$x_i = x_j + \left[\frac{x_{j+1} - x_j}{p_{j+1} - p_j} \right] (p_i - p_j). \tag{4.37}$$

This equation is of the same form as Equation 4.36, and therefore the same procedure can be used for any probability distribution. To demonstrate the

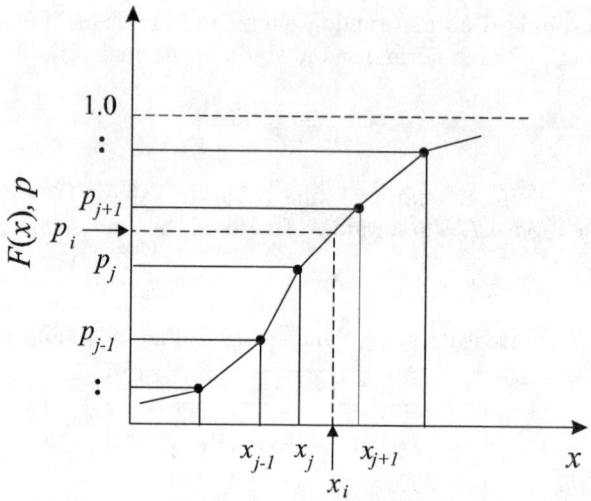

Figure 4.19: Discretized cumulative distribution for a general Monte Carlo method.

procedure, consider the following density function for random variable X,

$$f_X(x) = \begin{cases} 0, & x \leq 25 \\ 1/20, & 25 < x \leq 35 \\ 0, & 35 < x \leq 50 \\ 1/20, & 50 < x \leq 60. \end{cases} \quad (4.38)$$

In order to generate the random value realizations for random variable X according to the given probability density, the respective cumulative distribution is needed. It can be derived using the definition $F_X(x) = \int_{-\infty}^{x} f_X(x)\,dx$ and is shown in Figure 4.20.

Example 4.17 Random-Number Generation using Discretization

Find three random numbers that are distributed according to the cumulative distribution shown in Figure 4.20.

Solution Use the first three standard uniform random numbers listed in Table 4.2 and enter these into Equation 4.37. The points (x_j, p_j) and (x_{j+1}, p_{j+1}) are the end points of each line segment. The results are summarized in Table 4.8. Such a technique is useful for simulating experimentally determined distributions in a simple way, by using straight-line approximations.

4.4. MONTE CARLO METHODS

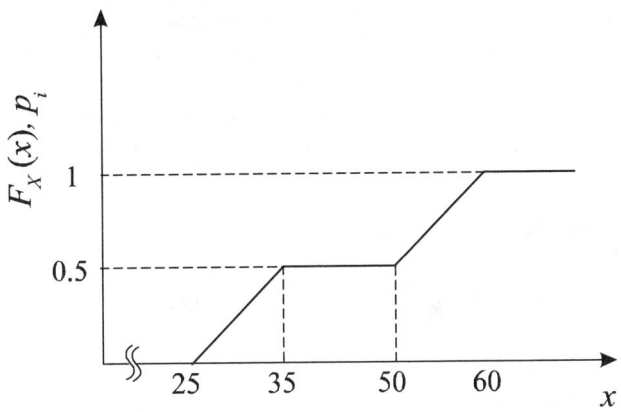

Figure 4.20: Cumulative distribution function $F_X(x)$.

Table 4.8: Random Numbers Using the Discrete Transform

p_i	$x_i = x_j + \left(\frac{x_{j+1}-x_j}{p_{j+1}-p_j}\right)(p_i - p_j)$	x_i
0.42742	$x_i = 25 + \left(\frac{35-25}{0.5-0}\right)(0.42742 - 0)$	33.5480
0.25981	$x_i = 25 + \left(\frac{35-25}{0.5-0}\right)(0.25981 - 0)$	30.1962
0.95105	$x_i = 50 + \left(\frac{60-50}{1.0-0.5}\right)(0.95105 - 0.5)$	59.0210

4.5 Concluding Summary

This chapter takes probabilistic modeling to the next level of complexity. Engineering and science depend on functions between variables for modeling physical processes. If some or all of these variables are random variables, then procedures are necessary to derive the density function of a variable that is related by this function to a random variable. For simple functions this is possible. For more complicated functions, approximate and numerical methods are needed. Such approaches are introduced and explained.

4.6 Problems

Section 4.1: Exact Functions of One Variable

1. Derive the probability density function for Y, given that $Y = X^4$, for the following densities. Sketch all density functions.

   (i) $f_X(x) = \frac{1}{2}$, $\quad 0 < x < 2$

   (ii) $f_X(x) = c \exp(-x)$, $\quad 0 < x < \infty$.

2. For the function $Y = aX^2$, find $f_Y(y)$ for the case where

   (i) $f_X(x) = c \exp(-x)$, $x \geq 0$,

   (ii) $f_X(x)$ is a Rayleigh density,

   (iii) $f_X(x)$ is a lognormal density.

   Sketch all density functions.

3. Derive the probability density function for Y, given $Y = a + X^3$. Consider two cases:

   (i) $f_X(x) = c \exp(-x)$, $x \geq 0$,

   (ii) $f_X(x)$ is a lognormal density.

   Sketch all density functions.

4. Derive the probability density function for Y, given $Y = X^3$ for $f_X(x) = 1/c^2$, $2 < x < 4$. Sketch all density functions.

5. The random variables X and Y are related by the equation,

$$Y = e^X.$$

   (i) Suppose that X is uniformly distributed as $0 \leq X \leq 5$. Find $f_Y(y)$. Sketch this density function.

4.6. PROBLEMS

(ii) Suppose now that X is governed by the density function $f_X = ce^{-x}$, $0 \leq X \leq 5$, where c is a constant. Find $f_Y(y)$. Sketch this density function.

6. Derive the probability density function for Y, given $Y = |X|$, and

$$f_X(x) = \begin{cases} \frac{1}{4} & \text{if } -2 \leq x \leq 0, \\ \frac{1}{2}\exp(-x) & \text{if } x \geq 0. \end{cases}$$

Sketch all density functions.

7. Given the following: $Y = 3X - 4$ and $f_X(x) = 0.5$, where $-1 \leq x \leq 1$, derive $f_Y(y)$ and sketch both density functions.

8. Given the following: $Y = 3X - 4$ and $f_X(x) = N(0, 0.33)$, derive $f_Y(y)$ and plot both density functions.

9. Given the fluid drag equation: $F_D = C_D V^2$, where C_D is a constant and $f_V(v) = 0.1$, for the range $10 \leq v \leq 20$, derive f_{F_D} and sketch both density functions.

10. For the same drag equation, $F_D = C_D V^2$, V is standard normal $N(0,1)$. Note that $F_D \geq 0$. Derive f_{F_D} and sketch both density functions.

11. For the function $Y = a \tan X$, $a > 0$, derive the general relation for $f_Y(y)$. Then, assume X is uniformly distributed over $[-\pi, \pi]$ and derive $f_Y(y)$. Sketch the density functions for X and Y.

Section 4.2: Functions of Two or More RVs

12. For the function $Z = XY$, find $f_Z(z)$ for the cases:
    (i) $f_{XY}(x,y) = [(b-a)(d-c)]^{-1}$. Sketch.
    (ii) $f_{XY}(x,y) = \exp[-(x+y)]$. Sketch.

13. For the function $Z = X + Y$, find $f_Z(z)$ for the cases:
    (i) $f_{XY}(x,y) = c\exp[-(x+y)]$ over the unit square. Plot.
    (ii) $f_{XY}(x,y)$ is jointly uniform over the unit square. Plot.

14. List the sequence of transformations needed to derive the general probability density function f_{Y_5} for

$$Y_5 = \left(X_1^2 + X_2\right)^2 + X_3 X_4,$$

assuming that all f_{X_i} are known.

15. Derive the general probability density function f_{Y_i}, $i = 1, \ldots, 5$, leading to
$$Y_5 = \left(X_1^2 + X_2\right)^2 + X_3 X_4,$$
assuming that all
$$f_{X_i} = 1, \quad 0 \leq x_i \leq 1.$$

Section 4.3: Approximate Analysis

16. Suppose $Y = a_1 X_1 + a_2 X_2$ with X_1 and X_2 uniform with mean values $\mu_{X_1} = 2$ and $\mu_{X_2} = 3$. Both probability density functions have magnitude $1/2$.

    (i) Calculate μ_Y from the given information directly.

    (ii) Calculate μ_Y using only the respective probability density functions.

    (iii) Use the method of transformation of RVs of Section 4.2 to derive f_Y assuming the given probability density functions. Then calculate μ_Y.

    (iv) Compare all results and discuss.

17. Reconsider Example 4.8, supposing that the angles are random variables. The 30° angle now is a random variable with mean of 30° and variance of 3° and the 45° angle now is a random variable with mean of 45° and variance of 4°. The applied force has the same statistics as in the example. Calculate the mean and variance of the tensions.

18. Derive Equation 4.26.

19. Find the general expression for the first-order approximation for μ and σ for the following:

    (i)
    $$\Delta = \frac{PL}{AE},$$
    with all variables uncorrelated and
    $$\begin{array}{llll}
    \mu_P = 1000 & \sigma_P = 10 & \mu_L = 35 & \sigma_L = 15 \\
    \mu_A = 0.1 & \sigma_A = 0.01 & \mu_E = 10 \times 10^6 & \sigma_E = 0.01 \times 10^6
    \end{array}$$

    (ii) solve *(i)* except that $\rho_{LA} = 0.5$ and all other correlations zero

    (iii)
    $$V = \pi L^3 + ar^3,$$

4.6. PROBLEMS

where π and a are constants and

$$\mu_L = 100 \quad \sigma_L = 5 \quad \mu_r = 40 \quad \sigma_r = 2,$$

(iv) solve (iii) for the cases: $\rho_{Lr} = 1, -1, 0$. Discuss.

20. Reconsider the projectile Example 4.14, and now assume that the speed and angle are correlated with $\rho_{V_o a} = 0.5$. Compare the results with those of the uncorrelated case. Draw any conclusions you can.

21. Obtain the mean and variance of Y as expanded in the Taylor series Equation 4.27.

22. Recalculate the numerical results of Example 4.11 for the following values of $\rho_{bh} = 0, 0.2, 0.7, 1.0$. Compare and discuss the results.

Section 4.4: Monte Carlo Method

23. Using a calculator or chart, generate 10 random numbers between 0 and 1.

24. Suppose $f = 1/(b-a)$ where $a = 2$ and $b = 4$. Use the results of the previous problem to specify 5 random numbers.

25. For the cantilever beam of Figure 4.18, with $R = 30$ lb and W a uniform random variable with $\mu_W = 100$ lb/ft and coefficient of variation $\delta_W = 0.10$, estimate the mean and standard deviation of the shear V and bending moment M.

26. For the cantilever beam of Figure 4.18, both parameters R and L are random variables. L is Gaussian with $\mu_L = 25$ ft and $\sigma_L = 0.1$ ft; R is a uniform random variable between the values $9800 - 10,200$ lb. Estimate the mean and standard deviation of the shear V and bending moment M.

27. Draw the density function, Equation 4.38. Generate an additional three values of x_i.

4.7 The Standard Normal Table

s	0.00	0.01	0.02	0.03	0.04	0.05	0.06	0.07	0.08	0.09
-3.4	0.0003	0.0003	0.0003	0.0003	0.0003	0.0003	0.0003	0.0003	0.0003	0.0002
-3.3	0.0005	0.0005	0.0005	0.0004	0.0004	0.0004	0.0004	0.0004	0.0004	0.0003
-3.2	0.0007	0.0007	0.0006	0.0006	0.0006	0.0006	0.0006	0.0005	0.0005	0.0005
-3.1	0.0010	0.0009	0.0009	0.0009	0.0008	0.0008	0.0008	0.0008	0.0007	0.0007
-3.0	0.0013	0.0013	0.0013	0.0012	0.0012	0.0011	0.0011	0.0011	0.0010	0.0010
-2.9	0.0019	0.0018	0.0017	0.0017	0.0016	0.0016	0.0015	0.0015	0.0014	0.0014
-2.8	0.0026	0.0025	0.0024	0.0023	0.0023	0.0022	0.0021	0.0021	0.0020	0.0019
-2.7	0.0035	0.0034	0.0033	0.0032	0.0031	0.0030	0.0029	0.0028	0.0027	0.0026
-2.6	0.0047	0.0045	0.0044	0.0043	0.0041	0.0040	0.0039	0.0038	0.0037	0.0036
-2.5	0.0062	0.0060	0.0059	0.0057	0.0055	0.0054	0.0052	0.0051	0.0049	0.0048
-2.4	0.0082	0.0080	0.0078	0.0075	0.0073	0.0071	0.0069	0.0068	0.0066	0.0064
-2.3	0.0107	0.0104	0.0102	0.0099	0.0096	0.0094	0.0091	0.0089	0.0087	0.0084
-2.2	0.0139	0.0136	0.0132	0.0129	0.0125	0.0122	0.0119	0.0116	0.0113	0.0110
-2.1	0.0179	0.0174	0.0170	0.0166	0.0162	0.0158	0.0154	0.0150	0.0146	0.0143
-2.0	0.0228	0.0222	0.0217	0.0212	0.0207	0.0202	0.0197	0.0192	0.0188	0.0183
-1.9	0.0287	0.0281	0.0274	0.0263	0.0262	0.0256	0.0250	0.0244	0.0283	0.0233
-1.8	0.0359	0.0352	0.0344	0.0336	0.0329	0.0322	0.0314	0.0307	0.0300	0.0294
-1.7	0.0446	0.0436	0.0427	0.0418	0.0406	0.0401	0.0392	0.0384	0.0375	0.0367
-1.6	0.0548	0.0537	0.0526	0.0516	0.0505	0.0495	0.0485	0.0475	0.0465	0.0455
-1.5	0.0668	0.0655	0.0643	0.0630	0.0618	0.0606	0.0594	0.0582	0.0570	0.0559
-1.4	0.0808	0.0793	0.0778	0.0764	0.0749	0.0735	0.0722	0.0708	0.0694	0.0681
-1.3	0.0968	0.0951	0.0934	0.0918	0.0901	0.0885	0.0869	0.0853	0.0838	0.0823
-1.2	0.1151	0.1131	0.1112	0.1093	0.1075	0.1056	0.1038	0.1020	0.1003	0.0985
-1.8	0.0359	0.0352	0.0344	0.0336	0.0329	0.0322	0.0314	0.0307	0.0300	0.0294
-1.7	0.0446	0.0436	0.0427	0.0418	0.0406	0.0401	0.0392	0.0384	0.0375	0.0367
-1.6	0.0548	0.0537	0.0526	0.0516	0.0505	0.0495	0.0485	0.0475	0.0465	0.0455
-1.5	0.0668	0.0655	0.0643	0.0630	0.0618	0.0606	0.0594	0.0582	0.0570	0.0559
-1.4	0.0808	0.0793	0.0778	0.0764	0.0749	0.0735	0.0722	0.0708	0.0694	0.0681
-1.3	0.0968	0.0951	0.0934	0.0918	0.0901	0.0885	0.0869	0.0853	0.0838	0.0823
-1.2	0.1151	0.1131	0.1112	0.1093	0.1075	0.1056	0.1038	0.1020	0.1003	0.0985
-1.1	0.1357	0.1335	0.1314	0.1292	0.1271	0.1251	0.1230	0.1210	0.1190	0.1170
-1.0	0.1587	0.1562	0.1539	0.1515	0.1492	0.1469	0.1446	0.1423	0.1401	0.1379
-0.9	0.1841	0.1814	0.1788	0.1762	0.1736	0.1711	0.1685	0.1660	0.1635	0.1611
-0.8	0.2119	0.2090	0.2061	0.2033	0.2005	0.1977	0.1949	0.1922	0.1894	0.1867
-0.7	0.2420	0.2389	0.2358	0.2327	0.2296	0.2266	0.2236	0.2206	0.2177	0.2148
-0.6	0.2743	0.2709	0.2676	0.2643	0.2611	0.2578	0.2546	0.2514	0.2483	0.2451
-0.5	0.3085	0.3050	0.3015	0.2981	0.2946	0.2912	0.2877	0.2843	0.2810	0.2776
-0.4	0.3446	0.3409	0.3372	0.3336	0.3300	0.3264	0.3228	0.3192	0.3156	0.3121
-0.3	0.3821	0.3783	0.3745	0.3707	0.3669	0.3632	0.3594	0.3557	0.3520	0.3483
-0.2	0.4207	0.4168	0.4129	0.4090	0.4052	0.4013	0.3974	0.3936	0.3897	0.3859
-0.1	0.4602	0.4562	0.4522	0.4483	0.4443	0.4404	0.4364	0.4325	0.4286	0.4247
-0.0	0.5000	0.4960	0.4920	0.4880	0.4840	0.4801	0.4761	0.4721	0.4681	0.4641
0.0	0.5000	0.5040	0.5080	0.5120	0.5160	0.5199	0.5239	0.5279	0.5319	0.5359
0.1	0.5398	0.5438	0.5478	0.5517	0.5557	0.5596	0.5636	0.5675	0.5714	0.5753
0.2	0.5793	0.5832	0.5871	0.5910	0.5948	0.5987	0.6026	0.6064	0.6103	0.6141
0.3	0.6179	0.6217	0.6255	0.6293	0.6331	0.6368	0.6406	0.6443	0.6480	0.6517
0.4	0.6554	0.6591	0.6628	0.6664	0.6700	0.6736	0.6772	0.6808	0.6844	0.6879
0.5	0.6915	0.6950	0.6985	0.7019	0.7054	0.7088	0.7123	0.7157	0.7190	0.7224

4.7. THE STANDARD NORMAL TABLE

s	00.0	0.01	0.02	0.03	0.04	0.05	0.06	0.07	0.08	0.09
0.6	0.7257	0.7291	0.7324	0.7357	0.7389	0.7422	0.7454	0.7486	0.7517	0.7549
0.7	0.7580	0.7611	0.7642	0.7673	0.7704	0.7734	0.7764	0.7794	0.7823	0.7852
0.8	0.7881	0.7910	0.7939	0.7967	0.7995	0.8023	0.8051	0.8078	0.8106	0.8133
0.9	0.8159	0.8186	0.8212	0.8238	0.8264	0.8289	0.8315	0.8340	0.8365	0.8389
1.0	0.8413	0.8438	0.8461	0.8485	0.8508	0.8531	0.8554	0.8577	0.8599	0.8621
1.1	0.8643	0.8665	0.8686	0.8708	0.8729	0.8749	0.8770	0.8790	0.8810	0.8830
1.2	0.8849	0.8869	0.8888	0.8907	0.8925	0.8944	0.8962	0.8980	0.8997	0.9015
1.3	0.9032	0.9049	0.9066	0.9082	0.9099	0.9115	0.9131	0.9147	0.9162	0.9177
1.4	0.9192	0.9207	0.9222	0.9236	0.9251	0.9265	0.9278	0.9292	0.9306	0.9319
1.5	0.9332	0.9345	0.9357	0.9370	0.9382	0.9394	0.9406	0.9418	0.9429	0.9441
1.6	0.9452	0.9463	0.9474	0.9484	0.9495	0.9505	0.9515	0.9525	0.9535	0.9545
1.7	0.9554	0.9564	0.9573	0.9582	0.9591	0.9599	0.9608	0.9616	0.9625	0.9633
1.8	0.9641	0.9649	0.9656	0.9664	0.9671	0.9678	0.9686	0.9693	0.9699	0.9706
1.9	0.9713	0.9719	0.9726	0.9732	0.9738	0.9744	0.9750	0.9756	0.9761	0.9767
2.0	0.9772	0.9778	0.9783	0.9788	0.9793	0.9798	0.9803	0.9808	0.9812	0.9817
2.1	0.9821	0.9826	0.9830	0.9834	0.9838	0.9842	0.9846	0.9850	0.9854	0.9857
2.2	0.9861	0.9864	0.9868	0.9871	0.9875	0.9878	0.9881	0.9884	0.9887	0.9890
2.3	0.9893	0.9896	0.9898	0.9901	0.9904	0.9906	0.9909	0.9911	0.9913	0.9916
2.4	0.9918	0.9920	0.9922	0.9925	0.9927	0.9929	0.9931	0.9932	0.9934	0.9936
2.5	0.9938	0.9940	0.9941	0.9943	0.9945	0.9946	0.9948	0.9949	0.9951	0.9952
2.6	0.9953	0.9955	0.9956	0.9957	0.9959	0.9960	0.9961	0.9962	0.9963	0.9964
2.7	0.9965	0.9966	0.9967	0.9968	0.9969	0.9970	0.9971	0.9972	0.9973	0.9974
2.8	0.9974	0.9975	0.9976	0.9977	0.9977	0.9978	0.9979	0.9979	0.9980	0.9981
2.9	0.9981	0.9982	0.9982	0.9983	0.9984	0.9984	0.9985	0.9985	0.9986	0.9986
3.0	0.9987	0.9987	0.9987	0.9988	0.9988	0.9989	0.9989	0.9989	0.9990	0.9990
3.1	0.9990	0.9991	0.9991	0.9991	0.9992	0.9992	0.9992	0.9992	0.9993	0.9993
3.2	0.9993	0.9993	0.9994	0.9994	0.9994	0.9994	0.9994	0.9995	0.9995	0.9995

Chapter 5

Random Processes

A random process can informally be thought of as a random variable that varies with time in a probabilistic way. With this in mind, it will be straightforward to extend the earlier definitions for random variables to those for random processes.

The discussion in the following sections is necessarily brief and is meant to be introductory and of an applied nature. This basic introduction provides us with the basic concepts needed to start understanding the study of random vibration. Examples where random processes are applied include earthquake engineering, offshore structural dynamics, wind engineering, turbulent forces on ocean and aerospace structures, machine dynamics, and crack propagation in materials.

5.1 Basic Random Process Descriptors

When the behavior of a dynamic process becomes so complicated that it cannot be described deterministically, probabilistic concepts are useful to quantify the behavior. In particular, by complicated behavior or motion we mean that, given the initial conditions of a function, $x(t_0)$ and $\dot{x}(t_0)$, the value $x(t_0 + \varepsilon)$ cannot be exactly predicted regardless of how small is ε. We will discover that complicated motion is comprised of energy at many frequencies, as compared to the simple harmonic motion that operates at one frequency. Such complicated dynamic behavior is interpreted as being random in nature since it is too complicated to model deterministically, and probabilistic concepts are useful to help organize the complexity.

A *random process* $X(t)$ may be understood to be a time-dependent random variable. For a specific time t, $X(t)$ is a random variable with distri-

bution function,
$$F_{X(t)}(x;t) = \Pr\left(X\left(t\right) \leq x\right).$$

This is the *first-order distribution* of the process $X(t)$. It describes how the probability characteristics of the random process change with time. That is, how do the mean value and variance change with time? It is customary to separate the time variable from the other variables by a semicolon. The corresponding *first-order density* is given by

$$f_{X(t)}(x;t) = \frac{\partial F_{X(t)}(x;t)}{\partial x}.$$

The random process at any specified time is a random variable. For example, $X(t_1)$, $X(t_2)$, and $X(t_3)$ are random variables. $X(t_1)$ and $X(t_2)$ may be independent of each other or correlated just like static random variables X and Y in Chapter 3. Correlation between two random variables is based on their joint probability density. The *second-order distribution* for $X(t_1)$ and $X(t_2)$ is the joint probability distribution function,

$$F_{X(t_1)X(t_2)}(x_1, x_2; t_1, t_2) = \Pr\left(X(t_1) \leq x_1, X(t_2) \leq x_2\right), \quad (5.1)$$

with corresponding *joint density*,

$$f_{X_1 X_2}(x_1, x_2; t_1, t_2) = \frac{\partial^2 F_{X_1 X_2}}{\partial x_1 \partial x_2}. \quad (5.2)$$

A simplified the notation is useful: $X_1 \equiv X(t_1)$ and $X_2 \equiv X(t_2)$.

Equation 5.1 is the probability that at time t_1, random process $X(t)$ will be less than or equal to the value x_1, *and* that at time t_2, it will be less than or equal to the value x_2. The density Equation 5.2 contains the same information but in a form that may be easier for calculation: the volume under the second-order density is a measure of the respective probability. Equations 5.1 and 5.2 are the starting points for the mathematical modeling of the probabilistic evolution of $X(t)$ in time; for random oscillations it is the input force that is modeled as a random process. The question considered next is: *If a random function is represented by many possible time histories, how can averages be determined?*

5.2 Ensemble Averaging

The random function of time $X(t)$ with density function $f(x;t)$ is representative of many possible time histories known as a *sample population*. Theoretically, there are an infinite number of elements in the sample $X_i(t)$ with statistical properties governed by the density function $f(x;t)$. Subscript i denotes the ith element in the sample population.

5.2. ENSEMBLE AVERAGING

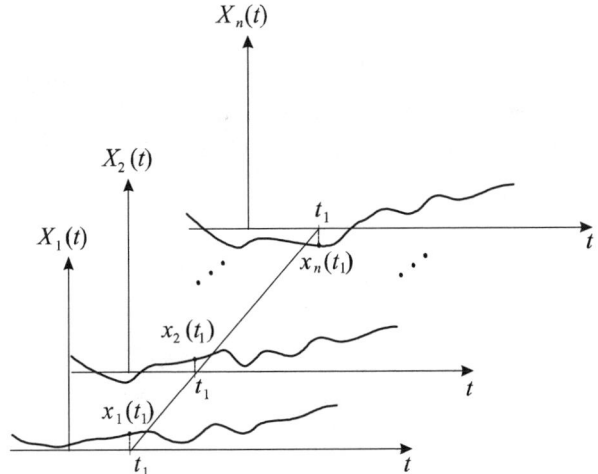

Figure 5.1: Ensemble averaging at $t = t_1$.

Figure 5.1 depicts that for any time t_1 there exist n possible values: $x_1(t_1)$, $x_2(t_1)$, $x_3(t_1)$, $\cdots x_n(t_1)$, where n can be a very large number. To utilize all these values, they are averaged to obtain the most likely value for the function at time t_1. This process is known as *ensemble averaging* since the group of samples is known as an *ensemble*. The averaging procedure uses the mathematical expectation introduced by Equation 3.7,

$$\mu_X(t) = E\{X(t)\} = \int_{-\infty}^{\infty} x(t) f(x;t) dx, \tag{5.3}$$

except that the statistical parameters $\mu_X(t)$ and $f(x;t)$ are now functions of time. Note that x is the dummy variable in the integration, which is performed over the ensemble, resulting in a mean value function of time. If $X(t)$ takes on discrete values, the mean value is given by

$$\mu_X(t) = \lim_{n \to \infty} \frac{1}{n} \sum_{k=1}^{n} x_k(t).$$

In general, the random process is governed by a time-dependent density function, and has a time-dependent mean value. An example of a density function is $f(x;t) = ce^{-xt}$, where c is a normalization constant. Similarly, it is necessary to evaluate the second-order averages that are very important in any random process model. A motivation for such averages is the question: *How does the value of the process $X(t)$ at $t = t_1$ affect its value at later time*

$t = t_2$? Knowing this helps us understand how rapidly a process varies. For a slowly varying function, we expect that if t_1 and t_2 are not *too far* apart, then $X(t_2)$ can be estimated given $X(t_1)$, as shown in Figure 5.2*(a)*. Thus, the values are correlated. On the other hand, if $X(t)$ varies rapidly, as in Figure 5.2*(b)*, any estimate of future values is not accurate and the values are much less correlated.

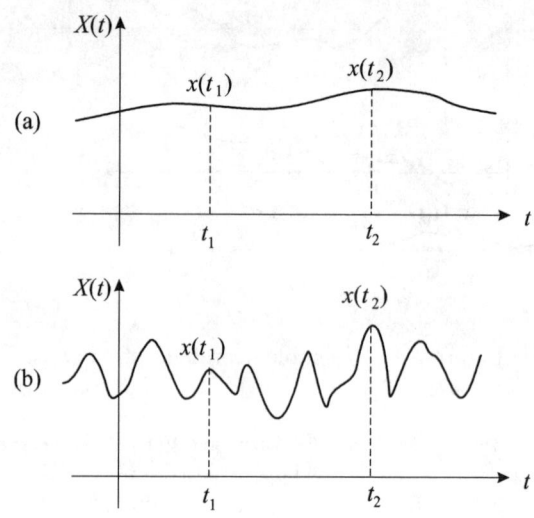

Figure 5.2: Second-order averages for *(a)* slowly varying and *(b)* rapidly varying random processes.

To address this question quantitatively, consider the random process $X(t)$ at any two time instances, $X(t_1)$ and $X(t_2)$. The second-order average is given by

$$E\{X(t_1)X(t_2)\} = \int_{-\infty}^{\infty} \int_{-\infty}^{\infty} x_1 x_2 f(x_1, x_2; t_1, t_2) dx_1 dx_2, \quad (5.4)$$

where the joint density function $f(x_1, x_2; t_1, t_2)$ is required. If $X(t)$ is a discrete random process, $E\{X(t_1)X(t_2)\}$ is given by

$$E\{X(t_1)X(t_2)\} = \lim_{n \to \infty} \frac{1}{n}\frac{1}{n} \sum_{k=1}^{n} \sum_{l=1}^{n} x_k(t_1) x_l(t_2). \quad (5.5)$$

The second-order average defined by Equation 5.4, and shown in Figure 5.3, is called the *autocorrelation function* and is given the notation $R_{XX}(t_1, t_2)$,

$$R_{XX}(t_1, t_2) = E\{X(t_1)X(t_2)\}, \quad (5.6)$$

5.2. ENSEMBLE AVERAGING

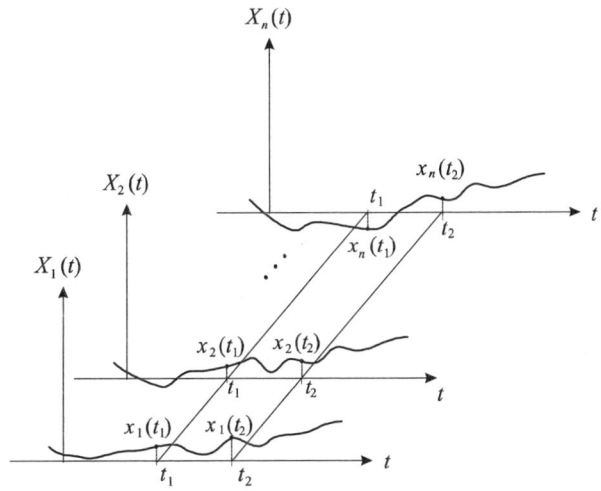

Figure 5.3: Second-order ensemble averaging.

where it is clear that the end result of the double integral is a function of t_1 and t_2.

In Equation 5.4, it is necessary to emphasize that at a specific time, a random process is nothing more than a random variable. So at each instant of time, the values a random process can take are governed by a probability density. This is exactly what the density $f(x;t)$ implies; the density changes with time, but once a specific time is chosen, the density function is only a function of the variable x, that is, $f(x;t_1) = f(x)$. This is shown schematically in Figure 5.4.

This is why in Equation 5.4 the double integral is over x_1 and x_2. The value of the random process $X(t)$ at $t = t_1$ becomes random variable X_1, represented by the dummy variable x_1 inside the integral. Similarly, the value of the random process $X(t)$ at $t = t_2$ becomes random variable X_2 represented by the dummy variable x_2. For the second-order average, then, the joint density function $f(x_1, x_2; t_1, t_2)$ is needed.

The notation for random process $X(t)$ can be interpreted as follows. The capital $X(t)$ represents the ensemble. The random process becomes a random variable at a particular time, say t_1, and is denoted as $X(t_1)$. Each realization of random variable $X(t_1)$ is a deterministic quantity and is denoted by the lower case $x_i(t_1)$.

The correlation function is a measure of the *similarity between the values of one stochastic process at two instants of time*, or *between different stochastic processes at two instances of time*. For different stochastic processes a

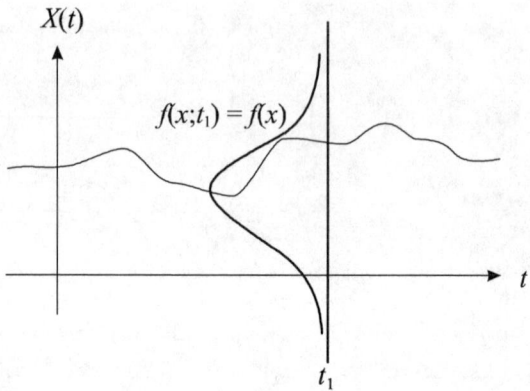

Figure 5.4: The density of a random process at specific time t_1.

cross-correlation function, $R_{XY}(t_1, t_2) = E\{X(t_1)Y(t_2)\}$, is defined.

Equations 5.4 and 5.6 show that the correlation is calculated by multiplying the corresponding values of the functions and then averaging these products using the expectation. If the two functions have similar shapes, then it is expected that a larger correlation will be found, since otherwise some of the products will be smaller or negative, leading to a smaller average. The largest correlation is found for $t_1 = t_2$, $R_{XX}(t_1, t_1) = E\{X(t_1)X(t_1)\}$. Possible correlation functions are subsequently discussed.

The *autocovariance* of $X(t)$ is defined as

$$C_{XX}(t_1, t_2) = E\{[X(t_1) - \mu_X(t_1)][X(t_2) - \mu_X(t_2)]\},$$

which can be expanded to yield

$$C_{XX}(t_1, t_2) = R_{XX}(t_1, t_2) - \mu_X(t_1)\mu_X(t_2).$$

Generally, functions such as $R_{XX}(t_1, t_2)$ are derived experimentally. But in many preliminary analyses one assumes a reasonable function for $R_{XX}(t_1, t_2)$ that is representative of the physical process under study. Predictions using this model are then verified using experimental data.

The *variance* is defined for $t_1 = t_2 = t$,

$$\sigma_X^2(t) = C_{XX}(t, t) = R_{XX}(t, t) - \mu_X^2(t).$$

For a normal random process, for example, the time-dependent probability density function is

$$f_X(x; t) = \frac{1}{\sqrt{2\pi C_{XX}(t, t)}} \exp\left[-\frac{(x - \mu_X(t))^2}{2C_{XX}(t, t)}\right].$$

5.3. STATIONARITY

Table 5.1: Sample Values of Discrete Random Process $X(t)$

$X(t_1)$	1	3	2	5	6	7	3	2	9
$X(t_2)$	4	1	2	5	3	6	7	10	2

We can imagine the Gaussian density function evolving in time due to the evolution of $\mu_X(t)$ and $C_{XX}(t,t)$ in time.

Example 5.1 Discrete Random Process

Consider a discrete random process $X(t)$. At two time instances t_1 and t_2, $X(t)$ can take on the respective values given in Table 5.1. Calculate the mean value and the autocorrelation function.

Solution The mean value is obtained by summing the discrete values and dividing the sum by the number of samples,

$$\mu_X(t_1) = \frac{1}{9}\sum_{i=1}^{9} x_i(t_1) = \frac{38}{9} = 4.22$$

$$\mu_X(t_2) = \frac{1}{9}\sum_{i=1}^{9} x_i(t_2) = \frac{40}{9} = 4.44.$$

Clearly, the mean value varies with time. The autocorrelation of $X(t)$ between the two time instances is obtained using the discrete version of Equation 5.4,

$$\begin{aligned} R_{XX}(t_1, t_2) &= \frac{1}{n}\frac{1}{n}\sum_{k=1}^{n}\sum_{l=1}^{n} x_k(t_1) x_l(t_2) \\ &= \frac{1}{9}\frac{1}{9}[1 \times (4 + 1 + ... + 10 + 2) + \ ... \\ &\quad + 9 \times (4 + 1 + ... + 10 + 2)] \\ &= 18.77. \end{aligned}$$

⊛

5.3 Stationarity

It can be observed that Equations 5.3 and 5.4 are difficult to evaluate, not only mathematically, but also due to the difficulty of obtaining the necessary data to define the joint density function. Sometimes it is appropriate to make the assumption of *stationarity*. The process is *strictly stationary* if the joint probability distribution of $X(t_1), X(t_2), \cdots, X(t_n)$ is identical to the joint distribution of $X(t_1 + \Delta t), X(t_2 + \Delta t), \cdots, X(t_n + \Delta t)$. That is, for any Δt and n,

$$f_{X_1 X_2 \cdots X_n}(x_1, x_2, \cdots, x_n; t_1, t_2, \cdots, t_n)$$
$$= f_{X_1 X_2 \cdots X_n}(x_1, x_2, \cdots, x_n; t_1 + \Delta t, t_2 + \Delta t, \cdots, t_n + \Delta t). \quad (5.7)$$

If the random process $X(t)$ is *weakly stationary*,[1] Equation 5.7 is only valid for $n = 1$ and 2. Therefore, a process that is strictly stationary is also weakly stationary whereas the reverse is not true. We often refer to the weakly stationary process as just a stationary process.

While the assumption of stationarity may appear to limit the applicability of the following models, in fact, with proper care and understanding, stationarity is a viable assumption for numerous practical applications. To use vibration terminology, *when we have assumed stationarity, we have assumed steady-state behavior in a statistical sense.*

Now let us discuss what the implications of a stationary process are. For $n = 1$, Equation 5.7 becomes

$$f_X(x; t) = f_X(x; t + \Delta t),$$

valid for any Δt. Therefore, *the first-order density function is independent of time*, or

$$f_X(x; t) = f_X(x),$$

and the first-order statistics such as mean value, variance or higher-order moments are also independent of time,

$$E\{X^m(t)\} = E\{X^m\} = \int_{-\infty}^{\infty} x^m f_X(x)\, dx.$$

For $n = 2$, Equation 5.7 can be written as

$$f_{X_1 X_2}(x_1, x_2; t_1, t_2) = f_{X_1 X_2}(x_1, x_2; t_1 + \Delta t, t_2 + \Delta t),$$

for any Δt. Thus, *the second-order density function only depends on the difference between t_1 and t_2*, or $t_2 - t_1$ or $t_1 - t_2$. If $\tau = t_2 - t_1$, then

$$f_{X_1 X_2}(x_1, x_2; t_1, t_2) = f_{X_1 X_2}(x_1, x_2; \tau). \quad (5.8)$$

Using Equation 5.8, it is possible to show that the second-order statistics

$$E\{X^m(t_1) X^p(t_2)\} \quad (5.9)$$

---

[1] It is also called *covariance stationary* or *stationary in the wide sense*.

5.3. STATIONARITY

are only functions of τ. For example, when $m = p = 1$, the expected value can be written as

$$\begin{aligned} E\left\{X\left(t_{1}\right) X\left(t_{2}\right)\right\} &= R_{XX}\left(t_{1}, t_{2}\right) \\ &= \int_{-\infty}^{\infty} \int_{-\infty}^{\infty} x_{1} x_{2} f_{X_{1} X_{2}}\left(x_{1}, x_{2} ; \tau\right) d x_{1} d x_{2} \\ &= R_{XX}\left(\tau\right), \end{aligned}$$

where $R_{XX}(\tau)$ is the autocorrelation function. Note that the second-order probability density and the second-order statistics are even functions of τ,

$$\begin{aligned} f_{X_1 X_2}\left(x_1, x_2; \tau\right) &= f_{X_1 X_2}\left(x_1, x_2; -\tau\right) \\ R_{XX}\left(\tau\right) &= R_{XX}\left(-\tau\right). \end{aligned} \quad (5.10)$$

Similarly, for two stationary processes $X(t)$ and $Y(t)$, the *cross-correlation* function is $R_{XY}(\tau)$,

$$R_{XY}(\tau) = E\left\{X(t) Y(t+\tau)\right\}.$$

Unlike the autocorrelation function, the cross correlation function is an odd function, $R_{XY}(\tau) \neq R_{YX}(\tau)$, but rather $R_{XY}(\tau) = R_{YX}(-\tau)$.

It is interesting to note that for $\tau = 0$ the *mean-square* value of $X(t)$ is obtained,

$$R_{XX}(\tau = 0) = E\{X^2(t)\} = \sigma_X^2 + \mu_X^2,$$

where the last equality is due to Equation 3.9. The mean square value is an important quantity in the following physical way. If $X(t)$ is a displacement of an elastic body, then $E\{X^2(t)\}$ is proportional to the average potential or strain energy. For instance, if the spring with constant k is stretched by $X(t)$, the average potential energy stored V is

$$E\{V_{potential}\} = \frac{1}{2} k E\{X^2(t)\}.$$

If $X(t)$ is the velocity of a mass, then $E\{X^2(t)\}$ is proportional to the average kinetic energy T of the mass,

$$E\{T_{kinetic}\} = \frac{1}{2} m E\{X^2(t)\}.$$

The displacement and the velocity can be normalized such that $X^2(t)$ can represent the energy itself. Therefore, $E\{X^2(t)\}$ can be viewed as the *average* energy.[2]

---

[2] The average energy of a stationary process is independent of time and equals the autocorrelation at $\tau = 0$. This knowledge will help us interpret the meaning of the *spectral density* in a subsequent section.

Stationarity also implies that the autocovariance is then only a function of τ and is given by

$$C_{XX}(\tau) = R_{XX}(\tau) - \mu_X^2,$$

where, for $\tau = 0$, we have $C_{XX}(0) = R_{XX}(0) - \mu_X^2 = \sigma_X^2$.
The *correlation coefficient* is defined as

$$\rho_{XX}(\tau) = C_{XX}(\tau)/C_{XX}(0).$$

The *correlation time* of a stochastic process can be defined as

$$\tau_c = \frac{1}{C_{XX}(0)} \int_0^\infty C_{XX}(\tau)\,d\tau.$$

The concept of correlation time becomes important in problems that must be solved approximately and simplification must be made on a physical basis. It is a measure of the rapidity of the oscillations of the dynamic process, and how far in time present values affect later values of the process.

The autocorrelation function can be written in terms of the correlation coefficient as follows,

$$\begin{aligned} R_{XX}(\tau) &= C_{XX}(\tau) + \mu_X^2 \\ &= \rho_{XX}(\tau)\sigma_X^2 + \mu_X^2. \end{aligned}$$

Since the correlation coefficient is between -1 and 1, the autocorrelation function is also bounded,

$$-\sigma_X^2 + \mu_X^2 \le R_{XX}(\tau) \le \sigma_X^2 + \mu_X^2.$$

$R_{XX}(\tau)$ reaches its maximum of $\sigma_X^2 + \mu_X^2$ when $\tau = 0$. For physical processes, as $\tau \to \infty$, the values $X(t)$ and $X(t+\tau)$ become uncorrelated. The correlation function is then

$$\begin{aligned} \lim_{\tau \to \infty} R_{XX}(\tau) &= \lim_{\tau \to \infty} E\{X(t_1)X(t_1+\tau)\} \\ &= E\{X(t_1)\}E\{X(t_1+\tau)\} \\ &= \mu_X^2. \end{aligned}$$

Figure 5.5 shows a sketch of an autocorrelation function for a stationary random process. Note the symmetry about the vertical axis: $R_{XX}(\tau) = R_{XX}(-\tau)$.

In most cases, the random processes are adjusted so that the mean value is zero, simplifying messy algebra so that $R_{XX}(\tau)$ approaches zero.

5.3. STATIONARITY

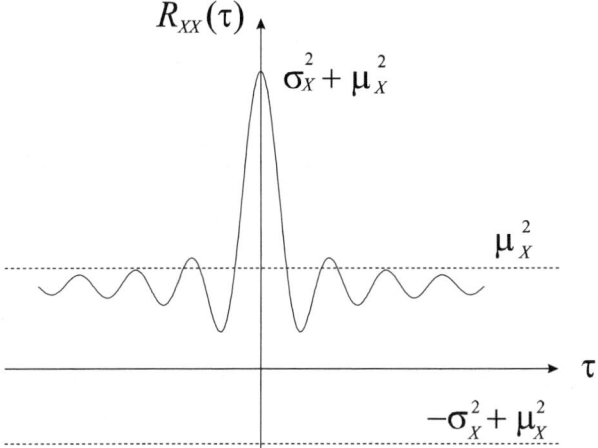

Figure 5.5: Sketch of autocorrelation function for a stationary random process.

Ergodicity

As a practical matter, one rarely has the benefit of numerous experiments, but usually must make the best use of *one* trial. This is especially true for expensive testing environments such as space or undersea. Utilizing one trial requires the introduction of the concept of *ergodicity*. A stationary random process is said to be an *ergodic* process if the *time average* of a *single* record is approximately equal to the ensemble average. This allows us to average a long *single* time history in time, as in Figure 5.6, rather than trying to obtain numerous records over which to perform an ensemble average. The mean value is then given by

$$\mu_X \simeq \overline{X} = \lim_{T\to\infty} \frac{1}{T} \int_{-T/2}^{T/2} X(t)dt,$$

where it is assumed that $X(t)$ is one particular realization of the random process. The overbar on X denotes time average.

Such an average makes sense only if μ_X is a constant, otherwise μ_X will be a function of T and the initial assumption is no longer valid. *An ergodic process is always stationary, but the opposite is not always true.*

The corresponding ergodic definition for the autocorrelation function is

$$R_{XX}(\tau) \simeq \lim_{T\to\infty} \frac{1}{T} \int_{-T/2}^{T/2} X(t)X(t+\tau)dt.$$

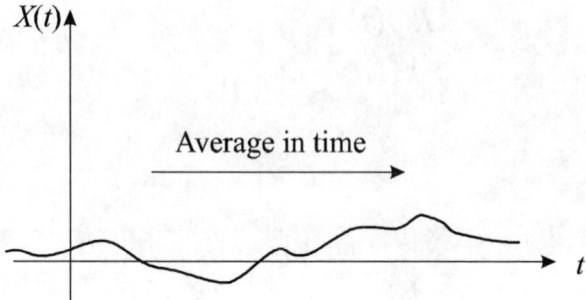

Figure 5.6: Ergodicity implies that averages of a stochastic process can be performed by averaging a single realization in time rather than averaging over an ensemble of realizations.

Example 5.2 Sample Correlation Functions

(i) Suppose the random process $X(t) = At$, where A is a random variable with μ_A and σ_A, and t is time. Then

$$\begin{aligned} E\{X(t)\} &= E\{At\} = \mu_A t \\ R_{XX}(t_1, t_2) &= E\{X(t_1) X(t_2)\} \\ &= E\{At_1 \cdot At_2\} \\ &= t_1 t_2 E\{A^2\} \\ &= t_1 t_2 \left(\sigma_A^2 + \mu_A^2\right). \end{aligned}$$

Both first- and second-order moments are functions of time and the process $X(t)$ is not stationary. It is sufficient that either moment be a function of time for the process to be nonstationarity.

(ii) Consider the random process $X(t) = A \sin \omega t$ where A is a random variable with density $f_A(a)$. Then the expected value and the correlation function are given by

$$\begin{aligned} E\{X(t)\} &= E\{A \sin \omega t\} = \mu_A \sin \omega t \\ R_{XX}(t_1, t_2) &= E\{A \sin \omega t_1 \cdot A \sin \omega t_2\} \\ &= \sin \omega t_1 \sin \omega t_2 E\{A^2\} \\ &= \sin \omega t_1 \sin \omega t_2 \left(\sigma_A^2 + \mu_A^2\right). \end{aligned}$$

The process is not stationary.

(iii) Consider the random process $X(t) = A \cos(\omega t - \Phi)$ where A is a constant and Φ is a random phase uniformly distributed between 0 and 2π.

5.3. STATIONARITY

The probability density function $f_\Phi(\phi)$ is given by

$$f_\Phi(\phi) = \frac{1}{2\pi}, \quad 0 < \phi \leq 2\pi.$$

The expected value and the correlation function are given by

$$\begin{aligned}
E\{X(t)\} &= 0 \\
R_{XX}(t_1, t_2) &= E\{A\cos(\omega t_1 - \Phi) A\cos(\omega t_2 - \Phi)\} \\
&= A^2 E\{\cos(\omega t_1 - \Phi)\cos(\omega t_2 - \Phi)\} \\
&= A^2 E\{(\cos\omega t_1 \cos\Phi + \sin\omega t_1 \sin\Phi) \\
&\quad \cdot (\cos\omega t_2 \cos\Phi + \sin\omega t_2 \sin\Phi)\} \\
&= A^2 (\cos\omega t_1 \cos\omega t_2 + \sin\omega t_1 \sin\omega t_2) \\
&= A^2 \cos\omega\tau,
\end{aligned}$$

where $\tau = t_2 - t_1$, and we utilized

$$\begin{aligned}
E\{\cos\Phi\} &= E\{\sin\Phi\} = E\{\sin\Phi\cos\Phi\} = 0 \\
E\{\cos^2\Phi\} &= E\{\sin^2\Phi\} = \frac{1}{2}.
\end{aligned}$$

The process is stationary.
⊛

Example 5.3 Ergodic Process

Consider a periodic step function $x(t)$ given by

$$x(t) = \begin{cases} 0 & \text{for } -\frac{T}{2} < t < 0 \\ 1 & \text{for } 0 < t < \frac{T}{2}, \end{cases}$$

where T is the period. Calculate the temporal mean value and autocorrelation function.

Solution Since the function is periodic, the averages can be calculated over one period instead of over a long time interval. The mean is given by

$$\mu_X = \frac{1}{T}\int_{-T/2}^{T/2} x(t)\,dt = \frac{1}{T}\int_0^{T/2} 1\,dt = \frac{1}{2}.$$

The autocorrelation function is given by

$$\begin{aligned}
R_{XX}(\tau) &= \frac{1}{T}\int_{-T/2}^{T/2} x(t)x(t+\tau)\,dt \\
&= \frac{1}{T}\int_0^{T/2} x(t)x(t+\tau)\,dt
\end{aligned}$$

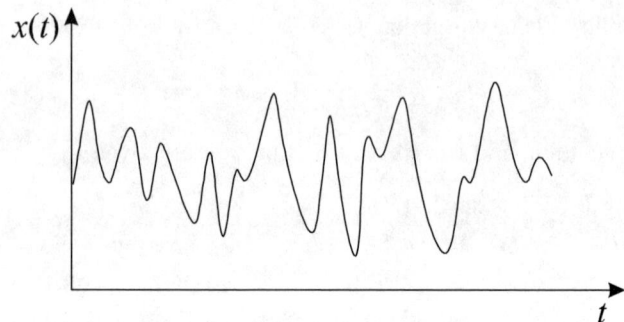

Figure 5.7: Variety of peaks in a sample realization $x(t)$ of a random process $X(t)$.

since $x(t) = 0$ for negative t. Then,

$$R_{XX}(\tau) = \frac{1}{T} \int_0^{T/2} x(t+\tau)dt,$$

since $x(t) = 1$ for positive t. The autocorrelation function is also periodic with a period of T. If τ is between 0 and $T/2$, $R_{XX}(\tau) = 1/2 - \tau/T$. If τ is between $-T/2$ and 0, $R_{XX}(\tau) = 1/2 + \tau/T$.

⊛

5.4 Derivatives of Stationary Processes

For certain important applications it is necessary to be able to count cycles. In fatigue life analysis, the number of cycles of a certain threshold magnitude is directly related to the amount of damage the structure has absorbed in that time. Certainly one way to count cycles is to count how many times the slope of a curve changes signs. However, counting cycles for a randomly varying dynamic process is not a simple proposition.

Consider the sample realization of Figure 5.7. Do we count each and every oscillation, regardless of how small the difference between peaks? These questions will be addressed in more detail in Chapter 9. Here, we provide the simple relations for calculating the correlation functions of the derivatives of random processes.

5.4. DERIVATIVES OF STATIONARY PROCESSES

Assume that random process $X(t)$ is differentiable. Then,

$$\begin{aligned}\frac{d}{d\tau}R_{XX}(\tau) &= \frac{d}{d\tau}E\{X(t)X(t+\tau)\} \\ &= E\left\{X(t)\frac{d}{d\tau}X(t+\tau)\right\} \\ &= E\left\{X(t)\dot{X}(t+\tau)\right\} \\ &= R_{X\dot{X}}(\tau).\end{aligned}$$

Similarly,

$$\begin{aligned}\frac{d^2}{d\tau^2}R_{XX}(\tau) &= \frac{d^2}{d\tau^2}E\{X(t)X(t+\tau)\} \\ &= E\left\{X(t)\frac{d^2}{d\tau^2}X(t+\tau)\right\} \\ &= E\left\{X(t)\ddot{X}(t+\tau)\right\} \\ &= R_{X\ddot{X}}(\tau).\end{aligned}$$

The second derivative can be written slightly differently by rewriting $R_{X\dot{X}}(\tau)$ as $E\left\{X(t-\tau)\dot{X}(t)\right\}$. This can be done since the correlation function only depends on the time difference. The second derivative is now written as

$$\begin{aligned}\frac{d^2}{d\tau^2}R_{XX}(\tau) &= \frac{d}{d\tau}R_{X\dot{X}}(\tau) \\ &= \frac{d}{d\tau}E\left\{X(t-\tau)\dot{X}(t)\right\} \\ &= -E\left\{\dot{X}(t-\tau)\dot{X}(t)\right\} \\ &= -R_{\dot{X}\dot{X}}(\tau).\end{aligned} \quad (5.11)$$

Therefore,

$$R_{X\ddot{X}}(\tau) = -R_{\dot{X}\dot{X}}(\tau). \quad (5.12)$$

Now reconsider the first derivative of the autocorrelation function at $\tau = 0$, $R_{X\dot{X}}(0)$. Since $R_{XX}(\tau)$ is an even function of τ, its slope at $\tau = 0$ equals zero. That is,

$$\left.\frac{dR_{XX}(\tau)}{d\tau}\right|_{\tau=0} = R_{X\dot{X}}(0) = 0. \quad (5.13)$$

Look ahead at Figure 5.13 at some examples of correlation functions to see that this is true. This is an important result and will be used later in Chapter 9.

Example 5.4 Derivatives of Autocorrelation Functions

Express $R_{\ddot{X}\ddot{X}}(\tau)$ in terms of derivatives of $R_{XX}(\tau)$.

Solution Start with $R_{X\ddot{X}}(\tau)$. The second derivative of $R_{XX}(\tau)$ has been shown to be

$$\frac{d^2}{d\tau^2}R_{XX}(\tau) = R_{X\ddot{X}}(\tau).$$

Then, the fourth derivative is given by

$$\begin{aligned}\frac{d^4}{d\tau^4}R_{XX}(\tau) &= \frac{d^2}{d\tau^2}E\left\{X(t)\ddot{X}(t+\tau)\right\} \\ &= R_{XX^{(4)}}(\tau).\end{aligned}$$

Since t is arbitrary, it can be adjusted so that

$$E\left\{X(t)\ddot{X}(t+\tau)\right\} = E\left\{X(t-\tau)\ddot{X}(t)\right\},$$

and then the fourth derivative can be written as

$$\begin{aligned}\frac{d^4}{d\tau^4}R_{XX}(\tau) &= \frac{d^2}{d\tau^2}E\left\{X(t-\tau)\ddot{X}(t)\right\} \\ &= R_{\ddot{X}\ddot{X}}(\tau).\end{aligned}$$

Therefore,

$$\frac{d^4}{d\tau^4}R_{XX}(\tau) = R_{\ddot{X}\ddot{X}}(\tau) = R_{XX^{(4)}}(\tau). \tag{5.14}$$

⊛

5.5 Fourier Series and Fourier Transforms

We learned in calculus that a periodic function $q(t)$ such as the one shown in Figure 5.8 can be expressed in a Fourier series as a sum of sines and cosines such that[3]

$$q(t) \Leftrightarrow a_0 + \sum_{n=1}^{\infty} a_n \cos\frac{2n\pi}{T}t + \sum_{n=1}^{\infty} b_n \sin\frac{2n\pi}{T}t,$$

---

[3] ⇔ means that $q(t)$ on the left-hand side has the Fourier series of $q(t)$ on the right-hand side. ⇔ is used rather than = in order to emphasize the fact that the Fourier series may not accurately represent the function in some places (known as the Gibbs phenomenon) and the series may even diverge. However, for convenience, we will use the equal sign in this text.

5.5. FOURIER SERIES AND FOURIER TRANSFORMS

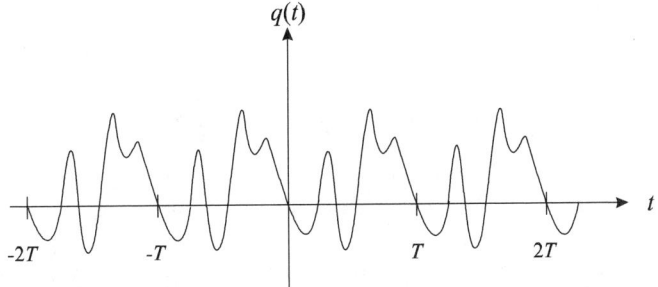

Figure 5.8: A periodic function with period T.

where the Fourier coefficients are given by

$$a_0 = \frac{1}{T} \int_{-T/2}^{T/2} q(t)\, dt$$

$$a_n = \frac{2}{T} \int_{-T/2}^{T/2} q(t) \cos \frac{2n\pi}{T} t\, dt$$

$$b_n = \frac{2}{T} \int_{-T/2}^{T/2} q(t) \sin \frac{2n\pi}{T} t\, dt, \tag{5.15}$$

for $n = 1, 2, 3, \cdots$. These relations can be written in the complex form,

$$q(t) \Leftrightarrow \sum_{n=-\infty}^{\infty} c_n \exp(i 2 n \pi t / T), \tag{5.16}$$

where

$$c_n = \frac{1}{T} \int_{-T/2}^{T/2} q(t) \exp(-i 2 n \pi t / T)\, dt, \tag{5.17}$$

$$n = \cdots -2, -1, 0, 1, 2, \cdots.$$

It can be found that the complex Fourier coefficient c_n equals

$$c_n = \frac{a_n + i b_n}{2}.$$

The Fourier representation of function $q(t)$ provides its *frequency composition*. The coefficients a_n and b_n are the amplitudes of the sinusoidal components, $\sin 2n\pi t/T$ and $\cos 2n\pi t/T$, of the function $q(t)$. The term,

$$\sqrt{a_n^2 + b_n^2} \text{ or } 2|c_n|,$$

is called the *spectrum of* $q(t)$.

If the function $q(t)$ is, for example,

$$q(t) = 2 + \sin\frac{\pi}{T}t + 3\cos\frac{2\pi}{T}t,$$

then the spectrum of $q(t)$ will have peaks at $\omega = 0$, π/T, and $2\pi/T$ with magnitude of 2, 1, and 3, respectively. The units of the spectrum are the same as the units of $q(t)$.

For this case where the function is periodic, the spectrum is discrete and the frequency is evenly spaced at $2\pi/T$ intervals. Noting that the terms $2n\pi/T$ are the angular frequencies ω_n (number of cycles in 2π s) and the angular frequencies are spaced at $2\pi/T$ intervals, we can let

$$\frac{2n\pi}{T} = \omega_n$$
$$\frac{2\pi}{T} = \Delta\omega.$$

Now, Equation 5.17 can be written as

$$c_n = \frac{\Delta\omega}{2\pi}\int_{-T/2}^{T/2} q(t)\exp(-i\omega_n t)\,dt. \quad (5.18)$$

It should be noted that if a segment of $q(t)$ from $-T/2$ to $T/2$ repeats itself, so does the segment from 0 to T. Therefore, the limits of the integral for the Fourier coefficients in Equations 5.15 and 5.17 can be changed such that

$$c_n = \frac{\Delta\omega}{2\pi}\int_{0}^{T} q(t)\exp(-i\omega_n t)\,dt.$$

The change in the limits does not change the values of the Fourier coefficients.

Substituting Equation 5.18 into Equation 5.16, we obtain

$$q(t) \Leftrightarrow \sum_{n=-\infty}^{\infty}\frac{\Delta\omega_n}{2\pi}\exp(i\omega_n t)\int_{-T/2}^{T/2} q(\hat{t})\exp(-i\omega_n\hat{t})\,d\hat{t},$$

where $\hat{t}$ is used to emphasize that it is a dummy variable and not to be confused with the time instant t when q is evaluated.

Consider a case where the period $T \to \infty$, and the function $q(t)$ does not repeat itself. As a consequence, the frequency interval of the spectrum $\Delta\omega$ can be replaced by $d\omega$, the discrete values ω_n limit to a continuous variable ω, and the summation is replaced by an integral. We can write

$$q(t) \Leftrightarrow \int_{-\infty}^{\infty}\left[\frac{1}{2\pi}\int_{-\infty}^{\infty} q(\hat{t})\exp(-i\omega\hat{t})\,d\hat{t}\right]\exp(i\omega t)\,d\omega. \quad (5.19)$$

5.5. FOURIER SERIES AND FOURIER TRANSFORMS

The term in the square brackets is denoted as $Q(\omega)$, and is called the Fourier transform of the function $q(t)$. The *Fourier transform*, therefore, is the continuous version of the Fourier coefficients. The well-known Fourier transform pair is then

$$Q(\omega) = \frac{1}{2\pi} \int_{-\infty}^{\infty} q(t) \exp(-i\omega t) \, dt \qquad (5.20)$$

$$q(t) = \int_{-\infty}^{\infty} Q(\omega) \exp(i\omega t) \, d\omega. \qquad (5.21)$$

The Fourier transform $Q(\omega)$ has the same units as $q(t) \times t$, whereas the Fourier coefficients have the same units as $q(t)$. Similar to the Fourier coefficients, the Fourier transform provides the frequency content of function $q(t)$. We can also denote the Fourier transform as $\mathcal{F}\{q(x)\}$ and the inverse transform as $\mathcal{F}^{-1}\{Q(\omega)\}$.

Note that a number of forms of Fourier transform pairs exist and they are equally valid. The Fourier transform pair can be defined as

$$Q(\omega) = A \int_{-\infty}^{\infty} q(t) \exp(-i\omega t) \, dt$$

$$q(t) = B \int_{-\infty}^{\infty} Q(\omega) \exp(i\omega t) \, d\omega, \qquad (5.22)$$

as long as $A \cdot B = 1/2\pi$ such that Equation 5.19 is still valid. The dummy variable can be changed in Equation 5.19: $\omega \rightarrow -\omega$. Then,

$$q(t) \Leftrightarrow \frac{1}{2\pi} \int_{-\infty}^{\infty} \left[\int_{-\infty}^{\infty} q(\hat{t}) \exp(i\omega \hat{t}) \, d\hat{t} \right] \exp(-i\omega t) \, d\omega,$$

with Fourier transform pair,

$$Q(\omega) = A \int_{-\infty}^{\infty} q(t) \exp(i\omega t) \, dt$$

$$q(t) = B \int_{-\infty}^{\infty} Q(\omega) \exp(-i\omega t) \, d\omega,$$

where $A \cdot B = 1/2\pi$. In this book, we use the Fourier transform pair as defined in Equations 5.20 and 5.21.

Fourier   The father of Joseph Fourier was a tailor in Auxerre. After the death of his first wife, with whom he had three children, he remarried and Joseph was the ninth of the twelve children of this second marriage. Joseph's mother died went he was nine years old and his father died the following year.

CHAPTER 5. RANDOM PROCESSES

Figure 5.9: Joseph Fourier (1768 - 1830).

Fourier's first schooling was at Pallais' school, run by the music master from the cathedral. There Joseph studied Latin and French and showed great promise. He proceeded in 1780 to the École Royale Militaire of Auxerre where at first he showed talent for literature but very soon, by the age of thirteen, mathematics became his real interest. By the age of 14 he had completed a study of the six volumes of Bézout's *Cours de Mathématique*. In 1783 he received the first prize for his study of Bossut's *Méchanique en Général*.

In 1787 Fourier decided to train for the priesthood and entered the Benedictine abbey of St. Benoit-sur-Loire. His interest in mathematics continued, however, and he corresponded with C.L. Bonard, the professor of mathematics at Auxerre. Fourier was unsure if he was making the right decision in training for the priesthood. He submitted a paper on algebra to Montucla in Paris and his letters to Bonard suggest that he really wanted to make a major impact in mathematics. In one letter Fourier wrote: *Yesterday was my 21st birthday, at that age Newton and Pascal had already acquired many claims to immortality.*

Fourier did not take his religious vows. Having left St. Benoit in 1789, he visited Paris and read a paper on algebraic equations at the Académie Royale des Sciences. In 1790 he became a teacher at the Benedictine college, École Royale Militaire of Auxerre, where he had studied. Up until this time Fourier suffered an inner conflict about whether he should follow a religious life or one of mathematical research. However, in 1793 a third element was added to this conflict when he became involved in politics and joined the local Revolutionary Committee. As he wrote: *As the natural ideas of equality developed it was possible to conceive the sublime hope of*

5.5. FOURIER SERIES AND FOURIER TRANSFORMS

establishing among us a free government exempt from kings and priests, and to free from this double yoke the long-usurped soil of Europe. I readily became enamored of this cause, in my opinion the greatest and most beautiful which any nation has ever undertaken.

Certainly Fourier was unhappy about the Terror that resulted from the French Revolution, and he attempted to resign from the committee. However this proved impossible and Fourier was now firmly entangled with the Revolution and unable to withdraw. The revolution was a complicated affair with many factions, with broadly similar aims, violently opposed to each other. Fourier defended members of one faction while in Orléans. A letter describing events relates that *Citizen Fourier, a young man full of intelligence, eloquence and zeal, was sent to Loiret. It seems that Fourier ... got up on certain popular platforms. He can talk very well and if he put forward the views of the Society of Auxerre he has done nothing blameworthy....*

This incident was to have serious consequences, but after it Fourier returned to Auxerre and continued to work on the revolutionary committee and continued to teach at the College. In July 1794 he was arrested, the charges relating to the Orléans incident, and he was imprisoned. Fourier feared the he would go to the guillotine but, after Robespierre himself went to the guillotine, political changes resulted in Fourier being freed.

Later in 1794 Fourier was nominated to study at the École Normale in Paris. This institution had been set up for training teachers and it was intended to serve as a model for other teacher-training schools. The school opened in January 1795 and Fourier was certainly the most able of the pupils, whose abilities ranged widely. He was taught by Lagrange, who Fourier described as the first among European men of science, and also by Laplace, who Fourier rated less highly, and by Monge, whom Fourier described as having a loud voice and as active, ingenious and very learned.

Fourier began teaching at the Collège de France and, having excellent relations with Lagrange, Laplace, and Monge, began further mathematical research. He was appointed to a position at the École Centrale des Travaux Publiques, the school being under the direction of Lazare Carnot and Gaspard Monge, which was soon to be renamed École Polytechnique. However, repercussions of his earlier arrest remained and he was arrested again and imprisoned. His release has been put down to a variety of different causes, pleas by his pupils, pleas by Lagrange, Laplace or Monge or a change in the political climate. In fact all three may have played a part.

By September 1, 1795 Fourier was back teaching at the École Polytechnique. In 1797 he succeeded Lagrange in being appointed to the chair of analysis and mechanics. He was renowned as an outstanding lecturer but he does not appear to have undertaken original research during this time.

In 1798 Fourier joined Napoleon's army in its invasion of Egypt as scientific adviser. Monge and Malus were also part of the expeditionary force. The expedition was at first a great success. Malta was occupied on June 10, 1798, Alexandria taken by storm on July 1, and the delta of the Nile quickly taken. However, on August 1, 1798 the French fleet was completely destroyed by Nelson's fleet in the Battle of the Nile, so that Napoleon found himself confined to the land that he was occupying. Fourier acted as an administrator as French-type political institutions and administration were set up. In particular, he helped establish educational facilities in Egypt and carried out archaeological explorations.

While in Cairo Fourier helped found the Cairo Institute and was one of the twelve members of the mathematics division, the others included Monge, Malus, and Napoleon himself. Fourier was elected secretary to the Institute, a position he continued to hold during the entire French occupation of Egypt. Fourier was also put in charge of collating the scientific and literary discoveries made during the time in Egypt.

Napoleon abandoned his army and returned to Paris in 1799, and he soon held absolute power in France. Fourier returned to France in 1801 with the remains of the expeditionary force, and resumed his post as Professor of Analysis at the École Polytechnique. However, Napoleon had other ideas about how Fourier might serve him and wrote: ... *the Prefect of the Department of Isère having recently died, I would like to express my confidence in citizen Fourier by appointing him to this place.*

Fourier was not happy at the prospect of leaving the academic world and Paris but could not refuse Napoleon's request. He went to Grenoble where his duties as Prefect were many and varied. His two greatest achievements in this administrative position were overseeing the operation to drain the swamps of Bourgoin and supervising the construction of a new highway from Grenoble to Turin. He also spent much time working on the *Description of Egypt*, which was not completed until 1810, when Napoleon made changes, rewriting history in places, to it before publication. By the time a second edition appeared every reference to Napoleon would have been removed.

It was during his time in Grenoble that Fourier did his important mathematical work on the theory of heat. His work on the topic began around 1804 and by 1807 he had completed his important memoir *On the Propagation of Heat in Solid Bodies*. The memoir was read to the Paris Institute on 21 December 1807 and a committee consisting of Lagrange, Laplace, Monge and Lacroix was set up to report on the work. Now this memoir is very highly regarded but at the time it caused controversy.

There were two reasons for the committee to feel unhappy with the work. The first objection, made by Lagrange and Laplace in 1808, was to Fourier's expansions of functions as trigonometric series, what we now

5.5. FOURIER SERIES AND FOURIER TRANSFORMS

call Fourier series. Further clarification by Fourier still failed to convince them. As is pointed out: *All these are written with such exemplary clarity - from a logical as opposed to calligraphic point of view, that their inability to persuade Laplace and Lagrange provides a good index of the originality of Fourier's views.*

The second objection was made by Biot against Fourier's derivation of the equations of heat transfer. Fourier had not made reference to Biot's 1804 paper on this topic but Biot's paper is certainly incorrect. Laplace, and later Poisson, had similar objections.

The Institute initiated a competition on the propagation of heat in solid bodies for the 1811 mathematics prize. Fourier submitted his 1807 memoir together with additional work on the cooling of infinite solids and terrestrial and radiant heat. Only one other entry was received and the committee set up to decide on the award of the prize, Lagrange, Laplace, Malus, Haüy and Legendre, awarded Fourier the prize. The report was not however completely favorable and states: ... *the manner in which the author arrives at these equations is not exempt of difficulties and that his analysis to integrate them still leaves something to be desired on the score of generality and even rigor.* With this rather mixed report there was no move in Paris to publish Fourier's work.

When Napoleon was defeated and on his way to exile in Elba, his route should have been through Grenoble. Fourier managed to avoid this difficult confrontation by sending word that it would be dangerous for Napoleon. When he learnt of Napoleon's escape from Elba and that he was marching towards Grenoble with an army, Fourier was extremely worried. He tried to persuade the people of Grenoble to oppose Napoleon and give their allegiance to the King. However as Napoleon marched into the town Fourier left in haste.

Napoleon was angry with Fourier, who he had hoped would welcome his return. Fourier was able to talk his way into favor with both sides and Napoleon made him Prefect of the Rhône. However Fourier soon resigned on receiving orders, possibly from Carnot, that he was to remove all administrators with royalist sympathies. He could not have completely fallen out with Napoleon and Carnot, however, for on 10 June 1815, Napoleon awarded him a pension of 6000 francs, payable from 1 July. However Napoleon was defeated on 1 July and Fourier did not receive any money. He returned to Paris.

Fourier was elected to the Académie des Sciences in 1817. In 1822 d'Alembert, who was the Secretary to the mathematical section of the Académie des Sciences, died and Fourier, together with Biot and Arago, applied for the post. After Arago withdrew, the election gave Fourier an easy win. Shortly after Fourier became Secretary, the Academy published his prize winning essay *Théorie Analytique de la Chaleur* in 1822. This was not a piece of political maneuvering by Fourier, however, since Delambre had arranged for the printing before he died. Fourier's work provided the impetus for later work on trigonometric series and the theory of functions of a real variable.

During Fourier's eight last years in Paris he resumed his mathematical researches and published a number of papers, some in pure mathematics while some were on applied mathematical topics. His life was not without problems however since his theory of heat still provoked controversy. Biot claimed priority over Fourier, a claim which Fourier had little difficulty showing to be false. Poisson, however, attacked both Fourier's mathematical techniques and also claimed to have an alternative theory. Fourier wrote *Historical Précis* as a reply to these claims but, although the work was shown to various mathematicians, it was never published.

Fourier's views on the claims of Biot and Poisson are given in the following: *Having contested the various results [Biot and Poisson] now recognize that they are exact but they protest that they have invented another method of expounding them and that this method is excellent and the true one. If they had illuminated this branch of physics by important and general views and had greatly perfected the analysis of partial differential equations, if they had established a principal element of the theory of heat by fine experiments ... they would have the right to judge my work and to correct it. I would submit with much pleasure .. But one does not extend the bounds of science by presenting, in a form said to be different, results which one has not found oneself and, above all, by forestalling the true author in publication.*

Example 5.5 Dirac Delta Function

Find the Fourier transform of the Dirac delta function $\delta(t)$. Use this result to find the Fourier transform of the number 1.

Solution   The Fourier transform of the delta function is given by

$$\Delta(\omega) = \frac{1}{2\pi} \int_{-\infty}^{\infty} \delta(t) \exp(-i\omega t)\, dt$$
$$= \frac{1}{2\pi}.$$

The Fourier transform of 1 is written as

$$\mathcal{F}\{1\} = \frac{1}{2\pi} \int_{-\infty}^{\infty} 1 \exp(-i\omega t)\, dt, \qquad (5.23)$$

5.5. FOURIER SERIES AND FOURIER TRANSFORMS

Figure 5.10: P.A.M. Dirac (1902-1984)

which is not easy to evaluate. To find $\mathcal{F}\{1\}$, we first use the fact that the inverse transform of $1/2\pi$ is a delta function,

$$\delta(t) = \int_{-\infty}^{\infty} \frac{1}{2\pi} \exp(i\omega t)\, d\omega.$$

With the following transformation of variables: $\omega \to -\tau$ and $t \to \Omega$, the integral can be written as

$$\delta(\Omega) = \int_{-\infty}^{\infty} \frac{1}{2\pi} \exp(-i\Omega\tau)\, d\tau.$$

Comparing with the definition of the Fourier transform of 1 in Equation 5.23, we find that the Fourier transform of 1 is simply the delta function.

⊛

Dirac Paul Dirac was born to Charles Adrien Ladislas Dirac and Florence Hannah Holten. Charles Dirac was a Swiss citizen born in Monthey, in the Valais Canton of Switzerland, while his mother came from Cornwall in England. Charles had been educated at the University of Geneva, then came to England in around 1888 and taught French in Bristol. There Dirac met Florence, whose father had moved to Bristol as Master Mariner on a Bristol ship, when she was working in the library there. Charles and

CHAPTER 5. RANDOM PROCESSES

Florence married in 1899 and they moved into a house in Bishopston, Bristol, which they named Monthey after the town of Charles' birth. By this time Charles was teaching French at the secondary school attached to the Merchant Venturers Technical College in Bristol.

Paul was one of three children, his older brother being Reginald Charles Felix Dirac and his younger sister being Beatrice Isabelle Marguerite Walla Dirac. Paul had a very strict family upbringing. His father insisted that only French be spoken at the dinner table and, as a result, Paul was the only one to eat with his father in the dining room. Paul's father was so strict with his sons that both were alienated and Paul was brought up in a somewhat unhappy home.

The first school that Paul attended was Bishop Primary school and already in this school his exceptional ability in mathematics became clear to his teachers. When he was twelve years old he entered secondary school, the one where his father taught, which was part of the Merchant Venturers Technical College. At about the time Paul entered this school, World War I had begun and this had a beneficial effect for Paul since the older boys in the school left for military service and the younger boys had more access to the science laboratories and other facilities.

Paul himself wrote about his school years: *The Merchant Venturers was an excellent school for science and modern languages. There was no Latin or Greek, something of which I was rather glad, because I did not appreciate the value of old cultures. I consider myself very lucky in having been able to attend the school. ... I was rushed through the lower forms, and was introduced at an especially early age to the basis of mathematics, physics, and chemistry in the higher forms. In mathematics I was studying from books which mostly were ahead of the rest of the class. This rapid advancement was a great help to me in my latter career.*

He completed his school education in 1918 and then studied electrical engineering at the University of Bristol. By this time the University had combined with the Merchant Venturers Technical College so Dirac remained in the same building as he had studied during his four years at secondary school. Although mathematics was his favorite subject he chose to study an engineering course at university since he thought that the only possible career for a mathematician was school teaching and he certainly wanted to avoid that profession. He obtained his degree in engineering in 1921 but following this, after an undistinguished summer job in engineering works, he did not find a permanent position. By this time he was developing a real passion for mathematics, but his attempts to study at Cambridge failed for rather strange reasons.

Taking the Cambridge scholarship examinations in June 1921 he was awarded a scholarship to study mathematics at St John's College, Cam-

5.5. FOURIER SERIES AND FOURIER TRANSFORMS

bridge, but it did not provide enough to support him. Additional support would have been expected from his local education authority, but he was refused support on the grounds that his father had not been a British citizen for long enough. Dirac was offered the chance to study mathematics at Bristol without paying fees and he did so being awarded first-class honors in 1923. Following this he was awarded a grant to undertake research at Cambridge and he began his studies there in 1923.

Dirac had been hoping to have his research supervised by Ebenezer Cunningham, for by this time Dirac had become fascinated by the general theory of relativity and wanted to undertake research on this topic. Cunningham already had as many research students as he was prepared to take on and so Dirac was supervised by Ralph Fowler: *Fowler was then the leading theoretician in Cambridge, well versed in the quantum theory of atoms; his own research was mostly on statistical mechanics. He recognized in Dirac a student of unusual ability. Under his influence Dirac worked on some problems in statistical mechanics. Within six months of arriving in Cambridge he wrote two papers on these problems.* No doubt Fowler aroused his interest in the quantum theory, and in May 1924 Dirac completed his first paper dealing with quantum problems. Four more papers were completed by November 1925.

Despite the obvious academic success Dirac enjoyed as a research student this was no easy time for him. His brother Reginald Dirac committed suicide during this period. No reason for the suicide seems to be known but Dirac's relations with his father, already strained, seemed almost to end completely after this, which does suggest that Dirac felt that his father carried at least some responsibility. Already a person who had few friends, this personal tragedy had the effect of making him even more withdrawn.

Although he had already made an excellent start to his research career, even more impressive work was to follow. This was as a result of Dirac being given proofs of a paper by Heisenberg to read in the summer of 1925. The significance of the algebraic properties of Heisenberg's commutators struck Dirac when he was out for a walk in the country. He realized that Heisenberg's uncertainty principle was a statement of the noncommutativity of the quantum mechanical observables. He realized the analogy with Poisson brackets in Hamiltonian mechanics. Higgs writes: *This similarity provided the clue which led him to formulate for the first time a mathematically consistent general theory of quantum mechanics in correspondence with Hamiltonian mechanics.*

The ideas were laid out in Dirac's doctoral thesis, *Quantum Mechanics*, for which he was awarded a Ph.D. in 1926. It is remarkable that Dirac had eleven papers in print before submitting his doctoral dissertation. Following the award of the degree he went to Copenhagen to work with Niels Bohr, moving on to Göttingen in February 1927 where he interacted with Robert Oppenheimer, Max Born, James Franck and the Russian Igor Tamm. Accepting an invitation from Ehrenfest, he spent a few weeks in Leiden on his way back to Cambridge. He was elected a Fellow of St John's College, Cambridge in 1927.

Dirac visited the Soviet Union in 1928. It was the first of many visits for he went again in 1929, 1930, 1932, 1933, 1935, 1936, 1937, 1957, 1965, and 1973. Also in 1928 he found a connection between relativity and quantum mechanics, his famous spin-1/2 Dirac equation. In 1929 he made his first visit to the United States, lecturing at the Universities of Wisconsin and Michigan. After the visit, along with Heisenberg, he crossed the Pacific and lectured in Japan. He returned via the trans-Siberian railway.

In 1930 Dirac published *The Principles of Quantum Mechanics* and for this work he was awarded the Nobel Prize for Physics in 1933. De Facio, reviewing, says of this book: *Dirac was not influenced by the feeding frenzy in experimental phenomenology of the time. This has given Dirac's book ... a lasting quality that few works can match*. Others comment that the book ... *reflects Dirac's very characteristic approach: abstract but simple, always selecting the important points and arguing with unbeatable logic*.

His lectures at Cambridge were closely modeled on *The Principles of Quantum Mechanics*, and they conveyed to generations of students a powerful impression of the coherence and elegance of quantum theory. They constituted his principal contribution to education, for he took very few research students.

Also in 1930 Dirac was elected a Fellow of the Royal Society. This honor came on the first occasion that his name was put forward, in itself quite an unusual event, which says much about the extremely high opinion that Dirac's fellow scientists had of him.

Dirac was appointed Lucasian professor of mathematics at the University of Cambridge in 1932, a post he held for 37 years. This was originally Newton's chair. In 1933 he published a pioneering paper on Lagrangian quantum mechanics which became the foundation on which Feynman later built his ideas of the path integral. In the same year, Dirac received the Nobel Prize for physics, which he shared with Schrödinger. It is an interesting comment on Dirac's nature that his first thought was to turn down the prize on the grounds that he hated publicity. However when it was pointed out to him that he would receive far more publicity if he turned down the prize, he accepted it. Another comment about this event is that Dirac was told that he could invite his parents to the award ceremony in Stockholm, but he chose to invite only his mother and not his father.

The academic year 1934-1935 was important for Dirac both for personal and professional reasons. He visited the Institute for Advanced Study

5.5. FOURIER SERIES AND FOURIER TRANSFORMS

at Princeton and there he became friendly with Wigner. While Dirac was there Wigner's sister Margit, who lived in Budapest, visited her brother. This chance meeting led, in January 1937, to Dirac marrying Margit in London. Margit had been married before and had two children, Judith and Gabriel Andrew, from her first marriage. Both children adopted the name Dirac. Gabriel Andrew Dirac went on to become a famous pure mathematician, particularly contributing to graph theory, becoming professor of pure mathematics at the University of Aarhus in Denmark.

In 1937, the same year that he married, Dirac published his first paper on large numbers and cosmological matters. We comment further on his ideas on cosmology below. He published his famous paper on classical electron theory, which included mass renormalization and radiative reaction in 1938. Dirac worked during World War II on uranium separation and nuclear weapons. In particular he acted as a consultant to a group in Birmingham working on atomic energy. This association led to Dirac being prevented by the British government from visiting the Soviet Union after the end of the war; he was not able to visit again until 1957.

We noted above that Dirac was elected a fellow of the Royal Society in 1930. He was awarded the Royal Society's Royal Medal in 1939 and the Society awarded him their Copley Medal in 1952 in recognition of his remarkable contributions to relativistic dynamics of a particle in quantum mechanics.

In 1969 Dirac retired from the Lucasian chair of mathematics at Cambridge and went with his family to Florida. He held visiting appointments at the University of Miami and at Florida State University. Then, in 1971, Dirac was appointed professor of physics at Florida State University where he continued his research.

In 1973 and 1975 Dirac lectured in the Physical Engineering Institute in Leningrad. In these lectures he spoke about the problems of cosmology or, to be more precise, to the problems of nondimensional combinations of world constants.

Although Dirac made vastly important contributions to physics, it is important to realize that he was always motivated by principles of mathematical beauty. Dirac unified the theories of quantum mechanics and relativity theory, but he also is remembered for his outstanding work on the magnetic monopole, fundamental length, antimatter, the d-function, and bra-kets, for example.

There is a standard folklore of Dirac stories, mostly revolving around Dirac saying exactly what he meant and no more. Once when someone, making polite conversation at dinner, commented that it was windy, Dirac left the table and went to the door, looked out, returned to the table and replied that indeed it was windy. It has been said in jest that his spoken vocabulary consisted of "Yes," "No," and "I don't know." Certainly when Chandrasekhar was explaining his ideas to Dirac he continually interjected "yes" then explained to Chandrasekhar that "yes" did not mean that he agreed with what he was saying, only that he wished him to continue. He once said: *I was taught at school never to start a sentence without knowing the end of it.* This may explain much about his conversation, and also about his beautifully written sentences in his books and papers.

Dirac received many honors for his work, some of which we have mentioned above. He refused to accept honorary degrees but he did accept honorary membership of academies and learned societies. The list of these is long but among them are USSR Academy of Sciences (1931), Indian Academy of Sciences (1939), Chinese Physical Society (1943), Royal Irish Academy (1944), Royal Society of Edinburgh (1946), Institut de France (1946), National Institute of Sciences of India (1947), American Physical Society (1948), National Academy of Sciences (1949), National Academy of Arts and Sciences (1950), Accademia delle Scienze di Torino (1951), Academia das Ciencias de Lisboa (1953), Pontifical Academy of Sciences, Vatican City (1958), Accademia Nazionale dei Lincei, Rome (1960), Royal Danish Academy (1962), and Académie des Sciences Paris (1963). He was appointed to the Order of Merit in 1973.

A memorial meeting was held at the University of Cambridge on April 19, 1985, and the papers presented at this meeting were published in *Tributes to Paul Dirac*, Cambridge, 1985 (Bristol, 1987). Achuthan, reviewing the volume, writes: ... *we vividly see everywhere the brilliant imprints of Dirac, unifier of quantum mechanics and relativity theory. Each of the pieces not only is in praise of an exceptionally gifted intellect but also places on record how deeply and abidingly the human mind can delve into the realms of mathematical insight and modeling, keeping intact the spirit of beauty and clarity of a creative genius. Only a few Nobel laureates ever can compare as well with this giant of mathematical sciences in whose demise the world of original thinking certainly has lost one of the most precious souls retaining fortunately still the glory for others to sing and emulate for a long time to come.*

In November 1995, a plaque was unveiled in Westminster Abbey commemorating Paul Dirac. The memorial address, of lectures presented to the Royal Society on this occasion, was presented by Stephen Hawking, who was Dirac's successor in the Lucasian chair of mathematics at Cambridge.

Time Differentiation

If one applies the Fourier transform to a differential equation, it is necessary to be able to take transforms of time derivatives. Consider then the Fourier transforms of dq/dt and d^2q/dt^2. Using the definition of the Fourier

5.5. FOURIER SERIES AND FOURIER TRANSFORMS

transform,
$$\mathcal{F}\left\{\frac{dq}{dt}\right\} = \frac{1}{2\pi}\int_{-\infty}^{\infty}\frac{dq}{dt}\exp(-i\omega t)\,dt.$$

Integrating by parts,
$$\mathcal{F}\left\{\frac{dq}{dt}\right\} = \frac{1}{2\pi}\exp(-i\omega t)q(t)\bigg|_{-\infty}^{\infty} + i\omega\frac{1}{2\pi}\int_{-\infty}^{\infty}q(t)\exp(-i\omega t)\,dt.$$

We assume that $q(t)$ is absolutely integrable[4] so that $q(t)$ must vanish at $-\infty$ and ∞. Then,

$$\begin{aligned}\mathcal{F}\left\{\frac{dq}{dt}\right\} &= i\omega\frac{1}{2\pi}\int_{-\infty}^{\infty}q(t)\exp(-i\omega t)\,dt \\ &= i\omega\mathcal{F}\{q(t)\}.\end{aligned} \quad (5.24)$$

Similarly, we find that the Fourier transform of the second derivative is

$$\begin{aligned}\mathcal{F}\left\{\frac{d^2q}{dt^2}\right\} &= -\omega^2\frac{1}{2\pi}\int_{-\infty}^{\infty}q(t)\exp(-i\omega t)\,dt \\ &= -\omega^2\mathcal{F}\{q(t)\}.\end{aligned} \quad (5.25)$$

Convolution Theorem

Another important property of the Fourier transform is the *convolution theorem*,

$$\begin{aligned}\mathcal{F}^{-1}\{P(\omega)Q(\omega)\} &= \frac{1}{2\pi}\int_{-\infty}^{\infty}p(\tau)q(t-\tau)\,d\tau \\ &= \frac{1}{2\pi}\int_{-\infty}^{\infty}q(\tau)p(t-\tau)\,d\tau,\end{aligned} \quad (5.26)$$

where $P(\omega)$ and $Q(\omega)$ are Fourier transforms of $p(t)$ and $q(t)$. This is proved next. The inverse Fourier transform of products of $P(\omega)$ and $Q(\omega)$ can be written as

$$\mathcal{F}^{-1}\{P(\omega)Q(\omega)\} = \int_{-\infty}^{\infty}P(\omega)Q(\omega)\exp(i\omega t)\,d\omega$$
$$= \int_{-\infty}^{\infty}P(\omega)\left\{\frac{1}{2\pi}\int_{-\infty}^{\infty}q(\hat{t})\exp(-i\omega\hat{t})\,d\hat{t}\right\}\exp(i\omega t)\,d\omega$$
$$= \frac{1}{2\pi}\int_{-\infty}^{\infty}\int_{-\infty}^{\infty}P(\omega)q(\hat{t})\exp\{-i\omega(\hat{t}-t)\}\,d\hat{t}\,d\omega,$$

---

[4] Absolutely integrable implies that the integral converges to a specific value and does not have a value at infinity.

where $\hat{t}$ is a dummy variable. Let $t - \hat{t} = \tau$,

$$\begin{aligned}\mathcal{F}^{-1}\{P(\omega)Q(\omega)\} &= \frac{1}{2\pi}\int_{-\infty}^{\infty}\int_{\infty}^{-\infty} P(\omega)q(t-\tau)\exp(i\omega\tau)(-d\tau)\,d\omega \\ &= \frac{1}{2\pi}\int_{-\infty}^{\infty}\int_{-\infty}^{\infty} P(\omega)q(t-\tau)\exp(i\omega\tau)\,d\tau\,d\omega.\end{aligned}$$

Change the order of integration,

$$\mathcal{F}^{-1}\{P(\omega)Q(\omega)\} = \frac{1}{2\pi}\int_{-\infty}^{\infty} q(t-\tau)\left[\int_{-\infty}^{\infty} P(\omega)\exp(i\omega\tau)\,d\omega\right]d\tau,$$

where the term in the square brackets is recognized as $p(t)$. Finally,

$$\mathcal{F}^{-1}\{P(\omega)Q(\omega)\} = \frac{1}{2\pi}\int_{-\infty}^{\infty} q(t-\tau)p(\tau)\,d\tau.$$

Changing variables, $\tau_2 = t - \tau$, we can also write

$$\begin{aligned}\mathcal{F}^{-1}\{P(\omega)Q(\omega)\} &= \frac{1}{2\pi}\int_{-\infty}^{\infty} q(\tau_2)p(t-\tau_2)\,d\tau_2 \\ &= \frac{1}{2\pi}\int_{-\infty}^{\infty} q(\tau)p(t-\tau)\,d\tau.\end{aligned}$$

Therefore, Equation 5.26 is proven. The integrals, $\int_{-\infty}^{\infty} q(\tau)p(t-\tau)\,d\tau$ and $\int_{-\infty}^{\infty} q(\tau)p(t-\tau)\,d\tau$, are often denoted as $p*q$.

Example 5.6 Convolution

Consider two rectangular functions,

$$\begin{aligned}p(t) &= 1, \quad -1 < t < 1 \\ q(t) &= 2, \quad -1 < t < 1.\end{aligned}$$

Find the convolution, $p*q$, and plot it as a function of t.

Solution The convolution of the two rectangular functions is given by

$$\begin{aligned}\int_{-\infty}^{\infty} q(\tau)p(t-\tau)\,d\tau &= \int_{-1}^{1} q(\tau)p(t-\tau)\,d\tau \\ &= \begin{cases} 0 & \text{for } t < -2 \\ 2(t+2) & \text{for } -2 < t < 0 \\ 2(-t+2) & \text{for } 0 < t < 2 \\ 0 & \text{for } t > 2. \end{cases}\end{aligned}$$

The convolution operation is shown in Figure 5.11.

⊛

5.5. FOURIER SERIES AND FOURIER TRANSFORMS

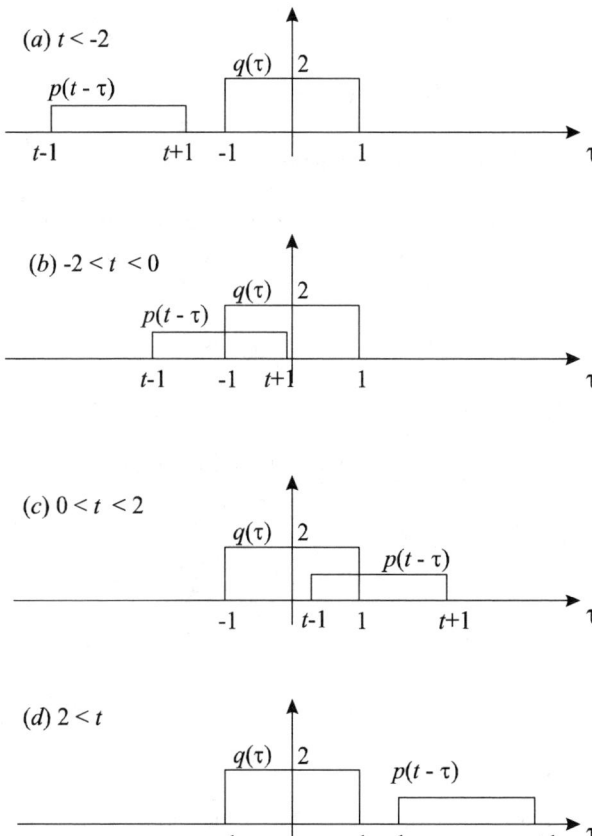

Figure 5.11: The convolution operation $q(\tau)p(t-\tau)$.

Shifting Theorem

If the Fourier transform of $p(t)$ is given by $P(\omega)$, then the Fourier transform of $p(t - t_o)$ is given by $P(\omega) \exp(-i\omega t_o)$. This can be shown as follows,

$$\begin{aligned}
\mathcal{F}\{p(t - t_o)\} &= \frac{1}{2\pi} \int_{-\infty}^{\infty} p(t - t_o) \exp(-i\omega t) \, dt \\
&= \frac{1}{2\pi} \int_{-\infty}^{\infty} p(t - t_o) \exp\{-i\omega(t - t_o)\} \exp(-i\omega t_o) \, dt \\
&= P(\omega) \exp(-i\omega t_o).
\end{aligned}$$

Example 5.7 Modulation theorem

Consider a function $p(t)$ with Fourier transform given by $P(\omega)$. Find the expression for the Fourier transform of $p(t) \cos \omega_o t$.

Solution Taking the Fourier transform of $p(t) \cos \omega_o t$, we can write

$$\mathcal{F}\{p(t) \cos \omega_o t\} = \frac{1}{2\pi} \int_{-\infty}^{\infty} p(t) \cos \omega_o t \exp(-i\omega t) \, dt.$$

Using the Euler identity, $\exp(i\theta) = \cos\theta + i\sin\theta$, the function $\cos \omega_o t$ becomes

$$\cos \omega_o t = \frac{\exp(i\omega_o t) + \exp(-i\omega_o t)}{2}.$$

Then,

$$\begin{aligned}
\mathcal{F}\{p(t) \cos \omega_o t\} &= \frac{1}{2\pi} \frac{1}{2} \int_{-\infty}^{\infty} p(t) [\exp\{-i(\omega - \omega_o)t\} \\
&\quad + \exp\{-i(\omega + \omega_o)t\}] \, dt \\
&= \frac{1}{2}[P(\omega - \omega_o) + P(\omega + \omega_o)].
\end{aligned}$$

Scaling Theorem

The scaling theorem is a statement that if the Fourier transform of $p(t)$ is $P(\omega)$, then the Fourier transform of $p(at)$ is $|a|^{-1} P(\omega/a)$, where a is a real and nonzero constant. This can be proved by taking the Fourier transform of $p(at)$, as follows,

$$\begin{aligned}
\mathcal{F}\{p(at)\} &= \frac{1}{2\pi} \int_{-\infty}^{\infty} p(at) \exp(-i\omega t) \, dt \\
&= \frac{1}{2\pi} \int_{-\infty}^{\infty} p(\tau) \exp\left(-i\frac{\omega}{a}\tau\right) \frac{1}{|a|} d\tau \\
&= \frac{1}{|a|} P(\omega/a).
\end{aligned}$$

5.6 Harmonic Processes

Our purpose in this section is to anticipate the subsequent derivation of the power spectrum. Knowing that a random process is composed of energy at many frequencies, we define a random process that is a sum of harmonics, similarly to the Fourier series. Begin first with a random function of one harmonic process, $X(t) = C \cos(\omega t - \phi)$, or equivalently,

$$X(t) = A \cos \omega t + B \sin \omega t,$$

where A and B are independent random variables. Assuming both to be zero mean and identically distributed, we have

$$\mu_A = \mu_B = 0$$
$$\sigma_A^2 = \sigma_B^2 = \sigma^2.$$

The autocorrelation is given by

$$\begin{aligned} R_{XX}(\tau) &= E\{X(t)X(t+\tau)\} \\ &= E\{(A \cos \omega t + B \sin \omega t)(A \cos \omega [t+\tau] + B \sin \omega [t+\tau])\}. \end{aligned}$$

Expanding the product and utilizing trigonometric identities results in

$$R_{XX}(\tau) = \sigma^2 \cos \omega \tau. \tag{5.27}$$

That is, the autocorrelation of a sinusoidal process is also sinusoidal.

Suppose that the frequency content of the random process is expanded,

$$\begin{aligned} X(t) &= \sum_{k=1}^{m} X_k(t) \\ &= \sum_{k=1}^{m} (A_k \cos \omega_k t + B_k \sin \omega_k t), \end{aligned}$$

where we make the same assumptions as were made above about A and B. Following the same procedure as for the above single-frequency process, we find

$$R_{XX}(\tau) = \sum_{k=1}^{m} R_{X_k X_k}(\tau) = \sum_{k=1}^{m} \sigma_k^2 \cos \omega_k \tau.$$

The total variance for the process is found by recalling that

$$\sigma^2 = E\{X^2(t)\} - \mu_X^2 = R_{XX}(0),$$

the last equality being true for the case where the mean equals zero. Then,

$$\sigma^2 = \sum_{k=1}^{m} \sigma_k^2.$$

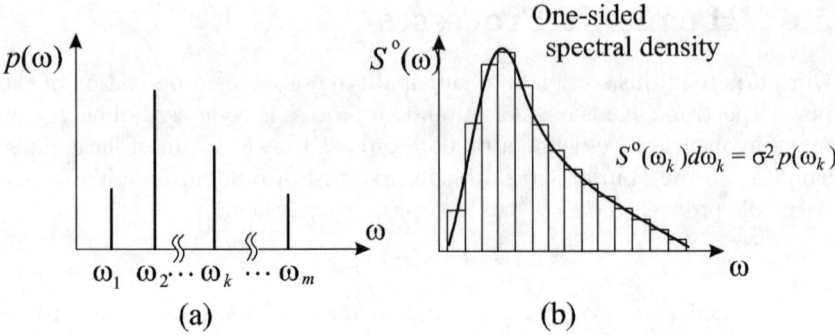

Figure 5.12: *(a)* Probability density $p(\omega)$ and *(b)* resulting one-sided spectral density $S^o(\omega)$.

Each frequency component ω_k contributes σ_k^2 to the total variance σ^2. The fraction of the total is given by the ratio σ_k^2/σ^2, which can be defined as $p(\omega_k)$ as shown in Figure 5.12(a). Note that $\sum_{k=1}^{m} p(\omega_k) = 1$. Then,

$$R_{XX}(\tau) = \sigma^2 \sum_{k=1}^{m} p(\omega_k) \cos \omega_k \tau,$$

where $p(\omega_k)$ acts as a weighting function. The above implies that $p(\omega_k)$ behaves like a probability density.

Suppose the frequency spectrum becomes very broad, including many frequencies, that is $m \to \infty$, resulting in a continuous frequency spectrum. Define $d\omega = \omega_{k+1} - \omega_k$. In an analogous manner to how we proceeded from a discrete to a continuous probability density function, replace $\sigma^2 p(\omega_k)$ by $S^o(\omega) d\omega$, and the sum above by an integral over the frequency range,

$$R_{XX}(\tau) = \int_0^\infty S^o(\omega) \cos \omega \tau \, d\omega. \qquad (5.28)$$

$S^o(\omega)$ is called the *one-sided spectral density* of the random process because it distributes the variance of the random process as a density across the frequency spectrum. The one-sided spectral density is shown in Figure 5.12(b).

5.7 Power Spectra

Our purpose in this section is to derive the Fourier transform pair,

$$S_{XX}(\omega) \Leftrightarrow R_{XX}(\tau),$$

5.7. POWER SPECTRA

also known as the *Wiener-Khinchine* relations or theorem. To do this, begin by taking the Fourier transform of the random process $X(t)$,

$$X(\omega) = \frac{1}{2\pi} \int_{-\infty}^{\infty} X(t) e^{-i\omega t} dt, \qquad (5.29)$$

which exists in the mean square sense if, and only if,

$$E\{X(\omega_1) X^*(\omega_2)\} = \frac{1}{(2\pi)^2} \int_{-\infty}^{\infty} \int_{-\infty}^{\infty} R_{XX}(t_1, t_2)$$
$$\cdot \exp\left[-i(\omega_1 t_1 - \omega_2 t_2)\right] dt_1 dt_2$$

is bounded for all values of ω_1 and ω_2. Superscript $*$ denotes complex conjugate.

If $X(t)$ is stationary, $X(\omega)$ does not exist.[5] To circumvent this difficulty, define the truncated Fourier transform,

$$X(\omega, T) = \frac{1}{2\pi} \int_{-T/2}^{T/2} X(t) e^{-i\omega t} dt.$$

As before, $X(\omega, T)$ exists in the mean square sense if, and only if, the following is bounded for all values of ω_1 and ω_2,

$$E\{X(\omega_1, T) X^*(\omega_2, T)\} = \frac{1}{(2\pi)^2} \int_{-T/2}^{T/2} \int_{-T/2}^{T/2} R_{XX}(t_1 - t_2)$$
$$\cdot \exp\left[-i(\omega_1 t_1 - \omega_2 t_2)\right] dt_1 dt_2.$$

Let $\tau = t_1 - t_2$, and $\omega_1 = \omega_2 = \omega$; then the last equation becomes

$$E\left\{|X(\omega, T)|^2\right\} = \frac{1}{(2\pi)^2} \int_{-T/2}^{T/2} \left[\int_{-T/2-t_2}^{T/2-t_2} R_{XX}(\tau) \exp(-i\omega\tau) d\tau\right] dt_2.$$

Reverse the order of integration on the right-hand side,

$$E\left\{|X(\omega, T)|^2\right\} = \frac{1}{(2\pi)^2} \left[\int_0^T R_{XX}(\tau) \exp(-i\omega\tau) \left(\int_{-T/2}^{T/2-\tau} dt_2\right) d\tau \right.$$
$$\left. + \int_{-T}^0 R_{XX}(\tau) \exp(-i\omega\tau) \left(\int_{-T/2-\tau}^{T/2} dt_2\right) d\tau\right]$$
$$= \frac{1}{(2\pi)^2} \int_{-T}^T (T - |\tau|) R_{XX}(\tau) \exp(-i\omega\tau) d\tau.$$

---

[5] It does not exist because a stationary process exists over all time and then the Fourier transform is indeterminate (infinity).

Multiply the last equation by $2\pi/T$ and take the limit as $T \to \infty$,

$$\lim_{T \to \infty} \frac{2\pi E\left\{|X(\omega,T)|^2\right\}}{T} = \lim_{T \to \infty} \frac{1}{2\pi} \int_{-T}^{T} \left(1 - \frac{|\tau|}{T}\right) R_{XX}(\tau) \exp(-i\omega\tau) \, d\tau$$

$$= S_{XX}(\omega) - \lim_{T \to \infty} \frac{1}{2\pi} \int_{-T}^{T} \left(\frac{|\tau|}{T}\right) R_{XX}(\tau) \exp(-i\omega\tau) \, d\tau. \quad (5.30)$$

Since $\lim_{\tau \to \infty} R_{XX}(\tau) \to 0$ for physical zero mean processes and $R_{XX}(\tau)$ is absolutely integrable such that it vanishes[6] at both $-\infty$ and ∞, the second term on the right-hand side of Equation 5.30 is arbitrarily small. Thus, we define the *power spectrum*, or *spectral density*, $S_{XX}(\omega)$, as the Fourier transform of its autocorrelation function,

$$S_{XX}(\omega) \equiv \frac{1}{2\pi} \int_{-\infty}^{\infty} R_{XX}(\tau) e^{-i\omega\tau} d\tau, \quad (5.31)$$

and

$$R_{XX}(\tau) = \int_{-\infty}^{\infty} S_{XX}(\omega) e^{i\omega\tau} d\omega. \quad (5.32)$$

Equation 5.30 also expresses the spectral density $S_{XX}(\omega)$ as

$$S_{XX}(\omega) = \lim_{T \to \infty} \frac{2\pi}{T} E\left\{|X(\omega,T)|^2\right\}.$$

Since $R_{XX}(\tau) = R_{XX}(-\tau)$, $S_{XX}(\omega)$ is not a complex function but a real even function,

$$S_{XX}(\omega) = S_{XX}(-\omega).$$

For $\tau = 0$,

$$\int_{-\infty}^{\infty} S_{XX}(\omega) \, d\omega = R_{XX}(0) = E\{X^2(t)\} \geq 0. \quad (5.33)$$

On physical grounds, then, it can be argued that since the area under the power spectral density equals σ_X^2 for a zero-mean random process, it must be a positive quantity for any $\Delta\omega$, that is, $S_{XX}(\omega) \geq 0$. Equation 5.33 implies that the integral of the power spectrum equals the "average or mean-square power" of the process $X(t)$.

For an ergodic process, the expected value can be written as

$$E\{X^2(t)\} \simeq \lim_{T \to \infty} \frac{1}{T} \int_{-T/2}^{T/2} X^2(t) dt,$$

---

[6] Y.K. Lin, *Probabilistic Theory of Structural Dynamics*, Robert E. Krieger Publishing Company, 1976, p. 57.

5.7. POWER SPECTRA

which is the total energy over the total time, or the average power. Therefore, the power spectrum is a measure of the "energy" of stochastic process $X(t)$. Specifically, it describes the distribution of the total mean-square value of the process over the frequency domain.

The Fourier transform pair in Equations 5.31 and 5.32 is known by the name *Wiener-Khinchine theorem*. This is a fundamental theorem in the theory of random processes. Note that this relation is valid for a weakly stationary random process. Figure 5.13 depicts some important pairs $S_{XX}(\omega) \Leftrightarrow R_{XX}(\tau)$. Where there is no chance of confusion, the subscripts used above can be omitted.

Cross-spectral densities are defined similarly,

$$S_{XY}(\omega) = \frac{1}{2\pi} \int_{-\infty}^{\infty} R_{XY}(\tau) e^{-i\omega\tau} d\tau = S_{YX}(\omega),$$

and

$$R_{XY}(\tau) = \int_{-\infty}^{\infty} S_{XY}(\omega) e^{i\omega\tau} d\omega.$$

Some examples of $R_{XX}(\tau)$ and $S_{XX}(\omega)$ that have broad physical application are presented in Examples 5.10 to 5.13.

Since the correlation function and the spectral density are even functions, we can use the Fourier cosine transform pair instead. Equations 5.31 and 5.32 can be written equivalently as

$$S_{XX}(\omega) = \frac{1}{\pi} \int_0^{\infty} R_{XX}(\tau) \cos\omega\tau \, d\tau, \qquad (5.34)$$

and

$$R_{XX}(\tau) = 2 \int_0^{\infty} S_{XX}(\omega) \cos\omega\tau \, d\omega. \qquad (5.35)$$

Example 5.8 Nonzero Mean Random Process

Consider a stationary random process $Y(t)$ with mean value μ_Y. First, define a random process $X(t) = Y(t) - \mu_Y$ such that the mean of $X(t)$ is zero. Compare the spectral densities of $X(t)$ and $Y(t)$.

Solution The autocorrelation function of $Y(t)$ in terms of $R_{XX}(\tau)$ is given by

$$\begin{aligned} R_{YY}(\tau) &= E\{Y(t)Y(t+\tau)\} \\ &= E\{(X(t) + \mu_Y)(X(t+\tau) + \mu_Y)\} \\ &= E\{X(t)X(t+\tau)\} + \mu_Y E\{X(t)\} + \mu_Y E\{X(t+\tau)\} + \mu_Y^2 \\ &= R_{XX}(\tau) + \mu_Y^2. \end{aligned}$$

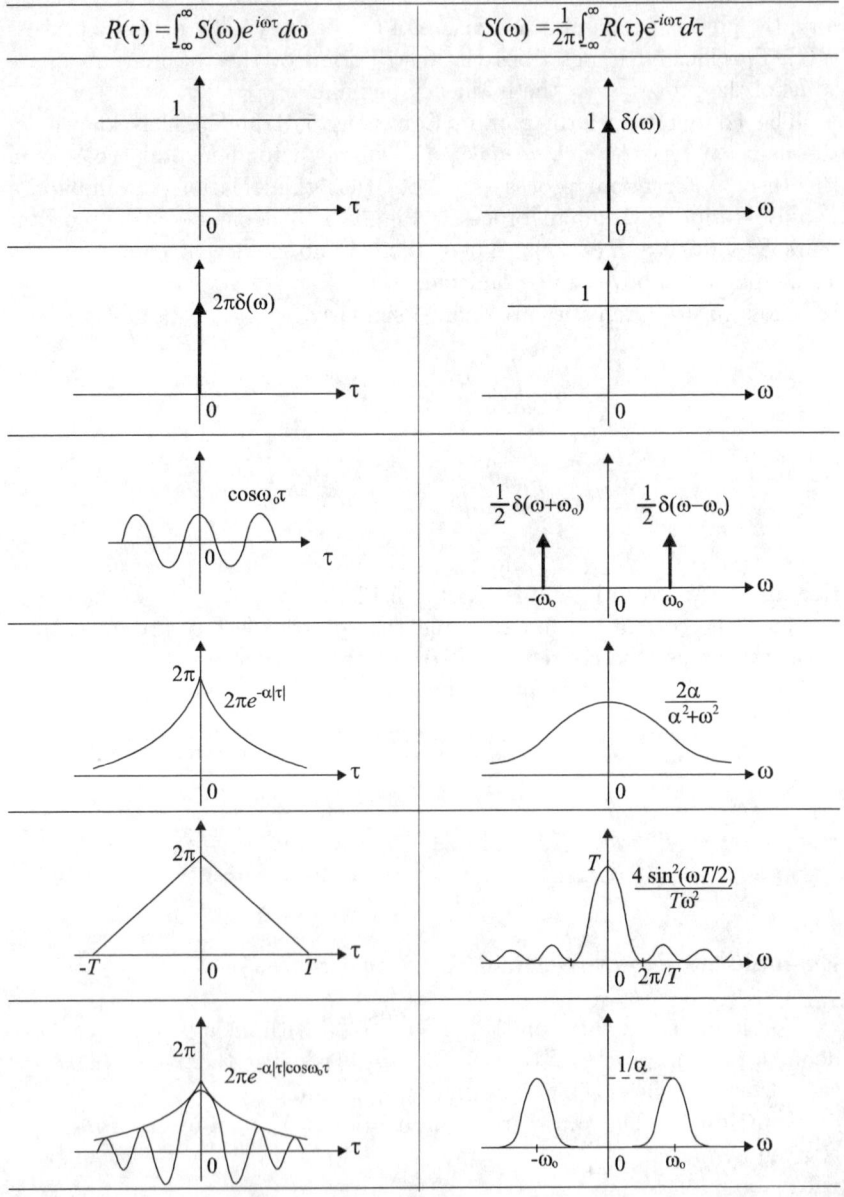

Figure 5.13: Relationship between the autocorrelation function and the power spectral density.

5.7. POWER SPECTRA

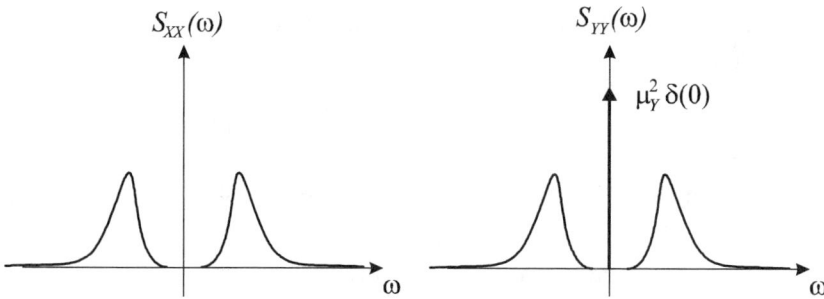

Figure 5.14: Spectral densities of a zero-mean random process $X(t)$, and of a nonzero-mean random process $Y(t)$.

Then, the spectral density is given by

$$\begin{aligned} S_{YY}(\omega) &= \frac{1}{2\pi} \int_{-\infty}^{\infty} R_{YY}(\tau) \exp(-i\omega\tau) \, d\tau \\ &= \frac{1}{2\pi} \int_{-\infty}^{\infty} \{R_{XX}(\tau) + \mu_Y^2\} \exp(-i\omega\tau) \, d\tau \\ &= S_{XX}(\omega) + \frac{1}{2\pi} \int_{-\infty}^{\infty} \mu_Y^2 \exp(-i\omega\tau) \, d\tau. \end{aligned}$$

The second term can be evaluated by considering the Fourier transform pair

$$\begin{aligned} \delta(\omega) &= \frac{1}{2\pi} \int_{-\infty}^{\infty} 1 \exp(-i\omega\tau) \, d\tau \\ 1 &= \int_{-\infty}^{\infty} \delta(\omega) \exp(i\omega\tau) \, d\omega, \end{aligned}$$

where $\delta(\omega)$ is the Dirac delta function and, therefore,

$$S_{YY}(\omega) = S_{XX}(\omega) + \mu_Y^2 \delta(\omega).$$

We have shown that the shape of the spectral densities are the same except at $\omega = 0$, where $\mu_Y^2 \delta(\omega)$ is added, as shown in Figure 5.14.
⊛

Example 5.9 Wiener-Khinchine Relationship

Derive the corresponding spectral densities for the following autocorrelation functions: *(i)* a pure sine function, $\sin \omega_o \tau$, *(ii)* $A \exp(-\alpha |\tau|) \cos \Omega \tau$, and *(iii)* $A \exp(-\alpha^2 \tau^2)$.

Solution The spectral density is related to the autocorrelation function by Equation 5.31,

$$S_{XX}(\omega) = \frac{1}{2\pi} \int_{-\infty}^{\infty} R_{XX}(\tau) e^{-i\omega\tau} d\tau.$$

(i) If the autocorrelation is a pure sine function, $S_{XX}(\omega)$ is given by the integral,

$$S_{XX}(\omega) = \frac{1}{2\pi} \int_{-\infty}^{\infty} \sin\omega_o\tau \, e^{-i\omega\tau} d\tau.$$

Using the Euler identity, $\sin\omega_o\tau = [\exp(i\omega_o\tau) - \exp(-i\omega_o\tau)]/2i$, the spectral density becomes

$$\begin{aligned}
S_{XX}(\omega) &= \frac{1}{2\pi} \int_{-\infty}^{\infty} \frac{1}{2i} [\exp(i\omega_o\tau) - \exp(-i\omega_o\tau)] e^{-i\omega\tau} d\tau \\
&= \frac{1}{2\pi} \int_{-\infty}^{\infty} \frac{1}{2i} \left(e^{-i(\omega-\omega_o)\tau} - e^{-i(\omega+\omega_o)\tau} \right) d\tau \\
&= \frac{1}{2i} [\delta(\omega - \omega_o) - \delta(\omega + \omega_o)].
\end{aligned}$$

(ii) If the autocorrelation is given by $A\exp(-\alpha|\tau|)\cos\Omega\tau$, the spectral density is given by

$$\begin{aligned}
S_{XX}(\omega) &\equiv \frac{1}{2\pi} \int_{-\infty}^{\infty} A\cos\Omega\tau \exp(-\alpha|\tau| - i\omega\tau) d\tau \\
&= \frac{A}{2\pi} \frac{1}{2} \left(\int_{-\infty}^{\infty} \exp(-\alpha|\tau| - i\omega\tau + i\Omega\tau) + \exp(-\alpha|\tau| - i\omega\tau - i\Omega\tau) d\tau \right) \\
&= \frac{A}{2\pi} \frac{1}{2} \int_0^{\infty} (\exp(-\alpha - i\omega + i\Omega)\tau + \exp(-\alpha - i\omega - i\Omega)\tau) d\tau \\
&\quad + \frac{A}{2\pi} \frac{1}{2} \int_{-\infty}^0 (\exp(\alpha - i\omega + i\Omega)\tau + \exp(\alpha - i\omega - i\Omega)\tau) d\tau \\
&= \frac{A}{2\pi} \frac{1}{2} \left(\frac{1}{\alpha + i\omega - i\Omega} + \frac{1}{\alpha - i\omega + i\Omega} + \frac{1}{\alpha + i\omega + i\Omega} + \frac{1}{\alpha - i\omega - i\Omega} \right) \\
&= \frac{A}{2\pi} \left(\frac{\alpha}{\alpha^2 + (\omega - \Omega)^2} + \frac{\alpha}{\alpha^2 + (\omega + \Omega)^2} \right).
\end{aligned}$$

For $\Omega = 0$, the spectral density is given by

$$S_{XX}(\omega) = \frac{A}{2\pi} \left(\frac{\alpha}{\alpha^2 + \omega^2} \right),$$

which coincides with the fourth plot in Figure 5.13. There are two distinct ranges of α. It can be shown that the spectral density has one peak at $\omega = 0$

5.7. POWER SPECTRA

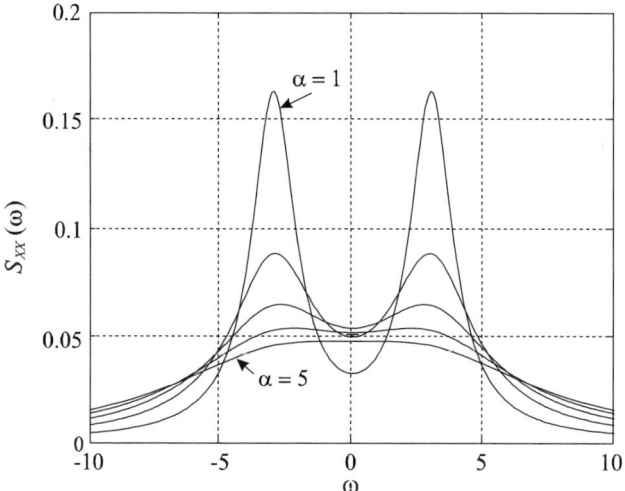

Figure 5.15: Spectral density for Example 5.9, with $A = 1$, $\Omega = 3$ rad/s and $\alpha = 1, 2, 3, 4,$ and 5.

when $\alpha^2 > 3\Omega$, with a peak value of

$$S_{XX}(0) = \frac{A}{\pi} \frac{\alpha}{\alpha^2 + \Omega^2}.$$

When $\alpha^2 < 3\Omega$, $S_{XX}(\omega)$ has two peaks, at

$$\omega_{peak} = \pm \left(\alpha^2 + \Omega^2\right)^{1/4} \left(2\Omega - \left(\alpha^2 + \Omega^2\right)^{1/2}\right)^{1/2},$$

with peak values

$$S_{XX}(\omega_{peak}) = \frac{A\alpha}{4\pi\Omega \left[(\alpha^2 + \Omega^2)^{1/2} - \Omega\right]}.$$

Figure 5.15 shows the spectral density for various values of α.

(iii) If the autocorrelation is given by $A\exp\left(-\alpha^2\tau^2\right)$, the spectral den-

sity is then

$$S_{XX}(\omega) \equiv \frac{1}{2\pi} \int_{-\infty}^{\infty} A \exp\left(-\alpha^2 \tau^2 - i\omega\tau\right) d\tau$$

$$= \frac{1}{2\pi} \int_{-\infty}^{\infty} A \exp\left(-\alpha^2 \left[\tau^2 + i\frac{\omega}{\alpha^2}\tau - \left(\frac{\omega}{2\alpha^2}\right)^2\right] - \alpha^2 \left(\frac{\omega}{2\alpha^2}\right)^2\right) d\tau$$

$$= \frac{1}{2\pi} \int_{-\infty}^{\infty} A \exp\left(-\alpha^2 \left(\tau + i\frac{\omega}{2\alpha^2}\right)^2 - \alpha^2 \left(\frac{\omega}{2\alpha^2}\right)^2\right) d\tau$$

$$= A \exp\left(-\alpha^2 \left(\frac{\omega}{2\alpha^2}\right)^2\right) \cdot \frac{1}{2\pi} \int_{-\infty}^{\infty} \exp\left(-\alpha^2 \left(\tau + i\frac{\omega}{2\alpha^2}\right)^2\right) d\tau.$$

The integral can be evaluated if we let $\alpha\left(\tau + i\omega/2\alpha^2\right) = t$,

$$\frac{1}{2\pi} \int_{-\infty}^{\infty} \exp\left(-\alpha^2 \left(\tau + i\frac{\omega}{2\alpha^2}\right)^2\right) d\tau = \frac{1}{2\pi} \int_{-\infty}^{\infty} \exp\left(-t^2\right) \frac{1}{\alpha} dt$$

$$= \frac{1}{2\pi} \frac{1}{\alpha} \sqrt{\pi},$$

where $\int_{-\infty}^{\infty} \exp\left(-t^2\right) dt = \sqrt{\pi}$. The spectral density is then given by

$$S_{XX}(\omega) = \frac{A}{2\alpha\sqrt{\pi}} \exp\left(-\frac{\omega^2}{4\alpha^2}\right).$$

⊛

Example 5.10 Ocean Wave Spectra

Ocean waves generated by wind are modeled as random processes. A much used spectral density of ocean wave elevation $\eta(t)$ is the Pierson-Moskowitz spectrum,

$$S_{\eta\eta}(\omega) = \frac{0.0081 g^2}{\omega^5} \exp\left[-0.74 \left(\frac{g}{V\omega}\right)^4\right] \text{ m}^2\text{s}, \quad (5.36)$$

where $\omega > 0$, g is the gravitational constant, and V is the wind speed at a height of 19.5 m above the still water level. Any consistent set of units for g and V can be used, with ω in rad/s. See Figure 5.16. The Pierson-Moskowitz is an experimentally determined spectrum, as are most if not all spectra used in applications.

Recall that the fluid particle velocities and accelerations can be obtained from the wave height elevation. There are other versions of this equation as well as other spectral densities in use. Each is specific to a particular part of the ocean for a particular time of year. See Chapter 13 for details

5.7. POWER SPECTRA

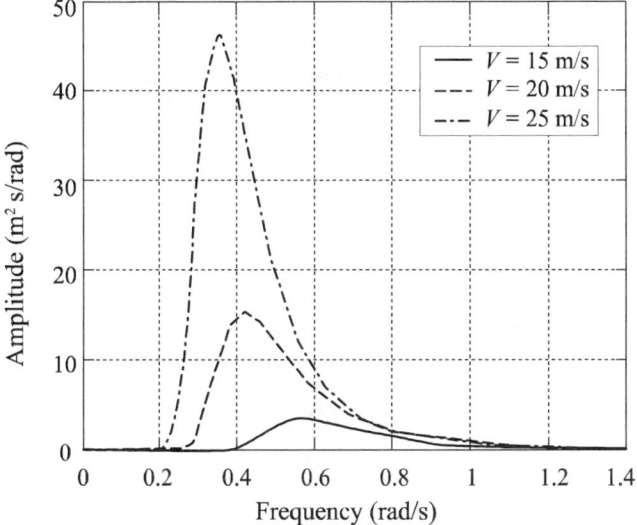

Figure 5.16: The Pierson-Moskowitz ocean wave height spectrum for three wind speeds V.

on such spectra as well as an introduction into the analysis of structures in the ocean environment.

❊

Example 5.11 Response to Ocean Waves

Consider the taut mooring shown in Figure 5.17, where the vertical member is a highly elastic synthetic rope with extensional rigidity, EA, and original length, L. Obtain the tension spectrum and the RMS tension in the rope assuming the buoy moves vertically with displacement $X(t)$ following the wave profile $\eta(t)$. The spectrum for the wave profile is the Pierson-Moskowitz in Equation 5.36 or, in terms of ω for $V = 10$ m/s,

$$S_{\eta\eta}^o(\omega) = \frac{0.78}{\omega^5} \exp\left[-\frac{0.69}{\omega^4}\right] \quad \text{m}^2\text{s}.$$

Solution Since the vertical motion of the buoy follows the wave profile, the spectral densities are also identical, $S_{XX}^o(\omega) = S_{\eta\eta}^o(\omega)$. The force is related to the displacement of the buoy by $F = (EA/L)X$, and thus the tension spectrum is given by $S_{FF}^o(\omega) = (EA/L)^2 S_{\eta\eta}^o(\omega)$. The

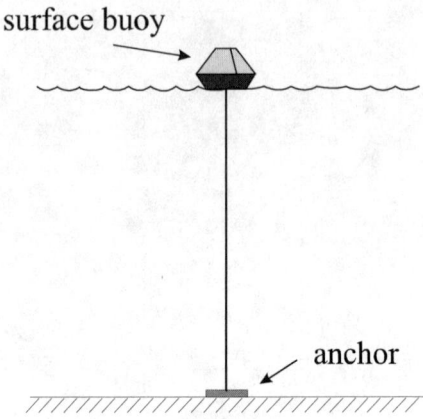

Figure 5.17: Taut mooring.

RMS tension is given by

$$E\{F^2\} = \int_0^\infty S_{FF}^o(\omega)\,d\omega$$
$$= \left(\frac{EA}{L}\right)^2 \int_0^\infty \frac{0.78}{\omega^5} \exp\left[-\frac{0.69}{\omega^4}\right] d\omega.$$

After integrating numerically using the trapezoidal rule, the RMS tension is found to equal

$$E\{F^2\} = \frac{0.28EA}{L} \text{N}^2.$$

This result is valuable for designing purposes in the following way. Knowing $E\{F^2\}$ and μ_F, the variance σ_F^2 can be evaluated and used to design the cable within prescribed sigma bounds. This procedure, usually iterative, can lead to modification of A or E.

⊛

Example 5.12 Wind Spectrum

All ground and mobile structures must be designed to withstand wind forces. One can classify structures subjected to wind as either streamlined or bluff bodies. Streamlined structures such as wings have a high aspect ratio (ratio of one dimension to another) and are shaped to optimally follow a streamline. They are designed so that the interaction of their shape with the wind results in a desirable configuration of forces. Bluff bodies such as tall buildings or chimneys are generally of low aspect ratio and may have

5.7. POWER SPECTRA

corners or sharp edges. They are designed more for strength than for fitting within the streamlines.

An example of a wind spectrum for horizontal velocity is[7]

$$S_v(f) = 4\kappa \overline{V}_{10}^2 \frac{L/\overline{V}_{10}}{\left(2+\overline{f}\right)^{5/6}} \left(\frac{\text{ft}}{\text{s}}\right)^2 \text{s}, \qquad (5.37)$$

where f is the frequency in Hz, $\overline{f} = fL/\overline{V}_{10}$ is a dimensionless frequency, L is a length scale[8] of approximately 4000 ft, $\overline{V}_{10}$ is the mean wind speed at 10 ft above the ground, and κ is a number in the range $0.005 \leq \kappa \leq 0.05$ that depends on wind elevation expected in the region. For $\kappa = 0.01$, $L = 4000$ ft, and $\overline{V}_{10} = 50, 100, 150$ ft/s, the wind spectrum is plotted in Figure 5.18.

As with the wave height spectra, wind characteristics have wide variability depending on location.

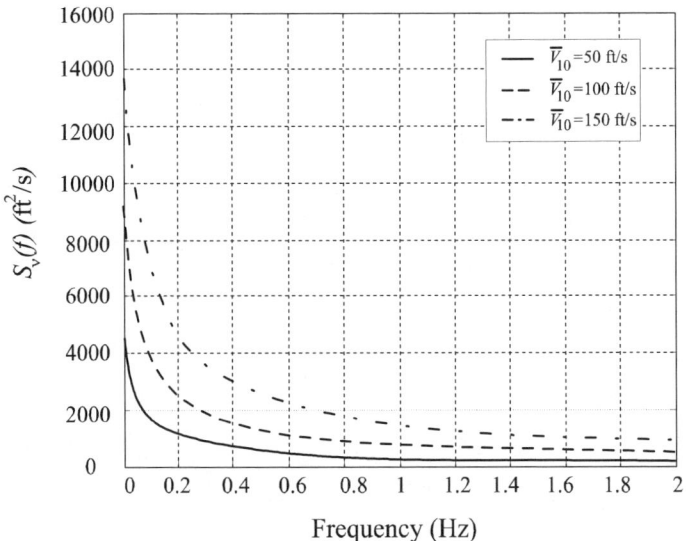

Figure 5.18: The wind spectral density, $S_V(f)$ for three different wind speeds $\overline{V}_{10}$ for $\kappa = 0.01$ and $L = 4000$ ft.

---

[7] See A.G. Davenport and M. Novak, in *Shock & Vibration Handbook*, Chapter 29, p. 23, C.M. Harris, Editor, McGraw-Hill, 1988.

[8] A length scale is a dimension that is representative of the process being modeled. For example, if the wavelength of a harmonic force is λ, then this is a length scale of the problem. Similarly, structural dimensions can be used as length scales.

242   CHAPTER 5. RANDOM PROCESSES

Example 5.13 Earthquake Spectrum

For earthquakes, the difficulty in specifying ground motion spectra for use as input to the structure is due to the significant variability in soil and geologic properties, even for two sites very near each other. We have all examined photos of earthquake-related devastation and noted how for two adjacent similar structures, one has survived relatively unscathed while the other has suffered severe damage. Differences can be due to even minor differences in the structures and the soil/foundation dynamic characteristics. The El Centro, California, earthquake[9] of May 18, 1940, has been used as input to numerous structural designs, to verify that they will survive anticipated ground motion.

The dynamic analysis of a structure in a seismic zone requires the estimation of the mechanical properties of the surrounding soil. Figure 5.19 is a schematic of such a structure, the displacement $x_i(t)$ of the incident wave, the resulting ground displacement $x_g(t)$ at the structure boundary condition, and the equivalent soil stiffness and damping properties, k_g and c_g, respectively.

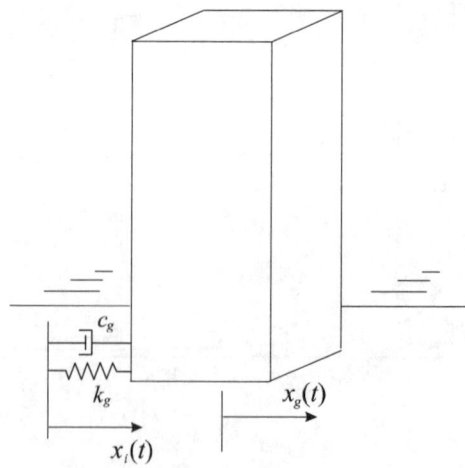

Figure 5.19: Model for structure and soil coupling.

An example of a quantitative relation between the spectrum of the incident disturbance $S_i(\omega)$ and the spectrum directly input to the structure

---
[9] See W.J. Hall, in Harris, Chapter 24, p. 5.

5.7. POWER SPECTRA

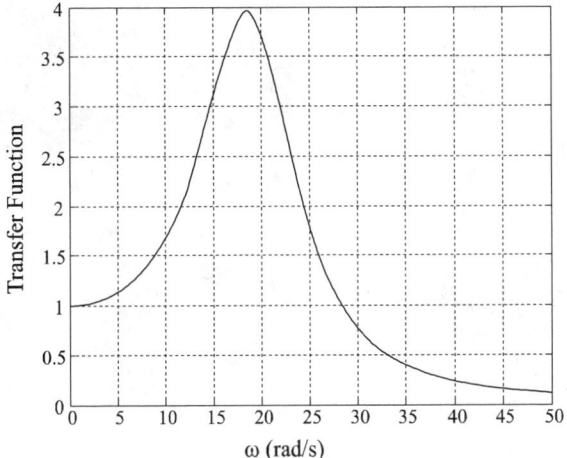

Figure 5.20: Transfer function using $\omega_g = 20$ rad/s and $\zeta_g = 0.3$.

through the ground directly beneath $S_g(\omega)$ is the Kanai-Tajimi spectrum,

$$S_g(\omega) = \left[\frac{\omega_g^4 + 4\zeta_g^2 \omega_g^2 \omega^2}{\left(\omega_g^2 - \omega^2\right)^2 + 4\zeta_g^2 \omega_g^2 \omega^2} \right] S_i(\omega),$$

where ω_g and ζ_g are the natural frequency and damping ratio for the soil layer at the site. Example mean values for these parameters are $\omega_g \simeq 20$ rad/s and $\zeta_g \simeq 0.3$. Figure 5.20 shows the transfer function (the term in the square brackets) using these mean values. The curve shows that energy in the frequency band $0 - 28$ Hz is amplified, with transfer function values greater than 1, and for frequencies above about 28 Hz the energy is filtered out.

Wiener Norbert Wiener was greatly influenced by his father, Leo Wiener, who was a Russian Jew, and so we should give some background regarding his father's education and career. Leo Wiener attended medical school at the University of Warsaw but was unhappy with the profession, so he went to Berlin where he began training as an engineer. This profession seemed only a little more interesting to him than the medical profession, and he emigrated to the United States having first landed in England. We should note that throughout his education Leo was interested

Figure 5.21: Norbert Wiener (1894-1964).

in mathematics and, although he never used his mathematical skills in any jobs he held, it was a deep amateur interest to him all through his life.

Arriving in New Orleans in 1880, Leo tried his hand at various jobs in factories and farms before becoming a school teacher in Kansas City. He progressed from being a language teacher in schools to becoming Professor of Modern Languages at the University of Missouri. While there he met and married Bertha Kahn, who was the daughter of a department store owner. Bertha, from a German-Jewish family, was ... *a small woman, healthy, vigorous and vivacious.* She joined her husband in the boarding house in Columbia, Missouri, where their son Norbert was born in the following year.

Not long after Norbert's birth a decision was taken to split the Modern Languages Department at the University of Missouri into separate departments of French and German. Leo was to join the German Department after the split but he lost out in some political maneuvering so the family left Columbia and moved to Boston. There Leo brought in money by taking a variety of teaching and other positions and eventually was appointed as an Instructor in Slavic Languages at Harvard. This did not pay well enough to provide for his family, so Leo kept various other positions to augment his salary. He remained at Harvard University for the rest of his career, being eventually promoted to professor.

As a young child Norbert had a nursemaid. When he was about four years old, a second child Constance was born; Wiener's second sister was

5.7. POWER SPECTRA

born in 1901. He writes about his upbringing: *I was brought up in a house of learning. My father was the author of several books, and ever since I can remember, the sound of the typewriter and the smell of the paste pot have been familiar to me.... I had full liberty to roam in what was the very catholic and miscellaneous library of my father. At one period or other the scientific interests of my father had covered most of the imaginable subjects of study.... I was an omnivorous reader....*

Wiener had problems regarding his schooling, partly because the reading which he had done at home had meant that he was advanced in certain areas but much less so in others. His parents sent him to the Peabody School when he was seven years old and, after worrying about which class he should enter, had him begin in the third grade. After a short time his parents and teachers felt he would be better suited to the fourth grade and he was moved up a year. However, he certainly did not fit into the school in either grade and his teacher had little sympathy with so young a boy in the fourth grade yet lacking certain skills that would be expected of the pupils at this stage in their education. He writes: *My chief deficiency was arithmetic. Here my understanding was far beyond my manipulation, which was definitely poor. My father saw quite correctly that one of my chief difficulties was that manipulative drill bored me. He decided to take me out of school and put me on algebra instead of arithmetic, with the purpose of offering a greater challenge and stimulus to my imagination.*

From this time on Wiener's father took over his education and he made rapid progress for so young a child. However, Wiener had problems relating to his movements and was obviously very clumsy. This stemmed partly from poor coordination but also partly from poor eyesight. Advised by a doctor to stop reading for six months to allow his eyes to recover, he still had regular lessons from his father, who now taught him to do mathematics in his head. After the six months were up Wiener went back to reading but he had developed some fine mental skills during this period that he retained all his life.

In the autumn of 1903, at age nine, he was sent to school again, this time to Ayer High School. The school agreed to experiment and to find the right level for Wiener, who was soon put into the senior-year class with pupils who were seven years older than he was. The school only formed part of his education, however, for his father continued to coach him. He graduated in 1906 from Ayer at the age of eleven and celebrated with his eighteen-year-old fellow students: *I owe a great deal to my Ayer friends. I was given a chance to go through some of the gawkiest stages of growing up in an atmosphere of sympathy and understanding.*

In September 1906, still only eleven years old, Wiener entered Tufts College. Socially a child, he was an adult in educational terms so his student days were not easy ones. Although taking various science courses, he took a degree in mathematics. Wiener's father continued to coach him in mathematics, leading to complete mastery of undergraduate-level topics. In 1909 Wiener graduated from Tufts at age fourteen and entered Harvard to begin graduate studies.

Rather against his father's advice, Wiener began graduate studies in zoology at Harvard. However, things did not go too well and by the end of a year a decision was taken, partly by Wiener, partly by his father, that he would change topics to philosophy. Having won a scholarship to Cornell he entered in 1910 to begin graduate studies in philosophy. Taking mathematics and philosophy courses, Wiener did not have a successful year and before it was finished his father had made the necessary arrangements to return to Harvard to continue philosophy.

Back at Harvard Wiener was strongly influenced by the fine teaching of Edward Huntington on mathematical philosophy. He received his Ph.D. from Harvard at the age of 18 with a dissertation on mathematical logic supervised by Karl Schmidt. From Harvard Wiener went to Cambridge, England, to study under Russell who told him that in order to study the philosophy of mathematics he needed to know more mathematics, so he attended courses by G. H. Hardy. In 1914 he went to Göttingen to study differential equations under Hilbert, and also attended a group theory course by Edmund Landau. He was influenced by Hilbert, Landau, and Russell but also, perhaps to an even greater degree, by Hardy. At Göttingen he learned that ... *mathematics was not only a subject to be done in the study but one to be discussed and lived with.*

Wiener returned to the United States a couple of days before the outbreak of World War I, but returned to Cambridge to study further with Russell. Back in the United States he taught philosophy courses at Harvard in 1915, worked for a while for the General Electric Company, then joined Encyclopedia Americana as a staff writer in Albany. While working there he received an invitation from Veblen to undertake war work on ballistics at the Aberdeen Proving Ground in Maryland. Taking about mathematics with his fellow workers while undertaking this war work revived his interest in mathematics. At the end of the war Osgood told him of a vacancy at MIT and he was appointed as an instructor in mathematics.

His first mathematical work at MIT led him to examine Brownian motion. In fact, as Wiener explained, this first work would provide a connecting thread through much of his later studies: ... *this study introduced me to the theory of probability. Moreover, it led me very directly to the periodogram, and to the study of forms of harmonic analysis more general than the classical Fourier series and Fourier integral. All these concepts have combined with the engineering preoccupations of a professor of the Mas-*

5.7. POWER SPECTRA

sachusetts Institute of Technology to lead me to make both theoretical and practical advances in the theory of communication, and ultimately to found the discipline of cybernetics, which is in essence a statistical approach to the theory of communication. Thus, varied as my scientific interests seem to be, there has been a single thread connecting all of them from my first mature work.

He attended the International Congress of Mathematicians at Strasbourg in 1920 and while there worked with Fréchet. He returned to Europe frequently in the next few years, visiting mathematicians in England, France, and Germany. Especially important was his contacts with Paul Lévy and with Göttingen where his work was seen to have important connections with quantum mechanics. This led to collaboration with Born.

In 1926 Wiener married Margaret Engemann, and after their marriage Wiener set off for Europe as a Guggenheim scholar. After visiting Hardy in Cambridge he returned to Göttingen where his wife joined him after completing her teaching duties in modern languages at Juniata College in Pennsylvania.

Another important year in Wiener's mathematical development was 1931-1932, which he spent mainly in England visiting Hardy at Cambridge. There he gave a lecture course on his own contributions to the Fourier integral, but Cambridge also provided a base from where he was able to visit many mathematical colleagues on the Continent. Among these were Blaschke, Menger, and Frank, who invited him to make a visit, while he also met Hahn, Artin, and Gödel.

Wiener's papers were hard to read. Sometimes difficult results appeared with hardly a proof as if they were obvious to Wiener, while at other times he would give a lengthy proof of a triviality. Freudenthal writes: *All too often Wiener could not resist the temptation to tell everything that cropped up in his comprehensive mind, and he often had difficulty in separating the relevant mathematics neatly from its scientific and social implications and even from his personal experiences. The reader to whom he appears to be addressing himself seems to alternate in a random order between the layman, the undergraduate student of mathematics, the average mathematician, and Wiener himself.*

Despite the style of his papers, Wiener contributed some ideas of great importance. We have already mentioned above his work in 1921 in Brownian motion. He introduced a measure in the space of one-dimensional paths, which brings in probability concepts in a natural way. From 1923 he investigated Dirichlet's problem, producing work that had a major influence on potential theory.

Wiener's mathematical ideas were very much driven by questions that were put to him by his engineering colleagues at MIT. These questions pushed him to generalize his work on Brownian motion to more general stochastic processes. This in turn led him to study harmonic analysis in 1930. His work on generalized harmonic analysis led him to study Tauberian theorems in 1932. His contributions on this topic won him the Bôcher Prize in 1933. He received the prize from the American Mathematical Society for his memoir Tauberian theorems published in *Annals of Mathematics* in the previous year. The work on Tauberian theorems naturally led him to study the Fourier transform and he published *The Fourier Integral, and Certain of Its Applications* (1933) and *Fourier Transforms* in 1934.

Wiener had an extraordinarily wide range of interests and contributed to many areas in addition to those we have mentioned above including communication theory, cybernetics (a term he coined), quantum theory and during World War II he worked on gunfire control. It is probably this latter work that motivated his invention of the new area of cybernetics, which he described in *Cybernetics: or, Control and Communication in the Animal and the Machine* (1948). Freudenthal writes: *While studying antiaircraft fire control, Wiener may have conceived the idea of considering the operator as part of the steering mechanism and of applying to him such notions as feedback and stability, which had been devised for mechanical systems and electrical circuits.... As time passed, such flashes of insight were more consciously put to use in a sort of biological research ... [Cybernetics] has contributed to popularizing a way of thinking in communication theory terms, such as feedback, information, control, input, output, stability, homeostasis, prediction, and filtering. On the other hand, it also has contributed to spreading mistaken ideas of what mathematics really means.*

Wiener himself was aware of these dangers and his wide dealings with other scientists led him to say: *One of the chief duties of the mathematician in acting as an adviser to scientists is to discourage them from expecting too much from mathematics.*

Some of Wiener's publications that we have not mentioned include *Nonlinear Problems in Random Theory* (1958), and *God and Golem, Inc.: A Comment on Certain Points Where Cybernetics Impinges on Religion* (1964).

We have mentioned above Freudenthal's comments on Wiener's poor writing style. His most famous work, *Cybernetics*, comes in for special criticism by Freudenthal: *Even measured by Wiener's standards "Cybernetics" is a badly organized work – a collection of misprints, wrong mathematical statements, mistaken formulas, splendid but unrelated ideas, and logical absurdities. It is sad that this work earned Wiener the greater part of his public renown, but this is an afterthought. At that time mathematical readers were more fascinated by the richness of its ideas than by its shortcomings.*

5.7. POWER SPECTRA

Freudenthal describes both Wiener's appearance and his character: *In appearance and behavior, Norbert Wiener was a baroque figure, short, rotund, and myopic, combining these and many qualities in extreme degree. His conversation was a curious mixture of pomposity and wantonness. He was a poor listener. His self-praise was playful, convincing, and never offensive. He spoke many languages but was not easy to understand in any of them. He was a famously bad lecturer.*

D. G. Kendall writes: *As a human being Wiener was above all stimulating. I have known some who found the stimulus unwelcome. He could offend publicly by snoring through a lecture and then asking an awkward question in the discussion, and also privately by proffering information and advice on some field remote from his own to an august dinner companion. I like to remember Wiener as I once saw him late at night in Magdalen College, Oxford, surrounded by a spellbound group of undergraduates, talking, endlessly talking.*

Figure 5.22: Aleksandr Yakovlevich Khinchine (1894-1959).

Khinchine Aleksandr Yakovlevich Khinchine, whose father was an engineer, attended the technical high school in Moscow where he became fascinated by mathematics. However, mathematics was certainly not his only interest when he was at secondary school, for he also had a passionate love of poetry and of the theatre. He completed his secondary education in 1911 and entered the Faculty of Physics and Mathematics of Moscow University in that year.

At university in Moscow, Khinchine worked with Luzin and others. He was an outstanding student being particularly interested in the metric theory of functions and before he graduated in 1916 he had already written his first paper on a generalization of the Denjoy integral. This first paper began a series of publications by Khinchine on properties of functions that are retained after deleting a set of density zero at a given point. He summarized his contributions to this area with the paper, *Recherches sur la structure des fonctions measurables*, in *Fundamanta mathematica* in 1927.

After graduating in 1916, Khinchine remained at Moscow University undertaking research for his dissertation, which would allow him to become a university teacher. After a couple of years he began teaching in a number of different colleges both in Moscow and Ivanovo. The town of Ivanovo, east of Moscow, was a center for the textile industry and it plays a surprisingly important part in the development of Russian mathematics with several of the major figures teaching in the town.

Around 1922 Khinchine took up new mathematical interests when he began to study the theory of numbers and probability theory. In the following year he strengthened results of Hardy and Littlewood with his introduction of the iterated logarithm, published in *Mathematische Zeitschrift*. With these ideas he also strengthened the law of large numbers due to Borel.

In 1927 Khinchine was appointed as a professor at Moscow University and, in the same year, he published *Basic Laws of Probability Theory*. Between 1932 and 1934 he laid the foundations for the theory of stationary random processes culminating in a major paper in *Mathematische Annalen* in 1934. Khinchine left Moscow in 1935 to spend two years at Saratov University but returned to Moscow University in 1937 to continue his role of building the school of probability theory there in partnership with Kolmogorov and others, including in particular their student Gnedenko. From the 1940s his work changed direction again and this time he became interested in the theory of statistical mechanics. In the last few years of his life his interests turned to developing Shannon's ideas on information theory.

We shall look at some of Khinchine's major publications and in this way get a feel for the large number of important contributions he made in a remarkably large range of topics. Some of these publications we have already mentioned in the brief description of his career that we gave above.

Khinchine first published the book, *Continued Fractions*, in 1936 with a second edition being published in 1949. The book consists of three chapters, the first two of which present the classical theory of continued fractions. The third chapter, the longest and most important, contains an account of Khinchine's own contributions to the topic of the metrical theory of Diophantine approximations. Another contribution by Khinchine to number theory is the short book, *Three Pearls of Number Theory*, which appeared in an English translation in 1952.

5.7. POWER SPECTRA

The book, *Eight Lectures on Mathematical Analysis*, by Khinchine ran to several editions. It was first published in 1943 and the eight lectures it contains are: continuum; limits; functions; series; derivative; integral; series expansions of functions; and differential equations. The book was designed to be used to supplement a standard course on the calculus and gives a careful treatment of some of the basic notions of mathematical analysis. Ivanov, reviewing the fourth edition, wrote: *The presentation is smooth, elegant, and interesting and makes very enjoyable reading.*

Khinchine published *Mathematical Principles of Statistical Mechanics* in 1943. It showed how to make classical statistical mechanics into a mathematically rigorous subject, developing a consistent presentation of the topic. In 1951 he extended the work of this 1943 book when he published *Mathematical Foundations of Quantum Statistics*. This new publication on the topic appeared in a German translation in 1956 and then in an English translation in 1960. The book was written in such a way as to be useful both to mathematicians who wanted to become better acquainted with some applications of analysis to physics, and also to physicists who wanted to understand more about the mathematical foundations for their subject. Topics covered included: local limit theorems for sums of identically distributed random variables; the foundations of quantum mechanics; general principles of quantum statistics; the foundations of the statistics of photons; entropy; and the second law of thermodynamics. The book has been rated as being equal in quality to von Neumann's masterpiece *Mathematical Foundations of Quantum Mechanics*.

Khinchine's book, *Mathematical Foundations of Information Theory*, translated into English from the original Russian in 1957, is important. It consists of English translations of two articles: "The entropy concept in probability theory," and "On the basic theorems of information theory," which were both published earlier in Russian. The second of these articles provides a refinement of Shannon's concepts of the capacity of a noisy channel and the entropy of a source. Khinchine generalized some of Shannon's results in this book ,which was written in an elementary style yet gave a comprehensive account with full details of all the results.

Among the many honors which Khinchine received for his work was election to the Soviet Academy of Sciences in 1939 and the award of a State Prize for scientific achievements in the following year.

Vere-Jones writes: *Khinchine was a fascinating figure ..., not least because of his early enthusiasms for poetry and acting, and his links with such figures of the revolution as the poet Mayakovsky and members of the Moscow Arts Theatre.*

Power Spectrum Units

At first sight, the power spectrum is a strange creation. We claim that it represents a real physical process, but Equation 5.32,

$$R_{XX}(\tau) = \int_{-\infty}^{\infty} S_{XX}(\omega) e^{i\omega\tau} d\omega,$$

is an integral over negative frequencies! Of course, these negative frequencies do not exist except mathematically, just as the complex exponential is used for mathematical ease with the understanding that only the real or imaginary parts are retained. Therefore, we often consider one-sided spectral densities, denoted as $S_{XX}^o(\omega)$.

One-sided spectra can be related to the standard spectral density by equating respective areas. The shaded areas shown in Figure 5.23(a) and (b) must be equivalent so that the total power is same in all cases, also that equal probabilities result. That is,

$$2S_{XX}(\omega)(\omega_o + d\omega - \omega_o) = S_{XX}^o(\omega)(\omega_o + d\omega - \omega_o).$$

Therefore,

$$S_{XX}^o(\omega) = 2S_{XX}(\omega).$$

The frequency ω has units of rad/s. In applications, frequencies f with units of cycles/s or Hz (Hertz) are more common. These two approaches can be simply accommodated by the relation[10] $f = \omega/2\pi$. Figure 5.23(c) shows the two-sided spectrum as a function of f, $S_{XX}(f)$. By equating shaded areas in Figure 5.23(a) and (c), we find,

$$S_{XX}(\omega) d\omega = S_{XX}(f) df.$$

Making use of $2\pi df = d\omega$,

$$S_{XX}(f) = 2\pi S_{XX}(\omega).$$

Figure 5.23(d) shows an equivalent one-sided spectral density function $W_{XX}(f)$. Equating shaded areas in Figure 5.23(a) and (d), we find

$$\begin{aligned} 2S_{XX}(\omega) d\omega &= W_{XX}(f) df \\ W_{XX}(f) &= 4\pi S_{XX}(\omega). \end{aligned} \quad (5.38)$$

In most applications, since these depend on experimental data, we are more likely to see the one-sided density function. The wave height spectrum

---

[10] 2π radians equals one Hz.

5.7. POWER SPECTRA

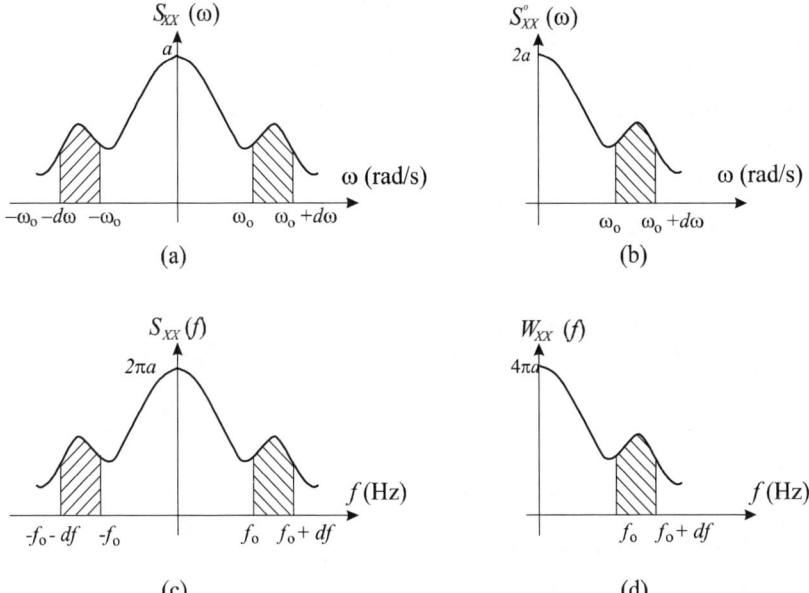

Figure 5.23: Two-sided spectra in rad/s and Hz, and their equivalent one-sided spectra.

(rad/s) and wind spectrum (Hz) in Figures 5.16 and 5.18 are in reality one-sided. It should be noted that $S_{XX}(\omega)$ has the units of X^2/ω, and $W_{XX}(f)$ of X^2/f.

The Wiener-Khinchine relations written as Fourier cosine transforms in Equations 5.34 and 5.35 can be written in terms of the one-sided density,

$$S_{XX}^o(\omega) = \frac{2}{\pi} \int_0^\infty R_{XX}(\tau) \cos \omega \tau \, d\tau$$

$$R_{XX}(\tau) = \int_0^\infty S_{XX}^o(\omega) \cos \omega \tau \, d\omega.$$

The latter relation coincides with Equation 5.28.

5.7.1 Narrow- and Broad-Band Processes

In engineering applications, certain types of random processes tend to be frequently found. Since the power spectrum is representative of the distribution of dynamic energy as a function of frequency, then it is possible to define general categories of vibration according to how energy is distributed. A structure vibrating at a single constant frequency ω_0 can be represented in the time and frequency domains as shown in Figure 5.24. $X(\omega)$ is the Fourier transform of the random process $X(t)$ given in Equation 5.29.

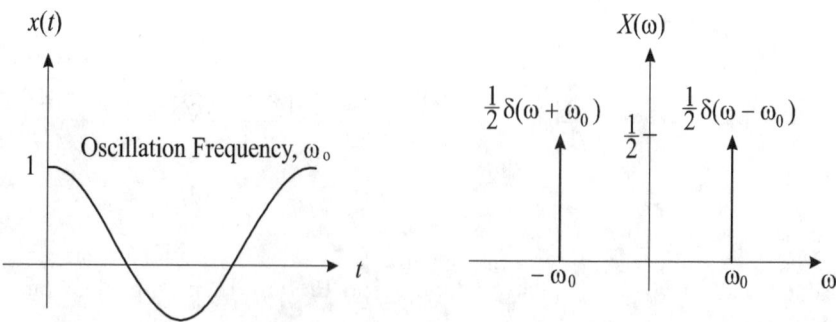

Figure 5.24: Deterministic process $x(t)$ in time and frequency $X(\omega)$ domains.

Two very important types of processes are *narrow-band* (Figure 5.25) and *broad-band* (Figure 5.26). Narrow and broad indicate the spread of the respective frequency bands.

A *narrow-band* process is an "almost" harmonic oscillator. Instead of vibrating at one distinct frequency as does a harmonic oscillator, it vibrates with frequencies in a narrow range: $\omega_1 \leq \omega \leq \omega_2$. The spectral density of

5.7. POWER SPECTRA

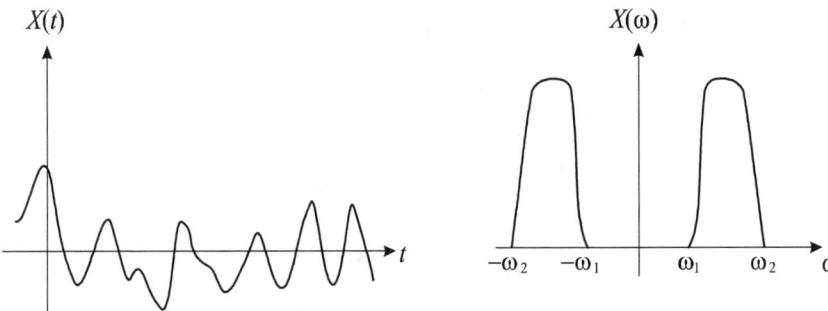

Figure 5.25: Narrow-band random process $X(t)$ in time and frequency $X(\omega)$ domains.

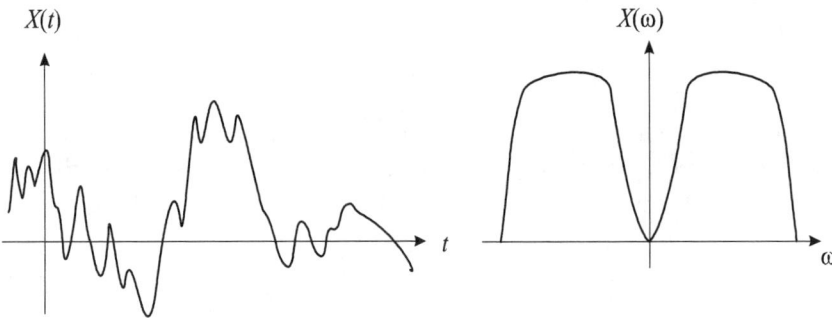

Figure 5.26: Broad-band random process $X(t)$ in time and frequency $X(\omega)$ domains.

a narrow-band process can be idealized by the spectrum in Figure 5.27. It has a flat spectrum in the frequency band, with constant magnitude S_0.

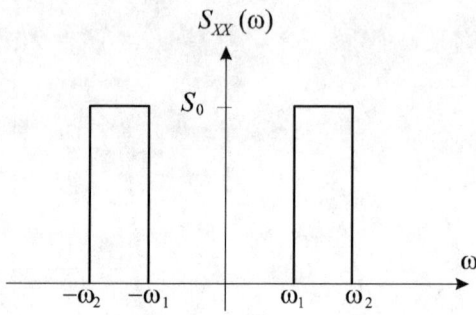

Figure 5.27: Spectral density of an idealized narrow-band process.

The autocorrelation function for such a process is evaluated as follows,

$$\begin{aligned} R_{XX}(\tau) &= \int_{-\infty}^{\infty} S_{XX}(\omega) e^{i\omega\tau} d\omega \\ &= 2 \int_{\omega_1}^{\omega_2} S_0 \cos \omega\tau \, d\omega, \end{aligned}$$

where the real part of the complex exponential is retained having made use of the symmetry of the power spectrum function. The integral is evaluated to give

$$R_{XX}(\tau) = 2\frac{S_0}{\tau} (\sin \omega_2 \tau - \sin \omega_1 \tau). \quad (5.39)$$

Note that the autocorrelation function consists of two harmonic functions at frequencies ω_1 and ω_2. When the frequencies are close to each other, beating is observed. This is clearer when Equation 5.39 is written as

$$R_{XX}(\tau) = 4\frac{S_0}{\tau} \cos\left\{\left(\frac{\omega_1 + \omega_2}{2}\right)\tau\right\} \sin\left\{\left(\frac{\omega_2 - \omega_1}{2}\right)\tau\right\}. \quad (5.40)$$

The mean-square value can be obtained by taking the following limit of Equation 5.40,

$$\begin{aligned} E\{X^2(t)\} &= R_{XX}(0) \\ &= \lim_{\tau \to 0} 4\frac{S_0}{\tau} \cos\left\{\left(\frac{\omega_1 + \omega_2}{2}\right)\tau\right\} \sin\left\{\left(\frac{\omega_2 - \omega_1}{2}\right)\tau\right\} \\ &= 2S_0(\omega_2 - \omega_1). \end{aligned}$$

5.7. POWER SPECTRA

Of course, in this case it is very easy to evaluate the mean square by just calculating the area under the spectral density curve,

$$E\{X^2(t)\} = \int_{-\infty}^{\infty} S_{XX}(\omega)\,d\omega = 2S_0(\omega_2 - \omega_1).$$

Figure 5.28 shows the autocorrelation function for an ideal narrow band process using $S_0 = 2$ m^2/s, $\omega_1 = 3$ rad/s, and $\omega_2 = 3.5$ rad/s.

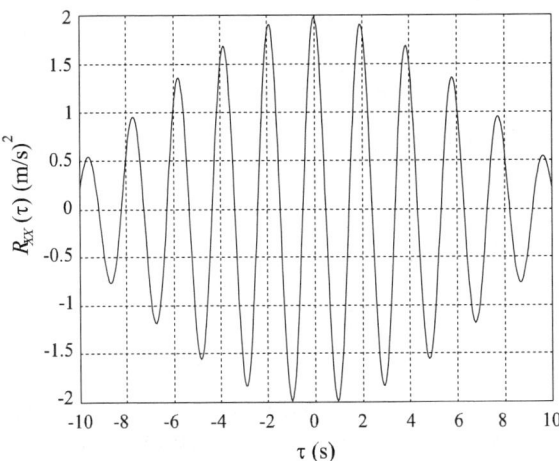

Figure 5.28: The autocorrelation function for an ideal narrow-band process. $S_o = 2$ m^2/s, $\omega_1 = 3$ rad/s, and $\omega_2 = 3.5$ rad/s.

A *broad-band* process is one that contains a significant range of frequencies, and therefore contains a wider range of amplitudes. The above results for the narrow-band process are valid here as well, except now $\omega_2 - \omega_1$ is a larger range. Figure 5.29 shows the autocorrelation function for $S_0 = 2$ m^2/s, $\omega_1 = 0$ rad/s, and $\omega_2 = 10$ rad/s.

Note that $R_{XX}(\tau)$ of a broad-band process has a distinguished peak at $\tau = 0$, but diminishes quickly as τ increases. This implies that the functions $X(t)$ and $X(t+\tau)$ have less and less in common, as τ increases.

5.7.2 White Noise Processes

A *white noise* process is an idealization made for mathematical expediency. Assuming that a process is white noise greatly simplifies the necessary algebra of an analysis. The term "white" is adopted from optics to signify that all frequencies are part of such a process, much like white light is composed of the whole color spectrum.

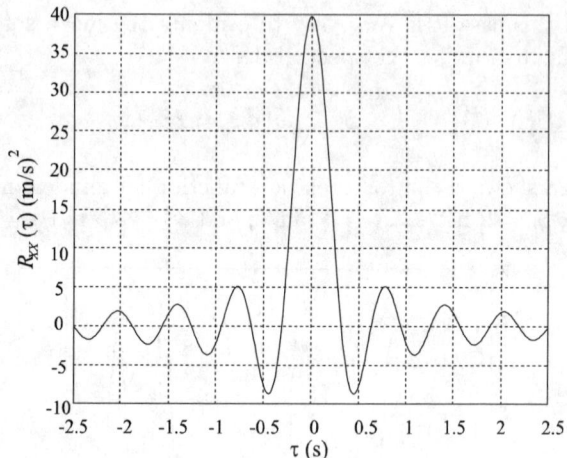

Figure 5.29: The autocorrelation function for an ideal narrow-band process. $S_0 = 2 \text{ m}^2/\text{s}$, $\omega_1 = 0$ rad/s, and $\omega_2 = 10$ rad/s.

The power spectrum ranges from $-\infty$ to ∞, and the autocorrelation function can be evaluated by setting $\omega_1 = 0$ and taking the limit $\omega_2 \to \infty$ in Equation 5.40,

$$\lim_{\omega_1 \to 0} 4\frac{S_0}{\tau} \cos\left\{\left(\frac{\omega_1 + \omega_2}{2}\right)\tau\right\} \sin\left\{\left(\frac{\omega_2 - \omega_1}{2}\right)\tau\right\} = 2S_0 \frac{\sin \omega_2 \tau}{\tau}$$

and then

$$\begin{aligned} R_{XX}(\tau) &= \lim_{\omega_2 \to \infty} 2S_0 \frac{\sin \omega_2 \tau}{\tau} \\ &= 2\pi S_0 \delta(\tau). \end{aligned} \quad (5.41)$$

Equation 5.41 can be confirmed using the definition of the spectral density given in Equation 5.31,

$$S_{XX}(\omega) = \frac{1}{2\pi} \int_{-\infty}^{\infty} 2\pi S_0 \delta(\tau) e^{-i\omega \tau} d\tau = S_0,$$

which is the spectral density that we started with. Figure 5.30 shows the spectral density of a white noise process and its equivalent one sided spectrum.

A random process that has a band-limited spectrum is called, following the optics analogy, *colored noise*.

5.7. POWER SPECTRA

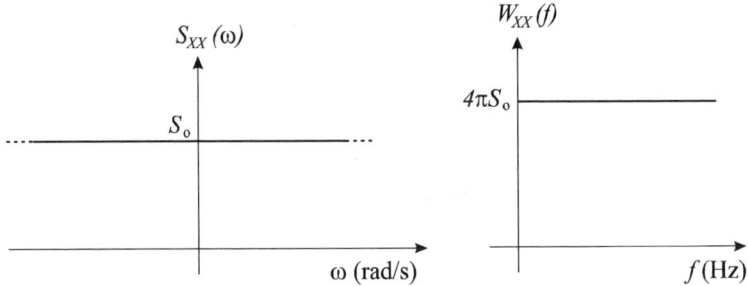

Figure 5.30: Two-sided and one-sided white noise spectra.

5.7.3 Spectral Densities of Derivatives of Stationary Random Processes

Previously in Section 5.4, we found the relationship between the derivatives of autocorrelation function and the cross-correlation function involving its derivatives. Namely, we showed

$$\frac{d}{d\tau}R_{XX}(\tau) = R_{X\dot{X}}(\tau)$$
$$\frac{d^2}{d\tau^2}R_{XX}(\tau) = R_{X\ddot{X}}(\tau)$$
$$= -R_{\dot{X}\dot{X}}(\tau).$$

In this section, similar relations are derived between their spectral densities, $S_{XX}(\omega)$, $S_{X\dot{X}}(\omega)$, $S_{X\ddot{X}}(\omega)$, and $S_{\dot{X}\dot{X}}(\omega)$.

The spectral density $S_{X\dot{X}}(\omega)$ is by definition,

$$S_{X\dot{X}}(\omega) = \frac{1}{2\pi}\int_{-\infty}^{\infty} R_{X\dot{X}}(\tau)\exp(-i\omega\tau)\,d\tau.$$

Using Equation 5.11,

$$S_{X\dot{X}}(\omega) = \frac{1}{2\pi}\int_{-\infty}^{\infty} \frac{d}{d\tau}R_{XX}(\tau)\exp(-i\omega\tau)\,d\tau.$$

By the property of the Fourier transform in Equation 5.24,

$$S_{X\dot{X}}(\omega) = i\omega\frac{1}{2\pi}\int_{-\infty}^{\infty} R_{XX}(\tau)\exp(-i\omega\tau)\,d\tau$$
$$= i\omega S_{XX}(\omega).$$

Similarly, the spectral density $S_{\dot{X}\dot{X}}(\omega)$ can be written as

$$\begin{aligned} S_{\dot{X}\dot{X}}(\omega) &= \frac{1}{2\pi} \int_{-\infty}^{\infty} R_{\dot{X}\dot{X}}(\tau) \exp(-i\omega\tau)\, d\tau \\ &= -\frac{1}{2\pi} \int_{-\infty}^{\infty} \frac{d^2}{d\tau^2} R_{XX}(\tau) \exp(-i\omega\tau)\, d\tau. \end{aligned}$$

Using Equation 5.25,

$$\begin{aligned} S_{\dot{X}\dot{X}}(\omega) &= \omega^2 \frac{1}{2\pi} \int_{-\infty}^{\infty} R_{XX}(\tau) \exp(-i\omega\tau)\, d\tau \\ &= \omega^2 S_{XX}(\omega). \end{aligned} \qquad (5.42)$$

Example 5.14 Spectral Densities of $X(t)$ and its Derivatives

Following the same procedure, derive $S_{X\ddot{X}}(\omega)$ and $S_{\ddot{X}\ddot{X}}(\omega)$ in terms of $S_{XX}(\omega)$.

Solution The spectral density $S_{X\ddot{X}}(\omega)$ is by definition,

$$S_{X\ddot{X}}(\omega) = \frac{1}{2\pi} \int_{-\infty}^{\infty} R_{X\ddot{X}}(\tau) \exp(-i\omega\tau)\, d\tau.$$

The cross-correlation function $R_{X\ddot{X}}(\tau)$ is

$$R_{X\ddot{X}}(\tau) = \frac{d^2}{d\tau^2} R_{XX}(\tau).$$

Then,

$$\begin{aligned} S_{X\ddot{X}}(\omega) &= \frac{1}{2\pi} \int_{-\infty}^{\infty} \frac{d^2}{d\tau^2} R_{XX}(\tau) \exp(-i\omega\tau)\, d\tau \\ &= -\omega^2 S_{XX}(\omega). \end{aligned}$$

The spectral density $S_{\ddot{X}\ddot{X}}$ is

$$S_{\ddot{X}\ddot{X}}(\omega) = \frac{1}{2\pi} \int_{-\infty}^{\infty} R_{\ddot{X}\ddot{X}}(\tau) \exp(-i\omega\tau)\, d\tau,$$

where the cross-correlation function $R_{\ddot{X}\ddot{X}}(\tau)$ is the fourth derivative of $R_{XX}(\tau)$ (Equation 5.14),

$$R_{\ddot{X}\ddot{X}}(\tau) = \frac{d^4}{d\tau^4} R_{XX}(\tau).$$

Then,
$$S_{\ddot{X}\ddot{X}}(\omega) = \frac{1}{2\pi}\int_{-\infty}^{\infty} \frac{d^4}{d\tau^4} R_{XX}(\tau)\exp(-i\omega\tau)\,d\tau.$$

Integrating by parts four times results in the density,

$$\begin{aligned}
S_{\ddot{X}\ddot{X}}(\omega) &= \frac{1}{2\pi}\left(\frac{d^3}{d\tau^3}R_{XX}(\tau) - (-i\omega)\frac{d^2}{d\tau^2}R_{XX}(\tau)\right.\\
&\quad \left.+(-i\omega)^2\frac{d}{d\tau}R_{XX}(\tau) - (-i\omega)^3 R_{XX}(\tau)\exp(-i\omega\tau)\right)\Big|_{-\infty}^{\infty}\\
&\quad +\frac{1}{2\pi}\int_{-\infty}^{\infty}(-i\omega)^4 R_{XX}(\tau)\exp(-i\omega\tau)\,d\tau.
\end{aligned}$$

If $R_{XX}(\tau)$ and its derivatives are absolutely integrable, the terms evaluated at $\pm\infty$ are each equal to zero. Then

$$\begin{aligned}
S_{\ddot{X}\ddot{X}}(\omega) &= \frac{1}{2\pi}\int_{-\infty}^{\infty}(-i\omega)^4 R_{XX}(\tau)\exp(-i\omega\tau)\,d\tau\\
&= \omega^4 S_{XX}(\omega).
\end{aligned}$$

⊛

5.8 Fourier Representation of a Stationary Random Process

If $X(t)$ is a result of many effects that are independent or nearly independent, then $X(t)$ is a Gaussian random process according to the central limit theorem.[11] If $X(t)$ is a stationary process on $[0,T]$, its realization can be represented using the Fourier series

$$X(t) = \sum_{n=1}^{N} a_n \cos\frac{2\pi n}{T}t + b_n \sin\frac{2\pi n}{T}t, \qquad (5.43)$$

where the coefficients a_n and b_n are random variables that have identical, independent normal distributions with zero mean,

$$\begin{aligned}
E\{a_n\} &= E\{b_n\} = 0\\
E\{a_n^2\} &= E\{b_n^2\} = \sigma_n^2.
\end{aligned}$$

---

[11] This section is based on Sections 1.7, 2.8, and 2.10 of S.O. Rice, "Mathematical Analysis of Random Noise," *Bell Syst. Tech. J.*, **23**, 282-332; **24**, 46-156; reprinted in N. Wax, *Selected Papers on Noise and Stochastic Processes*, Dover Publications, New York, 1954.

Independence implies

$$E\{a_n b_m\} = 0, \text{ for } 1 \le n, m \le N$$
$$E\{a_n a_m\} = 0, \text{ for } 1 \le n, m \le N \text{ and } n \ne m$$
$$E\{b_n b_m\} = 0, \text{ for } 1 \le n, m \le N \text{ and } n \ne m.$$

For example, a_1 is independent of all coefficients except for itself.

It should be noted that the number of Fourier components N should be large (at least 200) to duplicate the spectrum accurately.

The autocorrelation function is then given by

$$\begin{aligned}
R_{XX}(\tau) &= E\{X(t)X(t+\tau)\} \\
&= E\left\{\left(\sum_{n=1}^{N} a_n \cos\frac{2\pi n}{T}t + b_n \sin\frac{2\pi n}{T}t\right)\right. \\
&\quad \left.\cdot \left(\sum_{m=1}^{N} a_m \cos\frac{2\pi m}{T}(t+\tau) + b_m \sin\frac{2\pi m}{T}(t+\tau)\right)\right\} \\
&= \sum_{n=1}^{N}\sum_{m=1}^{N}\left[E\{a_n a_m\}\cos\frac{2\pi n}{T}t \cos\frac{2\pi m}{T}(t+\tau)\right. \\
&\quad + E\{a_n b_m\}\cos\frac{2\pi n}{T}t \sin\frac{2\pi m}{T}(t+\tau) \\
&\quad + E\{b_n a_m\}\sin\frac{2\pi n}{T}t \cos\frac{2\pi m}{T}(t+\tau) \\
&\quad \left.+ E\{b_n b_m\}\sin\frac{2\pi n}{T}t \sin\frac{2\pi m}{T}(t+\tau)\right].
\end{aligned}$$

Since $E\{a_n b_m\} = 0$ for any n and m, we can write

$$\begin{aligned}
R_{XX}(\tau) &= \sum_{n=1}^{N}\sum_{m=1}^{N}\left[E\{a_n a_m\}\cos\frac{2\pi n}{T}t \cos\frac{2\pi m}{T}(t+\tau)\right. \\
&\quad \left.+ E\{b_n b_m\}\sin\frac{2\pi n}{T}t \cos\frac{2\pi m}{T}(t+\tau)\right].
\end{aligned}$$

Using $E\{a_n a_m\} = 0$ for $n \ne m$ and $E\{a_n^2\} = E\{b_n^2\} = \sigma_n^2$ for $n = m$,

$$\begin{aligned}
R_{XX}(\tau) &= \sum_{n=1}^{N}\sigma_n^2\left(\cos\frac{2\pi n}{T}t \cos\frac{2\pi n}{T}(t+\tau)\right. \\
&\quad \left.+ \sin\frac{2\pi n}{T}t \cos\frac{2\pi n}{T}(t+\tau)\right) \\
&= \sum_{n=1}^{N}\sigma_n^2 \cos\frac{2\pi n}{T}\tau. \qquad (5.44)
\end{aligned}$$

5.8. FOURIER REPRESENTATION OF A RANDOM PROCESS

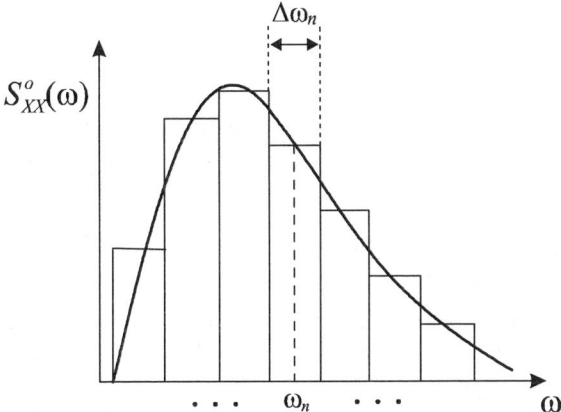

Figure 5.31: Area under the discretized spectral density curve.

The term $2\pi n/T$ is the angular frequency, denoted as ω_n. The interval between the angular frequencies is constant and is given by

$$\Delta\omega = \frac{2\pi(n+1)}{T} - \frac{2\pi n}{T} = \frac{2\pi}{T}.$$

The variance of $X(t)$, the autocorrelation function evaluated at $\tau = 0$, is simply the sum of individual variances associated with each frequency,

$$\sigma_X^2 = \sum_{n=1}^{N} \sigma_n^2,$$

since $X(t)$ is zero mean. σ_X^2 is the total area under the power spectral density curve, and thus each individual variance σ_n^2 is the area under the spectral density function between $\omega_n - \Delta\omega/2$ and $\omega_n + \Delta\omega/2$, as shown in Figure 5.31,

$$\sigma_n^2 = \int_{\omega_n - \Delta\omega/2}^{\omega_n + \Delta\omega/2} S_{XX}^o(\omega)\, d\omega, \tag{5.45}$$

or

$$\sigma_n^2 \simeq S_{XX}^o(\omega_n)\, \Delta\omega. \tag{5.46}$$

If the functional form of $S_{XX}^o(\omega)$ is given, the integral expression can be used for accuracy. If $S_{XX}^o(\omega)$ is given as discrete data, the approximation to the integral is used. Note that $S_{XX}^o(\omega)$ is one sided in this case.

This suggests that a sample time history of the Gaussian random process $X(t)$ can be expressed as a Fourier series with independent Gaussian random numbers a_n and b_n with variance σ_n^2.

However, this is not the only nor the most convenient representation of $X(t)$. Another representation is given by

$$X(t) = \sum_{n=1}^{N} \sqrt{2}\sigma_n \cos\left(\frac{2\pi n}{T}t - \varphi_n\right), \qquad (5.47)$$

where σ_n is given in either Equation 5.45 or 5.46, and φ_n are independent and distributed uniformly on $[0, 2\pi)$.

This representation does not satisfy the condition of a Gaussian process unless $N \to \infty$. However, there is no significant difference between two representations (Equations 5.43 and 5.47) for a large number of Fourier components ($N > 1000$).[12]

In this expression, $X(t)$ is represented by the sum of the sinusoids with fixed amplitude but uniform random phase shifts. Note that the representation in Equation 5.43 has $2 \times N$ normal random numbers (a_n and b_n), but here we only have N uniform random numbers (φ_n). Therefore, the latter representation is more convenient to use.

It can be shown that the autocorrelation function for this case is identical to Equation 5.44. Using Equation 5.47, the autocorrelation function is given by

$$\begin{aligned}
R_{XX}(\tau) &= E\{X(t)X(t+\tau)\} \\
&= E\left\{\sum_{n=1}^{N} \sqrt{2}\sigma_n \cos\left(\frac{2\pi n}{T}t - \varphi_n\right) \right. \\
&\qquad \left. \cdot \sum_{m=1}^{N} \sqrt{2}\sigma_m \cos\left(\frac{2\pi m}{T}(t+\tau) - \varphi_m\right)\right\} \\
\\
&= E\left\{\sum_{n=1}^{N} \sqrt{2}\sigma_n \left(\cos\varphi_n \cos\frac{2\pi n}{T}t + \sin\varphi_n \sin\frac{2\pi n}{T}t\right) \right. \\
&\qquad \left. \cdot \sum_{m=1}^{N} \sqrt{2}\sigma_m \left(\cos\varphi_m \cos\frac{2\pi m}{T}(t+\tau) + \sin\varphi_m \sin\frac{2\pi m}{T}(t+\tau)\right)\right\}
\end{aligned}$$

---

[12] This was shown in Elgar S., Guza, R.T, and Seymour R.J., "Wave Group Statistics from Numerical Simulations of a Random Sea," *Applied Ocean Research*, 7(2), 93-96, 1985.

5.8. FOURIER REPRESENTATION OF A RANDOM PROCESS

$$= \sum_{n=1}^{N}\sum_{m=1}^{N} 2\sigma_n\sigma_m \bigg[E\{\cos\varphi_n \cos\varphi_m\} \cos\frac{2\pi n}{T}t \cos\frac{2\pi m}{T}(t+\tau)$$
$$+ E\{\cos\varphi_n \sin\varphi_m\} \cos\frac{2\pi n}{T}t \sin\frac{2\pi m}{T}(t+\tau)$$
$$+ E\{\sin\varphi_n \cos\varphi_m\} \sin\frac{2\pi n}{T}t \cos\frac{2\pi m}{T}(t+\tau)$$
$$+ E\{\sin\varphi_n \sin\varphi_m\} \sin\frac{2\pi n}{T}t \cos\frac{2\pi m}{T}(t+\tau) \bigg].$$

Since φ_n are independent and uniformly distributed,

$$E\{\cos\varphi_n \sin\varphi_m\} = 0$$
$$E\{\sin\varphi_n \cos\varphi_m\} = 0.$$

For example,

$$E\{\sin\varphi_n \cos\varphi_m\} = \int_0^{2\pi}\int_0^{2\pi} \left(\frac{1}{2\pi}\right)^2 \sin\varphi_n \cos\varphi_m d\varphi_n d\varphi_m = 0.$$

Also, for $n \neq m$, the terms $E\{\cos\varphi_n \cos\varphi_m\}$ and $E\{\sin\varphi_n \sin\varphi_m\}$ are each equal to zero. For example,

$$E\{\sin\varphi_n \sin\varphi_m\} = \int_0^{2\pi}\int_0^{2\pi} \left(\frac{1}{2\pi}\right)^2 \sin\varphi_n \sin\varphi_m d\varphi_n d\varphi_m = 0.$$

For $n = m$,

$$E\{\sin\varphi_n \sin\varphi_n\} = \int_0^{2\pi}\int_0^{2\pi} \frac{1}{2\pi} \sin\varphi_n \sin\varphi_n d\varphi_n d\varphi_n$$
$$= \frac{1}{2\pi}\frac{1}{2}2\pi = \frac{1}{2}.$$

The autocorrelation function is then given by

$$R_{XX}(\tau) = \sum_{n=1}^{N} 2\sigma_n^2 \bigg[\frac{1}{2}\cos\frac{2\pi n}{T}t \cos\frac{2\pi n}{T}(t+\tau)$$
$$+ \frac{1}{2}\sin\frac{2\pi n}{T}t \sin\frac{2\pi n}{T}(t+\tau)\bigg]$$
$$= \sum_{n=1}^{N} \sigma_n^2 \cos\frac{2\pi n}{T}\tau,$$

which agrees with Equation 5.44.

Example 5.15 Fourier Representation of a Gaussian Process

Consider the Pierson-Moskowitz spectral density for the ocean wave elevation in Equation 5.36,

$$S_{\eta\eta}^o(\omega) = \frac{A}{\omega^5} \exp\left(-\frac{B}{\omega^4}\right) \text{ m}^2\text{s/rad, for } \omega > 0,$$

where the constants A and B are given by

$$\begin{aligned} A &= 0.0081g^2 \\ &= 0.7795 \text{ m}^2\text{rad}^4/\text{s}^4 \\ B &= 0.74\left(\frac{g}{V_{19.5}}\right)^4 \text{ (rad/s)}^4. \end{aligned}$$

g is the gravitational acceleration and $V_{19.5}$ is the wind speed evaluated at 19.5 m above the still water level. Plot a sample history for $T = 100$ s using 16 terms.

Solution Using the Fourier representation given by Equation 5.47, a sample time history for ocean wave elevation is given by

$$\eta(t) = \sum_{n=1}^{N} \sqrt{2}\sigma_n \cos\left(\frac{2\pi n}{T}t - \varphi_n\right),$$

where σ_n is given by Equation 5.45 or 5.46. Since the exact expression for $S_{\eta\eta}(\omega)$ is known, we can use Equation 5.46, or

$$\sigma_n = \sqrt{\int_{\omega_n - \Delta\omega/2}^{\omega_n + \Delta\omega/2} S_{\eta\eta}(\omega) \, d\omega},$$

and φ_n is distributed uniformly on $[0, 2\pi)$. First, we find that

$$\begin{aligned} \Delta\omega &= \frac{2\pi}{T} = \frac{2\pi}{100} = 0.0628 \text{ rad/s} \\ \omega_n &= \frac{2\pi n}{T} = \frac{2\pi n}{100}, \quad n = 1, \cdots, 16 \\ &= \begin{cases} 0.0628 & 0.126 & 0.188 & 0.251 \\ 0.314 & 0.377 & 0.440 & 0.502 \\ 0.565 & 0.628 & 0.691 & 0.754 \\ 0.816 & 0.879 & 0.942 & 1.00 \end{cases} \text{ rad/s,} \end{aligned}$$

5.8. FOURIER REPRESENTATION OF A RANDOM PROCESS

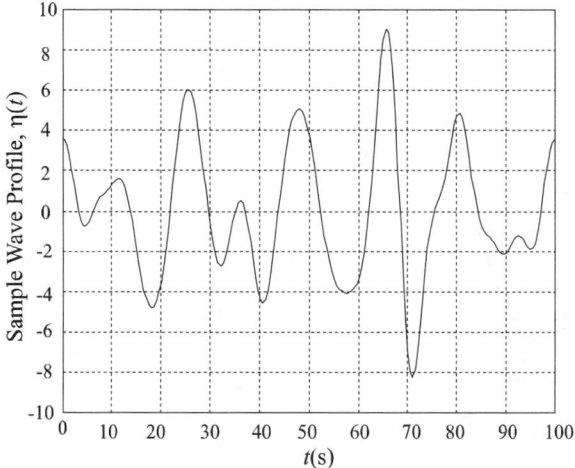

Figure 5.32: A sample time history for $\eta(t)$ from Example 5.15.

and

$$\sigma_1 = \sqrt{\int_{\omega_n - \Delta\omega/2}^{\omega_n + \Delta\omega/2} S_{XX}^o(\omega)\, d\omega} = 1.08 \times 10^{-242},$$

$\sigma_2 = 2.12 \times 10^{-14}$ $\sigma_3 = 1.39 \times 10^{-2}$ $\sigma_4 = 0.776$ $\sigma_5 = 1.63$
$\sigma_6 = 1.65$ $\sigma_7 = 1.26$ $\sigma_8 = 1.07$ $\sigma_9 = 0.846$
$\sigma_{10} = 0.668$ $\sigma_{11} = 0.537$ $\sigma_{12} = 0.437$ $\sigma_{13} = 0.360$
$\sigma_{14} = 0.301$ $\sigma_{15} = 0.254$ $\sigma_{16} = 0.217$.

Next, we need uniform random numbers between 0 and 2π for φ_n,

φ_n = 5.97, 3.05, 2.87, 1.45, 5.60, 0.116, 3.81, 4.79, 5.16, 2.79, 5.79, 2.55, 3.87, 4.64, 5.88, 4.98.

The procedure for obtaining random numbers with a given probability density is presented in Section 12.2. The sample time history is shown in Figure 5.32.
⊛

5.8.1 Borgman's Method of Frequency Discretization

When the frequencies ω_n are evenly spaced, the sample time history exhibits a frequency that depends on the discretization of the frequency band of the

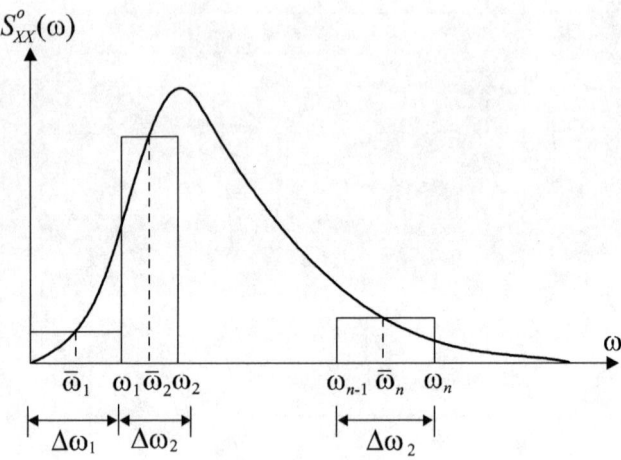

Figure 5.33: Discretization of a spectral density function using Borgman's method.

process. For example, the time history in the previous example repeats itself after 100 s.

In order to avoid the periodicity in the sample time history, Borgman[13] used a slightly different representation of a stationary Gaussian process. In his method, the frequencies ω_n are chosen such that the power spectrum $S^o_{XX}(\omega)$ is divided into N equal areas as shown in Figure 5.33. The individual variances σ_n^2 are now constant and equal to

$$\sigma_n^2 = \frac{\sigma^2}{N} \text{ for all } n,$$

where σ^2 is the total variance or the total area under the function $S_{XX}(\omega)$.

By replacing σ_n with $\sigma/\sqrt{N}$ in Equation 5.47, the random process is represented by

$$X(t) = \sum_{n=1}^{N} \sqrt{\frac{2}{N}} \sigma \cos(\bar{\omega}_n t - \varphi_n), \qquad (5.48)$$

where the phase angles φ_n are also independent and distributed uniformly on $[0, 2\pi)$. The frequencies $\bar{\omega}_n$ are the average of ω_{n-1} and ω_n,

$$\bar{\omega}_n = \frac{\omega_n + \omega_{n-1}}{2}. \qquad (5.49)$$

---

[13] L.E. Borgman, "Ocean Wave Simulation for Engineering Design," *Journal of Waterways and Harbors Division, ASCE*, 95 (WW4), 557-583, Nov. 1969.

5.8. FOURIER REPRESENTATION OF A RANDOM PROCESS 269

Example 5.16 Ocean Wave Elevation Simulation using Borgman's Method

The procedure developed here is by Borgman for ocean engineering applications. A primary characteristic of the ocean is the wave elevation $\eta(x,t)$, which is a function of position x and time t. The elevation $\eta(x,t)$ is modeled realistically as a random process in time and the possible wave heights are defined by a spectral density $S_{\eta\eta}(\omega)$. A commonly used one-sided spectrum is the Pierson-Moskowitz in Equation 5.36,

$$S_{\eta\eta}^o(\omega) = \frac{A}{\omega^5} \exp\left(-\frac{B}{\omega^4}\right) \text{ m}^2\text{s/rad, for } \omega > 0.$$

For computational purposes, discretize the integral over the range of ω where there is significant energy, say $\omega_0 < \omega < \omega_N$. The upper-limit frequency, ω_N, is chosen sufficiently large so that most of the area lies between 0 and ω_N. This discretization results in equal areas. The wave elevation is then given by

$$\eta(x,t) = \sum_{n=1}^{N} \sqrt{\frac{2}{N}} \sigma \cos\left(\bar{\omega}_n t - \bar{k}_n x - \varphi_n\right),$$

where $\bar{\omega}_n$ is given by Equation 5.49 and $\bar{k}_n$ is the wave number, related to the angular frequency by the dispersion relation,

$$\bar{k}_n = \bar{\omega}_n/g,$$

for deep water. The total variance σ^2 is the area under the function $S_{\eta\eta}^o(\omega)$,

$$\begin{aligned}
\sigma^2 &= \int_0^\infty S_{\eta\eta}^o(\omega)\, d\omega \\
&= \int_0^\infty \frac{A}{\omega^5} \exp\left(-\frac{B}{\omega^4}\right) d\omega \\
&= \left. \frac{A}{4B} \exp\left(-\frac{B}{\omega^4}\right) \right|_0^\infty \\
&= \frac{A}{4B}.
\end{aligned}$$

Then the wave elevation becomes

$$\begin{aligned}
\eta(x,t) &= \sum_{n=1}^{N} \sqrt{\frac{2}{N}} \sqrt{\frac{A}{4B}} \cos\left(\bar{\omega}_n t - \bar{k}_n x - \varphi_n\right) \\
&= \sum_{n=1}^{N} \sqrt{\frac{A}{2NB}} \cos\left(\bar{\omega}_n t - \bar{k}_n x - \varphi_n\right).
\end{aligned}$$

Now find ω_n for $n = 1, \cdots, N$. The area between 0 and ω_n equals n/N of the total area under the curve between 0 and ω_N,

$$\int_0^{\omega_n} S_{\eta\eta}^o(\omega)\, d\omega = \frac{n}{N}\int_0^{\omega_N} S_{\eta\eta}^o(\omega)\, d\omega.$$

$$\frac{A}{4B}\exp\left(-\frac{B}{\omega_n^4}\right) = \frac{n}{N}\frac{A}{4B}\exp\left(-\frac{B}{\omega_N^4}\right).$$

Solve this last equation for ω_n to find

$$\omega_n = \left(\frac{B}{\ln(N/n) + (B/\omega_N^4)}\right)^{1/4}, \qquad n = 1, 2, \cdots, N.$$

Figure 5.16 shows the Pierson-Moskowitz spectral density for several values of V, and Figure 5.34 shows the time history of a sample wave derived using the above procedure for wind velocity $V_{19.5} = 25$ m/s and $N = 10$ at $x = 0$.

※

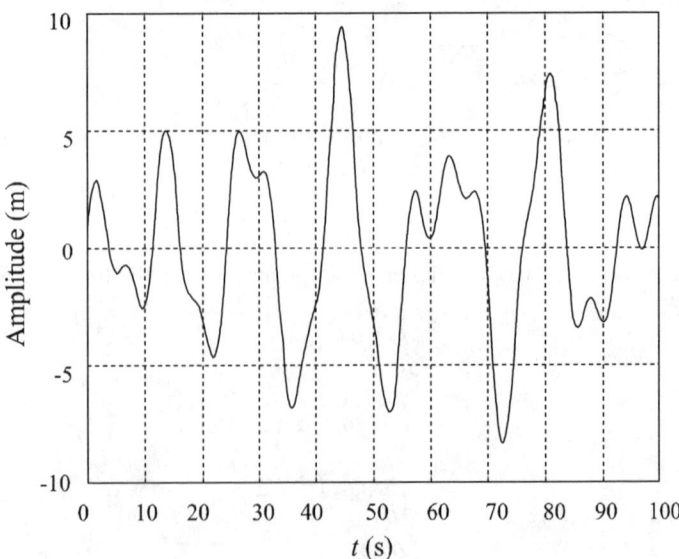

Figure 5.34: Sample wave profile derived using Borgman's method applied to the Pierson-Moskowitz wave height spectrum. $V_{19.5} = 25$ m/s, $N = 10$.

The disadvantage of Borgman's method is that the frequencies ω_n are chosen deterministically to simulate random processes. In Section 12.5.2, it will be shown how the frequencies can be selected randomly.

5.9 Concluding Summary

In this chapter, the first to consider time dependence of random variables, definitions and concepts for the analysis of random processes are introduced. Of particular importance are the concepts of stationarity, autocorrelation, and spectral density, and the Wiener-Khinchine relations. Various power spectra are introduced and applied. In addition, topics are introduced that are of importance in subsequent chapters, such as derivatives of stationary processes.

5.10 Problems

Section 5.1: Basic Random Process Descriptors

1. Let $X(t)$ be a random process with autocorrelation function,
$$R_{XX}(t_1, t_2) = b(t_1 + c)^{-2}(t_2 + c)^{-2},$$
where b and c are positive constants, and let $Y(t)$ be a random process defined by $Y(t) = \int_0^t X(s)\,ds$. Derive the cross-correlation and autocorrelation functions, $R_{XY}(t_1, t_2)$ and $R_{YY}(t_1, t_2)$.

2. The random process $X(t)$ is given by
$$X(t) = A\cos(\omega t - \Phi),$$
where A and Φ are random variables with the probability density function,
$$f_{A\Phi}(a, \phi) = \frac{1}{2\pi}(1 + (3a - 1)\cos\phi),$$
$$\text{for } 0 \leq \phi \leq 2\pi$$
$$\text{and } 0 \leq a \leq 1.$$
Derive *(i)* μ_X, *(ii)* σ_X^2, and *(iii)* $R_{XX}(t_1, t_2)$.

Section 5.2: Ensemble Averaging

3. Given the pair of random functions
$$X(t) = A\cos\omega_1 t$$
$$Y(t) = B\sin\omega_2 t,$$

where A and B are random variables with

$$E\{A\} = 2 \quad E\{B\} = 1$$
$$E\{A^2\} = 1 \quad E\{AB\} = 1 \quad E\{B^2\} = 4.$$

Evaluate the autocorrelation functions $R_{XX}(t_1, t_2)$, $R_{XY}(t_1, t_2)$, $R_{YX}(t_1, t_2)$, and $R_{YY}(t_1, t_2)$.

4. Let $X(t)$ be a discrete random process that equals the number of fish chosen randomly from a fish tank. Since fish breed freely in the tank, the total number of fish varies with time. The number of ways to choose x fish is given by the combination $_NC_x$, where N is the total number of fish. The probability that the number of fish chosen is x is

$$\Pr(X = x) = \frac{_NC_x}{\sum_{x=1}^{N} {_NC_x}}.$$

Assume that N varies exponentially in time such that

$$N = 1000 \exp(1.01t).$$

Find the mean $E\{X\}$ and the mean-square $E\{X^2\}$ values.

5. Derive the autocovariance function in the form

$$C_{XX}(t_1, t_2) = R_{XX}(t_1, t_2) - \mu_X(t_1)\mu_X(t_2).$$

6. Continuing Example 5.1, find $C_{XX}(t_1, t_2)$.

Section 5.3: Stationarity

7. Derive Equation 5.9.

8. Show that the cross-correlation function is an odd function, that is,

$$R_{XY}(\tau) = R_{YX}(-\tau).$$

9. Calculate the temporal mean and autocorrelation for the function

$$x(t) = \begin{cases} t & \text{for } 0 < t < 1 \\ 2 - t & \text{for } 1 < t < 2. \end{cases}$$

Plot the autocorrelation function.

5.10. PROBLEMS

Section 5.4: Derivatives of Stationary Processes

10. Let $X(t)$ be a stationary random process with autocorrelation function given by $R_{XX}(\tau)$. Find the covariance $\text{Cov}\left(X(t), \dot{X}(t)\right)$ where $X(t)$ is a zero-mean process. (Hint: Use the fact that $R_{XX}(\tau)$ is an even function and its derivative at $\tau = 0$ equals zero.)

Section 5.6: Harmonic Processes

11. Let $\dot{X}(t)$ be the random process for which the sample realizations are the derivatives of the sample realizations of

$$X(t) = A\cos(\omega t - \Phi),$$

where A is a constant and Φ is a uniform random variable between 0 and 2π. Show that $X(t_1)$ and $\dot{X}(t_2)$ are uncorrelated if $t_1 = t_2$.

12. Let the random process $X(t)$ be given by

$$X(t) = A\cos\omega t + B\sin\omega t,$$

where A and B are independent random variables. Assume that both have zero means and are identically distributed such that

$$\mu_A = \mu_B = 0$$
$$\sigma_A^2 = \sigma_B^2 = \sigma^2.$$

Show that the autocorrelation function is given by

$$R_{XX}(\tau) = \sigma^2 \cos\omega\tau.$$

13. Let $X(t) = A\sin(\omega_0 t + \Phi)$, where ω_0 is a constant and A and Φ are random variables with marginal probability density functions given by

$$f_A(a) = W_0/2 \quad \text{for} \quad -W_0 \leq a \leq W_0$$
$$f_\Phi(\phi) = 1/2\pi \quad \text{for} \quad 0 < \phi \leq 2\pi.$$

Find the autocorrelation function and the spectral density, $R_{XX}(\tau)$ and $S_{XX}(\omega)$.

Section 5.7: Power Spectra

14. Discuss whether the function

$$R_{XX}(\tau) = \begin{cases} R_o & \text{for} \quad |\tau| < \tau_o \\ 0 & \text{elsewhere} \end{cases}$$

can be an autocorrelation function. If not, explain. (Hint: The spectral density cannot be negative for any frequency.)

15. Consider a random signal with the constant spectral density,

$$W(f) = 0.002 \text{ cm}^2/\text{cps}$$
$$40 \leq f \leq 1600 \text{ cps},$$

and zero outside this frequency range. The mean of this signal is 2.0 cm. Find the variance of the response.

16. Consider Example 5.6, where $p(t)$ and $q(t)$ are square pulses.

    (i) Find the Fourier transforms of $p(t)$ and $q(t)$.

    (ii) Find the Fourier transform of the convolution $p * q$.

    (iii) Verify that the convolution theorem holds in this case, that is, $\mathcal{F}(p(t))\mathcal{F}(q(t)) = \mathcal{F}(p(t) * q(t))$.

17. Consider a random process with the autocorrelation function, $R_{XX}(\tau) = \sin \Omega \tau$. Find the corresponding spectral density.

18. A random process has the autocorrelation function,

$$R_{XX}(\tau) = A \exp(-\alpha |\tau|)(\cos \beta \tau + \gamma \sin \beta |\tau|).$$

    (i) Find the corresponding spectral density.

    (ii) Plot the spectral density for various values of α, β, and γ.

    (iii) Show that γ must be greater than α/β for the spectral density to be positive for all ω, and the given function for $R_{XX}(\tau)$ can indeed serve as an autocorrelation function.

19. Show that spectral density of a sum of two statistically independent processes is equal to the sum of the spectral densities of their respective processes.

20. Let

$$Z(t) = X(t)Y(t),$$

where $X(t)$ and $Y(t)$ are uncorrelated random processes. Derive the spectral density function of $Z(t)$ in terms of the spectral density functions of $X(t)$ and $Y(t)$.

Section 5.8: Fourier Representation of a Random Process

21. Consider a process with the spectral density

$$S_{XX}(\omega) = \frac{1}{2(\omega_2 - \omega_1)} \text{ m}^2\text{s, for } \omega_1 < |\omega| < \omega_2.$$

Find sample responses when (i) $\omega_1 = 10$ rad/s and $\omega_2 = 11$ rad/s, and (ii) $\omega_1 = 0$ rad/s and $\omega_2 = 20$ rad/s.

Chapter 6

Single-Degree-of-Freedom Dynamics

The main goal of the study of structures and systems in a random environment is to predict the response (output) statistics given the loading (input) statistics. *Statistics* is the discipline that organizes data in a form that is meaningful. It is important to distinguish between inherently random molecular forces due to Brownian motion, such as those experienced by atoms at that scale, and the environmental forces of concern here. Environmental forces in dynamic problems are not inherently random, but they undergo very complex cycles. Such complex phenomena cannot be effectively modeled using deterministic techniques. Therefore, the tools of probability and statistics are adopted for a framework to quantify uncertainties. In this way, uncertainties about the forces can be used to estimate structural response uncertainties.

Examples of significant applications of this theory include ocean engineering, aircraft design, earthquake engineering, wind engineering (skyscrapers and bridges), and spaceship design.

The reader is reminded that even though the discussions in these applied chapters is couched in the terminology of mechanical vibration and dynamics, the tools developed in this chapter and elsewhere in this book are applicable broadly to engineering and the physical sciences.

The convolution integral defines the linear dynamic response to a deterministic force $F(t)$. But what is to be done when the forcing function oscillates in such a complex manner as in Figure 6.1? One possibility is to carry out many experiments and gather data on $F(t)$ in the form of time histories.

Then, the time history with the largest amplitudes can be used for the

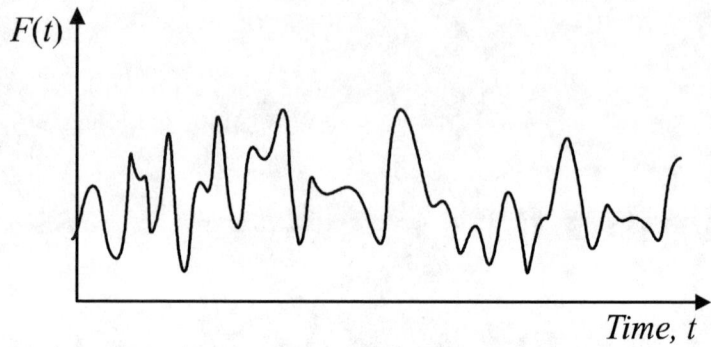

Figure 6.1: A sample realization of a random function $F(t)$.

deterministic analysis and design. This would work, but if the largest amplitude force occurs only infrequently, the structure would be overdesigned. This means that the structure is stronger than it has to be and is therefore uneconomical. What if all the time histories are averaged and this *averaged or mean value time history* is used as a deterministic force that is used in the convolution integral? This would be a good start, but of course, the response calculated in this manner would underestimate the actual response too often! How often depends on the scatter, or standard deviation, of possible time histories.

The next question to ask is then: *How much scatter is there above and below the mean value response?* Perhaps if we knew the mean value response as well as a measure of the scatter, this information could be used for a safe and economical design. This is indeed an approach that makes sense.

Finally: *How does the engineer know how often a very large amplitude force occurs?* If the very large force, such as an earthquake, occurs only once in 100 years, how is this fact used in a design?[1] What the designer needs is a way to give more *weight* to more likely events, without completely ignoring more severe but less likely events. The probability density function acts as such a weighting function.

All of the ideas described above are actually probabilistic concepts. We need to develop some introductory ideas with important motivating examples. This sets the stage for our efforts at random vibration modeling.

---

[1] A 100-year storm is one that will occur in any year with the probability 1/100. It does not mean that we can expect to see it once in 100 years.

6.1 Motivating Examples

Interesting and important vibrating systems include the bridge, satellite transport, rotating machinery, and the rocket ship; two of these are introduced below in detail. Relatively simple, single-degree-of-freedom idealizations, are introduced as preliminary analytical models. While it is true that for a complete detailed analysis one degree of freedom is insufficient, it is possible to estimate the gross behavior for preliminary design purposes.

6.1.1 Transport of a Satellite

Usually, vibration is studied with the primary goal of isolating a machine or structure from its potentially destructive dynamic and vibratory effects. If isolation is only partially successful, then the secondary goal is to use an understanding of vibration to design structures so that they are less susceptible to oscillation. Both goals are typically part of any analysis and design.

A specific application is the analysis, design, and testing of shipping containers. For structures such as satellites that cannot withstand the rigors of transport from the manufacturing site to the launch pad, it becomes necessary to design a container that, among many other purposes, isolates its contents from a spectrum of forcing. Satellites are designed to orbit the Earth in a microgravity, atmosphere-free environment. But the designer must also consider what the satellite will experience from its point of departure at the factory until it is operational in orbit. The satellite cannot be efficiently designed to simultaneously survive shipment as well as being blasted into orbit for space operations. Competing design constraints govern. Thus, the design requirement for a shipping container is to deliver the satellite in perfect condition to the launch pad. In this way, the satellite can be optimally designed for its final destination in orbit around Earth.

6.1.2 Rocket Ship

A rocket ship is an example of a system with a variable mass; mass varies with time. The rocket is propelled into space by thrust created by exhausting its fuel, resulting in a decreasing fuel mass within the fuel tanks.

Since most of us have not seen a rocket, much less been on one during flight, an analogy with a firing cannon is useful to help understand how thrust is generated. As the cannon is fired, it recoils in the direction opposite to that of the cannonball, satisfying the principle of conservation of linear momentum. If it is fired rapidly, it picks up speed with each shot. The cannon expels mass in one direction and moves in the opposite direction, moving faster if it shoots out more mass over a relatively short period of time.

The rocket is a container within which fuel is burned and ejected at high speed. The rocket accelerates in the direction opposite to that of the fuel ejection. The acceleration is a function of the rate of fuel consumption and the speed with which combustion gases are ejected. As this occurs, the total mass of the rocket varies (decreases) with time.

6.2 Deterministic SDoF Vibration

Dynamical systems are initially studied at the most simple level, where one coordinate completely describes the motion of the structure. This is called a single-degree-of-freedom (SDoF) model. The study of single-degree-of-freedom systems is very useful since many of the key principles developed here are of use throughout all of dynamics and vibration. One can explore many concepts with such simple models, and learn how to work with their governing equations.

Before considering random vibration, we briefly review some elements of deterministic vibration. This section provides an overview of the key ideas and formulations.

For a single-degree-of-freedom model, it is relatively straightforward to derive the equation of motion by utilizing an important concept known as the free body diagram. This tool allows the visualization of the forces acting on a free body. Newton's second law of motion for a body states that the sum of the external forces equals the rate of change of the momentum,

$$\sum_{i=1}^{n} F_i(t) = \frac{d(mv)}{dt}.$$

A sign convention needs to be established so that the vector sum in Newton's second law of motion properly accounts for the directions of the forces. For a body rotating in two dimensions, the corresponding equation of motion is $\sum_{i=1}^{n} M_i(t) = d(I\omega)/dt$, where externally applied moments are being summed. The inertial property is the mass moment of inertia I, and the rectilinear momentum mv is replaced by the angular momentum $I\omega$. Mathematically, though, the result is still a second-order linear differential equation.

Based on the above formulation, the general governing equation of motion for the single body shown in Figure 6.2(a), with stiffness k and damping c properties, undergoing rectilinear motion becomes

$$F_{external}(t) - c\dot{x} - k_s x = m\ddot{x},$$

6.2. DETERMINISTIC SDOF VIBRATION

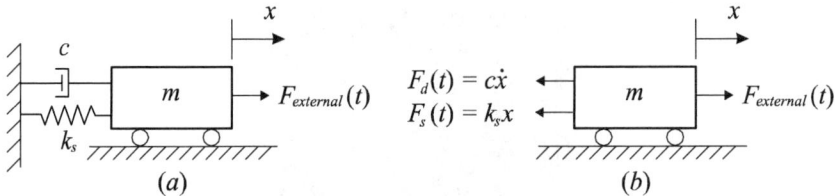

Figure 6.2: (a) Idealized system, (b) forces on a free body.

where $F_{external}(t)$ is the vectorial sum of external forces in one coordinate direction, an overdot denotes differentiation with respect to time. In more standard form, the governing differential equation becomes

$$m\ddot{x} + c\dot{x} + k_s x = F_{external}(t). \tag{6.1}$$

After dividing by the mass m, Equation 6.1 becomes

$$\ddot{x} + 2\zeta\omega_n\dot{x} + \omega_n^2 x = F(t),$$

where new parameters ζ and ω are introduced,

$$\zeta = \frac{c}{2m\omega_n}, \quad \omega_n^2 = \frac{k_s}{m}, \quad F(t) = \frac{F_{external}(t)}{m}.$$

ζ is the dimensionless viscous damping factor, and ω_n is the natural frequency of undamped oscillation with units of rad/s. This is the equation of motion for a structure idealized by a single-degree-of-freedom model, subjected to external forcing. The subscript s will be subsequently omitted from the stiffness constant.

The solution of the equation of motion leads to an understanding of the different classes of vibratory behavior. The character of the response turns out to be a function of the value of the parameter ζ. The next section begins to draw the basic outline of how this governing equation of motion is to be solved, and, as importantly, how to interpret the possible solutions as functions of damping, loading, and the initial conditions of the problem.

Example 6.1 Natural Frequency and Damping Factor

Consider a system with $m = 2.40$ kg, $k = 100$ N/mm, and damping factor $\zeta = 0.3$. Calculate the natural frequency ω_n and the damping coefficient c.

Figure 6.3: Sir Isaac Newton (1643-1727).

Solution   The natural frequency and the damping coefficient are given by

$$\omega_n = \sqrt{\frac{k}{m}} = \sqrt{\frac{100000}{2.40}} = 204 \text{ rad/s}$$
$$c = 2m\omega_n\zeta = 2 \cdot 2.4 \cdot 204 \cdot 0.3 = 294 \text{ Ns/m}.$$

⊛

Newton   The life of Isaac Newton can be divided into three quite distinct periods. The first is his boyhood days from 1643 up to his appointment to a chair in 1669. The second period from 1669 to 1687 was the highly productive period in which he was Lucasian professor at Cambridge. The third period (nearly as long as the other two combined) saw Newton as a highly paid government official in London with little further interest in mathematical research.

Isaac Newton was born in the manor house of Woolsthorpe, near Grantham in Lincolnshire. Although by the calendar in use at the time of his birth he was born on Christmas Day 1642, we give the date of January 4, 1643 in this biography, which is the "corrected" Gregorian calendar date bringing it into line with our present calendar. (The Gregorian calendar was not adopted in England until 1752.) Isaac Newton came from a family of farmers but never knew his father, also named Isaac Newton, who died in October 1642, three months before his son was born. Although Isaac's father owned property and animals, which made him quite a wealthy man, he was completely uneducated and could not sign his own name.

Isaac's mother Hannah Ayscough remarried Barnabas Smith, the minister of the church at North Witham, a nearby village, when Isaac was two years old. The young child was then left in the care of his grandmother Margery Ayscough at Woolsthorpe. Basically treated as an orphan, Isaac did not have a happy childhood. His grandfather James Ayscough was never mentioned by Isaac in later life and the fact that James left nothing to Isaac in his will, made when the boy was ten years old, suggests that there was no love lost between the two. There is no doubt that Isaac felt very bitter toward his mother and his stepfather Barnabas Smith. When examining his sins at age nineteen, Isaac listed: *Threatening my father and mother Smith to burn them and the house over them.*

Upon the death of his stepfather in 1653, Newton lived in an extended family consisting of his mother, his grandmother, one half-brother, and two half-sisters. From shortly after this time Isaac began attending the Free Grammar School in Grantham. Although this was only five miles from his home, Isaac lodged with the Clark family at Grantham. However, he seems to have shown little promise in academic work. His school reports described him as "idle" and "inattentive." His mother, by now a lady of reasonable wealth and property, thought that her eldest son was the right person to manage her affairs and her estate. Isaac was taken away from school but soon showed that he had no talent, or interest, in managing an estate.

An uncle, William Ayscough, decided that Isaac should prepare for entering university and, having persuaded his mother that this was the right thing to do, Isaac was allowed to return to the Free Grammar School in Grantham in 1660 to complete his school education. This time he lodged with Stokes, who was the headmaster of the school, and it would appear that, despite suggestions that he had previously shown no academic promise, Isaac must have convinced some of those around him that he had academic promise. Some evidence points to Stokes also persuading Isaac's mother to let him enter university, so it is likely that Isaac had shown more promise in his first spell at the school than the school reports suggest. Another piece of evidence comes from Isaac's list of sins referred to above. He lists one of his sins as ... *setting my heart on money, learning, and pleasure more than Thee* ..., which tells us that Isaac must have had a passion for learning.

We know nothing about what Isaac learned in preparation for university, but Stokes was an able man and almost certainly gave Isaac private coaching and a good grounding. There is no evidence that he learned any mathematics, but we cannot rule out Stokes introducing him to Euclid's Elements, which he was well capable of teaching (although there is evidence mentioned below that Newton did not read Euclid before 1663). Anecdotes abound about a mechanical ability that Isaac displayed at the school and stories are told of his skill in making models of machines, in particular of clocks and windmills. However, when biographers seek information about famous people there is always a tendency for people to report what they think is expected of them, and these anecdotes may simply be made up later by those who felt that the most famous scientist in the world ought to have had these skills at school.

Newton entered his uncle's old College, Trinity College Cambridge, on June 5, 1661. He was older than most of his fellow students but, despite the fact that his mother was financially well off, he entered as a sizar. A sizar at Cambridge was a student who received an allowance toward college expenses in exchange for acting as a servant to other students. There is certainly some ambiguity in his position as a sizar, for he seems to have associated with "better class" students rather than other sizars. Westfall has suggested that Newton may have had Humphrey Babington, a distant relative who was a Fellow of Trinity, as his patron. This reasonable explanation would fit well with what is known and mean that his mother did not subject him unnecessarily to hardship as some of his biographers claim.

Newton's aim at Cambridge was a law degree. Instruction at Cambridge was dominated by the philosophy of Aristotle but some freedom of study was allowed in the third year of the course. Newton studied the philosophy of Descartes, Gassendi, Hobbes, and in particular Boyle. The mechanics of the Copernican astronomy of Galileo attracted him and he also studied Kepler's Optics. He recorded his thoughts in a book which he entitled *Quaestiones Quaedam Philosophicae* (Certain Philosophical Questions). It is a fascinating account of how Newton's ideas were already forming around 1664. He headed the text with a Latin statement meaning "Plato is my friend, Aristotle is my friend, but my best friend is truth," showing himself a free thinker from an early stage.

How Newton was introduced to the most advanced mathematical texts of his day is slightly less clear. According to de Moivre, Newton's interest in mathematics began in the autumn of 1663 when he bought an astrology book at a fair in Cambridge and found that he could not understand the mathematics in it. Attempting to read a trigonometry book, he found that he lacked knowledge of geometry and so decided to read Barrow's edition of Euclid's Elements. The first few results were so easy that he almost gave up but he *changed his mind when he read that parallelograms upon the same base and between the same parallels are equal.*

Returning to the beginning, Newton read the whole book with a new respect. He then turned to Oughtred's *Clavis Mathematica* and Descartes' *La Géométrie*. The new algebra and analytical geometry of Viète was read by Newton from Frans van Schooten's edition of Viète's collected works published in 1646. Other major works of mathematics that he studied around

6.2. DETERMINISTIC SDOF VIBRATION

this time was the newly published major work by van Schooten, *Geometria a Renato Des Cartes,* which appeared in two volumes in 1659-1661. The book contained important appendices by three van Schooten disciples, Jan de Witt, Johan Hudde, and Hendrick van Heuraet. Newton also studied Wallis's Algebra and it appears that his first original mathematical work came from his study of this text. He read Wallis's method for finding a square of equal area to a parabola and a hyperbola that used indivisibles. Newton made notes on Wallis's treatment of series but also devised his own proofs of the theorems writing: *Thus Wallis doth it, but it may be done thus ...*

It would be easy to think that Newton's talent began to emerge on the arrival of Barrow to the Lucasian chair at Cambridge in 1663, when he became a Fellow at Trinity College. Certainly the date matches the beginnings of Newton's deep mathematical studies. However, it would appear that the 1663 date is merely a coincidence and that it was only some years later that Barrow recognized the mathematical genius among his students.

Despite some evidence that his progress had not been particularly good, Newton was elected a scholar on April 28, 1664 and received his bachelor's degree in April 1665. It would appear that his scientific genius had still not emerged, but it did so suddenly when the plague closed the University in the summer of 1665 and he had to return to Lincolnshire. There, in a period of less than two years, while Newton was still under 25 years old, he began revolutionary advances in mathematics, optics, physics, and astronomy.

While Newton remained at home he laid the foundations for differential and integral calculus, several years before its independent discovery by Leibniz. The "method of fluxions", as he termed it, was based on his crucial insight that the integration of a function is merely the inverse procedure to differentiating it. Taking differentiation as the basic operation, Newton produced simple analytical methods that unified many separate techniques previously developed to solve apparently unrelated problems such as finding areas, tangents, the lengths of curves and the maxima and minima of functions. Newton's *De Methodis Serierum et Fluxionum* was written in 1671 but Newton failed to get it published and it did not appear in print until John Colson produced an English translation in 1736.

When the University of Cambridge reopened after the plague in 1667, Newton put himself forward as a candidate for a fellowship. In October he was elected to a minor fellowship at Trinity College but, after being awarded his master's degree, he was elected to a major fellowship in July 1668, which allowed him to dine at the Fellows' Table. In July 1669 Barrow tried to ensure that Newton's mathematical achievements became known to the world. He sent Newton's text, *De Analysi,* to Collins in London writing: *[Newton] brought me the other day some papers, wherein he set down methods of calculating the dimensions of magnitudes like that of Mr. Mercator concerning the hyperbola, but very general; as also of resolving equations; which I suppose will please you; and I shall send you them by the next.*

Collins corresponded with all the leading mathematicians of the day so Barrow's action should have led to quick recognition. Collins showed Brouncker, the President of the Royal Society, Newton's results (with the author's permission) but after this Newton requested that his manuscript be returned. Collins could not give a detailed account but de Sluze and Gregory learned something of Newton's work through Collins. Barrow resigned the Lucasian chair in 1669 to devote himself to divinity, recommending that Newton (still only 27 years old) be appointed in his place. Shortly after this Newton visited London and twice met with Collins but, as he wrote to Gregory: ... *having no more acquaintance with him I did not think it becoming to urge him to communicate anything.*

Newton's first work as Lucasian Professor was on optics and this was the topic of his first lecture course begun in January 1670. He had reached the conclusion during the two plague years that white light is not a simple entity. Every scientist since Aristotle had believed that white light was a basic single entity, but the chromatic aberration in a telescope lens convinced Newton otherwise. When he passed a thin beam of sunlight through a glass prism Newton noted the spectrum of colors that was formed. He argued that white light is really a mixture of many different types of rays that are refracted at slightly different angles, and that each different type of ray produces a different spectral color. Newton was led by this reasoning to the erroneous conclusion that telescopes using refracting lenses would always suffer chromatic aberration. He therefore proposed and constructed a reflecting telescope.

In 1672 Newton was elected a fellow of the Royal Society after donating a reflecting telescope. Also in 1672 Newton published his first scientific paper on light and color in the Philosophical Transactions of the Royal Society. The paper was generally well received but Hooke and Huygens objected to Newton's attempt to prove, by experiment alone, that light consists of the motion of small particles rather than waves. The reception that his publication received did nothing to improve Newton's attitude to making his results known to the world. He was always pulled in two directions; there was something in his nature that wanted fame and recognition yet another side of him feared criticism and the easiest way to avoid being criticized was to publish nothing. Certainly one could say that his reaction to criticism was irrational, and certainly his aim to humiliate Hooke in public because of his opinions was abnormal. However, perhaps because of Newton's already high reputation, his corpuscular theory reigned until the wave theory was

6.2. DETERMINISTIC SDOF VIBRATION

revived in the 19th century.

Newton's relations with Hooke deteriorated further when, in 1675, Hooke claimed that Newton had stolen some of his optical results. Although the two men made their peace with an exchange of polite letters, Newton turned in on himself and away from the Royal Society, which he associated with Hooke as one of its leaders. He delayed the publication of a full account of his optical researches until after the death of Hooke in 1703. Newton's *Opticks* appeared in 1704. It dealt with the theory of light and color and with investigations of the colors of thin sheets, "Newton's rings," and diffraction of light. To explain some of his observations he had to use a wave theory of light in conjunction with his corpuscular theory.

Another argument, this time with the English Jesuits in Liège over his theory of color, led to a violent exchange of letters, and then in 1678 Newton appears to have suffered a nervous breakdown. His mother died in the following year and he withdrew further into his shell, mixing as little as possible with people for a number of years.

Newton's greatest achievement was his work in physics and celestial mechanics, which culminated in the theory of universal gravitation. By 1666 Newton had early versions of his three laws of motion. He had also discovered the law giving the centrifugal force on a body moving uniformly in a circular path. However he did not have a correct understanding of the mechanics of circular motion. Newton's novel idea of 1666 was to imagine that the Earth's gravity influenced the Moon, counterbalancing its centrifugal force. From his law of centrifugal force and Kepler's third law of planetary motion, Newton deduced the inverse-square law.

In 1679 Newton corresponded with Hooke who had written to Newton claiming ... *that the Attraction always is in a duplicate proportion to the Distance from the Center Reciprocall* ... . M. Nauenberg writes an account of the next events: *After his 1679 correspondence with Hooke, Newton, by his own account, found a proof that Kepler's areal law was a consequence of centripetal forces, and he also showed that if the orbital curve is an ellipse under the action of central forces then the radial dependence of the force is inverse square with the distance from the centre. This discovery showed the physical significance of Kepler's second law.*

In 1684 Halley, tired of Hooke's boasting [M. Nauenberg]: ... *asked Newton what orbit a body followed under an inverse square force, and Newton replied immediately that it would be an ellipse. However in De Motu ... he only gave a proof of the converse theorem that if the orbit is an ellipse the force is inverse square.* The proof that inverse square forces imply conic section orbits is sketched in Cor. 1 to Prop. 13 in Book 1 of the second and third editions of the *Principia*, but not in the first edition.

Halley persuaded Newton to write a full treatment of his new physics and its application to astronomy. Over a year later (1687) Newton published the *Philosophiae naturalis principia mathematica*, or *Principia*, as it is always known. The *Principia* is recognized as the greatest scientific book ever written. Newton analyzed the motion of bodies in resisting and nonresisting media under the action of centripetal forces. The results were applied to orbiting bodies, projectiles, pendulums, and free-fall near the Earth. He further demonstrated that the planets were attracted toward the Sun by a force varying as the inverse square of the distance and generalized that all heavenly bodies mutually attract one another. Further generalization led Newton to the law of universal gravitation: ... *all matter attracts all other matter with a force proportional to the product of their masses and inversely proportional to the square of the distance between them.*

Newton explained a wide range of previously unrelated phenomena: the eccentric orbits of comets, the tides and their variations, the precession of the Earth's axis, and motion of the Moon as perturbed by the gravity of the Sun. This work made Newton an international leader in scientific research. The Continental scientists certainly did not accept the idea of action at a distance and continued to believe in Descartes' vortex theory where forces work through contact. However this did not stop the universal admiration for Newton's technical expertise.

James II became king of Great Britain on February 6, 1685. He had become a convert to the Roman Catholic church in 1669 but when he came to the throne he had strong support from Anglicans as well as Catholics. However rebellions arose, which James put down, but he began to distrust Protestants and began to appoint Roman Catholic officers to the army. He then went further, appointing only Catholics as judges and officers of state. Whenever a position at Oxford or Cambridge became vacant, the king appointed a Roman Catholic to fill it. Newton was a staunch Protestant and strongly opposed to what he saw as an attack on the University of Cambridge.

When the King tried to insist that a Benedictine monk be given a degree without taking any examinations or swearing the required oaths, Newton wrote to the Vice-Chancellor: *Be courageous and steady to the Laws and you cannot fail.* The Vice-Chancellor took Newton's advice and was dismissed from his post. However Newton continued to argue the case strongly preparing documents to be used by the University in its defense. However William of Orange had been invited by many leaders to bring an army to England to defeat James. William landed in November 1688 and James, finding that Protestants had left his army, fled to France. The University of Cambridge elected Newton, now famous for his strong defense of the university, as one of their two members to the Convention Parliament

6.2. DETERMINISTIC SDOF VIBRATION

on January 15, 1689. This Parliament declared that James had abdicated and in February 1689 offered the crown to William and Mary. Newton was at the height of his standing, seen as a leader of the university and one of the most eminent mathematicians in the world. However, his election to Parliament may have been the event that let him see that there was a life in London that might appeal to him more than the academic world in Cambridge.

After suffering a second nervous breakdown in 1693, Newton retired from research. The reasons for this breakdown have been discussed by his biographers and many theories have been proposed: chemical poisoning as a result of his alchemy experiments; frustration with his researches; the ending of a personal friendship with Fatio de Duillier, a Swiss-born mathematician resident in London; and problems resulting from his religious beliefs. Newton himself blamed lack of sleep but this was almost certainly a symptom of the illness rather than the cause of it. There seems little reason to suppose that the illness was anything other than depression, a mental illness he must have suffered from throughout most of his life, perhaps made worse by some of the events we have just listed.

Newton decided to leave Cambridge to take up a government position in London becoming Warden of the Royal Mint in 1696 and Master in 1699. However, he did not resign his positions at Cambridge until 1701. As Master of the Mint, adding the income from his estates, we see that Newton became a very rich man. For many people a position such as Master of the Mint would have been treated as simply a reward for their scientific achievements. Newton did not treat it as such and he made a strong contribution to the work of the Mint. He led it through the difficult period of recoinage and he was particularly active in measures to prevent counterfeiting of the coinage.

In 1703 he was elected president of the Royal Society and was re-elected each year until his death. He was knighted in 1705 by Queen Anne, the first scientist to be so honored for his work. However the last portion of his life was not an easy one, dominated in many ways with the controversy with Leibniz over which had invented the calculus.

Given the rage that Newton had shown throughout his life when criticized, it is not surprising that he flew into an irrational temper directed against Leibniz. Perhaps all that is worth relating here is how Newton used his position as President of the Royal Society. In this capacity he appointed an "impartial" committee to decide whether he or Leibniz was the inventor of the calculus. He wrote the official report of the committee (although of course it did not appear under his name), which was published by the Royal Society, and he then wrote a review (again anonymously), which appeared in the *Philosophical Transactions of the Royal Society*. Newton's assistant

Whiston had seen his rage at first hand. He wrote: *Newton was of the most fearful, cautious and suspicious temper that I ever knew.*

6.2.1 Free Vibration With No Damping

The undamped forced oscillator is representative of a structure where damping plays a small role in the response. It is governed by the equation,

$$\ddot{x} + \omega_n^2 x = F(t), \qquad (6.2)$$

where the external force is set equal to zero in order to solve for the free vibration. Let $x(0)$ and $\dot{x}(0)$ be, respectively, the initial displacement and initial velocity of this system. As will be shown, the free response is only driven by the initial conditions of the system. Assume a harmonic solution of the form,

$$x(t) = C_1 \sin rt + C_2 \cos rt, \qquad (6.3)$$

where r is determined when this assumed solution is used to satisfy the differential equation of motion. Substituting Equation 6.3 into Equation 6.2 we find $-r^2 + \omega_n^2 = 0$ and, therefore, $r = \omega_n$ and

$$x(t) = C_1 \sin \sqrt{\frac{k}{m}} t + C_2 \cos \sqrt{\frac{k}{m}} t. \qquad (6.4)$$

The response of the oscillator is at its natural frequency, $\omega_n = \sqrt{k/m}$. The equation $-r^2 + \omega_n^2 = 0$ is called the *characteristic equation* and r is a root, or solution, of the equation. The negative root $r = -\omega_n$ is not considered for physical reasons; there is no negative frequency of oscillation.

To evaluate the constants in the response, the initial displacement $x(0)$ and initial velocity $\dot{x}(0)$ are needed. From Equation 6.4, set up equations for $x(0)$ and $\dot{x}(0)$ to find that $C_2 = x(0)$ and $C_1 = \dot{x}(0)/\omega_n$. The general response is then

$$x(t) = x(0) \cos \omega_n t + \frac{\dot{x}(0)}{\omega_n} \sin \omega_n t. \qquad (6.5)$$

For the undamped oscillator there is no decay; the oscillation continues without a decrease in peak amplitude. The response is a function of the initial conditions, with the frequency of oscillation a function of the square root of the ratio of stiffness to mass.

Equation 6.5 can be reformulated in terms of an amplitude, frequency, and phase of vibration, replacing the constants of integration $C_{1,2}$ by amplitude A and phase angle ϕ. To accomplish this, define the equivalent oscillation as

$$x(t) = A \cos(\omega_n t - \phi), \qquad (6.6)$$

6.2. DETERMINISTIC SDOF VIBRATION

where the amplitude A and the phase angle ϕ need to be evaluated. To do this, expand the cosine term and equate to Equation 6.5, to find

$$A = \sqrt{x^2(0) + \left(\frac{\dot{x}(0)}{\omega_n}\right)^2}, \quad \phi = \tan^{-1}\left(\frac{\dot{x}(0)}{x(0)\omega_n}\right). \tag{6.7}$$

This is a convenient form, allowing us to see the effects of the initial conditions on the response amplitude and phase, parameters about which we will learn much more in subsequent discussions. The phase angle in Equation 6.6 is shown with a negative sign in anticipation of the damped response, where there is a phase lag of the structure behind the forcing.

Example 6.2 Free Vibration Without Damping

Consider a single-degree-of-freedom oscillator with $\omega_n = 12$ rad/s and zero damping. Find the response to initial conditions $x(0) = -1$ m and $\dot{x}(0) = 2$ m/s. What are the amplitude and the phase angle for a solution of the form of a cosine function (Equation 6.6)?

Solution   The solution is given by Equation 6.5 or

$$x(t) = -\cos 12t + \frac{1}{6}\sin 12t.$$

The amplitude and the phase angle are given by Equation 6.7 or

$$A = \sqrt{(-1)^2 + \left(\frac{1}{6}\right)^2} = 1.014 \text{ m}$$

$$\phi = \tan^{-1}\left(-\frac{2}{12}\right).$$

When the inverse tangent function is calculated using either a calculator or numerically on a computer, the result one finds is the phase angle of -0.1651 rad. However, this is not correct. The correct phase angle can be found by recognizing that $x(0)$ is negative, which puts the phase angle in the second quadrant. Then, the phase angle is

$$\phi = \pi - 0.1651 = 3.3067 \text{ rad}.$$

❋

6.2.2 Harmonic Forced Vibration With No Damping

Begin with governing Equation 6.2 where $F(t) = (A/m)\cos\omega t$, ω is the frequency of the driving force $F(t)$ and $\omega \neq \omega_n$. There are two components to any forced vibration response: transient or free vibration, and steady state or forced vibration. In mathematical terms, the transient response is the homogeneous solution to the differential equation, and the steady state is the particular solution. By linear superposition, we can solve each problem separately, then add both solutions to obtain the complete response. It is the complete response that is used to satisfy the initial conditions. For a damped structure, the free vibration part of the solution is quickly "forgotten," while the forced vibration response continues as long as the force exists.

The forced response is assumed to be of the functional form of the loading,
$$x(t) = B\cos\omega t.$$
Substituting this assumed solution into the governing differential equation, B can be derived,
$$B = \frac{A/m}{-\omega^2 + \omega_n^2} = \frac{A/k}{1 - (\omega/\omega_n)^2}.$$
The forced or steady state response is therefore
$$x(t) = \frac{A/k}{1 - (\omega/\omega_n)^2}\cos\omega t,$$
and the complete response (free plus forced) is given by the sum,
$$x(t) = C_1 \sin\omega_n t + C_2 \cos\omega_n t + \frac{A/k}{1 - (\omega/\omega_n)^2}\cos\omega t, \qquad (6.8)$$
where C_1 and C_2 are established by satisfying the initial conditions.

Example 6.3 Beating Response

Consider a system with natural frequency $\omega_n = 1$ rad/s, subjected to harmonic forcing per unit mass of amplitude 2 N/kg and frequency 1.2 rad/s. *(i)* Find the response when the initial conditions are both equal to zero. *(ii)* Since the natural and forcing frequencies are close to each other, the response will exhibit beating. Write the envelope function and find the beat frequency. *(iii)* Find the initial conditions that will make C_1 and C_2 in Equation 6.8 equal to zero so that the response does not contain the natural frequency components.

Solution *(i)* It is given that $\omega_n = 1$ rad/s, $A/m = 2$ N/kg. Then, the solution takes the form,
$$\begin{aligned} x(t) &= C_1 \sin t + C_2 \cos t + \frac{2}{1 - 1.2^2}\cos 1.2t \\ &= C_1 \sin t + C_2 \cos t - 4.55 \cos 1.2t. \end{aligned}$$

6.2. DETERMINISTIC SDOF VIBRATION

When zero initial conditions are applied, the constant coefficients are found to be $C_1 = 0$ and $C_2 = 4.55$. The response can be written as

$$x(t) = 4.55(\cos t - \cos 1.2t)$$
$$= 9.1 \sin 1.1t \, \sin 0.1t.$$

(ii) From the solution given above, it follows that the envelope function is $9.10 \sin 0.1t$ and the beat frequency is $2|\omega_n + \omega_f| = 0.2$ rad/s.

(iii) In order for C_1 and C_2 to equal zero, the initial conditions must be

$$x(0) = -4.55 \text{ m}$$
$$\dot{x}(0) = 0 \text{ m/s}.$$

⊛

6.2.3 Free Vibration With Damping

It is important to consider the effects of viscous damping on the free vibration of the oscillator. The governing equation is then

$$\ddot{x} + 2\zeta\omega_n\dot{x} + \omega_n^2 x = 0. \tag{6.9}$$

All vibration is affected by the dissipation of energy. Proceed by assuming $x(t) = Ae^{rt}$, differentiate appropriately, and substitute into Equation 6.9. The two roots are determined to be

$$r_{1,2} = [-\zeta \pm \sqrt{\zeta^2 - 1}]\omega_n$$

and the solution for the response is

$$x(t) = A_1 e^{-[\zeta - \sqrt{\zeta^2 - 1}]\omega_n t} + A_2 e^{-[\zeta + \sqrt{\zeta^2 - 1}]\omega_n t}. \tag{6.10}$$

It is seen that the character of the solution depends on the value of the viscous damping factor ζ. The case where $\zeta = 1$ is known as *critical damping*, since it represents the boundary between aperiodic exponentially decaying motion ($\zeta > 1$), and exponentially decaying oscillatory motion ($\zeta < 1$). The critically damped system approaches equilibrium the fastest, but most structures will not have such a high viscous damping factor.

Vibration occurs where $\zeta < 1$, the *underdamped* case, and for such a damping factor, Equation 6.10 can be reduced to the response for a damped oscillator,

$$x(t) = (A_1 e^{i\omega_d t} + A_2 e^{-i\omega_d t})e^{-\zeta\omega_n t}$$
$$= Ce^{-\zeta\omega_n t}\cos(\omega_d t - \phi)$$
$$= e^{-\zeta\omega_n t}(C_1 \sin \omega_d t + C_2 \cos \omega_d t), \tag{6.11}$$

where $\omega_d = (1 - \zeta^2)^{1/2}\omega_n$ is the damped frequency of oscillation.

Example 6.4 Free Vibration With Damping

Consider a single-degree-of-freedom system with $m = 1$ kg, $c = 2$ Ns/m, and $k = 4$ N/m. *(i)* Determine whether the system is overdamped ($\zeta > 1$), underdamped ($\zeta < 1$), or critically damped ($\zeta = 1$), and *(ii)* find the response when $x(0) = 1.2$ m and $\dot{x}(0) = 2$ m/s.

Solution The natural frequency is given by

$$\omega_n = \sqrt{\frac{k}{m}} = 2 \text{ rad/s},$$

and the damping factor is given by

$$\zeta = \frac{c}{2\omega_n m} = 0.5.$$

Since the damping factor is less than 1, the system is underdamped, and the response takes the form of Equation 6.11.

The damped natural frequency is given by

$$\omega_d = \sqrt{1 - \zeta^2}\,\omega_n = 1.73 \text{ rad/s}.$$

Applying the initial conditions, the constant coefficients in Equation 6.11 are given by

$$C_1 = 0.0772 \text{ and } C_2 = 1.2.$$

The response is given by

$$x(t) = e^{-t}\left(0.0772 \sin \omega_d t + 1.20 \cos \omega_d t\right)$$

and is plotted in Figure 6.4. Due to the relatively high value of ζ, the system oscillation decays rapidly to zero.

⊛

6.2.4 Forced Vibration With Damping

Forces on structures are actually very complex functions of time and space. By first considering idealized versions of these functions, structural response to specific loads can be more readily studied and understood. Remarkably, for linear systems, it is possible to study a few types of forces and use these cases to build more general and complex forcing models.

The governing equation of motion is, in all these instances,

$$\ddot{x} + 2\zeta\omega_n \dot{x} + \omega_n^2 x = F(t), \tag{6.12}$$

6.2. DETERMINISTIC SDOF VIBRATION

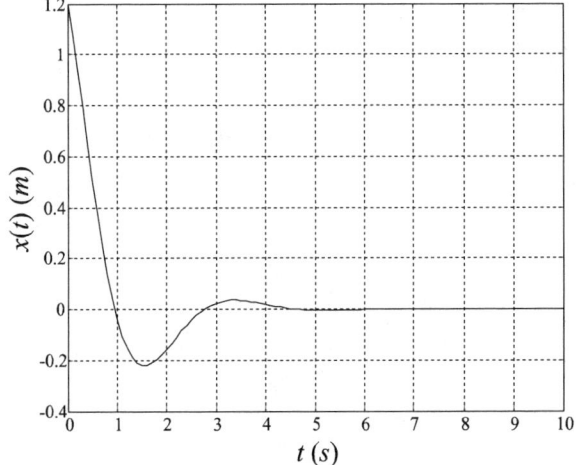

Figure 6.4: $x(t) = e^{-t}(0.0772\sin 1.73t + 1.20\cos 1.73t)$.

where specific forcing functions are substituted for $F(t)$, the force per unit mass. The governing equation of motion for harmonic forcing is given by

$$\ddot{x} + 2\zeta\omega_n\dot{x} + \omega_n^2 x = A\cos\omega t, \qquad (6.13)$$

where $A\cos\omega t$ is a harmonic load per unit mass with amplitude A oscillating at frequency ω. There is no loss in generality by having assumed that the phase of the load is set to zero, that is, let $A\cos(\omega t + \phi) = A\cos\omega t$.

The primary interest is to find the amplitude and the phase lag of the structural response. The reason why the introduction of damping causes this response lag is discussed below. In complex notation, the equivalent version of Equation 6.13 is

$$\ddot{x} + 2\zeta\omega_n\dot{x} + \omega_n^2 x = Ae^{i\omega t},$$

where the real part of the mathematical solution, $\text{Re}[x(t)]$, is the physical response.

Assume that the (particular) solution to Equation 6.13 has the form,

$$x(t) = B_1\cos\omega t + B_2\sin\omega t, \qquad (6.14)$$

where constants B_1 and B_2 are evaluated by requiring the assumed solution $x(t)$ to satisfy the governing equation, that is,

$$\left(-\omega^2 B_1 + 2\zeta\omega_n\omega B_2 + \omega_n^2 B_1 - A\right)\cos\omega t$$
$$+ \left(-\omega^2 B_2 - 2\zeta\omega_n\omega B_1 + \omega_n^2 B_2\right)\sin\omega t = 0.$$

This equation can be satisfied for all values of t only if the expressions in the parentheses vanish identically,

$$-\omega^2 B_1 + 2\zeta\omega_n\omega B_2 + \omega_n^2 B_1 = A$$
$$-\omega^2 B_2 - 2\zeta\omega_n\omega B_1 + \omega_n^2 B_2 = 0,$$

from which

$$B_1 = \frac{A(\omega_n^2 - \omega^2)}{(\omega_n^2 - \omega^2)^2 + 4\zeta^2\omega_n^2\omega^2}$$

$$B_2 = \frac{A(2\zeta\omega_n\omega)}{(\omega_n^2 - \omega^2)^2 + 4\zeta^2\omega_n^2\omega^2}.$$

Therefore, Equation 6.14 becomes

$$x(t) = \frac{A(\omega_n^2 - \omega^2)}{(\omega_n^2 - \omega^2)^2 + 4\zeta^2\omega_n^2\omega^2}\cos\omega t + \frac{A(2\zeta\omega_n\omega)}{(\omega_n^2 - \omega^2)^2 + 4\zeta^2\omega_n^2\omega^2}\sin\omega t. \tag{6.15}$$

This equation is in a form that makes the response difficult to visualize. To obtain a more useful form, Equation 6.15 can be equivalently written as

$$x(t) = D\cos(\omega t - \theta), \tag{6.16}$$

where, $D = \sqrt{B_1^2 + B_2^2}$, and $\theta = \tan^{-1}(B_2/B_1)$, or,

$$D = \frac{A}{\sqrt{(\omega_n^2 - \omega^2)^2 + (2\zeta\omega_n\omega)^2}}$$

$$= \frac{A/\omega_n^2}{\sqrt{(1 - \omega^2/\omega_n^2)^2 + (2\zeta\omega/\omega_n)^2}},$$

$$\theta = \tan^{-1}\frac{2\zeta\omega/\omega_n}{1 - \omega^2/\omega_n^2}.$$

Note that the phase lag is independent of the loading amplitude. It represents the lag in the peak structure displacement behind the load. Where $\zeta = 0$, there is no lag and the structure responds instantaneously.

A magnification factor β can be defined as

$$\beta = \frac{1}{\sqrt{(1 - \omega^2/\omega_n^2)^2 + (2\zeta\omega/\omega_n)^2}},$$

so that response Equation 6.16 simply becomes

$$x(t) = \frac{A}{\omega_n^2}\beta\cos(\omega t - \theta).$$

6.2. DETERMINISTIC SDOF VIBRATION

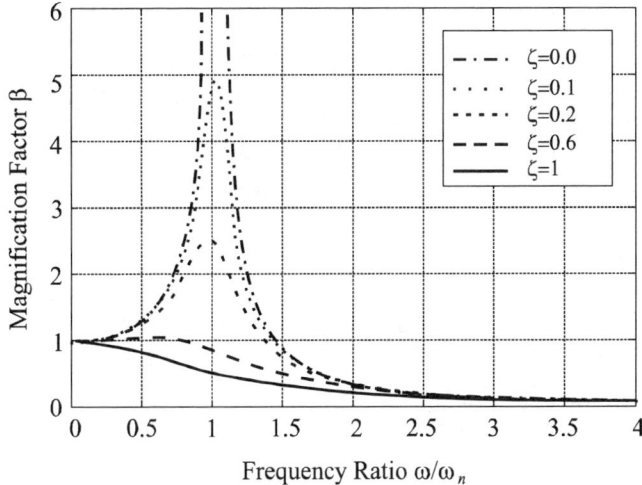

Figure 6.5: Magnification factor β as a function of ω/ω_n.

We are reminded that A incorporates the factor $1/m$, $(A = A_1/m)$, and therefore

$$D = \frac{A}{\omega_n^2}\beta = \frac{Am}{k}\beta = \frac{A_1}{k}\beta,$$

with A_1/k interpreted as the static deflection.

It is important and instructive to plot the magnification factor β and the phase lag θ as functions of the frequency ratio ω/ω_n for various values of the viscous damping factor ζ. The β curves in Figure 6.5 are a family, each curve represents a different value of ζ. An increase in the damping constant results in a decrease in the response amplitude. The maximum amplitude occurs, for $\zeta < 1$, at a frequency ratio which is slightly less[2] than 1.

The phase diagram, shown in Figure 6.6 as a function of frequency ratio, is equally interesting. For the case with no damping, for a load with frequency below resonance, the phase $\theta = 0$, and above resonance $\theta = \pi$. A discontinuity exists at the resonance point.

When damping is included, the sharp transition observed at resonance for the damped-free case is softened. For all cases regardless of damping, the

---

[2] β is maximum when

$$\frac{\omega}{\omega_n} = \sqrt{1-\zeta}.$$

Therefore, β reaches its maximum at a frequency ratio slightly less than 1 if and only if $\zeta < 1$.

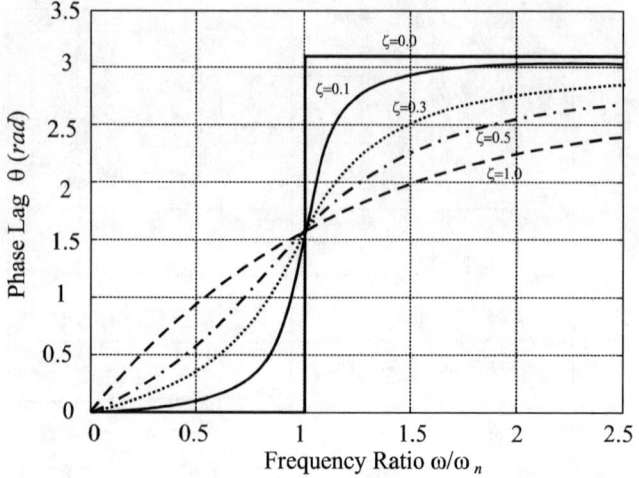

Figure 6.6: Phase lag θ as a function of ζ and ω/ω_n.

phase is $\pi/2$ radians or 90° at resonance. This property is useful in problems of structural identification and testing. If a vibration test is performed with phase data plotted as a function of ω/ω_n for a broad band of ω, then the point where this curve passes through the phase value $\pi/2$ is a resonant frequency, indicating that this is one of the natural frequencies of the structure. Structures have multiple natural frequencies, one for each degree of freedom.

6.2.5 Impulse Excitation

In this section, the response to an "impulse" at an arbitrary time $t = \tau$ is examined. Understanding the impulse response allows us to derive the general response solution. The impulse is defined as the change in momentum or

$$\text{Impulse} = \int d\,(\text{mass} \times \text{velocity})$$
$$= \int_t F(t)\,dt,$$

where Newton's second law of motion, $d\,(mv)/dt = F$, is utilized. If the force is applied for a short time duration, the impact of the force can be quantified using the concept of impulse. The Dirac delta function is often

6.2. DETERMINISTIC SDOF VIBRATION

used to represent the force in such cases, and is defined such that

$$\delta(t-\tau) = \begin{cases} 0, & t \neq \tau \\ \infty, & t = \tau, \end{cases}$$

where

$$\int_{-\infty}^{\infty} \delta(t-\tau)\,dt = 1$$

$$\int_{-\infty}^{\infty} \delta(t-\tau)f(t)\,dt = f(\tau).$$

For the force $F(t)$ given by the Dirac delta function, the impulse applied to the system at $t = \tau$ is

$$\begin{aligned} \text{Impulse} &= \int_{-\infty}^{\infty} F(t)\,dt \\ &= \int_{-\infty}^{\infty} \delta(t-\tau)\,dt \\ &= 1, \end{aligned}$$

and the system is said to be subjected to an impulse of magnitude 1, or unit impulse.

Consider the equation of motion,

$$m\ddot{x} + c\dot{x} + kx = \delta(t).$$

The solution $x(t)$ to a unit impulse applied at $t = 0$ with zero initial conditions is called the *impulse response function* and is denoted as $g(t)$. For this system, the impulse response function is given by[3]

$$g(t) = \begin{cases} (1/m\omega_d)\,e^{-\zeta\omega_n t}\sin\omega_d t, & t \geq 0 \\ 0, & t < 0. \end{cases} \qquad (6.17)$$

It makes sense that the response is zero before the impulse. By definition, the impulse response function $g(t)$ satisfies

$$m\ddot{g} + c\dot{g} + kg = \delta(t).$$

The response due to an impulse at $t = \tau$ is then

$$g(t-\tau) = \begin{cases} (1/m\omega_d)\,e^{-\zeta\omega_n(t-\tau)}\sin\omega_d(t-\tau), & t \geq \tau \\ 0, & t < \tau. \end{cases}$$

---

[3] H. Benaroya, *Mechanical Vibration*, Second Edition, Marcel Dekker, 2004.

If an impulse with magnitude f_o is applied at $t = \tau$, then $F(t) = f_o \delta(t - \tau)$ and the response is given by $f_o g(t - \tau)$.

For the governing equation,

$$\ddot{x} + 2\zeta\omega_n \dot{x} + \omega_n^2 x = \delta(t - \tau),$$

where $\delta(t - \tau)$ in this case is the force per unit mass, the impulse response function is given by

$$g(t) = \begin{cases} (1/\omega_d) e^{-\zeta\omega_n(t-\tau)} \sin \omega_d (t - \tau), & t \geq 0 \\ 0, & t < 0. \end{cases} \quad (6.18)$$

This is also the response when a unit impulse is applied to a unit mass at $t = \tau$.

Either definition of the impulse response function (Equation 6.17 or 6.18) is valid as long as we know which case we have considered.

The impulse response function can be simplified using the unit step function $u(t)$,

$$g(t) = \frac{1}{\omega_d} e^{-\zeta\omega_n t} \sin \omega_d t \cdot u(t),$$

where

$$u(t) = \begin{cases} 1, & t \geq 0 \\ 0, & t < 0. \end{cases}$$

Before proceeding to the derivations for arbitrary forcing, note that the unit impulse function $\delta(t)$ can be thought of as a force applied for a very short time duration Δt with a force magnitude of $1/\Delta t$, as shown in Figure 6.7. This idea is useful in the next section.

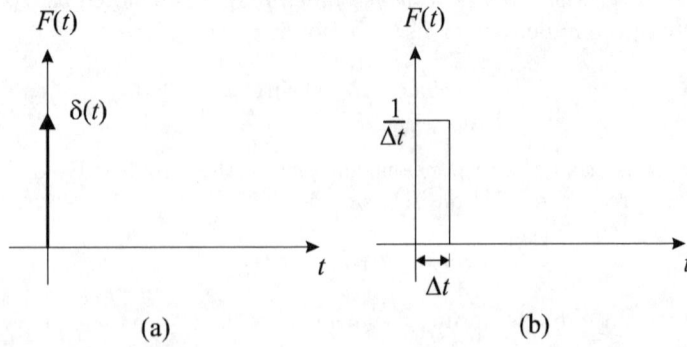

Figure 6.7: Unit impulse function, (a) idealized, and (b) closer to reality.

6.2.6 Arbitrary Loading: Convolution

In vibration, it is necessary to calculate the response of a structure to an arbitrary but defined load. Consider such an arbitrary deterministic load per unit mass, $F(t)$. Approximate the arbitrary function by a series of rectangles, vertical or horizontal, as shown in Figure 6.8. In the limit of zero thickness, both will approach the exact curve, but the approximating equations will be different.

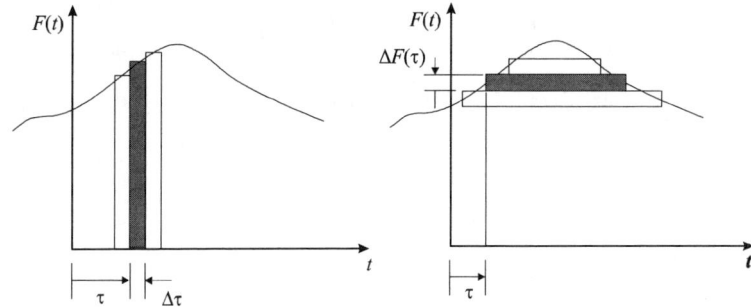

Figure 6.8: Two limiting cases for modeling an arbitrary load.

Consider the case where vertical impulses are used to approximate $F(t)$. As the figure shows, the impulse at time τ has a force magnitude $F(\tau)$ and a duration $\Delta\tau$. Each impulse has an intensity of $F(\tau)\Delta\tau$. The response due to the impulse applied at $t = \tau$ with intensity $F(\tau)\Delta\tau$ is

$$\Delta x(t, \tau) = F(\tau)\Delta\tau g(t - \tau).$$

Approximate the complete force by a sum of such impulses,

$$F(t) \simeq \sum_{\tau} F(\tau)\Delta\tau,$$

where τ is varied, resulting in the approximate response $x(t) \simeq \sum_{\tau} \Delta x(t, \tau)$, or

$$x(t) \simeq \sum_{\tau} F(\tau)\Delta\tau g(t - \tau).$$

Take the limit $\Delta\tau \to 0$,

$$x(t) = \int_{-\infty}^{t} F(\tau)g(t - \tau)d\tau, \qquad (6.19)$$

which is known as the convolution or Duhamel integral[4], and the force is applied at $t \to -\infty$. For the force applied at $t = 0$, the response is given by

$$x(t) = \int_0^t F(\tau)g(t-\tau)d\tau, \qquad (6.20)$$

since $g(\cdot)$ is zero if the argument is less than zero. Note that by letting $\tau^* = t - \tau$, and $\tau^* = -d\tau$, Equation 6.19 can be written as

$$x(t) = \int_0^\infty F(t-\tau^*)g(\tau^*)d\tau^*. \qquad (6.21)$$

Keep in mind that $F(t)$ and $g(t)$ are each zero for $t < 0$; the response can then be written as

$$\begin{aligned} x(t) &= \int_{-\infty}^\infty F(\tau)g(t-\tau)d\tau \\ &= \int_{-\infty}^\infty g(\tau)F(t-\tau)d\tau. \end{aligned} \qquad (6.22)$$

It is the superposition principle for linear systems that validates the idea of the convolution. Figure 6.9 visually depicts the meaning of the convolution of the two functions $g(t)$ and $F(t)$, which yields $x(t)$.

The convolution solution is due to force $F(t)$ with zero initial conditions. If the initial conditions are nonzero, then the full solution should include the result from the free response. That is, for the system governed by

$$\ddot{x} + 2\zeta\omega_n\dot{x} + \omega_n^2 x = F(t),$$

the complete solution is

$$\begin{aligned} x(t) &= e^{-\zeta\omega_n t}\left(\frac{\dot{x}(0) + \zeta\omega_n x(0)}{\omega_d}\sin\omega_d t + x(0)\cos\omega_d t\right) \\ &\quad + \frac{1}{\omega_d}\int_0^t F(\tau)e^{-\zeta\omega_n(t-\tau)}\sin\omega_d(t-\tau)d\tau, \end{aligned} \qquad (6.23)$$

where it is clear that the initial conditions can significantly affect the response at early time.

Duhamel   Jean-Marie Duhamel was a student at the École Polytechnique and then he became professor there in 1830. He was highly thought of as a teacher of mathematics and is reported to have given very fine lectures.

---

[4] It is called Duhamel when structural parameters are used.

6.2. DETERMINISTIC SDOF VIBRATION

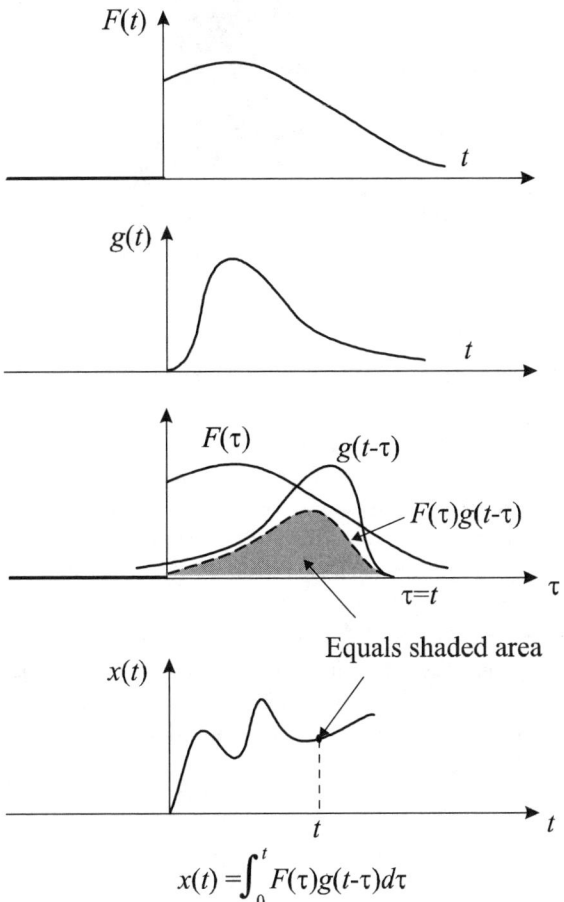

Figure 6.9: The convolution of $g(t)$ and $F(t)$.

Figure 6.10: Jean-Marie Duhamel (1797-1872).

During the period 1848 until 1851 Duhamel was Director of Studies at the École Polytechnique. From 1851, he again filled the analysis chair at the École Polytechnique. Also from 1851 he was professor at the Faculté des Sciences in Paris.

Duhamel worked on partial differential equations and applied his methods to the theory of heat, to rational mechanics and to acoustics. His acoustical studies involved vibrating strings and the vibration of air in cylindrical and conical pipes. His techniques in the theory of heat were mathematically similar to Fresnel's work in optics.

His theory of the transmission of heat in crystal structures was based on the work of Fourier and Poisson. "Duhamel's principle" in partial differential equations arose from his work on the distribution of heat in a solid with a variable boundary temperature. Duhamel was elected to the Académie des Sciences in 1840.

Example 6.5 Response Using the Impulse Response Function

Consider a mass-spring system subjected to a harmonic force with frequency ω_f. The equation of motion is given by

$$m\ddot{x} + kx = \cos\omega_f t.$$

Find the displacement $x(t)$ using the convolution integral assuming that the force is applied for $t \geq 0$ and the initial conditions equal zero. Compare this result with the general solution of Section 6.2.2.

6.2. DETERMINISTIC SDOF VIBRATION

Solution The unit impulse response function for the undamped system is given by

$$g(t) = \frac{1}{m\omega_n} \sin \omega_n t, \quad t \geq 0.$$

Then, response Equation 6.20 becomes

$$\begin{aligned} x(t) &= \frac{1}{m\omega_n} \int_0^t \cos \omega_f \tau \sin \omega_n (t - \tau) \, d\tau \\ &= \frac{1}{m} \frac{\cos \omega_f t - \cos \omega_n t}{\omega_n^2 - \omega_f^2}. \end{aligned}$$

Note that the lower limit of the integral is zero since the force is not applied for $t < 0$.

The solution obtained by solving the differential equation directly is

$$x(t) = c_1 \sin \omega_n t + c_2 \cos \omega_n t + \frac{\cos \omega_f t}{m\left(\omega_n^2 - \omega_f^2\right)}. \tag{6.24}$$

To compare the two solutions, apply the zero initial conditions, resulting in the constants of integration,

$$\begin{aligned} c_1 &= 0 \\ c_2 &= -\frac{1}{m\left(\omega_n^2 - \omega_f^2\right)}. \end{aligned}$$

Substitute these into Equation 6.24 and find

$$x(t) = \frac{1}{m} \frac{\cos \omega_f t - \cos \omega_n t}{\omega_n^2 - \omega_f^2},$$

which agrees with the solution obtained using the convolution integral.
⊛

6.2.7 Frequency Response Function

Approach the solution to Equation 6.12 from the frequency domain using Fourier transforms. First, take the Fourier transform of the equation of motion[5] to obtain

$$\left[(i\omega)^2 + i2\zeta\omega_n\omega + \omega_n^2\right] X(\omega) = F(\omega),$$

---
[5] The Fourier transform pair are defined in Equations 5.20 and 5.21.

where the Fourier transforms of time derivatives are available in Equations 5.24 and 5.25, and $F(\omega)$ is the Fourier transform of $F(t)$, the force per unit mass. Solve for $X(\omega)$,

$$X(\omega) = \frac{F(\omega)}{(i\omega)^2 + i2\zeta\omega_n\omega + \omega_n^2}. \qquad (6.25)$$

The inverse Fourier transform is the response $x(t)$.

The term $1/\left[(i\omega)^2 + i2\zeta\omega_n\omega + \omega_n^2\right]$ is called the *frequency response function* and is denoted as $H(i\omega)$,

$$H(i\omega) = \frac{1}{(i\omega)^2 + i2\zeta\omega_n\omega + \omega_n^2}. \qquad (6.26)$$

Equation 6.25 becomes

$$X(\omega) = H(i\omega) F(\omega).$$

$H(i\omega)$ is also called the *transfer function* for this system since it relates the input $F(\omega)$ to the output $X(\omega)$. Note that the transfer function is sometimes written as $H(i\omega)$ because i and ω always appear together. The notation $H(i\omega)$ is also useful because it is easier to see that $H(-\omega)$ is the complex conjugate of $H(\omega)$, that is,

$$H(-\omega) = H^*(\omega).$$

Here the simpler notation $H(\omega)$ is utilized.

Now consider the same equation of motion with $F(t) = \delta(t)$. By definition, the response to a unit impulse load is called the impulse response function and is denoted as $g(t)$. The equation of motion is given by

$$\ddot{g} + 2\zeta\omega_n\dot{g} + \omega_n^2 g = \delta(t).$$

Take the Fourier transform of the equation of motion,

$$G(\omega) = \frac{1}{2\pi\left[(i\omega)^2 + i2\zeta\omega_n\omega + \omega_n^2\right]},$$

where $G(\omega)$ is the Fourier transform of $g(t)$ and the Fourier transform of $\delta(t)$ is $1/2\pi$.

Then, Equation 6.25 can be written as

$$X(\omega) = 2\pi G(\omega) F(\omega), \qquad (6.27)$$

and $G(\omega)$ is related to $H(\omega)$ by

$$H(\omega) = 2\pi G(\omega)$$

6.2. DETERMINISTIC SDOF VIBRATION

or
$$H(\omega) = \int_{-\infty}^{\infty} g(t) \exp(-i\omega t)\, dt. \tag{6.28}$$

Using the Fourier transform pair in Equation 5.22 with $A = 1$ and $B = 1/2\pi$, we can immediately write the impulse response function as the transform of $H(\omega)$,
$$g(t) = \frac{1}{2\pi} \int_{-\infty}^{\infty} H(\omega) \exp(i\omega t)\, d\omega. \tag{6.29}$$

Using the convolution theorem, the inverse Fourier transform of Equation 6.27 is given by
$$\begin{aligned} x(t) &= 2\pi \mathcal{F}^{-1}\{G(\omega)F(\omega)\} \\ &= \int_{-\infty}^{\infty} F(\tau)g(t-\tau)d\tau \tag{6.30} \\ &= \int_{-\infty}^{\infty} g(\tau)F(t-\tau)d\tau. \tag{6.31} \end{aligned}$$

Since $g(t) = 0$ for $t < 0$, Equation 6.30 can be written as
$$x(t) = \int_{-\infty}^{t} F(\tau)g(t-\tau)d\tau. \tag{6.32}$$

Similarly, Equation 6.31 can be written as
$$x(t) = \int_{0}^{\infty} g(\tau)F(t-\tau)d\tau.$$

Equation 6.32 is identical to the previous result in Equation 6.19. Further assuming that the force is applied at $t = 0$,
$$\begin{aligned} x(t) &= \int_{0}^{t} F(\tau)g(t-\tau)d\tau \\ &= \int_{0}^{t} g(\tau)F(t-\tau)d\tau, \end{aligned}$$

which is also identical to Equation 6.20.

Example 6.6 Relative Motion

Consider a package containing sensitive equipment transported by the vehicle shown in Figure 6.11. The packing of the equipment can be modeled by a spring and a damper. The displacement of the vehicle is being recorded and is given by $y(t)$, and the displacement of the equipment with respect to the truck is given by $x(t)$. (*i*) Obtain the equation of motion for the equipment, (*ii*) find the transfer function $H(\omega)$ for the relative displacement, $x - y$, per unit acceleration of the vehicle, and (*iii*) find the transfer function for the relative velocity, $\dot{x} - \dot{y}$.

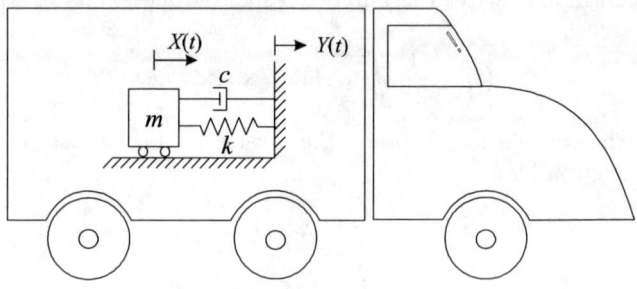

Figure 6.11: Truck transporting a sensitive instrument.

Solution (*i*) The sum of the forces in the horizontal direction are given by

$$\sum F_x = m\ddot{x}$$
$$= -k(x-y) - c(\dot{x}-\dot{y}).$$

Let the relative displacement be denoted as $z = x - y$, and the equation of motion can be rewritten as

$$m\ddot{z} + c\dot{z} + kz = -m\ddot{y}.$$

(*ii*) Take the Fourier transform of the equation of motion,

$$(-m\omega^2 + k + i\omega c)\mathcal{F}(z) = -m\mathcal{F}(\ddot{y})$$
$$\frac{\mathcal{F}(z)}{\mathcal{F}(\ddot{y})} = \frac{-m}{-m\omega^2 + k + i\omega c}$$
$$H(\omega) = -\frac{1}{\omega_n^2 - \omega^2 + i\omega 2\zeta\omega_n}.$$

(*iii*) Since $\mathcal{F}(\dot{z}) = i\omega\mathcal{F}(z)$,

$$\frac{\mathcal{F}(\dot{z})}{\mathcal{F}(\ddot{y})} = H(\omega) = -\frac{i\omega}{\omega_n^2 - \omega^2 + i\omega 2\zeta\omega_n}.$$

6.3 SDoF: The Response to Random Loads

Consider the second-order differential equation[6] governing the linear motion of an oscillator,

$$\ddot{X} + 2\zeta\omega_n\dot{X} + \omega_n^2 X = F(t), \tag{6.33}$$

where the input force per unit mass is given by stationary random process $F(t)$ and the output displacement by random process $X(t)$.

The results are presented *a priori* so that the reader will be able to better follow the mathematical manipulations that follow,

$$\mu_X = H(0)\mu_F$$
$$S_{XX}(\omega) = |H(\omega)|^2 S_{FF}(\omega)$$
$$|H(\omega)|^2 = \left[\frac{1/\omega_n^2}{\sqrt{(1-\omega^2/\omega_n^2)^2 + (2\zeta\omega/\omega_n)^2}}\right]^2.$$

These equations indicate that the mean value of the response, μ_X, is proportional to the mean value of the force μ_F, and the response spectral density, $S_{XX}(\omega)$, is proportional to the force spectral density $S_{FF}(\omega)$. For both results the proportionality constants depend on the structural or system parameters through the frequency response function.

6.3.1 Formulation

Consider the linear system defined by Equation 6.33 and assume $F(t)$ to be stationary, with mean μ_F and power spectrum $S_{FF}(\omega)$. The stationarity assumption for the forcing means that transient dynamic behavior cannot be directly considered here.[7] Thus, the initial loading transients of an earthquake, a wind gust, or an extreme ocean wave cannot be considered as stationary. Assuming that the character of the loading does not change, steady-state behavior can be assumed to be statistically stationary.

---

[6] In this instance, system parameters ζ and ω_n are deterministic, and, therefore, so is the governing equation of motion. If ζ and ω_n are either random variables or random processes then the governing equation is random. The case of random parameters is much more complicated because the system itself is random rather than just the forcing. We would have to solve the problem for the "many randomly prescribed systems" rather than just for one system with randomly prescribed forces.

[7] There are, however, clever ways by which stationary solutions can be utilized in nonstationary cases. One possibility is to multiply the stationary process by a deterministic time function such that the product is an *evolutionary* or nonstationary process. For example, use $A(t)F(t)$ as the forcing function, where $A(t)$ is a deterministic transient function and $F(t)$ is stationary. Nonstationary models are studied in Chapter 11.

6.3.2 Derivation of Equations

Begin with the convolution equation for the response of a linear oscillator, Equation 6.22, now written for random process $F(t)$,

$$X(t) = \int_{-\infty}^{\infty} g(\tau) F(t-\tau) d\tau, \qquad (6.34)$$

where $g(t)$ is the impulse response function given by

$$g(t) = \frac{1}{\omega_d} e^{-\zeta \omega_n t} \sin \omega_d t \cdot u(t),$$

the unit step function $u(t) = 1$ for $t \geq 0$ and zero otherwise, and stationary random load per unit mass $F(t)$ is applied at $t = -\infty$, that is, long before the present time to ensure stationarity. Beginning with Equation 6.34, take the expected value of both sides, and use the linear property of mathematical expectation to interchange it with the integral:

$$\begin{aligned}
E\{X(t)\} &= \int_{-\infty}^{\infty} g(\tau) E\{F(t-\tau)\} d\tau \\
&= E\{F(t)\} \int_{-\infty}^{\infty} g(\tau) d\tau \\
&= \mu_F \int_{-\infty}^{\infty} g(\tau) d\tau, \qquad (6.35)
\end{aligned}$$

where the stationarity of $F(t)$ was utilized in the second and third equations,[8] and the definition of the frequency response function is given in Equation 6.28. The integral in Equation 6.35 is recognized as $H(0) = H(\omega)|_{\omega=0}$. Using Equation 6.26, we arrive at the first important result,

$$\mu_X = H(0) \mu_F = \frac{1}{\omega_n^2} \mu_F, \qquad (6.36)$$

that since μ_F is time independent, then so must be μ_X. In addition, Equation 6.36 is identical to the static relation,

$$F_{static} = \frac{k}{m} x_{static},$$

where F_{static} is the static force per unit mass. Therefore, Equation 6.36 is an important result since it says that the dynamic problem can be treated separately and can be added later to the mean response μ_X.

In order to derive the output spectral density, intermediate results involving the correlation function need to be derived first.

---

[8] The force is stationary and has a constant mean value.

6.3. SDOF: THE RESPONSE TO RANDOM LOADS

6.3.3 Response Correlations

Since $f_{X_1 X_2}$ is not known, $R_{XX}(\tau)$ needs to be derived in an indirect way. First, derive the cross-correlation between $F(t)$ and $X(t)$. Multiply both sides of Equation 6.34 by $F(t - \alpha_1)$, and take expected values of both sides,

$$E\{X(t)F(t-\alpha_1)\} = \int_{-\infty}^{\infty} g(\tau_1) E\{F(t-\tau_1)F(t-\alpha_1)\}d\tau_1,$$

where

$$E\{F(t-\tau_1)F(t-\alpha_1)\} = R_{FF}(\tau_1 - \alpha_1)$$

is the autocorrelation of the force, and

$$E\{X(t)F(t-\alpha_1)\} = R_{XF}(-\alpha_1) = R_{XF}(\alpha_1)$$

is the cross-correlation between loading $F(t)$ and response $X(t)$. Thus,

$$R_{XF}(\alpha_1) = \int_{-\infty}^{\infty} g(\tau_1) R_{FF}(\tau_1 - \alpha_1) d\tau_1, \qquad (6.37)$$

where $R_{FF}(\tau)$ is known from experimental data. Next, multiply both sides of Equation 6.34 by $X(t + \alpha_2)$ and take expected values of both sides,

$$E\{X(t)X(t+\alpha_2)\} = \int_{-\infty}^{\infty} g(\tau_2) E\{F(t-\tau_2)X(t+\alpha_2)\}d\tau_2$$

$$R_{XX}(\alpha_2) = \int_{-\infty}^{\infty} g(\tau_2) R_{XF}(\tau_2 + \alpha_2) d\tau_2. \qquad (6.38)$$

Substitute Equation 6.37 into Equation 6.38 to find[9]

$$R_{XX}(\tau) = \int_{-\infty}^{\infty}\int_{-\infty}^{\infty} g(\alpha) g(\beta) R_{FF}(\tau + \beta - \alpha) d\alpha d\beta, \qquad (6.39)$$

which is a double convolution. To evaluate the variance, use

$$\sigma_X^2 = E\{X^2(t)\} - E^2\{X(t)\}$$
$$= R_{XX}(0) - \left[\frac{\mu_F}{\omega_n^2}\right]^2. \qquad (6.40)$$

While this information on the correlation is of interest, a more important result is the response spectral density, which is derived in the next section, after two examples.

---

[9] Keep careful track of the *dummy variables* so that appropriate arguments are maintained. Here, let $\alpha_2 \equiv \tau$, $\tau_1 \equiv \alpha$ and $\tau_2 \equiv \beta$ in order to simplify the notation.

Example 6.7 Response Mean and Variance

Following the previous discussion, examine how the response mean and variance (or standard deviation) can be very useful in a design. Suppose that an analysis results in the following response statistics: μ_X and σ_X, where stationarity has been assumed. Stationarity implies that the mean value is not a function of time and that the correlation is only a function of time difference τ.

Equation 6.40 was obtained by setting $\tau = 0$ and by substituting the response mean value. The designer needs both the mean value and the variance to establish bounds on the possible response. Example response bounds are: $\mu_X \pm \sigma_X$, $\mu_X \pm 2\sigma_X$, or $\mu_X \pm 3\sigma_X$. Of course, the larger the *sigma bounds* the more likely that all possible responses are covered. Along with a higher probability, that is, a safer structure, comes this broader band with its vagueness. There is no way around this uncertainty-type principle. These upper and lower bounds are used to define the least likely and most likely range of responses. If designing for strength, then the upper sigma bound can be used to size the structural components.

How wide or narrow the sigma bounds are depends on the underlying density function. For the Gaussian, there is an approximately 1:3 chance of being outside the one sigma bounds, but only a 1:200 chance of being outside the two sigma bounds. Different densities are structured differently.

Therefore, the designer must study the data in order to better understand the underlying density. There is no easy or clear-cut answer regarding how many sigma bounds to use in a design. As a practical matter, by retaining larger sigma bounds in the design, the design becomes more conservative, leading to a more costly structure or product.

Example 6.8 Nonstationary Transient Response

Consider the transient response of a single-degree-of-freedom system with equation of motion given by

$$\ddot{X}(t) + 2\zeta\omega_n \dot{X}(t) + \omega_n^2 X(t) = F(t),$$

where $F(t)$ is a zero mean stationary random process and $X(t)$ has zero initial conditions. Find the mean value and the variance of the response. Assume that the mass, stiffness, and damping are deterministic.

6.3. SDOF: THE RESPONSE TO RANDOM LOADS

Solution The general displacement response is given by

$$X(t) = X(0) e^{-\zeta \omega_n t} \left(\cos \omega_d t + \zeta \frac{\omega_n}{\omega_d} \sin \omega_d t \right) + \frac{\dot{X}(0)}{\omega_d} e^{-\zeta \omega_n t} \sin \omega_d t$$
$$+ \int_0^t g(t-\tau) F(\tau) d\tau,$$

where

$$g(t) = \frac{1}{\omega_d} e^{-\zeta \omega_n t} \sin \omega_d t.$$

The mean displacement is given by

$$E\{X(t)\} = E\{X(0)\} e^{-\zeta \omega_n t} \left(\cos \omega_d t + \zeta \frac{\omega_n}{\omega_d} \sin \omega_d t \right)$$
$$+ E\{\dot{X}(0)\} \frac{1}{\omega_d} e^{-\zeta \omega_n t} \sin \omega_d t + \int_0^t g(t-\tau) E\{F(\tau)\} d\tau,$$

where

$$\int_0^t g(t-\tau) E\{F(\tau)\} d\tau$$
$$= \frac{\mu_F}{\omega_n^2} \left(1 - e^{-\zeta \omega_n t} \cos \omega_d t - \frac{\zeta}{\sqrt{1-\zeta^2}} e^{-\zeta \omega_n t} \cos \omega_d t \right). \quad (6.41)$$

As $t \to \infty$, the terms due to the initial conditions approach zero and the mean value response approaches μ_F/ω_n^2, which agrees with Equation 6.36. This shows that the transient response to a stationary force approaches the theoretical stationary response with increasing time. How fast it approaches the stationary value depends on the system damping ζ.

Consider next the mean-square response,

$$E\{X^2(t)\} = \int_0^t \int_0^t g(t-\tau_1) g(t-\tau_2) R_{FF}(\tau_2 - \tau_1) d\tau_1 d\tau_2.$$

Substitute $R_{FF}(t) = \int_{-\infty}^{\infty} S_{FF}(\omega) \cos \omega t \, d\omega$, to find

$$E\{X^2(t)\} = \int_0^t \int_0^t \int_{-\infty}^{\infty} S_{FF}(\omega) \cos \omega (\tau_2 - \tau_1)$$
$$\cdot g(t-\tau_1) g(t-\tau_2) \, d\omega \, d\tau_1 d\tau_2.$$

The order of integration can be changed so that

$$E\{X^2(t)\}$$
$$= \frac{1}{\omega_d^2} \int_{-\infty}^{\infty} S_{FF}(\omega) \left[\int_0^t \int_0^t \cos \omega (\tau_2 - \tau_1) \exp(-\zeta \omega_n (2t - \tau_1 - \tau_2)) \right.$$
$$\left. \cdot \sin \omega_d (t-\tau_1) \sin \omega_d (t-\tau_2) \, d\tau_1 d\tau_2 \right] d\omega.$$

The integrations in τ_1 and τ_2 are performed to give

$$\begin{aligned}E\left\{X^2(t)\right\} &= \int_{-\infty}^{\infty} \frac{1}{\left(\omega_n^2 - \omega^2\right)^2 + 4\zeta^2\omega_n^2\omega^2} S_{FF}(\omega)\, d\omega \\ &\quad \cdot \left\{1 + e^{-2\zeta\omega_n t} + 2\frac{\zeta}{\sqrt{1-\zeta^2}} e^{-2\zeta\omega_n t} \sin\omega_d t \cos\omega_d t \right. \\ &\quad - 2e^{-\omega_n\zeta t}\left(\cos\omega_d t + \frac{\zeta}{\sqrt{1-\zeta^2}}\sin\omega_d t\right)\cos\omega t \\ &\quad - 2e^{-\omega_n\zeta t}\frac{\omega}{\omega_d}\sin\omega_d t \sin\omega t \\ &\quad \left. + e^{-2\zeta\omega_n t}\frac{(\omega_n\zeta)^2 - \omega_d^2 + \omega^2}{\omega_d^2}\sin^2\omega_d t\right\}. \end{aligned} \qquad (6.42)$$

This result was first obtained by Caughey and Stumpf.[10] As $t \to \infty$, the transient mean square response approaches the stationary value,

$$\lim_{t\to\infty} E\left\{X^2(t)\right\} = \int_{-\infty}^{\infty} \frac{1}{\left(\omega_n^2 - \omega^2\right)^2 + 4\zeta^2\omega_n^2\omega^2} S_{FF}(\omega)\, d\omega.$$

The transient response approaches the stationary response with increasing time. Equation 6.42 can be further reduced assuming that the forcing is represented by ideal white noise, $S_{FF}(\omega) = S_0$. Then,

$$\begin{aligned}E\left\{X^2(t)\right\} &= \frac{S_0\pi}{2\zeta\omega_n^3}\left(1 - e^{-2\omega_n\zeta t}\left(1 + \frac{\zeta}{\sqrt{1-\zeta^2}}\sin 2\omega_d t\right.\right. \\ &\quad \left.\left. + 2\frac{\zeta^2}{1-\zeta^2}\sin^2\omega_d t\right)\right),\end{aligned}$$

where the first term is the stationary value. This result is confirmed in Equation 6.45. Figure 6.12 shows the transient mean and the mean-square responses for various values of ζ.

Chapter 11 provides a fuller introduction to nonstationary processes and vibration.

⊛

6.3.4 Response Spectral Density

Begin with the Fourier transform relation between power spectrum and correlation function, $S_{XX}(\omega) = \frac{1}{2\pi}\int_{-\infty}^{\infty} R_{XX}(\tau)e^{-i\omega\tau}d\tau$ and substitute

---
[10] Caughey, T.K. and Stumpf, H.J., 1961, "Transient Response of a Dynamic System under Random Excitation," ASME *J. Applied Mechanics*, 28, pp. 563-566.

6.3. SDOF: THE RESPONSE TO RANDOM LOADS

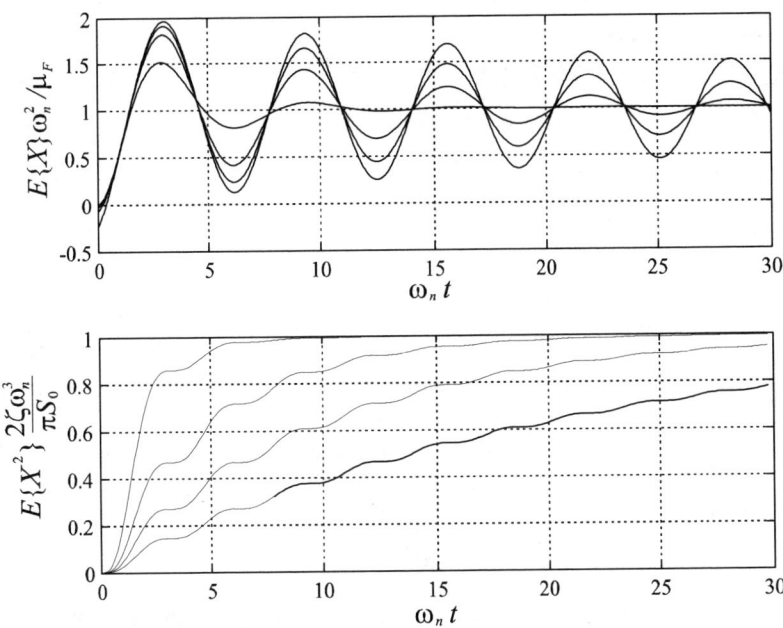

Figure 6.12: Transient mean and mean-square responses for $\zeta = 0.025, 0.05,$ 0.1, and 0.3.

Equation 6.39 for $R_{XX}(\tau)$, with $\lambda = \tau + \beta - \alpha$,

$$\begin{aligned}S_{XX}(\omega) &= \frac{1}{2\pi}\int_{-\infty}^{\infty} e^{-i\omega\tau}\left[\int_{-\infty}^{\infty}\int_{-\infty}^{\infty} g(\alpha)g(\beta)R_{FF}(\lambda)\,d\alpha d\beta\right]d\tau \\ &= \int_{-\infty}^{\infty} g(\alpha)e^{-i\omega\alpha}d\alpha \cdot \int_{-\infty}^{\infty} g(\beta)e^{+i\omega\beta}d\beta \\ &\quad \cdot \frac{1}{2\pi}\int_{-\infty}^{\infty} R_{FF}(\lambda)e^{-i\omega\lambda}d\lambda \\ &= H(\omega)H^*(\omega)S_{FF}(\omega),\end{aligned}$$

where $*$ denotes complex conjugate. The last equality is obtained by applying the appropriate definitions. Therefore,

$$S_{XX}(\omega) = |H(\omega)|^2 S_{FF}(\omega). \tag{6.43}$$

This is the **fundamental** result for random vibration and linear systems theory that relates the output spectral density to the known input spectral density and the system frequency response. It is emphasized here that the derivation of Equation 6.43 was based on the validity of the convolution equation, true for linear systems and structures. Any generalization for nonlinear behavior requires problem-specific approaches.[11] See Chapter 10.

Example 6.9 Oscillator Response to White Noise

Consider a simple application of the above ideas to an oscillator. What is the response of a damped oscillator to a force with white noise density?
Solution   The governing equation of motion is

$$\ddot{X} + 2\zeta\omega_n \dot{X} + \omega_n^2 X = F(t),$$

where $F(t)$ is the external force per unit mass, the frequency response function is given by

$$H(\omega) = \frac{1}{\omega_n^2 + i2\zeta\omega_n\omega + (i\omega)^2}.$$

The magnitude squared of $H(\omega)$ is given by[12]

$$|H(\omega)|^2 = \frac{1}{(\omega_n^2 - \omega^2)^2 + (2\zeta\omega_n\omega)^2}.$$

---

[11] Two widely used techniques for nonlinear problems are *stochastic linearization*, which leads to the use of linear theory, and *perturbation methods*, which transform a nonlinear equation into an infinite sequence of linear equations, again leading to linear equations. Therefore, the linear theory derived here is broadly useful.

[12] Multiply numerator and denominator by the complex conjugate,

$$\omega_n^2 - i2\zeta\omega_n\omega + (i\omega)^2.$$

6.3. SDOF: THE RESPONSE TO RANDOM LOADS

Therefore, given any input spectral density $S_{FF}(\omega)$, the response spectral density is

$$S_{XX}(\omega) = |H(\omega)|^2 S_{FF}(\omega) = \frac{S_{FF}(\omega)}{\left(\omega_n^2 - \omega^2\right)^2 + (2\zeta\omega_n\omega)^2}.$$

Suppose, for mathematical simplicity, that the forcing is white noise with $S_{FF}(\omega) = S_0$. Then,

$$S_{XX}(\omega) = \frac{S_0}{\left(\omega_n^2 - \omega^2\right)^2 + (2\zeta\omega_n\omega)^2}, \tag{6.44}$$

and the mean square response is given by

$$E\{X^2(t)\} = \int_{-\infty}^{\infty} S_{XX}(\omega)\,d\omega = \frac{\pi S_0}{2\omega_n^3 \zeta} = \frac{\pi m^2 S_0}{kc}. \tag{6.45}$$

This integral is not standard, but can be found in texts on random vibration. For example, the integral of this example problem is a specialized version of

$$\int_{-\infty}^{\infty} \left|\frac{B_0 + i\omega B_1}{A_0 + i\omega A_1 - \omega^2 A_2}\right|^2 d\omega = \frac{\pi \left(A_0 B_1^2 + A_2 B_0^2\right)}{A_0 A_1 A_2}, \tag{6.46}$$

where

$$A_0 = \omega_n^2 \quad A_1 = 2\zeta\omega_n \quad A_2 = 1 \quad B_0 = 1 \quad B_1 = 0.$$

Even though infinite mean-square energy is input to the system,[13] it responds with a finite mean-square energy. See Figure 6.13 for plots of the components of Equation 6.44. (Only the positive frequencies are shown, but the plots are symmetric about $\omega = 0$.) White noise is useful and frequently used, even though it is nonphysical, because it leads to good approximate results.

The mean-square response can also be written in terms of a one-sided spectrum using Equation 5.38,

$$E\{X^2(t)\} = \frac{\pi m^2 S_0}{kc} = \frac{\pi^2 m^2 W_0}{4kc},$$

using the one-sided density, $S_0 = \pi W_0/4$.

⊛

---

[13] The energy input equals the area under the spectral density, which for white noise is

$$\int_{-\infty}^{\infty} S_0\,d\omega = \infty.$$

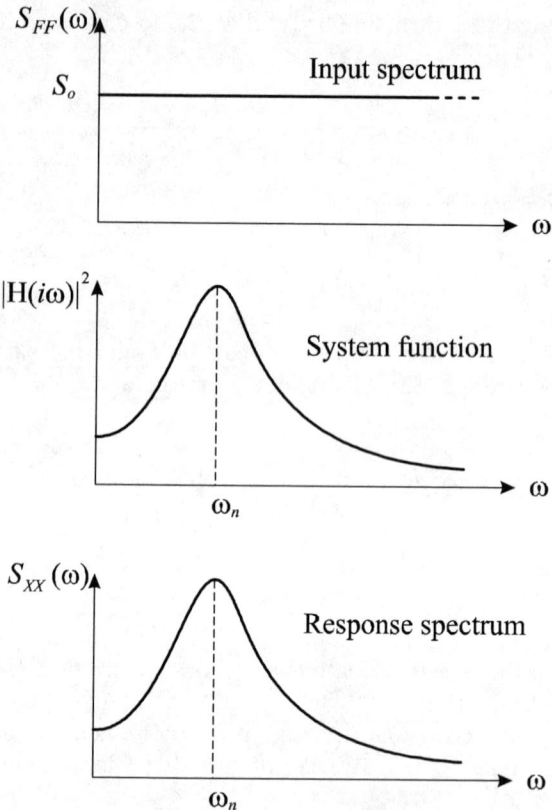

Figure 6.13: The input spectrum $S_{FF}(\omega)$ is filtered by the frequency response function $|H(i\omega)|^2$, resulting in the response spectrum $S_{XX}(\omega)$. The structure or system function acts as a filter allowing more energy to pass through in the frequency range around the natural frequency.

6.3. SDOF: THE RESPONSE TO RANDOM LOADS

Example 6.10 An SDoF System Subjected to White Noise Excitation

Consider a single-degree-of-freedom system subjected to white noise excitation with intensity S_0 N²s. Determine the RMS force in the spring.

Solution The response spectrum is related to the force spectrum by Equation 6.43,

$$S_{XX}(\omega) = |H(\omega)|^2 S_{FF}(\omega).$$

In this case, F is the input force rather than the force per unit mass. Therefore, the frequency response function is given by

$$H(\omega) = \frac{1}{-m\omega^2 + k + i\omega c}$$

and

$$|H(\omega)|^2 = \frac{1}{\left(-m\omega^2 + k\right)^2 + (\omega c)^2}.$$

Then, the response spectrum is given by

$$S_{XX}(\omega) = \frac{S_0}{\left(-m\omega^2 + k\right)^2 + (\omega c)^2}.$$

The force in the spring equals $F_s(t) = kX(t)$, so that $S_{F_s F_s}(\omega) = k^2 S_{XX}(\omega)$, or

$$\begin{aligned} S_{F_s F_s}(\omega) &= k^2 \frac{S_0}{\left(-m\omega^2 + k\right)^2 + (\omega c)^2} \\ &= \frac{S_0 \omega_n^4}{\left(-\omega^2 + \omega_n^2\right)^2 + (2\zeta\omega_n\omega)^2}. \end{aligned}$$

The mean-square value of the force in the spring is given by

$$\begin{aligned} E\{F_s^2\} &= \int_{-\infty}^{\infty} S_{F_s F_s}(\omega)\, d\omega \\ &= \int_{-\infty}^{\infty} \frac{S_0 \omega_n^4}{\left(-\omega^2 + \omega_n^2\right)^2 + (2\zeta\omega_n\omega)^2}\, d\omega = S_0 \frac{\pi \omega_n}{2\zeta}. \end{aligned}$$

This result can be used to estimate the probability that the force will exceed the design strength of the spring. If it is assumed that the processes are Gaussian, then it is possible to tie the mean-square value to a probability of exceedance.

⊛

Example 6.11 Response to Colored Noise

Suppose the same system as in the last example is subjected to more complex loading, where the spectral density of the forcing is not a constant, but a function of ω. How would the above analysis change?

The output spectral density becomes a more complex function of frequency, for example, if the loading density is something such as those found for wind loads. The mean-square response must then be evaluated numerically.

The applied problems are always solved numerically, although hopefully after some significant analytical exposition.

❊

Example 6.12 Cross-Spectral Density

Consider a single-degree-of-freedom system with a mass, spring and damper subjected to random force $F(t)$. Find the expression for the cross-spectral density $S_{XF}(\omega)$.

Solution The autocorrelation function and the spectral density are related by the Wiener-Khinchine relations,

$$S_{XF}(\omega) = \frac{1}{2\pi} \int_{-\infty}^{\infty} R_{XF}(\alpha_1) \exp(-i\omega\alpha_1) \, d\alpha_1,$$

where $R_{XF}(\alpha_1)$ is given by Equation 6.37,

$$R_{XF}(\alpha_1) = \int_{-\infty}^{\infty} g(\tau_1) R_{FF}(\tau_1 - \alpha_1) \, d\tau_1.$$

Then,

$$\begin{aligned} S_{XF}(\omega) &= \frac{1}{2\pi} \int_{-\infty}^{\infty} \int_{-\infty}^{\infty} g(\tau_1) R_{FF}(\tau_1 - \alpha_1) \exp(-i\omega\alpha_1) \, d\tau_1 d\alpha_1 \\ &= \frac{1}{2\pi} \int_{-\infty}^{\infty} g(\tau_1) \int_{-\infty}^{\infty} R_{FF}(\tau_1 - \alpha_1) \exp(-i\omega\alpha_1) \, d\alpha_1 d\tau_1, \end{aligned}$$

where the order of integration has changed. The integrand,

$$\int_{-\infty}^{\infty} R_{FF}(\tau_1 - \alpha_1) \exp(-i\omega\alpha_1) \, d\alpha_1,$$

can be rewritten using the transformation $\tau = \tau_1 - \alpha_1$ with $d\alpha_1 = -d\tau$.

6.3. SDOF: THE RESPONSE TO RANDOM LOADS

Then

$$\int_{-\infty}^{\infty} R_{FF}(\tau_1 - \alpha_1) \exp(-i\omega\alpha_1) \, d\alpha_1$$
$$= \int_{\tau_1+\infty}^{\tau_1-\infty} R_{FF}(\tau) \exp(-i\omega\tau_1 + i\omega\tau)(-d\tau)$$
$$= \int_{-\infty}^{\infty} R_{FF}(\tau) \exp(-i\omega\tau_1 + i\omega\tau) \, d\tau.$$

The cross-spectral density is then

$$S_{XF}(\omega) = \int_{-\infty}^{\infty} g(\tau_1) \exp(-i\omega\tau_1) \, d\tau_1 \cdot \frac{1}{2\pi} \int_{-\infty}^{\infty} R_{FF}(\tau) \exp(i\omega\tau) \, d\tau,$$

where the first part is recognized as $H(\omega)$ and the second part as $S_{FF}(-\omega)$. Recalling that $S_{FF}(\omega)$ is an even function,

$$S_{XF}(\omega) = H(\omega) S_{FF}(\omega).$$

Similarly, one can find

$$S_{FX}(\omega) = H^*(\omega) S_{FF}(\omega). \tag{6.47}$$

⊛

Example 6.13 SDoF Vibration, White Noise Approximation

Consider an electronic component that weighs 30 g subjected to a zero-mean random force with spectral density given by

$$W_{FF}(f) = \frac{118000}{100000 + (x - 820)^2} \text{ N}^2\text{s},$$

plotted in Figure 6.14. It is known that the natural frequency of this component is 300 Hz, and that damping is 20% of the critical damping. Obtain the probability that the amplitude of the vibrating component is greater than 0.1 mm assuming that the random force has a Gaussian density. Also find that probability using the white noise spectral approximation, that is, $W_{FF}(f) = W_{FF}(f_n)$. The force is still assumed Gaussian.

Solution Since f_n and m are known, the stiffness can be calculated by solving

$$f_n = \frac{1}{2\pi}\sqrt{\frac{k}{m}}$$

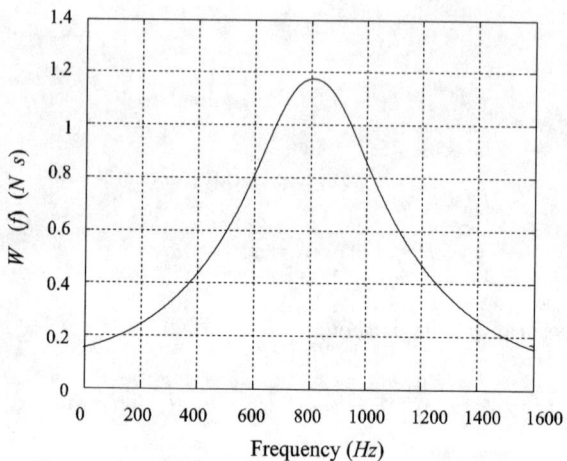

Figure 6.14: Force spectral density, $W_{FF}(f)$.

for $k = 106.5$ kN/m. It is also given that $\zeta = 0.2$. The transfer function is then given by

$$H(f) = \frac{1}{m(2\pi)^2} \frac{1}{(f_n^2 - f^2) + i2\zeta f f_n}$$

$$= 0.845 \frac{1}{(90000 - f^2) + i120f},$$

and its magnitude squared $|H(f)|^2$ is plotted in Figure 6.15.

The spectral density of the response, given by

$$W_{XX}(f) = |H(f)|^2 W_{FF}(f),$$

is plotted using a solid line in Figure 6.16. The area under the response spectral density is calculated numerically, and equals 0.0378 mm^2, which equals the variance σ_X^2.

It is known that for a linear system, if the forcing is Gaussian, then the response is also Gaussian. Then the probability that the response is greater than 0.1 mm is

$$\begin{aligned}
\Pr(|X| > 0.1) &= 2\Pr(X > 0.1) \\
&= 2\Pr\left(\frac{X}{\sigma_X} > \frac{0.1}{\sigma_X}\right) \\
&= 2\{1 - \Phi(0.514)\} \\
&= 0.607.
\end{aligned}$$

6.3. SDOF: THE RESPONSE TO RANDOM LOADS

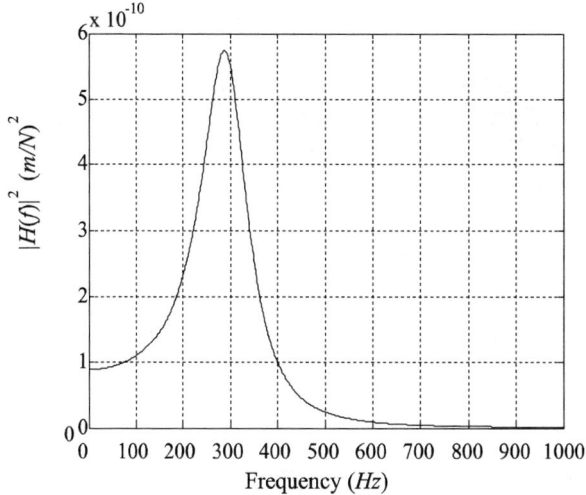

Figure 6.15: Transfer function for $m = 0.03$ kg, $f_n = 300$ Hz, and $\zeta = 0.2$.

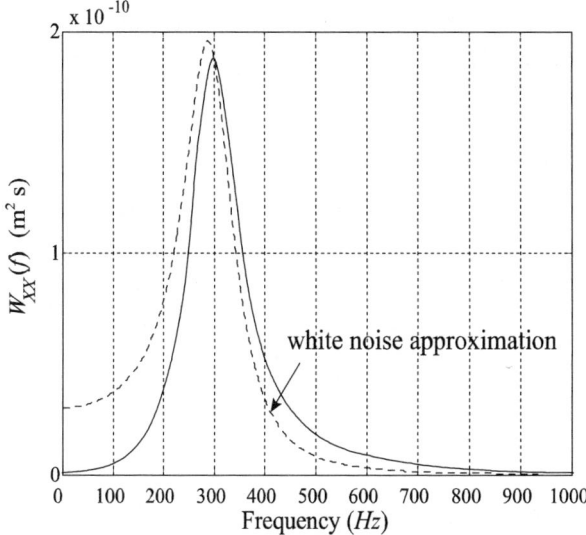

Figure 6.16: Response spectral density.

In this calculation the required probability is put into a form that can be looked up in the Gaussian probability table.

For the white noise approximation, the force spectral density is assumed to be of constant intensity evaluated at $W_{FF}(300) = 0.319$ N²/Hz. The response spectral density is plotted using a dotted line in Figure 6.16, which is quite similar to the actual response spectral density. The variance is numerically evaluated to be 0.0391 mm². The probability is then

$$\Pr(|X| > 0.1) = 2\{1 - \Phi(0.506)\}$$
$$= 0.613.$$

This is another example of how well the white noise approximation works in applications.

Example 6.14 Vibration of Leaf Spring with End Mass

Consider a leaf spring of length $L = 20$ cm with a square cross-section of area $(0.028 \text{ cm})^2$, as shown in Figure 6.17. The spring is made of steel with Young's modulus of 200 GPa. Assume that a white noise zero-mean Gaussian force of intensity 1 N²s is applied at the end of the spring. The point mass is 100 g and $\zeta = 0.1$.

What is the probability that the maximum displacement of the spring is less than 0.5 cm? If a probability of 0.95 is required, what should be the dimension of the cross-section?

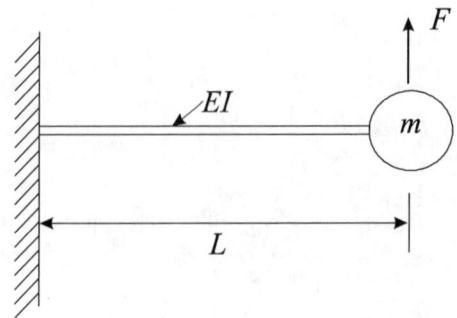

Figure 6.17: Leaf spring with a point mass.

Solution The equation of motion for the displacement at the tip is

$$m\ddot{X} + c\dot{X} + kX = F(t).$$

6.3. SDOF: THE RESPONSE TO RANDOM LOADS

The transfer function is given by

$$H(\omega) = \frac{1}{k + ic\omega - m\omega^2}.$$

The spring constant of a leaf spring is $3EI/L^3$, where I for a square cross-section is $b^4/12$. Then, $k = 384$ N/m. It is given in the problem statement that $\zeta = 0.1$. Then $c = 2\zeta\sqrt{km} = 1.24$ kg/s. The response spectral density is given by

$$\begin{aligned} S_{XX}(\omega) &= |H(\omega)|^2 S_0 \\ &= S_0 \left| \frac{1}{k + ic\omega - m\omega^2} \right|^2. \end{aligned}$$

The variance equals the area under the spectral density curve, so that

$$\sigma_X = \sqrt{S_0 \frac{\pi}{ck}} = 0.081 \text{ m},$$

where Equation 6.46 is used. The probability that the displacement is less than 0.05 m is given by the following calculations,

$$\begin{aligned} \Pr(|X| < 0.05) &= \Pr(-0.05 < X < 0.05) \\ &= 1 - \Pr(X > 0.05) - \Pr(X < -0.05) \\ &= 1 - 2\Pr(X > 0.05) \\ &= 1 - 2[1 - \Pr(X < 0.05)] \\ &= 2\Pr(X < 0.05) - 1 \\ &= 2\Pr\left(\frac{X}{0.081} < \frac{0.05}{0.081}\right) - 1 \\ &= 2\Phi(0.62) - 1 = 0.46. \end{aligned}$$

Since this probability of 0.46 does not meet the design criteria, it is necessary to increase the cross-section of the beam.

The required probability is 0.95 when

$$2\Phi\left(\frac{0.05}{\sigma_X}\right) - 1 = 0.95,$$

or

$$\Phi\left(\frac{0.05}{\sigma_X}\right) = 0.975.$$

From the standard normal distribution table, $0.05/\sigma_X = 1.96$. Then, $\sigma_X = 0.0255$ m. Since $\sigma_X = \sqrt{S_0\pi/\left(2\zeta\sqrt{kmk}\right)}$, with $S_0 = 1$ N^{2}s, $m = 0.1$ kg,

and $\zeta = 0.1$, we find that $k = 1800$ N/m and the new value for $b = 0.0313$ cm, which is about a 12% increase from the original value.
❋

Example 6.15 Response to a Band-Limited White Noise Process

Consider a spring-damper and massless system with parameters k and c subjected to external force as shown in Figure 6.2 with $m = 0$. Assume that $F_{external}(t)$ is band-limited white noise,

$$S_{FF}(\omega) = S_0, \quad |\omega| < \omega_c.$$

(i) Find the mean-square value, $R_{XX}(0)$, for various values of ω_c. What are the most rational nondimensional variables? *(ii)* Obtain the mean-square velocity as functions of normalized frequency ω_c.

Solution *(i)* The equation of motion for this system is given by

$$c\dot{X} + kX = F(t).$$

Taking the Fourier transform, we find the transfer function,

$$H(\omega) = \frac{1}{ci\omega + k}.$$

The response spectral density is given by

$$\begin{aligned} S_{XX}(\omega) &= |H(\omega)|^2 S_{FF}(\omega) \\ &= \frac{S_0}{k^2 + \omega^2 c^2} \quad \text{for } |\omega| < \omega_c. \end{aligned}$$

The autocorrelation function equals the Fourier transform of $S_{XX}(\omega)$,

$$\begin{aligned} R_{XX}(\tau) &= \int_{-\infty}^{\infty} \frac{S_0}{k^2 + \omega^2 c^2} \exp(i\omega\tau) \, d\omega \\ &= S_0 \int_{-\omega_c}^{\omega_c} \frac{1}{k^2 + \omega^2 c^2} \exp(i\omega\tau) \, d\omega. \end{aligned}$$

The mean-square value is given by

$$\begin{aligned} R_{XX}(0) &= S_0 \int_{-\omega_c}^{\omega_c} \frac{1}{k^2 + \omega^2 c^2} d\omega \\ &= \frac{2S_0}{kc} \tan^{-1}\left(\frac{c\omega_c}{k}\right). \end{aligned}$$

From this expression, the normalized mean-square displacement and the normalized frequency can be defined as

$$R_{XX}^*(0) = R_{XX}(0) \frac{kc}{2S_o} \quad \text{and} \quad \omega_c^* = \frac{c\omega_c}{k}.$$

Then,
$$R^*_{XX}(0) = \tan^{-1}(\omega^*_c).$$

(ii) The velocity autocorrelation function $R_{\dot{X}\dot{X}}(\tau)$ is related to the displacement autocorrelation by

$$R_{\dot{X}\dot{X}}(\tau) = -\frac{d^2 R_{XX}(\tau)}{d\tau^2}.$$

Then,
$$R_{\dot{X}\dot{X}}(\tau) = S_o \int_{-\omega_c}^{\omega_c} \frac{\omega^2}{k^2 + \omega^2 c^2} \exp(i\omega\tau)\, d\omega.$$

The mean-square velocity is given by

$$\begin{aligned} R_{\dot{X}\dot{X}}(0) &= S_o \int_{-\omega_c}^{\omega_c} \frac{\omega^2}{k^2 + \omega^2 c^2} d\omega \\ &= \frac{2S_o}{c}\left[\omega_c - \frac{k}{c}\tan^{-1}\left(\frac{c\omega_c}{k}\right)\right]. \end{aligned}$$

The mean-square velocity can be normalized by $2S_o k/c^2$ so that

$$R^*_{\dot{X}\dot{X}}(0) = \left[\omega^*_c - \tan^{-1}(\omega^*_c)\right].$$

⊛

6.4 Response to Two Random Loads

Previously, system response was due to a single randomly varying force. In general, the situation is more complicated due to the possibility that more than one force may act on a system, and the resulting response depends not only on the properties of each force but also on the correlation between the two forces.

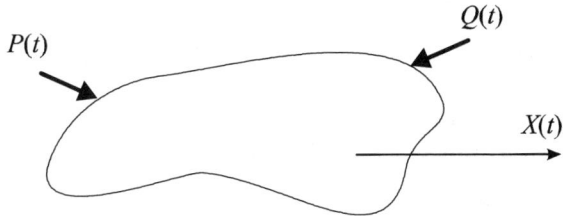

Figure 6.18: Response $X(t)$ to the two random loads $P(t)$ and $Q(t)$.

Consider the response of the system to two random forces, $P(t)$ and $Q(t)$, acting simultaneously but at different points on the system, as shown in Figure 6.18. We are interested in calculating $R_{XX}(\tau)$ and its Fourier transform $S_{XX}(\omega)$ at an arbitrary point on the system. Assume that $E\{P(t)\} = 0$ and $E\{Q(t)\} = 0$. Also, by utilizing available data, we are able to estimate $R_{PP}(\tau)$ and $R_{QQ}(\tau)$. Using linear superposition, and the convolution integral, the response due to both forces is given by the sum,

$$X(t) = \int_{-\infty}^{\infty} [g_{XP}(\tau_1) P(t-\tau_1) + g_{XQ}(\tau_1) Q(t-\tau_1)] d\tau_1.$$

Similarly for $X(t+\tau)$,

$$X(t+\tau) = \int_{-\infty}^{\infty} [g_{XP}(\tau_2) P(t+\tau-\tau_2) + g_{XQ}(\tau_2) Q(t+\tau-\tau_2)] d\tau_2,$$

where $g_{XP}(t)$ is the impulse response function at coordinate X due to force $P(t)$, and $g_{XQ}(t)$ is the impulse response function at X due to $Q(t)$. Then,

$$R_{XX}(\tau) = E\{X(t)X(t+\tau)\}$$
$$= E\left\{ \int_{-\infty}^{\infty} [g_{XP}(\tau_1) P(t-\tau_1) + g_{XQ}(\tau_1) Q(t-\tau_1)] d\tau_1 \right.$$
$$\left. \cdot \int_{-\infty}^{\infty} [g_{XP}(\tau_2) P(t+\tau-\tau_2) + g_{XQ}(\tau_2) Q(t+\tau-\tau_2)] d\tau_2 \right\}.$$

Expand the product and move the expectation operator to the random processes,

$$R_{XX}(\tau)$$
$$= \int_{-\infty}^{\infty} g_{XP}(\tau_1) \left[\int_{-\infty}^{\infty} g_{XP}(\tau_2) E\{P(t-\tau_1) P(t+\tau-\tau_2)\} d\tau_2 \right] d\tau_1$$
$$+ \int_{-\infty}^{\infty} g_{XP}(\tau_1) \left[\int_{-\infty}^{\infty} g_{XQ}(\tau_2) E\{P(t-\tau_1) Q(t+\tau-\tau_2)\} d\tau_2 \right] d\tau_1$$
$$+ \int_{-\infty}^{\infty} g_{XQ}(\tau_1) \left[\int_{-\infty}^{\infty} g_{XP}(\tau_2) E\{Q(t-\tau_1) P(t+\tau-\tau_2)\} d\tau_2 \right] d\tau_1$$
$$+ \int_{-\infty}^{\infty} g_{XQ}(\tau_1) \left[\int_{-\infty}^{\infty} g_{XQ}(\tau_2) E\{Q(t-\tau_1) Q(t+\tau-\tau_2)\} d\tau_2 \right] d\tau_1.$$

In this expression,

$$E\{P(t-\tau_1) P(t+\tau-\tau_2)\} = R_{PP}(\tau+\tau_1-\tau_2)$$
$$E\{Q(t-\tau_1) Q(t+\tau-\tau_2)\} = R_{QQ}(\tau+\tau_1-\tau_2).$$

6.4. RESPONSE TO TWO RANDOM LOADS

The expectations in the second and third terms are cross-correlations of the form $R_{PQ}(\tau) = E\{P(t)Q(t+\tau)\}$. Therefore, the autocorrelation of the response becomes

$$R_{XX}(\tau) = \int_{-\infty}^{\infty} g_{XP}(\tau_1) \left[\int_{-\infty}^{\infty} g_{XP}(\tau_2) R_{PP}(\tau + \tau_1 - \tau_2) d\tau_2\right] d\tau_1$$

$$+ \int_{-\infty}^{\infty} g_{XP}(\tau_1) \left[\int_{-\infty}^{\infty} g_{XQ}(\tau_2) R_{PQ}(\tau + \tau_1 - \tau_2) d\tau_2\right] d\tau_1$$

$$+ \int_{-\infty}^{\infty} g_{XQ}(\tau_1) \left[\int_{-\infty}^{\infty} g_{XP}(\tau_2) R_{QP}(\tau + \tau_1 - \tau_2) d\tau_2\right] d\tau_1$$

$$+ \int_{-\infty}^{\infty} g_{XQ}(\tau_1) \left[\int_{-\infty}^{\infty} g_{XQ}(\tau_2) R_{QQ}(\tau + \tau_1 - \tau_2) d\tau_2\right] d\tau_1.$$

The importance of this result is primarily in the observation that $R_{XX}(\tau)$ cannot be derived unless the cross-correlations $R_{PQ}(\tau)$ and $R_{QP}(\tau)$ are also known. Using the Fourier transform relation between $R_{XX}(\tau)$ and $S_{XX}(\omega)$,

$$S_{XX}(\omega) = \frac{1}{2\pi} \int_{-\infty}^{\infty} R_{XX}(\tau) e^{-i\omega\tau} d\tau,$$

we obtain, as in Section 6.3.4,

$$S_{XX}(\omega) = H_{XP}^*(\omega) H_{XP}(\omega) S_{PP}(\omega)$$
$$+ H_{XP}^*(\omega) H_{XQ}(\omega) S_{PQ}(\omega) + H_{XQ}^*(\omega) H_{XP}(\omega) S_{QP}(\omega)$$
$$+ H_{XQ}^*(\omega) H_{XQ}(\omega) S_{QQ}(\omega), \qquad (6.48)$$

where

$$H_{XP}^*(\omega) H_{XP}(\omega) = |H_{XP}(\omega)|^2$$
$$H_{XQ}^*(\omega) H_{XQ}(\omega) = |H_{XQ}(\omega)|^2$$

and

$$S_{PQ}(\omega) = \frac{1}{2\pi} \int_{-\infty}^{\infty} R_{PQ}(\tau) e^{-i\omega\tau} d\tau.$$

As expected, the evaluation of the output spectral density requires knowledge about the cross-spectra $S_{PQ}(\omega)$ and $S_{QP}(\omega)$. If there are more than two forces, then additional cross-spectra between each pair of forces are required.

Examining Equation 6.48 closely, we find that $S_{XX}(\omega)$ can be written in matrix form,

$$S_{XX}(\omega)$$
$$= \left\{ H_{XP}^*(\omega) \quad H_{XQ}^*(\omega) \right\} \begin{bmatrix} S_{PP}(\omega) & S_{PQ}(\omega) \\ S_{QP}(\omega) & S_{QQ}(\omega) \end{bmatrix} \left\{ \begin{array}{c} H_{XP}(\omega) \\ H_{XQ}(\omega) \end{array} \right\}.$$
$$(6.49)$$

Example 6.16 Conjugates of Cross-Spectra

It was briefly mentioned that $R_{PQ}(\tau) = R_{QP}(-\tau)$. Then, how are $S_{PQ}(\omega)$ and $S_{QP}(\omega)$ related?

Solution The cross-spectral density is given by

$$S_{PQ}(\omega) = \frac{1}{2\pi} \int_{-\infty}^{\infty} R_{PQ}(\tau) \exp(-i\omega\tau) d\tau.$$

Replacing $R_{PQ}(\tau)$ with $R_{QP}(-\tau)$,

$$S_{PQ}(\omega) = \frac{1}{2\pi} \int_{-\infty}^{\infty} R_{QP}(-\tau) \exp(-i\omega\tau) d\tau.$$

Letting $-\tau = t$,

$$S_{PQ}(\omega) = \frac{1}{2\pi} \int_{-\infty}^{\infty} R_{QP}(t) \exp(i\omega t) dt.$$

Then, we can write

$$S_{PQ}(\omega) = S_{QP}(-\omega) = S_{QP}^*(\omega).$$

That is, $S_{PQ}(\omega)$ and $S_{QP}(\omega)$ are complex conjugates.

⊛

Example 6.17 Response Spectrum due to Two Random Loads

Consider a mass-spring-damper system in Figure 6.19 subject to two random forces $P(t)$ and $Q(t)$. Find the response spectrum $S_{XX}(\omega)$ assuming that

$$\begin{aligned} S_{PP}(\omega) &= S_P \\ S_{PQ}(\omega) &= 0 \\ S_{QQ}(\omega) &= S_Q. \end{aligned}$$

Solution The equation of motion for this system is given by

$$m\ddot{X} + c\dot{X} + kX = P(t) + Q(t).$$

First assume that $Q(t) = 0$ in order to first obtain $H_{XP}(\omega)$. Taking the Fourier transform, the equation of motion is given by

$$(-m\omega^2 + ci\omega + k) X(\omega) = P(\omega).$$

6.4. RESPONSE TO TWO RANDOM LOADS

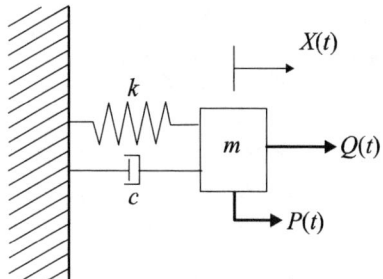

Figure 6.19: A single-degree-of-freedom system subjected to two random loads.

Then, the frequency response function $H_{XP}(\omega)$ is

$$H_{XP}(\omega) = \frac{1}{(-m\omega^2 + ci\omega + k)},$$

or

$$H_{XP}(\omega) = \frac{1}{m(-\omega^2 + 2\omega_n \zeta i\omega + \omega_n^2)}.$$

Similarly, $H_{XQ}(\omega)$ is obtained by setting $P(t) = 0$ and is also given by

$$H_{XQ}(\omega) = \frac{1}{m(-\omega^2 + 2\omega_n \zeta i\omega + \omega_n^2)}.$$

Then, the spectral density of the response is given by

$$\begin{aligned}
S_{XX}(\omega) &= \left\{ [m(\omega_n^2 - \omega^2 - 2\omega_n \zeta i\omega)]^{-1} \quad [m(\omega_n^2 - \omega^2 - 2\omega_n \zeta i\omega)]^{-1} \right\} \\
&\quad \cdot \begin{bmatrix} S_{PP}(\omega) & S_{PQ}(\omega) \\ S_{QP}(\omega) & S_{QQ}(\omega) \end{bmatrix} \left\{ \begin{matrix} [m(\omega_n^2 - \omega^2 + 2\omega_n \zeta i\omega)]^{-1} \\ [m(\omega_n^2 - \omega^2 + 2\omega_n \zeta i\omega)]^{-1} \end{matrix} \right\} \\
&= \frac{S_{PP}(\omega) + S_{PQ}(\omega) + S_{QP}(\omega) + S_{QQ}(\omega)}{m^2 \left[(\omega_n^2 - \omega^2)^2 + (2\omega_n \zeta)^2 \right]}.
\end{aligned} \qquad (6.50)$$

In this case, the spectral density is reduced to

$$S_{XX}(\omega) = \frac{S_P + S_Q}{\left[m^2(\omega_n^2 - \omega^2)^2 + (2\omega_n \zeta)^2 \right]}.$$

⊛

The spectral densities are rather complicated expressions to evaluate in general. A number of special cases is considered to better understand the effects of the cross-terms:

1. $P(t)$ and $Q(t)$ arise from independent sources and therefore are uncorrelated.[14] Then, $R_{PQ}(\tau) = 0$, $R_{QP}(\tau) = 0$, and $S_{PQ}(\omega) = 0$, $S_{QP}(\omega) = 0$.

2. $P(t)$ and $Q(t)$ are directly correlated, that is, $Q(t) = kP(t)$ where k is a constant.

3. $P(t)$ and $Q(t)$ are exponentially correlated,
$$E\{P(t)Q(t+\tau)\} = k_{PQ}\exp\{-\alpha\tau\},$$
where k_{PQ} is a constant.

4. $P(t)$ and $Q(t)$ are correlated in a "simplified" exponential, that is, with a triangular correlation defined by
$$E\{P(t)Q(t+\tau)\} = \overline{k}_{PQ}(1 - \tau/\tau_1), \quad -\tau_1 \leq \tau \leq \tau_1.$$

We consider in more detail the first two cases listed above. Where the loads are independent, because the cross-correlations are identically zero, the output spectral density is just the sum of the two respective spectral densities obtained with the forces acting separately,

$$\begin{aligned} S_{XX}(\omega) &= H_{XP}(\omega)H^*_{XP}(\omega)S_{PP}(\omega) \\ &\quad + H_{XQ}(\omega)H^*_{XQ}(\omega)S_{QQ}(\omega) \\ &= |H_{XP}(\omega)|^2 S_{PP}(\omega) + |H_{XQ}(\omega)|^2 S_{QQ}(\omega). \quad (6.51) \end{aligned}$$

Note that the output spectral density for a linear system subjected to more than one force follows a strict interpretation of the principle of linear superposition only when the forces are *uncorrelated*.

---

[14] Independence implies that
$$E\{P(t_1)Q(t_2)\} = E\{P(t_1)\}E\{Q(t_2)\}.$$
They are uncorrelated if
$$Cov\{P(t_1)Q(t_2)\} = E\{P(t_1)Q(t_2)\} - E\{P(t_1)\}E\{Q(t_2)\} = 0.$$
Independent processes are always uncorrelated whereas uncorrelated processes may not be independent. This was discussed in Section 3.5.1.

6.4. RESPONSE TO TWO RANDOM LOADS

For the case where $Q(t) = kP(t)$,

$$\begin{aligned}
R_{PQ}(\tau) &= E\{P(t)kP(t+\tau)\} \\
&= kR_{PP}(\tau) \\
R_{QP}(\tau) &= E\{kP(t)P(t+\tau)\} \\
&= kR_{PP}(\tau) \\
R_{QQ}(\tau) &= E\{kP(t)kP(t+\tau)\} \\
&= k^2 R_{PP}(\tau).
\end{aligned}$$

Then, the respective spectral densities are

$$\begin{aligned}
S_{PQ}(\omega) &= S_{QP}(\omega) = k S_{PP}(\omega) \\
S_{QQ}(\omega) &= k^2 S_{PP}(\omega),
\end{aligned}$$

leading to the spectral density of the response,

$$\begin{aligned}
S_{XX}(\omega) &= H_{XP}^*(\omega) H_{XP}(\omega) S_{PP}(\omega) + H_{XP}^*(\omega) H_{XQ}(\omega) k S_{PP}(\omega) \\
&\quad + H_{XQ}^*(\omega) H_{XP}(\omega) k S_{PP}(\omega) \\
&\quad + H_{XQ}^*(\omega) H_{XQ}(\omega) k^2 S_{PP}(\omega) \\
&= (H_{XP} + k H_{XQ})(H_{XP}^* + k H_{XQ}^*) S_{PP}(\omega) \\
&= |H_{XP} + k H_{XQ}|^2 S_{PP}(\omega). \quad (6.52)
\end{aligned}$$

This last expression is related to the relative phase between the two functions $H_{XP}(\omega)$ and $H_{XQ}(\omega)$. The addition of two frequency response functions $H_{XP} + k H_{XQ}$ is shown graphically in Figure 6.20.

Suppose $S_{PP}(\omega) = S_{QQ}(\omega) = S(\omega)$. Then, from Equation 6.51, for uncorrelated loadings,

$$S_{XX}(\omega) = \left[|H_{XP}|^2 + |H_{XQ}|^2\right] S(\omega). \quad (6.53)$$

If the forces are directly correlated with parameter $k = 1$, Equation 6.52 yields

$$\begin{aligned}
S_{XX}(\omega) &= |H_{XP} + H_{XQ}|^2 S(\omega) \\
&= \left[|H_{XP}|^2 + |H_{XQ}|^2 + 2|H_{XP}||H_{XQ}|\cos\phi\right] S(\omega),
\end{aligned} \quad (6.54)$$

where ϕ is the phase difference between H_{XP} and H_{XQ} as shown in Figure 6.20. The law of cosines is used for the second relation. Therefore, a comparison of Equation 6.53 with Equation 6.54 shows that the results of an uncorrelated loading will be identical to those that are correlated

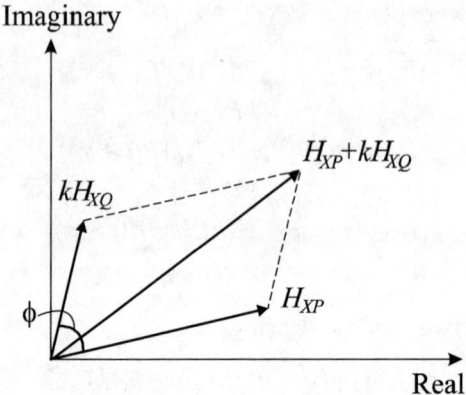

Figure 6.20: Addition of frequency response functions.

where $\cos\phi = 0$, that is, $\phi = \pm\pi/2$. This is when the two vectors in Figure 6.20 are perpendicular to each other. For other values of ϕ the spectral density in the correlated case may have any value in the range defined by $\left[|H_{XP}|^2 \pm |H_{XQ}|^2\right] S(\omega)$, depending on the value of ϕ.

If at some frequency $H_{XP} = -H_{XQ}$, the spectral density at that frequency for the correlated case with $k = 1$ will be zero. For any case where $H_{XP} = H_{XQ}$, the spectral density with correlation will be twice that obtained without correlation.

Another specialized result is where $Q(t)$ follows $P(t)$ after a lag of τ_0, so that $Q(t) = P(t + \tau_0)$. Then,

$$\begin{aligned} R_{PQ}(\tau) &= E\left\{P(t)P(t+\tau_0+\tau)\right\} \\ &= R_{PP}(\tau_0+\tau), \end{aligned}$$

with the respective spectral density,

$$\begin{aligned} S_{PQ}(\omega) &= \frac{1}{2\pi}\int_{-\infty}^{\infty} R_{PQ}(\tau) e^{-i\omega\tau} d\tau \\ &= \frac{1}{2\pi}\int_{-\infty}^{\infty} R_{PP}(\tau_0+\tau) e^{-i\omega\tau} d\tau \\ &= \frac{1}{2\pi} e^{i\omega\tau_0}\int_{-\infty}^{\infty} R_{PP}(\tau_0+\tau) e^{-i\omega(\tau_0+\tau)} d(\tau_0+\tau) \\ &= e^{i\omega\tau_0} S_{PP}(\omega). \end{aligned}$$

Since $S_{PQ}(\omega)$ and $S_{QP}(\omega)$ are complex conjugates,

$$S_{QP}(\omega) = e^{-i\omega\tau_0} S_{PP}(\omega).$$

6.5. CONCLUDING SUMMARY

Also, since it has been assumed above that $S_{QQ}(\omega) = S_{PP}(\omega)$,

$$S_{XX}(\omega) = \left[H_{XP}(\omega)H_{XP}^*(\omega) + e^{i\omega\tau_0}H_{XP}(\omega)H_{XQ}^*(\omega)\right.$$
$$\left. + e^{-i\omega\tau_0}H_{XQ}(\omega)H_{XP}^*(\omega) + H_{XQ}(\omega)H_{XQ}^*(\omega)\right]S_{PP}(\omega).$$

6.5 Concluding Summary

This chapter has provided a review of deterministic single-degree-of-freedom vibration, with a derivation of the convolution integral, the frequency response function and the impulse response function. These key equations are needed for the derivation of the general expressions for the response statistics of an oscillator to a random force. In addition, the fundamental input-output relation $S_{XX}(\omega) = |H(\omega)|^2 S_{FF}(\omega)$ is derived based on the impulse response and convolution equations.

6.6 Problems

Section 6.2: Deterministic SDoF Vibration

1. A single-degree-of-freedom system with $\omega_n = 1$ rad/s and $\zeta = 0.24$ is excited by an external force per unit mass, $F(t) = 2\sin\omega_f t$. Derive the response if the forcing frequency coincides with the damped natural frequency.

2. Derive the response of a single-degree-of-freedom system to a step function given by

$$F(t) = \begin{cases} 1 \text{ N/m} & \text{for } 0 < t < 1 \\ 0, & \text{elsewhere.} \end{cases}$$

    Assume that $\omega_n = 2.1$ rad/s and $\zeta = 0.7$. Use the convolution Equation 6.23.

3. A 4 lb electronic unit is mounted on top of a base with harmonic acceleration of amplitude $Y_0 = 37g$ and a frequency of 53 Hz. The electronic unit will fail if the acceleration exceeds $20g$, that is, it will fail if the unit is hard mounted on the base. Therefore, the electronic unit is isolated from the base using a spring and a damper. Determine the largest allowable value of the spring stiffness, k, assuming that the isolation will have a damping factor of $\zeta = 0.2$.

Section 6.3: SDoF: Response to Random Loads

4. A single-degree-of-freedom system with $m = 2$ kg and $k = 100$ N/m is harmonically forced by $F(t) = 2\sin 10t$ N. Determine the power spectral density and the corresponding autocorrelation function of the response.

5. Consider equipment to be used in a vibration environment with a white noise acceleration process having a spectral density $S_0 = 49g^2$s. An electronic unit is modeled as a single-degree-of-freedom system, has a natural frequency of 3 rad/s, and a damping ratio of 0.1. Determine

    (i) the acceleration spectrum, and

    (ii) the root-mean-square acceleration of the unit.

6. Derive Equation 6.47.

7. Assume that mass m is added in Example 6.15.

    (i) How will these plots change if a mass m is added?

    (ii) How will the plots change with variations in the natural frequency and damping ratio? The integrations that are needed to be performed, although tedious, can be done analytically.

8. A vehicle is traveling with constant velocity v_o on a rough surface. The vehicle is modeled as a single-degree-of-freedom system with stiffness and damping. The roughness of the road is recorded and has a known autocorrelation function,

$$R_{YY}(x_2 - x_1) = A\exp\left(-\alpha\left|x_2 - x_1\right|\right).$$

    Obtain

    (i) the response spectral density, $S_{XX}(\omega)$, and

    (ii) the response autocorrelation function, $R_{XX}(\omega)$.

9. Let the random force $F(t)$ have the autocorrelation function,

$$R_{FF}(\tau) = A\exp\left(-\alpha\left|\tau\right|\right)\left(\cos\beta\tau + \frac{\alpha}{\beta}\sin\beta\left|\tau\right|\right).$$

    Find the force spectrum, response spectrum, and the mean-square response.

6.6. PROBLEMS

10. A rocket contains sensitive equipment with $m = 0.324$ lb·s^2/in, $k = 72.4$ lb/in, and $\zeta = 0.0054$, as shown in Figure 6.21. The displacement of the sensitive equipment is denoted by $X(t)$, and the displacement of the rocket is denoted by $Y(t)$, where $Y(t)$ is a white noise-based random process of intensity $0.04g^2$/Hz, and g is the gravitational acceleration. Find the RMS of the acceleration $\ddot{X}$ in terms of g.

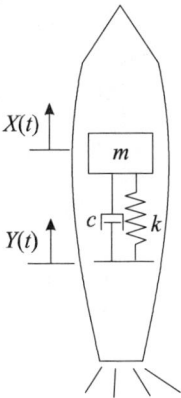

Figure 6.21: Schematic of a rocket and an electronic unit.

11. Let us revisit Example 6.13, where an small electronic element is subjected to a Gaussian random force with spectrum

$$W_{FF}(f) = \frac{118000}{100000 + (f - f_f)^2} \text{ N}^2\text{s},$$

where f_f is the dominant forcing frequency. Determine the probability that the amplitude of the vibration will exceed the specified tolerance of 1 mm when the dominant forcing frequency is varied from 50 Hz to 800 Hz. Solve using the exact form of the $W_{FF}(f)$, and then using the white noise approximation, $W_{FF}(f) = W_{FF}(f_n)$. Plot the percent error that results from using the white noise approximation. For what frequency is this error greatest?

12. Consider the system shown in Figure 6.22. The mass is driven by a random force, with the mean-value 2 N, given by

$$F(t) = 2 + G(t),$$

where $G(t)$ is a zero-mean white noise Gaussian random force whose spectral density is shown in Figure 6.23. The mass equals 1 kg, and the stiffness and damping coefficients equal 100 N/m and 1 kg/s, respectively. The spring and damper are 10 cm long when undeformed. Estimate the probability that the distance between the wall and the mass is greater than 5 cm, assuring that the damper and the spring do not become damaged during the vibration.

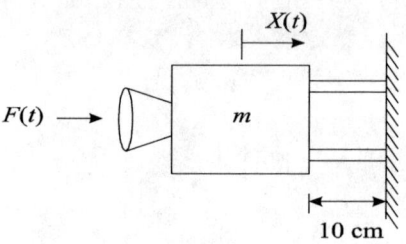

Figure 6.22: Mass driven by a random force with nonzero mean.

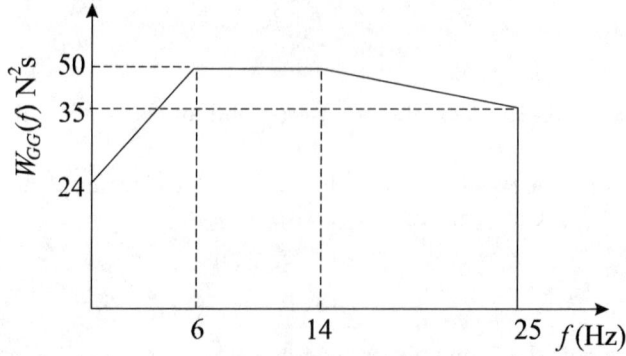

Figure 6.23: Spectral density for $G(t)$.

13. Consider an SDoF system with mass m, damping coefficient c, and no spring element. The mass is subjected to an excitation force given by $mX(t)$, where $X(t)$ is a stationary white noise random process of intensity S_0 m^2/s^3. Determine

    (i) the complex frequency response function $H(\omega)$ for the response velocity V, and

    (ii) the power spectral density $S_{VV}(\omega)$ and the mean-square value $E\{V^2\}$ of the stationary velocity response.

6.6. PROBLEMS

14. The mechanical system in Figure 6.24 is driven by a massless cart with random displacement given by $Y(t)$. Derive

    (i) the equation of motion for this system, and

    (ii) the frequency response function $H(\omega)$.

    (iii) Determine the response spectrum if $Y(t)$ is a white noise-based random process with intensity S_0 m^{2}s. Assume that $c_1 = c_2 = c/2$, $k_1 = k_2 = k/3$, and $m = 3c^2/4k$.

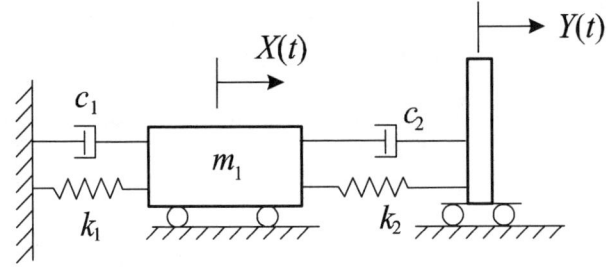

Figure 6.24: An SDoF system driven by a massless cart.

Section 6.4: Response to Two Random Loads

15. Consider the case where N random forces are applied to the single-degree-of-freedom system shown in Figure 6.19. The natural frequency of the system is ω_n, the damping factor is ζ, and the mass is m. Show that the response spectral density is given by

$$S_{XX}(\omega) = |H(\omega)|^2 \sum_{i=1}^{N} \sum_{i=j}^{N} S_{F_i F_j}(\omega),$$

where

$$|H(\omega)|^2 = \frac{1}{m^2 \left[(\omega_n^2 - \omega^2)^2 + (2\omega_n \zeta)^2\right]}.$$

Chapter 7

Multidegree-of-Freedom Vibration

In this chapter we begin to model more complex dynamic systems, where at least two degrees of freedom are required to describe the system behavior. New concepts related to coupling and correlation come into play for systems with at least two degrees of freedom. These concepts and analytical tools can be applied to systems of N degrees of freedom.

7.1 Deterministic Vibration

Consider a linear system with more than one-degree-of-freedom. The equations of motion can be written in matrix form as

$$[m]\{\ddot{x}(t)\} + [c]\{\dot{x}(t)\} + [k]\{x(t)\} = \{F(t)\}. \tag{7.1}$$

For an N-degree-of-freedom system, the dimension $N \times N$ matrices are the mass $[m]$, damping $[c]$, and stiffness $[k]$, and the response $\{x(t)\}$ and force $\{F(t)\}$ vectors are of dimension $N \times 1$.

For purposes of demonstration and discussion, the necessary concepts will be introduced by primarily working through the solution of a two-degree-of-freedom system. All the ideas transfer to larger systems, but with the two-degree-of-freedom model the key ideas can be explained without the complications of the major algebraic and numerical demands made by the larger systems.

For the system shown in Figure 7.1, the coupled equations of motion can be derived using either Newton's second law of motion applied to a free body diagram for each mass as shown in Figure 7.2, or by Lagrange's

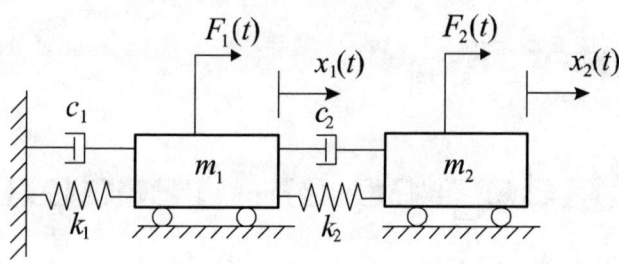

Figure 7.1: Two-degree-of-freedom mass-spring-damper system.

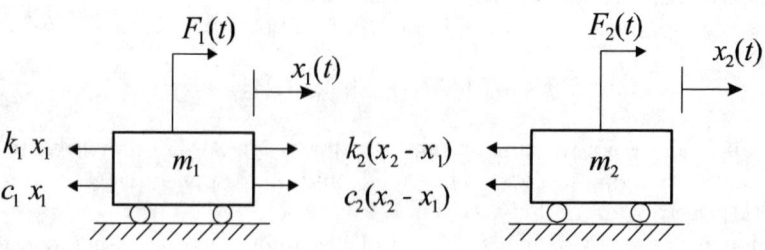

Figure 7.2: Free body diagrams for a two-degree-of-freedom system.

7.1. DETERMINISTIC VIBRATION

equation. In either case, the governing equations are

$$m_1 \ddot{x}_1 + (c_1 + c_2) \dot{x}_1 + (k_1 + k_2) x_1 - c_2 \dot{x}_2 - k_2 x_2 = F_1(t)$$
$$m_2 \ddot{x}_2 + c_2 \dot{x}_2 - c_2 \dot{x}_1 - k_2 x_1 + k_2 x_2 = F_2(t),$$

or in matrix form,

$$\begin{bmatrix} m_1 & 0 \\ 0 & m_2 \end{bmatrix} \begin{Bmatrix} \ddot{x}_1 \\ \ddot{x}_2 \end{Bmatrix} + \begin{bmatrix} c_1 + c_2 & -c_2 \\ -c_2 & c_2 \end{bmatrix} \begin{Bmatrix} \dot{x}_1 \\ \dot{x}_2 \end{Bmatrix}$$
$$+ \begin{bmatrix} k_1 + k_2 & -k_2 \\ -k_2 & k_2 \end{bmatrix} \begin{Bmatrix} x_1 \\ x_2 \end{Bmatrix} = \begin{Bmatrix} F_1(t) \\ F_2(t) \end{Bmatrix}. \quad (7.2)$$

In this case, the mass, damping, and stiffness matrices are given by

$$[m] = \begin{bmatrix} m_1 & 0 \\ 0 & m_2 \end{bmatrix}, \quad [c] = \begin{bmatrix} c_1 + c_2 & -c_2 \\ -c_2 & c_2 \end{bmatrix}, \quad [k] = \begin{bmatrix} k_1 + k_2 & -k_2 \\ -k_2 & k_2 \end{bmatrix}.$$

The property matrices are symmetric.

7.1.1 Solution by Frequency Response Function

In this section, we will follow a procedure similar to that in Section 6.2.7 for the single-degree-of-freedom system. Start by taking the Fourier transform of the equations of motion,

$$\left(-\omega^2 [m] + i\omega [c] + [k]\right) \{X(\omega)\} = \{F(\omega)\},$$

where $\{X(\omega)\}$ and $\{F(\omega)\}$ are Fourier transforms of $\{x(t)\}$ and $\{F(t)\}$, respectively. The matrix $\left(-\omega^2 [m] + i\omega [c] + [k]\right)$ is denoted by $[Z(\omega)]$, thus,

$$\{X(\omega)\} = [Z(\omega)]^{-1} \{F(\omega)\}.$$

The matrix $[Z(\omega)]^{-1}$ is identical to the frequency response matrix denoted as $[H(\omega)]$,

$$\{X(\omega)\} = [H(\omega)] \{F(\omega)\}. \quad (7.3)$$

For the two-degree-of-freedom system in Figure 7.1,

$$[H(\omega)]$$
$$= \frac{1}{\det[Z]} \begin{bmatrix} -m_2\omega^2 + i\omega c_2 + k_2 & i\omega c_2 + k_2 \\ i\omega c_2 + k_2 & -m_1\omega^2 + i\omega(c_1 + c_2) + (k_1 + k_2) \end{bmatrix}$$

$$\det[Z]$$
$$= \omega^4 m_1 m_2 - i\omega^3 (m_1 c_2 + c_1 m_2 + c_2 m_2) + i\omega (c_1 k_2 + k_1 c_2)$$
$$- \omega^2 (m_1 k_2 + c_1 c_2 + k_1 m_2 + k_2 m_2) + k_1 k_2. \quad (7.4)$$

Note that the frequency response function of a two-degree-of-freedom system is given as a matrix with dimension 2×2. To clarify the meaning of each element, expand Equation 7.3,

$$X_1(\omega) = H_{11}(\omega) F_1(\omega) + H_{12}(\omega) F_2(\omega)$$
$$X_2(\omega) = H_{21}(\omega) F_1(\omega) + H_{22}(\omega) F_2(\omega).$$

Now the meaning of each element is clear. Each element $H_{ij}(\omega)$ is the frequency response function for coordinate i due to a force at j. In general, $[H(\omega)]$ is a symmetric matrix since $[m]$, $[c]$, and $[k]$ are symmetric.

For a deterministic response, recall that the impulse response function is related to the frequency response function by (Equation 6.29)

$$g_{ij}(t) = \frac{1}{2\pi} \int_{-\infty}^{\infty} H_{ij}(\omega) \exp(i\omega t) \, d\omega.$$

For the two-degree-of-freedom system, the impulse response functions are

$$g_{11}(t) = \frac{1}{2\pi} \int_{-\infty}^{\infty} \frac{-m_2 \omega^2 + i\omega c_2 + k_2}{\det[Z]} \exp(i\omega t) \, d\omega$$

$$g_{12}(t) = \frac{1}{2\pi} \int_{-\infty}^{\infty} \frac{i\omega c_2 + k_2}{\det[Z]} \exp(i\omega t) \, d\omega$$

$$g_{21}(t) = g_{12}(t) = \frac{1}{2\pi} \int_{-\infty}^{\infty} \frac{i\omega c_2 + k_2}{\det[Z]} \exp(i\omega t) \, d\omega$$

$$g_{22}(t) = \frac{1}{2\pi} \int_{-\infty}^{\infty} \frac{-m_1 \omega^2 + i\omega (c_1 + c_2) + (k_1 + k_2)}{\det[Z]} \exp(i\omega t) \, d\omega,$$

where $[g(t)]$ is a symmetric matrix.

Using the convolution integral, the response $x_i(t)$ due to $F_j(t)$ is

$$x_i(t) = \int_{-\infty}^{\infty} g_{ij}(\tau) F_j(t-\tau) d\tau.$$

The total response $x(t)$ of mass m_i equals the sum of the individual responses to each of the forces. For the two-degree-of-freedom system, the responses are

$$x_1(t) = \int_{-\infty}^{\infty} [g_{11}(\tau) F_1(t-\tau) + g_{12}(\tau) F_2(t-\tau)] \, d\tau$$

$$x_2(t) = \int_{-\infty}^{\infty} [g_{21}(\tau) F_1(t-\tau) + g_{22}(\tau) F_2(t-\tau)] \, d\tau.$$

This analysis can be generalized to a system with N degrees of freedom. The equations of motion are written in matrix form as

$$[m]\{\ddot{x}(t)\} + [c]\{\dot{x}(t)\} + [k]\{x(t)\} = \{F(t)\},$$

7.1. DETERMINISTIC VIBRATION

where $\{x(t)\}$ and $\{F(t)\}$ are response and force vectors and the frequency response matrix is given by

$$[H(\omega)] = \left(-\omega^2 [m] + i\omega [c] + [k]\right)^{-1}. \tag{7.5}$$

The impulse response matrix $[g(t)]$ is given by

$$[g(t)] = \frac{1}{2\pi} \int_{-\infty}^{\infty} [H(\omega)] \exp(i\omega t) \, d\omega, \tag{7.6}$$

and the response matrix is given by

$$\{x(t)\} = \int_{-\infty}^{\infty} [g(\tau)] \{F(t-\tau)\} \, d\tau. \tag{7.7}$$

The impulse response vector is not trivial to evaluate, much less the convolution integrals. More importantly, this method becomes much more complex for each additional degree-of-freedom. Therefore, this method is rarely used, and we often rely on *modal analysis*.

7.1.2 Modal Analysis

First consider an undamped system in free vibration governed by the matrix equation of motion,

$$[m]\{\ddot{x}(t)\} + [k]\{x(t)\} = \{0\}. \tag{7.8}$$

The response $\{x(t)\}$ is harmonic at frequency ω,

$$\{x(t)\} = A_o \{u\} e^{i\omega t}.$$

Substituting this response into the equation of motion reduces them to the eigenvalue problem,

$$\left(-\omega^2 [m] + [k]\right) \{u\} = \{0\}. \tag{7.9}$$

To guarantee that the vector $\{u\}$ satisfies this equation requires the determinant of $(-\omega^2 [m] + [k])$ to be equal to zero,

$$\det\left[-\omega^2 [m] + [k]\right] = 0.$$

There are several values of ω^2 that can make the determinant equal to zero, and these values of ω^2 are called *eigenvalues*. For each nonrepeating eigenvalue, there is a corresponding vector $\{u\}$ that satisfies Equation 7.9. This vector is called the *eigenvector*. Physically, the eigenvalues of this problem are the *natural frequencies* squared, and the eigenvectors are the *mode shapes* since they represent the respective motion of each modal coordinate.

Solving this eigenvalue problem results in N sets of eigenvalues squared ω_i^2 with corresponding eigenvectors $\{u\}_i$. (An eigenvalue problem will be demonstrated in the next example.) It is customary that the eigenvalues are arranged in ascending order,

$$\omega_1^2 < \omega_2^2 < \cdots < \omega_N^2.$$

The eigenvectors can be *normalized with respect to the mass matrix*,

$$\{u\}_i^T [m] \{u\}_i = 1, \qquad i = 1, 2, \cdots, N, \qquad (7.10)$$

where

$$\{u\}_i = \left\{ \begin{array}{c} u_{1i} \\ \vdots \\ u_{Ni} \end{array} \right\}.$$

The eigen properties of ω_i^2 and $\{u\}_i$ are developed next. For the ith and jth set of eigenvalues and eigenvectors,

$$\begin{array}{rcl} -\omega_i^2 [m] \{u\}_i + [k] \{u\}_i & = & \{0\} \\ -\omega_j^2 [m] \{u\}_j + [k] \{u\}_j & = & \{0\}. \end{array}$$

Multiply the first equation by $\{u\}_j^T$ and the second equation by $\{u\}_i^T$ to obtain

$$\begin{array}{rcll} -\omega_i^2 \{u\}_j^T [m] \{u\}_i + \{u\}_j^T [k] \{u\}_i & = & \{0\} & (7.11) \\ -\omega_j^2 \{u\}_i^T [m] \{u\}_j + \{u\}_i^T [k] \{u\}_j & = & \{0\}. & (7.12) \end{array}$$

Take the transpose of the second equation,

$$-\omega_j^2 \{u\}_j^T [m]^T \{u\}_i + \{u\}_j^T [k]^T \{u\}_i = \{0\}.$$

Since $[m]$ and $[k]$ are symmetric, $[m] = [m]^T$ and $[k] = [k]^T$, this equation becomes

$$-\omega_j^2 \{u\}_j^T [m] \{u\}_i + \{u\}_j^T [k] \{u\}_i = \{0\}. \qquad (7.13)$$

Subtracting Equation 7.13 from Equation 7.11 results in

$$\left(\omega_j^2 - \omega_i^2\right) \{u\}_j^T [m] \{u\}_i = 0.$$

If the eigenvalues are unique $(\omega_j^2 \neq \omega_i^2)$, then the orthogonality of the eigenvectors with respect to the mass matrix is shown,

$$\{u\}_j^T [m] \{u\}_i = 0, \qquad i \neq j.$$

7.1. DETERMINISTIC VIBRATION

From Equation 7.13, by a similar procedure, the orthogonality of the eigenvectors with respect to the stiffness matrix is shown,

$$\{u\}_j^T [k] \{u\}_i = 0, \qquad i \neq j.$$

If the eigenvectors are normalized with respect to the mass matrix,

$$\{u\}_i^T [m] \{u\}_i = 1,$$

Equation 7.13 with $j = i$ can be written as

$$-\omega_i^2 \{u\}_i^T [m] \{u\}_i + \{u\}_i^T [k] \{u\}_i = \{0\},$$

and, therefore,

$$\{u\}_i^T [k] \{u\}_i = \omega_i^2.$$

In summary, the eigenvectors have the following properties:

$$\{u\}_i^T [m] \{u\}_j = \begin{cases} 1, & i = j \\ 0, & i \neq j \end{cases}$$

and

$$\{u\}_i^T [k] \{u\}_j = \begin{cases} \omega_i^2, & i = j \\ 0, & i \neq j. \end{cases}$$

Constructing a matrix composed of eigenvectors,

$$[P] = [\ \{u\}_1 \ \cdots \ \{u\}_N\ ], \qquad (7.14)$$

then

$$[P]^T [m] [P] = [I]$$

$$[P]^T [k] [P] = \begin{bmatrix} \omega_1^2 & 0 & 0 \\ 0 & \ddots & 0 \\ 0 & 0 & \omega_N^2 \end{bmatrix} = [\text{diag}(\omega^2)],$$

where $[I]$ is the identity matrix and $[P]$ is called the *modal matrix*.
If we let $\{x(t)\} = [P]\{z(t)\}$, Equation 7.8 can be written as

$$[m][P]\{\ddot{z}(t)\} + [k][P]\{z(t)\} = \{0\}.$$

Multiplying both sides by $[P]^T$, we obtain

$$\{\ddot{z}(t)\} + [\text{diag}(\omega^2)]\{z(t)\} = \{0\},$$

and the initial conditions are transformed to

$$\{z(0)\} = [P]^T [m]\{x(0)\} \text{ and } \{\dot{z}(0)\} = [P]^T [m]\{\dot{x}(0)\}.$$

The differential equations in $z_i(t)$ are decoupled and can be solved easily. Once $\{z(t)\}$ is obtained, $\{x(t)\}$ can be obtained using $\{x(t)\} = [P]\{z(t)\}$.

Example 7.1 Modal Matrix

Find the eigenvalues and modal matrix for the system with the property matrices,

$$[m] = \begin{bmatrix} 2 & 0 \\ 0 & 1 \end{bmatrix} \quad \text{and} \quad [k] = \begin{bmatrix} 2 & -1 \\ -1 & 1 \end{bmatrix}.$$

Solution The eigenvalue problem in Equation 7.9 becomes

$$\left(-\omega^2 \begin{bmatrix} 2 & 0 \\ 0 & 1 \end{bmatrix} + \begin{bmatrix} 2 & -1 \\ -1 & 1 \end{bmatrix} \right) \{u\} = \begin{Bmatrix} 0 \\ 0 \end{Bmatrix}$$

$$\begin{bmatrix} 2 - 2\omega^2 & -1 \\ -1 & 1 - \omega^2 \end{bmatrix} \{u\} = \begin{Bmatrix} 0 \\ 0 \end{Bmatrix}. \quad (7.15)$$

Since $\{u\}$ is not zero,[1] the equations must be linearly dependent, or

$$\det \begin{bmatrix} 2 - 2\omega^2 & -1 \\ -1 & 1 - \omega^2 \end{bmatrix} = 0$$

$$2\omega^4 - 4\omega^2 + 1 = 0.$$

The eigenvalues are then

$$\omega_1^2 = 0.293 \quad \text{and} \quad \omega_2^2 = 1.707 \; (\text{rad/s})^2.$$

For $\omega_1^2 = 0.293 \; (\text{rad/s})^2$, the corresponding eigenvector is obtained by substituting ω_1^2 into Equation 7.15,

$$\begin{bmatrix} 2 - 2(0.293) & -1 \\ -1 & 1 - (0.293) \end{bmatrix} \{u\}_1 = \{0\}$$

$$\begin{bmatrix} 1.414 & -1 \\ -1 & 0.707 \end{bmatrix} \{u\}_1 = \{0\}.$$

The eigenvector $\{u\}_1$ is given by

$$\{u\}_1 = c \begin{Bmatrix} 1 \\ 1.414 \end{Bmatrix},$$

where c is the normalization constant. Using Equation 7.10,

$$c^2 \begin{Bmatrix} 1 & 1.414 \end{Bmatrix} \begin{bmatrix} 2 & 0 \\ 0 & 1 \end{bmatrix} \begin{Bmatrix} 1 \\ 1.414 \end{Bmatrix} = 1,$$

---

[1] If $\{u\}$ is zero, this is called the trivial solution.

7.1. DETERMINISTIC VIBRATION

we find that

$$c = 0.5$$
$$\{u\}_1 = \begin{Bmatrix} 0.5 \\ 0.707 \end{Bmatrix}.$$

Similarly, the eigenvector $\{u\}_2$ that corresponds to $\omega_2^2 = 1.707$ (rad/s)^2 is given by

$$\{u\}_2 = \begin{Bmatrix} 0.5 \\ -0.707 \end{Bmatrix}.$$

The modal matrix is then

$$[P] = \begin{bmatrix} 0.5 & 0.5 \\ 0.707 & -0.707 \end{bmatrix}.$$

⊛

Consider the effect of damping on modal analysis with the original matrix equation of motion,

$$[m]\{\ddot{x}\} + [c]\{\dot{x}\} + [k]\{x\} = \{F(t)\}.$$

Restrict the problem to the proportional damping case, where $[c]$ is a linear combination of $[m]$ and $[k]$,

$$[c] = \alpha[m] + \beta[k],$$

with constant α and β.

Define a new set of coordinates $\{z(t)\}$ so that

$$\{x(t)\} = [P]\{z(t)\}, \tag{7.16}$$

where $[P]$ is the modal matrix defined in Equation 7.14. The matrix equation of motion becomes

$$[m][P]\{\ddot{z}\} + [c][P]\{\dot{z}\} + [k][P]\{z\} = \{F(t)\}.$$

Multiply by $[P]^T$,

$$[P]^T[m][P]\{\ddot{z}\} + [P]^T[c][P]\{\dot{z}\} + [P]^T[k][P]\{z\} = [P]^T\{F(t)\},$$

which reduces to

$$\{\ddot{z}\} + [\alpha[I] + \beta \times \text{diag}(\omega^2)]\{\dot{z}\} + \text{diag}(\omega^2)\{z\} = [P]^T\{F(t)\},$$

or

$$\ddot{z}_1 + (\alpha + \beta\omega_1^2)\dot{z}_1 + \omega_1^2 z = \{u\}_1^T\{F(t)\}$$
$$\vdots \qquad \vdots$$
$$\ddot{z}_N + (\alpha + \beta\omega_N^2)\dot{z}_N + \omega_N^2 z = \{u\}_N^T\{F(t)\}.$$

It is clear now why proportional damping is considered. The damping matrix $[c]$ becomes diagonalized so that the equations of motion are *decoupled*. The coordinates $z_i(t)$ are called the *modal coordinates* and $\{u\}_i^T\{F(t)\}$ are called the *modal forces*. Define

$$(\alpha + \beta\omega_i^2) = 2\omega_i\zeta_i$$
$$\{u\}_i^T\{F(t)\} = q_i(t)$$

to simplify the equations,

$$\ddot{z}_1 + 2\omega_1\zeta_1\dot{z}_1 + \omega_1^2 z_1 = q_1(t)$$
$$\vdots \qquad \vdots$$
$$\ddot{z}_N + 2\omega_N\zeta_N\dot{z}_N + \omega_N^2 z_N = q_N(t). \qquad (7.17)$$

The solution to these equations can be obtained using the impulse response function and the convolution integral for each coordinate (see Sections 6.2.5 and 6.2.6),

$$z_i(t) = \int_{-\infty}^{\infty} g_i(\tau)q_i(t-\tau)d\tau, \quad i = 1, 2, \cdots, N,$$

where

$$g_i(t) = \begin{cases} (1/\omega_{d_i})e^{-\zeta\omega_i t}\sin\omega_{d_i}t, & t \geq 0 \\ 0, & t < 0, \end{cases}$$

and

$$\omega_{d_i} = \omega_i\sqrt{1 - \zeta_i^2}.$$

The corresponding transfer function is given by

$$H_j(\omega) = \frac{1}{(\omega_j^2 - \omega^2) + i2\omega_j\zeta_j\omega}. \qquad (7.18)$$

Once all the $z_i(t)$ are evaluated, the physical coordinates are recovered by using the transformation in Equation 7.16, $\{x(t)\} = [P]\{z(t)\}$.

7.1.3 Advantages of Modal Analysis

Engineering structures, such as turbomachinery, bridges, or aircraft, have thousands of degrees of freedom. For such systems, it is very difficult if not impossible to obtain the response using the frequency response method. It would be necessary to obtain the frequency response matrix $[H(\omega)]$ in Equation 7.5 by inverting an $N \times N$ matrix, the impulse response matrix $[g(t)]$ in Equation 7.6 by evaluating the inverse Fourier transform of each element of $[H(\omega)]$, and the response $\{x(t)\}$ in Equation 7.7 by evaluating the convolution integral with N terms. However, the advantage of this method is that it does not require a special form for the damping as does the modal analysis.

In modal analysis, the response can be approximated using a few modes. Suppose that only the first *three* modes of a *one thousand*-degree-of-freedom model are the primary contributors to the response of the structure. Then the response can be approximated by

$$\{x(t)\}_{1000 \times 1} \simeq [P]_{1000 \times 3} \{z(t)\}_{3 \times 1},$$

where $[P]_{1000 \times 3}$ consists of the first three eigenvectors,

$$[P]_{1000 \times 3} = [\{u\}_1 \{u\}_2 \{u\}_3],$$

and

$$\{z(t)\}_{3 \times 1} = \begin{bmatrix} z_1(t) \\ z_2(t) \\ z_3(t) \end{bmatrix}.$$

Example 7.2 Approximation Using a Few Modes

Consider the 30-degree-of-freedom system shown in Figure 7.3. The mass and stiffness matrices are given by

$$[m] = \begin{bmatrix} m & 0 & 0 & 0 \\ 0 & \ddots & 0 & 0 \\ 0 & 0 & \ddots & 0 \\ 0 & 0 & 0 & m \end{bmatrix}, \quad [k] = \begin{bmatrix} 2k & -k & 0 & 0 \\ -k & \ddots & \ddots & 0 \\ 0 & \ddots & \ddots & -k \\ 0 & 0 & -k & 2k \end{bmatrix},$$

where $m = 0.01$ kg and $k = 1$ N/m. Evaluate the response for the initial displacement and the external force given by

$$x_n(t) = \begin{cases} 0.1 & \text{for } n = 1 \\ 0 & \text{for } n = 2, \ldots, 30 \end{cases}$$

$$F(t) = \sin 2.5 t.$$

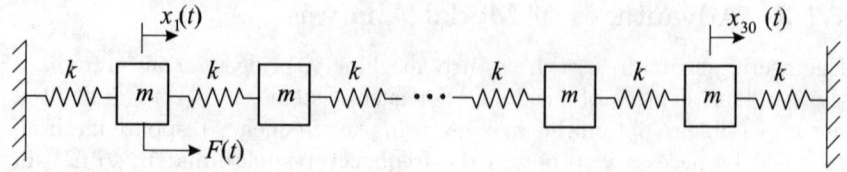

Figure 7.3: 30-degree-of-freedom discretized system.

Assume zero initial conditions. Approximate $x_{30}(t)$ using the first three modes.

Solution The natural frequencies and the mode shapes can be found using available software such as MATLAB. The first three natural frequencies in rad/s are found to be

$$\omega_1^2 = 1.0261, \quad \omega_2^2 = 4.0940, \quad \omega_3^2 = 9.1721.$$

The first three mode shapes are plotted in Figure 7.4. They are normalized with respect to the mass matrix.

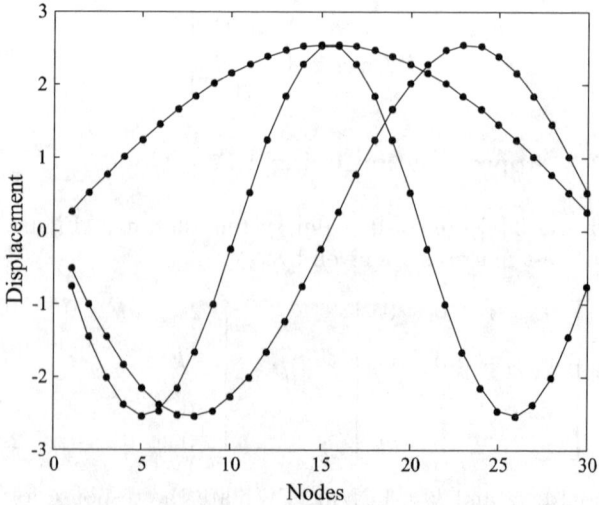

Figure 7.4: The first three mode shapes of the 30-DoF system.

The equations of motion for the first three modal coordinates are given

7.2. RESPONSE TO RANDOM LOADS

by

$$\ddot{z}_1 + 1.0261 z_1 = P_{11} \sin 2.5t$$
$$\ddot{z}_2 + 4.0940 z_2 = P_{12} \sin 2.5t$$
$$\ddot{z}_3 + 9.1721 z_3 = P_{13} \sin 2.5t,$$

where $P_{11} = 0.2570$, $P_{12} = -0.5113$, and $P_{13} = -0.7604$. Then, the general solution for z_i is given by

$$z_i(t) = -\frac{\omega_f}{\omega_i} \frac{P_{1i}}{\omega_i^2 - \omega_f^2} \sin \omega_i t + \frac{P_{1i}}{\omega_i^2 - \omega_f^2} \sin \omega_f t,$$

where $\omega_f = 2.5$ rad/s. $\{x(t)\}$ is then given by

$$\{x(t)\}_{30 \times 1} \simeq [P]_{30 \times 3} \{z(t)\}_{3 \times 1}$$
$$x_i(t) = P_{i1} z_1(t) + P_{i2} z_2(t) + P_{i3} z_3(t).$$

Figure 7.5 shows the exact solution and the approximation for $x_{30}(t)$. It is shown that the error is quite minimal. Note in the figure that it takes about 3 s for the pulse at $x_1(t)$ to reach the 30th mass.

⊛

7.2 Response to Random Loads

From the analysis of the single-degree-of-freedom system, knowledge of the frequency response function $H(\omega)$ is crucial to obtaining the response statistics to random loads. For a system with more than one-degree-of-freedom, there is more than one frequency response function $H(\omega)$. For an N-degree-of-freedom system, the frequency response functions are grouped as an $N \times N$ symmetric matrix $[H(\omega)]$.

In the single-degree-of-freedom system, the response statistics were given in terms of the mean value response μ_X and the spectral density of the response $S_{XX}(\omega)$. For the N-degree-of-freedom system, the response statistics of an N-degree-of-freedom system are given in terms of the vector mean response, $\{\mu_{X_1}, \cdots, \mu_{X_N}\}^T$, and a matrix of spectral densities of the response $[S(\omega)]$, which is also of dimension $N \times N$.

A priori, we present the results that are derived in the subsequent section; the mean response vector $\{\mu_X\}$ and the matrix spectral density of the output $[S_{XX}(\omega)]$:

$$\{\mu_X\} = [H(0)] \{\mu_F\}$$
$$[S_{XX}(\omega)] = [H^*(\omega)] [S_{FF}(\omega)] [H(\omega)]^T.$$

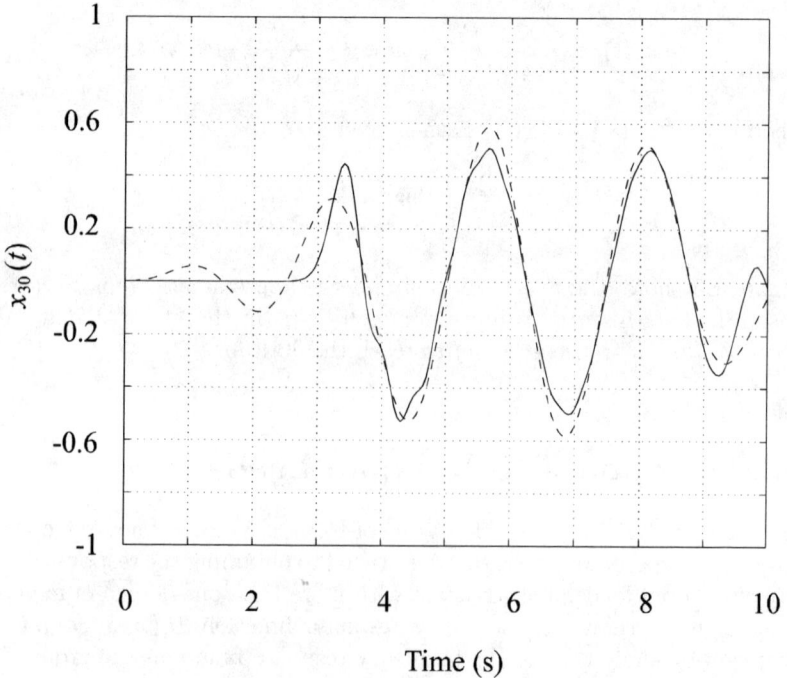

Figure 7.5: The exact solution for $x_{30}(t)$ using all thirty modes is plotted in the solid line. The approximation using the first three modes is plotted in the dotted line.

7.2. RESPONSE TO RANDOM LOADS

7.2.1 Response due to a Single Random Force

When an elastic body vibrates, points on the body move relative to each other. To describe this relative motion of two points on a body subjected to random loading, it is necessary to know their cross-spectral densities and cross-correlations in addition to the spectral densities and autocorrelation function of the individual motions of each point. This requirement is true even though there is only one force acting on the system. This is similar to the problem of the single mass loaded by two random loads of the last chapter.

Consider a body excited by a stationary randomly varying force $P(t)$ with autocorrelation function $R_{PP}(\tau)$ and spectral density $S_{PP}(\omega)$. Due to this loading, two points on the body have displacements $X(t)$ and $Y(t)$ with mean values μ_X and μ_Y, autocorrelations $R_{XX}(\tau)$ and $R_{YY}(\tau)$, and spectral densities $S_{XX}(\omega)$ and $S_{YY}(\omega)$, respectively.

From our earlier studies,

$$E\{X(t)\} = H_{XP}(0)\mu_P$$
$$E\{Y(t)\} = H_{YP}(0)\mu_P,$$

and

$$S_{XX}(\omega) = |H_{XP}(\omega)|^2 S_{PP}(\omega) \quad (7.19)$$
$$S_{YY}(\omega) = |H_{YP}(\omega)|^2 S_{PP}(\omega), \quad (7.20)$$

where

$$H_{XP}(\omega) = \int_{-\infty}^{\infty} g_{XP}(\tau) e^{-i\omega\tau} d\tau$$
$$H_{YP}(\omega) = \int_{-\infty}^{\infty} g_{YP}(\tau) e^{-i\omega\tau} d\tau,$$

and

$$X(t) = \int_{-\infty}^{\infty} g_{XP}(\tau) P(t-\tau) d\tau$$
$$Y(t) = \int_{-\infty}^{\infty} g_{YP}(\tau) P(t-\tau) d\tau.$$

With these relations, we can now find the expressions for the cross-correlation function $R_{XY}(\tau)$, and cross-spectral density $S_{XY}(\omega)$. The cross-correlation function is expressed as

$$
\begin{aligned}
R_{XY}(\tau) &= E\{X(t)Y(t+\tau)\} \\
&= E\left\{\int_{-\infty}^{\infty} g_{XP}(\tau_1) P(t-\tau_1) d\tau_1 \int_{-\infty}^{\infty} g_{YP}(\tau_2) P(t+\tau-\tau_2) d\tau_2\right\} \\
&= \int_{-\infty}^{\infty} g_{XP}(\tau_1) \left[\int_{-\infty}^{\infty} g_{YP}(\tau_2) E\{P(t-\tau_1)P(t+\tau-\tau_2)\} d\tau_2\right] d\tau_1 \\
&= \int_{-\infty}^{\infty} g_{XP}(\tau_1) \left[\int_{-\infty}^{\infty} g_{YP}(\tau_2) R_{PP}(\tau+\tau_1-\tau_2) d\tau_2\right] d\tau_1.
\end{aligned}
$$

Even though we have an explicit expression, it is generally not easy to interpret.

The cross-spectral density can be derived by using the Fourier transform relation between $R_{XY}(\tau)$ and $S_{XY}(\omega)$,

$$
\begin{aligned}
S_{XY}(\omega) &= \frac{1}{2\pi} \int_{-\infty}^{\infty} R_{XY}(\tau) e^{-i\omega\tau} d\tau \\
&= \frac{1}{2\pi} \int_{-\infty}^{\infty} \int_{-\infty}^{\infty} g_{XP}(\tau_1) \\
&\quad \cdot \int_{-\infty}^{\infty} g_{YP}(\tau_2) R_{PP}(\tau+\tau_1-\tau_2) d\tau_2 d\tau_1 e^{-i\omega\tau} d\tau \\
&= \frac{1}{2\pi} \int_{-\infty}^{\infty} g_{XP}(\tau_1) e^{+i\omega\tau_1} \int_{-\infty}^{\infty} g_{YP}(\tau_2) e^{-i\omega\tau_2} \\
&\quad \cdot \int_{-\infty}^{\infty} R_{PP}(\tau+\tau_1-\tau_2) e^{-i\omega(\tau+\tau_1-\tau_2)} d\tau d\tau_2 d\tau_1.
\end{aligned}
\tag{7.21}
$$

In order to integrate this last expression, it is easier to proceed by transforming the variables in the innermost integral according to

$$\lambda \equiv \tau + \tau_1 - \tau_2 \quad \text{and} \quad d\lambda = d\tau,$$

where τ_1 and τ_2 are dummy variables. Then, Equation 7.21 becomes

$$
\begin{aligned}
S_{XY}(\omega) &= \frac{1}{2\pi} \int_{-\infty}^{\infty} g_{XP}(\tau_1) e^{+i\omega\tau_1} d\tau_1 \int_{-\infty}^{\infty} g_{YP}(\tau_2) e^{-i\omega\tau_2} d\tau_2 \\
&\quad \cdot \int_{-\infty}^{\infty} R_{PP}(\lambda) e^{-i\omega\lambda} d\lambda \\
&= H_{XP}^*(\omega) H_{YP}(\omega) S_{PP}(\omega).
\end{aligned}
\tag{7.22}
$$

7.2. RESPONSE TO RANDOM LOADS

Where points X and Y coincide, this result reduces to the classical fundamental relation $S_{XX}(\omega) = |H_{XP}|^2 S_{PP}(\omega)$.

Equations 7.19, 7.20, and 7.22 can be written in matrix form,

$$\begin{bmatrix} S_{XX}(\omega) & S_{XY}(\omega) \\ S_{YX}(\omega) & S_{YY}(\omega) \end{bmatrix} = \begin{bmatrix} H_{XP}^*(\omega) \\ H_{YP}^*(\omega) \end{bmatrix} S_{PP}(\omega) \begin{bmatrix} H_{XP}(\omega) & H_{YP}(\omega) \end{bmatrix}, \quad (7.23)$$

or

$$\begin{bmatrix} S_{XX}(\omega) & S_{XY}(\omega) \\ S_{YX}(\omega) & S_{YY}(\omega) \end{bmatrix}$$
$$= \begin{bmatrix} H_{XP}^*(\omega) S_{PP}(\omega) H_{XP}(\omega) & H_{XP}^*(\omega) S_{PP}(\omega) H_{YP}(\omega) \\ H_{YP}^*(\omega) S_{PP}(\omega) H_{XP}(\omega) & H_{YP}^*(\omega) S_{PP}(\omega) H_{YP}(\omega) \end{bmatrix}.$$

Example 7.3 Response Spectra for a Two-Degree-of-Freedom System Subjected to a Single Random Force

Consider the mass-spring-damper system shown in Figure 7.6. Assume that the random force $F_1(t)$ is stationary white noise with $S_{F_1 F_1}(\omega) = S_o$. *(i)* Derive the frequency response functions $H_{11}(\omega)$ and $H_{21}(\omega)$, and *(ii)* and the spectral densities $S_{X_1 X_1}(\omega)$, $S_{X_1 X_2}(\omega)$, and $S_{X_2 X_2}(\omega)$.

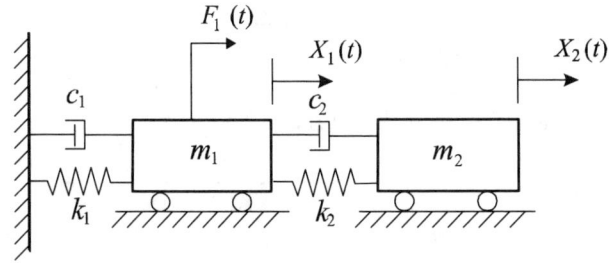

Figure 7.6: Two-degree-of-freedom system excited by a single random force.

Solution The equation of motion for this system has been obtained in Section 7.1, Equation 7.2, with $F_2(t) = 0$. We found previously that the frequency response function for this system is given by

$$[H(\omega)]$$
$$= \frac{1}{\det[Z]} \begin{bmatrix} -m_2\omega^2 + i\omega c_2 + k_2 & i\omega c_2 + k_2 \\ i\omega c_2 + k_2 & -m_1\omega^2 + i\omega(c_1 + c_2) + (k_2 + k_1) \end{bmatrix},$$

where

$$\det[Z] = \omega^4 m_1 m_2 - i\omega^3 (m_1 c_2 + c_1 m_2 + c_2 m_2) + i\omega (c_1 k_2 + k_1 c_2) \\ -\omega^2 (m_1 k_2 + c_1 c_2 + k_1 m_2 + k_2 m_2) + k_1 k_2.$$

Since $F_2 = 0$, the frequency response functions due to F_2, $H_{12}(\omega)$ and $H_{22}(\omega)$, equal zero. The frequency response functions due to F_1 are given by

$$H_{11}(\omega) = \frac{-m_2\omega^2 + i\omega c_2 + k_2}{\det[Z]}$$

$$H_{21}(\omega) = \frac{i\omega c_2 + k_2}{\det[Z]}.$$

Using Equation 7.23, the response spectra are given by

$$S_{X_1 X_1}(\omega) = H_{11}^*(\omega) H_{11}(\omega) S_{F_1 F_1}(\omega)$$
$$= S_o \frac{(k_2 - m_2\omega^2)^2 + (\omega c_2)^2}{(\det[Z])^2}$$

$$S_{X_2 X_2}(\omega) = H_{21}^*(\omega) H_{21}(\omega) S_{F_1 F_1}(\omega)$$
$$= S_o \frac{k_2^2 + (\omega c_2)^2}{(\det[Z])^2}$$

$$S_{X_1 X_2}(\omega) = H_{11}^*(\omega) H_{21}(\omega) S_{F_1 F_1}(\omega)$$
$$= S_o \frac{(-m_2\omega^2 - i\omega c_2 + k_2)(i\omega c_2 + k_2)}{(\det[Z])^2}$$

$$S_{X_2 X_1}(\omega) = H_{21}^*(\omega) H_{11}(\omega) S_{F_1 F_1}(\omega)$$
$$= S_o \frac{(-m_2\omega^2 + i\omega c_2 + k_2)(-i\omega c_2 + k_2)}{(\det[Z])^2}.$$

Note that $S_{X_2 X_1}(\omega) = S_{X_1 X_2}(-\omega) = S_{X_1 X_2}^*(\omega)$.
⊛

7.2.2 Response to Multiple Random Forces

Consider a multidegree-of-freedom system subjected to multiple random forces. The goal is to obtain the relation equivalent to Equation 7.23 for

7.2. RESPONSE TO RANDOM LOADS

the general response of N degrees of freedom. First consider the response of a single-degree-of-freedom to N forces, and then generalize to N degrees of freedom.

This will be done in two ways. First by extending the impulse response method, leading to the convolution integral, and then by utilizing the modal analysis approach where the coupled physical equations of motion are decoupled in a new *modal* coordinate system.

Impulse Response Approach

Start by expressing the response $X_k(t)$ using the impulse response functions $g_{ki}(t)$. Recall that $g_{ki}(t)$ is the impulse response of mass m_k to force $F_i(t)$. By linear superposition, the total response $X_k(t)$ of mass m_k is equal to the sum of the individual responses to each of the N forces,

$$\begin{aligned} X_k(t) &= \sum_{i=1}^{N} X_{ki}(t) \\ &= \sum_{i=1}^{N} \int_{-\infty}^{\infty} g_{ki}(\tau) F_i(t-\tau) d\tau. \end{aligned}$$

Assuming that the forces are stationary with respective mean values μ_{F_i} and cross-correlations $R_{F_i F_j}(\tau)$, the mean and correlation of the response can be found as follows,

$$\begin{aligned} E\{X_k(t)\} &= \sum_{i=1}^{N} \int_{-\infty}^{\infty} g_{ki}(\tau) E\{F_i(t-\tau)\} d\tau \\ \mu_{X_k} &= \sum_{i=1}^{N} \mu_{F_i} \int_{-\infty}^{\infty} g_{ki}(\tau) d\tau \\ &= \sum_{i=1}^{N} \mu_{F_i} H_{ki}(0), \end{aligned}$$

where Equation 6.28 is used. In matrix form,

$$\{\mu_X\} = [H(0)]\{\mu_F\}.$$

This is a static response and, therefore, can be ignored here, and added on at the end of the computations. Next evaluate the expressions for the response correlations and spectral densities, from which we can evaluate the mean-square response. By definition, the correlations are given by

$$R_{X_k X_j}(\tau)$$
$$= E\{X_k(t) X_j(t+\tau)\}$$
$$= E\left\{\sum_{m=1}^{N} X_{km}(t) \sum_{n=1}^{N} X_{jn}(t+\tau)\right\}$$
$$= E\left\{\sum_{m=1}^{N} \int_{-\infty}^{\infty} g_{km}(\zeta) F_m(t-\zeta) d\zeta \sum_{n=1}^{N} \int_{-\infty}^{\infty} g_{jn}(\xi) F_n(t+\tau-\xi) d\xi\right\}$$
$$= \sum_{m=1}^{N}\sum_{n=1}^{N} \int_{-\infty}^{\infty}\int_{-\infty}^{\infty} g_{km}(\zeta) g_{jn}(\xi) E\{F_m(t-\zeta) F_n(t+\tau-\xi)\} d\zeta d\xi$$
$$= \sum_{m=1}^{N}\sum_{n=1}^{N} \int_{-\infty}^{\infty}\int_{-\infty}^{\infty} g_{km}(\zeta) g_{jn}(\xi) R_{F_m F_n}(\tau-\xi+\zeta) d\zeta d\xi.$$

The response spectral density is by definition,

$$S_{X_k X_j}(\omega) = \frac{1}{2\pi} \int_{-\infty}^{\infty} R_{X_k X_j}(\tau) \exp(-i\omega\tau) d\tau.$$

Before substituting the correlation function into this equation, multiply it by $\exp(-i\omega[\zeta-\xi]) \cdot \exp(i\omega[\zeta-\xi])$. Also, define $\nu = \tau - \xi + \zeta$, with $d\nu = d\tau$. All of these manipulations allow us to put the spectral density in the following form,

$$\begin{aligned} S_{X_k X_j}(\omega) &= \sum_{m=1}^{N}\sum_{n=1}^{N} \int_{-\infty}^{\infty} g_{km}(\zeta) \exp(i\omega\zeta) d\zeta \int_{-\infty}^{\infty} g_{jn}(\xi) \exp(-i\omega\xi) d\xi \\ &\quad \cdot \frac{1}{2\pi} \int_{-\infty}^{\infty} R_{F_m F_n}(\nu) \exp(-i\omega\nu) d\nu \\ &= \sum_{m=1}^{N}\sum_{n=1}^{N} H_{km}^*(\omega) H_{jn}(\omega) S_{F_m F_n}(\omega), \end{aligned} \qquad (7.24)$$

where the star denotes complex conjugate. In matrix form,

$$S_{X_k X_j}(\omega) = \{H_k^*(\omega)\} [S_{FF}(\omega)] \{H_j(\omega)\}^T. \qquad (7.25)$$

In this notation, $\{H_k^*(\omega)\}$ is a row vector of dimension $1 \times N$,

$$\{H_k^*(\omega)\} = \{H_{k1}^*(\omega) \quad \cdots \quad H_{kN}^*(\omega)\},$$

$\{H_j(\omega)\}^T$ is a column vector of dimension $N \times 1$,

$$\{H_j(\omega)\}^T = \left\{\begin{array}{c} H_{j1}(\omega) \\ \vdots \\ H_{jN}(\omega) \end{array}\right\},$$

7.2. RESPONSE TO RANDOM LOADS

and $[S_{FF}(\omega)]$ is a matrix of dimension $N \times N$,

$$[S_{FF}(\omega)] = \begin{bmatrix} S_{F_1 F_1}(\omega) & \cdots & S_{F_1 F_N}(\omega) \\ \vdots & \ddots & \vdots \\ S_{F_N F_1}(\omega) & \cdots & S_{F_N F_N}(\omega) \end{bmatrix}.$$

Equation 7.25 can now be generalized for any X_k and X_j,

$$[S_{XX}(\omega)] = [H^*(\omega)][S_{FF}(\omega)][H(\omega)]^T, \qquad (7.26)$$

where

$$[H(\omega)] = \begin{bmatrix} H_{11}(\omega) & \cdots & H_{1N}(\omega) \\ \vdots & \ddots & \vdots \\ H_{N1}(\omega) & \cdots & H_{NN}(\omega) \end{bmatrix}$$

and

$$[S_{XX}(\omega)] = \begin{bmatrix} S_{X_1 X_1}(\omega) & \cdots & S_{X_1 X_N}(\omega) \\ \vdots & \ddots & \vdots \\ S_{X_N X_1}(\omega) & \cdots & S_{X_N X_N}(\omega) \end{bmatrix}.$$

Recall Equations 6.43, 6.49, and 7.23. Equation 6.43 is the response spectral density of a single-degree-of-freedom system to a single random force, Equation 6.49 is the response spectral density of a single-degree-of-freedom system to two random forces, and Equation 7.23 is the response spectral density matrix of a two-degree-of-freedom system due to a single force. Comparing them with Equation 7.26, we realize that they are all special cases of this equation.

For zero-mean response, the mean-square value of the jth coordinate is given by the relation,

$$\sigma_{X_j}^2 = \int_{-\infty}^{\infty} S_{X_j X_j}(\omega)\, d\omega.$$

Modal Analysis Approach

The modal analysis of a multidegree-of-freedom structure is developed in this section.[2] Start the analysis at the point where the modal equations of motion have been formulated, and the assumption of proportional damping has been made. From Equation 7.17, for a random vibration, the modal equations in indicial notation are

$$\ddot{Z}_i + 2\zeta_i \omega_i \dot{Z}_i + \omega_i^2 Z_i = Q_i(t), \qquad i = 1, 2, \cdots, N, \qquad (7.27)$$

---

[2] The approach of this section follows that in Chapter 23 of *Dynamics of Structures*, by R.W. Clough and J. Penzien, McGraw–Hill, Second Edition, 1993.

where the parameters are familiar. It is further assumed that the modal forces $Q_i(t)$ are *ergodic* random excitations. The transformation between physical and modal spaces is

$$X_j(t) = \sum_{i=1}^{N} u_{ji} Z_i(t),$$

and the transformation between physical and modal forces is

$$Q_j(t) = \sum_{i=1}^{N} u_{ij} F_i(t) = \{u\}_j^T \{F(t)\}, \qquad (7.28)$$

where $\{u\}_j$ is the modal vector for the jth degree of freedom and

$$\{u\}_j = \left\{ \begin{array}{c} u_{1j} \\ \vdots \\ u_{Nj} \end{array} \right\}.$$

In matrix form,

$$\begin{aligned} \{X(t)\} &= [P]\{Z(t)\} \\ \{Q(t)\} &= [P]^T \{F(t)\}, \end{aligned}$$

where $[P]$ is the modal matrix given by

$$[P] = \begin{bmatrix} \{u\}_1 & \cdots & \{u\}_N \end{bmatrix}.$$

For a two-degree-of-freedom system, we can be specific,

$$X_j(t) = \sum_{i=1}^{2} u_{ji} Z_i(t) = u_{j1} Z_1(t) + u_{j2} Z_2(t), \quad j=1,2 \quad (7.29)$$

$$Q_j(t) = \sum_{i=1}^{2} u_{ij} F_i(t) = u_{1j} F_1(t) + u_{2j} F_2(t), \quad j=1,2, \quad (7.30)$$

for each degree of freedom j of the two-degree-of-freedom structure. The goal in this analysis is to evaluate the statistics of the response, that is, to derive autocorrelations and cross-correlations $R_{X_k X_j}(\tau)$, and their Fourier transforms, the power spectra $S_{X_k X_j}(\omega)$.

Begin with the definition of the cross-correlation and substitute Equation 7.29 for $X_j(t)$,

$$\begin{aligned} R_{X_k X_j}(\tau) &= E\{X_k(t) X_j(t+\tau)\} \\ &= E\left\{\sum_{l=1}^{n} \sum_{m=1}^{n} u_{kl} u_{jm} Z_l(t) Z_m(t+\tau)\right\}, \quad (7.31) \end{aligned}$$

7.2. RESPONSE TO RANDOM LOADS

where $Z_i(t)$ is the solution to Equation 7.27,

$$Z_i(t) = \int_0^t Q_i(\tau) g_i(t-\tau) d\tau \qquad (7.32)$$

$$g_i(\tau) = \frac{1}{\omega_{d_i}} \exp(-\zeta_i \omega_i t) \sin \omega_{d_i} t \qquad (7.33)$$

$$\omega_{d_i} = \omega_i \left(1 - \zeta_i^2\right)^{1/2}. \qquad (7.34)$$

Since the impulse response function $g(t)$ equals zero for $t < 0$, the lower limit on the integral defining $Z(t)$ can be made $-\infty$ without changing the value of the integral. It should be noted that the single-indexed impulse response functions used here, $g_i(\tau)$, are the impulse response functions of the modal equations and are different from the double-indexed impulse response functions, $g_{ij}(\tau)$, used earlier.

Substituting Equations 7.32-7.34 into Equation 7.31, and letting the expectation operate only on the stochastic terms, results in the relation,

$$R_{X_k X_j}(\tau) = \sum_{l=1}^{n} \sum_{m=1}^{n} \int_{-\infty}^{t+\tau} \int_{-\infty}^{t} u_{kl} u_{jm} E\{Q_l(\theta_1) Q_m(\theta_2)\}$$
$$\cdot g_l(t-\theta_1) g_m(t+\tau-\theta_2) d\theta_1 d\theta_2, \qquad (7.35)$$

where θ_1 and θ_2 are dummy time variables, and

$$R_{Q_l Q_m}(\theta_2 - \theta_1) = E\{Q_l(\theta_1) Q_m(\theta_2)\},$$

due to the assumed ergodicity (and thus stationarity) of the forcing.

For a lightly damped system with well-separated modal frequencies, as is the case in many engineering structures, the response due to $Q_l(t)$ is almost statistically independent of the response due to $Q_m(t)$. The cross-correlation terms that arise in Equation 7.35 are then almost zero, with the only nonzero terms arising for $m = l$,

$$R_{Q_l Q_l}(\theta_2 - \theta_1) = E\{Q_l(\theta_1) Q_l(\theta_2)\}.$$

Now that we have the correlation function for the response in terms of the correlation function for the random forcing, we can proceed to evaluate the response spectral density, from which probabilities of occurrence can be evaluated. First transform variables[3] according to

$$y_1 \equiv t - \theta_1 \quad y_2 \equiv t + \tau - \theta_2$$
$$dy_1 = -d\theta_1 \quad dy_2 = -d\theta_2,$$

---
[3] Do not forget to transform the integration limits when you transform the variables.

resulting in the response correlation function,

$$R_{X_k X_j}(\tau) = \sum_{l=1}^{n} \sum_{m=1}^{n} \int_0^\infty \int_0^\infty u_{kl} u_{jm} R_{Q_l Q_m}(y_1 - y_2 + \tau)$$
$$\cdot g_l(y_1) g_m(y_2) \, dy_1 dy_2.$$

The power spectral density for response $X(t)$ equals the Fourier transform of the correlation function,

$$S_{X_k X_j}(\omega) = \frac{1}{2\pi} \int_{-\infty}^{\infty} R_{X_k X_j}(\tau) e^{-i\omega\tau} d\tau$$

$$= \frac{1}{2\pi} \int_{-\infty}^{\infty} \left\{ \sum_{l=1}^{n} \sum_{m=1}^{n} \int_0^\infty \int_0^\infty u_{kl} u_{jm} R_{Q_l Q_m}(y_1 - y_2 + \tau) \right.$$
$$\left. \cdot g_l(y_1) g_m(y_2) \, dy_1 dy_2 \right\} \exp(-i\omega\tau) \, d\tau,$$

or, recalling and utilizing the assumption that the processes are ergodic, and, therefore, by averaging in time,

$$S_{X_k X_j}(\omega) = \sum_{l=1}^{n} \sum_{m=1}^{n} u_{kl} u_{jm} \left\{ \lim_{T \to \infty} \frac{1}{2T} \int_{-T}^{T} g_l(y_1) \, dy_1 \right.$$
$$\cdot \lim_{T \to \infty} \frac{1}{2T} \int_{-T}^{T} g_m(y_2) \, dy_2$$
$$\left. \cdot \lim_{T \to \infty} \frac{1}{2T} \int_{-T}^{T} R_{Q_l Q_m}(y_1 - y_2 + \tau) \exp(-i\omega\tau) \, d\tau \right\}.$$

The lower limits on the integrals were set to $-T$ since $g(t)$ is zero for $t < 0$, and thus the changes in lower limits do not affect the values of the integrals. Using the change of variables,

$$\gamma \equiv y_1 - y_2 + \tau, \qquad d\gamma = d\tau,$$

the expression for the spectral density becomes

$$S_{X_k X_j}(\omega) = \sum_{l=1}^{n} \sum_{m=1}^{n} u_{kl} u_{jm} \left\{ \lim_{T \to \infty} \frac{1}{2T} \int_{-T}^{T} g_l(y_1) \exp(i\omega y_1) \, dy_1 \right.$$
$$\cdot \lim_{T \to \infty} \frac{1}{2T} \int_{-T}^{T} g_m(y_2) \exp(-i\omega y_2) \, dy_2$$
$$\left. \cdot \lim_{T \to \infty} \frac{1}{2T} \int_{-T-y_2+y_1}^{T-y_2+y_1} R_{Q_l Q_m}(\gamma) e^{-i\omega\gamma} d\gamma \right\}.$$

7.2. RESPONSE TO RANDOM LOADS

In the last integral, we make the physical argument that $R_{Q_l Q_m}(\gamma) \longrightarrow 0$ as $|\gamma|$ increases, and, therefore, the limits can be replaced by $-T$ and T, respectively.[4] Then,

$$H_l(-\omega) = \lim_{T \to \infty} \frac{1}{2T} \int_{-T}^{T} g_l(y_1) \exp(i\omega y_1) \, dy_1$$

$$H_m(\omega) = \lim_{T \to \infty} \frac{1}{2T} \int_{-T}^{T} g_m(y_2) \exp(-i\omega y_2) \, dy_2$$

$$S_{Q_l Q_m}(\omega) = \lim_{T \to \infty} \frac{1}{2T} \int_{-T-y_2+y_1}^{T-y_2+y_1} R_{Q_l Q_m}(\gamma) e^{-i\omega \gamma} d\gamma,$$

with the resulting response spectral density,

$$S_{X_k X_j}(\omega) = \sum_{l=1}^{n} \sum_{m=1}^{n} u_{kl} u_{jm} H_l(-\omega) H_m(\omega) S_{Q_l Q_m}(\omega).$$

Note that the single-indexed transfer functions, $H_l(\omega)$, are the transfer functions between the modal forces and the modal coordinates and they are given by Equation 7.18. They are different from the double-indexed transfer functions in Equation 7.4, $H_{lm}(\omega)$, which are the transfer functions between the actual forces and the displacements.

$$\begin{aligned} S_{X_k X_j}(\omega) &= \begin{bmatrix} u_{k1} & u_{k2} \end{bmatrix} \begin{bmatrix} H_1^*(\omega) & 0 \\ 0 & H_2^*(\omega) \end{bmatrix} \begin{bmatrix} S_{Q_1 Q_1}(\omega) & S_{Q_1 Q_2}(\omega) \\ S_{Q_2 Q_1}(\omega) & S_{Q_2 Q_2}(\omega) \end{bmatrix} \\ &\quad \cdot \begin{bmatrix} H_1(\omega) & 0 \\ 0 & H_2(\omega) \end{bmatrix} \begin{bmatrix} u_{j1} \\ u_{j2} \end{bmatrix} \\ &= u_{k1} u_{j1} H_1^*(\omega) H_1(\omega) S_{Q_1 Q_1}(\omega) \\ &\quad + u_{k1} u_{j2} H_1^*(\omega) H_2(\omega) S_{Q_1 Q_2}(\omega) \\ &\quad + u_{k2} u_{j1} H_2^*(\omega) H_1(\omega) S_{Q_2 Q_1}(\omega) \\ &\quad + u_{k2} u_{j2} H_2^*(\omega) H_2(\omega) S_{Q_2 Q_2}(\omega). \end{aligned}$$

To find $[S_{XX}]$ for the two-degree-of-freedom system, perform the following matrix product,

$$\begin{aligned}[] [S_{XX}] &= \begin{bmatrix} u_{11} & u_{12} \\ u_{21} & u_{22} \end{bmatrix} \begin{bmatrix} H_1^*(\omega) & 0 \\ 0 & H_2^*(\omega) \end{bmatrix} \begin{bmatrix} S_{Q_1 Q_1}(\omega) & S_{Q_1 Q_2}(\omega) \\ S_{Q_2 Q_1}(\omega) & S_{Q_2 Q_2}(\omega) \end{bmatrix} \\ &\quad \cdot \begin{bmatrix} H_1(\omega) & 0 \\ 0 & H_2(\omega) \end{bmatrix} \begin{bmatrix} u_{11} & u_{21} \\ u_{12} & u_{22} \end{bmatrix}. \end{aligned}$$

---

[4] Physically, as time difference γ increases, there will be an exponentially decaying correlation. This is borne out by experiments on physical systems.

In general, for N degrees of freedom,

$$[S_{XX}(\omega)] = [P][\mathcal{H}^*(\omega)][S_{QQ}(\omega)][\mathcal{H}(\omega)][P]^T, \qquad (7.36)$$

where

$$[\mathcal{H}(\omega)] = \begin{bmatrix} H_1(\omega) & 0 & 0 \\ 0 & \ddots & 0 \\ 0 & 0 & H_N(\omega) \end{bmatrix}$$

$$H_i(\omega) = \frac{1}{-\omega^2 + i2\zeta_i\omega_i\omega + \omega_i^2}.$$

The spectral densities of the modal forces $S_{Q_lQ_m}(\omega)$ can be obtained from $S_{FF}(\omega)$ using Equation 7.28. The cross-correlations $R_{Q_lQ_m}(\tau)$ are defined as

$$\begin{aligned} R_{Q_lQ_m}(\tau) &= E\{Q_l(t)Q_m(t+\tau)\} \\ &= E\left\{\sum_{i=1}^N u_{il}F_i(t) \sum_{j=1}^N u_{jm}F_j(t+\tau)\right\} \\ &= \sum_{i=1}^N \sum_{j=1}^N u_{il}u_{jm} E\{F_i(t)F_j(t+\tau)\} \\ &= \sum_{i=1}^N \sum_{j=1}^N u_{il}u_{jm} R_{F_iF_j}(\tau). \end{aligned}$$

Taking the Fourier transform,

$$S_{Q_lQ_m}(\omega) = \sum_{i=1}^N \sum_{j=1}^N u_{il}u_{jm} S_{F_iF_j}(\omega).$$

In indicial form,

$$S_{Q_lQ_m}(\omega) = \{u\}_l^T [S_{FF}(\omega)]\{u\}_m,$$

or in matrix form,

$$[S_{QQ}(\omega)] = [P]^T[S_{FF}(\omega)][P]. \qquad (7.37)$$

Substituting Equation 7.37 into Equation 7.36, we obtain

$$[S_{XX}(\omega)] = [P][\mathcal{H}^*(\omega)][P]^T[S_{FF}(\omega)][P][\mathcal{H}(\omega)][P]^T. \qquad (7.38)$$

7.2. RESPONSE TO RANDOM LOADS

Comparing this result with Equation 7.26, the following relations hold:

$$[H^*(\omega)] = [P][\mathcal{H}^*(\omega)][P]^T$$
$$[H(\omega)] = [P][\mathcal{H}(\omega)][P]^T.$$

$[H(\omega)]$ is a fully populated transfer function matrix between $\{F(t)\}$ and $\{X(t)\}$.

For lightly damped systems with well-spaced modal frequencies, the cross terms in the double summation above, those where $l \neq m$, contribute very little to the mean-square response given by $\int_{-\infty}^{\infty} S_{X_j X_j}(\omega) d\omega$. In this case, use the approximation,

$$S_{X_j X_j}(\omega) \simeq \sum_{l=1}^{n} u_{jl}^2 |H_l(\omega)|^2 S_{Q_l Q_l}(\omega), \qquad (7.39)$$

where

$$|H_l(\omega)|^2 = \left[\frac{1/\omega_l^2}{\sqrt{(1-\omega^2/\omega_l^2)^2 + (2\zeta_l \omega/\omega_l)^2}}\right]^2.$$

More details are available in specialized texts.[5]

Suppose that only the first three modes out of n degrees of freedom are utilized, where n is much larger than 3. Then, the spectral density $[S_{XX}]$ based on three modes can be obtained by

$$[S_{XX}(\omega)]_{n \times n} =$$
$$[P]_{n \times 3} [\mathcal{H}^*(\omega)]_{3 \times 3} [P]^T_{3 \times n} [S_{FF}(\omega)]_{n \times n} [P]_{n \times 3} [\mathcal{H}(\omega)]_{3 \times 3} [P]^T_{3 \times n}, \quad (7.40)$$

where

$$[P]_{n \times 3} = [\ \{u\}_1 \ \{u\}_2 \ \{u\}_3 \ ]$$

$$[\mathcal{H}(\omega)]_{3 \times 3} = \begin{bmatrix} H_1(\omega) & 0 & 0 \\ 0 & H_2(\omega) & 0 \\ 0 & 0 & H_3(\omega) \end{bmatrix}.$$

Such a procedure is computationally efficient.

Example 7.4 Response to Multiple Random Forces

Consider a two-degree-of-freedom system with mass, damping, and stiffness matrices $[m]$, $[c]$ and $[k]$. Assume that the first mass is subjected to a

---
[5] For example, see P. Wirsching, T. Paez, and H. Ortiz, *Random Vibration: Theory and Practice*, Wiley-Interscience, 1995.

single random force, $F_1(t)$, and the second mass is subjected to two random forces, $F_2(t)$ and $F_3(t)$. All the forces and displacements are along a single line. The force spectrum $[S_{FF}(\omega)]$ has a dimension of 3×3. Find the response spectra, $[S_{XX}(\omega)]$.

Solution The equation of motion is given by

$$[m]\{\ddot{X}\} + [c]\{\dot{X}\} + [k]\{X\} = \{F(t)\},$$

where $\{F(t)\} = [F_1(t) \quad F_2(t) + F_3(t)]^T$. The transfer function matrix can be found by taking the Fourier transform of the equation of motion and solving for $\{X(\omega)\}$,

$$([k] - \omega^2[m] + i\omega[c])\{X(\omega)\} = \{F(\omega)\}$$
$$\{X(\omega)\} = [H(\omega)]\{F(\omega)\},$$

where $[H(\omega)]$ is a square matrix. We can rewrite $\{X(\omega)\}$ as

$$\{X(\omega)\}_{2\times 1} = [H^*(\omega)]_{2\times 3}\{F(\omega)\}_{3\times 1}.$$

where

$$\{F(\omega)\}_{3\times 1} = \{F_1(\omega) \quad F_2(\omega) \quad F_3(\omega)\}^T$$
$$[H^*(\omega)]_{2\times 3} = \begin{bmatrix} H_{11}(\omega) & H_{12}(\omega) & H_{12}(\omega) \\ H_{21}(\omega) & H_{22}(\omega) & H_{12}(\omega) \end{bmatrix},$$

and $H_{ij}(\omega)$ is the frequency response function for coordinate i due to a force at j and is an element of the square matrix $[H(\omega)]$. Then, the response spectrum matrix is given by

$$[S_{XX}(\omega)]_{2\times 2} = [H^*(\omega)]_{2\times 3}[S_{FF}(\omega)]_{3\times 3}[H(\omega)]^T_{3\times 2}.$$

This is an extension of the result obtained for the single-degree-of-freedom system obtained in the previous chapter. Recall that for a single mass subjected to two random forces $P(t)$ and $Q(t)$, we found that the response spectrum is given by Equation 6.49 or

$$[S_{XX}(\omega)] = \{\ H^*_{XP}(\omega) \quad H^*_{XQ}(\omega)\ \} \begin{bmatrix} S_{PP}(\omega) & S_{PQ}(\omega) \\ S_{QP}(\omega) & S_{QQ}(\omega) \end{bmatrix} \begin{Bmatrix} H_{XP}(\omega) \\ H_{XQ}(\omega) \end{Bmatrix}.$$

Example 7.5 Approximation to the Response Spectral Density

Consider a two-degree-of-freedom system with equation of motion given by

$$[m]\{\ddot{X}\} + [c]\{\dot{X}\} + [k]\{X\} = \{F\},$$

7.2. RESPONSE TO RANDOM LOADS

where

$$[m] = \begin{bmatrix} m_1 & 0 \\ 0 & m_2 \end{bmatrix} \text{ kg}, \quad [k] = \begin{bmatrix} k_1+k_2 & -k_2 \\ -k_2 & k_2+k_3 \end{bmatrix} \text{ N/m},$$

$$[c] = [m] + 0.2[k] \text{ Ns/m}$$

$$[S_{FF}(\omega)] = \begin{bmatrix} S_{11}(\omega) & S_{12}(\omega) \\ S_{12}(\omega) & S_{22}(\omega) \end{bmatrix} \text{ N}^2\text{s}.$$

Obtain the response spectrum via modal analysis and the mean-square value. Assume $m_1 = 1$ kg, $m_2 = 1$ kg, $k_1 = 2$ N/m, $k_2 = 0.2$ N/m, $k_3 = 2$ N/m, $S_{11}(\omega) = 1$ N^{2}s, $S_{12}(\omega) = S_{21}(\omega) = 0$, and $S_{22}(\omega) = 2$ N^{2}s.

Solution It can be found that the natural frequencies are $\omega_1 = 1.20$ rad/s and $\omega_2 = 2.36$ rad/s, and the corresponding modal matrix is given by

$$[P] = \begin{bmatrix} 0.189 & 0.982 \\ 0.982 & -0.189 \end{bmatrix}.$$

The transfer functions $H_1(\omega)$ and $H_2(\omega)$ are given by

$$H_1(\omega) = \frac{1}{-\omega^2 + 1.162 + i\,1.232\omega}$$

$$H_2(\omega) = \frac{1}{-\omega^2 + 2.239 + i\,1.448\omega}.$$

Then, the response spectrum is given by Equation 7.38 or

$$S_{X_1X_1}(\omega) = 0.070 H_1^2(\omega) + 0.999 H_2^2(\omega)$$
$$- 0.035 \left[H_1(\omega) H_2^*(\omega) + H_1^*(\omega) H_2(\omega) \right]$$

$$S_{X_1X_2}(\omega) = 0.365 H_1^2(\omega) - 0.192 H_2^2(\omega)$$
$$- 0.179 H_1^*(\omega) H_2(\omega) + 0.007 H_1(\omega) H_2^*(\omega)$$

$$S_{X_2X_1}(\omega) = 0.365 H_1^2(\omega) - 0.192 H_2^2(\omega)$$
$$- 0.179 H_1(\omega) H_2^*(\omega) + 0.007 H_1^*(\omega) H_2(\omega)$$

$$S_{X_2X_2}(\omega) = 1.894 H_1^2(\omega) + 0.037 H_2^2(\omega)$$
$$+ 0.035 \left[H_1(\omega) H_2^*(\omega) + H_1^*(\omega) H_2(\omega) \right].$$

If we had ignored the cross-terms, as per Equation 7.39, then

$$S_{X_1X_1}(\omega) = u_{11}^2 |H_1(\omega)|^2 S_{Q_1Q_1}(\omega) + u_{12}^2 |H_2(\omega)|^2 S_{Q_2Q_2}(\omega)$$
$$= 0.070 |H_1(\omega)|^2 + 0.999 |H_2(\omega)|^2$$

$$S_{X_2X_2}(\omega) = u_{21}^2 |H_1(\omega)|^2 S_{Q_1Q_1}(\omega) + u_{22}^2 |H_2(\omega)|^2 S_{Q_2Q_2}(\omega)$$
$$= 1.894 |H_1(\omega)|^2 + 0.037 |H_2(\omega)|^2.$$

The mean-square values are given by

$$[R_{XX}(0)] = \int_{-\infty}^{\infty} [S_{XX}(\omega)]\, d\omega$$
$$= \begin{bmatrix} 1.033 & 0.392 \\ 0.392 & 4.281 \end{bmatrix},$$

where the integration can be performed using Equation 6.46. The approximation based on Equation 7.39 yields

$$R_{X_1 X_1}(0) = 1.122 \quad \text{and} \quad R_{X_2 X_2}(0) = 4.192.$$

Figure 7.7 shows the relative errors of the mean-square estimates for varying values of k_2. The relative error is defined as

$$\text{Relative Error} = \frac{\text{Approximation} - \text{Exact Value}}{\text{Exact Value}}.$$

The relative errors equal zero when there is no mechanical coupling between the two masses ($k_2 = 0$). Maximum relative errors occur for intermediate value of k_2 and decrease steadily for increasing k_2. The dynamics of the two masses may be correlated simply because the external forces are correlated. Figure 7.8 shows the relative errors when $k_2 = 0.2$ and $S_{12}(\omega)$ is varied. It is assumed that $S_{12}(\omega) = S_{21}(\omega)$. It is shown that the error is the smallest when $S_{12}(\omega) = S_{21}(\omega) = 0.2$ m²s. The errors increase as $S_{12}(\omega)$ increases. It should be noted that the errors equal zero for any values of $S_{12}(\omega)$ if the mechanical coupling between the two masses equals zero ($k_2 = 0$).

✱

Example 7.6 Approximation to the Response Spectral Density Using a Few Modes

Consider a continuous beam that vibrates in the longitudinal direction as shown in Figure 7.9. The length of the beam is 1 m, the cross-section is circular, and the radius of the cross-section varies along the beam according to

$$r(x) = -0.0005x + 0.001.$$

The beam is subjected to distributed load $f(x,t)$. The load is given by $f(x,t) = x^2 G(t)$ where $G(t)$ is a white noise-based random process with intensity 1. The beam is made of steel with density of $\rho = 7830$ kg/m³ and Young's modulus of 200 GPa Approximate this system as an $N = 10$-degree-of-freedom system connected by elastic springs. Find the natural frequencies and mode shapes. Find the approximate mean-square values using only the first three degrees of freedom.

7.2. RESPONSE TO RANDOM LOADS

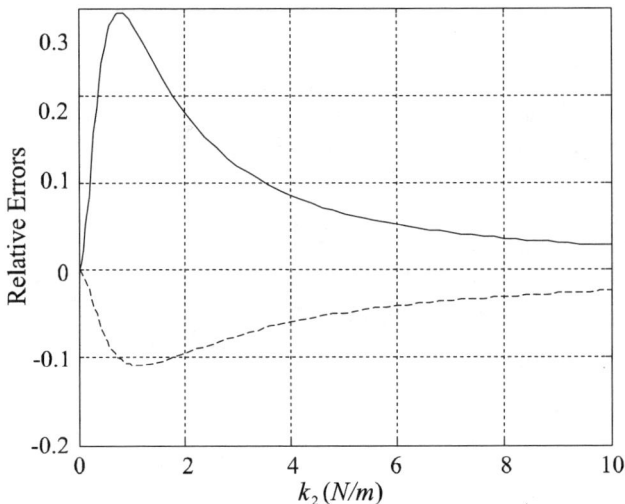

Figure 7.7: Relative errors for the mean-square values when k_2 is varied. The solid line is for $R_{X_1X_1}(0)$ and the dotted line is for $R_{X_2X_2}(0)$.

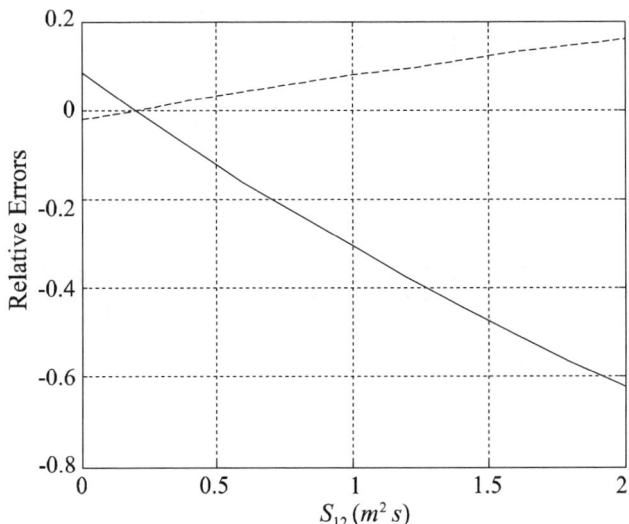

Figure 7.8: Relative errors for the mean-square values when S_{12} is varied. The solid line is for $R_{X_1X_1}(0)$ and the dotted line is for $R_{X_2X_2}(0)$.

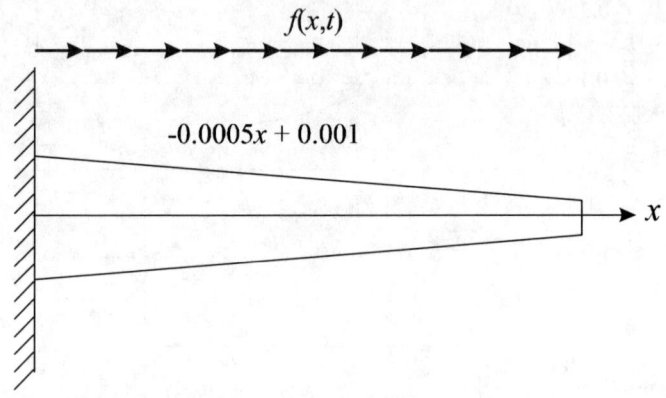

Figure 7.9: Longitudinally vibrating beam with cross-section radius $r(x) = -0.0005x + 0.001$.

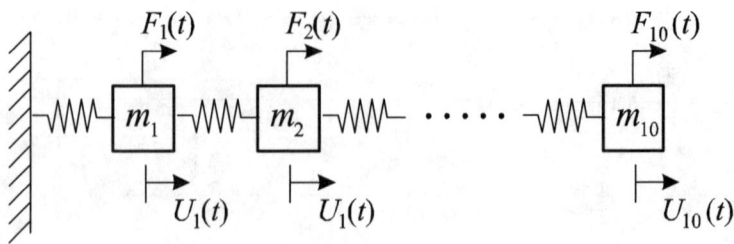

Figure 7.10: 10-degree-of-freedom approximation of a continuous beam undergoing longitudinal motion.

7.2. RESPONSE TO RANDOM LOADS

Solution First discretize the beam into N elements of equal length such that $x_i = (L/N)i$ for $i = 0, ..., N$. Figure 7.10 is a discretization into 10 elements. The lumped mass is given by

$$m_i = \int_{x_{i-1}}^{x_i} \rho A(x)\, dx$$

$$= \int_{x_{i-1}}^{x_i} \rho \pi (0.001 - 0.0005x)^2 dx \text{ for } i = 1, \cdots, N.$$

The location of the center of mass is given by

$$\bar{x}_i = \frac{\int_{x_{i-1}}^{x_i} \rho A(x)\, x\, dx}{\int_{x_{i-1}}^{x_i} \rho A(x)\, dx}.$$

From Hooke's law, the spring constant is defined by $k = $ force/change-in-length. For a longitudinal beam,

$$\text{force} = EA(x) \frac{du}{dx},$$

where dx is the original length of the incremental beam element and du is the change in length of the incremental beam element. For the ith beam element, we find that

$$\text{force} = \frac{E}{\int_{\bar{x}_{i-1}}^{\bar{x}_i} \frac{dx}{A(x)}} \Delta u_i$$

$$\frac{\text{force}}{\Delta u_i} = \frac{E}{\int_{\bar{x}_{i-1}}^{\bar{x}_i} \frac{dx}{A(x)}}.$$

The stiffness between the $(i-1)$th and ith lumped mass is then given by

$$k_i = \frac{E}{\int_{\bar{x}_{i-1}}^{\bar{x}_i} \frac{dx}{A(x)}}.$$

Note that the stiffness between the first mass and the wall is given by

$$k_1 = \frac{E}{\int_{\bar{x}_{i-1}}^{\bar{x}_i} \frac{dx}{A(x)}}.$$

The random force applied to ith beam element can be obtained by integrating the distributed force over that element,

$$F_i(t) = \int_{x_{i-1}}^{x_i} f(x,t)\,dx$$
$$= G(t)\int_{x_{i-1}}^{x_i} x^2\,dx,$$

where $\int_{x_{i-1}}^{x_i} x^2\,dx$ can be denoted as C_i. Then, the spectral density element is given by

$$S_{F_i F_j}(\omega) = C_i C_j S_{GG} = C_i C_j.$$

The lumped mass, stiffness, and C_i can be collected in matrices,

$$[m] = \begin{bmatrix} m_1 & 0 & 0 & 0 \\ 0 & m_i & 0 & 0 \\ 0 & 0 & \ddots & 0 \\ 0 & 0 & 0 & m_{10} \end{bmatrix},$$

$$[k] = \begin{bmatrix} k_1+k_2 & -k_2 & 0 & \\ -k_2 & k_2+k_3 & -k_3 & \\ 0 & -k_3 & \ddots & -k_{10} \\ 0 & 0 & -k_{10} & k_{10} \end{bmatrix}$$

$$[S_{FF}(\omega)] = S_{GG} \begin{bmatrix} C_1^2 & C_1 C_2 & \cdots & C_1 C_{10} \\ C_2 C_1 & C_2^2 & \cdots & \vdots \\ \vdots & \vdots & \ddots & \vdots \\ C_{10} C_1 & \cdots & \cdots & C_{10}^2 \end{bmatrix},$$

from which the natural frequencies and modal matrix can be obtained. The transfer functions $H_i(\omega)$ are given by

$$H_i(\omega) = \frac{1}{\omega_i^2 - \omega^2},$$

where ω_i is the ith natural frequency.

If five beam elements are used, the lumped masses, stiffnesses, and C_i are given by

$$m_1 = 0.00444, \quad m_2 = 0.00356, \quad m_3 = 0.00277,$$
$$m_4 = 0.00167208, \quad m_5 = 0.00149,$$
$$k_1 = 6.19 \times 10^6, \quad k_2 = 2.55 \times 10^6, \quad k_3 = 2.02 \times 10^6,$$
$$k_4 = 1.55 \times 10^6, \quad k_5 = 1.138 \times 10^6$$

7.3. PERIODIC STRUCTURES

$C_1 = 0.00267, \quad C_2 = 0.0187, \quad C_3 = 0.0507, \quad C_4 = 0.0987, \quad C_5 = 0.163.$

The natural frequencies are given by

$$\omega = 0.104, \quad 0.573, \quad 1.32, \quad 2.07, \quad 2.53 \; (\times 10^9) \; \text{rad/s}.$$

As the number of degrees of freedom used increases, the natural frequencies and the eigenvectors approach those of the continuous beam. If ten elements are used, we find that the first three natural frequencies are given by

$$\omega = 0.105, \quad 0.606, \quad 1.545 \; (\times 10^9) \; \text{rad/s}.$$

If twenty elements are used, the first three natural frequencies are given by

$$\omega = 0.105, \quad 0.614, \quad 1.605 \; (\times 10^9) \; \text{rad/s}.$$

Using the first three natural frequencies and eigenvectors of a 20-degree-of-freedom model, the response spectrum is given by Equation 7.38,

$$\begin{aligned}
&[S_{XX}(\omega)]_{20 \times 20} \\
&= [P]_{20 \times 3} \, [\mathcal{H}^*(\omega)]_{3 \times 3} \, [P]^T_{3 \times 20} \, [S_{FF}(\omega)]_{20 \times 20} \, [P]_{20 \times 3} \, [\mathcal{H}(\omega)]_{3 \times 3} \, [P]^T_{3 \times 20} \\
&= [P]_{20 \times 3} \begin{bmatrix} H_1^* H_1 S_{Q_1 Q_1} & H_1^* H_2 S_{Q_1 Q_1} & H_1^* H_3 S_{Q_1 Q_1} \\ H_2^* H_1 S_{Q_1 Q_1} & H_2^* H_2 S_{Q_1 Q_1} & H_2^* H_3 S_{Q_1 Q_1} \\ H_3^* H_1 S_{Q_1 Q_1} & H_3^* H_2 S_{Q_1 Q_1} & H_3^* H_3 S_{Q_1 Q_1} \end{bmatrix} [P]^T_{3 \times 20}.
\end{aligned}$$

The spectral densities can also be simplified using Equation 7.39,

$$S_{X_j X_j} = \sum_{j=1}^{20} u_{jl}^2 \, |H_l(\omega)|^2 \, S_{Q_l Q_l}(\omega).$$

Figure 7.11 shows the exact response spectrum $S_{X_1 X_1}(\omega)$ in the solid line. The dotted line is the approximation for the 20 degree-of-freedom system using the first three modes, the dashed center line is the approximation using the simplification of Equation 7.39. Note that the approximation of the spectral density using the first three modes is very close to the true solution.

7.3 Periodic Structures

Periodic structures are those with a repetitive pattern, where each *bay* is designed to be identical and joined to the next bay in the same manner. Examples of such structures are sections of aircraft fuselage that have repetitive

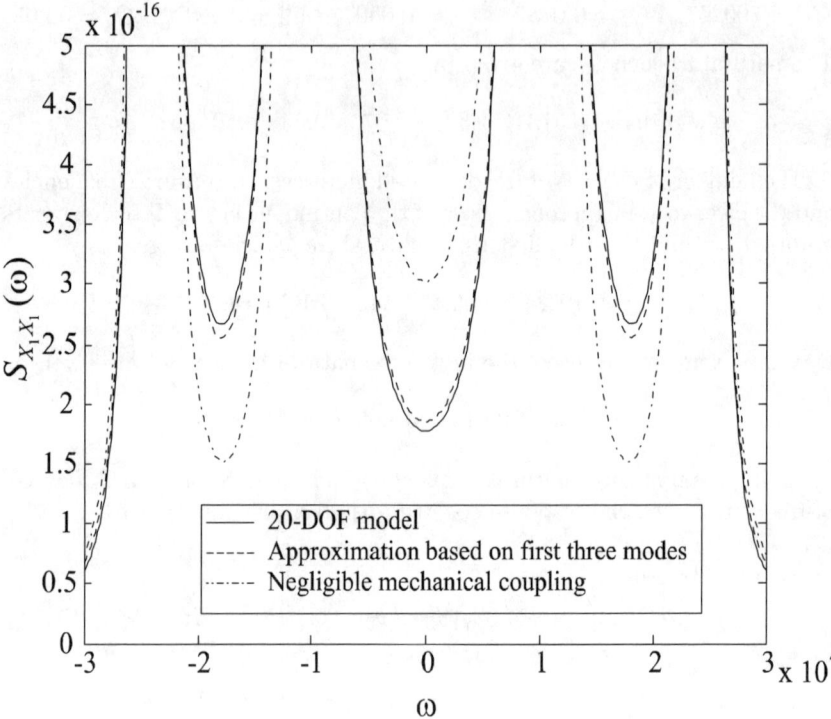

Figure 7.11: Approximation of the response spectral density: the solid line is the exact solution to the 20 DoF model; the dotted line is the approximation to the 20 DoF model based on the first three modes, and the dashed center line is the approximation to the 20 DoF model using the simplification of Equation 7.39.

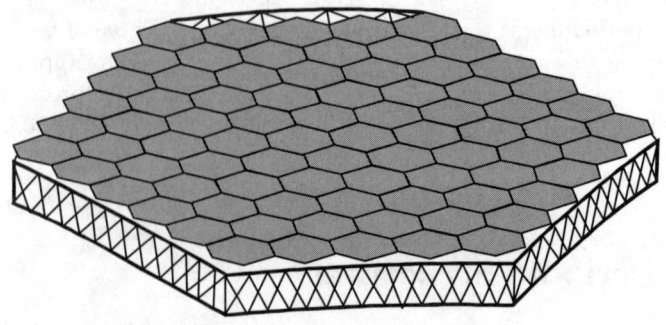

Figure 7.12: A structure with periodic properties.

7.3. PERIODIC STRUCTURES

stiffeners on a shell, turbine blades that have a circular periodicity, and antenna dishes. A schematic of a periodic structure is provided in Figure 7.12.

It is generally assumed that each bay of a repetitive structure is identical to the next one. This assumption tends to considerably simplify the analysis; only one bay plus the boundary conditions needs to be analyzed if the loading is also symmetric.

However, in actual structures, the periodic nature can never be exact and there are at least very small differences in material properties and geometry when moving from one bay to the next. Recent research has shown that even small imperfections can result in significant changes in structural response for such *near-periodic* structures. We will first examine the behavior of an exactly periodic structure, and then examine how an imperfection in the periodicity affects structural response.

7.3.1 Perfect Lattice Models

Perfectly periodic discrete structures are sometimes called *lattice* models because, historically, such spring–mass systems looked like lattices to the physicists who used them to model the interactions of atoms in a solid. Here, a 10-degree-of-freedom structure undergoing longitudinal motion (along the axis of the structure) is formulated, and some numerical results are presented and discussed. In this section, the structure is assumed to be perfectly periodic, and in the next section an imperfection is introduced so that we can examine its effects.

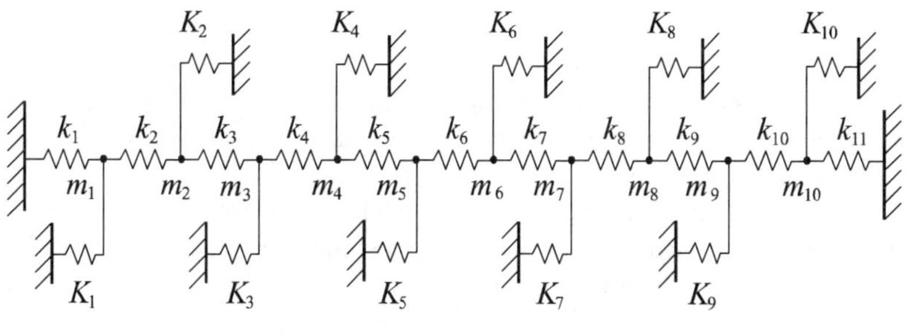

Figure 7.13: Ten-bay structure in longitudinal motion.

Consider the ten-mass structure of Figure 7.13. Each mass represents the inertial properties of a *substructure* or *bay*. A mass is attached to a neighboring mass by a coupling spring k_i that represents the coupling stiffness between substructures. To represent the stiffness of a substructure, the mass is also attached to a spring that is fixed to some immovable point. This is meant to be a conceptual model of certain classes of structures that are weakly coupled internally, but are attached to a much stiffer base structure.

Additional examples include space frame structures such as the space station, solar arrays attached to a satellite by highly stiff supports, and rotating machinery or other circular symmetric systems, where flexible blades are attached to a very stiff shaft. In Figure 7.13, if k_1 and k_{11} are made to be one and the same spring, then this model can be used for circular symmetric structures as well.

The matrix equation of motion for the periodic structure is

$$[m]\{\ddot{x}\} + [k]\{x\} = \{0\},$$

where

$$[m] = \begin{bmatrix} m_1 & 0 & \cdots & 0 \\ 0 & m_2 & \cdots & 0 \\ & \cdots & \ddots & 0 \\ 0 & \cdots & & m_{10} \end{bmatrix}, \quad \{x(t)\} = \begin{Bmatrix} x_1(t) \\ x_2(t) \\ \vdots \\ x_{10}(t) \end{Bmatrix},$$

and

$$[k] = \begin{bmatrix} k_1 + K_1 + k_2 & -k_2 & \cdots & & 0 \\ -k_2 & k_2 + K_2 + k_3 & & & 0 \\ \vdots & & 0 & \ddots & 0 \\ 0 & & & & -k_{10} \\ 0 & & \cdots & -k_{10} & k_{10} + K_{10} + k_{11} \end{bmatrix}.$$

The stiffness matrix is *tridiagonal*, meaning that nonzero elements appear only on the main and the two adjacent diagonals. The main diagonal is of the form $k_i + K_i + k_{i+1}$.

The procedures we have learned for free vibration response apply here for the evaluation of natural frequencies, modes, and response. Our purpose is to examine the time history response for each mass where, in this discussion, $m_i = 10$ kg and $K_i = 100$ N/m for all i.

A parameter found to be important in the behavior of such systems is the *coupling stiffness ratio*, defined as

$$\text{CSR} = \frac{k_i}{K_i}.$$

Once the CSR is prescribed, k_i is determined since K_i is already known. For example, a *weakly-coupled* structure may have a coupling stiffness ratio of CSR = 0.01 or 1%, and therefore, $k_i = 1$ N/m. The degree of coupling between bays, k_i, affects how fast energy can propagate from one bay to the

7.3. PERIODIC STRUCTURES

next. This can be physically understood by recognizing that energy propagates due to the compression and elongation of the spring during oscillation. For larger k_i, energy from one mass is transferred faster to the next mass. This coincides with our studies of coupled pendula.

Figure 7.14 shows the response of each of the ten masses due to a unit initial velocity applied at mass m_1. The time history is 600 s long, and we can see how the wave travels from position one to ten and then reflects back from the right end. Since the periodic system is perfectly periodic with no discontinuities or imperfections, there are no locations where a *mismatch* between the properties of adjacent cells or bays would result in some reflected energy.

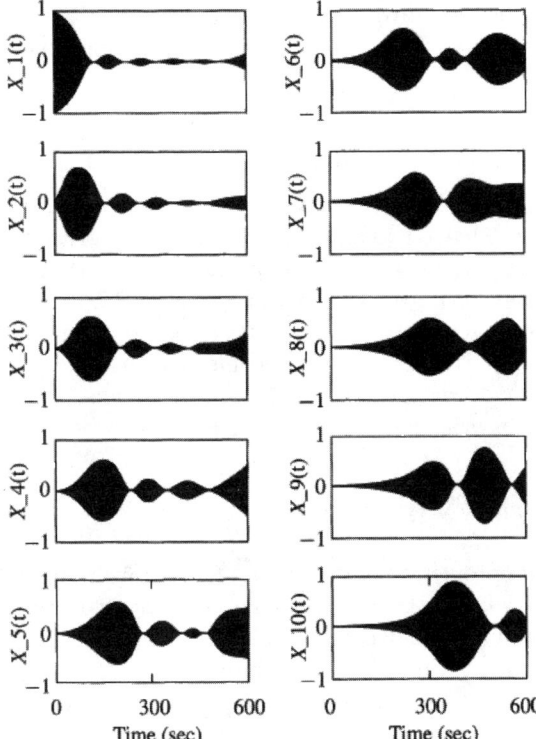

Figure 7.14: Response of ideal structure to a unit initial velocity at mass position one. (Impulse response.)

The next section discusses the effects of an imperfection on the character of the response.

7.3.2 Effects of Imperfection

The effects of imperfection can be studied by introducing a parameter that is a measure of the physical differences between adjacent bays.[6] Assume that imperfection is introduced due to differences between bay stiffnesses K_i. This *stiffness imperfection ratio* is defined as

$$\text{SIR} = \frac{K_d - K}{K},$$

where K_d is the introduced disordered bay stiffness and K is the ideal bay stiffness. For example, if SIR $= +10\%$ for an imperfectly periodic structure with ideal bay stiffness $K = 100$ N/m, then this implies that

$$\begin{aligned} K_d &= K(SIR+1) \\ &= 100(0.10+1) = 110 \text{ N/m}. \end{aligned}$$

If SIR $= -10\%$, then $K_d = 90$ N/m. Performing a standard free vibration analysis for the ten-bay structure with 10% stiffness imperfection in the fifth bay has the effect of *localizing* vibrational energy about the fifth mass. With CSR $= 1\%$, a unit initial velocity at the first mass is used to initiate a free vibration of the system with the resulting responses shown in Figure 7.15.

As an example of how the modes become distorted due to imperfections, see Figure 7.16. We see the change from a smooth mode curve to an irregular or distorted mode with the addition of imperfections.

7.4 Inverse Vibration

Inverse vibration problems come in many forms. Such problems are called inverse because what is known and what is evaluated are reversed. For example, in previous problems the system mass, damping, and stiffness were known quantities, and the known input force was used to solve for the response. In an inverse problem, the force and response are known and are used to evaluate the system mass, damping, and stiffness. Such problems are more difficult to solve than the usual *forward* problems because there may be more than one solution. That is, there may be more than one combination of system properties that satisfies the force-response relation.

In this section the eigenvalue data is used to calculate the properties of a linear dynamic system. Suppose a set of experiments is run to estimate the

---

[6] See S. Mester, H. Benaroya, "A Parameter Study of Localization," *Shock and Vibration*, Vol. 3, No. 1, pp. 1-10 (1996).

7.4. INVERSE VIBRATION

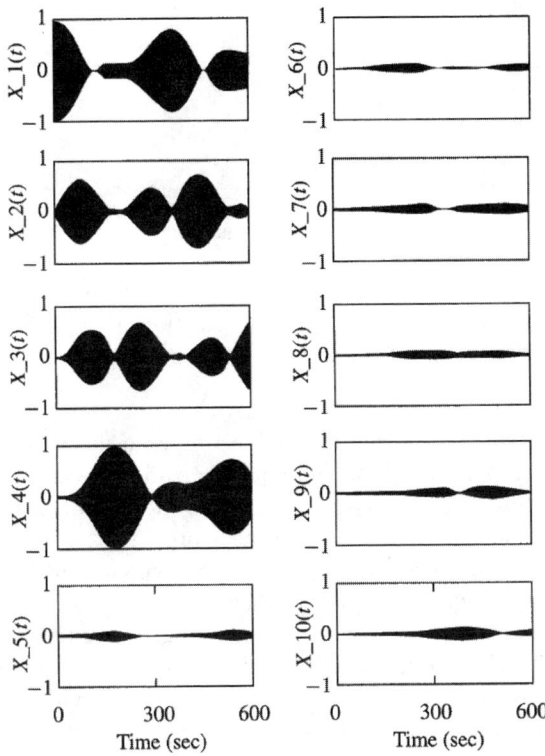

Figure 7.15: Response of a structure with SIR=10% located at the fifth bay, CSR=1%, with first bay loaded by unit initial velocity. (Impulse response.)

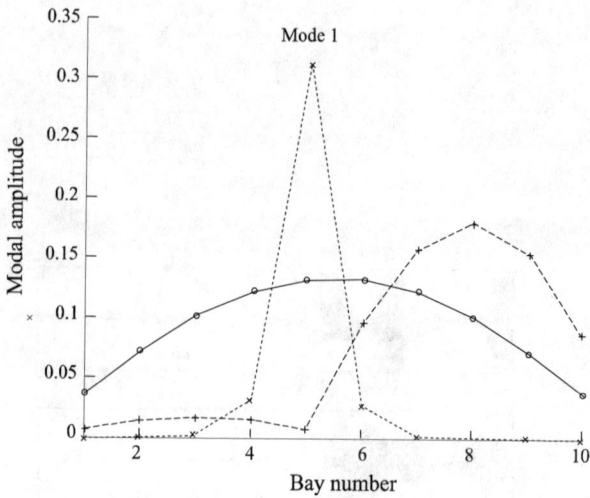

Figure 7.16: Distorted mode due to imperfection; [- × -] for negative disorder, [- ○ -] for no imperfection, [- + -] for positive disorder.

natural frequencies of a structure. For example, a multidegree-of-freedom system can be driven by a variable frequency load. At each resonance, there is a peak response and the phase angle is $\pi/2$ rad. Can this frequency data be used to evaluate the mass and stiffness properties of the structure? If not, then what additional information is necessary? This is a very simple statement of the inverse vibration problem.

This fascinating problem is examined using two approaches. The first is based on the work of Gladwell.[7] This is a deterministic approach that assumes all frequency data is exact. Such a study provides us with a new way of thinking about the relationships between structural properties and their respective free vibration characteristics.

In the second approach, it is more realistically assumed that the data has some small errors, regardless of the sophistication of the experimental setup. We are interested in finding out the mass and stiffness sensitivities to uncertainties in the frequency data.

A possible application of this type of work includes a method for nondestructive testing and evaluating structural integrity. Measurements at regular time intervals would be able to detect shifts in the spectral (frequency) properties of a given structure over time. In particular, such techniques

---

[7] G.M.L. Gladwell, *Inverse Problems in Vibration*, Martinus Nijhoff Publishers, 1986, and G.M.L. Gladwell, "Inverse Problems in Vibration," *Applied Mechanics Reviews*, Vol. 39, No. 7, July 1986, pp. 1013-1018.

7.4. INVERSE VIBRATION

could be utilized to estimate and locate changes in structural stiffness due to structural aging, and signify a need for repair.

7.4.1 Deterministic Inverse Vibration Problem

In a typical vibration problem, the physical parameters of the system are known at least approximately. These parameters are the masses and spring constants for a discrete system, or, density, modulus of elasticity, and physical dimensions for a continuous system. From an analysis of these parameters, the natural frequencies or the response to a particular excitation can be determined.

In an *inverse* vibration problem, the physical parameters of a system are determined from the spectral data, that is, frequencies and mode shapes, or eigenvalues and eigenvectors. Consider the simple spring-mass system shown in Figure 7.17. From vibration theory, it is known that this system has two distinct positive eigenvalues, λ_1 and λ_2, that are the roots of the characteristic equation,

$$\lambda^2 - \left[\frac{k_1 + k_2}{m_1} + \frac{k_2}{m_2}\right]\lambda + \frac{k_1 k_2}{m_1 m_2} = 0.$$

The respective natural frequencies equal $\sqrt{\lambda_1}$ and $\sqrt{\lambda_2}$. This equation yields the following relations between the eigenvalues:

$$\lambda_1 + \lambda_2 = \frac{k_1 + k_2}{m_1} + \frac{k_2}{m_2}, \quad \lambda_1 \lambda_2 = \frac{k_1 k_2}{m_1 m_2}. \tag{7.41}$$

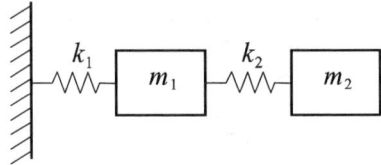

Figure 7.17: A two-degree-of-freedom system.

In inverse vibration problems, the goal is to use the eigenvalue data to reconstruct the physical system properties. For this example, there are two equations for the four unknown values k_1, k_2, m_1, and m_2. This implies that there are an infinite number of two-degree-of-freedom models that have the eigenvalues λ_1 and λ_2. It is, therefore, necessary to introduce two more equations so that the system can be completely specified.

In order to obtain more equations, consider the system shown in Figure 7.18. This system is identical to the previous one, except that the right end has been fixed, restricting it to a single degree of freedom. This constrained system has a single known eigenvalue, λ_3, given by

$$\lambda_3 = \frac{k_1 + k_2}{m_1}. \qquad (7.42)$$

Constraining an end is one way to obtain an additional equation.

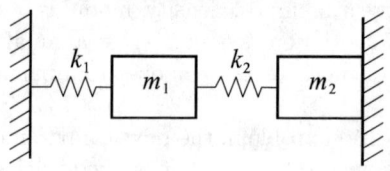

Figure 7.18: A constrained two-degree-of-freedom system.

By algebraically manipulating Equations 7.41 and 7.42, the ratios between the system properties can be obtained,

$$R_1 \equiv \frac{k_2}{m_2} = \lambda_1 + \lambda_2 - \lambda_3 \qquad (7.43)$$

$$R_2 \equiv \frac{k_1}{m_1} = \frac{\lambda_1 \lambda_2}{\lambda_1 + \lambda_2 - \lambda_3} \qquad (7.44)$$

$$R_3 \equiv \frac{k_2}{m_1} = \frac{(\lambda_3 - \lambda_1)(\lambda_2 - \lambda_3)}{\lambda_1 + \lambda_2 - \lambda_3}. \qquad (7.45)$$

Ratios R_1, R_2, and R_3 must all be positive if there is to be a corresponding physical system, since all masses and stiffnesses are positive quantities. The ratios are physically the squares of frequencies. This requires that the eigenvalues satisfy

$$0 < \lambda_1 < \lambda_3 < \lambda_2, \qquad (7.46)$$

which is predicted for this type of system by the *inclusion principle*.[8]

These ratios obviously reveal a great deal about the dynamic properties of the system, but they do not uniquely identify it. In order to do this, some

---

[8] The inclusion principle, sometimes called the *Sturmian separation theorem*, is a statement of how the natural frequencies of a system decrease as the number of degrees of freedom increase. For example, assume that there are two mathematical models of the same structure. One model is three degrees of freedom, and the other is two degrees of freedom. The first will have eigenvalues $\lambda_1 \leq \lambda_2 \leq \lambda_3$ while the other has eigenvalues $\Lambda_1 \leq \Lambda_2$. The inclusion principle can be used to show that $\lambda_1 \leq \Lambda_1 \leq \lambda_2 \leq \Lambda_2 \leq \lambda_3$.

This makes physical sense since the structure becomes less stiff, or more flexible, with more degrees of freedom, resulting in lower frequencies of oscillation. Such an understanding has implications on how we mathematically model and approximate a physical system.

7.4. INVERSE VIBRATION

further information is needed. For example, if the total mass of the system, $m = m_1 + m_2$, is known, then the parameters can be uniquely solved using Equations 7.43-7.45, resulting in

$$k_1 = \frac{R_1 R_2}{R_1 + R_3} m \qquad (7.47)$$

$$k_2 = \frac{R_1 R_3}{R_1 + R_3} m \qquad (7.48)$$

$$m_1 = \frac{R_1}{R_1 + R_3} m \qquad (7.49)$$

$$m_2 = \frac{R_3}{R_1 + R_3} m. \qquad (7.50)$$

The system's total mass serves only to scale the results, so in many cases it may be sufficient to assume a value if it is not known explicitly. Note that, from Equations 7.44 and 7.45, this approach will work if the eigenvalues are distinct and $\lambda_1 \neq \lambda_3 - \lambda_2$. Such degenerate cases require other techniques.

Example 7.7 A Two-Degree-of-Freedom Inverse Vibration Problem

In the above development, assume that the following data were obtained from two experiments,

$$\omega_1 = \sqrt{\lambda_1} = 2 \text{ Hz}$$
$$\omega_2 = \sqrt{\lambda_2} = 22 \text{ Hz}$$
$$\omega_3 = \sqrt{\lambda_3} = 5 \text{ Hz},$$

and that $m_1 + m_2 = 11$ kg. The first experiment provided ω_1 and ω_2, while the second experiment provided ω_3. Solve for k_1, k_2, m_1 and m_2.

Solution Note that ω_3 is the natural frequency of the structure when the second mass is fixed. See Equation 7.46. Using Equations 7.43-7.45, we find

$$R_1 = 463, \quad R_2 = 4.2, \quad R_3 = 20.8,$$

and using Equations 7.47-7.50,

$$k_1 = 44.2 \text{ N/m}, \quad k_2 = 218.9 \text{ N/m}$$
$$m_1 = 10.5 \text{ kg}, \quad m_2 = 0.47 \text{ kg}.$$

These results make physical sense since we expect a large disparity between mass and/or stiffness properties if there is a significant difference between the natural frequencies.

While it has been shown that it is possible to derive closed-form solutions to the inverse vibration problem associated with a two-degree-of-freedom discrete system, for larger systems more intricate numerical approaches are necessary. Gladwell's book is a good start to examine such problems. Here, we restrict ourselves to problems that can be demonstrated analytically.

⊛

7.4.2 Effect of Uncertain Data

Now introduce some uncertainty into the experimentally determined parameters. Quantities involved in the design and analysis of engineering systems generally exhibit some degree of randomness. This can be attributed to several sources. One such source is the uncertainty involved in measurements. A measurement may be made, in some cases, as precisely as is needed for a particular application. In other situations, a measurement can be made only as precisely as the measuring system technology will allow. In either case, there is some uncertainty in the resulting values.

Even if quantities could be exactly measured, the inherent statistical nature of material properties and production techniques suggest a need for probabilistic methods. Two seemingly identical components will, in general, exhibit slight characteristic differences that may affect their respective performances. Assemblies of such components are even more likely to differ from one to another. Finally, the modeling of engineering systems usually requires some approximation. Reasons for this include a lack of understanding of the particular system, or a need to simplify a particularly complex equation. Such assumptions may introduce some form of uncertainty into the solution. This last form of randomness obviously depends on the particular system. The analysis of this type of uncertainty can be quite difficult.

Here, we consider only randomness of the first two types. The system can be mathematically modeled using deterministic equations, and randomness is introduced in the variables. It is important to note that a probabilistic analysis of the type shown here not only provides a better model of the system, but also provides the analyst with a tool for quantifying statistical confidence in the analytical results.

From the discussion of the previous section, Equations such as 7.43-7.45 must be solved. λ_i are random variables and, therefore, we need to be able to work with a function of random variables. In order to work analytically, it is necessary to approximate ratios such as R_1, R_2, and R_3 using the venerable Taylor series representation.

Consider[9] a function R of random variables λ_i, $i = 1, 2, \ldots, n$. Each of

---

[9]See D. Moss, H. Benaroya, "A Discrete Inverse Vibration Problem with Parameter Uncertainties," *Applied Mathematics and Computation*, Vol. 69, 313-333 (1995).

7.4. INVERSE VIBRATION

these variables can be written as

$$\lambda_i = \mu_{\lambda_i} + \epsilon_i,$$

where μ_{λ_i} is the mean value of λ_i and ϵ_i is a (small) random parameter signifying some uncertainty about the actual (mean) value of the frequency. Therefore, $E[\epsilon_i] = 0$ and $E[\epsilon_i^2] = \sigma_{\lambda_i}^2$ since

$$\begin{aligned}\sigma_{\lambda_i}^2 &= E\left\{(\lambda_i - \mu_{\lambda_i})^2\right\} \\ &= E\left\{\epsilon_i^2\right\}.\end{aligned}$$

Before proceeding with the general expansion, consider the case of a two-degree-of-freedom structure, with two distinct roots (frequency squared), each having uncertainties,

$$\begin{aligned}\lambda_1 &= \mu_{\lambda_1} + \epsilon_1 \\ \lambda_2 &= \mu_{\lambda_2} + \epsilon_2.\end{aligned}$$

For a general general nonlinear function R_1 of both λ_1 and λ_2, the Taylor series expansion about the mean values of λ_1 and λ_2 is

$$\begin{aligned}R_1(\lambda_1, \lambda_2) &= \\ R_1(\mu_{\lambda_1}, \mu_{\lambda_2}) &+ \frac{\partial R_1(\mu_{\lambda_1}, \mu_{\lambda_2})}{\partial \lambda_1}(\lambda_1 - \mu_{\lambda_1}) + \frac{\partial R_1(\mu_{\lambda_1}, \mu_{\lambda_2})}{\partial \lambda_2}(\lambda_2 - \mu_{\lambda_2}) \\ &+ \frac{1}{2}\left[\frac{\partial^2 R_1(\mu_{\lambda_1}, \mu_{\lambda_2})}{\partial \lambda_1^2}(\lambda_1 - \mu_{\lambda_1})^2 + \frac{\partial^2 R_1(\mu_{\lambda_1}, \mu_{\lambda_2})}{\partial \lambda_1 \partial \lambda_2}(\lambda_1 - \mu_{\lambda_1})(\lambda_2 - \mu_{\lambda_2}) \right. \\ &\left. + \frac{\partial^2 R_1(\mu_{\lambda_1}, \mu_{\lambda_2})}{\partial \lambda_2^2}(\lambda_2 - \mu_{\lambda_2})^2\right] + \cdots,\end{aligned}$$

where $(\lambda_1 - \mu_{\lambda_1}) = \epsilon_1$ and $(\lambda_2 - \mu_{\lambda_2}) = \epsilon_2$. A similar expression can be derived for $R_2(\lambda_1, \lambda_2)$. It is important to observe that all terms in these expressions are evaluated at the respective mean values of λ_1 and λ_2, known quantities.

The function R_k for an n-degree-of-freedom structure is a function of all λ_i, and can be expanded into a Taylor series as

$$\begin{aligned}R_k(\lambda_1, \lambda_2, \ldots, \lambda_n) &= \\ R_k(\mu_{\lambda_1}, \mu_{\lambda_2}, \ldots, \mu_{\lambda_n}) &+ \sum_{i=1}^{n} \frac{\partial R_k}{\partial \lambda_i}\epsilon_i + \frac{1}{2}\sum_{i=1}^{n}\sum_{j=1}^{n} \frac{\partial^2 R_k}{\partial \lambda_i \partial \lambda_j}\epsilon_i \epsilon_j + \cdots,\end{aligned}$$

for $k = 1, 2, \ldots, n$. If the frequencies exhibit only a small degree of randomness, that is, if $\lambda_i - \mu_{\lambda_i} = \epsilon_i \ll 1$, the expansion can be truncated after only a few terms with little error due to small terms such as ϵ_i^2 and $\epsilon_i \epsilon_j$.

We only take the first two terms of the Taylor series to demonstrate this procedure. This is a linear approximation for the actual value of R,

$$R_k(\lambda_1, \lambda_2, \ldots, \lambda_n) \simeq R_k(\mu_{\lambda_1}, \mu_{\lambda_2}, \ldots, \mu_{\lambda_n}) + \sum_{i=1}^{n} \frac{\partial R_k}{\partial \lambda_i} \epsilon_i, \quad k = 1, 2, \ldots, n.$$

Since $E\{\epsilon_i\} = 0$, taking the expected value of R_k leads to the approximate result

$$E\{R_k\} \simeq R_k(\mu_{\lambda_1}, \mu_{\lambda_2}, \ldots, \mu_{\lambda_n}), \quad k = 1, 2, \ldots, n. \quad (7.51)$$

Thus, the linear or first-order approximation of the mean value of a complicated function can be obtained by substituting the mean values of all random variables in the function.

To obtain an estimate of the standard deviation of R_k, assume the variables to be statistically independent,[10] so that $E[\epsilon_i \epsilon_j] = E[\epsilon_i]E[\epsilon_j] = 0$ for $i \neq j$. The standard deviation of R_k is then estimated by

$$\sigma_{R_k}^2 = E\{R_k^2\} - E^2\{R_k\} = \sum_{i=1}^{n} \left(\frac{\partial R_k}{\partial \lambda_i}\right)^2 \sigma_{\lambda_i}^2, \quad k = 1, 2 \ldots, n, \quad (7.52)$$

where the partial derivatives on the right hand side are evaluated at the respective mean values. It should be noted that Equation 7.52 depends only on the mean values and standard deviations of the random variables. It is independent of the particular distribution of these variables, with the only assumption being that they are independent. The method can therefore prove useful in cases where little is known about the probabilistic nature of the random variables, but that estimates of their means and variances can be obtained. Example 7.8 below demonstrates the procedure just developed.

A more accurate prediction for the statistics of random variable R_k is obtained by retaining the second-order term of the Taylor series,

$$R_k(\lambda_1, \lambda_2, \ldots, \lambda_n) \simeq$$
$$R_k(\mu_{\lambda_1}, \mu_{\lambda_2}, \ldots, \mu_{\lambda_n}) + \sum_{i=1}^{n} \frac{\partial R_k}{\partial \lambda_i} \epsilon_i + \frac{1}{2} \sum_{i=1}^{n} \sum_{j=1}^{n} \frac{\partial^2 R_k}{\partial \lambda_i \partial \lambda_j} \epsilon_i \epsilon_j,$$

for $k = 1, 2, \ldots, n$. However, the convergence of approximate series such as these cannot be taken for granted.

It is necessary to verify solutions obtained using these approximate expansions since accuracy depends on the smallness of ϵ. To verify the accuracy of the truncated Taylor series, we have two options. The first is

---

[10] Such assumptions are generally made as a first approximation to the actual situation. For cases where this is not a valid assumption, it is necessary to somehow estimate, usually with experiments, what is the correlation so that $E\{\epsilon_i \epsilon_j\}$ can be evaluated.

7.4. INVERSE VIBRATION

to build an experiment that duplicates the vibrating structure and test it under various conditions. The other option is the powerful Monte Carlo simulation technique, which is discussed in a couple of places in this text. The key point to be made here is that there must be an effort at verification of any and every approximation made in an analysis.

Example 7.8 An Uncertain Two-Degree-of-Freedom System

Use the previous two-term Taylor series approximation and the following data to demonstrate this procedure. The eigenvalues,

$$\lambda_1 = 0.382 \text{ Hz}^2, \quad \lambda_2 = 2.618 \text{ Hz}^2$$

are for the original system shown in Figure 7.17, and

$$\lambda_3 = 1.000 \text{ Hz}^2$$

for the system with fixed end shown in Figure 7.18.

Use the following information: $m_1 + m_2 = 20$ kg exactly (zero standard deviation), where the mean values of m_1 and m_2 are each 10 kg, the stiffnesses are $k_1 = k_2 = 10$ kg/cm, and all λ values are approximate and assumed random with coefficients of variation $\delta = \sigma/\mu = 0.01$, a 1% variation. We are interested in estimating the mean values and variances of k_i and m_i, given the mean values and variances of the masses and eigenvalues.

Solution Utilize Equation 7.51 for the mean-value calculations and Equation 7.52 for the standard deviation calculations. The procedure is in two parts:

1. Given the mean values and standard deviations of λ_i, and using Equations 7.43-7.45, derive the estimated mean values and variances of each ratio R_i.

2. With these results, Equations 7.47-7.50 are used to derive the estimated mean values and variances of each stiffness and mass.

The procedure will be demonstrated only for some of the variables since the algebra becomes very long. Begin with ratio $R_1 = \lambda_1 + \lambda_2 - \lambda_3$. The mean value of R_1 is estimated by

$$\mu_{R_1} = E\{R_1\} = R_1(\mu_{\lambda_1}, \mu_{\lambda_2}, \mu_{\lambda_3})$$
$$= \mu_{\lambda_1} + \mu_{\lambda_2} - \mu_{\lambda_3}.$$

The variance is estimated by

$$\sigma^2_{R_1} = \left(\frac{\partial R_1}{\partial \lambda_1}\right)^2 \sigma^2_{\lambda_1} + \left(\frac{\partial R_1}{\partial \lambda_2}\right)^2 \sigma^2_{\lambda_2} + \left(\frac{\partial R_1}{\partial \lambda_3}\right)^2 \sigma^2_{\lambda_3}$$
$$= (1)^2 \sigma^2_{\lambda_1} + (1)^2 \sigma^2_{\lambda_2} + (-1)^2 \sigma^2_{\lambda_3}.$$

Next, follow the same procedure for ratio $R_2 = \lambda_1\lambda_2/(\lambda_1 + \lambda_2 - \lambda_3)$. The mean is estimated by

$$\mu_{R_2} = E\{R_2\} = \frac{\mu_{\lambda_1}\mu_{\lambda_2}}{\mu_{\lambda_1} + \mu_{\lambda_2} - \mu_{\lambda_3}},$$

and the variance by

$$\begin{aligned}
\sigma_{R_2}^2 &= \left(\frac{\partial R_2}{\partial \lambda_1}\right)^2 \sigma_{\lambda_1}^2 + \left(\frac{\partial R_2}{\partial \lambda_2}\right)^2 \sigma_{\lambda_2}^2 + \left(\frac{\partial R_2}{\partial \lambda_3}\right)^2 \sigma_{\lambda_3}^2 \\
&= \left(\frac{(\mu_{\lambda_1} + \mu_{\lambda_2} - \mu_{\lambda_3})\mu_{\lambda_2} - \mu_{\lambda_1}\mu_{\lambda_2}(1)}{(\mu_{\lambda_1} + \mu_{\lambda_2} - \mu_{\lambda_3})^2}\right)^2 \sigma_{\lambda_1}^2 \\
&+ \left(\frac{(\mu_{\lambda_1} + \mu_{\lambda_2} - \mu_{\lambda_3})\mu_{\lambda_1} - \mu_{\lambda_1}\mu_{\lambda_2}(1)}{(\mu_{\lambda_1} + \mu_{\lambda_2} - \mu_{\lambda_3})^2}\right)^2 \sigma_{\lambda_2}^2 \\
&+ \left(\frac{(\mu_{\lambda_1} + \mu_{\lambda_2} - \mu_{\lambda_3})(0) - \mu_{\lambda_1}\mu_{\lambda_2}(-1)}{(\mu_{\lambda_1} + \mu_{\lambda_2} - \mu_{\lambda_3})^2}\right)^2 \sigma_{\lambda_3}^2,
\end{aligned}$$

which can be algebraically simplified. The same procedure can be used to estimate the mean value and variance of ratio R_3.

Now that the statistics of each ratio R_i has been estimated, proceed with step two to use these in the estimation of the statistics of real interest here, those of k_i and m_i. Beginning with the relation $k_2 = R_1 R_3 m/(R_1 + R_3)$, estimate the mean value of k_2 by

$$E\{k_2\} = \frac{\mu_{R_1}\mu_{R_3}}{\mu_{R_1} + \mu_{R_3}}m,$$

where the total mass m is assumed to be an exact value with no variance. For the estimated value of the variance of k_1, we have

$$\begin{aligned}
\sigma_{k_2}^2 &= \left(\frac{\partial k_2}{\partial R_1}\right)^2 \sigma_{R_1}^2 + \left(\frac{\partial k_2}{\partial R_2}\right)^2 \sigma_{R_2}^2 + \left(\frac{\partial k_2}{\partial R_3}\right)^2 \sigma_{R_3}^2 \\
&= \left(\frac{(\mu_{R_1} + \mu_{R_3})\mu_{R_3}m - \mu_{R_1}\mu_{R_3}m(1)}{(\mu_{R_1} + \mu_{R_3})^2}\right)^2 \sigma_{R_1}^2 \\
&+ \left(\frac{(\mu_{R_1} + \mu_{R_3})(0) - \mu_{R_1}\mu_{R_3}m(0)}{(\mu_{R_1} + \mu_{R_3})^2}\right)^2 \sigma_{R_2}^2 \\
&+ \left(\frac{(\mu_{R_1} + \mu_{R_3})\mu_{R_1}m - \mu_{R_1}\mu_{R_3}m(1)}{(\mu_{R_1} + \mu_{R_3})^2}\right)^2 \sigma_{R_3}^2,
\end{aligned}$$

where it is noted that the second expression on the right-hand side equals zero since R_2 is not in the equation for k_2. The same procedure can then be

Table 7.1: Taylor Expansion Results Compared to Monte Carlo Simulations. (The values under the columns labeled MC $\star$ $\Delta\%$ show the percent differences between the perturbation results and the Monte Carlo results for μ and σ, respectively.)

Variable	μ	σ	$\delta = \sigma/\mu$	(MC μ $\Delta\%$)	(MC σ $\Delta\%$)
m_1	10.00	0.235	0.024	0.07	0.54
m_2	10.00	0.235	0.024	0.07	0.54
k_1	10.00	0.212	0.021	0.21	2.45
k_2	10.00	0.158	0.016	0.25	3.05

used to estimate the mean values and variances of the remaining parameters, k_1, m_1 and m_2. Substituting the mean values and variances given to us at the beginning of the problem statement, we find the results of the first-order expansion as listed in Table 7.1. Comparisons are made between the expansion values and a Monte Carlo (MC) simulation that is considered to be essentially exact.

We see larger errors for the approximate stiffness values than for the mass values when comparing with the Monte Carlo results. This is due to the greater complexity of the stiffness expressions, as seen in Equations 7.43–7.45.

⊛

For many engineering applications, procedures such as those presented in this section can suitably model uncertainties.

7.5 Random Eigenvalues

It is of interest to consider problems where parameters are defined as random variables. The ideas presented in this section can be of interest if exact solutions are desired in low order systems. For more complex problems, methods for nonlinear systems and Monte Carlo methods are better suited.

The procedure discussed here[11] is based on the ability to transform probability density functions, discussed in Section 4.1. The approach can be viewed as part of the general algebraic theory of random variables. As we know, there is no single more important descriptor of a linear system than its eigenstructure, its eigenvalues and eigenvectors or eigenfunctions. System characteristics, properties, and thus behavior, are all embedded in system eigenproperties. The effects of mass, damping, and stiffness distri-

---

[11] This section is based on the paper, "Random Eigenvalues and Structural Dynamics Models," H. Benaroya, in *Stochastic Structural Dynamics 1: New Theoretical Developments*, Eds. Y.K. Lin, I. Elishakoff, Springer-Verlag 1990.

butions determine the magnitude and distribution of the eigenvalues, as do structural imperfections, stress concentrations, and physical constraints.

Inherent inaccuracies in manufacturing, measurements, uncertainties of geometry and material, and other factors make engineering systems uncertain in the sense that their parameter values cannot be specified exactly, rather, via a probability density function. Such systems may be called disordered structural systems, random systems, or random eigenvalue problems.

Consider the discretized model of a structural system,

$$[M]\{\ddot{Y}\} + [C]\{\dot{Y}\} + [K]\{Y\} = \{F(t)\},$$

where $\{F(t)\}$ is a stochastic vector function of time. The elements of matrices $[M]$, $[C]$, $[K]$ contain random variables. For example, randomly distributed stiffness properties can be written as

$$[K] = [K_0] + [\kappa],$$

where $E\{[K]\} = [K_0]$, and $[\kappa]$ is a matrix of zero-mean random variables.

As an example, consider a two-degree-of-freedom system with no damping and no mass coupling. The random stiffness matrix is composed of random variable stiffnesses K_{ij},

$$[K] = \begin{bmatrix} K_{11} & K_{12} \\ K_{21} & K_{22} \end{bmatrix},$$

with system dynamic characteristics associated with the random eigenvalue problem,

$$|[K] - \lambda[I]| = 0.$$

λ are the squares of the modal frequencies, that are found by solving the random quadratic polynomial (the characteristic function),

$$A_2 \lambda^2 + A_1 \lambda + A_0 = 0,$$

where

$$\begin{aligned} A_2 &= 1 \\ A_1 &= -(K_{11} + K_{22}) \\ A_0 &= K_{11}K_{22} - K_{12}K_{21}. \end{aligned}$$

It has been assumed that all mass elements have numerical value 1. $\lambda_{1,2}$ are the two solutions of the eigenvalue problem and both are random variables. The joint density function of the random eigenvalues as a function of the density of the matrix elements is of interest. Three possible solution sets need to be considered:

7.5. RANDOM EIGENVALUES

1. $\lambda_{1,2}$ are real and unequal
2. $\lambda_{1,2}$ are real and equal
3. $\lambda_{1,2}$ are complex conjugates

If there is a more general system with inertial as well as stiffness coupling, the characteristic values are defined by

$$\det \begin{bmatrix} K_{11} - \lambda M_{11} & K_{12} - \lambda M_{12} \\ K_{21} - \lambda M_{21} & K_{22} - \lambda M_{22} \end{bmatrix} = 0.$$

Early work by Hamblen[12] considered random algebraic equations of the form,

$$\lambda^2 - A_1 \lambda + A_0 = 0,$$

which has roots,

$$\lambda_{1,2} = \frac{A_1}{2} \pm \sqrt{\frac{A_1^2}{4} - A_0},$$

or

$$A_1 = \lambda_1 + \lambda_2$$
$$A_0 = \lambda_1 \lambda_2.$$

The roots are either real or complex conjugates. The probabilities of these events are

$$\Pr(R) = \Pr\left(A_0 \leq \frac{A_1^2}{4}\right)$$
$$\Pr(C) = \Pr\left(A_0 > \frac{A_1^2}{4}\right),$$

where R is defined as an event when the roots are real and C as an event that the roots are complex.

Given $f_{A_1 A_0}(a_1, a_0)$, one can evaluate $\Pr(R)$ and $\Pr(C)$ by integrating the joint density over the appropriate domains. Also, the following conditional densities can be evaluated,

$$f_{A_1 A_0}(a_1, a_0 | C) = f_{A_1 A_0}(a_1, a_0) / \Pr(C)$$
$$f_{A_1 A_0}(a_1, a_0 | R) = f_{A_1 A_0}(a_1, a_0) / \Pr(R).$$

For example, suppose the joint density is given by the gamma distribution,

$$f_{A_1 A_0}(a_1, a_0) = \exp(-a_1 - a_0).$$

---

[12] *Distribution of Roots of Algebraic Equations with Variable Coefficients*, J.W. Hamblen, Ph.D. Dissertation, Purdue, 1955.

Then $\Pr(R) = 0.24$. and the marginal densities for the roots are found to be

$$g_1(\lambda_1|R) = \frac{1}{0.24}\frac{\lambda_1^2 + \lambda_1 - 1}{(1+\lambda_1)^2}\exp(-\lambda_1) + \frac{1}{0.24}\frac{1}{(1+\lambda_1)^2}\exp(-\lambda_1^2 - 2\lambda_1),$$
$$0 \leq \lambda_1 < \infty$$
$$g_2(\lambda_2|R) = \frac{1}{0.24}\frac{1}{(1+\lambda_2)^2}\exp(-\lambda_2^2 - 2\lambda_2), \quad 0 \leq \lambda_2 < \infty,$$

where $\lambda_{1,2}$ are the real roots and $g_{1,2}$ are their marginal densities. See Figure 7.19.

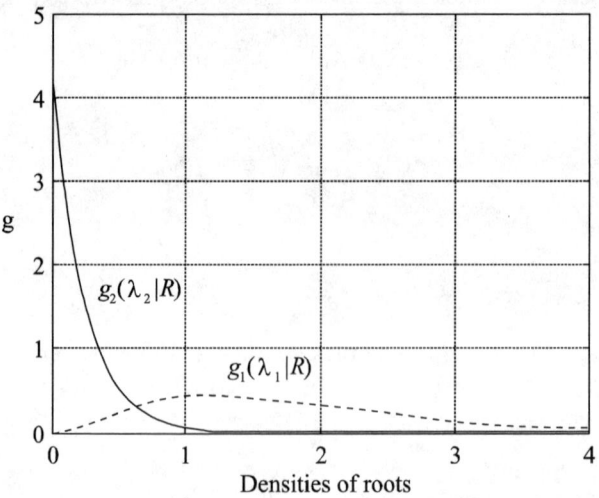

Figure 7.19: $g_1(\lambda_1|R)$ and $g_2(\lambda_2|R)$.

Example 7.9 Random Eigenvalues

Consider an undamped two-degree-of-freedom system with mass and stiffness matrices given by

$$[M] = \begin{bmatrix} 1 & 0 \\ 0 & 1 \end{bmatrix} \text{ kg} \quad [K] = \begin{bmatrix} k+1 & -1 \\ -1 & 1 \end{bmatrix} \text{ N/m},$$

where k is random, distributed uniformly between 0.5 and 1.5 N/m. Obtain the marginal densities for the roots.

Solution The characteristic equation is given by

$$\lambda^2 - (2+k)\lambda + k = 0.$$

7.5. RANDOM EIGENVALUES

The roots of the characteristic equation are given by

$$\lambda_{1,2} = \frac{1}{2}\left(2 + k \pm \sqrt{k^2 + 4}\right),$$

which are always real. The probability density $f_{\lambda_1}(\lambda_1)$ is related to $f_k(k)$ by

$$f_{\lambda_1}(\lambda_1) = f_k(k) \frac{dk}{d\lambda_1}.$$

After algebraic manipulations,

$$\frac{dk}{d\lambda} = \frac{(\lambda^2 - 2\lambda + 2)}{(1+\lambda)^2}.$$

The marginal densities are then given by

$$f_{\lambda_1}(\lambda_1) = \frac{(\lambda_1^2 - 2\lambda_1 + 2)}{(1+\lambda_1)^2} \quad 4.5616 < \lambda_1 < 6$$

$$f_{\lambda_2}(\lambda_2) = \frac{(\lambda^2 - 2\lambda_2 + 2)}{(1+\lambda_2)^2} \quad 0.4384 < \lambda_2 < 1.$$

⊛

7.5.1 A Two-Degree-of-Freedom Model

Consider the two-degree-of-freedom system governed by the governing equation,

$$\begin{bmatrix} 1 & 0 \\ 0 & 1 \end{bmatrix} \begin{Bmatrix} \ddot{X}_1 \\ \ddot{X}_2 \end{Bmatrix} + \begin{bmatrix} K_1 + K_2 & -K_2 \\ -K_2 & K_2 + K_3 \end{bmatrix} \begin{Bmatrix} X_1 \\ X_2 \end{Bmatrix} = \begin{Bmatrix} 0 \\ 0 \end{Bmatrix}.$$

For the characteristic equation $\lambda^2 - A_1\lambda + A_0 = 0$, let $\lambda = \omega^2$, and

$$\begin{aligned} A_1 &= K_1 + K_3 + 2K_2 > 0 \\ A_0 &= K_1 K_3 + K_2(K_1 + K_3) > 0. \end{aligned}$$

The stiffnesses are governed by $f_{K_1}(k_1)$ and $f_{K_2}(k_2)$, and $K_3 = k_3$ is deterministic. Our goal is to derive the joint density function of the roots $\lambda_{1,2}$. This is a two stage process:

1. $f_{A_1 A_0}(a_1, a_0) \Longleftarrow f_{K_1 K_2}(k_1, k_2)$
2. $f_{\lambda_1 \lambda_2}(\lambda_1, \lambda_2) \Longleftarrow f_{A_1 A_0}(a_1, a_0)$.

Define
$$K_{11} = A_1 - k_3 - 2K_{21}$$
$$K_{12} = A_1 - k_3 - 2K_{22}$$
$$K_{21} = -\frac{a}{2} + \frac{1}{2}\sqrt{a^2 - 4b}$$
$$K_{22} = -\frac{a}{2} - \frac{1}{2}\sqrt{a^2 - 4b}$$
$$a = -\frac{1}{2}(-A_1 + 2k_3)$$
$$b = -\frac{1}{2}(A_0 - k_3 A_1 + k_3^2).$$

The derived density is then
$$f_{A_1 A_0}(a_1, a_0) = \frac{f_{K_1 K_2}(k_{11}, k_{21})}{|J(k_{11}, k_{21})|} + \frac{f_{K_1 K_2}(k_{12}, k_{22})}{|J(k_{12}, k_{22})|},$$
where $|J|$ denotes the absolute value of the Jacobians, respectively,
$$J(k_{11}, k_{21}) = \det(k_{11} - k_3 - 2k_{21})$$
$$J(k_{12}, k_{22}) = \det(k_{12} - k_3 - 2k_{22}).$$

Marginal densities f_{A_1} and f_{A_0} can be obtained by integrating out the other variable.

Given $f_{A_1 A_0}(a_1, a_0)$, the second part of the solution is to derive the joint density of the roots. As before,
$$\lambda_{1,2} = \frac{A_1}{2} \pm \sqrt{\frac{A_1^2}{4} - A_0},$$
or
$$A_1 = \lambda_1 + \lambda_2$$
$$A_0 = \lambda_1 \lambda_2.$$

The probability of real and complex roots are given by
$$\Pr(R) = \Pr\left(A_0 \leq \frac{A_1^2}{4}\right) = \int\int_{A_0 \leq A_1^2/4} f_{A_1 A_0}(a_1, a_0) da_1 da_0$$
$$\Pr(C) = \Pr\left(A_0 > \frac{A_1^2}{4}\right) = \int\int_{A_0 > A_1^2/4} f_{A_1 A_0}(a_1, a_0) da_1 da_0.$$

Also,
$$f_{A_1 A_0}(a_1, a_0 | R) = \frac{f_{A_1 A_0}(a_1, a_0)}{\Pr(R)}$$
$$f_{A_1 A_0}(a_1, a_0 | C) = \frac{f_{A_1 A_0}(a_1, a_0)}{\Pr(C)}.$$

The complex roots are of interest in vibration. Let $\lambda_{1,2} = \alpha \pm i\beta$. The joint density $f_{\alpha\beta}(\alpha, \beta)$ can be found to be[13]

$$f_{\alpha\beta}(\alpha, \beta | C) = \frac{f_{A_1 A_0}(2\alpha, \alpha^2 + \beta^2) |J(\alpha, \beta)|}{\Pr(C)},$$

where

$$J(\alpha, \beta) = \det \begin{bmatrix} \partial A_1/\partial \alpha & \partial A_1/\partial \beta \\ \partial A_2/\partial \alpha & \partial A_2/\partial \beta \end{bmatrix} = 4\beta, \quad \beta > 0.$$

For $\beta < 0$, $J(\alpha, \beta) = -4\beta$.

This procedure is an interesting one that depends on the ability of performing analysis of the density and functional equations.

7.6 Concluding Summary

We introduced methods for the analysis of multidegree-of-freedom systems subjected to random forces. General relations were derived relating the matrices of forcing and response spectra. Special topics of periodic structures, inverse vibration and random eigenvalues were introduced and discussed.

7.7 Problems

Section 7.1: Deterministic Vibration

1. Consider an undamped three-degree-of-freedom system with property matrices:

$$[m] = \begin{bmatrix} 1 & 0.5 & 0 \\ 0.5 & 2 & 0.3 \\ 0 & 0.3 & 3 \end{bmatrix} \text{ kg}, \quad [k] = \begin{bmatrix} 1 & 0 & 0 \\ 0 & 1 & 0 \\ 0 & 0 & 2 \end{bmatrix} \text{ N/m}.$$

Obtain the response using modal analysis. Assume that the initial conditions are given by

$$\{x(0)\} = \begin{bmatrix} 0 \\ 0.5 \\ 0 \end{bmatrix} \text{ m}, \quad \{\dot{x}(0)\} = \begin{bmatrix} 0 \\ 0 \\ 0.1 \end{bmatrix} \text{ m/s}.$$

---

[13] Write

$$\alpha = A_1/2$$
$$\beta = \pm\sqrt{A_0 - A_1^2/4},$$

or $A_1 = 2\alpha$ and $A_0 = \alpha^2 + \beta^2$.

2. Consider a one-meter-long continuous longitudinal beam with a circular cross-section. This beam is fixed at one end and free at the other. The radius of the cross-section is given by

$$r(x) = 0.0001 + 0.00005x^2 \text{ m}.$$

The beam is made of steel of density 7830 kg/m^3 and Young's modulus is 200 GPa. Approximate this system by 5-, 10-, 15- and 20-degree-of-freedom systems. Compare the first natural frequencies. Follow the procedure in Example 7.6.

3. A transient excitation $f_1(t)$ is applied to the system shown in Figure 7.20. Assuming $k_1 = 1$ N/m, $k_2 = 2$ N/m, $k_3 = 3$ N/m, $m_1 = m_2 = 1$ kg, $f_2(t) = 0$, and $f_1(t)$ is given by

$$\begin{aligned} f_1(t) &= 1, \text{ for } 0 < t < 1 \\ &= 0 \text{ elsewhere.} \end{aligned}$$

Determine:

(i) the equations of motion

(ii) the frequency response functions $H_{ij}(\omega)$ $(i, j = 1, 2)$

(iii) the responses $y_1(t)$ and $y_2(t)$

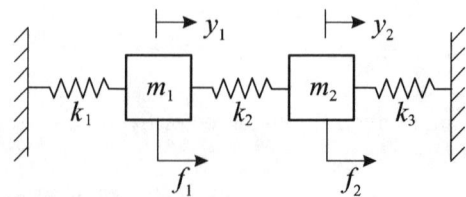

Figure 7.20: Two-degree-of-freedom system.

Section 7.2: Response to Random Loads

4. Consider the two-degree-of-freedom system shown in Figure 7.1 where $F_1 = 0$ and in addition to F_2, random force F_3 is applied to the second mass. The force spectrum is given by

$$[S_{FF}(\omega)] = \begin{bmatrix} S_{F_2 F_2}(\omega) & S_{F_2 F_3}(\omega) \\ S_{F_2 F_3}(\omega) & S_{F_3 F_3}(\omega) \end{bmatrix}.$$

Derive an expression for the response spectrum $[S_{XX}(\omega)]$ in terms of the force spectrum, the modal matrix, and the transfer function $[H(\omega)]$.

7.7. PROBLEMS

5. Derive Equation 7.24.

6. Consider a two-degree-of-freedom system in Figure 7.20 where the force spectrum is given by

$$S_{FF}(\omega) = S_0 \begin{bmatrix} 1 & 0 \\ 0 & 0 \end{bmatrix},$$

where $S_0 = 1$ N²s, $m_1 = 1$ kg, $m_2 = 2$ kg, $k_1 = 2$ N/m, $k_2 = 1$ N/m and $k_3 = 4$ N/m. Assume that there is a damper with $c = 1$ Ns/m between the wall and the first mass.

(i) Derive the expressions for the mean-squares, $R_{X_1 X_1}(0)$, $R_{X_1 X_2}(0)$ and $R_{X_2 X_2}(0)$.

(ii) Obtain the natural frequencies and the modal matrix.

(iii) Evaluate the mean-square values.

(iv) Find the relative error if the mean-square values are approximated using Equation 7.39.

7. Consider a three-degree-of-freedom system with property matrices:

$$[m] = \begin{bmatrix} m & 0 & 0 \\ 0 & m & 0 \\ 0 & 0 & m \end{bmatrix}, \quad [c] = \begin{bmatrix} c & 0 & 0 \\ 0 & c & 0 \\ 0 & 0 & c \end{bmatrix},$$

$$[k] = \begin{bmatrix} 3k & -k & -k \\ -k & 3k & -k \\ -k & -k & 3k \end{bmatrix}.$$

Determine the natural frequencies and the modal matrix. Assume that mass 1 is subjected to a white noise random force of intensity S_0. Find the mean-square value $R_{X_1 X_1}(0)$. What is the percent error if the autocorrelation functions are approximated using Equation 7.39?

8. Consider the mechanical system in Figure 7.21, where the massless driving cart is subjected to a random force $F(t)$. Find:

(i) the equation of motion for this system

(ii) the frequency response function $[H(\omega)]$

(iii) the response spectrum if $F(t)$ is a white noise-based random process of intensity S_0 N²s. Assume that $c_1 = c_2 = c/2$, $k_1 = k_2 = k/3$ and $m = 3c^2/4k$

9. Consider the simply supported massless beam of length L shown in Figure 7.22. The two point masses, m, and the springs with stiffness

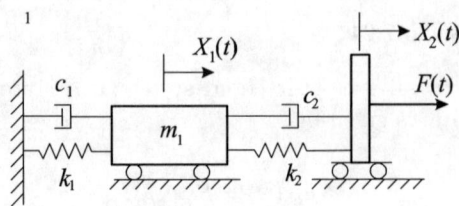

Figure 7.21: A multidegree-of-freedom system driven by a massless cart.

k, are attached at $x = L/3$ and $2L/3$. Mass 1 is subjected to a random load with correlation function,

$$R_{FF}(\tau) = 2\pi \left(1 - \frac{|\tau|}{T}\right).$$

Find the force spectrum, the response spectrum, and the response autocorrelation functions. Assuming that the beam has a flexural rigidity of EI, and the deflections are small so that the small angle assumption is valid, derive the matrix equation of motion for the two masses, the stiffness matrix, the natural frequencies, and the modal matrix. In addition, obtain the response spectra. Use $L = 1$ m, $T = 2$ s, $EI = 2$ Nm^2.

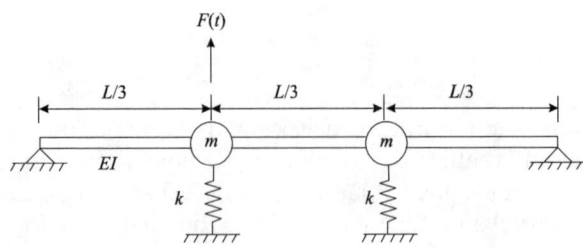

Figure 7.22: Simply supported beam subjected to random force.

10. Two disks with mass polar moments of inertia of I_1 and I_2 are mounted on a circular massless shaft as shown in Figure 7.23.

(i) Derive the equations of motion in terms of $\Theta_1(t)$ and $\Theta_2(t)$, the angles of twist.

(ii) Find the response spectrum $[S_{\Theta\Theta}(\omega)]$ if a white noise-based random torque T_1 with intensity S_0 N^{2}m^{2}s is applied to the first disk.

7.7. PROBLEMS

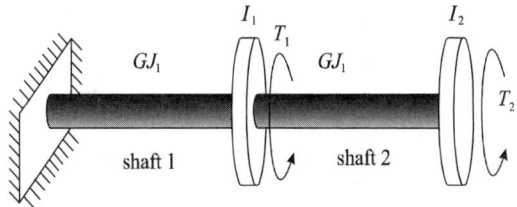

Figure 7.23: Two disks monted on a shaft.

11. A rigid bar with mass per unit length m carries a point mass M at its end. The bar is supported by two springs, as shown in Figure 7.24.

    (i) Derive the equations of motion in terms of the translation and rotation of the center of mass of the entire system.

    (ii) Determine the RMS responses if the system is subjected to a white noise random force $F(t)$ with $S_{FF} = 0.005$ N²s. Assume that the motion is small so that the small angle assumption is valid, and use parameter values $m = 0.2$ kg/m, $M = 1$ kg, $k_1 = 300$ N/m, $k_2 = 400$ N/m, and $L = 1$ m.

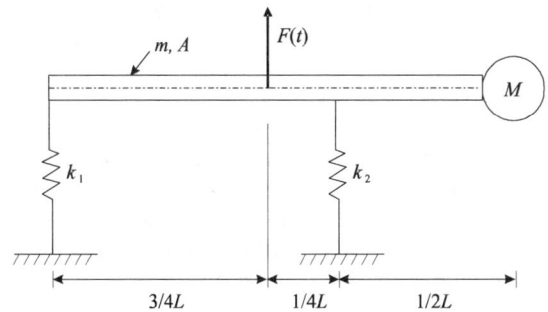

Figure 7.24: A rigid rod and a point mass.

12. A point mass is suspended from a massless rigid bar as shown in Figure 7.25. The point mass is subjected to white noise random forcing having a constant spectral density 0.008 N²m.

    (i) How many degrees of freedom does this system have?

    (ii) Determine the RMS of the responses. Use parameter values $m = 3$ kg, $k_1 = 250$ N/m, $k_2 = 300$ N/m, $I_o = 4$ kg m², and $L = 6$ m.

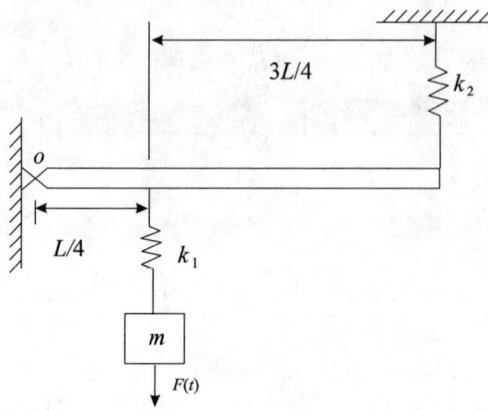

Figure 7.25: A rigid bar and a point mass.

Section 7.3: Periodic Structures

13. Give several examples of engineering periodic structures.

14. How significant do you think localization is in one-dimensional, two-dimensional and three-dimensional structures? Discuss physically.

Section 7.4: Inverse Vibration

15. Consider the two-degree-of-freedom system shown in Figure 7.26. It is found that the two natural frequencies of this system are 5 Hz and 20 Hz. When the second mass is fixed the natural frequency becomes 7 Hz. Assuming the total mass is 10 kg, find m_1, m_2, k_1, and k_2.

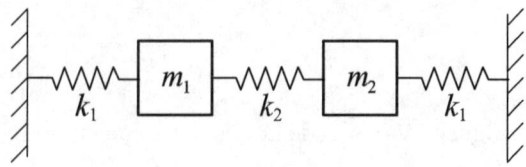

Figure 7.26: Two-degree-of-freedom system with unknown mass and stiffness.

16. Consider the two-degree-of-freedom system in the previous problem assuming now that the frequency measurements are random. When

7.7. PROBLEMS

the measurements are taken repeatedly, it is found that the mean natural frequencies are 5 Hz and 20 Hz for the original system, and 7 Hz when the second mass is fixed. The coefficients of variation are 0.02 for both first and second natural frequencies of the original system, and 0.01 for the natural frequency of the system with the fixed second mass. Estimate the mean values and variances of k_i and m_i. Assume that the total mass is an exact value.

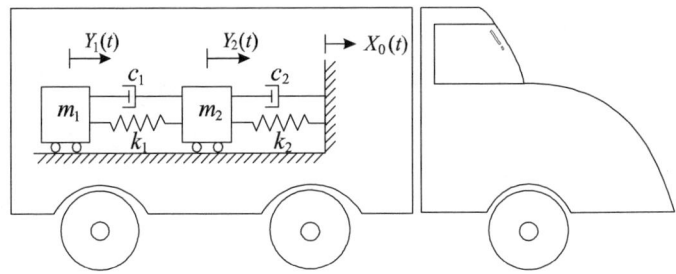

Figure 7.27: Truck transporting sensitive equipment.

17. A truck is transporting sensitive equipment as shown in Figure 7.27. The system is set in vibration through the acceleration of the truck $\ddot{X}_0(t)$. Obtain the equations of motion of the system in terms of the absolute displacement $\{Y(t)\}$ and also in terms of relative displacements $\{Z(t)\}$. Determine the complex frequency responses that correspond to the steady state excitation $\ddot{X}_0$.

Section 7.5: Random Eigenvalues

18. Repeat Example 7.9 assuming that k is normally distributed with a mean $\mu_k = 1$ and $\sigma_k = 0.1$.

19. The springs k_1 and k_2 in the two-degree-of-freedom system Figure 7.20 are uncertain, and they are measured experimentally. The measurements show that k_1 is distributed normally with $\mu_1 = 10$ N/m, $\sigma_1 = 2$ N/m, $\mu_2 = 20$ N/m, $\sigma_1 = 3$ N/m. Find the marginal densities for the natural frequencies. Use $m_1 = 2$ kg, $m_2 = 1$ kg, and $k_3 = 5$ N/m.

Chapter 8

Continuous System Vibration

Continuous models of vibrating systems are viewed as more realistic because the structural properties are distributed rather than being concentrated at discrete points.[1] The price to pay for increased realism is increased complexity. The governing equations of motion go from being ordinary differential equations for the discrete models, to partial differential equations for the continuous ones, since displacement is a function of both time and position.

For the elementary models of this chapter, deriving and solving the partial differential equations does not pose a great challenge. But, when nonuniform structures with varying cross-sections and material properties need to be modeled, approximate techniques are utilized. Our introduction to continuous systems focuses on strings and vibrating beams. These are found in most structures and machines, and understanding how they behave is of great use to modeling more complex systems. We will approach these studies using the direct and modal solutions.

This chapter begins with an introduction to deterministic models for continuous systems. Random models are based on these methods for multidegree-of-freedom systems, because continuous systems are effectively discretized when solved via modal analysis. Strings and beams are considered for the purpose of demonstrating how one would proceed in such analyses. Membranes and plates are extensions of string and beams, albeit with much additional algebra, are left for more advanced studies.

---

[1] Although numerical studies require the discretization of continuous models.

8.1 Deterministic Continuous Systems

In this section, three continuous systems are considered. The first is a taut string, and the second is an axially vibrating beam, and the third is a transversely vibrating beam.

8.1.1 Strings

A string is a valuable model for understanding the dynamic behavior of continuous systems. Strings have been used for simplified models of telephone wires, conveyor belts, even models of human DNA. When a string is stretched from its equilibrium position, the tension that is created internally acts as the restoring force to bring the string back to its original undeformed position. The string cannot transmit a bending moment but can resist axial tension.

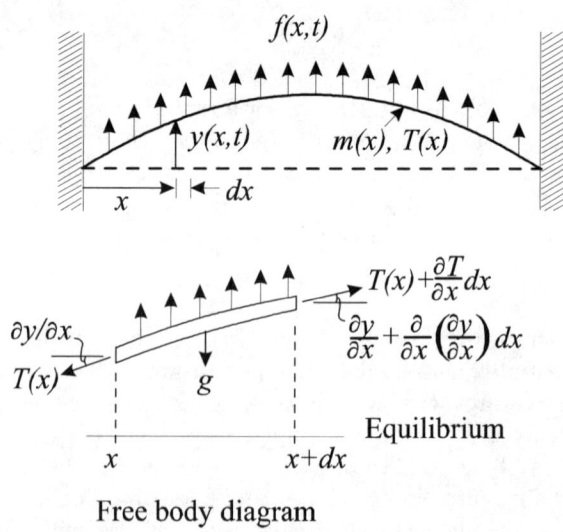

Figure 8.1: String and its free body diagram.

It is straightforward to derive the governing equation of motion using the free body diagram for a section of the displaced string, as per Figure 8.1. The variables introduced are $T(x)$, the tension in the string, $f(x,t)$, the applied transverse force per unit length, and $m(x)$, the mass per unit length. All are functions of position x. Using Newton's second law of motion, and assuming small displacements such that $\sin\theta \simeq \theta \simeq \tan\theta \simeq dy/dx$, the sum of the forces in the transverse direction y are set equal to the product of the

8.1. DETERMINISTIC CONTINUOUS SYSTEMS

mass and the acceleration of the string element in the y direction,

$$\left(T(x) + \frac{\partial T}{\partial x}dx\right)\left(\frac{\partial y}{\partial x} + \frac{\partial^2 y}{\partial x^2}dx\right) + f(x,t)dx - T(x)\frac{\partial y}{\partial x} = m(x)dx\frac{\partial^2 y}{\partial t^2}.$$

When expanding the products on the left-hand side, second-order terms $(dx)^2$ are ignored on the assumption that for linear vibration these terms are not significant. Divide by dx, with the result being the governing equation of linear motion,

$$\frac{\partial}{\partial x}\left[T(x)\frac{\partial y}{\partial x}\right] + f(x,t) = m(x)\frac{\partial^2 y}{\partial t^2}.$$

The free vibration problem used to solve for system frequencies and modes is obtained by setting $f(x,t) = 0$,

$$\frac{\partial}{\partial x}\left[T(x)\frac{\partial y}{\partial x}\right] = m(x)\frac{\partial^2 y}{\partial t^2}. \tag{8.1}$$

Note that for the case of constant tension, $T(x) = T$, the *wave equation* is obtained:

$$\frac{\partial^2 y}{\partial x^2} = \frac{1}{c^2}\frac{\partial^2 y}{\partial t^2},$$

where $c = \sqrt{T/m}$ is the velocity of wave propagation with units of length per unit time.

Possible Boundary Conditions

The equation of motion requires two boundary conditions and two initial conditions. The string in Figure 8.1 has fixed-fixed boundaries, and they can be prescribed mathematically as

$$\begin{aligned} y(0,t) &= 0 \\ y(L,t) &= 0. \end{aligned}$$

One possible set of initial conditions is

$$\begin{aligned} y(x,0) &= (x-L)x \\ \dot{y}(x,0) &= 0. \end{aligned}$$

These imply that the initial velocity is zero, and the initial displacement has a parabolic shape with its minimum at $x = L/2$. Note that the initial displacement must satisfy the boundary conditions.

Another possible boundary condition is that the ends of the string can be free to move up and down, as shown in Figure 8.2. Then the resisting boundary force in the vertical direction is zero, or

$$T(x)\frac{\partial y}{\partial x}\bigg|_{x=0 \text{ or } L} = 0.$$

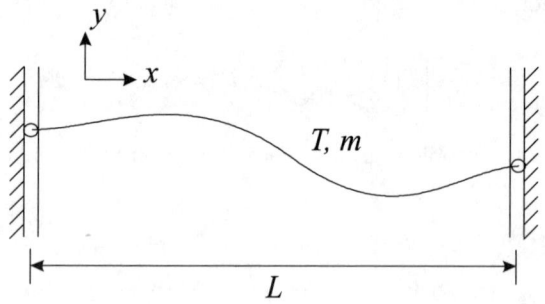

Figure 8.2: Free-free string.

8.1.2 Axial Vibration of Beams

Beams are fundamental components in structures and machines. They are also useful models for more complex structures.

Consider the schematic of the beam in Figure 8.3. Displacements, strains, and stresses are assumed uniform at a given cross-section. Newton's second law of motion is a statement of the force balance for any free body element of the beam. From the figure, force P acts to the left and this force plus an undetermined increment dP acts to the right. Had this been a static problem, $dP = 0$ for static equilibrium.

For the dynamic problem, the sum of the forces equals the product of mass and acceleration. Let the element have a mass per unit length of $m(x)$. Alternatively, $m(x) = \rho(x)A(x)$, where $\rho(x)$ is the density and $A(x)$ is the area of the cross-section at x. Then, by Newton's second law of motion for an element of length dx,

$$[P + dP](x,t) - P(x,t) = m(x)dx\frac{\partial^2 u(x,t)}{\partial t^2}.$$

From the strength of materials, $P = AE\varepsilon = AE\partial u(x,t)/\partial x$. The force

8.1. DETERMINISTIC CONTINUOUS SYSTEMS

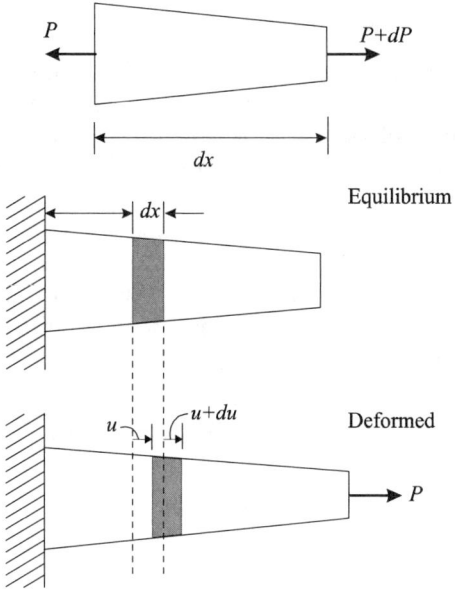

Figure 8.3: Schematic for the longitudinal vibration of a beam.

differential is then

$$dP(x,t) = \frac{\partial P(x,t)}{\partial x} dx$$
$$= \frac{\partial}{\partial x}\left(A(x)E\frac{\partial u}{\partial x}\right) dx,$$

and

$$\frac{\partial}{\partial x}\left(A(x)E\frac{\partial u}{\partial x}\right) = m(x)\frac{\partial^2 u}{\partial t^2}. \tag{8.2}$$

If the beam is uniform, then $m(x) = m = \rho A$ and the area A can be cancelled from both sides, leading to the wave equation governing $u(x,t)$,

$$c^2 \frac{\partial^2 u}{\partial x^2} = \frac{\partial^2 u}{\partial t^2},$$

where $c^2 = E/\rho$ has units of speed squared. Note that the equation of motion takes the same form as the equation of motion for the string, and therefore the general solutions to the free and forced vibrations are the same. However, they are two different physical systems, and the boundary conditions have different physical meaning. We will briefly discuss the possible boundary conditions for longitudinally vibrating beams.

Possible Boundary Conditions

For an axially vibrating beam, we can consider two end conditions: conditions on the displacement $u(x,t)$ or the longitudinal force $P = AE\, \partial u(x,t)/\partial x$. The displacement u is zero for a fixed end, and the longitudinal force $AE\partial u(x,t)/\partial x$ is zero for a free end. Other end conditions are shown in Figure 8.4.

These boundary conditions can be obtained using Newton's second law of motion. For example, the first boundary condition can be obtained by assuming that there is a body between the spring and the beam with zero mass. Then assume a positive displacement u and draw a free body diagram for this fictitious body as shown in Figure 8.5. Then,

$$\sum F = m_i \left.\frac{\partial^2 u}{\partial t^2}\right|_{(0,t)}$$

$$AE\left.\frac{\partial u}{\partial x}\right|_{(0,t)} - ku(0,t) = m_i \left.\frac{\partial^2 u}{\partial t^2}\right|_{(0,t)}.$$

Since $m_i = 0$, we have

$$AE\left.\frac{\partial u}{\partial x}\right|_{(0,t)} - ku(0,t) = 0.$$

8.1.3 Transversely Vibrating Beams

Consider a beam under pure bending as shown in Figure 8.6 where y is the displacement of the beam element in the transverse direction, M is the bending moment and Q is the shear force.

The force balance in the transverse direction gives

$$-[Q(x,t) + dQ] + Q(x,t) + f(x,t)\,dx = m(x)\,dx\frac{\partial^2 y(x,t)}{dt^2},$$

where m is the mass per unit length that may vary along the beam, dQ is the change in the shear force or

$$dQ = \frac{\partial Q}{\partial x}dx.$$

From strength of materials, the shear force and the moment can be approximated as

$$M = \left(EI(x)\frac{\partial^2 y}{\partial x^2}\right)$$

$$Q = \frac{\partial M}{\partial x} = \frac{\partial}{\partial x}\left(EI(x)\frac{\partial^2 y}{\partial x^2}\right).$$

8.1. DETERMINISTIC CONTINUOUS SYSTEMS

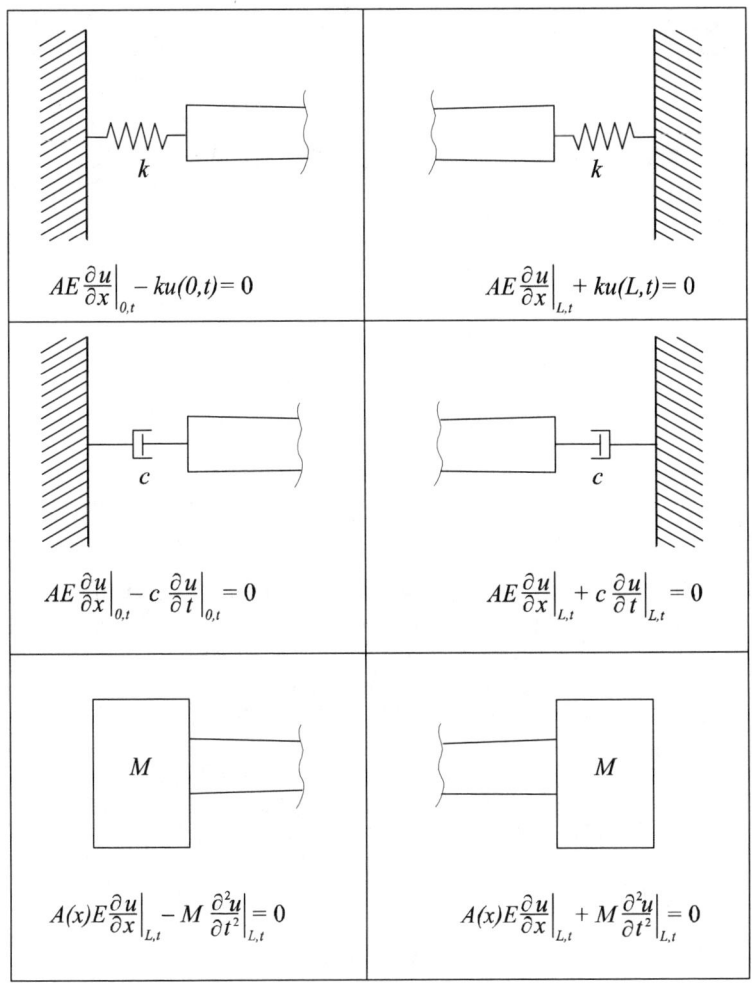

Figure 8.4: Sample boundary conditions.

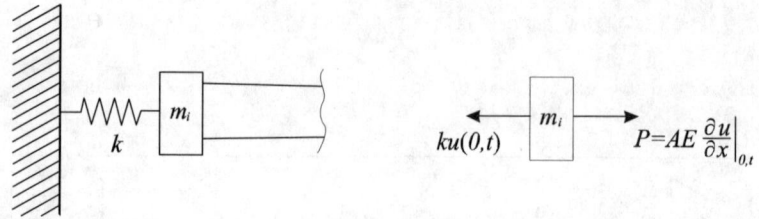

Figure 8.5: Derivation of boundary conditions.

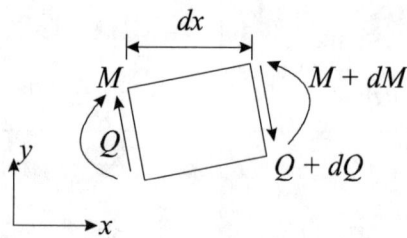

Figure 8.6: An incremental beam element in transverse vibration.

The shear and moment are drawn in the positive sense in Figure 8.6. The downward shear on the positive face of the incremental beam and the upward shear on the negative face are defined positive. For the moment, the counterclockwise moment on the positive face and the clockwise moment on the negative face are defined positive. Substituting the expression for the shear into the force balance equation, we find the equation of motion,

$$\frac{\partial^2}{\partial x^2}\left(EI(x)\frac{\partial^2 y}{\partial x^2}\right) + m(x)\frac{\partial^2 y}{dt^2} = f(x,t).$$

The possible boundary conditions are

clamped:	$y(0,t) = 0$	and	$y'(0,t) = 0$
hinged:	$y(0,t) = 0$	and	$EIy''(0,t) = 0$
free:	$EIy''(0,t) = 0$	and	$(EIy'')'(0,t) = 0$
sliding:	$y'(0,t) = 0$	and	$(EIy'')'(0,t) = 0.$

This is the Bernoulli-Euler beam model and is the simplest model for a transversely vibrating beam. This model assumes that the transverse motion is caused only by the bending of the beam. The Timoshenko beam model includes both bending and shear effects.

8.2 Sturm-Liouville Eigenvalue Problem

All the equations of motion considered here can be put in the following form:

$$L\left[y\left(x,t\right)\right] = m\frac{\partial^2 y}{\partial t^2},$$

with appropriate homogeneous boundary conditions. $L\left(\cdot\right)$ is a differential operator, and is listed below for string and beam,

$$\begin{aligned} \text{String} &: \quad L\left(y\right) = \partial\left(T\partial y/\partial x\right)/\partial x & (8.3)\\ \text{Axially vibrating beam} &: \quad L\left(y\right) = \partial\left(EA\partial y/\partial x\right)/\partial x & (8.4)\\ \text{Transversely vibrating beam} &: \quad L\left(y\right) = -\partial^2\left(EI\partial^2 y/\partial x^2\right)/\partial x. & (8.5) \end{aligned}$$

For the string and the axially vibrating beam, the boundary conditions at $x = 0$ and L must be of the form,

$$\begin{aligned} B_1 y\left(x,t\right) + B_2 \frac{\partial y\left(x,t\right)}{\partial x}\bigg|_{x=0} &= 0\\ C_1 y\left(x,t\right) + C_2 \frac{\partial y\left(x,t\right)}{\partial x}\bigg|_{x=L} &= 0. \end{aligned}$$

For the transversely vibrating beam, the boundary conditions must be of the form,

$$\begin{aligned} B_1 y\left(a,t\right) + B_2 \frac{\partial}{\partial x}\left(EI\frac{\partial y}{dx}\right)(a,t) &= 0\\ \text{and } B_3 \frac{\partial y\left(a,t\right)}{\partial x} + B_4 \left(EI\frac{\partial^2 y}{\partial x^2}\right)(a,t) &= 0, \end{aligned}$$

where a equals either 0 or L.

We make the usual and most elementary assumption regarding the solution of a partial differential equation, that its solution is *separable*,

$$y(x,t) = \eta(t)Y(x). \tag{8.6}$$

This form implies that behavior in time and space are independent. This is called the *method of separation of variables* and is successful when the governing partial differential equation and the corresponding boundary conditions are homogeneous.[2]

---

[2] In this case, it means that we are considering the free vibration case. In general, a differential equation is said to be homogeneous if every term in the equation contains the

Since Equation 8.6 must satisfy the governing equation, we take the appropriate derivatives, $y'(x,t) = \eta(t)Y'(x)$, $\ddot{y}(x,t) = \ddot{\eta}(t)Y(x)$, and substitute them into governing Equation 8.1 to find

$$L[Y(x)]\eta(t) = m(x)\ddot{\eta}(t)Y(x)$$
$$\frac{L[Y(x)]}{m(x)Y(x)} = \frac{\ddot{\eta}(t)}{\eta(t)} = -\omega^2.$$

In the last equation, time-dependent variables and space-dependent variables have been placed on opposite sides of the equal sign. Since the left-hand side is only a function of x, and the right-hand side is only a function of t, each must be equal to the same constant, say $-\omega^2$. The partial differential equation is then separated into two ordinary differential equations,

$$\frac{d^2\eta}{dt^2} + \omega^2 \eta(t) = 0$$
$$L[Y(x)] + m(x)\omega^2 Y(x) = 0. \tag{8.7}$$

The solution to the first ordinary differential equation in time is

$$\eta(t) = A\cos\omega t + B\sin\omega t,$$

where the constant ω is the oscillation frequency.

The second ordinary equation in x (Equation 8.7) must be satisfied on the domain $0 < x < L$. It can be found that, for a given set of boundary conditions, it is satisfied by more than one set of $Y(x)$ and ω^2, where $Y(x)$ is the eigenfunction and ω^2 the eigenvalue of the problem. This forms the basis for the modal analysis, developed in subsequent sections. In the next section, the orthogonality property of the eigenfunctions are discussed. This property is the fundamental basis for modal analysis.

---

dependent variable or its derivatives. For example,

$$\frac{d^2y}{dx^2} + 3\frac{dy}{dx} + x = 0$$

is not homogeneous because the third term does not contain y nor the derivatives of y.

Similarly, the boundary conditions are said to be homogeneous if every term contains the dependent variable or its derivatives. For example,

$$y(0) - 3\left(\frac{dy}{dx}\right)\bigg|_{x=0} = 0$$

is a homogeneous boundary condition while

$$y(0) + 3 = 0$$

is not due to the constant term 3.

8.2. STURM-LIOUVILLE EIGENVALUE PROBLEM

8.2.1 Orthogonality

From Equation 8.7, the spatial equation for nth and rth eigenfunctions and eigenvalues can be written as

$$L\left[Y_n(x)\right] + m(x)\omega_n^2 Y_n(x) = 0$$
$$L\left[Y_r(x)\right] + m(x)\omega_r^2 Y_r(x) = 0.$$

Multiplying the first equation by $Y_r(x)$, multiplying the second equation by $Y_n(x)$, and integrating over the domain, we obtain

$$\int_0^L Y_r(x)L\left[Y_n(x)\right]dx + \omega_n^2 \int_0^L m(x)Y_n(x)Y_r(x)dx = 0$$
$$\int_0^L Y_n(x)L\left[Y_r(x)\right]dx + \omega_r^2 \int_0^L m(x)Y_r(x)Y_n(x)dx = 0.$$

Subtracting the second from the first equation, we find

$$\int_0^L \{Y_r(x)L\left[Y_n(x)\right] - Y_n(x)L\left[Y_r(x)\right]\}dx$$
$$= \left(\omega_r^2 - \omega_n^2\right)\int_0^L m(x)Y_n(x)Y_r(x)dx. \tag{8.8}$$

The left-hand side integrates to zero when appropriate boundary conditions are applied. For the string problem, the left-hand side can be integrated by parts and becomes

$$LHS = Y_r(x)\left[T(x)Y_n(x)'\right] - Y_n(x)\left[T(x)Y_n(x)'\right]\big|_0^L,$$

which is reduced to zero due to the boundary conditions. For the transversely vibrating beam, the left-hand side can be integrated by parts and becomes

$$LHS = Y_r(x)\left(EI(x)Y_n''(x)\right)' - Y_n'(x)\left(EI(x)Y_n''(x)\right)\Big|_0^L,$$

which also reduces to zero due to the boundary conditions. Then, the right-hand side of Equation 8.8 must also equal zero,

$$\left(\omega_r^2 - \omega_n^2\right)\int_0^L m(x)Y_n(x)Y_r(x)dx = 0.$$

If the eigenvalues are unique, $\omega_n \neq \omega_r$, then,

$$\int_0^L m(x)Y_n(x)Y_r(x)dx = 0, \qquad n \neq r.$$

For $r = n$, we can *normalize* the eigenfunctions such that

$$\int_0^L m(x) Y_n(x) Y_n(x)\, dx = 1, \qquad n = 1, 2, \cdots.$$

This also implies another orthogonality condition,

$$\int_0^L Y_r(x) L\left[Y_n(x)\right] dx = -\omega_n^2 \delta_{nr}, \tag{8.9}$$

where δ_{nr} is the Kronecker delta. For the string, when the left-hand side is integrated by parts and the boundary conditions are applied, Equation 8.9 evaluates to

$$\int_0^L T(x) \frac{dY_r(x)}{dx} \frac{dY_n(x)}{dx}\, dx = -\omega_n^2 \delta_{nr},$$

and for the transversely vibrating beam,

$$\int_0^L EI(x) \frac{d^2Y_r(x)}{dx^2} \frac{d^2Y_n(x)}{dx^2}\, dx = -\omega_n^2 \delta_{nr}.$$

The value of these orthogonality conditions is shown next.

8.2.2 Natural Frequencies and Mode Shapes

In the previous section, the orthogonality property of the eigenfunction are discussed without actually obtaining the eigenfunctions for a specific system and its boundary conditions. The natural frequencies and mode shapes are a result of an eigenvalue-eigenvector analysis.

Consider the uniform vibrating string assuming constant tension, where the string is fixed at both ends. A uniform string is one where the geometry and material properties do not vary along its length. The constant tension approximation is valid for a string oscillating with small amplitudes. Equation 8.7 simplifies considerably for constant T and m, and the eigenvalue problem becomes

$$Y'' + \beta^2 Y = 0, \qquad \beta^2 = \frac{\omega^2 m}{T}, \tag{8.10}$$

with the boundary conditions $Y(0) = 0$ and $Y(L) = 0$. Recall that the wave speed c is given by $c = \sqrt{T/m}$ and therefore $\beta = \omega/c$. The second-order constant coefficient differential Equation 8.10 has a periodic solution in x,

$$Y(x) = C_1 \sin \beta x + C_2 \cos \beta x,$$

and satisfaction of the boundary conditions requires that the following two equations are solved,

$$\begin{aligned} Y(0) &= 0 = C_2 \\ Y(L) &= 0 = C_1 \sin \beta L. \end{aligned}$$

8.2. STURM-LIOUVILLE EIGENVALUE PROBLEM

This results in the requirement that $C_1 \neq 0$ and $\beta_r L = r\pi$, $r = 1, 2, \cdots$. The natural frequency is then obtained using Equation 8.10 or

$$\begin{aligned} \omega_r &= \sqrt{\frac{T}{m}} \beta_r \\ &= \sqrt{\frac{T}{m}} \frac{r\pi}{L}, \qquad r = 1, 2, \cdots. \end{aligned} \qquad (8.11)$$

Once we have the natural frequencies of the system, the mode shapes are given by[3]

$$Y_r(x) = \sqrt{\frac{2}{mL}} \sin \frac{r\pi x}{L}, \qquad r = 1, 2, \cdots. \qquad (8.12)$$

The mode shapes are plotted in Figure 8.7.

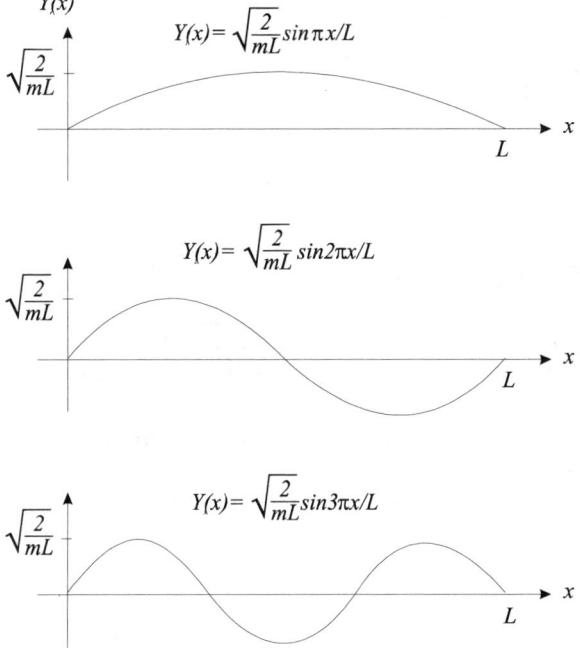

Figure 8.7: The first three mode shapes.

---

[3] The coefficient $\sqrt{2/mL}$ is arbitrary, and it is customarily obtained by a normalization with respect to m.

Example 8.1 Eigenvalues and Eigenfunctions of a Fixed-Fixed String

Consider a string fixed at $x = 0$ and 1 m. It is found that the string obeys the linear Hooke's law with a proportionality constant of 10 kN/m. The unstretched length of the string is 0.99 m and weights 0.1 N.

Solution The tension is given by

$$T = k\Delta x = 10000 \times 0.01 = 100 \text{ N},$$

and the total mass of the string is 0.1/9.8 kg or 0.0102 kg. The linear mass density is then

$$m = \text{mass/length} = 0.0102/1 = 0.010 \text{ kg/m}.$$

The natural frequencies are given by Equation 8.11, or

$$\omega_r = \sqrt{\frac{T}{m}} \frac{r\pi}{L}, \qquad r = 1, 2, \cdots.$$

Then, the first three natural frequencies are given by

$$\omega_r = 99\pi, \ 198\pi, \ 297\pi \text{ rad/s}.$$

⊛

Example 8.2 Eigenvalue Problem for Axially Vibrating Beam

Consider a uniform beam vibrating in the axial direction. The beam is fixed at one end and free at the other. Obtain the natural frequencies and mode shapes.

Solution The assumed solution for the fixed-free longitudinal vibration problem is $u(x,t) = U(x)F(t)$. $U(x)$ is the mode of vibration. Differentiate this assumed solution and substitute into Equation 8.2 to find

$$\frac{d}{dx}\left[EA(x)\frac{dU(x)}{dx}\right] = m(x)U(x)\frac{\ddot{F}}{F}$$
$$= -\omega^2 m(x)U(x).$$

If we assume that one end is fixed and the other end is free as shown in Figure 8.3, the boundary conditions in terms of $U(x)$ are given by

$$U(x)|_{x=0} = 0 \text{ and } EA(x)\frac{dU(x)}{dx}\bigg|_{x=L} = 0.$$

8.2. STURM-LIOUVILLE EIGENVALUE PROBLEM

For a uniform beam with constant properties along the beam, $A(x) = A$ and $m(x) = m$, the eigenvalue problem becomes

$$\frac{d^2U}{dx^2} + \frac{m}{EA}\omega^2 U = 0.$$

For the fixed-free beam, the natural frequencies and the eigenfunctions are given by the relations

$$\omega_r = \frac{(2r-1)\pi}{2L}\sqrt{\frac{EA}{m}}$$

$$U_r(x) = \sqrt{\frac{2}{mL}}\sin\frac{(2r-1)\pi}{2L}x, \quad r = 1, 2, \cdots.$$

The constant coefficient of the eigenfunctions is obtained using the normalization relation

$$\int_0^L mU_r^2(x)\,dx = 1.$$

⊛

Example 8.3 Longitudinal Beam with a Point End Mass

A longitudinal beam is fixed at one end and a point mass of M is attached to the other end. Assuming that the mass per unit length is m, the point mass is equal to the mass of the beam, and the constant extensional rigidity is EA, obtain the expressions for the eigenfunctions and the natural frequencies.

Solution   The spatial equation is given by

$$\frac{d^2U(x)}{dx^2} + \frac{m}{EA}\omega^2 U(x) = 0,$$

which is solved to yield

$$U(x) = c_1\sin\sqrt{\frac{m}{EA}}\omega x + c_2\cos\sqrt{\frac{m}{EA}}\omega x.$$

The boundary conditions are given by

$$U(x)|_{x=L} = 0,$$
$$EA\frac{dU(x)}{dx}\bigg|_{x=L} - M\omega^2 U(x)|_{x=L} = 0,$$

where $M = mL$. When the assumed solution is substituted into the boundary conditions, we find that the eigenfunctions are given by

$$Y_r = C\sin\beta_r x, \quad r = 1, 2, \cdots,$$

and the characteristic equation is given by

$$\cos \beta L - \beta L \sin \beta L = 0,$$

where $\beta = \sqrt{m/EA}\,\omega$. This transcendental equation can be solved numerically, and the first few roots are

$$\beta L = 0.860,\ 3.426,\ 6.437,\cdots.$$
$$\omega = 0.860\frac{1}{L}\sqrt{\frac{EA}{m}},\ 3.426\frac{1}{L}\sqrt{\frac{EA}{m}},\ 6.437\frac{1}{L}\sqrt{\frac{EA}{m}},\cdots.$$

⊛

Example 8.4 Transversely Vibrating Beam Supported by a Spring

Consider a beam that is simply supported (hinged support) at one end and is supported by a spring at the other end. Find the mode shapes and natural frequencies if the stiffness of the spring is $k = 0.001EI/L^3$.

Solution The mode shapes must satisfy the spatial equation given by

$$EI\frac{d^4 Y(x)}{dx^4} - m\omega^2 Y(x) = 0,$$

and the corresponding boundary conditions,

$$Y(x)|_{x=0} = 0$$
$$Y''(x)|_{x=0} = 0$$
$$Y''(x)|_{x=L} = 0$$
$$EIY'''(x) - kY(x)|_{x=L} = 0.$$

The last condition can be obtained by adding a zero mass where the spring is attached and writing a force balance as shown in Figure 8.8(a). The counterclockwise moment and the downward shear on the positive side of an incremental beam are defined positive as shown in Figure 8.6. Since the beam is attached on the left of the fictitious mass, the positive shear points upward as shown in Figure 8.8(b). For an incremental positive displacement, the spring exerts a downward force. The sum of the force in the vertical direction (Figure 8.8(c)) is then

$$EI\left.\frac{\partial^3 y(x,t)}{\partial x^3}\right|_{x=L} - k\,y(x,t)|_{x=L} = M\left.\frac{\partial^2 y(x,t)}{\partial t^2}\right|_{x=L}.$$

Since the end mass does not exist, we can let $M = 0$.

From the spatial equation, the form of the solution must be

$$Y(x) = c_1 \sin \beta x + c_2 \cos \beta x + c_3 \sinh \beta x + c_4 \cosh \beta x,$$

8.2. STURM-LIOUVILLE EIGENVALUE PROBLEM

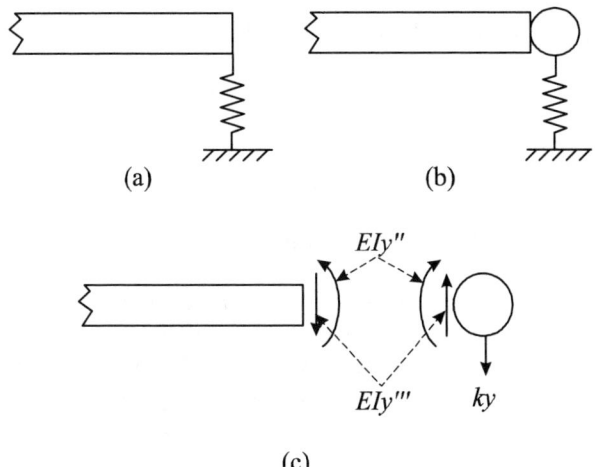

Figure 8.8: Free body diagram for a transverse beam supported by a spring at one end.

where $\beta^4 = m\omega^2/EI$. The first two boundary conditions render $c_2 = c_4 = 0$. Applying the third and fourth boundary conditions, we obtain

$$-c_1 \sin \beta L + c_3 \sinh \beta L = 0$$
$$c_1 \left(-\beta^3 EI \cos \beta L - k \sin \beta L\right) + c_3 \left(\beta^3 EI \cosh \beta L - k \sinh \beta L\right) = 0.$$

From the third boundary condition, the eigenfunction is given by

$$Y(x) = c_1 \left(\sin \beta x + \frac{\sin \beta L}{\sinh \beta L} \sinh \beta x\right).$$

From the third and the fourth boundary conditions, we obtain the characteristic equation,

$$2\frac{kL^3}{EI} \sin \beta L \sinh \beta L - \beta^3 L^3 \sin \beta L \cosh \beta L + \beta^3 L^3 \cos \beta L \sinh \beta L = 0.$$

Substituting $k = 0.001 EI/L^3$, the characteristic equation can be written as

$$0.002 \sin \beta L \sinh \beta L - \beta^3 L^3 \sin \beta L \cosh \beta L + \beta^3 L^3 \cos \beta L \sinh \beta L = 0.$$

The solutions to this equation are given by

$$\beta L = 3.927,\ 7.069,\ 10.21, \cdots.$$

The corresponding natural frequencies are given by

$$\omega = \sqrt{\frac{EI}{mL}}\,(\beta L)^2.$$

8.3 Deterministic Vibration

So far, we have obtained the natural frequencies and mode shapes for different continuous systems. In this section, the modes and natural frequencies are used to derive the response. The system response to nonzero initial conditions is considered first, and the system response to external forces is considered next.

8.3.1 Free Response

Recall that the solution to the partial differential equation is based on the assumption that it is separable and can be written as $y(x,t) = \eta(t)Y(x)$, where $\eta(t) = A\cos\omega t + B\sin\omega t$. We have subsequently found that there is more than a single product of ω and $Y(x)$ that satisfy the homogeneous differential equation and the boundary conditions. There is an infinite number of such products, $\eta_r(t)Y_r(x)$, or

$$\begin{aligned} y(x,t) &= \sum_{r=1}^{\infty} \eta_r(t)Y_r(x) \\ &= \sum_{r=1}^{\infty} Y_r(x)\left(A_r \sin\omega_r t + B_r \cos\omega_r t\right). \end{aligned}$$

The arbitrary constants A_r and B_r are functions of the initial conditions, which are specified next. For the initial conditions $y(x,0)$ and $\dot{y}(x,0)$, the constants A_r and B_r can be obtained using the orthogonality property of eigenfunctions. At $t = 0$,

$$y(x,0) = \sum_{r=1}^{\infty} B_r Y_r(x). \qquad (8.13)$$

$$\dot{y}(x,0) = \sum_{r=1}^{\infty} A_r \omega_r Y_r(x). \qquad (8.14)$$

Multiply Equation 8.13 by $mY_n(x)$ and integrate from 0 to L,

$$m\int_0^L y(x,0)Y_n(x)\,dx = \int_0^L \sum_{r=1}^{\infty} B_r m Y_r(x) Y_n(x)\,dx.$$

8.3. DETERMINISTIC VIBRATION

Using the orthonormality condition, only the term with $r = n$ will be nonzero. Then,

$$B_n = m \int_0^L y(x,0) Y_n(x)\, dx, \qquad n = 1, 2, \cdots.$$

Similarly, A_n can be obtained by multiplying Equation 8.14 by $mY_n(x)$ and integrating from 0 to L,

$$A_n = \frac{m}{\omega_n} \int_0^L \dot{y}(x,0) Y_n(x)\, dx, \qquad n = 1, 2, \cdots.$$

The complete solution is then

$$y(x,t) = \sum_{r=1}^{\infty} \left[Y_r \left(\frac{m}{\omega_r} \int_0^L \dot{y}(x,0) Y_r(x)\, dx \right) \sin \omega_r t \right.$$
$$\left. + Y_r \left(m \int_0^L y(x,0) Y_r(x)\, dx \right) \cos \omega_r t \right].$$

After the next example, we will develop the procedure for finding the solution when $f(x,t) \neq 0$.

Example 8.5 Free Response of a Fixed-Fixed String

Consider a 1 m long uniform spring with a wave propagation velocity equal to 100 m/s. The total mass of the string is 10 g. Find the free response if the initial displacement is given by

$$y(x,0) = \begin{cases} 0.2\, x, & 0 \le x < \tfrac{1}{2} \\ 0.2\,(1-x), & \tfrac{1}{2} < x \le 1. \end{cases} \qquad (8.15)$$

Assume zero initial velocity.

Solution The initial displacement describes a string that is initially pinched in the middle. The natural frequencies are given by Equation 8.11, or

$$\omega_r = \sqrt{\frac{T}{m}} \frac{r\pi}{L}, \qquad r = 1, 2, 3, \cdots,$$

where the wave propagation velocity is $\sqrt{T/m}$. Then, the natural frequencies are given by

$$\omega_r = 100\,\pi r, \qquad r = 1, 2, 3, \cdots.$$

The mass per unit length is $m = 0.01$ kg/m. The eigenfunctions (Equation 8.12) are given by

$$Y_r(x) = \sqrt{\frac{2}{mL}} \sin \frac{r\pi x}{L}$$
$$= \sqrt{\frac{2}{0.01}} \sin r\pi x.$$

The complete response is given by

$$y(x,t) = \sum_{r=1}^{\infty} Y_r(x) \left(A_r \sin \omega_r t + B_r \cos \omega_r t \right),$$

where $A_r = 0$ and

$$B_r = m \int_0^L y(x,0) Y_r(x)\, dx, \qquad r = 1, 2, \cdots.$$

It is found that B_n for even integer n are zeros and the first four nonzero coefficients are

$B_1 = 0.00573,\ B_3 = -0.000637,\ B_5 = 0.000229,\ B_7 = -0.000117,\cdots.$

Figure 8.9 shows the response for the first 0.01 s at 0.001 s intervals when the first eleven terms are used. The response is harmonic with a period of 0.02 s.

⊛

8.3.2 Forced Response via Eigenfunction Expansion

Consider a general case where damping and distributed external force are included. Then, the equation of motion can be written in the following form:

$$-L(y(x,t)) + c\frac{\partial y(x,t)}{\partial t} + m\frac{\partial^2 y(x,t)}{\partial t^2} = f(x,t). \qquad (8.16)$$

Again, $L(\cdot)$ is a differential operator such as those given in Equations 8.3-8.5. Homogeneous boundary conditions of the Sturm-Liouville problem are assumed. To solve this problem, we considered the associated eigenvalue problems in Section 8.2,

$$L(Y_r(x)) + \omega_r^2 m Y_r(x) = 0 \text{ for } 0 < x < L \text{ and } r = 1, 2, \cdots.$$

The subscript r is added to Y and ω^2 to signify that it is the rth eigenfunction and eigenvalue.

8.3. DETERMINISTIC VIBRATION

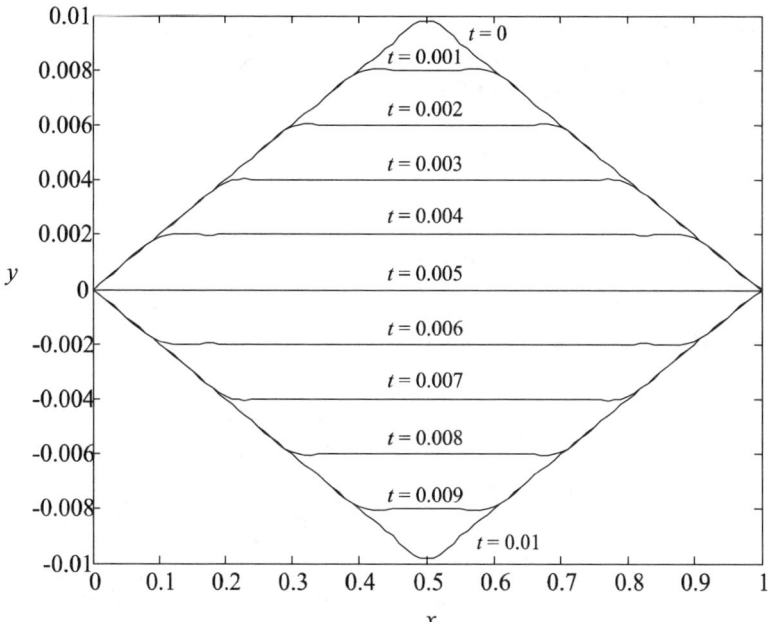

Figure 8.9: Response of a string to initial conditions Equations 8.15.

The solution can be found by expanding $y(x,t)$ in terms of the eigenfunctions of the problem, such that

$$y(x,t) = \sum_{r=1}^{\infty} \eta_r(t) Y_r(x), \qquad (8.17)$$

where the eigenfunctions $Y_r(x)$ are the solutions to the equivalent homogeneous problem in Section 8.2. The goal is to find $\eta_r(t)$. Knowing $\eta_r(t)$, we can then construct the response to the forced vibration using Equation 8.17.

This method is called the *eigenfunction expansion* because the solution to the forced system is expanded in terms of its eigenfunctions. It is also assumed that the distributed load $f(x,t)$ can be expanded in terms of the eigenfunctions,

$$f(x,t) = \sum_{r=1}^{\infty} q_r(t) m Y_r(x).$$

Substitute Equation 8.17 into Equation 8.16,

$$\sum_{r=1}^{\infty} -\eta_r(t) L(Y_r(x)) + c \frac{d\eta_r(t)}{dt} Y_r(x) + m \frac{d^2\eta_r(t)}{dt^2} Y_r(x) = \sum_{r=1}^{\infty} q_r(t) m Y_r(x),$$

where $L(Y_r(x))$ can be replaced by $-\omega_r^2 m Y_r(x)$. Then,

$$\sum_{r=1}^{\infty} \left(\frac{d^2\eta_r(t)}{dt^2} + \frac{c}{m} \frac{d\eta_r(t)}{dt} + \omega_r^2 \eta_r(t) \right) Y_r(x) = \sum_{r=1}^{\infty} q_r(t) Y_r(x). \qquad (8.18)$$

Multiply Equation 8.18 by $Y_n(x)$ and integrate from 0 to L, using modal orthogonality. Each $\eta_m(t)$ can be found by solving the ordinary differential equation,

$$\frac{d^2\eta_n(t)}{dt^2} + \frac{c}{m} \frac{d\eta_n(t)}{dt} + \omega_n^2 \eta_n(t) = q_n(t) \text{ for } n = 1, 2, \cdots. \qquad (8.19)$$

Equation 8.19 looks familiar since we first solved it for the single-degree-of-freedom system. The solution to any $q_n(t)$ can be found using the convolution integral Equation 6.23,

$$\eta_n(t) = \left(\frac{\dot{\eta}_n(0) + \zeta \omega_n \eta(0)}{\omega_{dn}} \sin \omega_{dn} t + \eta_n(0) \cos \omega_{dn} t \right) e^{-\zeta_n \omega_n t}$$

$$+ \int_0^t q_n(\tau) g_n(t-\tau) d\tau, \qquad (8.20)$$

8.3. DETERMINISTIC VIBRATION

where $\zeta = c/\left(2\sqrt{mk}\right)$ and $g_n(t)$ is the impulse response function,

$$g_n(t) = \frac{1}{\omega_{dn}} e^{-\zeta_n \omega_n t} \sin \omega_{dn} t, \text{ for } t > 0$$
$$= 0, \text{ for } t < 0,$$

where ω_{dn} is the nth damped natural frequency $\omega_{dn} = \omega_n \sqrt{1 - \zeta_n^2}$, and ζ_n is the damping ratio, equal to $c/(2m\omega_n)$.

The initial conditions for $\eta_m(t)$ are obtained from the initial conditions for $y(x,t)$. Given $y(x,0)$ and $\dot{y}(x,0)$, Equation 8.17 can be written as

$$y(x,0) = \sum_{r=1}^{\infty} \eta_r(0) Y_r(x)$$
$$\dot{y}(x,0) = \sum_{r=1}^{\infty} \dot{\eta}_r(0) Y_r(x).$$

Multiply these equations by $mY_m(x)$ and integrate from 0 to L,

$$\eta_m(0) = \int_0^L y(x,0) m Y_m(x)\, dx$$
$$\dot{\eta}_m(0) = \int_0^L \dot{y}(x,0) m Y_m(x)\, dx.$$

The solution to Equation 8.16 can be obtained by multiplying $\eta_m(t)$ by the mth eigenfunction $Y_m(t)$ for $m = 1, 2, 3, \cdots$, and summing them up using Equation 8.17.

Example 8.6 Forced Response of a String

A uniform string with $L = 1$ m, $m = 0.1$ kg/m and $T = 200$ N is forced by a traveling-point load

$$f(x,t) = \delta(x - 0.01t), \quad 0 \leq t \leq 100.$$

Find the response assuming zero initial conditions.

Solution The natural frequencies and the mode shapes of a fixed-fixed string are given by

$$\omega_r = \frac{r\pi}{1} \sqrt{\frac{200}{0.1}}$$
$$Y_r(x) = \sqrt{\frac{2}{0.1 \times 1}} \sin \frac{r\pi x}{1}, \quad r = 1, 2, 3 \cdots.$$

The response to the traveling load can be written in terms of eigenfunctions with time-dependent coefficients

$$y(x,t) = \sum_{r=1}^{\infty} \eta_r(t) Y_r(x),$$

where $\eta_r(t)$ is given by Equation 8.20,

$$\eta_r(t) = \frac{1}{\omega_r} \int_0^t q_r(\tau) \sin \omega_r (t - \tau) d\tau,$$

where

$$q_r(\tau) = \int_0^L f(x,\tau) Y_r(x) dx.$$

Note that the damping is zero so that $\omega_{d\,r} = \omega_r$ and $\zeta_r = 0$.
The generalized force $q_r(\tau)$ is given by

$$\begin{aligned} q_r(\tau) &= \int_0^L \delta(x - 0.01\tau) Y_r(x) dx = Y_r(0.01\tau) \\ &= \begin{cases} 4.472 \sin 0.01 r\pi\tau, & 0 \le \tau \le 100 \\ 0, & 100 < \tau, \end{cases} \end{aligned}$$

and the generalized coordinate $\eta_r(t)$ is given by

$$\eta_r(t) = \frac{4.472}{\omega_r} \int_0^t \sin 0.01 r\pi\tau \sin \omega_r (t - \tau) d\tau.$$

If $t < 100$,

$$\eta_r(t) = \frac{447.2}{\omega_r} \left(\frac{\pi r \sin \omega_r t - 100 \omega_r \sin 0.01 r\pi t}{(\pi r)^2 - 10000\omega_r^2} \right),$$

and if $t > 100$,

$$\begin{aligned} \eta_r(t) = \ &\frac{223.6}{\omega_r} \frac{1}{(\pi r)^2 - 10000\omega_r^2} \{(r\pi - 100\omega_r) \sin(r\pi + \omega_r(100 - t)) \\ &- (r\pi + 100\omega_r) \sin(r\pi - \omega_r(100 - t)) - 2r\pi \sin(\omega_r t)\}. \end{aligned} \quad (8.21)$$

✳

8.3. DETERMINISTIC VIBRATION

Example 8.7 Forced Response of a Hinged-Hinged Transversely Vibrating Beam

Consider a transversely vibrating cantilever beam governed by the equation of motion,

$$EI\frac{\partial^4 y}{\partial x^4} + c\frac{\partial y}{\partial t} + m\frac{\partial^2 y}{\partial t^2} = f(x,t).$$

Obtain the mode shapes, natural frequencies, and the response to an impulsive distributed load of $f(x,t) = f_0 x^2 \delta(t)$. Use $EI = 10^7$ Nm^2, $c = 1$ Ns/m^2, $m = 0.01$ kg/m and zero initial conditions.

Solution The eigenfunctions and the eigenvalues are the solutions of equation,

$$\frac{d^4 Y(x)}{dx^4} - \beta^4 Y(x) = 0,$$

and boundary conditions given by

$$Y(0) = \left.\frac{dY}{dx}\right|_{x=0} = 0$$

$$\left.\frac{d^2 Y}{dx^2}\right|_{x=L} = \left.\frac{d^3 Y}{dx^3}\right|_{x=L} = 0,$$

where $\beta^4 = m\omega^2/EI$ and L equals the beam length. The mode shapes and the natural frequencies are given by

$$\begin{aligned}
Y_r &= c_r \{(\sin\beta_r L - \sinh\beta_r L)(\sin\beta_r x - \sinh\beta_r x) \\
&\quad + (\cos\beta_r L + \cosh\beta_r L)(\cos\beta_r x - \cos\beta_r x)\} \\
\beta_r &= 1.88,\ 4.69,\ 7.86,\cdots \\
\omega_r &= 1.88^2\sqrt{\frac{EI}{m}},\ 4.69^2\sqrt{\frac{EI}{m}},\ 7.86^2\sqrt{\frac{EI}{m}},\cdots \\
c_r &= 0.329/\sqrt{mL},\ 0.0183/\sqrt{mL},\ 0.000776/\sqrt{mL},\cdots.
\end{aligned}$$

The forced response is given by

$$y(x,t) = \sum_{j=1}^{\infty} Y_j(x) \int_0^t \left[\int_0^L f(\xi,\tau) Y_j(\xi) d\xi\right] g_j(t-\tau) d\tau,$$

where the impulse response function is given by

$$g_j(t) = \frac{1}{\omega_{dj}} e^{-\zeta_j \omega_j t} \sin\omega_{dj} t.$$

The damping ratios and the damped natural frequencies are given by

$$\zeta_j = \frac{c}{2m\omega_j} \text{ and } \omega_{dj} = \omega_j\sqrt{1-\zeta_j^2}.$$

Evaluating the temporal part of the solution using the known eigenfunctions and impulse response functions, the response is given by

$$y(x,t) = f_o\sqrt{\frac{2}{mL}}L^3\left\{\frac{(\pi^2-4)}{\pi^3}Y_1(x)g_1(t) - \frac{1}{2\pi}Y_2(x)g_2(t)\right.$$
$$\left.+\frac{(9\pi^2-4)}{27\pi^3}Y_3(x)g_3(t) - \frac{1}{4\pi}Y_4(x)g_4(t) + ...\right\}. \quad (8.22)$$

❊

8.4 Random Vibration of Continuous Systems

From the analysis of the single-degree-of-freedom system, we know that knowledge of the frequency response function, $H(\omega)$, is crucial in obtaining the response statistics to random loads. For the multidegree-of-freedom system, the response spectrum matrix $[S_{XX}(\omega)]$ can be obtained if the matrix frequency response function $[H(\omega)]$ is known. In most cases, the matrix frequency response function is difficult to obtain or the expression can be very complicated, so that it is more sensible to use modal analysis.

The response is expanded in terms of eigenvectors with time-dependent coefficients called modal coordinates, and the force is expanded in terms of eigenvectors with time-dependent coefficients called modal forces. The matrix response spectral density is then expressed in terms of the modal force spectra, the frequency response functions relating the modal coordinates to the modal forces, and the collection of eigenvectors or modal matrix.

For a continuous system, an eigenfunction expansion method, which is the continuous version of the modal analysis, is used to derive the response spectrum. Recall that the response spectrum for multidegree-of-freedom systems is given as a matrix. For a system with three degrees of freedom, the response spectrum matrix has a dimension of 3×3. For example, $S_{X_1X_2}(\omega)$ is a cross-spectral density of the first and the second masses. Since the response spectrum is a continuous function of the spatial coordinate, the response will be expressed as a continuous function of x_1 and x_2, each denoting the spatial coordinate.

A priori, we present the results that are derived in the subsequent section. The cross-spectral density of the output $S_{yy}(x_1,x_2,\omega)$ relating the displacements at $x = x_1$ and $x = x_2$ is given by

$$S_{yy}(x_1,x_2;\omega) = \sum_{r=1}^{\infty}\sum_{k=1}^{\infty}Y_r(x_1)H_r^*(\omega)S_{Q_rQ_k}(\omega)H_k(\omega)Y_k(x_2),$$

8.4. RANDOM VIBRATION OF CONTINUOUS SYSTEMS

where $Y_r(x)$ is the rth eigenfunction, $H_r(\omega)$ is the transfer function between the generalized coordinate and generalized force and is given by

$$H_r(\omega) = \left(\omega_r^2 - \omega^2 + i2\zeta_r\omega_r\omega\right)^{-1},$$

ω_r is the rth natural frequency, ζ_r is the rth damping ratio, $S_{Q_rQ_k}(\omega)$ is the spectral density of the generalized force $Q_r(t)$ and is given by

$$S_{Q_rQ_k}(\omega) = \int_0^L \int_0^L S_{FF}(x_1,x_2;\omega)Y_r(x_1)Y_k(x_2)dx_1dx_2,$$

and $S_{FF}(x_1,x_2;\omega)$ is the spectral density of the loading.

8.4.1 Derivation of Response Spectral Density

Begin with the response to a distributed force in Equation 8.17 with $\eta_r(t)$ replaced by the convolution solution in Equation 8.20,

$$\mathcal{Y}(x,t) = \sum_{r=1}^{\infty} \left[\int_0^t Q_r(t)g_r(t-\tau)d\tau\right] Y_r(x),$$

where $Q_r(t)$ is the generalized force due to random force $F(x,t)$, given by

$$Q_r(t) = \int_0^L F(x,t)Y_r(x)\,dx.$$

Using change of variables, we can equivalently write

$$\mathcal{Y}(x,t) = \sum_{r=1}^{\infty} \left[\int_0^t Q_r(t-\tau)g_r(t)d\tau\right] Y_r(x).$$

The transient effects due to initial conditions have been neglected. The mean value of the response is then, in general,

$$E\{\mathcal{Y}(x,t)\} = \sum_{r=1}^{\infty} Y_r(x) \int_0^t \left[\int_0^L E\{F(x,t-\tau)\}Y_r(x)\,dx\right] g_r(\tau)\,d\tau.$$

Given the mean value of the forcing, we can calculate the mean value of the response. If the forcing is stationary, then

$$\begin{aligned}
E\{\mathcal{Y}(x,t)\} &= \mu_F \sum_{r=1}^{\infty} Y_r(x) \int_0^t \left[\int_0^L Y_r(x)\,dx\right] g_r(\tau)\,d\tau \\
&= \mu_F \sum_{r=1}^{\infty} Y_r(x) \left(\int_0^L Y_r(x)\,dx\right) \int_0^t g_r(\tau)\,d\tau \\
&= \mu_F \sum_{r=1}^{\infty} H_r(0) Y_r(x) \int_0^L Y_r(x)\,dx,
\end{aligned}$$

where
$$H_r(\omega) = \frac{1}{\omega_r^2 - \omega^2 + i2\zeta_r\omega_r\omega}.$$

For simplicity, the mean is taken as equal to zero. Then, the autocorrelation of the response can be derived as follows, initially without assuming stationarity,

$$\begin{aligned}
R_{\mathcal{YY}}(x_1, x_2,; t_1, t_2) &= E\{\mathcal{Y}(x_1, t_1)\mathcal{Y}(x_2, t_2)\} \\
&= \sum_{r=1}^{\infty}\sum_{k=1}^{\infty} E\{\eta_r(t_1)\eta_k(t_2)\} Y_r(x_1) Y_k(x_2) \\
&= \sum_{r=1}^{\infty}\sum_{k=1}^{\infty} R_{\eta_r\eta_k}(t_1, t_2) Y_r(x_1) Y_k(x_2), \quad (8.23)
\end{aligned}$$

where

$$\begin{aligned}
R_{\eta_r\eta_k}(t_1, t_2) &= \int_0^t \int_0^t E\{Q_r(t_1 - \theta) Q_k(t_2 - \kappa)\} g_r(\theta) g_k(\kappa)\, d\theta\, d\kappa \\
&= \int_0^t \int_0^t R_{Q_rQ_k}(t_1 - \theta, t_2 - \kappa) g_r(\theta) g_k(\kappa)\, d\theta\, d\kappa,
\end{aligned}$$

and, assuming stationarity,

$$R_{\eta_r\eta_k}(t_2 - t_1) = \int_0^t \int_0^t R_{Q_rQ_k}(t_2 - t_1 + \theta - \kappa) g_r(\theta) g_k(\kappa)\, d\theta\, d\kappa. \quad (8.24)$$

The expression for $R_{Q_rQ_k}(t_1, t_2)$ is defined by

$$\begin{aligned}
R_{Q_rQ_k}(t_1, t_2) &= E\{Q_r(t_1)Q_k(t_2)\} \\
&= \int_0^L \int_0^L E\{F(x_1, t_1)F(x_2, t_2)\} Y_r(x_1) Y_k(x_2)\, dx_1 dx_2 \\
&= \int_0^L \int_0^L R_{FF}(x_1, x_2; t_1, t_2) Y_r(x_1) Y_k(x_2)\, dx_1 dx_2.
\end{aligned}$$

For stationary loading, $R_{FF}(x_1, x_2; t_1, t_2) = R_{FF}(x_1, x_2; \tau)$ and, therefore, $R_{Q_rQ_k}(t_1, t_2) = R_{Q_rQ_k}(\tau)$, with $\tau = t_2 - t_1$,

$$R_{Q_rQ_k}(\tau) = \int_0^L \int_0^L R_{FF}(x_1, x_2; \tau) Y_r(x_1) Y_k(x_2)\, dx_1 dx_2.$$

The power spectrum of the response is found by taking the Fourier transform of Equation 8.23,

$$S_{\mathcal{YY}}(\omega) = \sum_{r=1}^{\infty}\sum_{k=1}^{\infty} \left[\frac{1}{2\pi}\int_{-\infty}^{\infty} R_{\eta_r\eta_k}(\tau)\exp(-i\omega\tau)d\tau\right] Y_r(x_1) Y_k(x_2).$$

8.4. RANDOM VIBRATION OF CONTINUOUS SYSTEMS

$R_{\eta_r \eta_k}(\tau)$ can be derived beginning with Equation 8.24. Take the Fourier transform of both sides, define $\lambda = \tau + \theta - \kappa$, to find

$$S_{\eta_r \eta_k}(\omega) = \frac{1}{2\pi} \int_{-\infty}^{\infty} e^{-i\omega(\lambda - \theta + \kappa)} \left[\int_{-\infty}^{\infty} \int_{-\infty}^{\infty} R_{Q_r Q_k}(\lambda) g_r(\theta) g_k(\kappa) d\theta d\kappa \right] d\lambda, \tag{8.25}$$

where τ has been replaced by $\lambda - \theta + \kappa$ and $d\tau$ by $d\lambda$. Rewrite Equation 8.25 in a more useful form by separating the integrals according to dummy variables,

$$S_{\eta_r \eta_k}(\omega) = \left(\int_{-\infty}^{\infty} g_r(\theta) e^{i\omega\theta} d\theta \right) \times \left(\int_{-\infty}^{\infty} g_k(\kappa) e^{-i\omega\kappa} d\kappa \right)$$
$$\times \left(\frac{1}{2\pi} \int_{-\infty}^{\infty} R_{Q_r Q_k}(\lambda) e^{-i\omega\lambda} d\lambda \right).$$

The Fourier transform of the impulse response function $g(t)$ is the frequency response function $H(\omega)$. Therefore,

$$S_{\eta_r \eta_k}(\omega) = H_r^*(\omega) H_k(\omega) S_{Q_r Q_k}(\omega),$$

where $S_{Q_r Q_k}(\omega)$, the spectral density of the modal force components, is derived assuming that the Fourier transform for $Q_j(t)$ exists, as it does for most physical processes,

$$Q_j(t) = \int_{-\infty}^{\infty} Q_j(\omega) e^{i\omega t} d\omega.$$

Then

$$S_{Q_r Q_k}(\omega) = \int_0^L \int_0^L S_{FF}(x_1, x_2; \omega) Y_r(x_1) Y_k(x_2) dx_1 dx_2, \tag{8.26}$$

where $S_{FF}(\omega)$ is the spectral density of the loading, a quantity that is estimated from data. Substituting appropriately,

$$S_{yy}(x_1, x_2; \omega) = \sum_{r=1}^{\infty} \sum_{k=1}^{\infty} H_r^*(\omega) H_k(\omega) S_{Q_r Q_k}(\omega) Y_r(x_1) Y_k(x_2). \tag{8.27}$$

One value of having such an equation is that the mean-square (MS) displacement can be evaluated,

$$\mathcal{Y}_{MS}(x) = R_{yy}(x, x; 0) = \int_{-\infty}^{\infty} S_{yy}(x, x; \omega) d\omega.$$

Recall that if $\mu_{\mathcal{Y}}(x) = 0$ then $\mathcal{Y}_{MS}(x) = \sigma_{\mathcal{Y}}^2(x)$, the variance.

Now, let us examine our results again by sampling the response spectral density at n discrete locations such that

$$S_{y_i y_j}(\omega) = S_{yy}(x = x_i, x = x_j; \omega).$$

Then, we can construct an $n \times n$ matrix of the response spectrum. Assume that we only consider first m eigenfunctions. Then, we can construct an $n \times m$ matrix Y_{ij} with elements consisting of jth eigenfunctions evaluated at $x = x_i$. Note that this matrix is equivalent to the modal matrix P_{ij}, which is the jth eigenvector at ith coordinate.

Let H_{ij} be an $m \times m$ matrix whose ith diagonal element is $H_i(\omega)$. $[S_{QQ}(\omega)]$ is an $m \times m$ matrix whose element is given by $S_{Q_i Q_j}(\omega)$. Then, Equation 8.27 can be written as

$$[S_{yy}(\omega)]_{n \times n}$$
$$= [Y(x)]_{n \times m} [H^*(\omega)]_{m \times m} [S_{QQ}(\omega)]_{m \times m} [H(\omega)]_{m \times m} [Y(x)]_{m \times n}^T.$$

Note that this result is remarkably similar to the result for the multidegree-of-freedom system in Equation 7.40 where the spectral density is approximated by the first m modes.

The derivations are now complete, but what do they mean, and what do they do for us? One of the functions of a probabilistic analysis is to help the designer bound uncertainties and understand how randomness in the forcing results in a scatter of possible structural responses. Furthermore, this scatter is not haphazard, but is defined by a standard deviation, and possibly a density function. It is the variance that is used to bound the mean value response. These are subjects for more advanced studies, which are now accessible to us.

Example 8.8 Random Vibration of a String

Consider a fixed-fixed uniform beam with $m = 0.01$ kg/m and $EA = 100$ N, and $L = 1$ m is forced by a white noise based random excitation with $S_0 = 1$ N^{2}s. Obtain the response spectra at the midpoint of the beam, $S_{yy}(L/2, L/2; \omega)$. Compare this with the response spectra when the beam is modeled as a discrete system. See Example 7.6.

Solution The eigenfunctions and the natural frequencies are found by solving

$$EA \frac{d^2 Y}{dx^2} + m\omega^2 Y = 0.$$

For the fixed-fixed beam, the eigenfunctions and the natural frequencies are

8.4. RANDOM VIBRATION OF CONTINUOUS SYSTEMS

given by

$$\begin{aligned} Y_r(x) &= \sqrt{\frac{2}{mL}} \sin \frac{r\pi x}{L}, \qquad r = 1, 2, \cdots \\ &= 14.14 \sin r\pi x \\ \omega_r &= \frac{r\pi}{L}\sqrt{\frac{EA}{m}} = 100\pi r, \qquad r = 1, 2, \cdots. \end{aligned}$$

Note that the eigenfunctions are normalized with respect to m so that they satisfy

$$\int_0^L m Y_r(x) Y_n(x) = \delta_{rn}$$

$$\int_0^L EA \frac{d^2 Y_r}{dx^2} Y_n(x) = -\omega_n^2 \delta_{rn}.$$

We now assume that the solution $y(x,t)$ and the distributed force $f(x,t)$ can be expressed in terms of the eigenfunctions such that

$$\begin{aligned} y(x,t) &= \sum_{r=1}^{\infty} \eta_r(t) Y_r(x) \\ f(x,t) &= \sum_{r=1}^{\infty} m q_r(t) Y_r(x). \end{aligned}$$

Substituting these expansions into the equation of motion, we find ordinary differential equations for $\eta_r(t)$

$$\ddot{\eta}_r(t) + \omega_r^2 \eta_r(t) = q_r(t), \text{ for } r = 1, 2, \cdots.$$

The rth frequency response function is then given by

$$H_r(\omega) = \frac{1}{\omega_r^2 - \omega^2}.$$

The response spectral density is given by Equation 8.27,

$$S_{yy}(x_1, x_2; \omega) = \sum_{r=1}^{\infty} \sum_{k=1}^{\infty} H_r^*(\omega) H_k(\omega) S_{Q_r Q_k}(\omega) Y_r(x_1) Y_k(x_2),$$

where $S_{Q_r Q_k}(\omega)$ is given by Equation 8.26,

$$S_{Q_r Q_k}(\omega) = \int_0^L \int_0^L S_{FF}(x_1, x_2; \omega) Y_r(x_1) Y_k(x_2) dx_1 dx_2.$$

For this problem, they are given by

$$S_{Q_rQ_k}(\omega) = \begin{cases} S_0 \dfrac{8L}{m\pi^2} \dfrac{1}{rk} & \text{for odd } r \text{ and } k \\ 0 & \text{for even } r \text{ or even } k. \end{cases}$$

For instance,

$$\begin{aligned} S_{Q_1Q_1}(\omega) &= S_0 \dfrac{8L}{m\pi^2} \\ S_{Q_1Q_2}(\omega) &= S_{Q_2Q_1}(\omega) = S_{Q_2Q_2}(\omega) = 0 \\ S_{Q_1Q_3}(\omega) &= S_{Q_3Q_1}(\omega) = S_0 \dfrac{8L}{3m\pi^2} \quad \cdots. \end{aligned}$$

Then, the response spectral density is given by

$$\begin{aligned} &S_{yy}(x_1, x_2; \omega) \\ =\ & H_1^*(\omega) H_1(\omega) S_{Q_1Q_1}(\omega) Y_1(x_1) Y_1(x_2) \\ & + (H_1^*(\omega) H_3(\omega) + H_3^*(\omega) H_1(\omega)) S_{Q_1Q_3}(\omega) Y_1(x_1) Y_3(x_2) \\ & + H_3^*(\omega) H_3(\omega) S_{Q_3Q_3}(\omega) Y_3(x_1) Y_3(x_2) \\ & + H_5^*(\omega) H_5(\omega) S_{Q_5Q_5}(\omega) Y_5(x_1) Y_5(x_2) \\ & + (H_1^*(\omega) H_5(\omega) + H_5^*(\omega) H_1(\omega)) S_{Q_1Q_5}(\omega) Y_1(x_1) Y_5(x_2) \\ & + (H_3^*(\omega) H_5(\omega) + H_5^*(\omega) H_3(\omega)) S_{Q_3Q_5}(\omega) Y_3(x_1) Y_5(x_2) \\ & + \cdots. \end{aligned}$$

The spectral density at $x = L/2$ is given by

$$\begin{aligned} & S_{yy}(L/2, L/2; \omega) \\ =\ & \dfrac{16 S_0}{m^2 \pi^2} \Bigg\{ \dfrac{1}{(\omega_1^2 - \omega^2)^2} + \dfrac{1}{9} \dfrac{1}{9(\omega_3^2 - \omega^2)^2} + \dfrac{1}{25} \dfrac{1}{(\omega_5^2 - \omega^2)^2} \\ & + \dfrac{2}{3} \dfrac{1}{(\omega_1^2 - \omega^2)(\omega_3^2 - \omega^2)} + \dfrac{2}{5} \dfrac{1}{(\omega_1^2 - \omega^2)(\omega_5^2 - \omega^2)} \\ & + \dfrac{2}{15} \dfrac{1}{(\omega_3^2 - \omega^2)(\omega_5^2 - \omega^2)} + \cdots \Bigg\}. \end{aligned}$$

Figure 8.10 shows $S_{yy}(L/2, L/2; \omega)$ around the first natural frequency. The response spectra at the midpoint when the system is modeled using 151 discrete masses is also plotted in dotted line. The discrepancy in the first natural frequency is due to the fact that the discrete model is an approximation to the continuous model.

The first natural frequency of the discrete model as a function of number of degrees of freedom is shown in Figure 8.11. Note that the first natural

8.4. RANDOM VIBRATION OF CONTINUOUS SYSTEMS

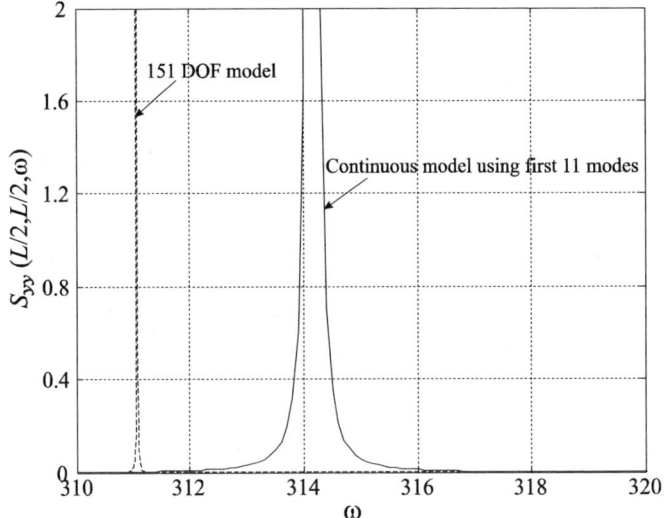

Figure 8.10: Response spectra at $x = L/2$.

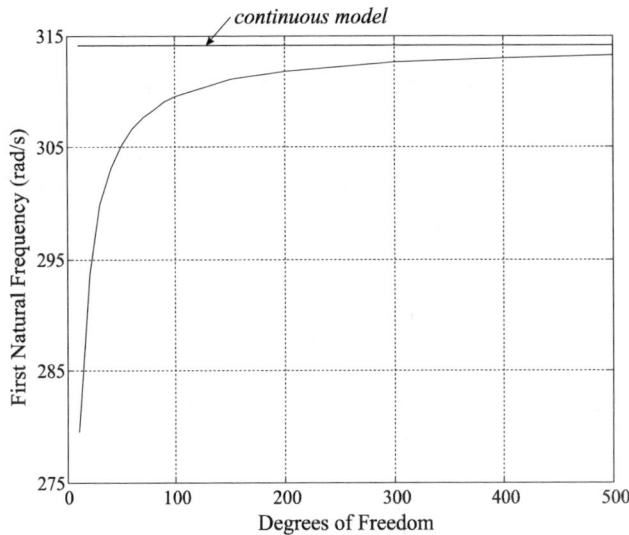

Figure 8.11: First natural frequency of the discrete multidegree-of-freedom system as a function of number of degrees of freedom. The exact fundamental frequency for the continuous model is shown as a straight line at the top.

frequency of the discrete system approaches that of the continuous system with an increasing number of degrees of freedom.

⊛

Example 8.9 Random Vibration of a Transversely Vibrating Beam

A simply supported beam has uniform mass per unit length m, damping per unit length c, and flexural rigidity of EI. The beam carries a stationary, random load $F(x,t)$ with cross-spectrum $S_{FF}(x_1, x_2; \omega)$ between the load intensities at two different positions x_1 and x_2 along the span of the beam, given by

$$S_{FF}(x_1, x_2; \omega) = \frac{q^2}{\omega_c L^2} \sin^2\left(\frac{\pi x_1}{L}\right) \sin^2\left(\frac{\pi x_2}{L}\right), \quad |\omega| < \omega_c$$
$$= 0 \quad \text{elsewhere,}$$

where q and ω_c are known constants. *(i)* Find the autocorrelation function $R_{PP}(\tau)$ of the total force $P(t)$ acting on the beam where

$$P(t) = \int_0^L F(x,t)dx,$$

(ii) find the mean-square value $E(F^2)$, and *(iii)* find the mean-square value of the response $E(\mathcal{Y}^2)$. Assume that $c = 0.1\sqrt{mEI}\pi^2/L^2$ and $\omega_c = 10\omega_1$.

Solution *(i)* The autocorrelation of the total force can be written in terms of the autocorrelation function of the distributed force,

$$R_{PP}(\omega) = E\left\{\int_0^L f(x_1, t)\,dx_1 \int_0^L f(x_2, t+\tau)\,dx_2\right\}$$
$$= \int_0^L \int_0^L R_{FF}(x_1, x_2; \omega)\,dx_1 dx_2.$$

It follows that the spectral density of the total force is related to the spectral density of the distributed force by

$$S_{PP}(\omega) = \int_0^L \int_0^L S_{FF}(x_1, x_2; \omega)\,dx_1 dx_2$$
$$= \frac{q^2}{\omega_c \pi}, \quad |\omega| < \omega_c.$$

(ii) The mean-square response equals the area under the density,

$$E\{F^2\} = \frac{2q^2}{\pi}.$$

8.4. RANDOM VIBRATION OF CONTINUOUS SYSTEMS

(iii) In order to find the mean-square response, we first start with the eigenvalue problem. The mode shapes must satisfy the spatial equation given by

$$EI\frac{d^4 Y(x)}{dx^4} - m\omega^2 Y(x) = 0$$

and corresponding boundary conditions,

$$\begin{aligned} Y(0) &= 0, \quad Y''(0) = 0, \\ Y(L) &= 0, \quad Y''(L) = 0. \end{aligned}$$

After applying the boundary conditions, we find that the rth eigenfunctions and eigenvalues are given by

$$Y_r(x) = \sqrt{\frac{2}{mL}} \sin\frac{r\pi}{L}x$$

$$\omega_r = \sqrt{\frac{EI}{m}}\left(\frac{r\pi}{L}\right)^2, \quad r = 1, 2, \cdots.$$

The frequency response functions are given by

$$H_r(\omega) = \frac{1}{\omega_r^2 - \omega^2 + i\omega c/m}.$$

The next step is obtain the spectral density for the generalized forces, $S_{Q_r Q_k}(\omega)$,

$$S_{Q_r Q_k}(\omega) = \int_0^L \int_0^L S_{FF}(x_1, x_2; \omega) Y_r(x_1) Y_k(x_2) dx_1 dx_2.$$

In this case, it evaluates to

$$S_{Q_r Q_k}(\omega) = \frac{q^2}{\omega_c L^2}\frac{2}{mL}\frac{L^2}{\pi^2}\frac{16}{kr(k^2-4)(r^2-4)}, \quad |\omega| < \omega_c,$$

for odd r and odd k. The spectral density for the generalized force becomes zero when r or k is even.

The response spectral density is given by

$$S_{yy}(x_1, x_2; \omega) = \sum_{r=1}^{\infty}\sum_{k=1}^{\infty} H_r^*(\omega) H_k(\omega) S_{Q_r Q_k}(\omega) Y_r(x_1) Y_k(x_2).$$

In our case,

$$S_{yy}(\omega) = \sum_{r=1}^{\infty}\sum_{k=1}^{\infty} H_r^*(\omega) H_k(\omega) S_{Q_r Q_k}(\omega) \int_0^L Y_r(x_1) dx_1 \int_0^L Y_k(x_2) dx_2,$$

where the integration of the eigenfunction is

$$\int_0^L Y_r(x_1)dx_1 = \sqrt{\frac{2}{mL}}\frac{2L}{r\pi}.$$

Then,

$$S_{yy}(\omega) = \sum_{\substack{r=1 \\ \text{odd } r}}^{\infty} \sum_{\substack{k=1 \\ \text{odd } k}}^{\infty} H_r^*(\omega) H_k(\omega) \frac{q^2}{\omega_c m^2} \frac{1}{\pi^4} \frac{256}{k^2 r^2 (k^2-4)(r^2-4)}, \quad |\omega| < \omega_c.$$

The mean-square value of the total force equals $R_{yy}(0)$, which is equivalent to the area under the spectral density function,

$$\begin{aligned} E(\mathcal{Y}^2) &= \int_{-\omega_c}^{\omega_c} S_{yy}(\omega)\,d\omega \\ &= \sum_{\substack{r=1 \\ \text{odd } r}}^{\infty} \sum_{\substack{k=1 \\ \text{odd } k}}^{\infty} \frac{q^2}{\omega_c m^2} \frac{1}{\pi^4} \frac{256}{k^2 r^2 (k^2-4)(r^2-4)} A_{rk}, \end{aligned}$$

where we let

$$A_{rk} = \int_{-\omega_c}^{\omega_c} H_r^*(\omega) H_k(\omega)\,d\omega.$$

In our case, the frequency response function can be written as

$$H_r(\omega^*) = \frac{1}{\omega_1^2 (r^4 - \omega^{*2} + i0.1\omega^*)},$$

where ω_1 is the first natural frequency, and ω^* is the frequency normalized by the first natural frequency, ω/ω_1. Then,

$$A_{rk}^* = \omega_1^3 A_{rk} = \int_{-\omega_c/\omega_1}^{\omega_c/\omega_1} H_r^*(\omega^*) H_k(\omega^*)\,d\omega^*,$$

which can be evaluated for various values of r and k. The imaginary part of this integral is an odd function of ω and becomes zero. A_{rk} are evaluated to

$$\begin{aligned} A_{11}^* &= 31.4 \\ A_{31}^* &= -0.00138 \\ A_{33}^* &= 0.383 \\ A_{51}^* &= 0.000269 \\ A_{53}^* &= 0.000541 \\ A_{55}^* &= 0.0000576. \end{aligned}$$

The mean-square response is then given by

$$E(\mathcal{Y}^2) = \sum_{\substack{r=1 \\ \text{odd } r}}^{\infty} \sum_{\substack{k=1 \\ \text{odd } k}}^{\infty} \frac{q^2}{\omega_c m^2} \frac{1}{\pi^4} \frac{256}{k^2 r^2 (k^2-4)(r^2-4)} \frac{1}{\omega_1^3} A_{rk}^*$$

$$= \frac{q^2}{m^2 \pi^4} \frac{256}{\omega_1^4} \left(\frac{31.4}{9} + \frac{-0.00138}{-135} + \frac{0.383}{2025} + \cdots \right).$$

Noting that the terms becomes progressively smaller, the mean-square response can be approximated by the first term,

$$E(\mathcal{Y}^2) \simeq 893 \frac{q^2}{m^2 \pi^4} \frac{1}{\omega_1^4}.$$

⊛

8.5 Beams with Complex Loading

Here we consider the generally complex problems of beams with more realistic boundary conditions and forces. A simplified theory is derived and applications from the literature discussed.

8.5.1 Transverse Vibration of Beam with Axial Force

Consider the lateral vibration of a beam, loaded both laterally and axially, a schematic of which appears in Figure 8.12. Our interest lies with establishing the added effect of the axial force on the response. Whenever a beam or column is compressed, there is concern about its buckling. As the axial load is increased, a critical value is reached where a new (buckled) deformation configuration is possible. Since this configuration is generally undesirable, structural failure is assumed to occur at this critical load. Design codes generally assume a failure at some load less than the buckling load.

To formulate this problem, use the free body diagram of an arbitrary section of the beam and draw all external forces, noting that there is an additional moment term Sy due to the constant axial force S, where y is the deflection at the section under consideration. This is shown in Figure 8.13.

Summing the forces in the y direction, we find

$$-(Q+dQ) + Q + (S+dS)(\theta + d\theta) - S\theta + p(x,t)\,dx = m(x)\,dx \frac{\partial^2 y}{\partial t^2},$$

where

$$Q = \frac{\partial}{\partial x}\left(EI \frac{\partial^2 y}{\partial x^2}\right), \quad \theta = \frac{\partial y}{\partial x}$$

$$dQ = \frac{\partial Q}{\partial x} dx, \quad dS = \frac{\partial S}{\partial x} dx, \quad d\theta = \frac{\partial \theta}{\partial x} dx.$$

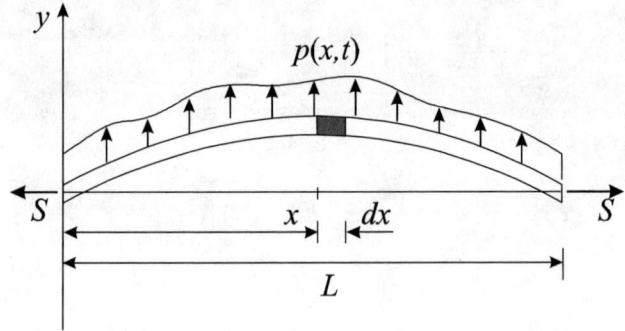

Figure 8.12: Beam with transverse and axial forces.

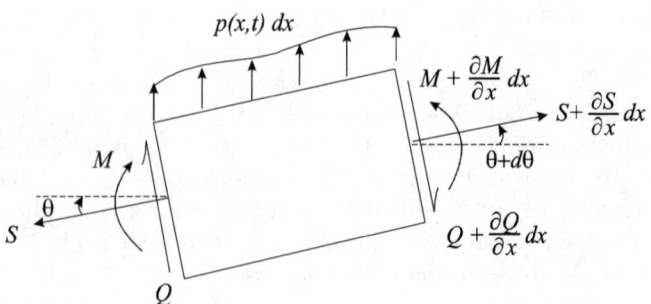

Figure 8.13: Free body for the transverse vibration of beam with axial force.

8.5. BEAMS WITH COMPLEX LOADING

Simplifying,

$$-\frac{\partial^2}{\partial x^2}\left(EI\frac{\partial^2 y}{\partial x^2}\right)dx + S\frac{\partial^2 y}{\partial x^2}dx + \frac{\partial y}{\partial x}\frac{\partial S}{\partial x}dx$$
$$+\frac{\partial^2 y}{\partial x^2}\frac{\partial S}{\partial x}dx^2 + p(x,t)\,dx = m(x)\,dx\frac{\partial^2 y}{\partial t^2}.$$

Dividing by dx and taking the limit as $dx \to 0$, the governing equation becomes

$$m(x)\frac{\partial^2 y}{\partial t^2} + \frac{\partial^2}{\partial x^2}\left(EI(x)\frac{\partial^2 y}{\partial x^2}\right) - \frac{\partial}{\partial x}\left(S(x)\frac{\partial y}{\partial x}\right) = p(x,t). \quad (8.28)$$

Making the uniformity assumption and setting $p(x,t) = 0$, $EI(x) = EI$ and $S(x) = S$ results in the eigenvalue problem,

$$EI\frac{d^4 Y}{dx^4} - S\frac{d^2 Y}{dx^2} = \omega^2 m Y(x), \quad (8.29)$$

where the effect of axial force S can be significant. For this problem, the modes are

$$Y_r(x) = C_r \sin\frac{r\pi x}{L}, \quad r = 1, 2, \cdots.$$

Take appropriate derivatives and substitute these into Equation 8.29, to find

$$EIC_r\left(\frac{r\pi}{L}\right)^4 \sin\frac{r\pi x}{L} + SC_r\left(\frac{r\pi}{L}\right)^2 \sin\frac{r\pi x}{L} = \omega_r^2 m C_r \sin\frac{r\pi x}{L},$$

where the coefficients C_r can be cancelled. Thus,

$$\omega_r = \left(\frac{r\pi}{L}\right)^2 \sqrt{\frac{EI}{m}}\sqrt{1 + \frac{S}{EI}\left(\frac{L}{r\pi}\right)^2}, \quad r = 1, 2, \cdots.$$

Note that for a tensile axial force $+S$, the effect is an increase in the frequencies of free vibration. For a compressive force $-S$, the frequencies are

$$\omega_r = \left(\frac{r\pi}{L}\right)^2 \sqrt{\frac{EI}{m}}\sqrt{1 - \frac{S}{EI}\left(\frac{L}{r\pi}\right)^2}, \quad (8.30)$$

resulting in lower natural frequencies. The question to ask here is, for what compressive load will the frequencies shift down so that the fundamental frequency becomes zero?

For $r = 1$, the term,

$$\frac{S}{EI}\left(\frac{L}{r\pi}\right)^2,$$

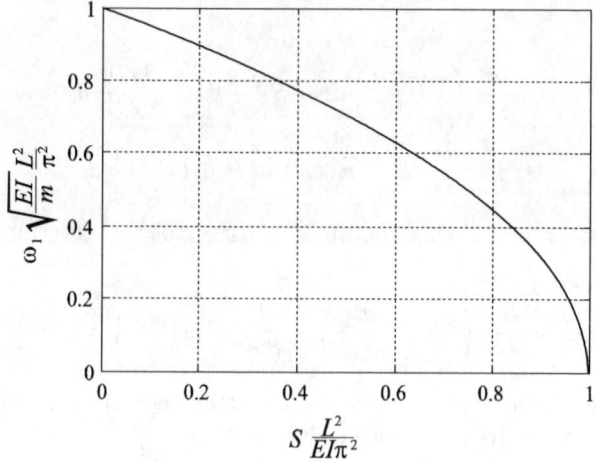

Figure 8.14: Natural frequency ω_1 as a function of axial force S.

is the ratio of S to the Euler buckling load. If $SL^2/EI\pi^2 \to 1$, the lowest mode of vibration approaches a zero frequency and transverse buckling occurs for $S = EI\pi^2/L^2$. A plot of $\omega_1\sqrt{EI/m}\,(L/\pi)^2$ as a function of $S\,(L/\pi)^2/EI$ using Equation 8.30 is shown in Figure 8.14.

8.5.2 Transverse Vibration of Beam on Elastic Foundation

All previous boundary conditions were at discrete points, at the ends of the beam. It is possible that a continuous boundary effect occurs along the length of the beam. In such cases, the boundary condition becomes part of the governing equation. An important example of such a problem is the beam on an elastic foundation,[4] a schematic of which appears in Figure 8.15. Applications include machine vibration, the vibration of a structure on a foundation, and structural response to ground shock such as earthquakes and explosives.

Here the elastic restraint against transverse motion is distributed continuously along the length of the beam and damping effects are ignored. For element of length dx, the differential equation of motion is

$$\frac{\partial^2}{\partial x^2}\left[EI(x)\frac{\partial^2 y}{\partial x^2}\right]dx = -k_f(x)y\,dx - \varrho A(x)dx\frac{\partial^2 y}{\partial t^2},$$

---

[4] The example of the rotating shaft that is immersed in a lubricated sleeve is one where there is continuous damping along the length instead of continuous stiffness.

8.5. BEAMS WITH COMPLEX LOADING

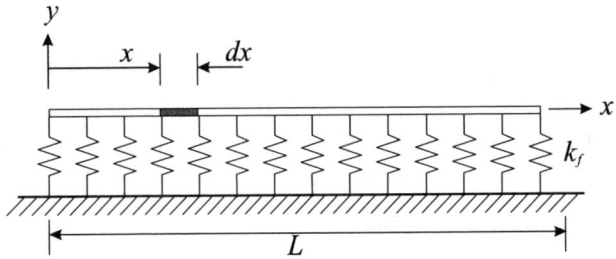

Figure 8.15: Beam on elastic foundation.

where the restraining force is given by the term $k_f(x) y \, dx$. The parameter $k_f(x)$ is the constant stiffness per unit length of foundation, and $\varrho A(x) = m(x)$, the mass of the beam per unit length. For a *prismatic beam* (uniform cross-section),

$$EI \frac{\partial^4 y}{\partial x^4} + k_f y = -\varrho A \frac{\partial^2 y}{\partial t^2}.$$

It is clear from this equation how the foundation stiffness alters the mathematical character of the governing equation. Assuming a product solution and performing the usual derivatives and substitutions, the equation governing the modal variable $Y(x)$ is found to be

$$EI \frac{d^4 Y}{dx^4} - (\varrho A \omega^2 - k_f) Y = 0,$$

or

$$\frac{d^4 Y}{dx^4} + \left(\frac{-\varrho A}{EI} \omega^2 + \frac{k_f}{EI} \right) Y = 0. \quad (8.31)$$

Let $m = \varrho A$, redefine β so as to include the foundation stiffness, $\beta^4 = (\omega^2 m - k_f)/EI$, and rewrite Equation 8.31 in standard form,

$$Y''''(x) - \beta^4 Y(x) = 0.$$

The equation for ω becomes

$$\omega_r = \left(\frac{r\pi}{L} \right)^2 \sqrt{\frac{EI}{m} + \frac{k_f}{m} \left(\frac{L}{r\pi} \right)^4}.$$

The importance of the term $k_f (L/r\pi)^4$ is problem-specific and by comparison to the value of EI. For most values of k_f, this term can be ignored when compared with EI. The importance of the elastic foundation is primarily on the lowest frequencies.

Example 8.10 Axially and Randomly Loaded Timoshenko Beam on an Elastic Foundation

This problem combines the loadings discussed above.[5] Many structures can be idealized as a beam on an elastic foundation. Examples include bridges, runways, rails, roadways, and pipelines. As a first analysis for a particular application, the understanding of how a beam on an elastic foundation vibrates can be very useful. The axial loading indicates that the beam is only a part of a larger structure, and that structure imparts a constraining force.

A Timoshenko beam is used, assumed to be uniform, damped, with generalized boundary conditions and subjected to an axial load. The Timoshenko beam includes the effects of shear deformation and rotary inertia, which are important if the span/depth ratio is relatively small. We will only outline the key steps and present the primary results since the details are ably developed in the source paper.

The coupled governing partial differential equations for transverse translation $u(x,t)$ and rotation $\phi(x,t)$ are

$$m\ddot{u} + c_1\dot{u} + k_l u + Su'' - kGA(u' + \phi)' = p(x,t)$$
$$I_o\ddot{\phi} + c_2\dot{\phi} + k_r\phi - EI\phi'' + kGA(u' + \phi) - k_s u' = q(x,t),$$

where the system is loaded by an external random force p and random moment q, and $I_o = mr^2$ is the rotary moment of inertia per unit length, S is the static axial load, I is the area moment of inertia. k is the effective area coefficient in shear, k_l, k_r, and k_s are stiffnesses, and the c values are damping coefficients.

The analysis is in three parts. The first step is to perform a free vibration analysis to derive the frequencies and modes of vibration. The second step is to perform a modal forced vibration analysis, and finally, a random vibration analysis.

For the free vibration analysis, the external forcing and the damping are set to zero, by definition. Harmonic solutions are assumed,

$$u = U_n(x)\exp(i\omega_n t)$$
$$\phi = \Phi_n(x)\exp(i\omega_n t),$$

$n = 1, 2, 3, \cdots$. Substituting these equations into the governing equations results in two simultaneous fourth-order ordinary differential equations for the eigenfunctions U_n and Φ_n. One of these two equations is, for example,

$$U_n'''' + \alpha_n^2 U_n'' - \beta_n^4 U_n = 0,$$

---

[5] This development is based on the paper, "Deterministic and Random Vibration of an Axially Loaded Timoshenko Beam Resting on an Elastic Foundation," T-P. Chang, *Journal of Sound and Vibration* (1994) 178(1), 55-66.

8.5. BEAMS WITH COMPLEX LOADING

with solution

$$U_n(x) = D_1 \sin \delta_n x + D_2 \cos \delta_n x + D_3 \sinh \epsilon_n x + D_4 \cosh \epsilon_n x.$$

δ_n and ϵ_n are functions of α_n^2 and β_n^4, which are functions of the physical parameters of the system. Similar equations are found for Φ_n.

For the forced vibration, expand the forced responses in terms of the eigenfunctions,

$$u(x,t) = \sum_{n=1}^{\infty} \eta_n(t) U_n(x)$$

$$\phi(x,t) = \sum_{n=1}^{\infty} \eta_n(t) \Phi_n(x).$$

Substitute these solutions into the governing equations, and then multiply by $U_m(x)$ and $\Phi_m(x)$, respectively. Then, the use of modal orthogonality results in the decoupled equations,

$$\ddot{\eta}_n + 2\zeta_n \omega_n \dot{\eta}_n + \omega_n^2 \eta_n = P_n(t) + Q_n(t), \qquad (8.32)$$

where the parameters are defined in a standard way, and it is assumed that the damping coefficients are related by $c_2 = r^2 c_1$ for convenience. The modal forcing terms are related to the external forcing in the usual way, for example,

$$P_n(t) = \frac{1}{\mu_m} \int_0^L U_n(x) p(x,t) dx,$$

where μ_m is a generalized mass. The moment is usually of greater interest than the rotation, and it can be obtained using the relation $M(x,t) = EI\phi'(x,t)$. Once $p(x,t)$ and $q(x,t)$ are defined, the equations can be solved using the Duhamel convolution integral.

The modal random vibration analysis begins with Equation 8.32. In terms of the frequency response function,

$$\eta_n(t) = \int_{-\infty}^{\infty} H_n(\omega) \exp(i\omega t) [P_n(t) + Q_n(t)] d\omega,$$

$$H_n(\omega) = \left[-\omega^2 + \omega_n^2 + 2i\zeta_n \omega_n \omega\right]^{-1}.$$

For the random vibration problem, the excitations $P(x,t)$ and $Q(x,t)$ are assumed random and stationary, and therefore so are $P_n(t)$ and $Q_n(t)$. The loads are assumed to be independent stochastic processes so that the cross-spectral densities between $P_n(t)$ and $Q_n(t)$ can be excluded. Of course, it may be necessary to include such correlations for particular applications. The sequence of calculations are as follows,

$$R_{\mathcal{U}_{x_1}\mathcal{U}_{x_2}}(x_1,x_2,\tau) = E\{\mathcal{U}(x_1,t)\mathcal{U}(x_2,t+\tau)\}$$

$$= \frac{1}{2\pi}\sum_{m=1}^{\infty}\sum_{n=1}^{\infty} U_m(x_1)U_n(x_2)$$

$$\cdot \int_{-\infty}^{\infty} H_m(\omega)H_n^*(\omega)[S_{P_mP_n}(\omega) + S_{Q_mQ_n}(\omega)]\exp(i\omega\tau)d\omega,$$

where

$$S_{P_mP_n}(\omega) = \frac{1}{\mu_m^2}\int_0^L\int_0^L U_m(x_1)U_n(x_2)S_{P_{x_1}P_{x_2}}(x_1,x_2,\omega)dx_1dx_2$$

$$S_{Q_mQ_n}(\omega) = \frac{1}{\mu_m^2}\int_0^L\int_0^L \Phi_m(x_1)\Phi_n(x_2)S_{Q_{x_1}Q_{x_2}}(x_1,x_2,\omega)dx_1dx_2.$$

$S_{P_{x_1}P_{x_2}}(x_1,x_2,\omega)$ and $S_{Q_{x_1}Q_{x_2}}(x_1,x_2,\omega)$ are spatial cross-spectral densities that are either assumed or based on data. Given these, the rest of the equations can be evaluated.

The above expressions are for two arbitrary points on the beam, that is, at $x = x_1$ and $x = x_2$. Setting $x = x_1 = x_2$, the autocorrelation function can be found, $R_{\mathcal{U}\mathcal{U}}(x,\tau)$, as can be the mean-square displacement by setting $\tau = 0$, as usual. Similarly, the mean-square bending moment is given by

$$E\{M^2(x)\} = \frac{1}{2\pi}(EI)^2\sum_{m=1}^{\infty}\sum_{n=1}^{\infty}\Phi'_m(x_1)\Phi'_n(x_2)$$

$$\cdot \int_{-\infty}^{\infty} H_m(\omega)H_n^*(\omega)[S_{P_mP_n}(\omega) + S_{Q_mQ_n}(\omega)]d\omega.$$

※

8.5.3 Response of a Beam to a Traveling Force

A brief introduction is provided here to the formulation of traveling force problems. A number of important applications can be cast in this form. Vehicle-structure *interaction problems*[6] play a significant role in modern engineering design. Familiar examples are automobile-bridge, jet-aircraft carrier, structure-ocean wave, and train-track interactions. High-speed ground transportation systems based on magnetically levitated (*maglev*) vehicles result in electromagnetic coupling between vehicle and track. Figures 8.16 and 8.17 hint at how preliminary models may be formulated.

---
[6] Interaction problems are generally those where two dissimilar systems are coupled in some way, and, therefore, their vibration characteristics must be solved simultaneously.

8.5. BEAMS WITH COMPLEX LOADING

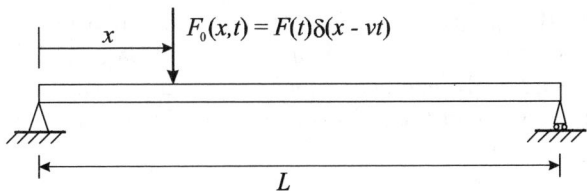

Figure 8.16: Traveling point force on a beam.

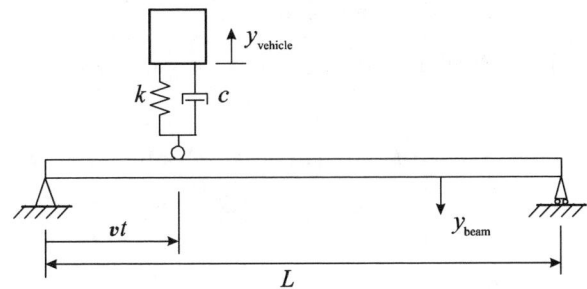

Figure 8.17: Traveling force due to single degree of freedom vehicle. Position of vehicle is $x = vt$ for constant v. For arbitrary speed, $x = w(t)$.

In the figures, the variable v denotes the velocity of the vehicle. $F_0(t)$ can be a harmonic function such as $\sin \Omega t$, and studies of the relation between beam vibration characteristics and parameters v and Ω are of importance.

In Figure 8.17, an n-degree-of-freedom model can be used for the vehicle, if necessary. The beam model may be as simple as the Bernoulli-Euler, or may include the additional effects of the Timoshenko beam. For a rail system, the beam may be on a foundation, and damping can be added. Many options exist, depending on application, and a realistic model can be built by adding some of the individual effects of the more basic models we have already studied.

Example 8.11 Beam Response Due to Moving Random Loads and Deterministic Axial Forces

This problem also combines two loading cases.[7] A Bernoulli-Euler beam is subjected to random loads moving with time-varying velocity. The problem of moving loads on a structure is broadly applicable, for example, to bridges, railways, piping systems subjected to two-phase flow, beams subjected to pressure waves, and machining operations where high axial speed may be employed. The deterministic axial forces may be due to prestressing or due to the effects of adjacent structures.

Equation 8.28 for $y(x, t)$ becomes

$$EI\frac{\partial^4 \mathcal{Y}}{\partial x^4} - S\frac{\partial^2 \mathcal{Y}}{\partial x^2} + m\frac{\partial^2 \mathcal{Y}}{\partial t^2} + c\frac{\partial \mathcal{Y}}{\partial t} = \mathcal{P}(x, t),$$

where $\mathcal{Y}(x, t)$ is the random response, and with simply supported boundary conditions,

$$\mathcal{Y}(0, t) = 0, \quad \mathcal{Y}(L, t) = 0$$
$$\left.\frac{\partial^2 \mathcal{Y}}{\partial x^2}\right|_{x=0} = 0, \quad \left.\frac{\partial^2 \mathcal{Y}}{\partial x^2}\right|_{x=L} = 0,$$

and the initial conditions

$$\mathcal{Y}(x, 0) = 0, \quad \left.\frac{\partial \mathcal{Y}(x, 0)}{\partial t}\right|_{t=0} = 0.$$

Note that the boundary and initial conditions are prescribed with probability 1.

---

[7] This development is based on the paper, "Stochastic Vibration of an Elastic Beam due to Random Moving Loads and Deterministic Axial Forces," H.S. Zibdeh, *Engineering Structures*, Vol. 17, No. 7, pp. 530-535, 1995.

8.5. BEAMS WITH COMPLEX LOADING

The positive sign of random force $\mathcal{P}(x,t)$ is taken to reflect that it acts downward, and c is the damping coefficient per unit length. To incorporate the moving load, let

$$\mathcal{P}(x,t) \equiv F_0(x,t) = F(t)\delta\left(x - w(t)\right),$$

where $w(t)$ is the position of the load, $F(t)$ is a random forcing,

$$F(t) = \mu_F + P(t),$$

μ_F is the mean value of the force and $P(t)$ is the zero-mean random part of the force. The forcing magnitude is random, but the location is given as deterministic.

In modal form, the vertical deflection is given by

$$\mathcal{Y}(x,t) = \sum_{n=1}^{\infty} Y_n(t) \sin \frac{n\pi x}{L}.$$

This expansion is substituted into the governing equation, then multiplied by $\sin(n\pi x/L)$ and integrated over the domain to utilize the orthogonality conditions, resulting in the nth mode of the generalized deflection,

$$EIY_n \frac{L}{2}\left(\frac{n\pi}{L}\right)^4 + SY_n \frac{L}{2}\left(\frac{n\pi}{L}\right)^2 + m\frac{L}{2}\ddot{Y}_n + c\frac{L}{2}\dot{Y}_n$$
$$= \int_0^L \mathcal{P}(x,t) \sin \frac{n\pi x}{L} dx.$$

This equation can be written in a simpler form as

$$\ddot{Y}_n + \frac{c}{n}\dot{Y}_n + \omega_n^2 Y_n = Q_n(t), \qquad (8.33)$$

where

$$Q_n(t) = \frac{2}{mL}\int_0^L \mathcal{P}(x,t) \sin \frac{n\pi x}{L} dx$$
$$\omega_n^2 = \frac{EI}{m}\left(\frac{\pi}{L}\right)^4 n^2 \left(n^2 + \left[\frac{S}{S_{critical}}\right]^2\right)$$
$$S_{critical} = \frac{\pi^2 EI}{L^2}.$$

$S_{critical}$ is the Euler buckling load. $Y_n(t)$ is given by the convolution integral

$$Y_n(t) = \int_0^t g_n(t-\tau) Q_n(\tau) d\tau, \qquad (8.34)$$

where $g_n(t)$ is the impulse response function. The solution to Equation 8.33, using Equation 8.34, is

$$\mathcal{Y}(x,t) = \frac{2}{mL} \sum_{n=1}^{\infty} \frac{1}{\omega_{d_n}} \sin \frac{n\pi x}{L}$$

$$\cdot \int_0^t \sin \omega_{d_n}(t-\tau) \exp\left(-\frac{c}{2m}[t-\tau]\right) \sin \frac{n\pi w(\tau)}{L} F(\tau) d\tau,$$

where the damped circular frequency is given by

$$\omega_{d_n}^2 = \omega_n^2 - \left(\frac{c}{2m}\right)^2.$$

The load position function $w(t)$ can be generally defined as

$$w(t) = a + bt + \frac{1}{2}ct^2,$$

where a is the point of application of the force, b is the initial speed, and c is the constant acceleration.

The mean-value response, which is also the deterministic part of the deflection, is given if the above equations are solved using $F(t) = \mu_F$. For the stochastic case, we use $F(t) = \mu_F + P(t)$. We already have the deterministic part of the solution. The stochastic part of the solution, also thought of as the variance about the mean-value response, requires an evaluation of the response covariance.

Begin by calculating the covariance of the force,

$$Cov_{FF}(t_1, t_2) = E\{P(t_1)P(t_2)\}.$$

Then

$$Cov_{PP}(x_1, x_2, t_1, t_2) = \delta(x_1 - w(t_1))\delta(x_2 - w(t_2)) Cov_{FF}(t_1, t_2).$$

If $P(t)$ is white noise of intensity (magnitude) S_{PP}, then

$$Cov_{FF}(t_1, t_2) = S_{PP}\delta(t_2 - t_1) = \begin{cases} S_{PP}, & t_2 = t_1 \\ 0, & \text{otherwise,} \end{cases}$$

and

$$Cov_{Q_l Q_n}(t_1, t_2)$$
$$= E\{Q_l Q_n\}$$
$$= \frac{4}{m^2 L^2} \int_0^L \int_0^L \delta(x_1 - w(t_1))\delta(x_2 - w(t_2)) Cov_{FF}(t_1, t_2)$$
$$\cdot \sin \frac{l\pi x_1}{L} \sin \frac{n\pi x_2}{L} dx_1 dx_2$$
$$= \frac{4}{m^2 L^2} \sin \frac{l\pi w(t_1)}{L} \sin \frac{n\pi w(t_2)}{L} S_{PP}\delta(t_2 - t_1),$$

where the second equality is obtained by utilizing the property of the delta function[8] that eliminates the integral. Similarly,

$$Cov_{Y_l Y_n}(t_1, t_2) = \frac{4S_{PP}}{m^2 L^2} \int_0^t g_l(t_1 - \tau_1) g_n(t_2 - \tau_1)$$
$$\cdot \sin \frac{l\pi w(\tau_1)}{L} \sin \frac{n\pi w(\tau_2)}{L} d\tau_1.$$

Finally, the covariance of the deflection is given by

$$Cov_{yy}(x_1, x_2, t_1, t_2) = \sum_{l=1}^{\infty} \sum_{n=1}^{\infty} \sin \frac{l\pi x_1}{L} \sin \frac{n\pi x_2}{L} Cov_{Y_l Y_n}(t_1, t_2).$$

From this expression, the variance of the deflection is found to be

$$\sigma_y^2(x,t) = Cov_{yy}(x, x, t, t)$$
$$= \frac{4S_{PP}}{m^2 L^2} \sum_{l=1}^{\infty} \sin^2 \frac{l\pi x}{L} \int_0^t g_l^2(t - \tau_1) \sin^2 \frac{l\pi w(\tau_1)}{L} d\tau_1,$$

which can be integrated. The coefficient of variation is defined as

$$\rho_y(x,t) = \frac{\sigma_y(x,t)}{y_{static}},$$

where y_{static} is the static deflection due to a concentrated force P at midspan, equal to $PL^3/48EI$. These statistics can be used in a design process based on specified mean displacements and a probability of being within a certain number of standard deviations of the mean value.

⊛

8.6 Concluding Summary

This chapter introduced elementary concepts for the modeling of dynamic continuous parametered structures, in particular strings and beams. A modal approach is demonstrated deterministically and then extended to include random forcing. Special boundary conditions are introduced to show how complex physical problems can be initially tackled using relatively simple physical models such as beams.

---

[8] $\int x \delta(x-y) dx = y$ since the only nonzero value exists where $x = y$.

8.7 Problems

Section 8.1: Continuous Systems

1. A transverse beam is supported by torsional springs at both ends with stiffness k. Find the equation of motion and corresponding boundary conditions.

Section 8.2: Sturm-Liouville Eigenvalue Problem

2. A string is fixed at one end and is free to move up and down at the other end. Obtain the eigenfunctions and eigenvalues.

3. Prove the orthogonality condition for the string in Equation 8.9.

4. Consider a uniform rod in axial vibration with mass per unit length m, length L, and axial stiffness EA. The one end at $x = 0$ is attached to a spring with stiffness k and is free at $x = L$. Obtain the eigenfunctions and eigenvalues. Use $k = EA/L$.

5. Consider a uniform string fixed at both ends. It has mass per unit length m, length L, and tension T. A spring with stiffness k is attached to the string at $x = L/2$. Find the eigenfunctions and eigenvalues.

6. Derive the boundary-value problem for a rod in axial vibration with a lumped mass M at $x = L$ and fixed at $x = 0$. And then, derive the eigenvalue problem. Obtain the eigenvalues, eigenfunctions, and the orthogonality condition for the eigenfunctions. Let $EA(x) = EA$, $m(x) = m$, and $M = 0.5mL$.

Section 8.3: Deterministic Vibration

7. Determine the response of a fixed-fixed string with uniform cross-section to the step initial displacement,

$$y(x,0) = \begin{cases} A, & 0.45L < x < 0.55L \\ 0, & \text{elsewhere}. \end{cases}$$

Assume that the initial velocity is zero.

8. A uniform cantilever beam vibrating in the transverse direction carries a point mass M at its free end. The beam has flexural rigidity of EI, mass per unit length of m, and the point mass, $M = 0.5mL$. The beam is subjected to a distributed load $f(x,t) = x \sin \omega_f t$, where ω_f does not coincide with one of the natural frequencies. Find the response of the point mass. Assume zero initial conditions.

9. Plot Equation 8.21 for the given parameter values for $r = 1$, for $r = 2$, and for $r = 3$..

10. Plot the solution $y(x,t)$ of Equation 8.22 using a one-term $(\eta_1 Y_1)$, a two-term $(\eta_1 Y_1 + \eta_2 Y_2)$, and a three-term $(\eta_1 Y_1 + \eta_2 Y_2 + \eta_3 Y_3)$ approximation. What conclusions can be drawn?

Section 8.4: Random Vibration of Continuous System

11. A uniform string that is fixed at both ends is subjected a distributed load $f(x,t)$ whose mean is not zero but a function of x such that $\mu_F = \sin x$. Find the mean response. Assume that the mass per unit length is m, the tension is T, and the length is L.

12. Consider a building that can be modeled as a uniform cantilever transversely vibrating beam with the flexural rigidity EI, mass per unit length m, and structural damping c. The building is subjected to the stationary random ground motion with zero-mean and a white spectral density $S_0(\omega) = S_0$. Estimate the spectral density function $S_{yy}(L, L; \omega)$ for the response deflection at the top of the building, $\mathcal{Y}(L,t)$. Also, find the mean-square response.

13. A uniform fixed-fixed string is subjected to a weakly stationary random excitation $f(x,t)$ that is characterized by

$$E\{F(x,t)\} = 0$$
$$R_{FF}(x_1, x_2; \tau) = \sigma^2 \exp(-\alpha |\tau|)\delta(x_1 - x_2),$$

where $\delta(\cdot)$ is the Dirac Delta function. Compute the cross-correlation function and the cross-spectral density of the response.

14. A transversely vibrating beam is fixed at one end and free at the other. The beam is uniform with constant m, c and EI throughout the entire span length L. A point load $F(t)$ is applied at the free end. Express the spectral density of the random normal force at the fixed end in terms of the spectral density of the concentrated load $F(t)$.

15. Consider a uniform bar in bending, simply supported at both ends and subjected to the excitation,

$$f(x,t) = F(t)\delta(x - L/2),$$

where $F(t)$ is an ergodic random process with ideal white noise power spectral density, and $\delta(x - L/2)$ a spatial Dirac delta function. Derive expressions for the cross-correlation function between the response at the points $x = L/4$ and $x = 3L/4$ and for the mean-square value of the response at $x = L/4$. Assume constant m, c, and EI along the beam.

Section 8.5: Beams with Complex Loading

16. A tower of a suspension bridge is subjected to earthquake excitations. The tower is idealized as a transverse beam that is under constant compression S due to the suspension cables. The ground motion in the horizontal direction is denoted by $Q(t)$ and the total transverse displacement of the beam column by $Y(x,t)$. The relative transverse displacement,
$$Y_0(x,t) = Y(x,t) + Q(t),$$
is directly related to the bending moment of the beam column. Assume that the ground motion can be modeled as
$$Q(t) = \int_0^t (e^{-\alpha \tau} - e^{-\beta \tau}) G(\tau) d\tau,$$
where $G(\tau)$ is a stationary Gaussian random process with a zero expectation and a spectral density $S_{GG}(\omega)$. In addition, the beam properties, EI, S, c, and m, are constants along the length of the beam. Find the expression for the cross-correlation function of the response.

Chapter 9

Reliability

9.1 Introduction

Suppose a component of a dynamic system is undergoing random vibration. This component may be a truss in a bridge, a steel column in a skyscraper, or the fuselage panel of an airplane, for example. These components can fail for many different reasons depending on the material, the geometry, and the types of loading. The truss in a bridge may fail due to fatigue, the steel column in a skyscraper may buckle, and the panel in an airplane may develop a crack around a point of stress concentration.

These components are designed so that they function properly for a certain period of time. The time at which a part fails, T_f, varies from realization to realization, and is a random variable. Empirical evidence clearly shows that T_f cannot be determined deterministically. "Identical components" subjected to "identical loads" will fail at different times, and the time to failure can only be described probabilistically. Of course, at some level the components and the loads are not identical. But such small differences are not quantifiable.

In this chapter, we consider two types of failure. The first is called the *first excursion failure* or *first passage failure*. In this case, a structure fails when a certain parameter exceeds a certain level for the first time, as shown in Figure 9.1. For example, a part may fail when the displacement or acceleration exceeds a threshold value, Z. If a part is designed to stay in the elastic range, it is considered to fail when the stress at a critical location (perhaps a location with high stress concentration) exceeds the yield strength. A rod that is in compression will fail when the axial load exceeds the critical load for buckling. A crack in a plate that is subjected to a tensile force will grow when the stress intensity factor exceeds its critical value and therefore the part is considered to fail.

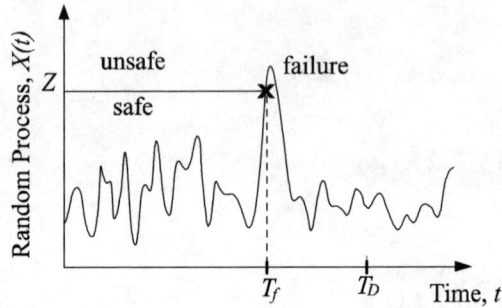

Figure 9.1: First excursion failure occurs when $X(t)$ exceeds Z. Here Z is taken to be a specific deterministic value, but it may also be a random value.

The other type of failure we consider is *fatigue failure*. When a part is subjected to cyclic loads, damage is done by repeated exposures or cycles. The part eventually fails when the accumulated damage reaches the total damage that a part can absorb. It can fail even at a much lower stress level than the yield strength.

The goal of an engineering design is to produce a part that will function properly during its design lifetime, T_D. If the time to failure, T_f, is greater than T_D, then the part has operated successfully. Figure 9.1 shows a first excursion failure, where $T_f < T_D$.

Since T_f is a random variable, even if the part functions properly throughout its design lifetime can only be expressed as a probability, called the *reliability*, and denoted as p_r,

$$p_r = \Pr\left(T_f > T_D\right).$$

The probability of failure, denoted as p_f, is then given by

$$\begin{aligned} p_f &= 1 - p_r \\ &= \Pr\left(T_f \leq T_D\right). \end{aligned}$$

T_D is prescribed by the engineer and is a deterministic quantity.

If the probability density function of T_f is known and denoted as $f_{T_f}(t_f)$, then the reliability and the probability to failure can be written as

$$\begin{aligned} p_r(T_D) &= \int_{T_D}^{\infty} f_{T_f}(t_f)\, dt_f \\ p_f(T_D) &= \int_0^{T_D} f_{T_f}(t_f)\, dt_f. \end{aligned}$$

9.2. FIRST EXCURSION FAILURE

Therefore, the probability to failure is also the cumulative distribution,

$$p_f(T_D) = F_{T_f}(T_D). \tag{9.1}$$

The reliability and the probability of time to failure are the areas under the probability density function of T_f, as shown in Figure 9.2. In this chapter it is assumed that the design lifetime T_D is a fixed number.

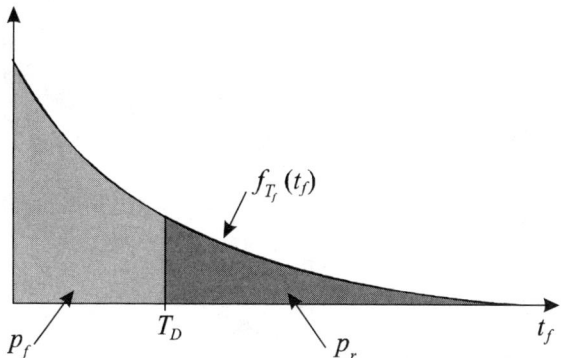

Figure 9.2: Reliability and probability of time to failure.

The general problem of obtaining a probability density function for the first arrival time, $f_{T_f}(t_f)$, is called the *first passage time problem*. There are numerous forms of $f_{T_f}(t_f)$ depending on the assumptions made. We will consider exponential, gamma, normal, and Weibull failure laws. The exponential failure law is the most common failure law, and we will show how it is derived in the next section.

9.2 First Excursion Failure

Suppose that a component is subjected to random forces, and parameters such as displacement, acceleration, stress at the critical location, compressive load, or the stress intensity factor at a crack opening are monitored. Define any one of these parameters $X(t)$ and assume that it is a weakly stationary random process. Every now and then, $X(t)$ may exceed a desirable range and be recorded as a "disturbance" as shown in Figure 9.3. The cause of the disturbance may be a sudden gust of wind, a bump on the road, or turbulence in the air.

Assume that these disturbances are independent of each other. That is, a bump on the road at point A does not imply a bump at point B and vice versa. Also assume that the component will fail when the first disturbance occurs. This type of failure is called a *first excursion failure*. Although a first excursion failure of a structure is often destructive so that $X(t)$ after the failure may not even exist, we assume for now that the structure can recover immediately after the failure so that $X(t)$ can still be monitored.

Let $N_Z(t, t + \Delta t)$ be the number of times that the sample realization crosses Z during the interval $(t, t+\Delta t)$. For example, $N_Z(0, a) = 1$ in Figure 9.3. The subscript Z is used to indicate that the threshold for $X(t)$ is Z. Since the random process is stationary, the number of crossings depends only on the length of the interval. Therefore, we write $N_Z(\Delta t)$ instead of $N_Z(t, t + \Delta t)$, where t is arbitrary. $N_Z(0, t)$ is denoted as $N_Z(t)$, which is a random process, whose probability density function $p_N(n, t)$ varies with time. The number of crossings over a fixed time interval $N_Z(t_o)$ is a random variable.

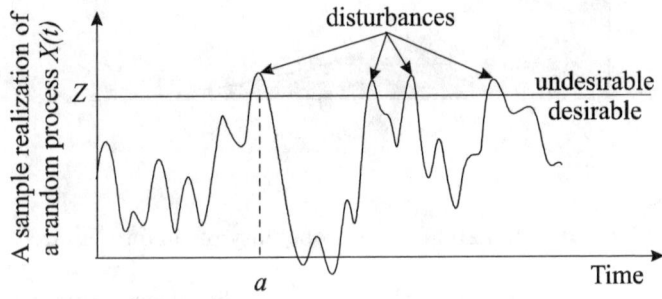

Figure 9.3: A sample realization of $X(t)$ as it wanders into undesirable behavior.

If we assume that each disturbance is independent of other disturbances and the arrival rate is constant, then the number of arrivals N is governed by the Poisson probability density function,

$$p_N(n, t) = \frac{(\nu t)^n e^{-\nu t}}{n!}, \qquad n \geq 0, \ t \geq 0,$$

where ν is the arrival rate, the number of arrivals per unit time. Note that the term νt is the number of arrivals in a given time and was denoted as λ in Section 3.4.6. $p_N(n, t)$ is the probability of exactly n arrivals during $(0, t)$.

For example, $p_N(0, t)$ is the probability of no arrival during $(0, t)$, and $p_N(1, t)$ is the probability of exactly 1 arrival during $(0, t)$. The probabilities of zero and one arrival ($n = 0$ and $n = 1$) are plotted as functions of time in Figure 9.4. The probability of zero arrival is 1 when $t = 0$, and 0 when $t \to \infty$, as they should be. On the other hand, the probability of one arrival

9.2. FIRST EXCURSION FAILURE

is zero for both $t = 0$ and $t \to \infty$. As $t \to \infty$, it is likely that there will be more than one arrival. There is an optimum time t when the probability of one arrival is maximum.

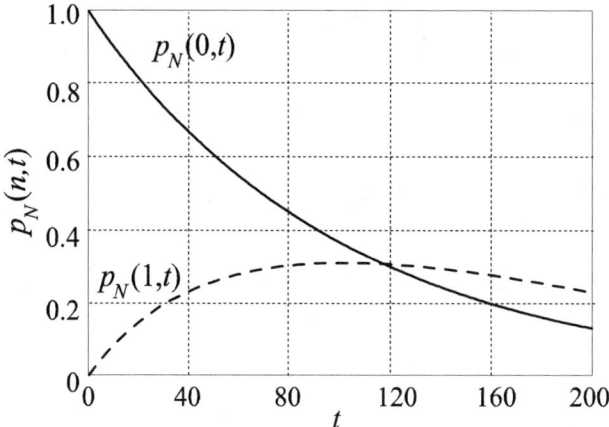

Figure 9.4: The probability of zero or one arrival in time t for $\lambda = 0.01$.

For any time interval t, the summation of the probabilities of 0 to ∞ arrivals must add up to one,

$$\begin{aligned}
\sum_{n=0}^{\infty} p_N(n,t) &= \sum_{n=0}^{\infty} \frac{(\nu t)^n e^{-\nu t}}{n!} \\
&= e^{-\nu t}\left(1 + \frac{(\nu t)^1}{1!} + \frac{(\nu t)^2}{2!} + \frac{(\nu t)^2}{3!} + ...\right) \\
&= e^{-\nu t}\left(e^{\nu t}\right) = 1.
\end{aligned}$$

The summation of the probabilities of 0 to M arrivals is the cumulative distribution, given by

$$\begin{aligned}
P_N(M,t) &= \sum_{n=0}^{M} p_N(n,t) \\
&= \sum_{n=0}^{M} \frac{(\nu t)^n e^{-\nu t}}{n!}, \qquad n \geq 0,\ t \geq 0.
\end{aligned}$$

To take advantage of the properties of the Poisson process, we realize that each disturbance can be counted each time $X(t)$ crosses Z with a positive slope. Therefore, the up-crossing rate, ν_Z^+, is the arrival rate ν in the Poisson process. The superscript $+$ is used to indicate that it is the up-crossing rate. Similarly, the down-crossing rate is denoted as ν_Z^-.

The probability density function for the number of up-crossings, N_Z^+, is given by

$$p_{N_Z^+}(n,t) = \frac{\left(\nu_Z^+ t\right)^n e^{-\nu_Z^+ t}}{n!}, \qquad n \geq 0,\ t \geq 0. \tag{9.2}$$

The mean value and the variance of N_Z^+ are given by

$$\mu_{N_Z^+} = \sum_{i=0}^{\infty} n p_{N_Z^+}(n,t) = \nu_Z^+ t \tag{9.3}$$

$$\sigma^2_{N_Z^+} = \sum_{n=0}^{\infty} n^2 p_{N_Z^+}(n,t) - \mu^2_{N_Z^+} = \nu_Z^+ t. \tag{9.4}$$

Reliability is the probability that the component does not fail for $t \leq T_D$. Equivalently, it is the probability that there is no up-crossing for $t \leq T_D$. Mathematically,

$$\begin{aligned} p_r &= p_{N_Z^+}(0, T_D) \\ &= e^{-\nu_Z^+ T_D}. \end{aligned}$$

The probability of failure is then

$$p_f = 1 - e^{-\nu_Z^+ T_D}.$$

Recalling that the probability of failure is the cumulative distribution as shown in Equation 9.1, then

$$\begin{aligned} F_{T_f}(T_D) &= \int_0^{T_D} f_{T_f}(t_f)\,dt_f \\ &= 1 - e^{-\nu_Z^+ T_D}, \end{aligned}$$

from which we deduce that

$$f_{T_f}(t_f) = \frac{dF_{T_f}(t_f)}{dt_f} = \nu_Z^+ e^{-\nu_Z^+ t_f}, \qquad t_f \geq 0, \tag{9.5}$$

which is the exponential failure law. The only required parameter is the up-crossing rate ν_Z^+. The expression for ν_Z^+ will be derived in Section 9.2.3. It should be noted that the crossing rate ν_Z^+ is typically a small number so that the first crossing occurs at a reasonably large t_f.

9.2. FIRST EXCURSION FAILURE

Example 9.1 Mean and Variance of the Poisson Process

Derive Equations 9.3 and 9.4.
Solution The mean value is given by

$$\mu_{N_Z^+} = \sum_{n=0}^{\infty} n p_{N_Z^+}(n,t)$$

$$= \sum_{n=0}^{\infty} n \frac{\left(\nu_Z^+ t\right)^n e^{-\nu_Z^+ t}}{n!}.$$

Note that the first term equals zero for $n = 0$, so the index can start from $n = 1$,

$$\mu_{N_Z^+} = e^{-\nu_Z^+ t} \sum_{n=1}^{\infty} \frac{\left(\nu_Z^+ t\right)^n}{(n-1)!} = \left(\nu_Z^+ t\right) e^{-\nu_Z^+ t} \sum_{n=1}^{\infty} \frac{\left(\nu_Z^+ t\right)^{n-1}}{(n-1)!}.$$

Let $k = n - 1$,

$$\mu_{N_Z^+} = \left(\nu_Z^+ t\right) e^{-\nu_Z^+ t} \sum_{k=0}^{\infty} \frac{\left(\nu_Z^+ t\right)^k}{k!}.$$

The summation $\sum_{k=0}^{\infty} \left(\nu_Z^+ t\right)^k / k!$ can be replaced by $\exp\left(\nu_Z^+ t\right)$,

$$\mu_{N_Z^+} = \left(\nu_Z^+ t\right) e^{-\nu_Z^+ t} e^{\nu_Z^+ t} = \nu_Z^+ t.$$

Similarly, the variance is given by

$$\sigma_{N_Z^+}^2 = \sum_{n=0}^{\infty} n^2 p_{N_Z^+}(n,t) - \mu_{N_Z^+}^2.$$

The first term can be written as

$$\sum_{n=0}^{\infty} n^2 p_{N_Z^+}(n,t) = \sum_{n=0}^{\infty} n \frac{\left(\nu_Z^+ t\right)^n e^{-\nu_Z^+ t}}{(n-1)!}$$

$$= \sum_{n=1}^{\infty} n \frac{\left(\nu_Z^+ t\right)^n e^{-\nu_Z^+ t}}{(n-1)!}.$$

Let $k = n - 1$,

$$\sum_{n=0}^{\infty} n^2 p_{N_Z^+}(n,t) = \left(\nu_Z^+ t\right) e^{-\nu_Z^+ t} \sum_{k=0}^{\infty} (k+1) \frac{\left(\nu_Z^+ t\right)^k}{k!}$$

$$= \left(\nu_Z^+ t\right) e^{-\nu_Z^+ t} \left[\sum_{k=0}^{\infty} k \frac{\left(\nu_Z^+ t\right)^k}{k!} + \sum_{k=0}^{\infty} \frac{\left(\nu_Z^+ t\right)^k}{k!}\right],$$

and $l = k - 1$,

$$\sum_{n=0}^{\infty} n^2 p_{N_Z^+}(n,t) = (\nu_Z^+ t) e^{-\nu_Z^+ t} \left[(\nu_Z^+ t) \sum_{l=0}^{\infty} \frac{(\nu_Z^+ t)^l}{l!} + \sum_{k=0}^{\infty} \frac{(\nu_Z^+ t)^k}{k!} \right]$$
$$= (\nu_Z^+ t) e^{-\nu_Z^+ t} \left((\nu_Z^+ t) e^{\nu_Z^+ t} + e^{\nu_Z^+ t} \right)$$
$$= (\nu_Z^+ t)(\nu_Z^+ t + 1).$$

Then,

$$\sigma_{N_Z^+}^2 = (\nu_Z^+ t)^2 + (\nu_Z^+ t) - (\nu_Z^+ t)^2$$
$$= \nu_Z^+ t.$$

⊛

9.2.1 Exponential Failure Law

We found that the time to failure is distributed according to the exponential law when the disturbances of a weakly stationary process are distributed according to a Poisson process. The probability density function of T_f is given by Equation 9.5,

$$f_{T_f}(t_f) = \nu_Z^+ e^{-\nu_Z^+ t_f}, \qquad t_f \geq 0.$$

Note that the exponential failure law is entirely characterized by this constant rate, ν_Z^+.

The exponential density was previously discussed in Section 3.4.2. The mean value and the standard deviation of time to failure are given by

$$E\{T_f\} = \int_0^\infty \nu_Z^+ t_f e^{-\nu_Z^+ t_f} dt_f = \frac{1}{\nu_Z^+}$$
$$Var\{T_f\} = \frac{1}{(\nu_Z^+)^2}.$$

The mean time to failure is the inverse of the rate, which makes sense intuitively. For example, the disturbance rate of 0.01 means 0.01 disturbances per second. Therefore, in 100 s, we can expect to have one disturbance. It should be noted that the variance is relatively large, which indicates that the time to failure is spread over a large range.

The probability density function of T_f, $f_{T_f}(t_f)$, is shown in Figure 9.5. The mean time to failure is $1/\nu_Z^+ = 100$ s and the variance is $1/(\nu_Z^+)^2 = 10000$ s^2.

9.2. FIRST EXCURSION FAILURE

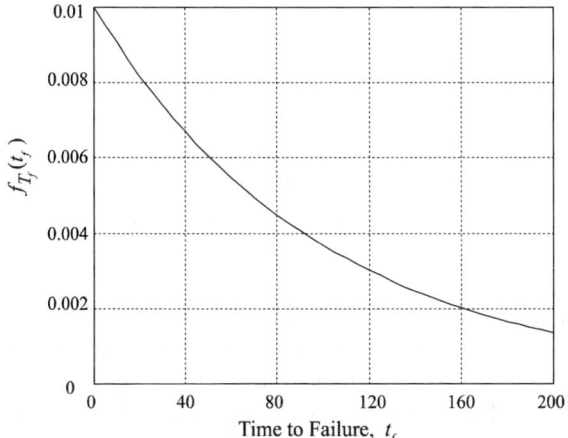

Figure 9.5: The probability density function for time to failure, $f_{T_f}(t_f) = \nu_Z^+ e^{-\nu_Z^+ t_f}$ with $\nu_Z^+ = 0.01$.

The constant failure rate ν_Z^+ is interpreted to mean that after the item has been in use, its reliability has not changed. That is, there is no 'wearing-out' effect and the probability is "memoryless." For example, if a steel beam in a skyscraper has not yet buckled, we can assume that it is as good as new. Many component follow the exponential failure law. This is demonstrated in the next example.

Example 9.2 Memoryless Probability

Express the following probabilities: *(i)* the probability that a component will work for at least t hours, *(ii)* the probability that a component will work t more hours given that it has already worked successfully for τ hr, and *(iii)* what conclusion can be drawn from *(ii)*?

Solution *(i)* The probability that a component will work for at least t hr is, by definition, the reliability $p_r(t)$,

$$\begin{aligned} \Pr(T_f \geq t) &= \int_t^\infty f_{T_f}(t_f)\, dt_f \\ &= \int_t^\infty \nu_Z^+ \exp\left(-\nu_Z^+ t_f\right) dt_f \\ &= \exp\left(-\nu_Z^+ t\right). \end{aligned}$$

(ii) The probability that a component will work t more hours given that it has already worked successfully for τ hr is the conditional probability of

event $(T_f \geq t + \tau | T_f \geq \tau)$,

$$\Pr(T_f \geq t + \tau | T_f \geq \tau) = \frac{\Pr([T_f \geq t + \tau] \cap [T_f \geq \tau])}{\Pr(T_f \geq \tau)}$$

$$= \frac{\Pr(T_f \geq t + \tau)}{\Pr(T_f \geq \tau)}$$

$$= \frac{\exp\left(-\nu_Z^+ t - \nu_Z^+ \tau\right)}{\exp\left(-\nu_Z^+ \tau\right)} = \exp\left(-\nu_Z^+ t\right).$$

(iii) Comparing the two probabilities, we find that the exponential distribution is a memoryless distribution. The next t hr will be like the last t hr. If a component has an exponential distribution, then the probability that it will operate for at least t more hours is the same for an old component as for a new component. Consider 100 new components that are designed to last 10 hr with reliability 0.90. After 10 hr, we can expect that 90 components are still working. After another 10 hr, we can expect that 90% of the previous 90 components or 81 components are functioning.

While the rate of component failure is constant, eventually, as a practical matter, all components will fail.

⊛

Now, formally define the failure rate, ν_Z^+. First consider the probability that a component will fail within dt given that it has already worked successfully for t. In terms of disturbances, this is the probability that the disturbance will occur between t and $t + dt$ given that no disturbance occurred up to t_f, or

$$\Pr(t \leq T_f \leq t + dt | T_f \geq t).$$

Using the definition of conditional probability we can write

$$\Pr(t \leq T_f \leq t + dt | T_f \geq t) = \frac{\Pr(t \leq T_f \leq t + dt)}{\Pr(T_f \geq t)}$$

$$= \frac{1}{p_r(t)} \int_t^{t+dt} f_{T_f}(t_f) \, dt_f,$$

where the denominator is recognized as the reliability, the probability that the component will function until time t. As $dt \to 0$, the numerator can be approximated by

$$\int_t^{t+dt} f_{T_f}(t_f) \, dt_f \simeq f_{T_f}(t) \, dt.$$

Divide by dt,

$$\frac{1}{dt} \Pr(t \leq T_f \leq t + dt | T_f \geq t) = \frac{f_{T_f}(t)}{p_r(t)},$$

9.2. FIRST EXCURSION FAILURE

and define the instantaneous failure rate by

$$\nu(t) \equiv \frac{1}{dt} \Pr\left(t \leq T_f \leq t + dt \mid T_f \geq t\right)$$
$$= \frac{f_{T_f}(t)}{p_r(t)}.$$

Then,

$$\nu(t) = \frac{f_{T_f}(t)}{1 - F_{T_f}(t)}. \tag{9.6}$$

Integrate Equation 9.6,

$$\int_0^t \nu(s)\,ds = -\ln\left(1 - F_{T_f}(s)\right)\Big|_0^t$$
$$= -\ln\left(1 - F_{T_f}(t)\right) + \ln\left(1 - F_{T_f}(0)\right).$$

If $F_{T_f}(0) = 0$, then the following relationships hold,

$$F_{T_f}(t) = 1 - \exp\left(-\int_0^t \nu(s)\,ds\right)$$
$$f_{T_f}(t) = \frac{d}{dt} F_{T_f}(t)$$
$$= \nu(t) \exp\left(-\int_0^t \nu(s)\,ds\right). \tag{9.7}$$

For the exponential failure law, we can verify that the failure rate is

$$\nu(t) = \frac{f_{T_f}(t)}{1 - F_{T_f}(t)}$$
$$= \frac{\nu_Z^+ e^{-\nu_Z^+ t}}{e^{-\nu_Z^+ t}}$$
$$= \nu_Z^+.$$

9.2.2 Modified Exponential Failure Law

Suppose that disturbances appear according to the Poisson law. Also suppose that at the first encounter with a disturbance, the part will survive with a probability p instead of failing right away. That is, a component will function properly until time t while no disturbance occurs, or one disturbance with no failure, or two disturbances with no failures from both, and so on. The probability of zero disturbance in time t is $p_N(0, t)$, where $p_N(0, t)$ is given by Equation 9.2. The probability of one disturbance with no failure is $p_N(1,t)p$, and the probability of two disturbances with no failure is $p_N(2,t)p^2$. Then the probability that a component will function properly during $[0,t]$ is

$$\begin{aligned}\Pr(T_f > t) &= p_N(0,t) + p_N(1,t)p + p_N(2,t)p^2 + \cdots \\ &= \sum_{n=0}^{\infty} \frac{(\nu_Z^+ t)^n e^{-\nu_Z^+ t}}{n!} p^n \\ &= \exp\left[-\nu_Z^+ (1-p)t\right],\end{aligned}$$

and the probability that the component will fail during $[0,t]$ is

$$\begin{aligned}\Pr(T_f \leq t) &= 1 - \Pr(T_f > t) \\ &= 1 - \exp\left[-\nu_Z^+ (1-p)t\right].\end{aligned}$$

The probability density of T_f is given by

$$\begin{aligned}f_{T_f}(t_f) &= \frac{dF_{T_f}(t_f)}{dt_f} = \frac{d}{dt_f} \Pr(T_f \leq t_f) \\ &= \nu_Z^+ (1-p) \exp\left[-\nu_Z^+ (1-p)t_f\right], \qquad t_f \geq 0,\end{aligned}$$

which is an exponential density with failure rate $\nu_Z^+ (1-p)$. If the component fails with probability 1 when it encounters a disturbance ($p=0$), then the failure rate becomes ν_Z^+.

9.2.3 Calculation of Up-Crossing Rate

In this section, the expression for the up-crossing rate ν_Z^+ of a weakly stationary process $X(t)$ is obtained. First, let A be the event that a sample realization of $X(t)$ crosses Z with a positive slope in dt,

$$A = \{X(t) \text{ crosses } Z \text{ with a positive slope during } (t, t+dt)\},$$

with probability p,

$$p = \Pr(A).$$

Consider a limiting case where the interval becomes infinitesimal so that there can be either one or zero up-crossing in dt as shown in Figure 9.6. Therefore, the number of up-crossings in dt, $N_Z^+(dt)$, can be either one or zero. That is,

$$\begin{aligned}\Pr(N_Z^+(dt) = 1) &= p \\ \Pr(N_Z^+(dt) = 0) &= 1-p.\end{aligned}$$

9.2. FIRST EXCURSION FAILURE

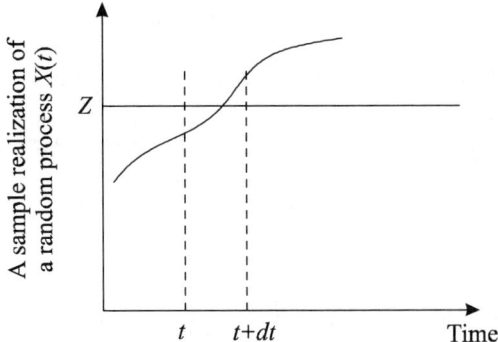

Figure 9.6: Level crossing with positive slope in dt.

The expected value of $N_Z^+ dt$ is given by

$$E\left\{N_Z^+ (dt)\right\} = p \times 1 + (1-p) \times 0 = p.$$

Assuming that the expected value is proportional to the length of the interval with the proportionality constant equal to the crossing rate,

$$E\left\{N_Z^+ (dt)\right\} = \nu_Z^+ dt,$$

the probability p is then

$$p = \nu_Z^+ dt. \tag{9.8}$$

Given p, the up-crossing rate ν_Z^+ can be obtained.

Go back to the event A to find p. Event A is composed of three events. They are:

1. $X(t) < Z$ in the beginning of the interval
2. $\dot{X}(t) > 0$ in the beginning of the interval, and
3. $X(t+dt) > Z$ at the end of the interval.

$X(t+dt)$ can be expanded about t using the Taylor series,

$$X(t+dt) = X(t) + \dot{X}(t)\,dt + \ddot{X}(t)\frac{dt^2}{2!} + \cdots .$$

Taking the first two terms of the Taylor series, the event $\{X(t+dt) > Z\}$, can be written as

$$\left\{X(t) + \dot{X}(t)\,dt > Z\right\} \quad \text{or} \quad \left\{X(t) > Z - \dot{X}(t)\,dt\right\}.$$

Combining all three events, event A can be written as

$$A = \left\{(X(t) < Z) \cap \left(Z - \dot{X}(t)\, dt < X(t)\right) \cap \dot{X}(t) > 0\right\}$$
$$= \left\{\left(Z - \dot{X}(t)\, dt < X(t) < Z\right) \cap \dot{X}(t) > 0\right\}.$$

That is,

$$\Pr(A) = p$$
$$= \int_0^\infty \int_{Z-\dot{x}dt}^Z f_{X\dot{X}}(x, \dot{x})\, dx\, d\dot{x}.$$

This integral can be simplified if dt tends to zero. As $dt \to 0$, the lower limit approaches the upper limit. Therefore, the integrand can be evaluated at $x = Z$. Replace dx by the difference between the upper and lower limits, $Z - (Z - \dot{x}\, dt)$. Then,

$$\int_{Z-\dot{x}dt}^Z f_{X\dot{X}}(x, \dot{x})\, dx \simeq f_{X\dot{X}}(Z, \dot{x})\, \dot{x}\, dt,$$

so that

$$p = \int_0^\infty f_{X\dot{X}}(Z, \dot{x})\, \dot{x}\, d\dot{x}\, dt. \qquad (9.9)$$

Comparing Equations 9.8 and 9.9, the up-crossing rate is given by

$$\nu_Z^+ = \int_0^\infty f_{X\dot{X}}(Z, \dot{x})\, \dot{x}\, d\dot{x}. \qquad (9.10)$$

Similarly, the down-crossing rate is given by

$$\nu_Z^- = -\int_{-\infty}^0 f_{X\dot{X}}(Z, \dot{x})\, \dot{x}\, d\dot{x}, \qquad (9.11)$$

and the total crossing rate is

$$\nu_Z = \int_{-\infty}^\infty f_{X\dot{X}}(Z, \dot{x})\, |\dot{x}|\, d\dot{x}.$$

Note that the derivation of the crossing rate does not require the crossings to be independent of each other. They are valid as long as $X(t)$ is weakly stationary. The Poisson model only applies when the crossings are independent.

9.2. FIRST EXCURSION FAILURE

Example 9.3 Calculation of Crossing Rates

Assume that a machine part will fail due to both excessive tension and compression. *(i)* Find the total crossing rate if the part fails when the stress $|X(t)| \geq Z$. *(ii)* Also find the total crossing rate if the part fails when the tensile yield strength is Z_1 and the compressive yield strength is $-Z_2$.

Solution   *(i)* The total rate is the summation of the rates when $X(t)$ up-crosses Z and down-crosses $-Z$. Using Equations 9.10 and 9.11, we find that

$$\nu_Z^+ = \int_0^\infty f_{X\dot{X}}(Z,\dot{x})\,\dot{x}\,d\dot{x}$$

$$\nu_{-Z}^- = -\int_{-\infty}^0 f_{X\dot{X}}(-Z,\dot{x})\,\dot{x}\,d\dot{x},$$

and the total rate equals

$$\nu = \int_0^\infty f_{X\dot{X}}(Z,\dot{x})\,\dot{x}\,d\dot{x} - \int_{-\infty}^0 f_{X\dot{X}}(-Z,\dot{x})\,\dot{x}\,d\dot{x}$$

$$= \int_0^\infty [f_{X\dot{X}}(Z,\dot{x}) - f_{X\dot{X}}(-Z,-\dot{x})]\,\dot{x}\,d\dot{x}.$$

(ii) Similarly, the total rate is the summation of the rates when $X(t)$ up-crosses Z_1 and down-crosses $-Z_2$. Using Equations 9.10 and 9.11, we find that

$$\nu_{Z_1}^+ = \int_0^\infty f_{X\dot{X}}(Z_1,\dot{x})\,\dot{x}\,d\dot{x}$$

$$\nu_{-Z_2}^- = -\int_{-\infty}^0 f_{X\dot{X}}(-Z_2,\dot{x})\,\dot{x}\,d\dot{x},$$

and the total rate equals

$$\nu = \int_0^\infty f_{X\dot{X}}(Z_1,\dot{x})\,\dot{x}\,d\dot{x} - \int_{-\infty}^0 f_{X\dot{X}}(-Z_2,\dot{x})\,\dot{x}\,d\dot{x}$$

$$= \int_0^\infty [f_{X\dot{X}}(Z_1,\dot{x}) - f_{X\dot{X}}(-Z_2,-\dot{x})]\,\dot{x}\,d\dot{x}.$$

Now given the joint density function $f_{X\dot{X}}(Z,\dot{x})$ the crossing rate ν can be calculated, either exactly or numerically. The following examples show how this is done for a Gaussian process.
⊛

Example 9.4 Up-Crossing Rate of a Gaussian Process

Obtain the up-crossing rate, ν_Z^+, of Gaussian process $X(t)$ with zero mean. Also find the zero up-crossing rate.

Solution   The joint probability density of $X(t)$ and $\dot{X}(t)$ is given by

$$f_{X\dot{X}}(x,\dot{x}) = \frac{1}{2\pi\sigma_X\sigma_{\dot{X}}} \exp\left[-\frac{1}{2}\left(\frac{x}{\sigma_X}\right)^2 - \frac{1}{2}\left(\frac{\dot{x}}{\sigma_{\dot{X}}}\right)^2\right],$$
$$-\infty < x < \infty, \quad -\infty < \dot{x} < \infty.$$

The up-crossing rate is given by

$$\begin{aligned} \nu_Z^+ &= \int_0^\infty f_{X\dot{X}}(Z,\dot{x})\, \dot{x}\, d\dot{x}. \\ &= \frac{1}{2\pi\sigma_X\sigma_{\dot{X}}} \exp\left[-\frac{1}{2}\left(\frac{Z}{\sigma_X}\right)^2\right] \int_0^\infty \exp\left[-\frac{1}{2}\left(\frac{\dot{x}}{\sigma_{\dot{X}}}\right)^2\right] \dot{x}\, d\dot{x} \\ &= \frac{\sigma_{\dot{X}}}{2\pi\sigma_X} \exp\left[-\frac{1}{2}\left(\frac{Z}{\sigma_X}\right)^2\right], \end{aligned}$$

and the zero up-crossing rate is found by setting $Z = 0$,

$$\nu_0^+ = \frac{\sigma_{\dot{X}}}{2\pi\sigma_X}. \tag{9.12}$$

Then, the up-crossing rate in terms of ν_0^+ can be written as

$$\nu_Z^+ = \nu_0^+ \exp\left[-\frac{1}{2}\left(\frac{Z}{\sigma_X}\right)^2\right].$$

※

Example 9.5 Up-Crossing Rate of a Gaussian Process: Vibrating Structure

Consider a package containing sensitive equipment transported by the truck shown in Figure 9.7. The packing materials can be modeled as a spring and a damper. Assume that the truck is moving at a constant velocity. The package is subject to vertical movement due to the irregular road surface. The irregularities $Y(t)$ can be modeled as a stationary Gaussian white noise process with intensity S_o. Assume that the equipment will be damaged if the relative acceleration exceeds a_{cr}. Find the up-crossing rate for a_{cr}.

9.2. FIRST EXCURSION FAILURE

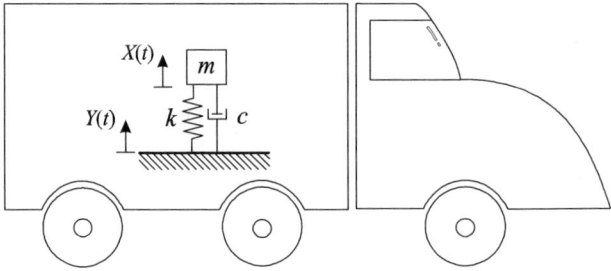

Figure 9.7: Relative motion of equipment in a moving truck.

Solution The quantity of interest is the relative acceleration A, given by
$$A = \ddot{X} - \ddot{Y}.$$
Since X and Y are Gaussian, then so is A, and
$$f_{A\dot{A}}(a, \dot{a}) = \frac{1}{2\pi\sigma_A\sigma_{\dot{A}}} \exp\left[-\frac{1}{2}\left(\frac{a}{\sigma_A}\right)^2 - \frac{1}{2}\left(\frac{\dot{a}}{\sigma_{\dot{A}}}\right)^2\right],$$
$$-\infty < a < \infty, \quad -\infty < \dot{a} < \infty.$$

From Equation 9.10, the up-crossing rate is given by
$$\begin{aligned}
\nu^+_{A=a_{cr}} &= \int_0^\infty f_{A\dot{A}}(a_{cr}, \dot{a})\, \dot{a}\, d\dot{a}. \\
&= \frac{1}{2\pi\sigma_A\sigma_{\dot{A}}} \exp\left[-\frac{1}{2}\left(\frac{a_{cr}}{\sigma_A}\right)^2\right] \int_0^\infty \exp\left[-\frac{1}{2}\left(\frac{\dot{a}}{\sigma_{\dot{A}}}\right)^2\right] \dot{a}\, d\dot{a} \\
&= \frac{\sigma_{\dot{A}}}{2\pi\sigma_A} \exp\left[-\frac{1}{2}\left(\frac{a_{cr}}{\sigma_A}\right)^2\right],
\end{aligned}$$
where
$$\sigma_A = \sqrt{\int_{-\infty}^\infty S_{AA}(\omega)\, d\omega}$$
$$\sigma_{\dot{A}} = \sqrt{\int_{-\infty}^\infty \omega^2 S_{AA}(\omega)\, d\omega}.$$

To find the spectral density for the relative acceleration, denote W as the relative displacement so that $\ddot{W} = A$ and $W = X - Y$. Then the equation of motion can be written as

$$m\ddot{X} + c\left(\dot{X} - \dot{Y}\right) + k(X - Y) = 0$$
$$m\ddot{W} + c\dot{W} + kW = -m\ddot{Y}.$$

Taking the Fourier transform, we find

$$\frac{W(\omega)}{Y(\omega)} = H(\omega) = \frac{\omega^2}{-\omega^2 + 2i\zeta\omega_n\omega + \omega_n^2},$$

where $W(\omega)$ and $Y(\omega)$ are the Fourier transforms of $W(t)$ and $Y(t)$ and

$$\omega_n^2 = \frac{k}{m}, \quad 2\zeta\omega_n = \frac{c}{m}.$$

The spectral density of $W(t)$ is then given by

$$S_{WW}(\omega) = |H(\omega)|^2 S_{YY}(\omega)$$
$$= \left|\frac{\omega^2}{-\omega^2 + 2i\zeta\omega_n\omega + \omega_n^2}\right|^2 S_o, \quad -\infty < \omega < \infty.$$

Using the result obtained in Example 5.14 in Chapter 5, the spectral density of $\ddot{W}(t)$ equals

$$S_{AA}(\omega) = S_{\ddot{W}\ddot{W}}(\omega) = \omega^4 S_{WW}(\omega)$$
$$= \left|\frac{\omega^2}{-\omega^2 + 2i\zeta\omega_n\omega + \omega_n^2}\right|^2 S_o.$$

The standard deviations σ_A and $\sigma_{\dot{A}}$ are then

$$\sigma_A = \sqrt{\int_{-\infty}^{\infty} \omega^8 \left|\frac{\omega^2}{-\omega^2 + 2i\zeta\omega_n\omega + \omega_n^2}\right|^2 S_o d\omega}$$

$$\sigma_{\dot{A}} = \sqrt{\int_{-\infty}^{\infty} \omega^{10} \left|\frac{\omega^2}{-\omega^2 + 2i\zeta\omega_n\omega + \omega_n^2}\right|^2 S_o d\omega}.$$

⊛

9.2.4 Narrow-Band Process – Envelope Function

When the process is narrow band, the crossings occur in clumps or packs as shown in Figure 9.8. That is, once threshold Z is crossed, it is highly likely that it will be crossed again. In this case, the crossings are not independent of each other. Therefore, it is not correct to model $N_{X=Z^+}(t)$ as a Poisson

9.2. FIRST EXCURSION FAILURE

process. Instead, Rice[1] has defined an *envelope function* shown in Figure 9.8. From the figure, $X(t)$ will cross the threshold Z approximately when

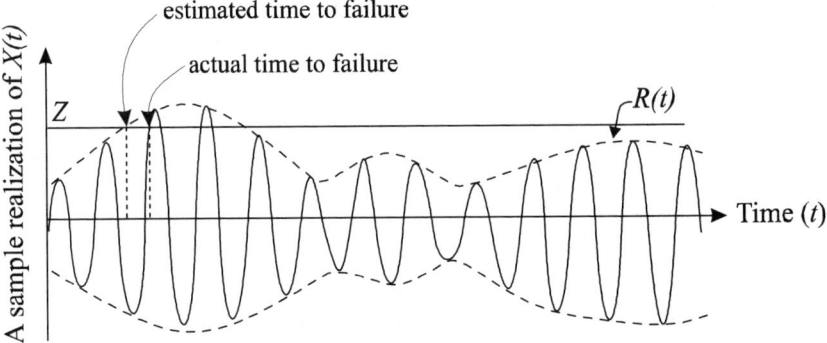

Figure 9.8: Narrow-band process $X(t)$ and its envelope function $R(t)$.

the envelope function $R(t)$ crosses Z. Assuming that the crossings are independent, the number of crossings of the envelope function, $N_{R=Z^+}$, can be modeled as a Poisson process. Then the probability density function of the number of up-crossings is given by Equation 9.2, or

$$p_{N_{R=Z^+}}(n,t) = \frac{\left(\nu_{R=Z}^+ t\right)^n \exp\left(-\nu_{R=Z}^+ t\right)}{n!}, \; n \geq 0, \; t \geq 0,$$

where the crossing rate is given by

$$\nu_{R=Z}^+ = \int_0^\infty f_{R\dot{R}}(Z,\dot{r})\dot{r}\,d\dot{r}. \tag{9.13}$$

The reliability is given by

$$\begin{aligned} p_r &= p_{N_{R=Z^+}}(0, T_D) \\ &= \exp\left(-\nu_{R=Z}^+ T_D\right), \end{aligned} \tag{9.14}$$

where T_D is the design lifetime. The average time to the first crossing is

$$E\{T_f\} = \frac{1}{\nu_{R=Z}^+}, \tag{9.15}$$

with variance

$$Var\{T_f\} = \frac{1}{\left(\nu_{R=Z}^+\right)^2}.$$

---

[1] S.O. Rice, "Mathematical Analysis of Random Noise," *Bell Syst. Tech. J.*, **23**, 282-332; **24**, 46-156; reprinted in N. Wax, *Selected Papers on Noise and Stochastic Processes*, Dover, New York, 1954. See pp. 213-216.

To obtain the up-crossing rate of the envelope, the joint probability density function is evaluated at $r = Z$, that is, $f_{R\dot{R}}(Z, \dot{r})$. In Section 9.2.5, we will find that $f_{R\dot{R}}(r, \dot{r})$ for a narrow-band Gaussian process $X(t)$ is given by

$$f_{R\dot{R}}(r, \dot{r}) = \frac{r}{\sqrt{2\pi}\sigma_X^2\sqrt{\sigma_{\dot{X}}^2 - \omega_m^2\sigma_X^2}} \exp\left(-\frac{1}{2}\left[\frac{r^2}{\sigma_X^2} + \frac{\dot{r}^2}{\sigma_{\dot{X}}^2 - \omega_m^2\sigma_X^2}\right]\right)$$

$$r \geq 0, \ -\infty < \dot{r} < \infty,$$

where

$$\omega_m = \frac{\int_0^\infty \omega S_{XX}^o(\omega)\, d\omega}{\int_0^\infty S_{XX}^o(\omega)\, d\omega}$$

$$\sigma_X^2 = \int_0^\infty S_{XX}^o(\omega)\, d\omega$$

$$\sigma_{\dot{X}}^2 = \int_0^\infty \omega^2 S_{XX}^o(\omega)\, d\omega,$$

and $S_{XX}^o(\omega)$ is a one-sided spectrum. Then the up-crossing rate for a narrow band process is given by

$$\nu_{R=Z}^+ = \int_0^\infty f_{R\dot{R}}(Z, \dot{r})\dot{r}\, d\dot{r}$$

$$= \frac{Z\sqrt{\sigma_{\dot{X}}^2 - \omega_m^2\sigma_X^2}}{\sqrt{2\pi}\sigma_X^2} \exp\left(-\frac{1}{2}\frac{Z^2}{\sigma_X^2}\right).$$

9.2.5 Rice's Envelope Function for Gaussian Narrow-Band Process $X(t)$

First assume that the narrow-band Gaussian process $X(t)$ can be written as

$$X(t) = R(t)\cos(\omega_m t - \Phi(t)),$$

where ω_m is the midband frequency, $R(t)$ is the envelope function, and $\Phi(t)$ is the phase angle.[2] The envelope function and the phase angle vary much more slowly than $X(t)$ with respect to t. We can assume $R(t)$ is positive and $\Phi(t)$ ranges between 0 and 2π without loss of generality. The goal is to obtain the joint probability density function $f_{R\dot{R}}(r, \dot{r})$. Once the joint

---

[2] Here, $\Phi(t)$ represents a random phase angle. The same variable is also sometimes used to represent a standard normal random variable. Since both are common usages of this variable, we retain that usage here in the hope that the context will help avoid confusion.

9.2. FIRST EXCURSION FAILURE

probability density function is found, the up-crossing rate of the envelope function is given by Equation 9.13. Then the reliability and the mean time to failure (Equations 9.14 and 9.15) can be obtained using the assumption that the up-crossings of the envelope function are distributed according to the Poisson process.

Random process $X(t)$ can also be written as

$$X(t) = C(t)\cos\omega_m t - S(t)\sin\omega_m t, \qquad (9.16)$$

where

$$\begin{aligned} C(t) &= R(t)\cos\Phi(t) \\ S(t) &= R(t)\sin\Phi(t), \end{aligned}$$

and the derivative of $X(t)$ is

$$\begin{aligned} \dot{X}(t) &= \dot{C}(t)\cos\omega_m t - \omega_m C(t)\sin\omega_m t \\ &\quad - \dot{S}(t)\sin\omega_m t - \omega_m S(t)\cos\omega_m t, \end{aligned}$$

where

$$\begin{aligned} \dot{C}(t) &= \dot{R}(t)\cos\Phi(t) - R(t)\dot{\Phi}(t)\sin\Phi(t) \\ \dot{S}(t) &= \dot{R}(t)\sin\Phi(t) + R(t)\dot{\Phi}(t)\cos\Phi(t). \end{aligned}$$

Given the joint probability density function $f_{CS\dot{C}\dot{S}}(c,s,\dot{c},\dot{s})$, the probability density $f_{R\dot{R}\Phi\dot{\Phi}}(r,\dot{r},\phi,\dot{\phi})$ is derived by transforming variables. The joint marginal probability density $f_{R\dot{R}}(r,\dot{r})$ can then be obtained from the joint density $f_{R\dot{R}\Phi\dot{\Phi}}(r,\dot{r},\phi,\dot{\phi})$. We first find $f_{CS\dot{C}\dot{S}}(c,s,\dot{c},\dot{s})$ for Gaussian processes $C(t)$, $S(t)$, $\dot{C}(t)$ and $\dot{S}(t)$.[3]

Derive $f_{CS\dot{C}\dot{S}}(c,s,\dot{c},\dot{s})$

Recall that a stationary Gaussian random process can be written as (Equation 5.47)

$$X(t) = \sum_{n=1}^{N}\sqrt{2}\sigma_n\cos(\omega_n t - \varphi_n), \qquad (9.17)$$

with variance

$$\sigma_X^2 = \sum_{n=1}^{N}\sigma_n^2. \qquad (9.18)$$

---
[3] If $X(t)$ is Gaussian, then so are $\dot{X}(t)$, $C(t)$, $S(t)$, $\dot{C}(t)$, and $\dot{S}(t)$.

σ_n is a portion of the area under the power spectral density, approximated by

$$\sigma_n \simeq S^o_{XX}(\omega_n) \Delta\omega$$
$$\simeq 2S_{XX}(\omega_n) \Delta\omega,$$

where $S^o_{XX}(\omega)$ is the one-sided and $S_{XX}(\omega)$ is the two-sided power spectral density. The phase angles φ_n are independent of each other and distributed uniformly on $[0, 2\pi)$.

Equation 9.17 can be rewritten as

$$X(t) = \sum_{n=1}^{N} \sqrt{2}\sigma_n \cos(\omega_n t - \omega_m t - \varphi_n + \omega_m t)$$
$$= \sum_{n=1}^{N} \sqrt{2}\sigma_n \cos(\omega_n t - \omega_m t - \varphi_n) \cos\omega_m t$$
$$- \sum_{n=1}^{N} \sqrt{2}\sigma_n \sin(\omega_n t - \omega_m t - \varphi_n) \sin\omega_m t.$$

A comparison of this equation with Equation 9.16 results in $C(t)$ and $S(t)$ given by

$$C(t) = \sum_{n=1}^{N} \sqrt{2}\sigma_n \cos(\omega_n t - \omega_m t - \varphi_n)$$
$$S(t) = \sum_{n=1}^{N} \sqrt{2}\sigma_n \sin(\omega_n t - \omega_m t - \varphi_n). \qquad (9.19)$$

$C(t)$ and $S(t)$ are identically distributed stationary Gaussian processes with zero mean values, $\mu_c = \mu_s = 0$. They are also uncorrelated,

$$Cov\{C(t), S(t)\}$$
$$= E\{(C(t) - \mu_c)(S(t) - \mu_s)\}$$
$$= E\{C(t)S(t)\}$$
$$= \left(\frac{1}{2\pi}\right)^N \int_0^{2\pi}\int_0^{2\pi}\cdots\int_0^{2\pi} C(t)S(t)\, d\varphi_1 d\varphi_2 \cdots d\varphi_N$$
$$= \left(\frac{1}{2\pi}\right)^N \sum_{n=1}^{N}\sum_{l=1}^{N}\int_0^{2\pi}\int_0^{2\pi}\cdots\int_0^{2\pi} [2\sigma_n\sigma_l \cos(\omega_n t - \omega_m t - \varphi_n)$$
$$\cdot \sin(\omega_l t - \omega_m t - \varphi_l)]\, d\varphi_1 d\varphi_2 \cdots d\varphi_N$$
$$= 0.$$

9.2. FIRST EXCURSION FAILURE

We know that independent processes are also uncorrelated while the reverse is not generally true. However, if the processes are Gaussian, uncorrelated Gaussian processes are also independent.[4] Therefore, $C(t)$ and $S(t)$ are uncorrelated and therefore independent of each other.

The variance of $C(t)$ is derived as follows,

$$\begin{aligned}
\sigma_C^2 &= E\left\{(C(t) - \mu_c)^2\right\} \\
&= \left(\frac{1}{2\pi}\right)^N \int_0^{2\pi}\int_0^{2\pi}\cdots\int_0^{2\pi} C^2(t)\,d\varphi_1 d\varphi_2\cdots d\varphi_N \\
&= \left(\frac{1}{2\pi}\right)^N \sum_{n=1}^{N}\sum_{l=1}^{N}\int_0^{2\pi}\int_0^{2\pi}\cdots\int_0^{2\pi}[2\sigma_n\sigma_l\cos(\omega_n t - \omega_m t - \varphi_n) \\
&\quad \cdot \cos(\omega_l t - \omega_m t - \varphi_l)]\,d\varphi_1 d\varphi_2\cdots d\varphi_N \\
&= \frac{1}{2\pi}\sum_{n=1}^{N}\int_0^{2\pi} 2\sigma_n^2 \cos^2(\omega_n t - \omega_m t - \varphi_n)\,d\varphi_n \\
&= \sum_{n=1}^{N}\sigma_n^2 = \sigma_X^2.
\end{aligned}$$

The variance of $S(t)$ can be obtained in a similar fashion. We find that the variances of $C(t)$ and $S(t)$ are equal to each other and to the variance of the random process $X(t)$, σ_X^2.

So far, we have established that $C(t)$ and $S(t)$ are independent Gaussian processes with variances equal to σ_X^2. Now let us turn our attention to $\dot{C}(t)$ and $\dot{S}(t)$.

The covariance of $C(t)$ and $\dot{C}(t)$ is given by

$$\begin{aligned}
Cov\left\{C(t),\dot{C}(t)\right\} &= E\left\{(C(t) - \mu_C)\left(\dot{C}(t) - \mu_{\dot{C}}\right)\right\} \\
&= R_{C\dot{C}}(0) - \mu_C\mu_{\dot{C}}.
\end{aligned}$$

We showed in Equation 5.13 that the autocorrelation function of a stationary process and its time derivative equals zero at $\tau = 0$, or

$$R_{C\dot{C}}(0) = 0.$$

For a zero-mean process ($\mu_C = 0$), C and $\dot{C}$ are uncorrelated Gaussian processes. Therefore, they are also independent. Similarly, we can find that $S(t)$ and $\dot{S}(t)$ are independent. So far, we showed that $C(t)$ and $S(t)$, $C(t)$ and $\dot{C}(t)$, and $S(t)$ and $\dot{S}(t)$ are independent. We still need to show

---

[4] $\rho = 0$ implies $f(y|x) = f(y)$ for Gaussian X and Y.

that the pairs $\left(\dot{C}(t), \dot{S}(t)\right)$, $\left(C(t), \dot{S}(t)\right)$, and $\left(S(t), \dot{C}(t)\right)$ are independent so that $C(t)$, $S(t)$, $\dot{C}(t)$, and $\dot{S}(t)$ are independent among themselves. We will do so by showing that $E\left\{\dot{C}(t)\dot{S}(t)\right\} = E\left\{C(t)\dot{S}(t)\right\} = E\left\{S(t)\dot{C}(t)\right\} = 0$.

The expected value of $\dot{C}(t)\dot{S}(t)$ is found to equal zero, as follows,

$$\begin{aligned}
E\left\{\dot{C}(t)\dot{S}(t)\right\} &= \left(\frac{1}{2\pi}\right)^N \int_0^{2\pi} \cdots \int_0^{2\pi} \dot{C}(t)\dot{S}(t)\, d\varphi_1 \cdots d\varphi_N \\
&= \left(\frac{1}{2\pi}\right)^N \sum_{n=1}^N \sum_{l=1}^N \int_0^{2\pi} \cdots \int_0^{2\pi} [2\sigma_n \sigma_l (\omega_l - \omega_m)(\omega_n - \omega_m)] \\
&\quad \cdot \sin(\omega_n t - \omega_m t - \varphi_n) \cos(\omega_l t - \omega_m t - \varphi_l)]\, d\varphi_1 \cdots d\varphi_N = 0.
\end{aligned}$$

The expected value of $C(t)\dot{S}(t)$ is given by

$$\begin{aligned}
E\left\{C(t)\dot{S}(t)\right\} &= \left(\frac{1}{2\pi}\right)^N \int_0^{2\pi} \cdots \int_0^{2\pi} C(t)\dot{S}(t)\, d\varphi_1 \cdots d\varphi_N \\
&= \left(\frac{1}{2\pi}\right)^N \sum_{n=1}^N \sum_{l=1}^N \int_0^{2\pi} \cdots \int_0^{2\pi} [2\sigma_n \sigma_l (\omega_l - \omega_m) \cos(\omega_n t - \omega_m t - \varphi_n) \\
&\quad \cdot \cos(\omega_l t - \omega_m t - \varphi_l)]\, d\varphi_1 \cdots d\varphi_N.
\end{aligned}$$

For $n \neq l$, the integral becomes zero, and we are left with

$$\begin{aligned}
E\left\{C(t)\dot{S}(t)\right\} &= \frac{1}{2\pi} \sum_{n=1}^N \int_0^{2\pi} 2\sigma_n^2 (\omega_n - \omega_m) \cos^2(\omega_n t - \omega_m t - \varphi_n)\, d\varphi_n \\
&= \sum_{n=1}^N \sigma_n^2 (\omega_n - \omega_m).
\end{aligned}$$

By choosing ω_m in a special way, we can make $E\left\{C(t)\dot{S}(t)\right\} = 0$. Also, $E\left\{C(t)\dot{S}(t)\right\}$ will be zero if

$$\sum_{n=1}^N \sigma_n^2 (\omega_n - \omega_m) = 0,$$

9.2. FIRST EXCURSION FAILURE

or

$$\omega_m = \frac{\sum_{n=1}^{N} \omega_n \sigma_n^2}{\sum_{n=1}^{N} \sigma_n^2}. \qquad (9.20)$$

If this is the case, $C(t)$, $S(t)$, $\dot{C}(t)$, and $\dot{S}(t)$ are independent among themselves. The joint probability density function can be written as

$$f_{CS\dot{C}\dot{S}}(c,s,\dot{c},\dot{s}) = \frac{1}{4\pi^2 \sigma_C^2 \sigma_{\dot{C}}^2} \exp\left(-\frac{c^2+s^2}{2\sigma_C^2} - \frac{\dot{c}^2+\dot{s}^2}{2\sigma_{\dot{C}}^2}\right),$$

$$-\infty < c, s, \dot{c}, \dot{s} < \infty,$$

where σ_C^2 is the variance of $C(t)$ or $S(t)$, and $\sigma_{\dot{C}}^2$ is the variance of $\dot{C}(t)$ or $\dot{S}(t)$.

Find $f_{R\Phi\dot{R}\dot{\Phi}}\left(r,\phi,\dot{r},\dot{\phi}\right)$

The joint probability density function of R, Φ, $\dot{R}$, and $\dot{\Phi}$ is given by

$$f_{R\Phi\dot{R}\dot{\Phi}}\left(r,\phi,\dot{r},\dot{\phi}\right) = J f_{CS\dot{C}\dot{S}}(c,s,\dot{c},\dot{s}),$$

where the Jacobian J is defined as the following determinant,

$$J = \begin{vmatrix} \partial c/\partial r & \partial c/\partial \phi & \partial c/\partial \dot{r} & \partial c/\partial \dot{\phi} \\ \partial s/\partial r & \partial s/\partial \phi & \partial s/\partial \dot{r} & \partial s/\partial \dot{\phi} \\ \partial \dot{c}/\partial r & \partial \dot{c}/\partial \phi & \partial \dot{c}/\partial \dot{r} & \partial \dot{c}/\partial \dot{\phi} \\ \partial \dot{s}/\partial r & \partial \dot{s}/\partial \phi & \partial \dot{s}/\partial \dot{r} & \partial \dot{s}/\partial \dot{\phi} \end{vmatrix}.$$

Recalling that

$$\begin{aligned} c &= r\cos\phi \\ s &= r\sin\phi \\ \dot{c} &= \dot{r}\cos\phi - r\dot{\phi}\sin\phi \\ \dot{s} &= \dot{r}\sin\phi + r\dot{\phi}\cos\phi, \end{aligned}$$

the Jacobian is given by

$$J = \begin{vmatrix} \cos\phi & -r\sin\phi & 0 & 0 \\ \sin\phi & r\cos\phi & 0 & 0 \\ -\dot{\phi}\sin\phi & -\dot{r}\sin\phi - r\dot{\phi}\cos\phi & \cos\phi & -r\sin\phi \\ \dot{\phi}\cos\phi & \dot{r}\cos\phi - r\dot{\phi}\sin\phi & \sin\phi & r\cos\phi \end{vmatrix} = r^2.$$

The joint probability density function $f_{R\Phi \dot{R}\dot{\Phi}}\left(r,\phi,\dot{r},\dot{\phi}\right)$ is then

$$f_{R\Phi \dot{R}\dot{\Phi}}\left(r,\phi,\dot{r},\dot{\phi}\right) = \frac{r^2}{4\pi^2 \sigma_C^2 \sigma_{\dot{C}}^2} \exp\left(-\frac{r^2}{2\sigma_C^2} - \frac{\dot{r}^2 + r^2\dot{\phi}^2}{2\sigma_{\dot{C}}^2}\right),$$

$$r \geq 0$$
$$-\infty < \dot{r}, \dot{\phi} < \infty$$
$$0 < \phi < 2\pi.$$

Find $f_{R\dot{R}}(r,\dot{r})$

The joint probability density function of R and $\dot{R}$ is then given by

$$\begin{aligned} f_{R\dot{R}}(r,\dot{r}) &= \int_{-\infty}^{\infty}\int_{0}^{2\pi} f_{R\Phi\dot{R}\dot{\Phi}}\left(r,\phi,\dot{r},\dot{\phi}\right) d\phi\, d\dot{\phi} \\ &= \frac{r}{\sqrt{2\pi}\sigma_C^2 \sigma_{\dot{C}}} \exp\left(-\frac{r^2}{2\sigma_C^2} - \frac{\dot{r}^2}{2\sigma_{\dot{C}}^2}\right), \qquad r \geq 0, -\infty < \dot{r} < \infty. \end{aligned}$$

Above is the result for a stationary narrow-band Gaussian process $X(t)$.

We previously established that σ_C^2 equals σ_X^2. What about $\sigma_{\dot{C}}^2$? In order to find the expression for $\sigma_{\dot{C}}^2$, first start with $\dot{X}(t)$. Using Equation 9.17, $\dot{X}(t)$ is given by the sum

$$\dot{X}(t) = \sum_{n=1}^{N} \sqrt{2}\sigma_n \omega_n \cos(\omega_n t - \varphi_n).$$

The mean of $\dot{X}(t)$ is given by

$$E\left\{\dot{X}(t)\right\} = \sum_{n=1}^{N} \sqrt{2}\sigma_n \omega_n E\left\{\cos(\omega_n t - \varphi_n)\right\} = 0,$$

and the variance of $\dot{X}(t)$ is given by

$$\sigma_{\dot{X}}^2 = \sum_{n=1}^{N} \sigma_n^2 \omega_n^2. \qquad (9.21)$$

From Equation 9.19, the derivatives of $C(t)$ and $S(t)$ are given by

$$\begin{aligned} \dot{C}(t) &= \sum_{n=1}^{N} -\sqrt{2}\sigma_n(\omega_n - \omega_m)\sin(\omega_n t - \omega_m t - \varphi_n) \\ \dot{S}(t) &= \sum_{n=1}^{N} \sqrt{2}\sigma_n(\omega_n - \omega_m)\cos(\omega_n t - \omega_m t - \varphi_n). \end{aligned}$$

9.2. FIRST EXCURSION FAILURE

The variance of $\dot{C}(t)$ is

$$\begin{aligned}
\sigma_{\dot{C}}^2 &= E\left\{\left(\dot{C}(t) - \mu_{\dot{c}}\right)^2\right\} \\
&= \frac{1}{2\pi}\sum_{n=1}^{N}\int_0^{2\pi} 2\sigma_n^2(\omega_n - \omega_m)^2 \sin^2(\omega_n t - \omega_m t - \varphi_n)\, d\varphi_n \\
&= \sum_{n=1}^{N}\sigma_n^2(\omega_n - \omega_m)^2.
\end{aligned}$$

It can be shown that the variance of $\dot{S}(t)$ is identical to the variance of $\dot{C}(t)$. Using Equations 9.18, 9.20, and 9.21, the variance $\sigma_{\dot{C}}^2$ can be written as

$$\begin{aligned}
\sigma_{\dot{C}}^2 &= \sum_{n=1}^{N}\sigma_n^2\left(\omega_n^2 - 2\omega_n\omega_m + \omega_m^2\right) \\
&= \sum_{n=1}^{N}\sigma_n^2\omega_n^2 - 2\omega_m\sum_{n=1}^{N}\sigma_n^2\omega_n + \omega_m^2\sum_{n=1}^{N}\sigma_n^2 \\
&= \sigma_{\dot{X}}^2 - 2\omega_m\omega_m\sigma_X^2 + \omega_m^2\sigma_X^2 \\
&= \sigma_{\dot{X}}^2 - \omega_m^2\sigma_X^2.
\end{aligned}$$

Finally, the joint probability density function of R and $\dot{R}$ is given by

$$f_{R\dot{R}}(r,\dot{r}) = \frac{r}{\sqrt{2\pi}\sigma_X^2\sqrt{\sigma_{\dot{X}}^2 - \omega_m^2\sigma_X^2}}\exp\left(-\frac{1}{2}\left[\frac{r^2}{\sigma_X^2} + \frac{\dot{r}^2}{\sigma_{\dot{X}}^2 - \omega_m^2\sigma_X^2}\right]\right),$$

$$r \geq 0,\ -\infty < \dot{r} < \infty, \qquad (9.22)$$

where

$$\omega_m = \frac{\sum_{n=1}^{N}\omega_n\sigma_n^2}{\sum_{n=1}^{N}\sigma_n^2} \qquad (9.23)$$

$$\sigma_n^2 = S_{XX}^o(\omega_n)\Delta\omega. \qquad (9.24)$$

The continuous-parametered version of Equation 9.23 is

$$\omega_m = \frac{\int_0^\infty \omega S_{XX}^o(\omega)\, d\omega}{\int_0^\infty S_{XX}^o(\omega)\, d\omega}. \qquad (9.25)$$

Now, the procedure of obtaining $f_{R\dot{R}}(r,\dot{r})$ is clear. If $S_{XX}^o(\omega)$ is given in a functional form, the midband frequency ω_m is the average frequency defined by Equation 9.25. The variances σ_X^2 and $\sigma_{\dot{X}}^2$ are obtained by integrating $S_{XX}^o(\omega)$ over their entire range,

$$\sigma_X^2 = \int_0^\infty S_{XX}^o(\omega)\,d\omega$$

$$\sigma_{\dot{X}}^2 = \int_0^\infty S_{\dot{X}\dot{X}}^o(\omega)\,d\omega,$$

where $S_{\dot{X}\dot{X}}^o(\omega)$ is simply (Equation 5.42)

$$S_{\dot{X}\dot{X}}^o(\omega) = \omega^2 S_{XX}^o(\omega).$$

Find $\nu_{R(t)=Z}^+$

The up-crossing rate of the envelope $R(t)$ is then given by (Equation 9.13):

$$\nu_{R(t)=Z}^+$$
$$= \int_0^\infty f_{R\dot{R}}(Z,\dot{r})\,\dot{r}\,d\dot{r}$$
$$= \int_0^\infty \frac{Z}{\sqrt{2\pi}\sigma_X^2 \sqrt{\sigma_{\dot{X}}^2 - \omega_m^2 \sigma_X^2}} \exp\left(-\frac{1}{2}\left[\frac{Z^2}{\sigma_X^2} + \frac{\dot{r}^2}{\sigma_{\dot{X}}^2 - \omega_m^2 \sigma_X^2}\right]\right)\dot{r}\,d\dot{r}$$
$$= \frac{Z\sqrt{\sigma_{\dot{X}}^2 - \omega_m^2 \sigma_X^2}}{\sqrt{2\pi}\sigma_X^2} \exp\left(-\frac{1}{2}\frac{Z^2}{\sigma_X^2}\right). \tag{9.26}$$

Defining $\omega_{0+} \equiv 2\pi\nu_0^+$, Equation 9.12 can be written as

$$\frac{\sigma_{\dot{X}}}{\sigma_X} = \omega_{0+}.$$

Then Equation 9.26 simplifies to

$$\nu_{R(t)=Z}^+ = \frac{Z\sqrt{\omega_{0+}^2 - \omega_m^2}}{\sqrt{2\pi}\sigma_X} \exp\left(-\frac{1}{2}\frac{Z^2}{\sigma_X^2}\right). \tag{9.27}$$

Note that

$$\omega_{0+}^2 = \frac{\sigma_{\dot{X}}^2}{\sigma_X^2} = \frac{\int_0^\infty \omega^2 S_{XX}^o(\omega)\,d\omega}{\int_0^\infty S_{XX}^o(\omega)\,d\omega}. \tag{9.28}$$

The ratio of the up-crossing rate of $X(t)$ to that of $R(t)$, $\nu_{X=Z}^+/\nu_{R=Z}^+$, is called the *average clump size*. For example, if $\nu_{X=Z}^+/\nu_{R=Z}^+ = 3$, then

9.2. FIRST EXCURSION FAILURE

$X(t)$ crosses Z three times while its envelope crosses once. The average clump size, using Equations 9.10 and 9.26, is

$$\text{average clump size} = \frac{\nu^+_{X=Z}}{\nu^+_{R=Z}}$$

$$= \frac{\sigma_X}{\sqrt{2\pi} Z \sqrt{1 - \omega_m^2 \sigma_X^2 / \sigma_{\dot X}^2}}$$

$$= \frac{\sigma_X}{\sqrt{2\pi} Z \sqrt{1 - \omega_m^2 / \omega_{0+}^2}}. \tag{9.29}$$

Example 9.6 Narrow-Band Gaussian Process Crossing Parameters

Consider a case where $X(t)$ is narrow-band Gaussian process with one-sided spectral density given by

$$S(\omega) = \begin{cases} S_o \omega, & \omega_1 \leq \omega \leq \omega_2 \\ 0, & \text{elsewhere.} \end{cases}$$

Find the midband frequency ω_m, zero up-crossing frequency ω_{0+}, σ_X, $\sigma_{\dot X}$ and the clump size.

Solution The midband frequency, given by Equation 9.25, is reduced to

$$\omega_m = \frac{\int_{\omega_1}^{\omega_2} S_o \omega^2 d\omega}{\int_{\omega_1}^{\omega_2} S_o \omega d\omega}$$

$$= \frac{2(\omega_2^3 - \omega_1^3)}{3(\omega_2^2 - \omega_1^2)}$$

$$= \frac{2(\omega_2^2 + \omega_2 \omega_1 + \omega_1^2)}{3(\omega_2 + \omega_1)},$$

and the zero up-crossing rate in Equation 9.28 is given by

$$\omega_{0+} = \sqrt{\frac{\int_0^\infty \omega^2 S_{XX}(\omega) d\omega}{\int_0^\infty S_{XX}(\omega) d\omega}}$$

$$= \sqrt{\frac{\int_{\omega_1}^{\omega_2} S_o \omega^3 d\omega}{\int_{\omega_1}^{\omega_2} S_o \omega d\omega}} = \sqrt{\frac{(\omega_2^2 + \omega_1^2)}{2}}.$$

Then,

$$\frac{\omega_m^2}{\omega_{0+}^2} = \frac{8}{9} \frac{(\omega_2^2 + \omega_2 \omega_1 + \omega_1^2)^2}{(\omega_1 + \omega_2)^2 (\omega_2^2 + \omega_1^2)}.$$

The standard deviations are given by

$$\sigma_X = \sqrt{\int_{\omega_1}^{\omega_2} S_o \omega \, d\omega} = \sqrt{\frac{S_o}{2}(\omega_2^2 - \omega_1^2)}$$

$$\sigma_{\dot{X}} = \sqrt{\int_{\omega_1}^{\omega_2} S_o \omega^3 \, d\omega} = \sqrt{\frac{S_o}{4}(\omega_2^4 - \omega_1^4)}.$$

The average clump size using Equation 9.29 is given by:

$$\frac{\sigma_X}{\sqrt{2\pi}Z\sqrt{1 - \omega_m^2/\omega_{0+}^2}} = \frac{\sqrt{\frac{S_o}{2}(\omega_2^2 - \omega_1^2)}}{\sqrt{2\pi}Z\sqrt{1 - \frac{8}{9}\frac{(\omega_2^2 + \omega_2\omega_1 + \omega_1^2)^2}{(\omega_1 + \omega_2)^2(\omega_2^2 + \omega_1^2)}}}$$

$$= \frac{1}{Z}\sqrt{\frac{S_o}{4\pi}\frac{(\omega_2 + \omega_1)}{(\omega_2 - \omega_1)}}\sqrt{\frac{(\omega_2^4 - \omega_1^4)}{(\omega_1^2 + 4\omega_2\omega_1 + \omega_2^2)}}.$$

⊛

Example 9.7 Probability Density Function of the Envelope Process

Determine the probability density function $f_R(r)$ of the envelope process for a narrow-band Gaussian process with zero mean.

Solution We have

$$f_{R\dot{R}}(r, \dot{r}) = \frac{r}{\sqrt{2\pi}\sigma_X^2\sqrt{\sigma_{\dot{X}}^2 - \omega_m^2\sigma_X^2}} \exp\left(-\frac{1}{2}\left[\frac{r^2}{\sigma_X^2} + \frac{\dot{r}^2}{\sigma_{\dot{X}}^2 - \omega_m^2\sigma_X^2}\right]\right)$$

$$r \geq 0, \ -\infty < \dot{r}, \dot{\phi} < \infty.$$

Then $f_R(r)$ is the marginal density,

$$f_R(r)$$
$$= \int_{-\infty}^{\infty} f_{R\dot{R}}(r, \dot{r}) \, d\dot{r}$$
$$= \frac{r}{\sqrt{2\pi}\sigma_X^2\sqrt{\sigma_{\dot{X}}^2 - \omega_m^2\sigma_X^2}} \exp\left(-\frac{1}{2}\frac{r^2}{\sigma_X^2}\right) \int_{-\infty}^{\infty} \exp\left(-\frac{1}{2}\frac{\dot{r}^2}{\sigma_{\dot{X}}^2 - \omega_m^2\sigma_X^2}\right) d\dot{r}.$$

Using the integral

$$\int_{-\infty}^{\infty} \exp\left(-\frac{1}{2}x^2\right) dx = \sqrt{2\pi},$$

9.2. FIRST EXCURSION FAILURE

we find that
$$f_R(r) = \frac{r}{\sigma_X^2} \exp\left(-\frac{1}{2}\frac{r^2}{\sigma_X^2}\right), \qquad r \geq 0.$$

Note that $R(t)$ has a Rayleigh density function. The mean and the variance of $R(t)$ is then $\mu_R = 1.253\sigma_X$ and $\sigma_R = 0.655\sigma_X$. See Section 3.4.5.

It should be noted that $f_R(r)$ also tells us how the peaks are distributed since R, the envelope function, is the magnitude of the peaks by definition. It will be shown independently that the peaks are distributed according to the Rayleigh density function in Section 9.3.2.

⊛

Example 9.8 First Passage for a Narrow-Band Gaussian Process

Consider a random force in a machine component. The force is denoted by $X(t)$, and it is considered to be an ideal narrow-band Gaussian process. The force has frequency components in the range $10 - 11$ rad/s, and the magnitude of the power spectrum is 2 N²s. The component will fail when it first exceeds 3.3 N, and the design life for this part is 30 s. What is the reliability? Also, what is the calculated reliability if we did not realize that the process is narrow band?

Solution Find $f_{R\dot{R}}(r,\dot{r})$, the crossing rate ν_Z^+, and the reliability. We are given that
$$S_{XX}(\omega) = S_o, \qquad \omega_1 < \omega < \omega_2.$$

The midband frequency is then
$$\omega_m = \frac{\int_{\omega_1}^{\omega_2} \omega S_o d\omega}{\int_{\omega_1}^{\omega_2} S_o d\omega}$$
$$= \frac{\omega_2 + \omega_1}{2},$$

and the variances are given by
$$\sigma_X^2 = \int_{\omega_1}^{\omega_2} S_o d\omega = S_o(\omega_2 - \omega_1)$$
$$\sigma_{\dot{X}}^2 = \int_{\omega_1}^{\omega_2} S_o \omega^2 d\omega = S_o\frac{(\omega_2^3 - \omega_1^3)}{3}.$$

For $S_o = 2$ N²s, $\omega_1 = 10$ rad/s and $\omega_2 = 11$ rad/s,
$$\omega_m = 10.5 \text{ rad/s}$$
$$\sigma_X^2 = 2 \text{ N}^2$$
$$\sigma_{\dot{X}}^2 = 220.667 \text{ (N/s)}^2$$
$$\omega_{0+} = 10.504 \text{ rad/s}.$$

The joint probability density function using Equation 9.22 is

$$f_{R\dot{R}}(r,\dot{r}) = \frac{r}{\sqrt{2\pi}2\sqrt{221-(10.5)^2 \cdot 2}} \exp\left(-\frac{1}{2}\left[\frac{r^2}{2} + \frac{\dot{r}^2}{221-(10.5)^2 \cdot 2}\right]\right)$$
$$= 0.484\, r \exp\left(-\left[\frac{r^2}{4} + \frac{\dot{r}^2}{0.825}\right]\right),$$
$$r > 0$$
$$-\infty < \dot{r} < \infty,$$

as shown in Figure 9.9.

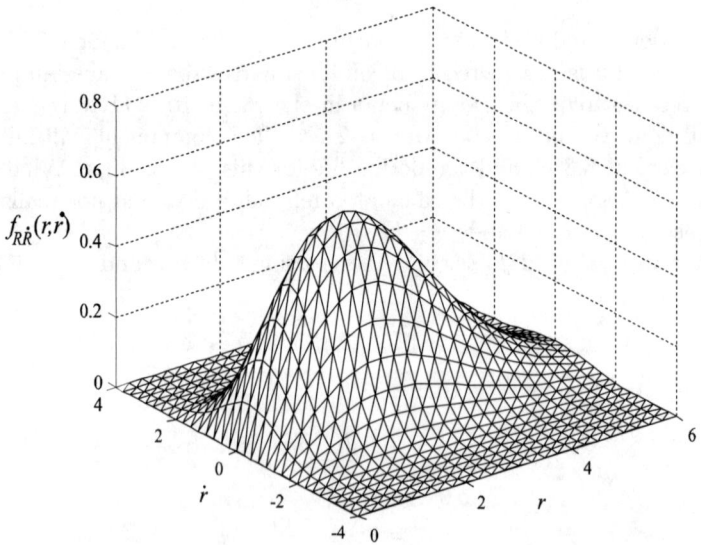

Figure 9.9: The joint probability density function $f_{R\dot{R}}(r,\dot{r})$.

The up-crossing rate (Equation 9.27) is given by

$$\nu^+_{R=Z} = \frac{Z\sqrt{\omega_{0+}^2 - \omega_m^2}}{\sqrt{2\pi}\sigma_X} \exp\left(-\frac{1}{2}\frac{Z^2}{\sigma_X^2}\right)$$
$$= 8.14 \times 10^{-2} Z \exp\left(-\frac{Z^2}{4}\right).$$

For $Z = 3.3$ N,
$$\nu^+_{R=Z} = 0.0177.$$

9.2. FIRST EXCURSION FAILURE

The average time to the first crossing is

$$E\{T_f\} = \frac{1}{\nu^+_{R=Z}} = 56.5\,\text{s}$$

and the variance is

$$Var\{T_f\} = \frac{1}{\left(\nu^+_{R=Z}\right)^2} = 3190\,\text{s}^2.$$

The reliability of this component is then (Equation 9.14)

$$\begin{aligned} p_r &= e^{-\nu^+_{R=Z} T_D} \\ &= e^{-0.531} = 0.588. \end{aligned}$$

This is the probability that the component will not fail during its design life.

If we did not realize that the process is narrow banded, then we would calculate the up-crossing rate using Equation 9.10. Then,

$$\begin{aligned} \nu^+_{X=Z} &= \frac{\sigma_{\dot{X}}}{2\pi \sigma_X} \exp\left[-\frac{1}{2}\left(\frac{Z}{\sigma_X}\right)^2\right]. \\ &= 0.1099, \end{aligned}$$

and the reliability would have been

$$\begin{aligned} p_r &= \exp\left(-\nu^+_{R=Z} T_D\right) \\ &= 0.0404, \end{aligned}$$

which is much lower than the actual reliability of 0.588.

For this problem, the average clump size is 6.22 (Equation 9.29). Figure 9.10 shows a sample response obtained using the Fourier representation in Equation 5.48 in Section 5.8.1. The sample response is given by

$$X(t) = \sum_{n=1}^{N} \sqrt{\frac{2}{N}} \sigma \cos\left(\bar{\omega}_n t - \varphi_n\right),$$

where $N = 150$, $\sigma = \sqrt{S_o(\omega_2 - \omega_1)} = \sqrt{2}$, $\bar{\omega}_n = 2\pi(2n-1)/T$, and $T = 100$. For this response, the part fails at 45.4 s. We can expect that the average failure time will approach the theoretical value of 56.5 s if a sufficient number of sample responses are averaged. The first three clump sizes in this response are 3, 6, and 7. Again, the average should be close to 6.22.

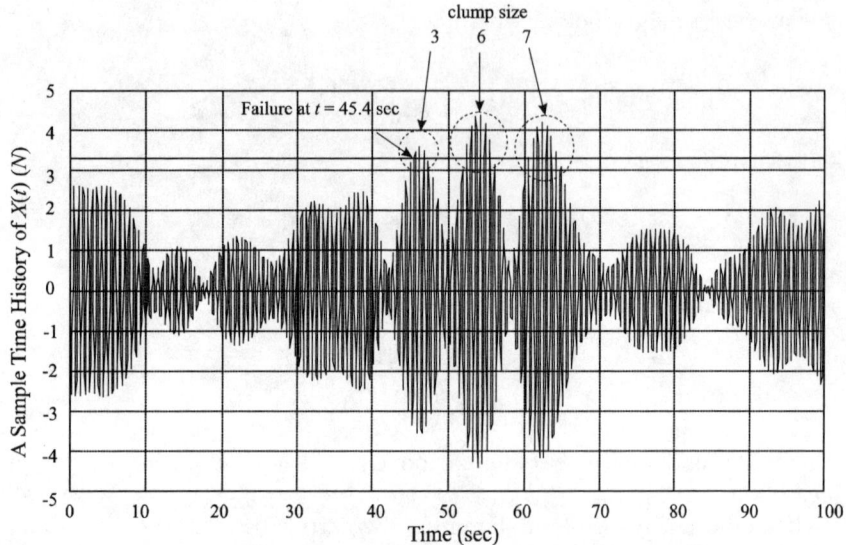

Figure 9.10: Sample response of Example 9.8.

9.2.6 Other Failure Laws

Here, we discuss the gamma, normal, and Weibull failure laws. These failure laws include *wear-out effects*.

Gamma Failure Law

The time to failure has a gamma probability distribution if its probability density is given by

$$f_{T_f}(t_f) = \frac{\nu}{\Gamma(r)} (\nu t_f)^{r-1} e^{-\nu t_f}, \qquad t_f \geq 0,$$

where ν is the positive disturbance rate or failure rate, r is a positive integer, and $\Gamma(r)$ is the gamma function, defined by

$$\Gamma(r) = \int_0^\infty t^{r-1} e^{-t} dt. \tag{9.30}$$

The probability density functions for three values of r are plotted in Figure 9.11. For positive integer r, the gamma function is reduced to

$$\Gamma(r) = (r-1)!.$$

9.2. FIRST EXCURSION FAILURE

This can be verified by integrating the gamma function $r - 1$ times. The probability density function is then written as

$$f_{T_f}(t_f) = \frac{\nu}{(r-1)!}(\nu t_f)^{r-1} e^{-\nu t_f}, \qquad t_f \geq 0.$$

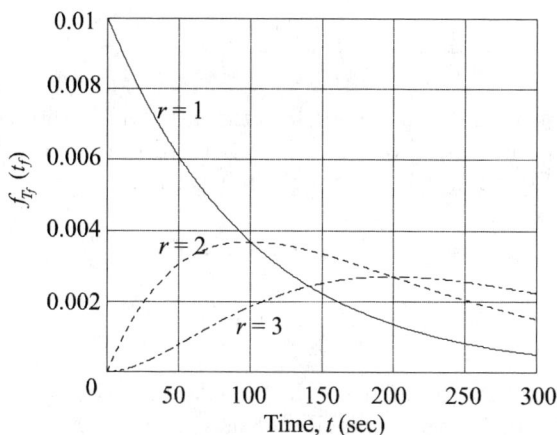

Figure 9.11: Gamma probability density function for $\nu = 0.01$ and $r = 1, 2, 3$.

The reliability can be written as

$$p_r = \int_{T_D}^{\infty} f_{T_f}(t_f)\, dt_f = \int_{T_D}^{\infty} \frac{\nu(\nu t_f)^{r-1}}{(r-1)!} e^{-\nu t_f}\, dt_f. \tag{9.31}$$

The gamma failure law arises when failure occurs if r or more disturbances occur in time T_D. Suppose that disturbances appear according to a Poisson process. The reliability, the probability that the component will not fail in time t, is the probability of $0, 1, \cdots$, or $r - 1$ disturbances occurring in time T_D,

$$\begin{aligned} p_r &= p_N(0, T_D) + p_N(1, T_D) + \cdots + p_N(r-1, T_D) \\ &= \sum_{n=0}^{r-1} \frac{(\nu T_D)^n e^{-\nu T_D}}{n!}. \end{aligned} \tag{9.32}$$

Equations 9.31 and 9.32 are equivalent, verifiable by integrating Equation 9.31 by parts $(r - 1)$ times. For example, integrating by parts once, we obtain

$$\int_{T_D}^{\infty} \frac{\nu (\nu t_f)^{r-1}}{(r-1)!} e^{-\nu t_f} dt_f$$

$$= \left[\frac{\nu (\nu t_f)^{r-1}}{(r-1)!} \frac{e^{-\nu t_f}}{-\nu} \right]_{T_D}^{\infty} - \int_{T_D}^{\infty} \frac{\nu^2 (r-1)(\nu t_f)^{r-2}}{(r-1)!} \frac{e^{-\nu t_f}}{-\nu} dt_f$$

$$= \frac{(\nu T_D)^{r-1} e^{-\nu T_D}}{(r-1)!} + \int_{t}^{\infty} \frac{\nu (\nu t_f)^{r-2}}{(r-2)!} e^{-\nu t_f} dt_f, \qquad (9.33)$$

where the first term in Equation 9.33 is the last term in the series in Equation 9.32. By integrating by parts $(r-2)$ more times, we will recover all the terms in Equation 9.32. The probability of failure is then

$$p_f = 1 - \int_t^{\infty} \frac{\nu}{(r-1)!} (\nu t_f)^{r-1} e^{-\nu t_f} dt_f$$

$$= \int_0^t \frac{\nu}{(r-1)!} (\nu t_f)^{r-1} e^{-\nu t_f} dt_f.$$

Since the probability of failure is the cumulative distribution of T_f, $p_f = F_{T_f}(t_f)$, we find that the probability density is given by

$$f_{T_f}(t_f) = \frac{\nu}{(r-1)!} (\nu t_f)^{r-1} e^{-\nu t_f}$$

$$= \frac{\nu}{\Gamma(r)} (\nu t_f)^{r-1} e^{-\nu t_f}, \qquad t_f \geq 0.$$

The mean value and the variance are found as follows,

$$E\{T_f\} = \int_0^{\infty} \frac{1}{(r-1)!} (\nu t_f)^r e^{-\nu t_f} dt_f = \frac{r}{\nu}$$

$$Var\{T_f\} = \int_0^{\infty} \frac{\nu t_f^2}{(r-1)!} (\nu t_f)^{r-1} e^{-\nu t_f} dt_f - \left(\frac{r}{\nu}\right)^2 = \frac{r}{\nu^2}.$$

Note that the gamma failure law is reduced to the exponential failure law if the failure occurs at the first disturbance $(r = 1)$.

Normal Failure Law

Another failure law, not as common as the exponential failure law, is the normal failure law. The normal law states that the time to failure is governed by

$$f_{T_f}(t_f) = \frac{1}{\sqrt{2\pi}\sigma} \exp\left(-\frac{1}{2}\left[\frac{t_f - \mu}{\sigma}\right]^2\right), \qquad t_f \geq 0.$$

9.2. FIRST EXCURSION FAILURE

Note that this distribution is truncated at $t_f = 0$ since it makes no sense for a part to fail before $t_f = 0$. The truncation is acceptable assuming μ is sufficiently far removed from zero.

The normal failure law indicates that 68.27% of failures occur within one standard deviation of the mean, 95.45% within two standard deviations of the mean, and 99.73% within three standard deviations of the mean.

The Weibull Failure Law

The Weibull failure law is similar to the exponential failure law except that the failure rate is no longer constant. Instead, the failure rate is given by

$$\nu(t_f) = (\alpha\beta) t_f^{\beta-1},$$

where α and β are positive constants. The probability density (Equation 9.7) is then given by

$$\begin{aligned} f_{T_f}(t_f) &= \nu(t_f) \exp\left(-\int_0^{t_f} \nu(s)\, ds\right) \\ &= (\alpha\beta) t_f^{\beta-1} \exp\left(-\int_0^{t_f} (\alpha\beta) s^{\beta-1} ds\right) \\ &= (\alpha\beta) t_f^{\beta-1} \exp\left(-\alpha t_f^{\beta}\right), \qquad t_f > 0. \end{aligned} \qquad (9.34)$$

Figure 9.12 shows the probability density function for $\alpha = 1$, $\beta = 1, 2,$ and 3.

Example 9.9 Mean Value and Variance of the Weibull Distribution

Show that the mean value and the variance of Weibull distribution is given by

$$\begin{aligned} E\{T_f\} &= \alpha^{-1/\beta} \Gamma\left(\frac{1}{\beta}+1\right) \\ Var\{T_f\} &= \alpha^{-2/\beta} \left\{\Gamma\left(\frac{2}{\beta}+1\right) - \left[\Gamma\left(\frac{1}{\beta}+1\right)\right]^2\right\}. \end{aligned}$$

Solution The mean value is given by

$$\begin{aligned} E\{T_f\} &= \int_0^{\infty} t_f f_{T_f}(t_f)\, dt_f \\ &= \int_0^{\infty} (\alpha\beta) t_f^{\beta} \exp\left(-\alpha t_f^{\beta}\right) dt_f. \end{aligned}$$

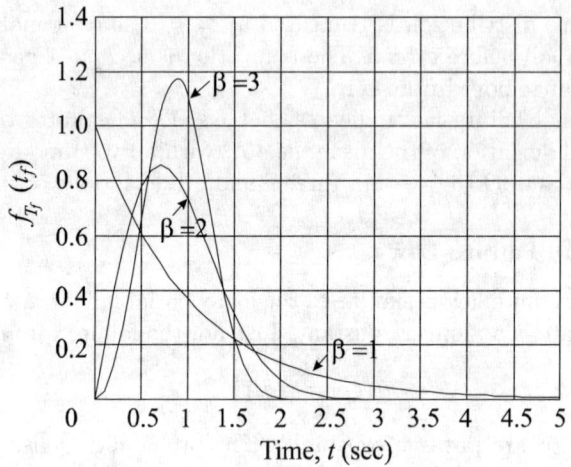

Figure 9.12: Weibull probability density function for $\alpha = 1$ and $\beta = 1, 2, 3$.

Let $\alpha t_f^\beta = x$ so that

$$t_f = \left(\frac{x}{\alpha}\right)^{-1/\beta}$$
$$dt_f = \beta^{-1}\alpha^{-1/\beta}x^{1/\beta-1}dx.$$

Simplifying and using the definition of the gamma function (Equation 9.30), we can write

$$E\{T_f\} = \alpha^{-1/\beta}\int_0^\infty x^{1/\beta+1-1}\exp(-x)\,dx$$
$$= \alpha^{-1/\beta}\Gamma\left(\frac{1}{\beta}+1\right).$$

The mean-square value is given by

$$E\{T_f^2\} = \int_0^\infty t_f^2 f_{T_f}(t_f)\,dt_f$$
$$= \int_0^\infty (\alpha\beta)\,t_f^{\beta+1}\exp\left(-\alpha t_f^\beta\right)dt_f.$$

9.3. FATIGUE LIFE PREDICTION

Using the same change of variables, we obtain

$$\begin{aligned} E\{T_f^2\} &= \int_0^\infty t_f^2 f_{T_f}(t_f)\, dt_f \\ &= \int_0^\infty \alpha^{1-(\frac{1}{\beta}+1)-\frac{1}{\beta}} x^{(\frac{1}{\beta}+1)+1/\beta-1} \exp(-x)\, x^{1/\beta-1} dx \\ &= \int_0^\infty \alpha^{-2/\beta}(x)^{2/\beta} \exp(-x)\, dx \\ &= \int_0^\infty \alpha^{-2/\beta}(x)^{(2/\beta+1)-1} \exp(-x)\, dx \\ &= \alpha^{-2/\beta}\Gamma\left(\frac{2}{\beta}+1\right). \end{aligned}$$

The variance is given by

$$\begin{aligned} Var\{T_f\} &= E\{T_f^2\} - (E\{T_f\})^2 \\ &= \alpha^{-2/\beta}\left\{\Gamma\left(\frac{2}{\beta}+1\right) - \left[\Gamma\left(\frac{1}{\beta}+1\right)\right]^2\right\}. \end{aligned}$$

⊛

9.3 Fatigue Life Prediction

When machines or structural members are subjected to repeated dynamic stresses even below the yield strength, they may exhibit diminished strength and ductility. When the cyclic stresses are continued, cracks in the material start to propagate and the parts may eventually *fracture*. This phenomenon is called fatigue, and the number of stress cycles prior to fracture is called the *fatigue life*. Before failure, a fatigue crack spreads from a location with high stress concentration, which is due to imperfections in material, surface smoothness, and structural geometry.[5]

It is generally difficult to predict fatigue life due to inherent scatter in experimental data, and difficulties in translating laboratory data for real-life situations. The goal of this section is to introduce the reader to damage prediction and fatigue life using Miner's rule. Although it is unsatisfactory in some cases, Miner's rule is the simplest and most widely used stress-based fatigue failure rule. It assumes that

- the mean stress is zero

---

[5] This is by no means the complete description of fatigue failure. The interested reader should refer to *Metal Fatigue*, N.E. Frost, K.J. Marsh, and L.P. Pook, Oxford University Press, 1974, reprinted by Dover Publications, 1999.

- the stress levels or amplitudes are discrete and can be defined, and

- the order in which the stress levels are applied has no effect.

In Miner's rule, the damage done to a component is expressed as a summation of ratios of number of cycles applied, n_i, to the number of cycles to failure at a given stress level, N_i. That is, the total damage done is given by

$$D = \sum_i \frac{n_i}{N_i}. \qquad (9.35)$$

The component fails when the damage D reaches 1. The total number of cycles to failure at a given stress level is obtained experimentally. The number of cycles can be obtained by counting the number of peaks at a certain stress level. Mathematically, the stress level s is a local maximum or a peak in a given cycle. The form of Miner's rule in Equation 9.35 is applicable when a discrete number of stress levels is applied.

However, a random process may reach any stress level, and it is not possible to count the number of cycles at a discrete stress level. Instead, the number of cycles can be counted for a stress range. For instance, the number of cycles between stress levels s_1 and s_2 is

$$n = \int_{s_1}^{s_2} n(s)\, ds,$$

where $n(s)$ is the number of cycles per unit stress range and indicates how the number of applied cycles is distributed over the stress level. The continuous version of Miner's rule is then given by

$$D = \int_s \frac{n(s)}{N(s)} ds.$$

$N(s)$ is the number of cycles to failure at the stress level s and obtained from the so-called failure curve, Wölher curve, or S-N curve, introduced in the next section. In practice, such curves are determined by experiments for various materials and geometry.

The term $n(s)$ can be normalized by the total number of peaks n_t, resulting in the probability density function of the applied stress levels,

$$f_S(s) = \frac{n(s)}{n_t}.$$

Let the random process $X(t)$ be the applied stress. The stress levels s are the peaks of $X(t)$. Therefore, the probability density function of peak

9.3. FATIGUE LIFE PREDICTION

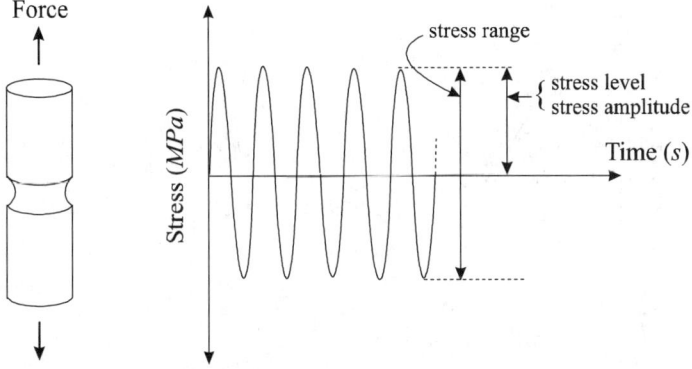

Figure 9.13: Uniaxial fatigue test to generate S-N curves.

heights, $f_H(h)$, is also the probability density function of stress levels, $f_S(s)$. The total damage predicted by Miner's rule is then

$$D = n_t \int_s \frac{f_S(s)}{N(s)} ds. \tag{9.36}$$

For fatigue failure problems, we first obtain the probability density of peak heights $f_H(h)$, or equivalently $f_S(s)$. Using $f_S(s)$ and an empirical law (S-N curve) obtained from experimental data, the total damage is obtained. Unlike first excursion failure problems, it is often difficult to obtain analytical expressions for the time to failure probability density.

9.3.1 Failure Curves

Fatigue behavior of structural components are often described by S-N curves. These curves are generated by classical uniaxial fatigue tests, as shown in Figure 9.13. A specimen is subjected to constant amplitude stresses S, and the number of cycles to failure N for each stress level is recorded and plotted. Each dot in Figure 9.14 corresponds to a single experiment. The result is plotted in logarithmic scale for both S and N. As the stress level decreases, there is a stress level at which the specimen will not fail regardless of the number of load cycles. The stress level at which this occurs is called the *fatigue limit* or *endurance limit strength*, and we denote this limit as S_e.

In most cases, the number of cycles to failure corresponding to a certain stress level will vary considerably even if specimens with the same material and same dimensions are subjected to the same loading patterns. That is,

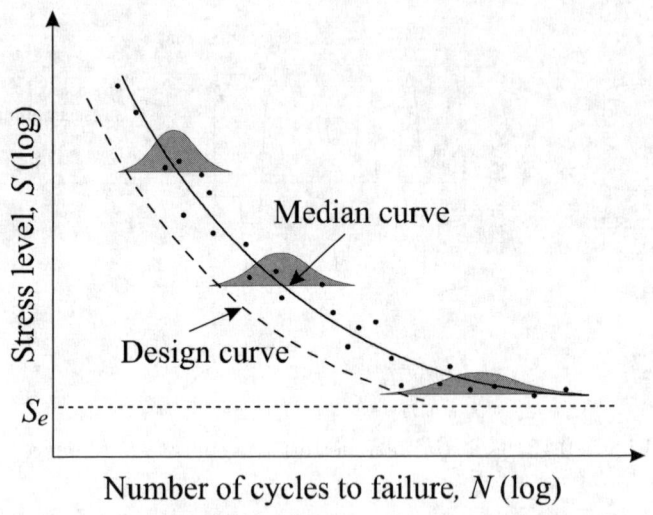

Figure 9.14: Results of fatigue tests and resulting S-N curve.

for a certain stress level, N is a random variable with a probability density function. This probability density function is called the *life distribution function*. The life distribution function varies with the stress level. As the stress level approaches the endurance limit strength, the scatter in the data increases. When a large enough number of specimens is tested, a probability density function that gives an adequate fit is the Gaussian or normal distribution. It is found that the Gaussian distribution can be a good approximation to the distribution of $\log N$.

S-N curves are used for design purposes. They are determined such that they give the lower bound. If we assume that $\log N$ is distributed normally and define the design curve to be mean-minus-two-standard-deviation curves, then the S-N curves are associated with a 97.7% probability of survival.

The Gaussian distribution is not the only nor the best approximation for the life distribution. The Weibull probability density function has been proposed to give a better fit.

In principle, Miner's rule applies to zero-mean processes. However, in practice, the applied load will not have a zero mean. When the applied load has a nonzero mean, modifications are often made in the S-N curve. One of the most important consequences of nonzero mean is the change in the fatigue life. Models that predict the fatigue life for nonzero-mean processes are modified Goodman, Gerber, and Soderberg models.

9.3. FATIGUE LIFE PREDICTION

For the remainder of this section, the S-N curves refer to the design S-N curves. We assume that they are definite and can be written in the form,
$$NS^m = A,$$
where m and A are determined from experiment. Take the logarithm of both sides to find
$$\log S = -\frac{1}{m}\log N + \frac{1}{m}\log A.$$
When $\log S$ is plotted against $\log N$, a line with slope $-1/m$ is obtained as shown in Figure 9.15(b). For this reason, this is called the linear model S-N curve. Often, a set of m and A are valid for a certain range of N. Therefore, the entire S-N curve may consist of more than one line, as shown in Figure 9.16.

Damage Equation 9.36 can be written as
$$\begin{aligned} D &= \frac{n_t}{A}\int_s s^m f_S(s)\,ds \\ &= \frac{n_t}{A}E\{S^m\}. \end{aligned} \tag{9.37}$$

In the next section, we will find the expression for the peak probability density function $f_S(s)$.

9.3.2 Peak Distribution for Stationary Random Process

Assume that the process $X(t)$ is weakly stationary with a zero mean. Appropriate modifications have been made so that the S-N curves take into account the effects of nonzero mean values.

Consider first the probability that $X(t)$ has a peak *with any magnitude* on $(t, t+dt)$. $X(t)$ will have a peak if the slope of $X(t)$ at the beginning of the interval is positive and negative at the end of the interval, $\dot{X}(t) > 0$ and $\dot{X}(t+dt) < 0$, and $X(t)$ is concave down as shown in Figure 9.17.

Mathematically, we can write
$$\begin{aligned} &\Pr\left[X(t)\ \text{has a peak on}\ (t, t+dt)\right] \\ &= \Pr\left(\dot{X}(t) > 0 \cap \dot{X}(t+dt) < 0 \cap \ddot{X}(t) < 0\right). \end{aligned}$$

Making the approximation,
$$\dot{X}(t+dt) \simeq \dot{X}(t) + \ddot{X}(t)\,dt,$$

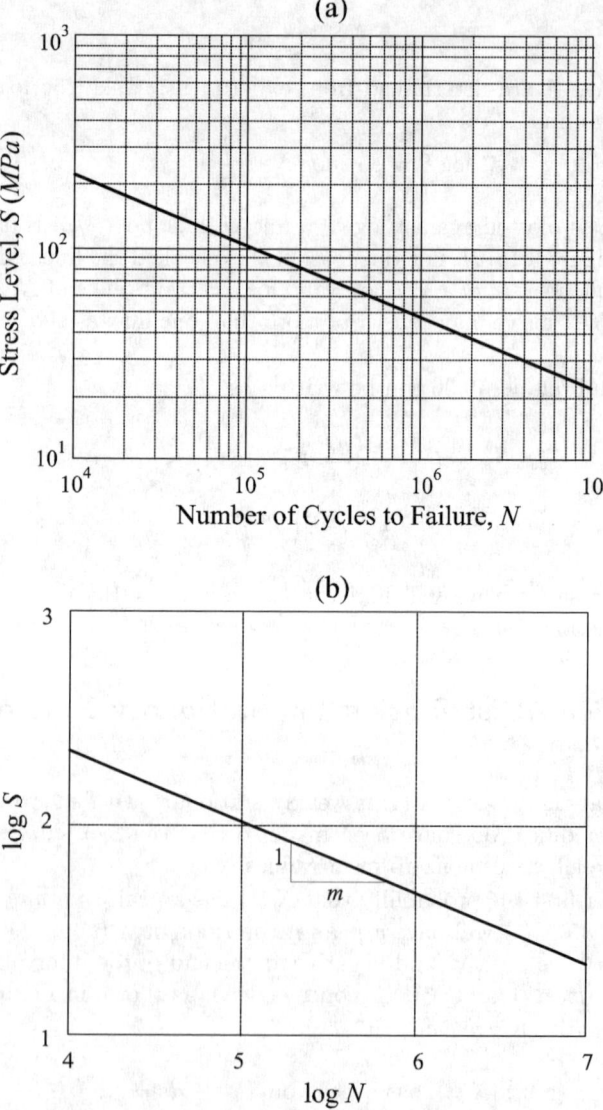

Figure 9.15: *(a)* S-N curve, *(b)* plot of $\log S$ vs. $\log N$.

9.3. FATIGUE LIFE PREDICTION

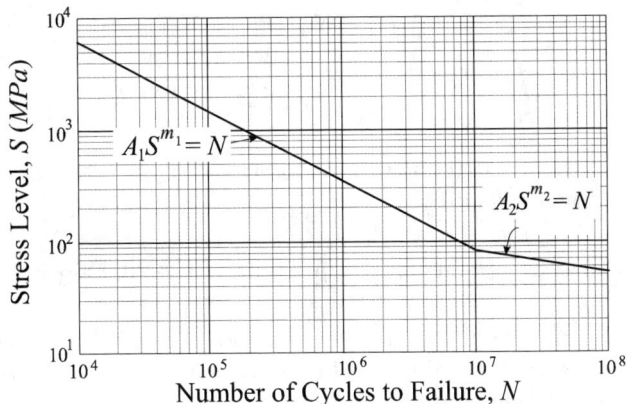

Figure 9.16: *S-N* curve for a different range of *N*.

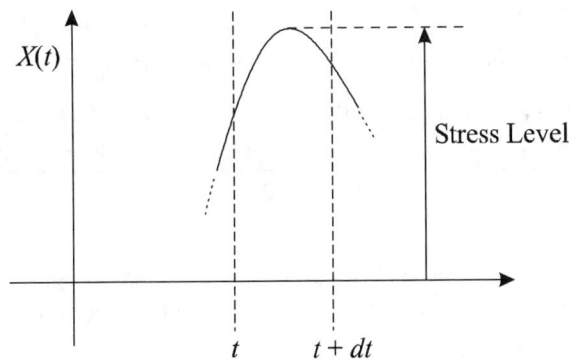

Figure 9.17: Peak of a zero-mean stationary random process.

we have

$$\Pr[X(t) \text{ has a peak on } (t, t+dt)]$$
$$= \Pr\left(0 < \dot{X}(t) < \ddot{X}(t)\,dt \;\cap\; \ddot{X}(t) < 0\right)$$
$$= \int_{-\infty}^{0} \int_{0}^{-\ddot{x}(t)dt} \int_{-\infty}^{\infty} f_{X\dot{X}\ddot{X}}(x,\dot{x},\ddot{x})\,dx\,d\dot{x}\,d\ddot{x}, \tag{9.38}$$

where $f_{X\dot{X}\ddot{X}}(x,\dot{x},\ddot{x})$ is the joint probability density function of X, $\dot{X}$, $\ddot{X}$. The probability is reduced to

$$\Pr[X(t) \text{ has a peak on } (t, t+dt)] = -\int_{-\infty}^{0} f_{\dot{X}\ddot{X}}(0,\ddot{x})\,\ddot{x}\,d\ddot{x}\,dt. \tag{9.39}$$

Note that the same result could have been obtained easily upon the realization that the desired probability is the probability that $\dot{X}(t)$ down-crosses zero on $(t, t+dt)$. This expression is given in Equation 9.11 with X replaced by $\dot{X}$, $\dot{X}$ replaced by $\ddot{X}$, and $Z = 0$.

We then ask what is the probability that the peak lies in the range $(s, s+ds)$ provided that $X(t)$ indeed has a peak on $(t, t+dt)$. That is,

$$\Pr[\text{peak occurs on } (s, s+ds) \mid X(t) \text{ has a peak on } (t, t+dt)].$$

Note that this probability can be expressed in terms of the probability density of peak magnitude as

$$\Pr[\text{peak occurs on } (s, s+ds) \mid X(t) \text{ has a peak on } (t, t+dt)]$$
$$= f_S(s)\,ds. \tag{9.40}$$

Recall that the conditional probability can be written as

$$= \frac{\Pr[\text{peak on } (s, s+ds) \mid X(t) \text{ has a peak on } (t, t+dt)]}{\Pr[X(t) \text{ has a peak on } (t, t+dt)]}.$$
$$\tag{9.41}$$

The numerator is the probability that the $X(t)$ has a peak on $(t, t+dt)$ with magnitudes on $(s, s+ds)$. The lower and upper limits of the first integration in Equation 9.38 is replaced by s and $s+ds$,

$$\Pr[\text{peak occurs on } (s, s+ds) \cap X(t) \text{ has a peak on } (t, t+dt)]$$
$$= \int_{-\infty}^{0} \int_{0}^{-\ddot{x}(t)dt} \int_{s}^{s+ds} f_{X\dot{X}\ddot{X}}(x,\dot{x},\ddot{x})\,dx\,d\dot{x}\,d\ddot{x}$$
$$= -\int_{-\infty}^{0} f_{X\dot{X}\ddot{X}}(s,0,\ddot{x})\,\ddot{x}\,d\ddot{x}\,ds\,dt. \tag{9.42}$$

9.3. FATIGUE LIFE PREDICTION

Substituting Equations 9.39 and 9.42 into Equation 9.41, we obtain

$$\Pr[\text{peak on } (s, s+ds) \mid X(t) \text{ has a peak on } (t, t+dt)]$$
$$= \frac{\int_{-\infty}^{0} f_{X\dot{X}\ddot{X}}(s, 0, \ddot{x}) \ddot{x} \, d\ddot{x}}{\int_{-\infty}^{0} f_{\dot{X}\ddot{X}}(0, \ddot{x}) \ddot{x} \, d\ddot{x}}.$$

Equating the above with Equation 9.40, the desired probability density of peaks is given by

$$f_S(s) = \frac{\int_{-\infty}^{0} f_{X\dot{X}\ddot{X}}(s, 0, \ddot{x}) \ddot{x} \, d\ddot{x}}{\int_{-\infty}^{0} f_{\dot{X}\ddot{X}}(0, \ddot{x}) \ddot{x} \, d\ddot{x}}. \tag{9.43}$$

9.3.3 Peak Distribution of a Gaussian Process

If $X(t)$ is a stationary Gaussian process, the joint probability density function is given by

$$f_{X\dot{X}\ddot{X}}(x, \dot{x}, \ddot{x})$$
$$= \frac{1}{(2\pi)^{3/2} |M|^{1/2}} \exp\left(-\frac{1}{2}(\{x\} - \{\mu_x\})^T [M]^{-1} (\{x\} - \{\mu_x\})\right),$$

where

$$[M] = \begin{bmatrix} Var\{X\} & 0 & Cov\{\ddot{X}, X\} \\ 0 & Var\{\dot{X}\} & 0 \\ Cov\{X, \ddot{X}\} & 0 & Var\{\ddot{X}\} \end{bmatrix}$$

$$\{x\} - \{\mu_x\} = \begin{Bmatrix} x - \mu_x \\ \dot{x} - \mu_{\dot{x}} \\ \ddot{x} - \mu_{\ddot{x}} \end{Bmatrix}.$$

$[M]$ is the covariance matrix and $|M|^{1/2}$ is the square root of the determinant of $[M]$. The inverse is denoted by $[M]^{-1}$. Note that

$$-Cov\{X, \ddot{X}\} = -Cov\{\ddot{X}, X\} = Var\{\dot{X}\}.$$

This can be shown as follows. The variance and the covariance are written as

$$Var\{\dot{X}\} = R_{\dot{X}\dot{X}}(0) - \mu_{\dot{X}}^2$$
$$Cov\{X, \ddot{X}\} = R_{X\ddot{X}}(0) - \mu_X \mu_{\ddot{X}}.$$

Using the fact that the mean values of the derivatives of a stationary process equal zero,[6] we can write

$$Var\left\{\dot{X}\right\} = R_{\dot{X}\dot{X}}(0)$$
$$Cov\left\{X,\ddot{X}\right\} = R_{X\ddot{X}}(0).$$

Using Equation 5.12, we have

$$Var\left\{\dot{X}\right\} = -Cov\left\{X,\ddot{X}\right\}.$$

Let

$$Var\left\{X\right\} = \sigma_X^2$$
$$Var\left\{\dot{X}\right\} = \sigma_{\dot{X}}^2$$
$$Var\left\{\ddot{X}\right\} = \sigma_{\ddot{X}}^2,$$

then the probability density function of a stationary Gaussian process $X(t)$ is given by

$$f_{X\dot{X}\ddot{X}}(x,\dot{x},\ddot{x}) = \frac{1}{(2\pi)^{3/2}\left|\sigma_{\dot{X}}^2\left(\sigma_X^2\sigma_{\ddot{X}}^2 - \sigma_{\dot{X}}^4\right)\right|^{1/2}}$$

$$\cdot \exp\left(-\frac{\sigma_{\dot{X}}^2\sigma_{\ddot{X}}^2 x^2 + \left(\sigma_X^2\sigma_{\ddot{X}}^2 - \sigma_{\dot{X}}^4\right)\dot{x}^2 + \sigma_X^2\sigma_{\dot{X}}^2\ddot{x}^2 + 2x\ddot{x}\sigma_{\dot{X}}^4}{2\left|\sigma_{\dot{X}}^2\left(\sigma_X^2\sigma_{\ddot{X}}^2 - \sigma_{\dot{X}}^4\right)\right|}\right).$$

The joint probability density function of $\dot{X}(t)$ and $\ddot{X}(t)$ is given by

$$f_{\dot{X}\ddot{X}}(\dot{x},\ddot{x}) = \int_{-\infty}^{\infty} f_{X\dot{X}\ddot{X}}(x,\dot{x},\ddot{x})\,dx$$

$$= \frac{1}{2\pi\sigma_{\dot{X}}\sigma_{\ddot{X}}}\exp\left(-\frac{1}{2}\left[\frac{\dot{x}^2}{\sigma_{\dot{X}}^2} + \frac{\ddot{x}^2}{\sigma_{\ddot{X}}^2}\right]\right).$$

---

[6] For instance, $dE\{X(t)\}/dt$ is zero for stationary $X(t)$ because $E\{X(t)\}$ is constant. $dE\{X(t)\}/dt$ can also be written as

$$\frac{d}{dt}E\{X(t)\} = E\left\{\frac{d}{dt}X(t)\right\}$$
$$= E\{\dot{X}(t)\}.$$

Therefore,
$$E\{\dot{X}(t)\} = 0.$$

9.3. FATIGUE LIFE PREDICTION

Then, the denominator of Equation 9.43 is given by

$$\int_{-\infty}^{0} f_{\dot{X}\ddot{X}}(0,\ddot{x})\,\ddot{x}\,d\ddot{x} = \int_{-\infty}^{0} \frac{1}{2\pi\sigma_{\dot{X}}\sigma_{\ddot{X}}} \exp\left(-\frac{1}{2}\frac{\ddot{x}^2}{\sigma_{\ddot{X}}^2}\right)\ddot{x}\,d\ddot{x}$$

$$= \frac{1}{2\pi}\frac{\sigma_{\ddot{X}}}{\sigma_{\dot{X}}}, \qquad (9.44)$$

and the numerator is given by

$$\int_{-\infty}^{0} f_{X\dot{X}\ddot{X}}(s,0,\ddot{x})\,\ddot{x}\,d\ddot{x}$$

$$= \int_{-\infty}^{0} \frac{1}{(2\pi)^{3/2}\,C} \exp\left(-\frac{\left[\sigma_{\dot{X}}^2\sigma_{\ddot{X}}^2 s^2 + \sigma_{X}^2\sigma_{\dot{X}}^2\ddot{x}^2 + 2s\ddot{x}\sigma_{\dot{X}}^4\right]}{2\left|\sigma_{\dot{X}}^2\left(\sigma_{X}^2\sigma_{\ddot{X}}^2 - \sigma_{\dot{X}}^4\right)\right|}\right)\ddot{x}\,d\ddot{x}$$

$$= \frac{1}{(2\pi)^{3/2}\sigma_{X}^2\sigma_{\dot{X}}^2}C\exp\left(\frac{-\sigma_{\dot{X}}^2\sigma_{\ddot{X}}^2 s^2}{2\left|\sigma_{\dot{X}}^2\left(\sigma_{X}^2\sigma_{\ddot{X}}^2 - \sigma_{\dot{X}}^4\right)\right|}\right)$$

$$+ \sigma_{\dot{X}}^4 s\sqrt{\frac{\pi}{2\sigma_{X}^2\sigma_{\dot{X}}^2}}\left(1 + \mathrm{erf}\left(\frac{\sigma_{\dot{X}}^3 s}{\sqrt{2}C\sigma_{X}}\right)\right)\exp\left(\frac{-s^2}{2\sigma_{X}^2}\right), \qquad (9.45)$$

where $C = \left|\sigma_{\dot{X}}^2\left(\sigma_{X}^2\sigma_{\ddot{X}}^2 - \sigma_{\dot{X}}^4\right)\right|^{1/2}$ and $\mathrm{erf}(s)$ is the error function,

$$\mathrm{erf}(s) = \int_{0}^{s} \frac{2}{\sqrt{\pi}} \exp\left(-t^2\right) dt.$$

Note that the error function is related to the cumulative standard normal distribution by

$$\Phi(x) = \int_{-\infty}^{x} \frac{1}{\sqrt{2\pi}} \exp\left(-\frac{1}{2}t^2\right) dt$$

$$= \frac{1}{2}\mathrm{erf}\left(\frac{x}{\sqrt{2}}\right) + \frac{1}{2}.$$

Using Equations 9.43, 9.44, and 9.45, $f_S(s)$ is given by

$$f_S(s)$$
$$= \frac{1}{\sqrt{2\pi}\sigma_X^2 \sigma_{\dot{X}} \sigma_{\ddot{X}}} \left[\sigma_{\dot{X}} \sqrt{\left(\sigma_X^2 \sigma_{\ddot{X}}^2 - \sigma_{\dot{X}}^4\right)} \exp\left(\frac{-\sigma_{\dot{X}}^2 \sigma_{\ddot{X}}^2 s^2}{2\sigma_{\dot{X}}^2 \left(\sigma_X^2 \sigma_{\ddot{X}}^2 - \sigma_{\dot{X}}^4\right)}\right) \right.$$
$$\left. + 2\sigma_{\dot{X}}^4 s \sqrt{\frac{\pi}{2\sigma_X^2 \sigma_{\dot{X}}^2}} \Phi\left(\frac{\sigma_{\dot{X}}^3 s}{\sqrt{\sigma_{\dot{X}}^2 \left(\sigma_X^2 \sigma_{\ddot{X}}^2 - \sigma_{\dot{X}}^4\right)}\sigma_X}\right) \exp\left(\frac{-s^2}{2\sigma_X^2}\right) \right].$$

The expression can be simplified if we let

$$\alpha = \frac{\Pr(X(t) \text{ up-crosses zero on } (t, t+dt))}{\Pr(X(t) \text{ has a peak on } (t, t+dt))}$$
$$= \left(\frac{1}{2\pi} \frac{\sigma_{\dot{X}}}{\sigma_X} dt\right) / \left(\frac{1}{2\pi} \frac{\sigma_{\ddot{X}}}{\sigma_{\dot{X}}} dt\right)$$
$$= \frac{\sigma_{\dot{X}}^2}{\sigma_X \sigma_{\ddot{X}}}. \qquad (9.46)$$

The probability that $X(t)$ up-crosses zero on $(t, t+dt)$ is obtained from Equations 9.8 and 9.12, and the probability that $X(t)$ has a peak on $(t, t+dt)$ is obtained from Equations 9.39 and 9.44. The probability density for the peak magnitude for a Gaussian process $X(t)$ is given by

$$f_S(s) = \frac{\sqrt{1-\alpha^2}}{\sqrt{2\pi}\sigma_X} \exp\left(\frac{-s^2}{2\sigma_X^2(1-\alpha^2)}\right)$$
$$+ 2s \frac{\alpha}{2\sigma_X^2} \Phi\left(\frac{s\alpha}{\sigma_X\sqrt{1-\alpha^2}}\right) \exp\left(\frac{-s^2}{2\sigma_X^2}\right). \qquad (9.47)$$

This is also called the Rice distribution.

Special Cases

The factor α in Equation 9.47 is called the *irregularity factor*. Using the definition of α in Equation 9.46, $\alpha \simeq 1$ describes a narrow-band process. That is, every time $X(t)$ up-crosses zero, it is likely to have a peak. Then, the probability density for the peak magnitude for a narrow-band Gaussian process is reduced to

$$f_S(s) = \frac{s}{\sigma_X^2} \exp\left(\frac{-s^2}{2\sigma_X^2}\right), \qquad s \geq 0. \qquad (9.48)$$

9.3. FATIGUE LIFE PREDICTION

The random variable S is said to have a Rayleigh distribution, which is discussed in Section 3.4.5. It should be noted that σ_X is *not* the standard deviation of S. It is the standard deviation of the random process $X(t)$.

Note that the probability density for the peak magnitude for a narrow-band Gaussian process obtained above is identical to the probability density for Rice's envelope function $R(t)$ obtained for a narrow-band Gaussian $X(t)$ in Example 9.7. This is because the envelope function describes the magnitude of the peaks.

Consider a case where α is a very small number such that $\alpha \simeq 0$. This is the case when the number of peaks is much larger than the number of zero crossings. It also means that $X(t)$ is broad band. Figure 9.18 shows such a case. Then the probability density for the peak magnitude for a broad-band Gaussian process approaches Gaussian or

$$f_S(s) \simeq \frac{1}{\sqrt{2\pi}\sigma_X} \exp\left(\frac{-s^2}{2\sigma_X^2}\right), \qquad -\infty < s < \infty. \tag{9.49}$$

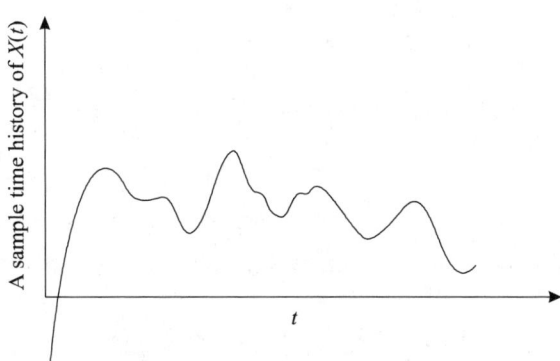

Figure 9.18: A sample time history of $X(t)$ when $\alpha \simeq 0$.

Miner's rule applies to narrow-band processes where damage is done by the stress cycles. For the time history of stress $X(t)$ shown in Figure 9.18, it is hard to say how much damage is done to the system since each cycle is not as definite. The rainflow method proposed by Dowling[7] is the most widely accepted method to count the number of cycles in a wide band process.

Except in these extreme conditions, the probability density for the peak magnitude is neither Rayleigh nor Gaussian, but between those. Also, the input may not be stationary, and nor is $X(t)$. In practice, the Weibull distribution is used for long-term stress level distribution due to nonstationary

---

[7] Dowling, N.E., "Fatigue Failure Predictions of Complicated Stress-Strain Histories," *ASTM J. Mater.*, Vol. 7, No. 1, 1972, pp. 71-87.

environmental loading. The Weibull distribution is given by

$$f_S(s) = \zeta \left(\frac{1}{\delta}\right)^\zeta s^{\zeta-1} \exp\left(-(s/\delta)^\zeta\right), \qquad s \geq 0, \qquad (9.50)$$

where ζ and δ are constants and are related to the Weibull constants in Equation 9.34 by

$$\beta = \zeta, \qquad \alpha = \left(\frac{1}{\delta}\right)^\zeta.$$

Here, ζ and δ are preferred to α and β when the Weibull distribution describes the stress level distribution. ζ is called the Weibull shape parameter, and δ is called the Weibull scale parameter.

Suppose that $S_{\max}$ is the maximum stress that we expect to see once during the entire operation with N_s cycles. In other words, the probability that the stress exceeds $S_{\max}$ is $1/N_s$. We can write

$$\int_{S_{\max}}^\infty f_S(s)\,ds = \frac{1}{N_s}$$

$$\exp\left(-(S_{\max}/\delta)^\zeta\right) = \frac{1}{N_s}$$

$$\frac{1}{\delta} = \frac{1}{S_{\max}} \left(\ln \frac{1}{N_s}\right)^{1/\zeta}.$$

δ can be eliminated, and the Weibull probability distribution becomes

$$f_S(s) = -\frac{\ln N_s}{N_s} \zeta \frac{s^{\zeta-1}}{S_{\max}^\zeta} \exp\left(-\left(\frac{s}{S_{\max}}\right)^\zeta\right), \qquad s \geq 0.$$

Example 9.10 Cumulative Damage of a System Subjected to Gaussian Input Using Miner's Rule

Find the total damage done when $S(t)$ can be characterized by *(i)* Rayleigh and *(ii)* Weibull distributions. Assume that the S-N curve is given by $NS^m = A$.

Solution Damage is given by (Equation 9.37)

$$D = \frac{n_t}{A}\int_s s^m f_S(s)\,ds = \frac{n_t}{A} E\{S^m\}.$$

The damage can be found by substituting $f_S(s)$ for each case.

9.3. FATIGUE LIFE PREDICTION

(i) This is the case where $X(t)$ is a narrow-band Gaussian process. The probability density function $f_S(s)$ is given by Equation 9.48. The damage is given by

$$D = \frac{n_t}{A} \int_0^\infty \frac{s^{m+1}}{\sigma_X^2} \exp\left(\frac{-s^2}{2\sigma_X^2}\right) ds.$$

Let $s^2/2\sigma_X^2 = t$,

$$s = \sqrt{2t}\sigma_X$$
$$ds = \sqrt{2}\sigma_X \frac{1}{2} t^{-1/2} dt.$$

The damage is given by

$$\begin{aligned} D &= \frac{n_t}{A}\left(\sqrt{2}\sigma_X\right)^m \int_0^\infty t^{m/2} \exp(-t)\, dt \\ &= \frac{n_t}{A}\left(\sqrt{2}\sigma_X\right)^m \int_0^\infty t^{\left(\frac{m}{2}+1\right)-1} \exp(-t)\, dt \\ &= \frac{n_t}{A}\left(\sqrt{2}\sigma_X\right)^m \Gamma\left(\frac{m}{2}+1\right), \end{aligned} \tag{9.51}$$

where the gamma function is defined as

$$\Gamma(r) = \int_0^\infty t^{r-1} \exp(-t)\, dt.$$

(ii) The probability density $f_S(s)$ for this case is given by Equation 9.50, and the damage is given by

$$D = \frac{n_t}{A} \int_0^\infty s^m \zeta \left(\frac{1}{\delta}\right)^\zeta s^{\zeta-1} \exp\left(-(s/\delta)^\zeta\right) ds.$$

Let $(s/\delta)^\zeta = t$, then

$$s = \delta t^{\frac{1}{\zeta}}$$
$$ds = \frac{\delta}{\zeta} t^{\frac{1}{\zeta}-1} dt.$$

Damage is given by

$$\begin{aligned} D &= \frac{n_t}{A} \int_0^\infty \zeta \left(\frac{1}{\delta}\right)^\zeta \left(\delta t^{\frac{1}{\zeta}}\right)^{m+\zeta-1} \exp(-t) \frac{\delta}{\zeta} t^{\frac{1}{\zeta}-1} dt \\ &= \frac{n_t}{A} \int_0^\infty \delta^m t^{\frac{m}{\zeta}+1-1} \exp(-t)\, dt \\ &= \frac{n_t}{A} \delta^m \Gamma\left(\frac{m}{\zeta}+1\right). \end{aligned}$$

Table 9.1: Summary of Example 9.10

Peak Distributions	$E\{S^m\}$
Rayleigh Distribution	$\left(\sqrt{2}\sigma_X\right)^m \Gamma\left(\frac{m}{2}+1\right)$
Gaussian Distribution	$\frac{1}{2\sqrt{\pi}}\left(\sqrt{2}\sigma_X\right)^m \Gamma\left(\frac{m}{2}+\frac{1}{2}\right)$
Weibull Distribution	$\delta^m \Gamma\left(\frac{m}{\zeta}+1\right)$

The results obtained in this example problem are summarized in Table 9.1.
⊛

Example 9.11 Application of Miner's rule

Find the total damage done to a component if the stress $X(t)$ is a stationary narrow-band Gaussian process with zero-mean value and variance of 10 N/m², and the total number of stress cycles is 100,000. The stress cycles are applied over 10,000 s. Assume that the failure curve obeys

$$N(s) = \frac{1.08 \times 10^{14}}{s^{3.5}}, \qquad s > 0,$$

where S has units of MPa.
Solution   We identify that

$$A = 1.08 \times 10^{14}$$
$$m = 3.5.$$

The total damage done to the component is given by (Equation 9.51)

$$\begin{aligned}
D &= \frac{n_t}{A}\left(\sqrt{2}\sigma_X\right)^m \Gamma\left(\frac{m}{2}+1\right) \\
&= \frac{100000}{1.08 \times 10^{14}} \left(\sqrt{2 \times 10}\right)^{3.5} \Gamma\left(\frac{3.5}{2}+1\right) \\
&= \frac{100000}{1.08 \times 10^{14}} \left(\sqrt{2 \times 10}\right)^{3.5} \times 1.6084 \\
&= 2.8169 \times 10^{-7}.
\end{aligned}$$

Usually, we are concerned with the fatigue life left in the part. That is, what is the estimated failure time T_f? If the process is stationary, we can write the total number of cycles as

$$n_t = \nu t.$$

9.3. FATIGUE LIFE PREDICTION

Then, the mean-value damage is given by

$$E\{D\} = \nu E\{t\} \int_s \frac{f_S(s)}{N(s)} ds,$$

where the term,

$$\nu \int_s \frac{f_S(s)}{N(s)} ds,$$

can be thought of as the damage rate. The estimated time until the damage reaches one is then

$$E\{T_f\} = \left[\nu \int_s \frac{f_S(s)}{N(s)} ds\right]^{-1}.$$

Since 100,000 stress cycles are applied in 10,000 s, the peak rate ν equals 0.1, or 10 peaks per second. The estimated time to failure is then

$$\begin{aligned}
E\{T_f\} &= \frac{1}{\frac{\nu}{A}\left(\sqrt{2}\sigma_X\right)^m \Gamma\left(\frac{m}{2}+1\right)} \\
&= \left(\frac{0.1}{1.08 \times 10^{14}} \left(\sqrt{2 \times 10}\right)^{3.5} \times 1.6084\right)^{-1} \\
&= 3.550 \times 10^{12} \text{ s.}
\end{aligned}$$

✳

In Equation 9.37, the damage done to the system is given by

$$D = \frac{n_t}{A} E\{S^m\},$$

where n_t is the total number of cycles, and A and m are the parameters of the S-N curve. Consider a case where the same number of constant amplitude stress cycles is applied to a new specimen with the same physical and geometrical properties as the previous one. The damage by this constant amplitude stress S_e is

$$D = \frac{n_t}{A} S_e^m.$$

Then what is the magnitude of S_e such that the same amount of damage is done to the system? This can be found by equating damage expressions,

$$\begin{aligned}
\frac{n_t}{A} &= \frac{n_t}{A} S_e^m \\
S_e &= (E\{S^m\})^{1/m},
\end{aligned}$$

and S_e is called the equivalent constant amplitude stress.

Example 9.12 Equivalent Constant Amplitude Stress

Consider a spring-mass system in Figure 9.19. A Gaussian random force $MF(t)$ with power spectral density of

$$S^o_{FF}(\omega) = \begin{cases} S_o, & \omega_1 < \omega < \omega_2 \\ 0, & \text{elsewhere} \end{cases}$$

is applied to the mass. The spring is likely to fail at the point of attachment to the wall due to fatigue. The stress felt at the wall is $X(t) = KY(t)$, where K is the spring constant.

Find *(i)* $S^o_{XX}(\omega)$, *(ii)* probability density of peaks $f_S(s)$, *(iii)* damage done to the system assuming that the total number of cycles is $n_t = 100,000$, and *(iv)* the equivalent stress level S_e.

Use $M = 1$ kg, $K = 10$ N/m, $S_o = 20$ m/s, $\omega_1 = 4$ rad/s and $\omega_2 = 5$ rad/s. Use the failure law,

$$N(s) = \frac{1.08 \times 10^{14}}{s^{3.5}}, \qquad s > 0.$$

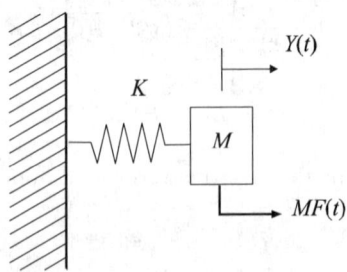

Figure 9.19: Spring-mass system subjected to Gaussian force.

Solution The equation of motion for this system is given by

$$M\ddot{Y} + KY = MF(t).$$

The frequency response function for this system is

$$H(\omega) = \frac{\mathcal{F}\{Y(t)\}}{\mathcal{F}\{F(t)\}} = \frac{1}{\omega_n^2 - \omega^2},$$

where $\mathcal{F}\{Y(t)\}$ denotes the Fourier transform of $Y(t)$, and ω_n is the natural frequency, defined by

$$\omega_n = \sqrt{\frac{K}{M}}.$$

9.3. FATIGUE LIFE PREDICTION

The spectral density of $Y(t)$ is given by

$$S_{YY}^o(\omega) = |H(\omega)|^2 S_{FF}^o(\omega)$$
$$= \left(\frac{1}{\omega_n^2 - \omega^2}\right)^2 S_o, \quad \omega_1 < \omega < \omega_2.$$

The quantity of interest is the stress KY or X

$$X = KY.$$

The spectral density of $X(t)$ can be found using the relation between the autocorrelation functions,

$$R_{XX}(\tau) = E\{X(t)X(t+\tau)\}$$
$$= K^2 E\{Y(t)Y(t+\tau)\} = K^2 R_{YY}(\tau).$$

Taking the Fourier transform of the autocorrelations, we have

$$S_{XX}^o(\omega) = K^2 S_{YY}^o(\omega) = \left(\frac{K}{\omega_n^2 - \omega^2}\right)^2 S_o.$$

The standard deviations are given by

$$\sigma_X = \sqrt{\int_{\omega_1}^{\omega_2} S_{XX}^o(\omega)\, d\omega} = 4.79$$

$$\sigma_{\dot{X}} = \sqrt{\int_{\omega_1}^{\omega_2} \omega^2 S_{XX}^o(\omega)\, d\omega} = 20.9$$

$$\sigma_{\ddot{X}} = \sqrt{\int_{\omega_1}^{\omega_2} \omega^4 S_{XX}^o(\omega)\, d\omega} = 91.9,$$

and the irregularity factor is given by

$$\alpha = \frac{\sigma_{\dot{X}}^2}{\sigma_X \sigma_{\ddot{X}}} = 0.992.$$

Using Equation 9.47, the probability density function of the stress level is given by

$$f_S(s) = \frac{1}{\sqrt{2\pi}\, 4.79 \sqrt{1 - (0.992)^2}} \exp\left(\frac{-s^2}{2(4.79^2)\left(1 - (0.992)^2\right)}\right)$$

$$+ s\frac{0.992}{(4.79)^2} \Phi\left(\frac{s}{4.79\sqrt{(1/(0.992)^2 - 1)}}\right) \exp\left(\frac{-s^2}{2(4.79)^2}\right)$$

$$= 0.468 \exp\left(-0.688 s^2\right) + 0.0216\, s\, \Phi(1.16s) \exp\left(-0.0109 s^2\right).$$

Since the irregularity factor is close to 1, we can approximate $f_S(s)$ by the Rayleigh distribution,

$$f_S(s) \simeq \frac{s}{\sigma_X^2} \exp\left(\frac{-s^2}{2\sigma_X^2}\right)$$
$$= 0.0416 s \exp\left(-0.0218 s^2\right), \qquad s \geq 0.$$

Using Equation 9.51, the damage is given by

$$D = \frac{n_t}{A} \left(\sqrt{2}\sigma_X\right)^m \Gamma\left(\frac{m}{2}+1\right).$$

⊛

9.4 Concluding Summary

This chapter introduced the concept of mechanical damage and fatigue. A variety of rules for counting cycles are derived, and various mechanisms for failure are outlined. This introduction is quite sophisticated and shows one of the most important applications of random process/vibration theory. Based on this chapter the reader can continue more advanced studies in the literature.

9.5 Problems

Section 9.2: First Excursion Failure

1. Consider a car that is being hit by hail. The number of times that hail hits the car is modeled as a Poisson process with the arrival rate of $\nu = 1$ hail/s. Find the probability that the car will be hit more than 100 times in 20 minutes. It is estimated that only one tenth of the hail is big enough to cause permanent damage to the car. What is the probability that the car will have more than 5 such dents in 20 minutes.

2. An electronic component is subjected to a random force that is a banded Gaussian process with spectrum given by

$$S_{FF}(\omega) = \frac{S_0}{\omega_2 - \omega_1}, \qquad \omega_1 < |\omega| < \omega_2.$$

When can we expect that the force will exceed its threshold of F_0 for the first time? The electronic component is protected by a vibration isolator with damping and stiffness of c and k, respectively. The electronic component will fail if the acceleration of the component exceeds F_0/m. How does the expected time of first excursion change?

9.5. PROBLEMS

3. In the previous problem, assume that $\omega_2 - \omega_1$ is small so that the force can be considered a narrow-band process. What is the expected time that the force will exceed F_0 for the first time with or without the vibration isolator? How do the results change from the previous problem?

4. A delicate system is enclosed in a box, which is also protected by a spring and damper system, as shown in Figure 9.20. The delicate system will fail if its acceleration exceeds a certain threshold, a_c m/s². The box itself is subjected to a Gaussian random force $F(t)$ with a white noise spectrum at level S_0 N²s. What is the expected time of failure?

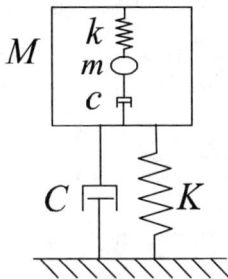

Figure 9.20: Two-degree-of-freedom system subjected to random force.

5. Determine the probability density function $f_{\dot{R}}(\dot{r})$ of the time derivative of the envelope process for a narrow-band Gaussian process with zero mean.

6. Find the average clump size of a narrow band Gaussian process $X(t)$ with one-sided spectral density given by

$$S^o(\omega) = S_o, \quad \omega_1 < \omega < \omega_2$$
$$= 0 \text{ elsewhere.}$$

7. Show that uncorrelated Gaussian processes are also independent.

Section 9.3: Fatigue Life Prediction

8. $X(t)$ is a narrow-band Gaussian process whose envelope function is distributed according to the Rayleigh distribution. What is the probability that the peaks will exceed 5 times the standard deviation of $X(t)$? Also, what is the value of the peak if the probability of being exceeded is 0.1?

9. Useful statistical parameters that describe the random ocean wave heights are mean-square height, mean height, and significant wave height, where the significant wave height is the mean of the one-third highest waves. Let $\eta(t)$ be the narrow-band random wave elevation measured from still water level, with wave amplitude, A that follow the Rayleigh density,

$$f_A(a) = \frac{a}{\sigma_\eta^2} \exp\left(-\frac{1}{2}\frac{a^2}{\sigma_\eta^2}\right) \quad \text{for } 0 < a < \infty.$$

The wave height is defined by the distance between the trough and the crest, or $2A$. Find the mean square height, mean height and the significant wave height in terms of σ_η.

10. Obtain the irregularity factor α that is used in the Rice distribution if the random process has the spectral density,

$$S_{XX}(\omega) = \frac{S_0}{\omega_0}, \quad |\omega| < \omega_0.$$

11. Determine the total damage done when the stress of the system, $S(t)$, can be characterized by the Gaussian distribution. Assume that the S-N curve is given by $NS^m = A$.

Chapter 10

Nonlinear Dynamic Models

Nonlinearities in governing equations of motion can occur due to a variety of physical causes, material or geometric, or loading. By nonlinearity we mean functions for which the principle of linear superposition does not apply, for example, x^p, $\cos\phi$, $\exp(-xt)$. In all these instances, the functional relationship between displacement and damping and/or spring force is not linear. Because of the complexity of nonlinear behavior, there is no single overarching principle that governs the solution of nonlinear equations. There are, however, general approaches that can be utilized for the solution of certain classes of nonlinear differential equations. An excellent and readable monograph on nonlinear vibration is by Stoker.[1]

The study of nonlinear system behavior can be broadly categorized as either qualitative or quantitative. Qualitative approaches are concerned less with response time histories and more with the stability characteristics of the system in the neighborhood of an equilibrium. Quantitative approaches, on the other hand, are devoted to the derivation of usually approximate solutions to the governing nonlinear equation of motion. Perturbation techniques[2,3] are a set of quantitative methods used to approximate the response of systems with small nonlinearity.

Of particular interest in nonlinear vibration are systems that have peri-

---

[1] *Nonlinear Vibrations in Mechanical and Electrical Systems*, J.J. Stoker, original edition 1950, Wiley Classics Library Edition, reprinted 1992.

[2] *Perturbation Techniques in Mathematics, Physics, and Engineering*, R. Bellman, Holt, Rinehart and Winston, New York, 1966.

[3] *Nonlinear Oscillations*, A.H. Nayfeh, D.T. Mook, Wiley-Interscience, New York, 1979.

odic solutions.[4] A nonlinear equation may have periodic solutions as well as nonperiodic solutions. Where nonlinear effects are large, the analyst must use numerical methods to find the response time history. The origin of the quantitative method lies in the early history of astronomical calculations, where it received the name *perturbation method*.

Stochasticity usually enters the problem due to complexity in the loading environment. This complexity, such as turbulence, is modeled using probabilistic tools. Sometimes the randomness is in the system itself, for example, in the material or the geometry. Solutions of stochastic nonlinear models, just as stochastic linear models, depend on the availability of deterministic nonlinear solutions.

In this chapter, the focus is on single-degree-of-freedom oscillators where the key concepts can be introduced. Higher-order nonlinear systems are certainly important, but they are primarily tackled using numerical methods. We introduce and explore deterministic solutions and their stochastic extensions.

10.1 Examples of Nonlinear Vibration

Some examples of physical systems governed by nonlinear equations are given next. First consider systems where dissipation is ignored:

- *motion in a gravitational field:* A pendulum oscillates in a gravitational field. The governing equation of motion is

$$\ddot{\theta} + \frac{g}{l}\sin\theta = 0.$$

  In linearized theory, for small oscillations, θ is small, and as an approximation we let $\sin\theta \simeq \theta$, resulting in the equation of the simple harmonic oscillator, $\ddot{\theta} + (g/l)\theta = 0$. Figure 10.1 shows the range over which the approximation looks reasonable.

- *restoring moments for floating bodies:* A floating body, such as a ship, oscillates in response to wave, current, and wind loads that act on it. In general the motion is governed by a nonlinear equation.

- *elastic restoring forces:* The restoring force for a moored body is nonlinear as is the force-displacement relation for a shallow arch that can undergo snap-through buckling. While we generally may assume a linear elastic model for the relation between an applied force and the subsequent deformation, cases such as those mentioned above are elastic but nonlinear.

---

[4] *Nonlinear Ordinary Differential Equations*, D.W. Jordan, P. Smith, Oxford University Press, Second Edition, 1988.

10.1. EXAMPLES OF NONLINEAR VIBRATION

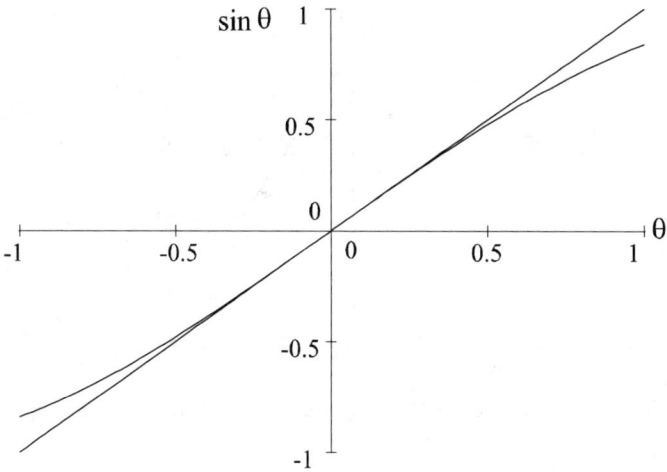

Figure 10.1: $\sin\theta$ versus its approximation θ rad.

- *geometric nonlinearities:* As a body moves, nonlinear effects become significant. This can be true of a continuous process or a discontinuous process. The latter may occur when additional forces come into play at discrete locations. For example, heat exchanger tubes with stops and ships with mooring lines and fenders have on-off constraints.

Consider the following nonlinear dissipative forces:

- *internal damping:* As a structure cycles through inelastic ranges of its constitutive relation, there is an energy loss through permanent deformation and the structure is said to be hysteretic.

- *interface damping or friction:* A body sliding on a surface experiences Coulomb friction.

- *flow-induced forces:* Drag between a fluid and structure results in a force $\sim v\left|v\right|$, where v is the relative flow velocity.

Example 10.1 Simple Pendulum

Consider the simple pendulum of Figure 10.2. The equation of motion is given as
$$ml^2\ddot{\theta} + mgl\sin\theta = 0.$$

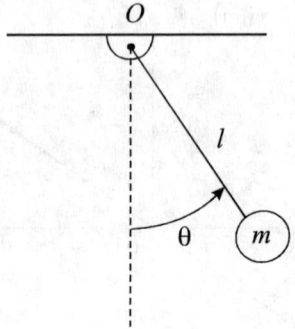

Figure 10.2: Simple pendulum.

Using the linear approximation $\sin\theta \simeq \theta$, the response is given by the harmonic function,

$$\theta = A\sin(\omega_n t + \phi),$$
$$\omega_n = \sqrt{g/l}.$$

It is relatively straightforward to estimate the angles for which the approximation is no longer acceptable. For those cases, it is possible to retain the first two terms of the series that the sine represents, $\sin\theta \simeq \theta - \theta^3/6$. The equation of motion is then

$$ml^2\ddot{\theta} + mgl\left(\theta - \frac{\theta^3}{6}\right) = 0,$$

or

$$\ddot{\theta} + \omega_n^2\left(\theta - \frac{\theta^3}{6}\right) = 0. \tag{10.1}$$

This is an oscillator with a nonlinear spring, $\ddot{\theta} + k(\theta) = 0$, where $k(\theta)$ is a general nonlinear function of θ. A linear spring behaves according to $k\theta$. The nonlinear spring can be categorized as a *soft spring*, where the slope decreases with increasing θ, or a *hard spring*, where the slope increases with θ, as seen in Figure 10.3.

10.2. FUNDAMENTAL NONLINEAR EQUATIONS

Therefore,

$$\text{if } \frac{dk}{d\theta} = k \text{ then the spring is linear}$$

$$\text{if } \frac{d^2k}{d\theta^2} > k \text{ then the spring is hard}$$

$$\text{if } \frac{d^2k}{d\theta^2} < k \text{ then the spring is soft.}$$

By these categories we imply that the hard spring is strictly increasing and the soft spring is strictly decreasing, and k is some arbitrary stiffness (slope) that describes how the spring behaves for small θ. Of course, hybrid behavior is possible. This is generalized next.

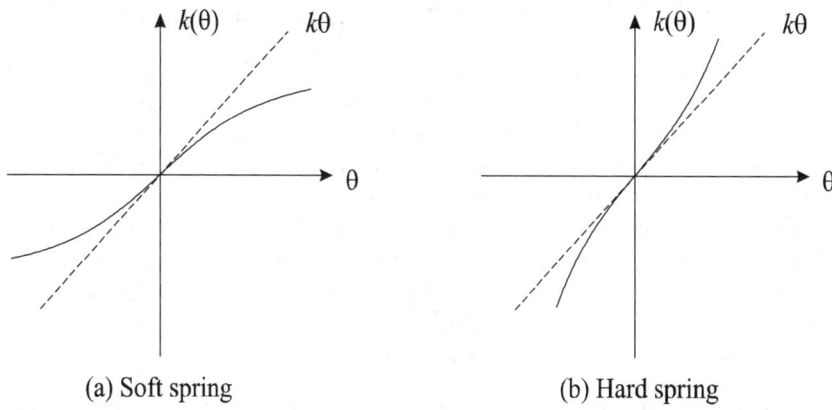

(a) Soft spring (b) Hard spring

Figure 10.3: *(a)* Soft spring results in larger displacement for same force increment. *(b)* Hard spring requires greater force for same displacement increment.

✽

10.2 Fundamental Nonlinear Equations

A general nonlinear oscillator, where x is the displacement, can be represented by

$$m\ddot{x} + g(\dot{x}) + h(x) = F(t).$$

The inertia force is given by $m\ddot{x}$, the damping force by $-g(\dot{x})$, and the spring or restoring force by $-h(x)$. The external force or excitation is represented by the function $F(t)$, which may be a random process. The reason for the negative signs is that by Newton's second law of motion, the sum of the forces on a free body is equated to the inertia force, that is, $m\ddot{x} = -g(\dot{x}) - h(x) + F(t)$. This general equation is specialized to two important classes of problems. One is the class where only the stiffness force is nonlinear, and the other where only the damping force is nonlinear.

The nonlinear stiffness class is governed by the equation,

$$m\ddot{x} + c\dot{x} + h(x) = F(t).$$

The stiffness force $-h(x)$ is a nonlinear function of x, signifying how the restoring force acts in the different displacement regimes. Equations of this form are classified as the *Duffing* equation.

A class of nonlinear damping equations is governed by

$$m\ddot{x} + g(\dot{x}) + kx = F(t), \quad \text{where} \begin{cases} \dot{x}g(\dot{x}) < 0 & \text{for small } \dot{x} \\ \dot{x}g(\dot{x}) > 0 & \text{for large } \dot{x}. \end{cases}$$

The damping force $-g(\dot{x})$ is a nonlinear function of velocity $\dot{x}$. Note the interesting behavior of this oscillator depending on whether the damping force acts in the direction of motion or acts opposite to the direction of velocity. These are called *self-excited* oscillators because the "damping" force is in the direction of velocity for small velocities so that the state of rest is unstable and motion develops from the rest position under the slightest disturbance even if the external force is zero. A special case of this equation is known as the *van der Pol* equation. We will study these equations in more detail later in this chapter.

We note two classes of equations:

Definition 1 *An **autonomous** equation is one where time t does not appear explicitly.*

Definition 2 *A **nonautonomous** equation is one where t appears explicitly, for example, in the forcing term.*

The difference between the nonautonomous and the autonomous systems is that the solutions of the former have the period of the external excitation, or more generally, are in a rational ratio to this period, while the period of the latter are determined by the parameters of the differential equation itself.

Also in an autonomous differential equation, t can be replaced by $t + t_0$, where t_0 is the phase, and still have the same solution. This means that the time axis can be arbitrarily translated and that the origin can be selected so that the initial velocity $dx/dt = 0$. Even so, sometimes for nonautonomous equations, additional conditions such as the zero initial velocity are imposed. The resulting analysis is less general, but greatly simplified and sometimes still useful. On occasion we make such additional assumptions in this chapter.

10.3 Statistical Equivalent Linearization

It is sometimes possible to create an equivalent linear system that reproduces the essential characteristics of the original nonlinear system. Consider the general nonlinear system,

$$\ddot{Y} + \alpha \dot{Y} + \omega^2 Y + g(Y, \dot{Y}) = f(t), \qquad (10.2)$$

where $\alpha = c/m$, $\omega^2 = k/m$, $g(Y,\dot{Y}) = g_0(Y,\dot{Y})/m$ is the nonlinear components, and $f(t) = F(t)/m$ is the random force per unit mass. An example of a nonlinear force is $g(Y,\dot{Y}) = \gamma \dot{Y}\left|\dot{Y}\right| + \beta Y^3$. Of course, there are numerous classes of nonlinear functions that arise in applications.

The deterministic equivalent linearization procedure attempts to linearize the governing equation while maintaining an energy equality between the linear approximation and the exact nonlinear equation. This can be accomplished by minimizing the error introduced by replacing Equation 10.2 by the following,

$$\ddot{Y} + \alpha_e \dot{Y} + \omega_e^2 Y = f(t) + \varepsilon(t).$$

Introduced are an equivalent damping parameter α_e and frequency parameter ω_e^2, and on the right-hand side a term $\varepsilon(t)$ that embodies the errors made due to using this simplified version of the governing equation.

Comparing the linearized equation with the original equation provides us with the expression for the error,

$$\varepsilon(t) = (\alpha_e - \alpha)\dot{Y} + \left(\omega_e^2 - \omega^2\right)Y - g(Y,\dot{Y}). \qquad (10.3)$$

Since this equation is assumed to have a stochastic forcing $f(t)$, the output process $Y(t)$ is also a stochastic process. The error is then random and can be minimized in one way by minimizing the mean-square error $E\{\varepsilon^2(t)\}$. The procedure will be to differentiate twice with respect to the two parameters α_e and ω_e^2 to verify that there is a minimum (second derivative is positive), and then to set the respective first derivatives to zero in order to have the two equations needed to find expressions for α_e and ω_e.

We can observe that the second derivative of $E\left\{\varepsilon^2(t)\right\}$ with respect to α_e equals $2E\left\{\dot{Y}^2\right\}$, and with respect to ω_e^2 equals $2E\{Y^2\}$. These results can be found purely by observation since we are looking for second derivatives. The only terms that will remain after a second derivative will have had to be multiplied by the square of the differentiated variable, that is, $(\alpha_e)^2$ and $\left(\omega_e^2\right)^2$.

Considering Equation 10.3, we see that only terms multiplied by α_e and ω_e^2 will survive a first derivative. Therefore, take the derivative with respect to α_e,

$$\frac{\partial}{\partial \alpha_e} E\left\{\varepsilon^2(t)\right\} = 0$$

$$(\alpha_e - \alpha) E\left\{\dot{Y}^2\right\} + (\omega_e^2 - \omega^2) E\left\{Y\dot{Y}\right\} - E\left\{\dot{Y}g(Y,\dot{Y})\right\} = 0$$

$$\alpha_e = \alpha - \frac{(\omega_e^2 - \omega^2) E\left\{Y\dot{Y}\right\} - E\left\{\dot{Y}g(Y,\dot{Y})\right\}}{E\left\{\dot{Y}^2\right\}}. \tag{10.4}$$

Then, the derivative with respect to ω_e^2 is

$$\frac{\partial}{\partial \omega_e^2} E\left\{\varepsilon^2(t)\right\} = 0$$

$$(\omega_e^2 - \omega^2) E\left\{Y^2\right\} + (\alpha_e - \alpha) E\left\{Y\dot{Y}\right\} - E\left\{Yg(Y,\dot{Y})\right\} = 0$$

$$\omega_e^2 = \omega^2 - \frac{(\alpha_e - \alpha) E\left\{Y\dot{Y}\right\} - E\left\{Yg(Y,\dot{Y})\right\}}{E\left\{Y^2\right\}}. \tag{10.5}$$

With these equivalent parameters in hand, we can proceed with a linearized analysis of the equation of motion,

$$\ddot{Y} + \alpha_e \dot{Y} + \omega_e^2 Y = f(t). \tag{10.6}$$

The approximation comes into Equation 10.6 by virtue of dropping the $\varepsilon(t)$ on the right hand side. We see from Equations 10.4 and 10.5 that if the system is linear, $g(Y,\dot{Y}) = 0$, $\alpha_e = \alpha$, and $\omega_e^2 = \omega^2$.

To proceed further we need additional information regarding the nonlinear function $g(Y,\dot{Y})$ as well as the probabilistic properties of $(Y,\dot{Y})$. As an example, assume that the nonlinear stiffness and nonlinear damping are separable, that is $g(Y,\dot{Y}) \equiv g_1(Y) + g_2(\dot{Y})$, and that excitation $f(t)$ is Gaussian. The second assumption implies that the response is also Gaussian and that $E\left\{Y\dot{Y}\right\} = 0$. Solving for the equivalent parameters, we find

$$\alpha_e = \alpha + \frac{E\left\{\dot{Y}g_2(\dot{Y})\right\}}{E\left\{\dot{Y}^2\right\}} \tag{10.7}$$

$$\omega_e^2 = \omega^2 + \frac{E\left\{Yg_1(Y)\right\}}{E\left\{Y^2\right\}}, \tag{10.8}$$

where it has been possible to separate out the nonlinear damping and nonlinear stiffness. If the system has only nonlinear damping, then $g_1(Y) = 0$,

10.3. STATISTICAL EQUIVALENT LINEARIZATION

and if only nonlinear stiffness then, $g_2(\dot{Y}) = 0$. It is clear from Equations 10.7 and 10.8 that equivalent parameters α_e and ω_e^2 are actually functions of time, since they are functions of $(Y, \dot{Y})$. Therefore, it is assumed that in an average sense the parameter variations do not affect the response statistics.

Example 10.2 Flow-Induced Vibration of a Cylindrical Structure

Offshore structural mechanics is a complex and fascinating discipline. The intricacies of how energy is transferred between structure and fluid, as the fluid flows past the structure, are many. Here we examine the in-line vibration, that is, vibration in the same direction as the randomly fluctuating flow velocity.[5]

A single-degree-of-freedom oscillator is used to represent a structural cylinder that is typical in offshore structures. The elastic restraint on the cylinder is modeled by a linear spring of stiffness k, and structural damping by a linear viscous damper with coefficient c. As the structure vibrates, it drags fluid with it, creating an added mass. The structural oscillation is a function of the total mass, structure plus added.

The assumed random equation of motion is

$$m\ddot{Y} + c\dot{Y} + kY = F(t), \qquad (10.9)$$

where Y is the absolute displacement of the cylinder. Morison's equation is used to relate the flow induced force $F(t)$ to the fluctuating flow velocity,

$$F(t) = \rho A \ddot{U} + C_I \rho A \left(\ddot{U} - \ddot{Y}\right) + \frac{1}{2} C_D \rho D \left|\dot{U} - \dot{Y}\right| \left(\dot{U} - \dot{Y}\right), \qquad (10.10)$$

where the motion of the cylinder has been taken into account by relative acceleration $\ddot{U} - \ddot{Y}$ and relative velocity $\left|\dot{U} - \dot{Y}\right|\left(\dot{U} - \dot{Y}\right)$ terms. ρ is the fluid density, A and D equal the cross-sectional area and diameter of the cylinder, respectively, and C_I and C_D are the inertia and drag coefficients, determined experimentally. The flow velocity $\dot{U}$ can be decomposed into a mean value component V and a zero mean fluctuating component $\dot{\xi}$,

$$\dot{U} = V + \dot{\xi}(t). \qquad (10.11)$$

The fluid displacement $\xi(t)$ is assumed to be stationary and Gaussian, from which it follows that its derivatives are also stationary zero-mean Gaussian.

---

[5]This example is based on Section 5.6.3 in *Random Vibration and Statistical Linearization*, J.B. Roberts, P.D. Spanos, John Wiley and Sons, 1990. Now published by Dover Publications.

A relative displacement coordinate is defined,

$$Q = \xi - Y. \tag{10.12}$$

Substitute Equations 10.10-10.12 into Equation 10.9, to find

$$M\ddot{Q} + c\dot{Q} + \frac{1}{2}C_D \rho D \left|V + \dot{Q}\right|\left(V + \dot{Q}\right) + kQ = (M_0 - \rho A)\ddot{U} + c\dot{U} + kU,$$

where

$$M = m + C_I \rho A.$$

Divide through by M to find

$$\ddot{Q} + \beta \dot{Q} + \omega_n^2 Q + \nu G\left(\dot{Q}\right) = \mathcal{F}(t), \tag{10.13}$$

where

$$\beta = \frac{c}{M}, \quad \omega_n^2 = \frac{k}{M}, \quad \nu = \frac{C_D \rho D}{2M}$$

$$G\left(\dot{Q}\right) = \left|V + \dot{Q}\right|\left(V + \dot{Q}\right)$$

$$\mathcal{F}(t) = \frac{1}{M}(M_0 - \rho A)\ddot{\xi} + \beta \dot{\xi} + \omega_n^2 \xi.$$

From this equation for $\mathcal{F}(t)$, we deduce that $E\{\mathcal{F}(t)\} = 0$ since ξ and its derivatives are zero mean. But since $G\left(\dot{Q}\right)$ is odd and nonlinear on the left-hand side of Equation 10.13, $E\{Q\} \neq 0$.

Using the above procedures, we search for values of β and ω_n^2 to define an equivalent linear system,

$$\ddot{Q}_0 + \beta_e \dot{Q}_0 + \omega_e^2 Q_0 = \mathcal{F}(t), \tag{10.14}$$

where $Q_0 = Q - E\{Q\}$. Since the nonlinearity is only in the damping, $\omega_e^2 = \omega_n^2$, and, assuming Q is Gaussian, it is found that

$$\beta_e = \beta + \nu \left[\left(\frac{8}{\pi}\right)^{1/2} \sigma_{\dot{Q}} \exp(-v^2) + 2V \operatorname{erf}(v)\right],$$

where $\sigma_{\dot{Q}}$ is the standard deviation of $\dot{Q}$ and

$$v = \frac{V}{\sqrt{2}\sigma_{\dot{Q}}}$$

$$\operatorname{erf}(v) = \frac{2}{\sqrt{\pi}} \int_0^v \exp\left(-t^2\right) dt.$$

10.3. STATISTICAL EQUIVALENT LINEARIZATION

We see that the equivalent parameter β_e depends on knowledge of the solution Q through $\sigma_{\dot{Q}}$. Clearly then additional equations and assumptions are necessary. For the linear system of Equation 10.14,

$$\sigma_{\dot{Q}}^2 = \int_{-\infty}^{\infty} |H(\omega)|^2 \omega^2 S_{\mathcal{F}\mathcal{F}}(\omega)\, d\omega,$$

where the variables are defined as usual and

$$H(\omega) = \frac{1}{\omega_e^2 - \omega^2 + i\beta_e \omega}.$$

The spectrum of the force $S_{\mathcal{F}\mathcal{F}}(\omega)$ can be related to the spectrum of the fluid displacement.

We also need an expression for $E\{Q\}$. Take the expected value of Equation 10.13,

$$E\{\ddot{Q}\} + \beta E\{\dot{Q}\} + \omega_n^2 E\{Q\} + \nu E\{G(\dot{Q})\} = E\{\mathcal{F}(t)\}$$

$$\omega_n^2 E\{Q\} + \nu E\{G(\dot{Q})\} = 0,$$

since

$$E\{\ddot{Q}\} = E\{\ddot{Q}_0\} = 0$$
$$E\{\dot{Q}\} = E\{\dot{Q}_0\} = 0.$$

These zero-mean values arise using linear Equation 10.14. In this equation, the mean value of the forcing is zero, therefore, the output and derivatives are zero. Again assuming that $Q(t)$ is Gaussian, we find

$$E\{Q\} = -\frac{\nu}{\omega_n^2}\left[\left(\sigma_{\dot{Q}}^2 + V^2\right)\operatorname{erf} v + \left(\frac{2}{\pi}\right)^{1/2} V\sigma_{\dot{Q}} \exp(-v^2)\right].$$

❋

Example 10.3 Nonlinear Random Response of a Circular Plate

This example examines the random dynamic response of a nonuniform orthotropic circular plate.[6] An *orthotropic* structure or material has at least 2 orthogonal planes of symmetry, where material properties are independent of direction within each plane. Such materials require 9 elastic constants in their constitutive matrices. In contrast, a material without any planes of symmetry is fully anisotropic and requires 21 elastic constants, whereas a

---

[6] This example is based on the paper "Nonlinear Dynamic Response of a Nonuniform Orthotropic Circular Plate Under Random Excitation," T-P. Chang, J-L. Ke, *Computers & Structures*, Vol. 60, No. 1, pp. 113-123, 1996.

material with an infinite number of symmetry planes (that is, every plane is a plane of symmetry) is isotropic, and requires only 2 elastic constants.

Plates are a fundamental component of many engineering structures. The problem under consideration is the large deflection of an axisymmetrically clamped orthotropic circular plate of radius a subjected to an axisymmetrically random dynamic load $P(r,t)$, applying the von Kármán plate theory. An outline of the formulation is given here. The interested reader is urged to study the details in the source paper.

The derivation of the governing equation requires the following for utilization in Hamilton's principle:

1. The strain-displacement relations

2. The stress-strain relations

3. The kinetic energy

4. The strain energy

5. Rayleigh's dissipation function of the damping force

Two coupled equations of motion are derived, one for u, the radial stretching, and one for w, the transverse deformation. Forming the Lagrangian function and applying Hamilton's principle results in two coupled governing partial differential equations. The nonuniform thickness of the plate is given by the relation,

$$h = h_0 \left[1 - \gamma \left(\frac{r}{a}\right)\right],$$

where γ is a constant that describes the variation of the linear nonuniform thickness of the circular plate. Separation of variables solutions are assumed,

$$u(r,t) = h_0 g(t) \left(\frac{r}{a}\right)^2 \left[1 - \left(\frac{r}{a}\right)^2\right]$$
$$w(r,t) = h_0 f(t) \left[1 - \left(\frac{r}{a}\right)^2\right],$$

that satisfy the clamped edge conditions. A Galerkin procedure is used to derive the equation of motion governing transverse deflection in nondimensional form,

$$f(\tau) + cf(\tau) + \omega^2 \left[f(\tau) + \varepsilon f^3(\tau)\right] = \theta(\tau)\psi(\tau). \qquad (10.15)$$

10.3. STATISTICAL EQUIVALENT LINEARIZATION

The random force is given as the product $\theta(\tau)\psi(\tau)$ of a stationary Gaussian random process $\psi(\tau)$ with zero mean and a nondimensional deterministic time function $\theta(\tau)$, where τ is nondimensional time. Equation 10.15 is then solved via the procedure of stochastic equivalent linearization.

Figure 10.4: Boris Grigorievich Galerkin (1871-1945).

Galerkin   Boris Grigorievich Galerkin came from a poor family and this was to mean that he had a harder time through his years of education than would otherwise have been the case. He attended secondary school in Minsk, then in 1893 he entered the Petersburg Technological Institute. Here he studied mathematics and engineering but he needed to make money to survive so at first he took on private tutoring, then from 1896 he worked as a designer.

After graduating from the Technological Institute in 1899 he got a job at the Kharkov Locomotive Plant. In 1903 Galerkin went to St. Petersburg and there he became engineering manager at the Northern Mechanical and Boiler Plant.

From 1909 Galerkin began to study building sites and construction works throughout Europe. In the same year he began teaching at the Petersburg Technological Institute. His first publication on longitudinal curvature also appeared in 1909, work which carried on from beginnings which had been laid by Euler. This paper was highly relevant to his study of construction sites since the results were applied to the construction of bridges and frames for buildings.

His visits around European construction sites ended around 1914 but his academic work then turned to the area for which he is today best known, namely the method of approximate integration of differential equations, known as the Galerkin method. He published his finite element method in 1915.

In 1920 Galerkin was promoted to Head of Structural Mechanics at the Petersburg Technological Institute. By this time he also held two chairs, one in elasticity at the Leningrad Institute of Communications Engineers and one in structural mechanics at Leningrad University.

In 1921 the St. Petersburg Mathematical Society was reopened (it had closed in 1917 due to the Russian Revolution) as the Petrograd Physical and Mathematical Society. Galerkin played a major role in the Society along with Steklov, Sergei Bernstein, Friedmann and others.

Other work for which Galerkin is famous is his work on thin elastic plates. His major monograph on this topic *Thin Elastic Plates* was published in 1937. From 1940 until his death, Galerkin was head of the Institute of Mechanics of the Soviet Academy of Sciences.

A.T. Grigorian describes his other work: *Galerkin's scientific research in the theory of casing (1934-1945) revealed its broad application in industrial construction. His works in the field constitute a new direction in this important area. Galerkin was a consultant in the planning and building of many of the Soviet Union's largest hydrostations. In 1929, in connection with the building of the Dnepr dam and hydroelectric station, Galerkin investigated stress in dams and retaining walls with trapezoidal profile. His results were used in planning the dam.*

Example 10.4 Seismic Vibration Control Using a Liquid Column Damper

Containers of liquid have been used as passive and active structural dampers to wind and earthquake loads. In this example[7] a liquid column damper is utilized for the seismic vibration control of short period structures. The structure is modeled as a linear, viscously damped single-degree-of-freedom system. The fluid nonlinear damping is stochastic equivalently linearized. The system is shown in Figure 10.5.

The system is a passive control device that provides consistent behavior across a wide range of excitation levels. The tube in the figure has cross-sectional area A, horizontal dimension B, and contains a liquid of mass density ρ and a column of length L. The coefficient of head loss is controlled by the opening ratio of the orifice(s) and is denoted by ξ. The mass of the

---

[7] This example is based on the paper "Seismic Vibration Control of Short Period Structures Using the Liquid Column Damper," A. Ghosh, B. Basu, *Engineering Structures* 26 (2004), 1905-1913.

10.3. STATISTICAL EQUIVALENT LINEARIZATION

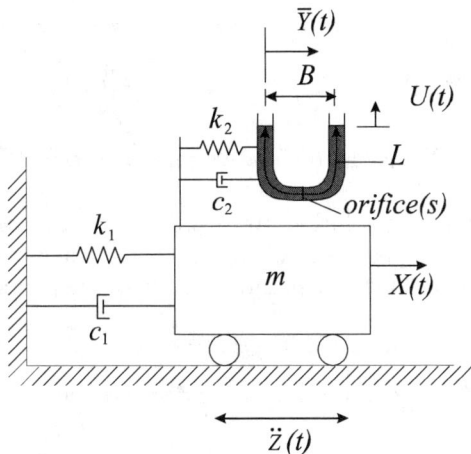

Figure 10.5: Schematic of liquid column damper for seismic vibration control.

enclosed fluid plus container is given by $\rho AL + M_c$. The structure fluid-damper system is subjected to base acceleration, $\ddot{Z}(t)$. The equation of motion of the liquid column is

$$\rho AL\ddot{U}(t) + \frac{1}{2}\rho A\xi \left|\dot{U}(t)\right|\dot{U}(t) + 2\rho AgU(t)$$
$$= -\rho AB\left[\frac{d^2\overline{Y}}{dt^2} + \ddot{X}(t) + \ddot{Z}(t)\right], \qquad (10.16)$$

where the parameters are defined in the figure. This equation for the fluid motion $U(t)$ is nonlinear. A statistical equivalent linearization results in the linear equation,

$$\rho AL\ddot{U}(t) + 2\rho AC_p\dot{U}(t) + 2\rho AgU(t) = -\rho AB\left[\frac{d^2\overline{Y}}{dt^2} + \ddot{X}(t) + \ddot{Z}(t)\right], \qquad (10.17)$$

where C_p represents the equivalent linearized damping coefficient, obtained by minimizing the mean square error between Equations 10.16 and 10.17, specifically,

$$C_p = \frac{\sigma_{\dot{U}}\xi}{\sqrt{2\pi}}.$$

Since C_p depends on $\sigma_{\dot{U}}$, and thus $U(t)$, an iterative solution procedure for C_p is required.

The equation of motion for the mass, m, is given by

$$m\left[\ddot{X}(t) + \ddot{Z}(t)\right] + c_1\dot{X}(t) + k_1 X(t) = c_2 \frac{d\overline{Y}}{dt} + k_2 \overline{Y}(t), \qquad (10.18)$$

where $c_2 Y(t) + k_2 Y(t)$ is the interactive force between structure and fluid damper.

The key steps in the solution are to normalize the above equations of motion, derive the respective frequency response functions, and then use the fundamental relation between input power spectrum and output power spectrum,

$$S_{XX}(\omega) = |H(\omega)|^2 S_{\ddot{Z}\ddot{Z}}(\omega),$$

to find the root mean-square value of the displacement.

⊛

10.3.1 Equivalent Nonlinearization

Efforts to improve the quality of linearizations has led to numerous other approximate techniques. One of these is the method of *equivalent nonlinearization*. As the name implies, the original nonlinear equation is replaced by an approximate nonlinear equation that is simpler to solve.[8]

The procedure is demonstrated using an offshore structural dynamics example. An offshore structure, such as the compliant structure in Figure 10.6, is modeled using a single-degree-of-freedom model. The equation of motion is given as

$$\ddot{X} + 2\zeta\omega_n \dot{X} + \omega_n^2 \left[X + BX^2 + CX^3\right] = \frac{c_e}{m} F_d(t),$$

where B and C are parameters that define the nonlinearity, m is the structural mass, $F_d(t)$ is the resultant of the Morison wave force $p_d(z,t)$, and c_e is a coefficient that transforms the Morison force into one acting at the level of the displacement $X(t)$. The resultant Morison force is given by

$$\begin{aligned} F_d(t) &= \int_0^{-d} p_d(z,t)\,dz \\ &= \int_0^{-d} \left[k_d \mathcal{U}(z,t)|\mathcal{U}(z,t)| + k_m \dot{\mathcal{U}}(z,t)\right] dz, \qquad (10.19) \end{aligned}$$

where $\mathcal{U}(z,t)$ is the water particle velocity, composed of a mean value added to a fluctuating component that is zero mean Gaussian,

$$\mathcal{U}(z,t) = \mu_{\mathcal{U}}(z) + U(z,t).$$

---

[8]This section is based on the paper, "Stochastic Response of Offshore Structures via Statistical Cubicization," C. Floris, R. Pulega, *Meccanica* **37**: 15-32, 2002.

10.3. STATISTICAL EQUIVALENT LINEARIZATION

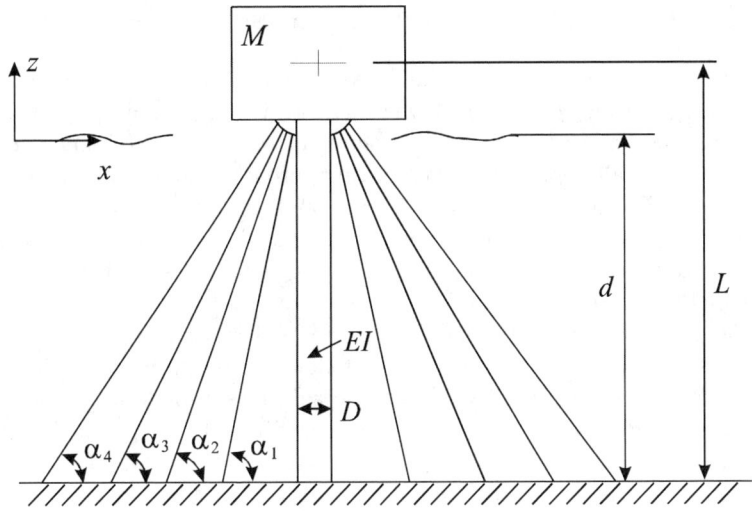

Figure 10.6: Compliant offshore tower with cable supports.

Clearly, $\dot{\mathcal{U}}(z,t) = \ddot{U}(z,t)$. k_d and k_m are drag and inertia coefficients. $U(z,t)$ is completely specified by the power spectrum of the wave elevation $\eta(t)$ multiplied by a transfer function,

$$\begin{aligned} S_{UU}(\omega, z) &= T^2(\omega, z) S_{\eta\eta}(\omega) \\ &= \omega^2 \frac{\cosh^2[\kappa(z+d)]}{\sinh^2(\kappa d)} S_{\eta\eta}(\omega), \end{aligned}$$

where $S_{\eta\eta}(\omega)$ can be taken to be the Pierson-Moskowitz spectrum, κ is the wave number, which for deep water is approximated by $\kappa \simeq \omega^2/g$, and

$$S_{\eta\eta}(\omega) = \alpha g^2 \omega^{-5} \exp\left[-\beta\left(\frac{U_{19.5}}{g\omega}\right)^4\right],$$

where $\alpha = 0.0081$, $\beta = 0.74$, and $U_{19.5}$ is the wind velocity at 19.5 m above still water. In an approach to simplify the Morison force defined in Equation 10.19, $k_d \mathcal{U}(z,t) |\mathcal{U}(z,t)|$ is replaced by a cubic polynomial, with coefficients obtained by minimizing the mean-square error. The resulting Morison force is

$$p_d(z,t) = b_0(z) + b_1(z) U(z,t) + b_2(z) U^2(z,t) + b_3(z) U^3(z,t) + k_m \dot{\mathcal{U}}(z,t).$$

At this point, given the functions $U(z,t)$ and $\dot{\mathcal{U}}(z,t)$, it is possible to integrate p_d to find $F_d(t)$, which then becomes the forcing in the nonlinear governing equation.

10.4 Perturbation or Expansion Methods

A variety of approximation techniques are available to model and solve nonlinear governing equations. One quantitative approach is by a series expansion method known as a *perturbation* of the system. Inherent in this approach, as in all approximate analytical approaches, is that the nonlinearity is small. Perturbation methods[9] are useful as first steps in a nonlinear analysis to get a general behavior of the system. The method becomes cumbersome for more realistic models. Once the general behavior is understood other methods are useful.

The essence of the perturbation method as it is used to locate periodic solutions of nonlinear oscillators is the following. Given a harmonic oscillator, there exists an infinity of periodic solutions depending on the two constants of integration. The question arises, what will happen if the linear system $\ddot{x} + x = 0$ is perturbed by adding a small term, say, $\varepsilon f(t, x, \dot{x})$? A general statement is not possible but depends on the function $f(t, x, \dot{x})$ and the smallness of ε.

For example, if $\varepsilon f(t, x, \dot{x}) = \varepsilon c \dot{x}$, then the trajectories become logarithmic spirals, stable if $c > 0$ and unstable if $c < 0$. The value of ε must be sufficiently small to ensure that the series solution converges. The perturbation method attempts to ascertain under what conditions the perturbed equation has periodic solutions. This is sometimes called the *problem of Poincaré*, who studied such problems rigorously.

Consider the unforced nonlinear pendulum as a motivating example.

Example 10.5 Nonlinear Pendulum

The approximate equation for an undamped nonlinear pendulum, Equation 10.1,

$$\ddot{\theta} + \omega_n^2 \left(\theta - \frac{\theta^3}{6}\right) = 0,$$

can be written in the more general form,

$$\ddot{\theta} + \omega_n^2 \theta + \varepsilon \theta^3 = 0, \qquad (10.20)$$

where $\omega_n^2 = \sqrt{g/l}$ and $\varepsilon = -\omega_n^2/6$. Equation 10.20 is known as an unforced *Duffing* equation. It is reasonable to assume that the nonlinearity is weak, that is, ε is small. In this way, the perturbation solution is given by

$$\theta(t) = \theta_0(t) + \varepsilon \theta_1(t) + \varepsilon^2 \theta_2(t) + \cdots,$$

---

[9] *Perturbations, Theory and Methods*, J.A. Murdock, Wiley-Interscience, 1991. This is an excellent book, the only one referenced here that provides a rigorous discussion of perturbation methods, their use, and their mathematical basis.

10.4. PERTURBATION METHODS

where $\theta_i(t)$ are to be determined for all i. Take a two-term approximation for $\theta(t)$, and substitute this into the equation of motion,

$$\left(\ddot{\theta}_0 + \varepsilon\ddot{\theta}_1\right) + \omega_n^2\left(\theta_0 + \varepsilon\theta_1\right) + \varepsilon\left(\theta_0 + \varepsilon\theta_1\right)^3 = 0,$$

and expand and group according to the order of the perturbation parameter ε,

$$\left(\ddot{\theta}_0 + \omega_n^2\theta_0\right) + \varepsilon\left(\ddot{\theta}_1 + \omega_n^2\theta_1 + \theta_0^3\right) + \varepsilon^2\left(3\theta_0^2\theta_1\right)$$
$$+ \varepsilon^3\left(3\theta_0\theta_1^2\right) + \varepsilon^4\theta_1^3 = 0.$$

Since ε is assumed to be small, terms of order ε^2, ε^3, and ε^4 are higher than the order of the assumed solution and therefore truncated. The solution obtained in this way is said to be correct to order ε. Following the general procedure, each order of the solution is satisfied independently,[10]

$$\varepsilon^0 \;:\; \ddot{\theta}_0 + \omega_n^2\theta_0 = 0 \tag{10.21}$$
$$\varepsilon^1 \;:\; \ddot{\theta}_1 + \omega_n^2\theta_1 = -\theta_0^3. \tag{10.22}$$

θ_0 is known as the *generating solution* for the sequence of equations.

The Duffing equation requires two initial conditions. These can be specified in full generality, but assume that $\theta(0) = C$ and $\dot{\theta}(0) = 0$. If the initial velocity is other than zero, only the phase is changed but not the character of the solution for autonomous Equation 10.20. It is easiest to stipulate that the ε^0 order equation satisfy the initial conditions. Then the remaining equations of higher order will satisfy zero initial conditions.

As we know, these equations are solved in sequence. Equation 10.21 is solved for θ_0, which is then substituted into the right-hand side of Equation 10.22 so that the solution θ_1 can be obtained. Therefore,

$$\theta_0(t) = A\sin(\omega_n t + \phi),$$

where satisfying the initial conditions results in $A = C$ and $\phi = \pi/2$. Equation 10.22 becomes[11]

$$\ddot{\theta}_1 + \omega_n^2\theta_1 = -C^3\sin^3(\omega_n t + \pi/2)$$
$$= -C^3\left[\frac{3}{4}\sin(\omega_n t + \pi/2) - \frac{1}{4}\sin 3(\omega_n t + \pi/2)\right].$$

---

[10] When substituting the expansion into the differential equation, we obtain a power series in ε that must vanish *identically* in ε. This is why each order in ε results in a differential equation that must be satisfied.

[11] Of course, $C\sin(\omega_n t + \pi/2) = C\cos\omega_n t$.

The solution to this equation is

$$\theta_1(t) = \frac{3}{8\omega_n} t C^3 \cos(\omega_n t + \pi/2) - \frac{C^3}{32\omega_n^2} \sin 3(\omega_n t + \pi/2). \quad (10.23)$$

The two-term approximate solution is then $\theta(t) = \theta_0(t) + \varepsilon\theta_1(t)$. We immediately see a problem with the solution so obtained in the first term on the right-hand side of Equation 10.23. This is a *secular term*, that is, an expression that grows without bounds, here due to the factor t multiplying the cosine. We know that the solution to Equation 10.20 should be periodic for small ε. The problem with the method as prescribed so far is that the truncation has removed terms that would balance, and in the limit, result in a periodic solution. This effect is demonstrated by expanding the following sine function,

$$\begin{aligned}\sin(\omega_n + \varepsilon)t &= \sin\omega_n t \cos\varepsilon t + \cos\omega_n t \sin\varepsilon t \\ &= \left(1 - \frac{1}{2!}\varepsilon^2 t^2 + \frac{1}{4!}\varepsilon^4 t^4 - \cdots\right)\sin\omega_n t \\ &\quad + \left(\varepsilon t - \frac{1}{3!}\varepsilon^3 t^3 + \frac{1}{5!}\varepsilon^5 t^5 - \cdots\right)\cos\omega_n t.\end{aligned}$$

Suppose we approximated the expression $\sin(\omega_n + \varepsilon)t$ by two terms of the expansion on the right-hand side. The sine is harmonic and therefore the secular terms should not be there and must be removed in a logical way. This problem is corrected with Lindstedt's method derived in the next section.

⊛

10.4.1 Lindstedt-Poincaré Method

An examination of the straightforward expansion of Example 10.5 shows that the solution is constrained to oscillate at the constant value ω_n. Linear harmonic systems are characterized by a constant period of oscillation, regardless of initial conditions. The linear response ($\varepsilon = 0$) is harmonic and of period $T_n = 2\pi/\omega_n$. Nonlinear quasiharmonic systems, however, have periods, and thus frequencies, that are functions of the nonlinearity and the initial conditions. The breakdown of the above method, leading to secular terms, is due to ignoring this nonlinearity. In the presence of nonlinear terms the response is periodic and of period $T = 2\pi/\omega$, where ω is an unknown fundamental frequency depending on ε and initial conditions.

The *Lindstedt-Poincaré* method gets around this problem by expanding ω, the response frequency, as well as θ, in powers of a small parameter. Recall that the initial conditions of linear and nonlinear systems are coupled

10.4. PERTURBATION METHODS

to the solution through the constants of integration. We therefore expect that the expansion in ω will be a function of an integration constant. Since ω does not appear in the equation of motion, only ω_n appears, the transformation $\tau = \omega t$ is first introduced. Then, $dt = d\tau/\omega$, and

$$\frac{d}{dt} = \omega \frac{d}{d\tau}$$

$$\frac{d}{dt}\left(\frac{d}{dt}\right) = \omega \frac{d}{d\tau}\left(\omega \frac{d}{d\tau}\right)$$

$$\text{or} \quad \frac{d^2}{dt^2} = \omega^2 \frac{d^2}{d\tau^2},$$

where τ is in radians. The essential feature of this method is emphasized by Minorsky.[12] There appears to be an arbitrariness in the approximations since two expansions have been introduced in the same differential equations. This arbitrariness enables us to dispose of the available constants so as to gradually eliminate the secular terms in the subsequent approximations.

Equation 10.20, $\ddot{\theta} + \omega_n^2 \theta + \varepsilon \theta^3 = 0$, becomes

$$\omega^2 \theta'' + \omega_n^2 \theta + \varepsilon \theta^3 = 0, \tag{10.24}$$

where primes denote differentiation with respect to τ, $\omega_n^2 = \sqrt{g/l}$, and $\varepsilon = -\omega_n^2/6$. For $\varepsilon < 0$, the spring is soft and for $\varepsilon > 0$, the spring is hard. One can assume the initial conditions $\theta(\varepsilon, 0) = a_0$ and $\theta'(\varepsilon, 0) = 0$. For an autonomous system, we can allow the first-order solution $\theta_0(\tau)$ to satisfy these conditions: $\theta_0(\varepsilon, 0) = a_0$ and $\theta_0'(\varepsilon, 0) = 0$ with the remaining orders satisfying zero initial conditions $\theta_i(\varepsilon, 0) = 0$ and $\theta_i'(\varepsilon, 0) = 0$, $i = 1, 2, \cdots$. For nonautonomous systems, the arbitrary constants of integration must be certain values so that periodic solutions exist. Not all initial conditions lead to periodic solutions.

We next expand θ and ω, and consider a linear approximation (only ε^1 terms are retained in addition to the linear terms),

$$\theta(\varepsilon, \tau) = \theta_0(\tau) + \varepsilon \theta_1(\tau) + \cdots$$
$$\omega = \omega_n + \varepsilon \omega_1 + \cdots.$$

Note that the oscillation frequency reduces to ω_n for $\varepsilon = 0$. Substitute these expansions into Equation 10.24, to find

$$(\omega_n + \varepsilon \omega_1)^2 (\theta_0 + \varepsilon \theta_1)'' + \omega_n^2 (\theta_0 + \varepsilon \theta_1) + \varepsilon (\theta_0 + \varepsilon \theta_1)^3 = 0.$$

---

[12] *Nonlinear Oscillations*, N. Minorsky, Krieger Publishing Company, 1987. Reprint of original edition of 1962.

This equation is expanded and grouped according to powers of ε,

$$\varepsilon^0 \; : \; \omega_n^2 \theta_0'' + \omega_n^2 \theta_0 = 0 \tag{10.25}$$
$$\varepsilon^1 \; : \; \omega_n^2 \theta_1'' + \omega_n^2 \theta_1 = -\theta_0^3 - 2\omega_n \omega_1 \theta_0'' \tag{10.26}$$
$$\vdots$$

We only retain the first two equations since a two-term expansion is utilized. Note how such sets of equations are solved iteratively. Equation 10.25 is solved for $\theta_0(\tau)$ and then its powers and derivatives substituted into Equation 10.26 which is then solved for $\theta_1(\tau)$. Also note that each equation is linear and the nonlinear effects are on the right-hand side, acting as inputs to the system. Linear theory is used to solve this sequence of equations. Simplify Equations 10.25 and 10.26,

$$\theta_0'' + \theta_0 = 0$$
$$\theta_1'' + \theta_1 = -\frac{1}{\omega_n^2}\theta_0^3 - 2\frac{\omega_1}{\omega_n}\theta_0''.$$

The solution to the first equation is $\theta_0(\tau) = a_0 \cos \tau$, after satisfying the initial conditions. The second equation becomes[13]

$$\theta_1'' + \theta_1 = \left(-\frac{3}{4}\frac{a_0^3}{\omega_n^2} + 2\frac{\omega_1 a_0}{\omega_n}\right) \cos \tau - \frac{a_0^3}{4\omega_n^2} \cos 3\tau.$$

To remove secular terms, the coefficient of the resonant loading $\cos \tau$ must be set to zero. This leads to an equation for ω_1,

$$\omega_1 = \frac{3}{8}\frac{a_0^2}{\omega_n}.$$

Therefore,

$$\theta_1(\tau) = a_1 \cos \tau + b_1 \sin \tau + \frac{a_0^3}{32\omega_n^2} \cos 3\tau.$$

a_1 and b_1 are found by applying the zero initial conditions for $\theta_1(\tau)$, leading to $a_1 = -a_0^3/32\omega_n^2$ and $b_1 = 0$. The approximate solution for $\theta(\varepsilon, \tau)$ is then

$$\theta(\varepsilon, \tau) = a_0 \cos \tau - \varepsilon \frac{a_0^3}{32\omega_n^2}(\cos \tau - \cos 3\tau) + O(\varepsilon^2). \tag{10.27}$$

---

[13] Use the trigonometric identity,

$$\cos^3 \tau = \frac{3}{4} \cos \tau + \frac{1}{4} \cos 3\tau.$$

10.4. PERTURBATION METHODS

Transforming back to the t domain,

$$\theta(\varepsilon, t) = a_0 \cos \omega t - \varepsilon \frac{a_0^3}{32\omega_n^2} (\cos \omega t - \cos 3\omega t) + O(\varepsilon^2)$$

$$\omega = \omega_n + \varepsilon \frac{3}{8} \frac{a_0^2}{\omega_n} + O(\varepsilon^2). \tag{10.28}$$

These equations provide a measure of the effect of the system nonlinearity on system response and frequency of response.

⊛

On When to Expand ω

There may be some ambiguity on when a second parameter expansion is needed. Sometimes we see the expansion in ω and other times not. Essentially, one expands parameters where their value is unknown. Of course, the oscillator response amplitude is unknown and expanded. For the autonomous oscillator the period and thus frequency are unknown, and ω is therefore expanded.

In the following sections, forced nonlinear oscillators are studied with the purpose of finding periodic oscillations at the forcing period or frequency. Thus there is no need to expand ω.

One of the interesting and challenging aspects of nonlinear equations is that they generally have numerous solutions. Depending on what the analyst needs to find out about the behavior of the nonlinear oscillator, different techniques may be appropriate. Here we are only interested in periodic solutions of slightly nonlinear oscillators.

Poincaré   Jules Henri Poincaré was born on April 29, 1854 in Nancy, Lorraine, France and died on July 17, 1912 in Paris, France. Poincaré entered the École Polytechnique in 1873, graduating in 1875. He was well ahead of all the other students in mathematics. His memory was remarkable and he retained much from all the texts he read but not in the manner of learning by rote, rather by linking the ideas he was assimilating in a visual way. His ability to visualize what he heard proved particularly useful when he attended lectures since his eyesight was so poor that he could not see properly what his lecturers were writing on the blackboard.

Upon graduating from the École Polytechnique Poincaré continued his studies at the École des Mines. After completing his studies there, Poincaré spent a short while as a mining engineer at Vesoul while completing his doctoral work. As a student of Charles Hermite, Poincaré received his doctorate in mathematics from the University of Paris in 1879. His thesis was on differential equations and the examiners were somewhat critical of the work. In 1886 Poincaré was nominated for the chair of mathematical

CHAPTER 10. NONLINEAR DYNAMIC MODELS

Figure 10.7: Jules Henri Poincaré (1854-1912).

physics and probability at the Sorbonne. The intervention and support of Hermite was to ensure that Poincaré was appointed to this chair and to a chair at the École Polytechnique.

Before looking briefly at the many contributions that Poincaré made to mathematics and to other sciences, we should say a little about his way of thinking and working. He is considered as one of the great geniuses of all time. Poincaré kept very precise working hours. He undertook mathematical research for four hours a day, between 10 am and noon then again from 5 pm to 7 pm. He would read articles in journals later in the evening. An interesting aspect of Poincaré's work is that he tended to develop his results from first principles. For many mathematicians there is a building process with more and more being built on top of the previous work. This was not the way that Poincaré worked and not only his research, but also his lectures and books, were all developed carefully from basics.

Poincaré was a scientist preoccupied by many aspects of mathematics, physics and philosophy, and he is often described as the last universalist in mathematics. He made contributions to numerous branches of mathematics, celestial mechanics, fluid mechanics, the special theory of relativity, and the philosophy of science. Much of his research involved interactions between different mathematical topics and his broad understanding of the whole spectrum of knowledge allowed him to attack problems from many different angles.

In applied mathematics he studied optics, electricity, telegraphy, capillarity, elasticity, thermodynamics, potential theory, quantum theory, theory of relativity, and cosmology. In the field of celestial mechanics he studied

10.4. PERTURBATION METHODS

the three-body problem, as well as the theories of light and of electromagnetic waves. He is acknowledged as a codiscoverer, with Albert Einstein and Hendrik Lorentz, of the special theory of relativity.

After Poincaré achieved prominence as a mathematician, he turned his superb literary gifts to the challenge of describing for the general public the meaning and importance of science and mathematics. Poincaré's popular works include *Science and Hypothesis* (1901), *The Value of Science* (1905), and *Science and Method* (1908).

Poincaré achieved the highest honors for his contributions of true genius. He was elected to the Académie des Sciences in 1887 and in 1906 was elected President of the Academy. The breadth of his research led to him being the only member elected to every one of the five sections of the Academy, namely the geometry, mechanics, physics, geography, and navigation sections. In 1908 he was elected to the Académie Francaise and was elected director in the year of his death. He was also made chevalier of the Légion d'Honneur and was honored by a large number of learned societies around the world. He won numerous prizes, medals and awards.

10.4.2 Forced Oscillations of Quasiharmonic Systems

Consider the forced Duffing equation,

$$\ddot{x} + \omega_n^2 x = \varepsilon \left[-\omega_n^2 \left(\alpha x + \beta x^3 \right) + F \cos \Omega t \right], \quad \varepsilon \ll 1, \qquad (10.29)$$

where α, β are given constant parameters, $\omega_n^2 = k/m$, and the harmonic forcing is $\varepsilon F \cos \Omega t$. In this case, the harmonic excitation is of small magnitude. It is expected therefore that the response will be almost or *quasiharmonic*. Therefore, it is of interest to determine the circumstances under which response $x(t)$ is periodic with period $T = 2\pi/\Omega$. An expansion in the frequency of response is therefore not necessary.

This problem is approached using the perturbation method, with the following change of variables,

$$\Omega t = \tau + \phi$$
$$\frac{d}{dt} = \Omega \frac{d}{d\tau},$$

where τ is the new time variable and ϕ is a phase angle. This transformation results in a time scale with a period of oscillation of 2π. With these transformations, Equation 10.29 becomes

$$\Omega^2 x'' + \omega_n^2 x = \varepsilon \left[-\omega_n^2 \left(\alpha x + \beta x^3 \right) + F \cos \left(\tau + \phi \right) \right], \quad \varepsilon \ll 1, \qquad (10.30)$$

where primes denote differentiation with respect to τ. Recalling that secular terms must be removed, as they are nonphysical, requires that the solution to Equation 10.30 be periodic, $x(\tau + 2\pi) = x(\tau)$. Assume also that $x(0) = C_0$, and for convenience that $x'(0) = 0$. The value of C_0 is not arbitrary, but is determined in terms of other system parameters below. The perturbation solution is based on the expansions of $x(\tau)$ and ϕ,

$$\begin{aligned} x(\tau) &= x_0(\tau) + \varepsilon x_1(\tau) + \varepsilon^2 x_2(\tau) + \cdots \\ \phi &= \phi_0 + \varepsilon \phi_1 + \varepsilon_2^2 \phi + \cdots, \end{aligned}$$

where $x_i(\tau + 2\pi) = x_i(\tau)$, $x_i'(0) = 0$, for all i, and $x_0(0) = C_0$. These expansions are substituted into Equation 10.30, and equating the coefficients of like powers of ε, we find

$$\Omega^2 x_0'' + \omega_n^2 x_0 = 0 \qquad (10.31)$$
$$\Omega^2 x_1'' + \omega_n^2 x_1 = -\omega_n^2 \left(\alpha x_0 + \beta x_0^3\right) + F \cos(\tau + \phi_0) \qquad (10.32)$$
$$\Omega^2 x_2'' + \omega_n^2 x_2 = -\omega_n^2 \left(\alpha x_1 + 3\beta x_0^2 x_1\right) + F \cos(\tau + \phi_1) \qquad (10.33)$$
$$\vdots$$

These equations are solved in sequence for $x_i(\tau)$, for all i, applying the periodicity and initial conditions. The solution to Equation 10.31 is

$$x_0(\tau) = C_0 \cos \frac{\omega_n}{\Omega} \tau,$$

where C_0 is a constant amplitude. 2π-periodicity in τ must be satisfied; this is possible only if $\Omega = \omega_n$. This substitution is made in the subsequent Equations 10.32 and 10.33. Substitute for $x_0(\tau)$ in Equation 10.32 to find

$$x_1'' + x_1 = -\left(\alpha C_0 \cos \tau + \beta C_0^3 \cos^3 \tau\right) + \frac{F}{\omega_n^2} \cos(\tau + \phi_0).$$

This expression can be simplified using the trigonometric relation $\cos^3 \tau = (3\cos\tau + \cos 3\tau)/4$,

$$\begin{aligned} x_1'' + x_1 &= -\alpha C_0 \cos\tau - \beta C_0^3 \frac{1}{4}(3\cos\tau + \cos 3\tau) \\ &\quad + \frac{F}{\omega_n^2}(\cos\tau\cos\phi_0 - \sin\tau\sin\phi_0) \\ &= -\frac{F}{\omega_n^2}\sin\phi_0 \sin\tau - \left(\alpha C_0 + \frac{3}{4}\beta C_0^3 - \frac{F}{\omega_n^2}\cos\phi_0\right)\cos\tau \\ &\quad - \frac{1}{4}\beta C_0^3 \cos 3\tau. \end{aligned}$$

To avoid secular terms requires that the coefficients of $\sin\tau$ and $\cos\tau$ be set to zero, yielding conditions and relations for the respective parameters

10.4. PERTURBATION METHODS

for periodicity. There are two sets of expressions that satisfy this condition: $\phi_0 = 0$, and $\phi_0 = \pi$,

$$\phi_0 = 0, \quad \alpha C_0 + \frac{3}{4}\beta C_0^3 - \frac{F}{\omega_n^2} = 0 \qquad (10.34)$$

$$\phi_0 = \pi, \quad \alpha C_0 + \frac{3}{4}\beta C_0^3 + \frac{F}{\omega_n^2} = 0, \qquad (10.35)$$

where C_0 is now determined, since α, β and F are all known. Equations 10.34 and 10.35 offer the same information. Equations 10.35 tell us that for phase $\phi_0 = \pi$, the response and the forcing are 180° out of phase. This is the same as being in phase with response of negative amplitude.

Utilizing Equations 10.34, we can solve for $x_1(\tau)$,

$$x_1(\tau) = C_1 \cos \tau + \frac{1}{32}\beta C_0^3 \cos 3\tau. \qquad (10.36)$$

The value of C_1 is determined based on the required periodicity of $x_2(\tau)$, just as C_0 was determined based on the periodicity of $x_1(\tau)$. Equation 10.36 is substituted into Equation 10.33, and utilizing a number of trigonometric identities, a set of expressions are derived that must be examined for the possibility of secular responses.

Suppose only $x_1(\tau)$ is retained, then to order ε, or $O(\varepsilon)$,

$$\begin{aligned} x(\tau) &\simeq x_0(\tau) + \varepsilon x_1(\tau) \\ &= C_0 \cos \tau + \varepsilon \left(C_1 \cos \tau + \frac{1}{32}\beta C_0^3 \cos 3\tau \right), \end{aligned}$$

where $\phi \simeq \phi_0 + \varepsilon \phi_1 = 0$, since the ε^2 solution yields $\phi_1 = 0$. All the phase angle terms are zero since there is no damping in the system. This is recalled from our studies of linear undamped oscillators, and we could have therefore avoided the ϕ expansion.

Equation 10.34 reminds us of the frequency response function $H(\omega)$. Both functions are relations between response amplitude and forcing amplitude. Equation 10.34 is the analogous nonlinear relation.

Introduce the relation,

$$\omega_0^2 \equiv (1 + \varepsilon \alpha)\omega_n^2, \qquad (10.37)$$

so that Equation 10.29 becomes

$$\ddot{x} + \omega_0^2 x + \varepsilon \omega_n^2 \beta x^3 = \varepsilon F \cos \Omega t, \quad \varepsilon \ll 1.$$

Solve Equation 10.37 for α and substitute the result into the second of Equations 10.34. Solve this equation for ω_n^2 to find

$$\omega_n^2 = \omega_0^2 \left(1 + \frac{3}{4}\varepsilon \beta C_0^2\right) - \varepsilon \frac{F}{C_0}. \qquad (10.38)$$

In deriving this equation, the following approximation is made

$$\left(1 - \frac{3}{4}\varepsilon\beta C_0^2\right)^{-1} \simeq \left(1 + \frac{3}{4}\varepsilon\beta C_0^2\right),$$

based on the Taylor expansion for $1/(1+x) \simeq 1 - x$ for small x, and also a term of order ε^2 is dropped. Equation 10.38 can be plotted as $|C_0|$ versus ω_n, where ω_n is measured in units of ω_0. The product $\varepsilon\beta$ is given and εF is a parameter. Representative curves are shown in Figure 10.8 for $\varepsilon\beta = \pm 0.1$. Note that the parametric curves for a single case appear in pairs, on either side of the $F = 0$ curve.

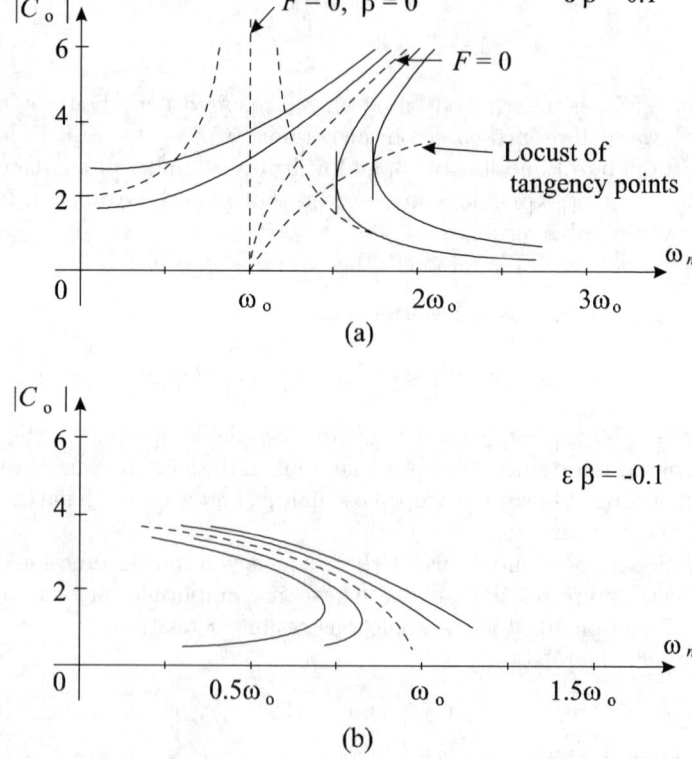

Figure 10.8: Amplitude curves for the Duffing equation for two values of $\varepsilon\beta$. In (a), the dotted lines for $\beta = 0$ represent the linear problem.

Quite a wealth of information can be discerned from this set of curves. The sign of $\varepsilon\beta$ determines whether the curves slant to the right, $\varepsilon\beta > 0$,

10.4. PERTURBATION METHODS

for a hardening spring, or to the left, $\varepsilon\beta < 0$, for a softening spring. Unlike for the linear case, there is no resonance condition where, for an undamped oscillator, the amplitudes grow ever larger around the natural frequency of the system.

Consider the line labeled as the locus of tangency points. A vertical line to the right of this line intersects the curves at two locations on one branch and at one location on the mirror branch. Any intersection represents a real root, and so this vertical line indicates that there are three possible amplitudes corresponding to a given excitation force amplitude. A vertical line to the left has only one real root, but also two complex roots. Following a representative curve as ω_0 increases results in ever larger amplitudes.

10.4.3 Jump Phenomenon

Of course, all systems have some damping, and it is interesting to consider a lightly damped Duffing oscillator to examine the significant changes that are predicted. Consider the governing equation,

$$\ddot{x} + \omega_n^2 x = \varepsilon\left[-2\zeta\omega_n\dot{x} - \omega_n^2\left(\alpha x + \beta x^3\right) + F\cos\Omega t\right], \quad \varepsilon \ll 1,$$

where the small damping term, $-2\varepsilon\zeta\omega_n\dot{x}$, has been added. (Note that α is dimensionless and β has units of x^{-2}.) Following the same procedure as before, the comparable equation to Equation 10.38 is found,

$$\left[\omega_0^2\left(1 + \frac{3}{4}\varepsilon\beta C_0^2\right) - \omega_n^2\right]^2 + \left(2\varepsilon\zeta\omega_0^2\right)^2 = \left(\frac{\varepsilon F}{C_0}\right)^2.$$

A plot of this equation shows similar, but significantly different curves. See Figure 10.9. The difference is that the two branches connect at some location. Compare with Figure 10.8.

This is significant because it means that an increase in the frequency does not always lead to an increase in amplitude. At some value, there will be a drop in amplitude to the other solution branch. This is clearly seen if one follows path 4 for increasing frequency, with a drop at point 1 to point 2. Similarly, for decreasing frequency, at location 3 there is a jump in amplitude to point 4. The path from point 1 to point 3 is unstable.

This behavior is called the *jump phenomenon*. It exists for nonlinear systems with damping. There are practical implications. For example, as a motor runs up to operating frequency, the designer will need to make sure that there are no jumps in amplitude.

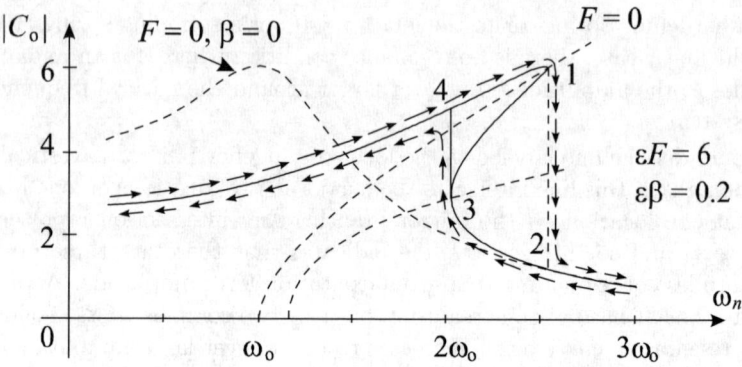

Figure 10.9: Jump phenomenon. The path between points 1 and 3 is unstable.

10.4.4 Periodic Solutions of Nonautonomous Systems

Consider the system governed by

$$\ddot{u} + \omega_n^2 u = F_0 \cos t + \varepsilon f(u, \dot{u}, t, \varepsilon),$$

where the force is not small but the nonlinearities are of order ε. When $\varepsilon = 0$ and $F_0 = 0$, the nonlinear equation reduces to the simple harmonic oscillator with frequency ω_n and period $T = 2\pi/\omega_n$. For the case of $\varepsilon = 0$, the first-order solution u_0 is

$$u_0 = C_0 \cos \omega_n (t - t_0) + \frac{F_0}{\omega_n^2 - 1} \cos t, \quad \omega_n^2 \neq 1.$$

C_0 and t_0 are arbitrary constants. There are two ways the solution can have the same 2π-period as the forcing term; ω_n must be an integer. Otherwise, if ω_n is an irrational number then the solution u_0 can be periodic only if $C_0 = 0$, in which case the period is automatically 2π, like the forcing.

Forced Pendulum

Nonlinear problems exhibit a coupling between the free and forced response and the initial conditions. The arbitrary constants of integration do not only specify the allowable initial conditions for periodic solutions, but are a result of enforcing periodicity.

Consider the *forced pendulum*, governed by the equation,

$$\ddot{\theta} + \omega_n^2 \sin \theta = F \cos \omega t.$$

10.4. PERTURBATION METHODS

An approximate nonlinear equation that is representative of the fully nonlinear pendulum equation is

$$\ddot{\theta} + \omega_n^2\left(\theta - \frac{\theta^3}{6}\right) = F\cos\omega t.$$

Consider a more general version of this equation. To do this first nondimensionalize the variables using the transformation $\tau = \omega t$. The nonlinear governing equation then becomes

$$\omega^2\theta'' + \omega_n^2\theta - \frac{1}{6}\omega_n^2\theta^3 = \Gamma\omega^2\cos\tau$$

$$\text{or} \quad \theta'' + \Omega^2\theta - \frac{1}{6}\Omega^2\theta^3 = \Gamma\cos\tau,$$

where $\Gamma\omega^2 = F$ and $\Omega = \omega_n/\omega$. To further generalize,[14] define $\varepsilon \equiv -\Omega^2/6$. Apply the perturbation expansion to

$$\theta'' + \Omega^2\theta + \varepsilon\theta^3 = \Gamma\cos\tau, \tag{10.39}$$

reducing to the linear problem for $\varepsilon = 0$. We look for solutions of the same period as the driving force, which has 2π-periodicity.

The expansion is taken to be a function of ε and τ in the following way,

$$\theta(\varepsilon, \tau) \equiv \theta_0(\tau) + \varepsilon\theta_1(\tau) + \varepsilon^2\theta_2(\tau) + \cdots, \tag{10.40}$$

where the periodicity condition $\theta_i(\tau + 2\pi) = \theta_i(\tau)$, $i = 0, 1, 2, \ldots$, removes any secular terms from the solution. If this approach is to work in a practical sense, it is necessary that $\varepsilon \ll 1$ such that $\theta_0(\tau) \gg \varepsilon\theta_1(\tau) \gg \varepsilon^2\theta_2(\tau)\ldots$. Then we can retain the expansion to order ε or ε^2 and obtain a good approximation. What has been discussed so far makes no assumptions about the character of the forcing. An expansion in ω is not necessary here since oscillations with known period are sought.

The procedure is to substitute the expression for $\theta(\varepsilon, \tau)$ and its second derivative with respect to τ, $\theta''(\varepsilon, \tau)$, into Equation 10.39, with the result

$$\left(\theta_0'' + \varepsilon\theta_1'' + \varepsilon^2\theta_2'' + \cdots\right) + \Omega^2\left(\theta_0 + \varepsilon\theta_1 + \varepsilon^2\theta_2 + \cdots\right)$$
$$+ \varepsilon\left(\theta_0 + \varepsilon\theta_1 + \varepsilon^2\theta_2 + \cdots\right)^3 = \Gamma\cos\tau.$$

Note that when the cubic term is expanded, expressions of order ε^3 and ε^4 and higher are found.[15] Retain a solution to order ε^2 and drop all higher order terms, thus,

$$\varepsilon\left(\theta_0 + \varepsilon\theta_1 + \varepsilon^2\theta_2\right)^3 \simeq \varepsilon\left(\theta_0^3 + 3\theta_0^2\varepsilon\theta_1\right).$$

---

[14] We can define ε as a positive quantity and retain the negative sign in the equation of motion. Of course the results are the same, but sometimes retaining the physical meaning of ε as a frequency is desirable and then $\varepsilon > 0$.

[15] The expansion of the cubic term is

Since the θ_i are independent of ε, and the above must be satisfied for all ε, the expanded differential equation is actually a sequence of differential equations that can be identified by equating terms of the same order of ε,

$$\varepsilon^0 \quad : \quad \theta_0'' + \Omega^2 \theta_0 = \Gamma \cos \tau \tag{10.41}$$
$$\varepsilon^1 \quad : \quad \theta_1'' + \Omega^2 \theta_1 = -\theta_0^3 \tag{10.42}$$
$$\varepsilon^2 \quad : \quad \theta_2'' + \Omega^2 \theta_2 = -3\theta_0^2 \theta_1. \tag{10.43}$$

The approximate solution is then given by Equation 10.40, with an error of $O(\varepsilon^3)$. We can immediately observe the major advantage to this expansion technique is the transformation of a nonlinear equation into a set of linear equations that are solved in sequence. The linear equations are all the same except for the input, and they can be solved using the convolution integral. Proceeding through this iterative solution, we will utilize our physical understanding of the system to help us in rejecting parts of the solution that do not fit our understanding. This will become clearer as we proceed to solve the example.

Solve Equations 10.41-10.43 in sequence. The solution to the ε^0 order governing equation is

$$\theta_{0h}(\tau) = a_0 \cos \Omega \tau + b_0 \sin \Omega \tau,$$

where the subscript h denotes homogeneous solution, and a_0 and b_0 are arbitrary constants, the values of which are established below by enforcing the required periodicity. The particular solution is given by

$$\theta_{0p}(\tau) = \frac{\Gamma}{\Omega^2 - 1} \cos \tau.$$

The complete solution is the sum of the homogeneous and particular solutions. The particular solution is not valid at or near the resonant condition $\Omega = 1$. This case is solved separately.[16] We also are interested here, for the sake of brevity, only at solutions that have the period of the loading, that is, $T = 2\pi/1$ s. Other solutions exist, some of which are examined in the next section.

$$()^3 = \theta_0^3 + 3\theta_0^2 \varepsilon \theta_1 + 3\theta_0^2 \varepsilon^2 \theta_2 + 3\theta_0 \varepsilon^2 \theta_1^2$$
$$+ 6\theta_0 \varepsilon^3 \theta_1 \theta_2 + 3\theta_0 \varepsilon^4 \theta_2^2 + \varepsilon^3 \theta_1^3 + 3\varepsilon^4 \theta_1^2 \theta_2 + 3\varepsilon^5 \theta_1 \theta_2^2 + \varepsilon^6 \theta_2^3.$$

[16] Assume $\Omega \simeq 1$, and
$$\Omega^2 = 1 + \varepsilon \beta, \tag{10.44}$$
with $\Gamma = \varepsilon \gamma$, and expand as before.

10.4. PERTURBATION METHODS

If the response must have the period 2π, then $a_0 = b_0 = 0$, and $\theta_0(\tau) = \theta_{0p}(\tau)$. This function is cubed and becomes the forcing for the equation governing $\theta_1(\tau)$,

$$\theta_0^3 = \left(\frac{\Gamma}{\Omega^2 - 1}\cos\tau\right)^3$$
$$= \left(\frac{\Gamma}{\Omega^2 - 1}\right)^3 \left(\frac{3}{4}\cos\tau + \frac{1}{4}\cos 3\tau\right).$$

The expansion of $\cos^3\tau$ is important because it helps identify the various harmonic components contained therein. Again ignoring the homogeneous solution, the particular solution is

$$\theta_{1p}(\tau) = a_1 \cos\tau + b_1 \cos 3\tau; \qquad \Omega \ne 1, 3,$$
$$a_1 = \frac{3\Gamma^3}{4(\Omega^2 - 1)^4}$$
$$b_1 = \frac{\Gamma^3}{4(\Omega^2 - 1)(\Omega^2 - 9)}.$$

For the next term in the series, it is necessary to evaluate the input $3\theta_0^2\theta_1$. This step is omitted here. The truncated solution correct to order ε is

$$\theta(\varepsilon, \tau) \simeq \theta_0(\tau) + \varepsilon\theta_1(\tau) + O(\varepsilon^2),$$

where expressions for $\theta_0(\tau)$ and $\theta_1(\tau)$ are known, and $\varepsilon = -\Omega^2/6$. The variables can now be transformed back to the physical ones.

Arbitrary Forcing

Now consider the possibility that the forcing is general, $f(t)$. Begin with the general nonlinear governing equation,

$$\ddot{\theta} + \alpha\dot{\theta} + \omega_n^2\theta + \varepsilon g(\theta, \dot{\theta}) = f(t),$$

with the expansion solution,

$$\theta(\varepsilon, t) = \theta_0(t) + \varepsilon\theta_1(t) + \varepsilon^2\theta_2(t) + \ldots,$$

where $\theta_0(t)$ satisfies the linear differential equation. Before substituting the expansion into the governing equation, the nonlinear function $g(\theta, \dot{\theta})$ is itself expanded, *about the linear solution* $\left(\theta_0, \dot{\theta}_0\right)$, so that we can keep track of the various orders in ε,

$$\begin{aligned}g(\theta,\dot{\theta}) &\equiv g\left(\theta_0+\varepsilon\theta_1+\varepsilon^2\theta_2+\ldots,\dot{\theta}_0+\varepsilon\dot{\theta}_1+\varepsilon^2\dot{\theta}_2+\ldots\right)\\ &= g(\theta_0,\dot{\theta}_0)+\left(\varepsilon\theta_1+\varepsilon^2\theta_2+\ldots\right)\frac{\partial}{\partial\theta}g(\theta_0,\dot{\theta}_0)\\ &\quad+\left(\varepsilon\dot{\theta}_1+\varepsilon^2\dot{\theta}_2+\ldots\right)\frac{\partial}{\partial\dot{\theta}}g(\theta_0,\dot{\theta}_0)\\ &\quad+\text{ higher-order terms.}\end{aligned}$$

Retaining terms to order ε^2, and gathering terms of equal power in ε, the following sequence is generated,

$$\begin{aligned}\varepsilon^0 &: \ddot{\theta}_0+\alpha\dot{\theta}_0+\omega_n^2\theta_0=f(t)\\ \varepsilon^1 &: \ddot{\theta}_1+\alpha\dot{\theta}_1+\omega_n^2\theta_1=-g(\theta_0,\dot{\theta}_0)\\ \varepsilon^2 &: \ddot{\theta}_2+\alpha\dot{\theta}_2+\omega_n^2\theta_2=-\theta_1\frac{\partial}{\partial x}g(\theta_0,\dot{\theta}_0)-\dot{\theta}_1\frac{\partial}{\partial\dot{x}}g(\theta_0,\dot{\theta}_0).\end{aligned}$$

As a practical matter, it becomes very difficult to solve more than a two- or three-term approximation. The solutions for θ_0 and θ_1 are now obtained in sequence using the convolution integral. Proceeding for a general nonlinear function $g(\theta,\dot{\theta})$,

$$\begin{aligned}\theta_0(t) &= \int_{-\infty}^{\infty}f(t-\tau)h(\tau)d\tau\\ \theta_1(t) &= -\int_{-\infty}^{\infty}g\left[\theta_0(t-\tau),\dot{\theta}_0(t-\tau)\right]h(\tau)d\tau\\ &\quad\vdots\end{aligned}$$

and then $\theta(t)\simeq\theta_0(t)+\varepsilon\theta_1(t)$. The response is approximately given by

$$\begin{aligned}\theta(t) &\simeq \int_{-\infty}^{\infty}f(t-\tau)h(\tau)d\tau\\ &\quad-\varepsilon\int_{-\infty}^{\infty}g\left[\theta_0(t-\tau),\dot{\theta}_0(t-\tau)\right]h(\tau)d\tau.\end{aligned}$$

The above expansion searches for oscillations at the forcing frequency. One can also transform the differential equation and introduce a response frequency that is also expanded as in the Lindstedt expansion.

10.4. PERTURBATION METHODS

Example 10.6 Stochastic Parametric Excitation via Perturbation Technique

This example is based on a series of papers by Benaroya and Rehak[17] that studied various aspects of stochastic parametric vibration. The term parametric[18] implies that some part of the external physical forcing appears on the left-hand side of the equation of motion, within one or more of the parameters such as damping or stiffness. One interesting by-product is that system stability is determined by the relative magnitudes of the system parameters. Therefore, parametric vibration must include an explicit study of system stability.

Begin with the generic governing equation of motion,

$$m\frac{d^2\chi}{dT^2} + c\frac{d\chi}{dT} + K_{\text{rand}}(T)\chi = F(T), \qquad (10.45)$$

where $K_{\text{rand}}(T)$ and $F(T)$ are random functions of time T. The "random" stiffness can be decomposed into a deterministic mean value added to a zero mean random fluctuation, $K_{\text{rand}}(T) = k_d + K_f$, and therefore $E\{K_{\text{rand}}(T)\} = k_d$. Nondimensionalize the governing equation using the following definitions,

$$\omega_n = k_d/m \qquad \zeta = c/2m\omega_n \qquad t = \omega_n T$$

$$X = \chi k_d/E\{F\} \qquad K = K_f/k_d \qquad F = F(T)/E\{F\}.$$

Making these substitutions, Equation 10.45 becomes

$$\frac{d^2X}{dt^2} + 2\zeta\frac{dX}{dt} + X + KX = F,$$

where it is assumed that K and F are statistically independent stationary processes, although this does not necessarily have to be true. All the random

---

[17] "Parametric Random Excitation. I: Exponentially Correlated Parameters," H. Benaroya, M. Rehak, *Journal of Engineering Mechanics*, p. 861-874, Vol. 113, No. 6, June, 1987.

"Parametric Random Excitation. II: White-Noise Parameters," H. Benaroya, M. Rehak, *Journal of Engineering Mechanics*, p. 875-884, Vol. 113, No. 6, June, 1987.

"Response and Stability of a Random Differential Equation: Part I – Moment Equation Method," H. Benaroya, M. Rehak, *Journal of Applied Mechanics*, p. 192-195, Vol. 111, March, 1989.

"Response and Stability of a Random Differential Equation: Part II – Expansion Method," H. Benaroya, M. Rehak, *Journal of Applied Mechanics*, p. 196-201, Vol. 111, March, 1989.

[18] See this fine review for more background. "Structural Dynamics with Parameter Uncertainties," R.A. Ibrahim, *Appl. Mech. Rev.*, Vol. 40, No. 3, Mar. 1987, pp. 309-328.

inputs are placed on the right-hand side of the governing equation, but this is solely to organize the solution more clearly,

$$\frac{d^2 X}{dt^2} + 2\zeta \frac{dX}{dt} + X = F - KX.$$

This equation is linear, but note that X appears on both sides of the equation and therefore the linearity is not pure. Rather, an iterative or sequential solution procedure is required, as follows. In the formal solution, ignore temporarily that X appears on both sides of the equation. The convolution integral is the solution or response to each of the forces that appears on the right-hand side, F and $-KX$, and by linearity, we can superpose those two components of the response,

$$\begin{aligned} X(t) &= \int_{-\infty}^{\infty} g(\tau) F(t-\tau) d\tau \\ &\quad - \int_{-\infty}^{\infty} g(\tau) K(t-\tau) X(t-\tau) d\tau, \end{aligned} \quad (10.46)$$

where $g(t)$ is the impulse response function,

$$g(t) = \frac{\exp(-\zeta t)}{\sqrt{1-\zeta^2}} \sin \sqrt{1-\zeta^2} t, \quad t \geq 0.$$

It is emphasized that since $X(t)$ appears on both sides of Equation 10.46, a recursive procedure is necessary. The solution can be organized by assuming the solution $X(t) = \sum_{j=0}^{\infty} X_j(t)$. Then

$$\begin{aligned} X(t) &= \int_{-\infty}^{\infty} g(\tau) F(t-\tau) d\tau \\ &\quad - \int_{-\infty}^{\infty} g(\tau) K(t-\tau) [X_0(t-\tau) + X_1(t-\tau) + \ldots] d\tau, \end{aligned}$$

where $X_0(t) \equiv \int_{-\infty}^{\infty} g(\tau) F(t-\tau) d\tau$ is the solution to the respective deterministic problem, and then

$$\begin{aligned} X_1(t) &\equiv -\int_{-\infty}^{\infty} g(t_1) K(t-t_1) X_0(t-t_1) dt_1 \\ X_2(t) &\equiv -\int_{-\infty}^{\infty} g(t_2) K(t-t_2) X_1(t-t_2) dt_2 \\ &\vdots \end{aligned} \quad (10.47)$$

Note that a perturbation expansion of the form $X(t) = \sum_{j=0}^{\infty} \varepsilon^j X_j(t)$ would lead to the same series of equations as above for $X_j(t)$. Making the appropriate substitutions in the right-hand sides of Equation 10.47 shows us

10.4. PERTURBATION METHODS

explicitly the hidden complexity of these equations,

$$X_0(t) = \int_{-\infty}^{\infty} g(\tau) F(t-\tau) d\tau$$

$$X_1(t) = -\int_{-\infty}^{\infty} dt_1 \, g(t_1) K(t-t_1) \int_{-\infty}^{\infty} dt_2 \, g(t_2) F(t-t_1-t_2)$$

$$X_2(t) = -\int_{-\infty}^{\infty} dt_1 \, g(t_1) K(t-t_1) \int_{-\infty}^{\infty} dt_2 \, g(t_2) K(t-t_1-t_2)$$

$$\cdot \int_{-\infty}^{\infty} dt_3 \, g(t_3) F(t-t_1-t_2-t_3)$$

$$\vdots$$

Since K and F are random processes, proceed by taking the mathematical expectations of each $X_j(t)$. Thus,

$$E\{X(t)\} = \sum_{j=0}^{\infty} E\{X_j(t)\},$$

where

$$E\{X_0(t)\} = \int_{-\infty}^{\infty} g(\tau) E\{F(t-\tau)\} d\tau$$

$$E\{X_1(t)\} = -\int_{-\infty}^{\infty} \int_{-\infty}^{\infty} dt_1 dt_2 \, g(t_1) g(t_2) E\{K(t-t_1)F(t-t_1-t_2)\}$$

$$E\{X_2(t)\} = -\int_{-\infty}^{\infty} \int_{-\infty}^{\infty} \int_{-\infty}^{\infty} dt_1 dt_2 dt_3 \, g(t_1) g(t_2) g(t_3)$$

$$\cdot E\{K(t-t_1)K(t-t_1-t_2)F(t-t_1-t_2-t_3)\}$$

$$\vdots$$

The evaluation of the correlation,

$$E\{K(t-t_1)F(t-t_1-t_2)\},$$

is based on the assumption that K and F are statistically independent,

$$E\{K(t-t_1)F(t-t_1-t_2)\} = E\{K(t-t_1)\} E\{F(t-t_1-t_2)\},$$

where it is recalled that $E\{K\} = 0$. For the expectation $E\{X_2(t)\}$ it is necessary to evaluate

$$E\{K(t-t_1)K(t-t_1-t_2)F(t-t_1-t_2-t_3)\}$$
$$= E\{K(t-t_1)K(t-t_1-t_2)\} E\{F(t-t_1-t_2-t_3)\}$$
$$= R_{KK}(t_2) E\{F\}.$$

The mean value of the response is a function of the correlation function of the input. Later terms in the series require the evaluation of higher-order statistics. To proceed further with the terms expressed above requires the autocorrelation of the stiffness process. Assume the physical correlation model,

$$R_{KK}(\tau) = \sigma_K^2 \exp(-\beta|\tau|), \tag{10.48}$$

where σ_K and β would depend on the particular application and its associated data. For various values of β we have the following plots for $\beta = 1, 2, 4$ in Figure 10.10.

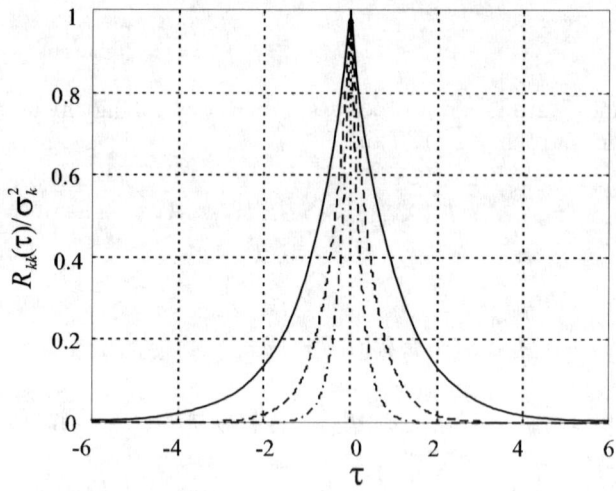

Figure 10.10: $R_{KK}(\tau)/\sigma_K^2 = \exp(-\beta|\tau|)$.

Lacking additional data, a parametric study becomes suitable, where numerical values are varied in order to explore the sensitivity of the output results to such variations.

⊛

10.4.5 Random Duffing Oscillator

Consider the Duffing equation subjected to a random force and apply the perturbation technique to derive estimates of the response correlation and spectral density. The governing equation is

$$\ddot{X} + 2\zeta\omega_n \dot{X} + \omega_n^2 \left(X + rX^3\right) = F(t).$$

Define $\varepsilon = \omega_n^2 r$. Recall that in customary oscillator notation $c/m = 2\zeta\omega_n$ and $k/m = \omega_n^2$. The excitation per unit mass $F(t)$ is a stationary Gaussian

10.4. PERTURBATION METHODS

random process with power spectrum $S_{FF}(\omega)$. The equations that govern the terms in the approximation $X \simeq X_0 + \varepsilon X_1 + \varepsilon^2 X_2$ are

$$\ddot{X}_0 + 2\zeta\omega_n \dot{X}_0 + \omega_n^2 X_0 = F(t) \tag{10.49}$$
$$\ddot{X}_1 + 2\zeta\omega_n \dot{X}_1 + \omega_n^2 X_1 = -X_0^3 \tag{10.50}$$
$$\ddot{X}_2 + 2\zeta\omega_n \dot{X}_2 + \omega_n^2 X_2 = -3X_0^2 X_1. \tag{10.51}$$

Equation 10.49 is a linear oscillator driven by a random load $F(t)$. Note again that this procedure takes a nonlinear oscillator and converts it into a series of linear oscillators. The nonlinearities have been moved to the right-hand side of each linear oscillator, but the system equation, that is the left-hand side, is always linear and amenable to solution via the convolution integral regardless of the form of the function driving it on the right-hand side.

The first three terms in the infinite sequence of solutions is then

$$X_0(t) = \int_{-\infty}^{\infty} g(t) F(t-\tau) d\tau \tag{10.52}$$

$$X_1(t) = -\int_{-\infty}^{\infty} g(t) X_0^3(t-\tau) d\tau \tag{10.53}$$

$$X_2(t) = -3\int_{-\infty}^{\infty} g(t) X_0^2(t-\tau) X_1(t-\tau) d\tau, \tag{10.54}$$

where

$$g(t) = (1/\omega_d)\exp(-\zeta\omega_d t)$$
$$\omega_d = \omega_n\sqrt{1-\zeta^2}.$$

The solution to Equation 10.52 is then cubed and substituted into Equation 10.53, and that equation is integrated to find the expression $X_1(t)$. This expression and the one for $X_0^2(t)$ are then appropriately substituted into Equation 10.54 for integration. The point has been made that this is a tedious process, the accuracy of which depends on the value of ε. Since $F(t)$ is a random function of time, only the statistics of the response can be estimated. For a three-term expansion, the mean value of the response is given by

$$E\{X\} = E\{X_0\} + \varepsilon E\{X_1\} + \varepsilon^2 E\{X_2\},$$

where

$$E\{X_0\} = \int_{-\infty}^{\infty} g(t) E\{F(t-\tau)\} d\tau$$

$$E\{X_1\} = -\int_{-\infty}^{\infty} g(t) E\{X_0^3(t-\tau)\} d\tau$$

$$E\{X_2\} = -3\int_{-\infty}^{\infty} g(t) E\{X_0^2(t-\tau) X_1(t-\tau)\} d\tau.$$

The evaluation of these expectations can be quite difficult analytically and generally requires some assumptions regarding the statistical properties of the variables.

Even more intricate is the estimation of the output correlation function $R_{XX}(\tau)$ and spectral density $S_{XX}(\omega)$,

$$\begin{aligned}R_{XX}(\tau) &= E\{X(t)X(t+\tau)\} \\ &\simeq E\{[X_0(t) + \varepsilon X_1(t) + \varepsilon^2 X_2(t)] \\ &\quad \cdot [X_0(t+\tau) + \varepsilon X_1(t+\tau) + \varepsilon^2 X_2(t+\tau)]\} \\ &= E\{X_0(t)X_0(t+\tau) + \varepsilon X_0(t)X_1(t+\tau) + \varepsilon^2 X_0(t)X_2(t+\tau) \\ &\quad + \varepsilon X_1(t)X_0(t+\tau) + \varepsilon^2 X_1(t)X_1(t+\tau) + \varepsilon^3 X_1(t)X_2(t+\tau) \\ &\quad + \varepsilon^2 X_2(t)X_0(t+\tau) + \varepsilon^3 X_2(t)X_1(t+\tau) + \varepsilon^4 X_2(t)X_2(t+\tau)\}.\end{aligned}$$

To evaluate this expression data is needed for numerous correlations and cross-correlations. It is reasonable to drop terms of order greater than ε^2 for consistency with the fact that the expansion for $X(t)$ is only to order ε^2, thus,

$$\begin{aligned}R_{XX}(\tau) &\simeq R_{X_0 X_0}(\tau) + \varepsilon [R_{X_0 X_1}(\tau) + R_{X_1 X_0}(\tau)] \\ &\quad + \varepsilon^2 [R_{X_1 X_1}(\tau) + R_{X_0 X_2}(\tau) + R_{X_2 X_0}(\tau)].\end{aligned}$$

The response spectral density approximation is the Fourier transform of $R_{XX}(\tau)$,

$$\begin{aligned}S_{XX}(\omega) &\simeq S_{X_0 X_0}(\omega) + \varepsilon [S_{X_0 X_1}(\omega) + S_{X_1 X_0}(\omega)] \\ &\quad + \varepsilon^2 [S_{X_1 X_1}(\omega) + S_{X_0 X_2}(\omega) + S_{X_2 X_0}(\omega)].\end{aligned}$$

While the math is relatively neat to derive, the reality of these kinds of equations is that the necessary data are rarely available to perform these computations. Usually, for an approximation the analyst might only retain terms to order ε, and even then, major simplifications are needed.

Another approach to this problem is to apply the fundamental theorem to Equation 10.49,

$$S_{X_0 X_0}(\omega) = |H(i\omega)|^2 S_{FF}(\omega).$$

10.5 The van der Pol Equation

Given $S_{X_0 X_0}(\omega)$, it is possible to find $S_{X_0^3 X_0^3}(\omega)$ for a Gaussian process, and then use the fundamental theorem again with Equation 10.50, and then with Equation 10.51. For non-Gaussian processes, this is more difficult, if not impossible analytically.

10.5 The van der Pol Equation

The unforced van der Pol equation is given by

$$\ddot{x} - \alpha\left(1 - x^2\right)\dot{x} + x = 0, \quad \alpha > 0. \tag{10.55}$$

If α is small in some sense, then the ε notation can be used and the van der Pol equation becomes

$$\ddot{x} + x = \varepsilon\left(1 - x^2\right)\dot{x}. \tag{10.56}$$

Expand the right-hand side to order ε^2, that is, $x = x_0 + \varepsilon x_1 + \varepsilon^2 x_2$,

$$\begin{aligned}
\left(1 - x^2\right)\dot{x} &\simeq \left[1 - \left(x_0 + \varepsilon x_1 + \varepsilon^2 x_2\right)^2\right]\left(\dot{x}_0 + \varepsilon \dot{x}_1 + \varepsilon^2 \dot{x}_2\right) \\
&\simeq \left(1 - x_0^2\right)\dot{x}_0 + \varepsilon\left[-2x_0 x_1 \dot{x}_0 + \left(1 - x_0^2\right)\dot{x}_1\right] \\
&\quad + \varepsilon^2\left[-x_1^2 \dot{x}_0 - 2x_0 x_2 \dot{x}_0 - 2x_0 x_1 \dot{x}_1 + \left(1 - x_0^2\right)\dot{x}_2\right],
\end{aligned}$$

and insert into Equation 10.56. Equate terms of same order in ε to find the following perturbation equations,

$$\begin{aligned}
\varepsilon^0 &: \ddot{x}_0 + x_0 = 0 \\
\varepsilon^1 &: \ddot{x}_1 + x_1 = \left(1 - x_0^2\right)\dot{x}_0 \\
\varepsilon^2 &: \ddot{x}_2 + x_2 = -2x_0 x_1 \dot{x}_0 + \left(1 - x_0^2\right)\dot{x}_1 \\
\varepsilon^3 &: \ddot{x}_3 + x_3 = -x_1^2 \dot{x}_0 - 2x_0 x_2 \dot{x}_0 - 2x_0 x_1 \dot{x}_1 + \left(1 - x_0^2\right)\dot{x}_2 \\
&\vdots
\end{aligned}$$

The equation of $O\left(\varepsilon^3\right)$ is ignored since the expansion is to order ε^2. Each equation is now solved sequentially and the approximate solution for $x(t)$ is given by

$$x(t) \simeq x_0 + \varepsilon x_1 + \varepsilon^2 x_2.$$

In the case initial conditions are random variables, the response becomes a random process, and the governing equation becomes

$$\ddot{X} - \alpha\left(1 - X^2\right)\dot{X} + X = 0, \quad \alpha > 0.$$

The expansion of the solution results in random functions $X_i(t)$. Then,

$$E\{X(t)\} \simeq E\{X_0 + \varepsilon X_1 + \varepsilon^2 X_2\}$$
$$= E\{X_0\} + \varepsilon E\{X_1\} + \varepsilon^2 E\{X_2\}.$$

The difficulties in evaluating these expectations might lead one to directly simulate the governing differential equation rather than pursue an expansion approximate solution.

10.5.1 Limit Cycles

In the above introduction to the stability of motion, all motion either tended toward the equilibrium point, or moved out into instability. Another possibility exists, and it most closely resembles the elliptical trajectories mentioned earlier. The difference between limit cycles and closed trajectories is that the periodic motion of a limit cycle is not a closed trajectory. Rather, it is a trajectory that can be approached from the equilibrium point or from other initial conditions beyond.

In such damped vibration problems, the trajectories can start either very close to the origin or far away from the origin, and approach the same closed curve about the origin. This curve represents a periodic, but not harmonic, solution of the governing equation. This curve is called the *limit cycle*. The classical equation that has such a limit cycle is the *van der Pol* oscillator, one mathematical form of which is Equation 10.55. The van der Pol equation can only be solved numerically.

This equation is applied to a number of disciplines, including nonlinear electrical and vibratory systems. Physically, for vibratory systems, the form of the damping force, $\alpha\left(1 - x^2\right)\dot{x}$, is such that the damping is negative for small amplitudes and positive for larger amplitudes; this depends on the sign of $\left(1 - x^2\right)$. Consider the phase plane for the van der Pol equation,

$$\dot{x} = y$$
$$\dot{y} = \alpha\left(1 - x^2\right)y - x,$$

with trajectories defined by

$$\frac{dy}{dx} = \frac{\alpha\left(1 - x^2\right)y - x}{y}.$$

Sketches of a variety of trajectories, regardless of initial conditions, approach the limit cycle asymptotically. An initial point inside the limit cycle follows an outwardly spiraling trajectory. An initial point outside the limit

10.5. THE VAN DER POL EQUATION

cycle follows an inwardly spiraling trajectory. An infinity of isoclines[19] pass through the origin, which is a singularity. The limit cycle has the interesting property that the maximum value of x is always close to 2 regardless of the value of α. Figure 10.11 shows the trajectories of two cases.

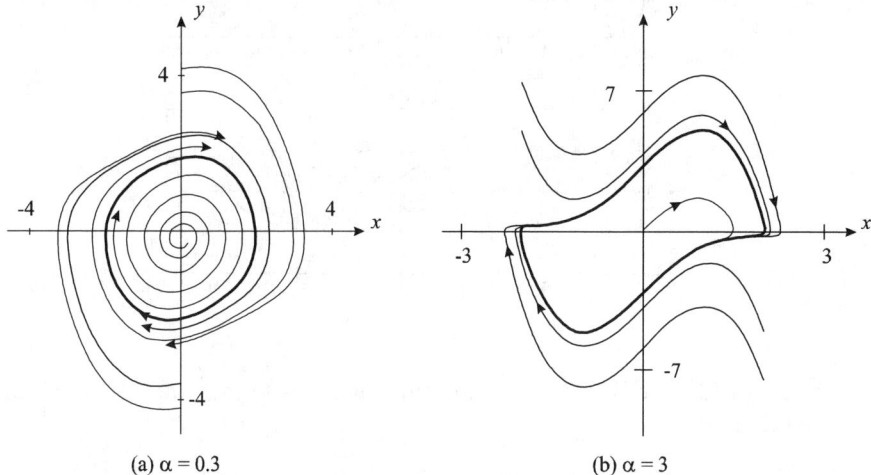

(a) $\alpha = 0.3$ (b) $\alpha = 3$

Figure 10.11: Trajectories for two values of α for the van der Pol equation for a number of initial conditions. The limit cycle is shown in a dark line.

Should the initial conditions be random variables, then it is expected that the crisply defines trajectories of Figure 10.11 are replaced by a band of possible trajectories defined by the probability density of the initial conditions.

Figure 10.12 shows the time histories of this oscillator for the same two cases, for several deterministic initial conditions. For random variable initial conditions, Figure 10.12 would be of the ensemble of possible time functions.

10.5.2 The Forced van der Pol Equation

The forced van der Pol equation is a model for a system that is capable of self-oscillation while acted upon by another oscillator. Rand[20] points to a biological application involving the human sleep-wake cycle in which

---

[19] An isocline is defined as the locus of points at which the trajectories passing through them have a constant slope. The method of isoclines is used to construct the trajectories of dynamical systems with one degree of freedom.

[20] *Lecture Notes on Nonlinear Vibrations*, R.H. Rand. Look up the latest version of lecture notes at http://www.tam.cornell.edu/randdocs.

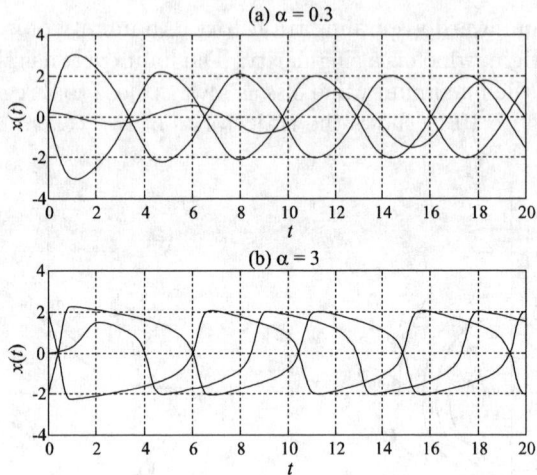

Figure 10.12: Van der Pol equation time histories for several sets of initial conditions.

a person's biological clock is modeled by a van der Pol oscillator, and the daily night-day cycle caused by the Earth's rotation is modeled as a periodic forcing term.

The phenomenon where the period of a forced oscillation is an integer multiple of the period of the forcing is called *frequency entrainment*. The frequency of the solution, say ν, is said to be entrained by the frequency of the forcing, say ω, so that $\nu = \omega/l$ for some positive integer l called the *entrainment index*. The van der Pol equation has such characteristics.

Consider the following forced van der Pol equation,

$$\ddot{x} + \varepsilon\left(x^2 - 1\right)\dot{x} + x = A\cos\omega t. \tag{10.57}$$

The unforced case has a limit cycle with a radius of approximately 2 and a period of approximately 2π. The limit cycle is generated by the balance between the internal energy loss and energy generation. A forcing term will alter this balance.

If A is small, the excitation is called weak, and its effect depends on whether ω is close to the natural frequency. If it is, an oscillation will be generated that is a perturbation of the limit cycle. If A is not small, the excitation is called hard. If the natural and imposed frequencies are not close, we expect that the natural oscillation will be damped out as happens with the corresponding linear equation.

10.5. THE VAN DER POL EQUATION

Stretch time according to $\omega t = \tau$, and Equation 10.57 becomes

$$\omega^2 x'' + \varepsilon\omega \left(x^2 - 1\right) x' + x = A\cos\tau.$$

A number of cases require further examination. Suppose hard excitation far from resonance and assume that ω is not close to one. Expand as usual,

$$x(\varepsilon, \tau) = x_0(\tau) + \varepsilon x_1(\tau) + \cdots, \qquad (10.58)$$

to find

$$\begin{aligned} \omega^2 x_0'' + x_0 &= A\cos\tau \\ \omega^2 x_1'' + x_1 &= -\omega\left(x_0^2 - 1\right) x_0', \end{aligned}$$

where $x_0(\tau)$ and $x_1(\tau)$ have period 2π. Therefore,

$$x_0(\tau) = \frac{A}{1 - \omega^2} \cos\tau.$$

We know the solution to $x_1(\tau)$ is of $O(\varepsilon)$, thus,

$$x(\varepsilon, \tau) = \frac{A}{1 - \omega^2} \cos\tau + O(\varepsilon).$$

The solution is a perturbation of the ordinary linear response and the limit cycle is suppressed as expected. If the excitation is soft and far from resonance, the procedure is similar as for the hard excitation but the response is usually unstable.

Suppose soft excitation near resonance. Then let $A = \varepsilon\gamma$, and near resonance $\omega = 1 + \varepsilon\omega_1$, then use expansion Equation 10.58. The resulting equations are

$$\begin{aligned} x_0'' + x_0 &= 0 \\ x_1'' + x_1 &= -2\omega_1 x_0'' - \left(x_0^2 - 1\right) x_0' + \gamma\cos\tau \\ &\vdots \end{aligned}$$

As usual, solutions are sought with period 2π, then

$$x_0(\tau) = a_0 \cos\tau + b_0 \sin\tau$$

and

$$\begin{aligned} x_1'' + x_1 &= \left[\gamma + 2\omega_1 a_0 - b_0\left(\frac{1}{4}r_0^2 - 1\right)\right]\cos\tau \\ &\quad + \left[2\omega_1 b_0 + a_0\left(\frac{1}{4}r_0^2 - 1\right)\right]\sin\tau \\ &\quad + \cdots \\ r_0 &= +\sqrt{a_0^2 + b_0^2}, \end{aligned}$$

where r_0 is the response amplitude for $x_0(\tau)$. For a periodic solution to exist, the coefficients of the harmonic functions must be set equal to zero,

$$2\omega_1 a_0 - b_0\left(\frac{1}{4}r_0^2 - 1\right) = -\gamma$$

$$2\omega_1 b_0 + a_0\left(\frac{1}{4}r_0^2 - 1\right) = 0.$$

These two equations can be combined into

$$r_0^2\left[4\omega_1^2 + \left(\frac{1}{4}r_0^2 - 1\right)^2\right] = \gamma^2,$$

which gives possible values of r_0. Note that there may be as many as 3 real solutions for $r_0 > 0$.

Van der Pol   Balthazar van der Pol (1889-1959) was a Dutch electrical engineer who initiated modern experimental dynamics in the laboratory during the 1920s and 1930s. Van der Pol investigated electrical circuits employing vacuum tubes and found that they have stable oscillations, now called limit cycles. When these circuits are driven with a signal whose frequency is near that of the limit cycle, the resulting periodic response shifts its frequency to that of the driving signal. That is to say, the circuit becomes "entrained" to the driving signal. The waveform, or signal shape, however, can be quite complicated and contain a rich structure of harmonics and subharmonics.

In the September 1927 issue of the British journal, *Nature*, he and his colleague van der Mark reported that an "irregular noise" was heard at certain driving frequencies between the natural entrainment frequencies. By reconstructing his electronic tube circuit, we now know that they had discovered deterministic chaos. Their paper is probably one of the first experimental reports of chaos, something that they failed to pursue in more detail.

Van der Pol built a number of electronic circuit models of the human heart to study the range of stability of heart dynamics. His investigations with adding an external driving signal were analogous to the situation in which a real heart is driven by a pacemaker. He was interested in finding out, using his entrainment work, how to stabilize a heart's irregular beating or "arrhythmias."

10.6 Markov Process-Based Models

In this section, a certain kind of random process called the *Markov* process[21] is introduced. It has some interesting properties, and it is linked to the *Fokker-Planck* equation that governs the evolutionary probability density of the response process. The Fokker-Planck equation is derived herein. To do this, the Markov process is defined and the *Chapman-Kolmogorov* equation derived. The Chapman-Kolmogorov equation is a differential equation whose solution is the probability density function of the response.

Unlike other methods that are used to solve nonlinear problems, such as perturbation technique discussed in the previous section, the Markov process approach has an advantage that it is not limited to systems with weak nonlinearity (Recall that the nonlinear terms seen previously were all of order ϵ.) In this section, we will apply this theory to the Duffing and van der Pol oscillators.

10.6.1 Probability Background

A stochastic process $X(t)$ is said to be a Markov process if the following conditional probability is satisfied,

$$\Pr\left(X(t_n) \leq x_n | X(t_{n-1}) = x_{n-1}, \ldots, X(t_2) = x_2, X(t_1) = x_1\right)$$
$$= \Pr\left(X(t_n) \leq x_n | X(t_{n-1}) = x_{n-1}\right), \qquad (10.59)$$

where $t_1 < t_2 < \ldots < t_{n-1} < t_n$. Equation 10.59 implies that the probability of a random process $X(t)$ to take on a value x_n at $t = t_n$ only depends on the most recent value, $X(t_{n-1})$, so that the probability is independent of the time history prior to $t = t_{n-1}$.

The random process $X(t)$ can be either a discrete or continuous process. A discrete Markov process is called the *Markov chain*, and a continuous Markov process is called a *diffusion process*.

A continuous random process $X(t)$ is a Markov process if the following conditional probability density function is satisfied,

$$f_{X_1 \ldots X_n}\left(x_n, t_n | x_{n-1}, t_{n-1}; \ldots; x_2, t_2; x_1, t_1\right)$$
$$= f_{X_1 \ldots X_n}\left(x_n, t_n | x_{n-1}, t_{n-1}\right), \qquad (10.60)$$

where $X_1 \equiv X(t_1), \ldots, X_n \equiv X(t_n)$. The Markov process is such that the probability that $X(t)$ lies in $(x_n, x_n + dx_n)$ at time t_n, given that X lies in $(x_{n-1}, x_{n-1} + dx_{n-1})$ at time t_{n-1}, in $(x_{n-2}, x_{n-2} + dx_{n-2})$ at time $t_{n-2}, \ldots$, and in $(x_1, x_1 + dx_1)$ at time t_1, is identical to the probability that X lies

---
[21] *Markov Processes, An Introduction for Physical Scientists*, D.T. Gillespie, Academic Press, 1992.

in the interval $(x_n, x_n + dx_n)$ at time t_n given that X lies in the interval $(x_{n-1}, x_{n-1} + dx_{n-1})$ at time t_{n-1}.

The notation in Equation 10.60 can be cumbersome. We can write the same equation as

$$f_{\{X\}}(x_n(t_n)|x_{n-1}(t_{n-1}),\ldots,x_1(t_1)) = f_{\{X\}}(x_n(t_n)|x_{n-1}(t_{n-1})),$$

where the subscript is replaced by the vector $\{X\}$, and $x_n(t_n)$ is used instead of x_n, t_n. The notation $x_n(t_n)$ does *not* mean that x_n is a function of t_n. It means that the random process $X(t)$ can take on a value x_n at time t_n. Therefore, the notation (t_n) seems redundant. In some cases, we may drop (t_n). Then, we can write

$$f_{\{X\}}(x_n|x_{n-1},\ldots,x_1) = f_{\{X\}}(x_n|x_{n-1}).$$

Recall the expression that defines the conditional density,

$$f(E_2|E_1) = \frac{f(E_2, E_1)}{f(E_1)}, \qquad (10.61)$$

which can generalized for any number of events, for example, for three events,

$$f(E_3|E_2, E_1) = \frac{f(E_3, E_2, E_1)}{f(E_2, E_1)}. \qquad (10.62)$$

Utilizing Equations 10.61 and 10.62,

$$\begin{aligned} f_{\{X\}}(x_3, x_2, x_1) &= f_{\{X\}}(x_3|x_2, x_1) f_{\{X\}}(x_2, x_1) \\ &= f_{\{X\}}(x_3|x_2, x_1) f_{\{X\}}(x_2|x_1) f_{\{X\}}(x_1). \end{aligned} \qquad (10.63)$$

If $X(t)$ is a Markov process, Equation 10.63 can be rewritten as

$$f_{\{X\}}(x_3, x_2, x_1) = f_{\{X\}}(x_3|x_2) f_{\{X\}}(x_2|x_1) f_{\{X\}}(x_1). \qquad (10.64)$$

In general,

$$f_{\{X\}}(x_n, x_{n-1}, \ldots, x_1) = f_{\{X\}}(x_1) \prod_{i=2}^{i=n} f(x_i|x_{i-1}), \qquad (10.65)$$

where $f(x_i|x_{i-1})$ is called the *transition probability density function*, or the *evolution density function*. Equation 10.65 implies that the complete probabilistic structure can be found if the probability density is known at the initial time, t_1, and its evolution mechanism is given by the transition probability density.

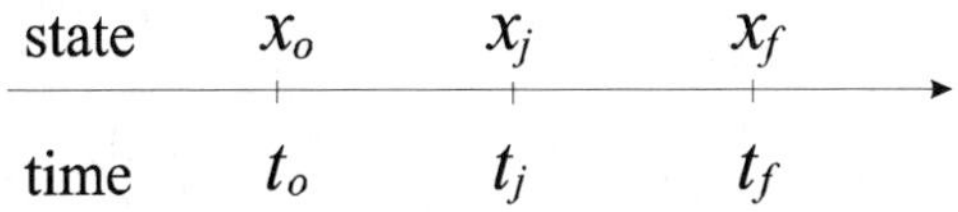

10.6. MARKOV PROCESS-BASED MODELS

```
state      x_o        x_j         x_f
         ─┼──────────┼───────────┼────────►
time       t_o        t_j         t_f
```

Figure 10.13: States of a Markov process.

Now, consider the schematic in Figure 10.13. In passing from state $x_0(t_0)$ to the present state $x_f(t_f)$ in time $(t_0 - t_f)$, the process $X(t)$ passes through some intermediate state x_j at t_j. The joint density function that defines these three states is $f(x_f(t), x_j(t_j), x_0(t_0))$, where x_j represents any state between the initial and final states. The marginal joint density of the initial and final states can be derived by integrating over all possible intermediate states,

$$f_{\{X\}}(x_f(t_f), x_0(t_0)) = \int f_{\{X\}}(x_f(t_f), x_j(t_j), x_0(t_0)) \, dx_j.$$

Utilize Equation 10.64 to find that

$$\begin{aligned} f_{\{X\}}(x_f(t_f), x_0(t_0)) &= \int f_{\{X\}}(x_f(t_f)|x_j(t_j), x_0(t_0)) \\ &\quad \cdot f_{\{X\}}(x_j(t_j)|x_0(t_0)) f_{\{X\}}(x_0(t_0)) \, dx_j. \end{aligned}$$

If $X(t)$ is Markovian, then

$$\begin{aligned} f_{\{X\}}(x_f(t_f), x_0(t_0)) &= \int f_{\{X\}}(x_f(t_f)|x_j(t_j)) \\ &\quad \cdot f_{\{X\}}(x_j(t_j)|x_0(t_0)) f_{\{X\}}(x_0(t_0)) \, dx_j. \end{aligned}$$

$f_{\{X\}}(x_0(t_0))$ is not a function of x_j and can therefore be pulled out of the integration. Then, divide both sides by $f_{\{X\}}(x_0(t_0))$, and make the following substitution,

$$\frac{f_{\{X\}}(x_f(t_f), x_0(t_0))}{f_{\{X\}}(x_0(t_0))} \equiv f_{\{X\}}(x_f(t_f)|x_0(t_0))$$

to find

$$f_{\{X\}}(x_f(t_f)|x_0(t_0)) = \int f_{\{X\}}(x_f(t_f)|x_j(t_j)) f_{\{X\}}(x_j(t_j)|x_0(t_0)) \, dx_j,$$

or in more concise notation,

$$f_{\{X\}}(x_f|x_0) = \int f_{\{X\}}(x_f|x_j) f_{\{X\}}(x_j|x_0) \, dx_j.$$

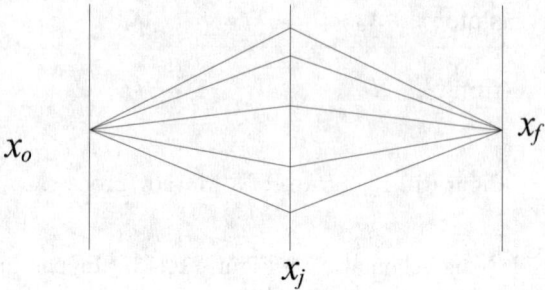

Figure 10.14: Possible transitions of a Markov process.

This is the *Chapman-Kolmogorov* equation for Markov process $X(t)$, which states that the conditional probability of the final state $x_f(t_f)$, given that the initial state is $x_0(t_0)$, is equal to the sum (integral) of all the possible transitions through all intermediate states $x_j(t_j)$. One can consider this to be a *law of conservation of probability*, and it can be depicted by the schematic in Figure 10.14. The equation is an accounting of the fact that in the transition between the initial and final state, the process $X(t)$, being random, can proceed through a variety of intermediate states x_j, and all this is governed by the respective density functions.

For discrete Markov process $X(t)$, the equivalent Chapman-Kolmogorov equation is

$$\Pr(X(t_f) = x_f | X(t_0) = x_0)$$
$$= \sum_j \Pr(X(t_f) = x_f | X(t_j) = x_j) \Pr(X(t_j) = x_j | X(t_0) = x_0).$$

We should keep in mind that if $X(t)$ is a Markov process, then the Chapman-Kolmogorov equation is satisfied. However, the converse is not necessarily true, all solutions to the Chapman-Kolmogorov equation are not necessarily Markov processes.

Markov Andrei A. Markov was a graduate of Saint Petersburg University (1878), where he began as professor in 1886. Markov's early work was mainly in number theory and analysis, continued fractions, limits of integrals, approximation theory and the convergence of series.

After 1900 Markov applied the method of continued fractions, pioneered by his teacher Pafnuty Chebyshev, to probability theory. He also studied sequences of mutually dependent variables, hoping to establish the limiting laws of probability in their most general form. He proved the central limit theorem under fairly general assumptions.

10.6. MARKOV PROCESS-BASED MODELS

Figure 10.15: Andrei A. Markov (1856-1922).

Markov is particularly remembered for his study of Markov chains, sequences of random variables in which the future variable is determined by the present variable but is independent of the way in which the present state arose from its predecessors. This work launched the theory of stochastic processes.

In 1923 Norbert Wiener became the first to treat rigorously a continuous Markov process. The foundation of a general theory was provided during the 1930s by Andrei Kolmogorov. Markov was also interested in poetry and he made studies of poetic style, interestingly Kolmogorov had similar interests. Markov had a son (of the same name) who was born on September 9, 1903, and followed his father in also becoming a renowned mathematician.

10.6.2 The Fokker-Planck Equation

We will now utilize the Chapman-Kolmogorov equation as a basis for deriving the Fokker-Planck equation.[22] This is accomplished by defining an integral that contains an arbitrary function as a factor. By working with this integral, it is possible to derive an equation that governs the evolutionary density. This equation is the Fokker-Planck.

For Markov process $X(t)$ define the following integral,

$$I = \int_{-\infty}^{\infty} R(y) \frac{\partial}{\partial t} f_{\{X\}} \left(y(t) | x_0(t_0) \right) dy, \qquad t > t_0, \tag{10.66}$$

---
[22] Also, Fokker-Planck-Kolmogorov.

where $R(y)$ is an *arbitrary function* of y, which goes to zero sufficiently fast as $y \to \pm\infty$,

$$\lim_{y \to \pm\infty} \frac{d^n}{dy^n} R(y) = 0,$$

for any n. Note that y is a dummy variable in the integration, and it can be interpreted as an arbitrary state governed by the probability density function $f_{\{X\}}(y(t)|x_0(t_0))$. In fact, $y(t)$ is a value that the random process $X(t)$ can take on at an arbitrary time t. We would normally denote it simply as $x(t)$. However, we write it as $y(t)$ so that the following derivation is clearer.

Make the following substitution,

$$\frac{\partial}{\partial t} f_{\{X\}}(y(t)|x_0(t_0)) = \lim_{\Delta t \to 0} \frac{1}{\Delta t} \left[f_{\{X\}}(y(t+\Delta t)|x_0(t_0)) - f_{\{X\}}(y(t)|x_0(t_0)) \right].$$

Then,

$$I = \int_{-\infty}^{\infty} R(y) \lim_{\Delta t \to 0} \frac{1}{\Delta t} \left[f_{\{X\}}(y(t+\Delta t)|x_0(t_0)) - f_{\{X\}}(y(t)|x_0(t_0)) \right] dy$$

$$= \lim_{\Delta t \to 0} \frac{1}{\Delta t} \int_{-\infty}^{\infty} R(y) \left[f_{\{X\}}(y(t+\Delta t)|x_0(t_0)) - f_{\{X\}}(y(t)|x_0(t_0)) \right] dy,$$

since it is assumed that the integral converges uniformly in the neighborhood of t. Now use the Chapman-Kolmogorov equation to make the following substitution,

$$f_{\{X\}}(y(t+\Delta t)|x_0(t_0)) = \int_{-\infty}^{\infty} f_{\{X\}}(y(t+\Delta t)|x(t)) f_{\{X\}}(x(t)|x_0(t_0)) \, dx.$$

Note that $x(t)$ is an arbitrary intermediate state that will be integrated out and thus does not appear in the result on the left hand side of the equation. Thus,

$$I = \lim_{\Delta t \to 0} \frac{1}{\Delta t} \int_{-\infty}^{\infty} R(y) \left[\int_{-\infty}^{\infty} f_{\{X\}}(y(t+\Delta t)|x(t)) f_{\{X\}}(x(t)|x_0(t_0)) \, dx \right.$$
$$\left. - f_{\{X\}}(y(t)|x_0(t_0)) \right] dy$$

$$= \lim_{\Delta t \to 0} \frac{1}{\Delta t} \left[\int_{-\infty}^{\infty} dx \left\{ f_{\{X\}}(x(t)|x_0(t_0)) \right. \right.$$
$$\left. \times \int_{-\infty}^{\infty} dy \left\{ R(y) f_{\{X\}}(y(t+\Delta t)|x(t)) \right\} \right\}$$
$$\left. - \int_{-\infty}^{\infty} dy \left\{ R(y) f_{\{X\}}(y(t)|x_0(t_0)) \right\} \right].$$

10.6. MARKOV PROCESS-BASED MODELS

The last integral is rewritten in the following way, recognizing that y is a dummy variable of integration,

$$\int_{-\infty}^{\infty} dy \left\{ R(y) f_{\{X\}}(y(t)|x_0(t_0)) \right\} \equiv \int_{-\infty}^{\infty} dx \left\{ R(x) f_{\{X\}}(x(t)|x_0(t_0)) \right\}.$$

Then,

$$I = \lim_{\Delta t \to 0} \frac{1}{\Delta t} \int_{-\infty}^{\infty} dx \left\{ f_{\{X\}}(x(t)|x_0(t_0)) \right.$$
$$\left. \cdot \left[\int_{-\infty}^{\infty} dy \left\{ R(y) f_{\{X\}}(y(t+\Delta t)|x(t)) \right\} - R(x) \right] \right\}. \tag{10.67}$$

This equation has, mathematically, two dummy variables x and y, one for each integration. Physically, we may interpret them as two arbitrary states of the Markov process with initial state $x_0(t_0)$. These two arbitrary states may be different or may be the same. One way to relate them generally is to expand one in a Taylor series about the other. Given the form of Equation 10.67, we choose to expand the function $R(y)$ about state x,

$$R(y) = R(x) + (y-x) R'(x) + \frac{(y-x)^2}{2} R''(x) + \cdots,$$

where any order may be retained. Equation 10.67 can then be written as

$$I = \lim_{\Delta t \to 0} \frac{1}{\Delta t} \int_{-\infty}^{\infty} dx \left\{ f_{\{X\}}(x(t)|x_0(t_0)) \left[\int_{-\infty}^{\infty} dy \left\{ R(x) + (y-x) R'(x) \right. \right. \right.$$
$$\left. \left. \left. + \frac{(y-x)^2}{2} R''(x) + \cdots \right\} f_{\{X\}}(y(t+\Delta t)|x(t)) - R(x) \right] \right\}.$$

Expand the above products, and retain up to the second-order term in the expansion,

$$\begin{aligned} I = \lim_{\Delta t \to 0} \frac{1}{\Delta t} & \left[\int_{-\infty}^{\infty} dx \left\{ f_{\{X\}}(x(t)|x_0(t_0)) \int_{-\infty}^{\infty} dy \left\{ A \right\} \right\} \right. \\ & + \int_{-\infty}^{\infty} dx \left\{ f_{\{X\}}(x(t)|x_0(t_0)) \int_{-\infty}^{\infty} dy \left\{ B \right\} \right\} \\ & + \int_{-\infty}^{\infty} dx \left\{ f_{\{X\}}(x(t)|x_0(t_0)) \int_{-\infty}^{\infty} dy \left\{ C \right\} \right\} \\ & \left. - \int_{-\infty}^{\infty} dx \left\{ f_{\{X\}}(x(t)|x_0(t_0)) R(x) \right\} \right], \tag{10.68} \end{aligned}$$

where

$$A = R(x) f_{\{X\}}(y(t+\Delta t)|x(t))$$
$$B = (y-x) R'(x) f_{\{X\}}(y(t+\Delta t)|x(t))$$
$$C = \frac{(y-x)^2}{2} R''(x) f_{\{X\}}(y(t+\Delta t)|x(t)).$$

The second-order expansion is exact for the case of a Gaussian white noise process.

We can take $R(x)$ and its derivatives outside the integrals with respect to y. Note that

$$\int_{-\infty}^{\infty} dy \left\{ f_{\{X\}}(y(t+\Delta t)|x(t)) \right\} = 1$$

and therefore the sum of the first and last terms in Equation 10.68 cancel

$$\int_{-\infty}^{\infty} dx \left\{ f_{\{X\}}(x(t)|x_0(t_0)) \int_{-\infty}^{\infty} dy \left\{ R(x) f_{\{X\}}(y(t+\Delta t)|x(t)) \right\} \right\}$$
$$- \int_{-\infty}^{\infty} dx \left\{ f_{\{X\}}(x(t)|x_0(t_0)) R(x) \right\} = 0.$$

For ease of notation, define the following,

$$A(x,t) = \lim_{\Delta t \to 0} \frac{1}{\Delta t} \int_{-\infty}^{\infty} dy \left\{ (y-x) f_{\{X\}}(y(t+\Delta t)|x(t)) \right\} \quad (10.69)$$

$$B(x,t) = \lim_{\Delta t \to 0} \frac{1}{\Delta t} \int_{-\infty}^{\infty} dy \left\{ (y-x)^2 f_{\{X\}}(y(t+\Delta t)|x(t)) \right\}. \quad (10.70)$$

These are limits of expectations. Then,

$$I = \int_{-\infty}^{\infty} dx \left\{ f_{\{X\}}(x(t)|x_0(t_0)) \left[R'(x) A(x,t) + \frac{R''(x)}{2} B(x,t) \right] \right\},$$

and $A(x,t)$ and $B(x,t)$ can be written as

$$A(x,t) = \lim_{\Delta t \to 0} \frac{1}{\Delta t} E\left\{ x(t+\Delta t) - x(t)|x(t) \right\}$$

$$B(x,t) = \lim_{\Delta t \to 0} \frac{1}{\Delta t} E\left\{ (x(t+\Delta t) - x(t))^2 |x(t) \right\}.$$

To put this expression into a useful form, we need to be able to factor out the arbitrary function $R(x)$. This can be accomplished by integrating each of the expressions on the right hand side by parts, once for the term with factor $A(x,t)$, and twice for the term with factor $B(x,t)$. In the process

10.6. MARKOV PROCESS-BASED MODELS

of integrating by parts we obtain boundary condition terms. In these cases above, the boundary terms will go to zero in the limit because of the given property for the arbitrary function $R(x)$. We are then left with the following expressions,

$$\int_{-\infty}^{\infty} dx \left\{ f_{\{X\}}\left(x(t)|x_0\left(t_0\right)\right) R'(x) A(x,t) \right\}$$
$$= - \int_{-\infty}^{\infty} dx \left\{ R(x) \frac{\partial}{\partial x} \left(f_{\{X\}}\left(x(t)|x_0\left(t_0\right)\right) A(x,t) \right) \right\}$$
$$\int_{-\infty}^{\infty} dx \left\{ f_{\{X\}}\left(x(t)|x_0\left(t_0\right)\right) \frac{R''(x)}{2} B(x,t) \right\}$$
$$= + \int_{-\infty}^{\infty} dx \left\{ \frac{R(x)}{2} \frac{\partial^2}{\partial x^2} \left(f_{\{X\}}\left(x(t)|x_0\left(t_0\right)\right) B(x,t) \right) \right\}.$$

From Equation 10.66, replace I by its equivalent, and therefore

$$\int_{-\infty}^{\infty} \left\{ R(x) \frac{\partial}{\partial t} f_{\{X\}}\left(x(t)|x_0\left(t_0\right)\right) \right\} dx$$
$$= - \int_{-\infty}^{\infty} \left\{ R(x) \frac{\partial}{\partial x} \left(f_{\{X\}}\left(x(t)|x_0\left(t_0\right)\right) A(x,t) \right) \right\} dx$$
$$+ \int_{-\infty}^{\infty} \left\{ \frac{R(x)}{2} \frac{\partial^2}{\partial x^2} \left(f_{\{X\}}\left(x(t)|x_0\left(t_0\right)\right) B(x,t) \right) \right\} dx.$$

Factoring the function $R(x)$ and combining terms, results in

$$\int_{-\infty}^{\infty} dx \left\{ R(x) \left[\frac{\partial f_{\{X\}}}{\partial t} + \frac{\partial}{\partial x} \left(A f_{\{X\}} \right) - \frac{1}{2} \frac{\partial^2}{\partial x^2} \left(B f_{\{X\}} \right) \right] \right\} = 0.$$

Since $R(x)$ is an arbitrary function, the expression in the square brackets must equal zero in order for the expression to be valid. Then,

$$\frac{\partial f_{\{X\}}}{\partial t} = -\frac{\partial}{\partial x} \left(A f_{\{X\}} \right) + \frac{1}{2} \frac{\partial^2}{\partial x^2} \left(B f_{\{X\}} \right),$$

which is the *Fokker-Planck* equation[23] governing the evolution of the probability density function $f_{\{X\}}$ for the Markov process $X(t)$. Note that the initial condition is

$$f_{\{X\}}\left(x(t)|x_0\left(t_0\right)\right)|_{t=t_0} = \delta\left(x - x_0\right).$$

---

[23] The Fokker-Planck equation is one of a class of equations known as the Kramers-Moyal equations. When we truncated after the second order, we made use of the fact that for a Markov process, the higher-order terms are all zero [Gillespie, p. 120].

Also, the characterizing functions A and B are the (Gaussian) mean and variance of a *propagator* density function, which is related to the Markov density f [Gillespie, p. 115].

Equations 10.69 and 10.70 are based on the particular governing differential equation of motion. For a multidimensional process, $\mathbf{x}(t)$ is a vector random process given by

$$\mathbf{x}(t) = [x_{(1)}(t)\ x_{(2)}(t) \ldots\ x_{(M)}(t)]^T,$$

and the respective Fokker-Planck equation is given by

$$\frac{\partial f_{\{X\}}(\mathbf{x}(t)|\mathbf{x}_0(t_0))}{\partial t} = -\sum_i \frac{\partial}{\partial x_i}\left[A_i(\mathbf{x},t) f_{\{X\}}(\mathbf{x}(t)|\mathbf{x}_0(t_0))\right]$$
$$+ \frac{1}{2}\frac{\partial^2}{\partial x_i \partial x_j}\sum_{i,j}\left[B_{ij}(\mathbf{x},t) f_{\{X\}}(\mathbf{x}(t)|\mathbf{x}_0(t_0))\right], \tag{10.71}$$

$$A_i(\mathbf{x},t) = \lim_{\Delta t \to 0}\frac{1}{\Delta t}E\{\Delta x_{(i)}|\mathbf{x}\}$$
$$B_{ij}(\mathbf{x},t) = \lim_{\Delta t \to 0}\frac{1}{\Delta t}E\{\Delta x_{(i)} \Delta x_{(j)}|\mathbf{x}\}.$$

The terms $\Delta x_{(i)}$ are interpreted to mean $x_{(i)}(t+\Delta t) - x_{(i)}(t)$, where Δt is a very small time interval. The conditioned variable implies $\mathbf{x}(t) = \mathbf{x}$, a particular value. Thus,

$$E\{\Delta x_{(i)}|\mathbf{x}\} \equiv E\{x_{(i)}(t+\Delta t) - x_{(i)}(t)|\mathbf{x}(t) = \mathbf{x}\}$$
$$E\{\Delta x_{(i)} \Delta x_{(j)}|\mathbf{x}\} \equiv E\{(x_{(i)}(t+\Delta t) - x_{(i)}(t))$$
$$\cdot (x_{(j)}(t+\Delta t) - x_{(j)}(t))|\mathbf{x}(t) = \mathbf{x}\}.$$

Then,

$$\{\Delta x_{(i)}|\mathbf{x}\} = \dot{x}_{(i)}\Delta t + O(\Delta t)^2$$

and

$$\lim_{\Delta t \to 0}\frac{1}{\Delta t}E\{\Delta x_{(i)}|\mathbf{x}\} \simeq \dot{x}_{(i)}$$

since the Δt cancels.

Example 10.7 The Random Duffing Oscillator

As an example, consider the general governing equation for a Duffing-type oscillator,
$$\ddot{X} + c\dot{X} + h(X) = F(t),$$
where $h(X) = X + X^3$, and where input $F(t)$ is taken to be zero mean and delta correlated,

$$E\{F\} = 0 \qquad E\{F(t)F(t+\tau)\} = 2\alpha\delta(\tau).$$

10.6. MARKOV PROCESS-BASED MODELS

Convert the second-order differential equation into two first-order differential equations. Define the state space by letting $x_1 = X$ and $x_2 = \dot{x}_1$. Then the system becomes

$$\left\{ \begin{array}{c} \dot{x}_1 \\ \dot{x}_2 \end{array} \right\} = \left\{ \begin{array}{c} x_2 \\ -cx_2 - h(x_1) + F(t) \end{array} \right\}.$$

By this set of definitions, x_1 represents the displacement and x_2 the velocity. $\dot{x}_2$ is then the acceleration. Now proceed to evaluate the necessary coefficients, a_1, a_2, b_{11}, b_{12}, b_{22} of Equation 10.71:

$$a_1 = \lim_{\Delta t \to 0} \frac{1}{\Delta t} E\{\Delta x_1\} = \lim_{\Delta t \to 0} \frac{E\{\dot{x}_1 \Delta t\}}{\Delta t} = \dot{x}_1 = x_2 = v$$

$$a_2 = \lim_{\Delta t \to 0} \frac{1}{\Delta t} E\{\Delta x_2\} = \lim_{\Delta t \to 0} \frac{E\{\dot{x}_2 \Delta t\}}{\Delta t} = \dot{x}_2 = -cx_2 - h(x_1)$$

$$b_{11} = \lim_{\Delta t \to 0} \frac{1}{\Delta t} E\{(\Delta x_1)^2\} = \lim_{\Delta t \to 0} \frac{E\{(\dot{x}_1 \Delta t)^2\}}{\Delta t} = \lim_{\Delta t \to 0} E\{\dot{x}_1^2 \Delta t\} = 0$$

$$b_{12} = \lim_{\Delta t \to 0} \frac{1}{\Delta t} E\{\Delta x_1 \Delta x_2\} = \lim_{\Delta t \to 0} \frac{E\{\dot{x}_1 \dot{x}_2 (\Delta t)^2\}}{\Delta t}$$
$$= \lim_{\Delta t \to 0} E\{\dot{x}_1 \dot{x}_2 \Delta t\} = 0.$$

The final term to be evaluated requires a bit more computation,

$$b_{22} = \lim_{\Delta t \to 0} \frac{1}{\Delta t} E\{(\Delta x_2)^2\} = \lim_{\Delta t \to 0} \frac{1}{\Delta t} E\{(\dot{x}_2 \Delta t)^2\}$$
$$= \lim_{\Delta t \to 0} \frac{1}{\Delta t} E\{(-cx_2 - h(x_1) + F(t))^2 (\Delta t)^2\}.$$

The squared expression needs to be expanded, resulting in the following terms:

$$c^2 x_2^2 + h^2 - 2cx_2 h$$
$$F^2(t)$$
$$-2cx_2 F(t) - 2hF(t).$$

Each term is multiplied by $(\Delta t)^2$. We need to rewrite $F(t)$ in the following way,

$$F(t) = \frac{d}{dt} \int F(u) du$$
$$= \lim_{\Delta t \to 0} \frac{1}{\Delta t} \int_t^{t+\Delta t} F(u) du;$$

and similarly,

$$F^2(t) = \lim_{\Delta t \to 0} \frac{1}{(\Delta t)^2} \int_t^{t+\Delta t} F(u)\,du \cdot \int_t^{t+\Delta t} F(v)\,dv.$$

By doing this we obtain the appropriate Δt factor in the expected value. Then,

$$\lim_{\Delta t \to 0} \frac{1}{\Delta t} E\left\{(-cx_2 - h(x_1))^2 (\Delta t)^2\right\} = 0$$

$$\lim_{\Delta t \to 0} \frac{1}{\Delta t} E\left\{(-2cx_2 - 2h)(\Delta t)^2 \cdot \lim_{\Delta t \to 0} \frac{1}{\Delta t} \int_t^{t+\Delta t} F(u)\,du\right\}$$

$$= (-2cx_2 - 2h) \lim_{\Delta t \to 0} \int_t^{t+\Delta t} E\{F(u)\}\,du = 0$$

$$\lim_{\Delta t \to 0} \frac{1}{\Delta t} E\left\{F^2(t)(\Delta t)^2\right\} = \lim_{\Delta t \to 0} \frac{1}{\Delta t} \int_t^{t+\Delta t}\int_t^{t+\Delta t} E\{F(u)F(w)\}\,du\,dw$$

$$= \lim_{\Delta t \to 0} \frac{1}{\Delta t} \int_t^{t+\Delta t}\int_t^{t+\Delta t} R_{FF}(u-w)\,du\,dw$$

$$= \lim_{\Delta t \to 0} \frac{1}{\Delta t} \int_t^{t+\Delta t}\int_t^{t+\Delta t} 2\alpha\delta(u-w)\,du\,dw$$

$$= \lim_{\Delta t \to 0} \frac{1}{\Delta t} \int_t^{t+\Delta t}\int_t^{t+\Delta t} 2\alpha\delta(\Delta t)\,du\,d(u-\Delta t)$$

$$= \lim_{\Delta t \to 0} \frac{1}{\Delta t} \int_t^{t+\Delta t} 2\alpha d(u-\Delta t)$$

$$= \lim_{\Delta t \to 0} \frac{1}{\Delta t} 2\alpha \cdot \Delta t$$

$$= 2\alpha.$$

Therefore, $b_{22} = 2\alpha$. Equation 10.71 can now be written as

$$\begin{aligned}\frac{\partial f}{\partial t} &= -\frac{\partial}{\partial x_1}(a_1 f) - \frac{\partial}{\partial x_2}(a_2 f) \\ &\quad + \frac{1}{2}\frac{\partial^2}{\partial x_1^2}(b_{11} f) + \frac{\partial^2}{\partial x_1 \partial x_2}(b_{12} f) + \frac{1}{2}\frac{\partial^2}{\partial x_2^2}(b_{22} f) \\ &= -\frac{\partial}{\partial x_1}(x_2 f) - \frac{\partial}{\partial x_2}\left([-cx_2 - h(x_1)]f\right) + \frac{1}{2}\frac{\partial^2}{\partial x_2^2}(2\alpha f),\end{aligned}$$

(10.72)

10.6. MARKOV PROCESS-BASED MODELS

where $f \equiv f(x_1, x_2)$, and substituting the appropriate coefficients with $h(x) = x_1 + x_1^3$ results in the equation,

$$\begin{aligned}\frac{\partial f}{\partial t} &= -\frac{\partial}{\partial x_1}(x_2 f) - \frac{\partial}{\partial x_2}\left(\left[-cx_2 - x_1 - x_1^3\right]f\right) + \frac{1}{2}\frac{\partial^2}{\partial x_2^2}(2\alpha f) \\ &= -\frac{\partial}{\partial x}(vf) - \frac{\partial}{\partial v}\left(\left[-cv - x - x^3\right]f\right) + \frac{1}{2}\frac{\partial^2}{\partial v^2}(2\alpha f).\end{aligned}$$

Expand the derivatives to find

$$\frac{\partial f}{\partial t} = -v\frac{\partial f}{\partial x} + c\frac{\partial(vf)}{\partial v} + (x + x^3)\frac{\partial f}{\partial v} + \alpha\frac{\partial^2 f}{\partial v^2}.$$

This equation has the initial density condition,

$$f(x, v; 0|x(0),)v(0)) = \delta(x - x(0))\delta(v - v(0))$$

for all (x, v) in the domain.

Due to the complexity of this differential equation, the "steady-state" solution for $f(x, v, t)$ is found by setting $\partial f/\partial t = 0$ and then solving for $f_s(x, v)$.

Consider Equation 10.72 in steady state,

$$-x_2\frac{\partial f_s}{\partial x_1} + \frac{\partial}{\partial x_2}\left(\left[cx_2 + h(x_1)\right]f_s\right) + \alpha\frac{\partial^2 f_s}{\partial x_2^2} = 0, \quad (10.73)$$

where $(x, v) \equiv (x_1, x_2)$ and we simplified the signs of the second expression and pulled x_2 outside the derivative with respect to x_1. In general, the solution of this partial differential equation would require some approximate analytical tools or numerical analysis. However, it is possible to rewrite Equation 10.73 in the following form,

$$\left[h(x_1)\frac{\partial f_s}{\partial x_2} - x_2\frac{\partial f_s}{\partial x_1}\right] + \frac{\partial}{\partial x_2}\left[cx_2 f_s + \alpha\frac{\partial f_s}{\partial x_2}\right] = 0. \quad (10.74)$$

Following Caughey,[24] Equation 10.74 can be solved for arbitrary $h(x_1)$ if the following separation is used,

$$\begin{aligned}\left[h(x_1)\frac{\partial f_s}{\partial x_2} - x_2\frac{\partial f_s}{\partial x_1}\right] &= 0 \\ \frac{\partial}{\partial x_2}\left[cx_2 f_s + \alpha\frac{\partial f_s}{\partial x_2}\right] &= 0.\end{aligned}$$

---

[24] For additional details, see "Nonlinear Theory of Random Vibrations," T.K. Caughey, *Advances in Applied Mechanics*, Vol. 11, 1971.

Another way is shown by Soong[25] assuming $f_s(x_1, x_2) = f_s(x_1) f_s(x_2)$. Rearrange Equation 10.73 as follows,

$$\frac{\partial}{\partial x_2}\left[h(x_1) f_s + \frac{\alpha}{c}\frac{\partial f_s}{\partial x_1}\right] + \left(c\frac{\partial}{\partial x_2} - \frac{\partial}{\partial x_1}\right)\left[x_2 f_s + \frac{\alpha}{c}\frac{\partial f_s}{\partial x_2}\right] = 0,$$

which can have a solution for arbitrary $h(x_1)$ only if each term in square brackets is identically zero. Therefore,

$$\begin{aligned}f_s(x_1,x_2) &= \text{const} \cdot \exp\left(-\frac{c}{\alpha}\left[\int_0^{x_1} h(x)\,dx + \frac{x_2^2}{2}\right]\right) \\ &= \text{const} \cdot \exp\left(-\frac{c}{\alpha}\int_0^{x_1} h(x)\,dx\right)\exp\left(-\frac{c}{\alpha}\frac{x_2^2}{2}\right).\end{aligned}$$

By the uniqueness of the solution, this is the only solution.

⊛

Example 10.8 The Random van der Pol Oscillator

As another example, consider the following van der Pol equation,

$$\ddot{X} + c\left(X^2 - a^2\right)\dot{X} + X = F(t).$$

The input forcing $F(t)$ has the same properties as in the previous example. Define $x_1 = X$ and $x_2 = \dot{x}_1$; then

$$\left\{\begin{array}{c}\dot{x}_1 \\ \dot{x}_2\end{array}\right\} = \left\{\begin{array}{c}x_2 \\ c\left(x_1^2 - a^2\right)x_2 - x_1 + F(t)\end{array}\right\}.$$

Following the same procedure as before, we have

$$\begin{aligned}a_1 &= \lim_{\Delta t \to 0}\frac{1}{\Delta t}E\{\Delta x_1\} = \lim_{\Delta t \to 0}\frac{1}{\Delta t}E\{\dot{x}_1 \Delta t\} = \dot{x}_1 = x_2 = v \\ a_2 &= \lim_{\Delta t \to 0}\frac{1}{\Delta t}E\{\Delta x_2\} = \lim_{\Delta t \to 0}\frac{1}{\Delta t}E\{\dot{x}_2 \Delta t\} = \dot{x}_2 = c\left(x_1^2 - a^2\right)x_2 - x_1 \\ b_{11} &= \lim_{\Delta t \to 0}\frac{1}{\Delta t}E\left\{(\Delta x_1)^2\right\} = \lim_{\Delta t \to 0}\frac{1}{\Delta t}E\left\{(\dot{x}_1 \Delta t)^2\right\} \\ &= \lim_{\Delta t \to 0} E\{\dot{x}_1^2 \Delta t\} = 0 \\ b_{12} &= \lim_{\Delta t \to 0}\frac{1}{\Delta t}E\{\Delta x_1 \Delta x_2\} = \lim_{\Delta t \to 0}\frac{1}{\Delta t}E\left\{\dot{x}_1 \dot{x}_2 (\Delta t)^2\right\} \\ &= \lim_{\Delta t \to 0} E\{\dot{x}_1 \dot{x}_2 \Delta t\} = 0 \\ b_{22} &= \lim_{\Delta t \to 0}\frac{1}{\Delta t}E\left\{\left(c\left(x_1^2 - a^2\right)x_2 - x_1 + F(t)\right)^2 (\Delta t)^2\right\} = 2\alpha,\end{aligned}$$

---

[25] See p. 197-200 in *Random Differential Equations in Science and Engineering*, T.T. Soong, Academic Press, New York, 1973.

10.6. MARKOV PROCESS-BASED MODELS

with the resulting Fokker-Planck equation,

$$\frac{\partial f}{\partial t} = -v\frac{\partial f}{\partial x} + c\left(X^2 - a^2\right)\frac{\partial(vf)}{\partial v} + X\frac{\partial f}{\partial v} + \alpha\frac{\partial^2 f}{\partial v^2}.$$

⊛

Example 10.9 Fokker-Planck Equation for a Duffing Oscillator Excited by Colored Noise

Suppose the system is driven by the colored noise process $y(t)$. Generate $y(t)$ using the following first-order linear differential filter of the Gaussian white noise (zero mean) process $w(t)$, with $E\left\{w(t)w(t+\tau)\right\} = 2\alpha\delta(\tau)$,

$$c_1\dot{y} = c_2 y + c_3 w(t),$$

where the c_i are constants. One form of this equation incorporates the correlation time τ_c,

$$\tau_c\dot{y} = -y + w(t).$$

Note that as the correlation time approaches zero, $y(t)$ approaches $w(t)$.

The second-order Duffing oscillator

$$\ddot{x} + \gamma\dot{x} + x + x^3 = y(t)$$

can now be written as a system of three first-order differential equations, with $\mu = 1/\tau_c$,

$$\begin{aligned} \dot{x} &= v \\ \dot{v} &= y - \gamma v - x - x^3 \\ \dot{y} &= -\mu y + \mu w(t). \end{aligned}$$

Define $f(x, v, y; t)$ to be the joint probability density function of x, v, and y at time t. To derive the Fokker-Planck equation we need to derive the coefficients as we have done in the previous two examples,

$$\begin{aligned} a_1 &= \lim_{\Delta t \to 0} \frac{1}{\Delta t} E\left\{\dot{x}\Delta t\right\} = \lim_{\Delta t \to 0} E\left\{v\right\} = v \\ a_2 &= \lim_{\Delta t \to 0} \frac{1}{\Delta t} E\left\{\dot{v}\Delta t\right\} \\ &= \lim_{\Delta t \to 0} E\{y - \gamma v - x - x^3\} = y - \gamma v - x - x^3 \\ a_3 &= \lim_{\Delta t \to 0} \frac{1}{\Delta t} E\left\{\dot{y}\Delta t\right\} \\ &= \lim_{\Delta t \to 0} E\{-\mu y + \mu w(t)\} = -\mu y. \end{aligned}$$

Continuing,

$$b_{11} = \lim_{\Delta t \to 0} \frac{1}{\Delta t} E\left\{(\dot{x}\Delta t)^2\right\} = \lim_{\Delta t \to 0} \Delta t E\left\{\dot{x}^2\right\} = 0$$

$$b_{12} = \lim_{\Delta t \to 0} \frac{1}{\Delta t} E\left\{\dot{x}\dot{v}(\Delta t)^2\right\} = \lim_{\Delta t \to 0} \Delta t E\left\{\dot{x}\dot{v}\right\} = 0$$

$$b_{22} = \lim_{\Delta t \to 0} \frac{1}{\Delta t} E\left\{(\dot{v}\Delta t)^2\right\} = \lim_{\Delta t \to 0} \Delta t E\left\{\dot{v}^2\right\} = 0.$$

Finally,

$$\begin{aligned}
b_{13} &= \lim_{\Delta t \to 0} \frac{1}{\Delta t} E\left\{\dot{x}\dot{y}(\Delta t)^2\right\} \\
&= \lim_{\Delta t \to 0} E\left\{v\left(-\mu y \Delta t + \mu \int_t^{\Delta t} w(u)du\right)\right\} = 0 \\
b_{23} &= \lim_{\Delta t \to 0} \frac{1}{\Delta t} E\left\{\dot{v}\dot{y}(\Delta t)^2\right\} \\
&= \lim_{\Delta t \to 0} E\left\{(y - \gamma v - x - x^3)\left(-\mu y \Delta t + \mu \int_t^{\Delta t} w(u)du\right)\right\} = 0 \\
b_{33} &= \lim_{\Delta t \to 0} \frac{1}{\Delta t} E\left\{(\dot{y}\Delta t)^2\right\} \\
&= \lim_{\Delta t \to 0} \frac{1}{\Delta t} E\left\{\left(-\mu y \Delta t + \mu \int_t^{\Delta t} w(u)du\right)^2\right\} \\
&= \lim_{\Delta t \to 0} \frac{1}{\Delta t} E\left\{(\mu y \Delta t)^2 - 2\mu^2 y \Delta t \int_t^{\Delta t} w(u)du \right. \\
&\qquad \left. + \mu^2 \int_t^{\Delta t}\int_t^{\Delta t} w(u_1)w(u_2)du_1 du_2\right\} \\
&= \lim_{\Delta t \to 0} \frac{1}{\Delta t} E\left\{\mu^2 \int_t^{\Delta t}\int_t^{\Delta t} w(u_1)w(u_2)du_1 du_2\right\} \\
&= 2\mu^2 \alpha.
\end{aligned}$$

The Fokker-Planck equation becomes

$$\begin{aligned}
\frac{\partial f}{\partial t} &= -\frac{\partial}{\partial x}(vf) - \frac{\partial}{\partial v}[(y - \gamma v - x - x^3)f] - \frac{\partial}{\partial y}(-\mu y f) + \frac{1}{2}\frac{\partial^2}{\partial y^2}(2\mu^2 \alpha f) \\
&= -v\frac{\partial f}{\partial x} - \frac{\partial}{\partial v}[(y - \gamma v - x - x^3)f] + \mu\frac{\partial}{\partial y}(yf) + \mu^2 \alpha \frac{\partial^2 f}{\partial y^2}.
\end{aligned}$$

⊛

10.6. MARKOV PROCESS-BASED MODELS

Example 10.10 Fokker-Planck Equation for Coupled Linear Duffing Oscillator

Sometimes more complicated dynamical systems require coupled oscillator models, where each oscillator is representative of a different subsystem. We may see this in multicomponent structures and machines, and we see this in fluid-structure interaction problems, for example.

We will again force the system with colored noise. In our assumed model, the colored noise drives the linear oscillator, with the coupled Duffing oscillator driven by the relative displacements between it and the linear oscillator. The assumed governing equations are as follows,

$$\dot{y} + \mu y = \mu w(t)$$
$$\ddot{x}_1 + \gamma_1 \dot{x}_1 + x_1 = y$$
$$\ddot{x}_2 + \gamma_2 \dot{x}_2 + x_2 + \varepsilon x_2^3 = k[x_1 - x_2],$$

where the parameters have the same properties and meanings as in the last example. Define $v_1 = \dot{x}_1$ and $v_2 = \dot{x}_2$, and rewrite the equations of motion in state space format, that is, as first-order differential equations,

$$\dot{y} = -\mu y + \mu w(t)$$
$$\dot{x}_1 = v_1$$
$$\dot{v}_1 = y - \gamma_1 v_1 - x_1$$
$$\dot{x}_2 = v_2$$
$$\dot{v}_2 = kx_1 - \gamma_2 v_2 - (1+k)x_2 - \varepsilon x_2^3.$$

We next proceed to derive the coefficients required for the Fokker-Planck equation,

$$a_1 = \lim_{\Delta t \to 0} \frac{1}{\Delta t} E\{\dot{y} \Delta t\} = \lim_{\Delta t \to 0} \mu E\{-y + w(t)\} = -\mu y$$

$$a_2 = \lim_{\Delta t \to 0} \frac{1}{\Delta t} E\{\dot{x}_1 \Delta t\} = \lim_{\Delta t \to 0} E\{v_1\} = v_1$$

$$a_3 = \lim_{\Delta t \to 0} \frac{1}{\Delta t} E\{\dot{v}_1 \Delta t\}$$
$$= \lim_{\Delta t \to 0} E\{y - \gamma_1 v_1 - x_1\} = y - \gamma_1 v_1 - x_1$$

$$a_4 = \lim_{\Delta t \to 0} \frac{1}{\Delta t} E\{\dot{x}_2 \Delta t\} = \lim_{\Delta t \to 0} E\{v_2\} = v_2$$

$$a_5 = \lim_{\Delta t \to 0} \frac{1}{\Delta t} E\{\dot{v}_2 \Delta t\}$$
$$= \lim_{\Delta t \to 0} E\{kx_1 - \gamma_2 v_2 - (1+k)x_2 - \varepsilon x_2^3\}$$
$$= kx_1 - \gamma_2 v_2 - (1+k)x_2 - \varepsilon x_2^3$$

$$b_{11} = \lim_{\Delta t \to 0} \frac{1}{\Delta t} E\left\{(\dot{y}\Delta t)^2\right\}$$

$$= \lim_{\Delta t \to 0} \frac{1}{\Delta t} E\left\{\left(-\mu y \Delta t + \mu \int_t^{\Delta t} w(u) du\right)^2\right\}$$

$$= \lim_{\Delta t \to 0} \frac{1}{\Delta t} E\left\{(\mu y \Delta t)^2 - 2\mu^2 y \Delta t \int_t^{\Delta t} w(u) du\right.$$

$$\left. + \mu^2 \int_t^{\Delta t} \int_t^{\Delta t} w(u_1) w(u_2) du_1 du_2\right\}$$

$$= \lim_{\Delta t \to 0} \frac{1}{\Delta t} E\left\{\mu^2 \int_t^{\Delta t} \int_t^{\Delta t} w(u_1) w(u_2) du_1 du_2\right\}$$

$$= 2\mu^2 \alpha.$$

Continuing,

$$b_{12} = \lim_{\Delta t \to 0} \frac{1}{\Delta t} E\left\{\dot{y}\dot{x}_1(\Delta t)^2\right\} = \lim_{\Delta t \to 0} \Delta t E\left\{\mu[-y + w(t)]v_1\right\} = 0$$

$$b_{13} = \lim_{\Delta t \to 0} \frac{1}{\Delta t} E\{\Delta y \Delta v_1\}$$

$$= \lim_{\Delta t \to 0} \frac{1}{\Delta t} E\left\{\dot{y}\dot{v}_1 (\Delta t)^2\right\}$$

$$= \lim_{\Delta t \to 0} \Delta t E\left\{\mu[-y + w(t)][y - \gamma_1 v_1 - x_1]\right\} = 0$$

$$b_{14} = \lim_{\Delta t \to 0} \frac{1}{\Delta t} E\left\{\dot{y}\dot{x}_2 (\Delta t)^2\right\} = \lim_{\Delta t \to 0} \Delta t E\left\{\mu[-y + w(t)]v_2\right\} = 0$$

$$b_{15} = \lim_{\Delta t \to 0} \frac{1}{\Delta t} E\left\{\dot{y}\dot{v}_2 (\Delta t)^2\right\}$$

$$= \lim_{\Delta t \to 0} E\left\{\mu[-y + w(t)][kx_1 - \gamma_2 v_2 - (1+k)x_2 - \varepsilon x_2^3]\right\} = 0$$

$$b_{22} = \lim_{\Delta t \to 0} \frac{1}{\Delta t} E\left\{(\dot{x}_1 \Delta t)^2\right\} = \lim_{\Delta t \to 0} \Delta t E\left\{v_1^2\right\} = 0$$

$$b_{23} = \lim_{\Delta t \to 0} \frac{1}{\Delta t} E\left\{\dot{x}_1 \dot{v}_1 (\Delta t)^2\right\} = \lim_{\Delta t \to 0} \Delta t E\{v_1[y - \gamma_1 v_1 - x_1]\} = 0$$

$$b_{24} = \lim_{\Delta t \to 0} \frac{1}{\Delta t} E\left\{\dot{x}_1 \dot{x}_2 (\Delta t)^2\right\} = \lim_{\Delta t \to 0} \Delta t E\{v_1 v_2\} = 0$$

10.6. MARKOV PROCESS-BASED MODELS

$$\begin{aligned}
b_{25} &= \lim_{\Delta t \to 0} \frac{1}{\Delta t} E\left\{\dot{x}_1 \dot{v}_2 \left(\Delta t\right)^2\right\} \\
&= \lim_{\Delta t \to 0} \Delta t E\{v_1[kx_1 - \gamma_2 v_2 - (1+k)x_2 - \varepsilon x_2^3]\} = 0 \\
b_{33} &= \lim_{\Delta t \to 0} \frac{1}{\Delta t} E\left\{(\dot{v}_1 \Delta t)^2\right\} \\
&= \lim_{\Delta t \to 0} \Delta t E\{[y - \gamma_1 v_1 - x_1]^2\} = 0 \\
b_{34} &= \lim_{\Delta t \to 0} \frac{1}{\Delta t} E\left\{\dot{v}_1 \dot{x}_2 \left(\Delta t\right)^2\right\} \\
&= \lim_{\Delta t \to 0} \Delta t E\{[y - \gamma_1 v_1 - x_1]v_2\} = 0
\end{aligned}$$

$$\begin{aligned}
b_{35} &= \lim_{\Delta t \to 0} \frac{1}{\Delta t} E\left\{\dot{v}_1 \dot{v}_2 \left(\Delta t\right)^2\right\} \\
&= \lim_{\Delta t \to 0} \Delta t E\{[y - \gamma_1 v_1 - x_1][kx_1 - \gamma_2 v_2 - (1+k)x_2 - \varepsilon x_2^3]\} = 0 \\
b_{44} &= \lim_{\Delta t \to 0} \frac{1}{\Delta t} E\left\{(\dot{x}_2 \Delta t)^2\right\} = \lim_{\Delta t \to 0} \Delta t E\left\{v_2^2\right\} = 0
\end{aligned}$$

$$\begin{aligned}
b_{45} &= \lim_{\Delta t \to 0} \frac{1}{\Delta t} E\left\{\dot{x}_2 \dot{v}_2 \left(\Delta t\right)^2\right\} \\
&= \lim_{\Delta t \to 0} \Delta t E\{v_2[kx_1 - \gamma_2 v_2 - (1+k)x_2 - \varepsilon x_2^3]\} = 0 \\
b_{55} &= \lim_{\Delta t \to 0} \frac{1}{\Delta t} E\left\{(\dot{v}_2 \Delta t)^2\right\} \\
&= \lim_{\Delta t \to 0} \Delta t E\{[kx_1 - \gamma_2 v_2 - (1+k)x_2 - \varepsilon x_2^3]^2\} = 0.
\end{aligned}$$

The resulting Fokker-Planck equation for the evolutionary density function $f \equiv f(y, x_1, v_1, x_2, v_2; t)$ is

$$\begin{aligned}
\frac{\partial f}{\partial t} &= \frac{\partial}{\partial y}(\mu y f) - \frac{\partial}{\partial x_1}(v_1 f) - \frac{\partial}{\partial v_1}\left[(y - \gamma_1 v_1 - x_1)f\right] - \frac{\partial}{\partial x_2}(v_2 f) \\
&\quad - \frac{\partial}{\partial v_2}\left[(kx_1 - \gamma_2 v_2 - (1+k)x_2 - \varepsilon x_2^3)f\right] + \mu^2 \alpha \frac{\partial^2 f}{\partial y^2}. \quad (10.75)
\end{aligned}$$

This equation can be simplified somewhat by expanding the derivatives,

resulting in

$$\frac{\partial f}{\partial t} = \mu \frac{\partial}{\partial y}(yf) - v_1 \frac{\partial f}{\partial x_1} - \frac{\partial}{\partial v_1}\left[(y - \gamma_1 v_1 - x_1)f\right] - v_2 \frac{\partial f}{\partial x_2}$$

$$- \frac{\partial}{\partial v_2}\left[(kx_1 - \gamma_2 v_2 - (1+k)x_2 - \varepsilon x_2^3)f\right]$$

$$+ \mu^2 \alpha \frac{\partial^2 f}{\partial y^2}.$$

This equation is generally solved numerically. If $\partial f/\partial t = 0$, the solution is a steady-state solution.

Figure 10.16: Max Karl Ernst Ludwig Planck (1858-1947).

Planck Max Planck came from an academic family, his father being professor of law at Kiel and both his grandfather and great-grandfather had been professors of theology at Göttingen. In 1867 Planck's family moved to Munich and he attended school there. He did well at school, but not brilliantly, usually coming somewhere between third and eighth in his class.

In 1874, at the age of 16, he entered the University of Munich. Before he began his studies he discussed the prospects of research in physics with Philipp von Jolly, the professor of physics there, and was told that physics was essentially a complete science with little prospect of further developments. Fortunately Planck decided to study physics despite the bleak future for research that was presented to him.

10.6. MARKOV PROCESS-BASED MODELS 581

Planck describes why he chose physics: *The outside world is something independent from man, something absolute, and the quest for the laws which apply to this absolute appeared to me as the most sublime scientific pursuit in life.*

Planck then studied at Berlin where his teachers included Helmholtz and Kirchhoff. He later wrote that he admired Kirchhoff greatly but found him dry and monotonous as a teacher. Planck returned to Munich and received his doctorate at the age of 21 with a thesis on the second law of thermodynamics. He was then appointed to a teaching post at the University of Munich in 1880 and he taught there until 1885.

In 1885 Planck was appointed to a chair in Kiel and held this chair for four years. After the death of Kirchhoff in 1887, Planck succeeded him to the chair of theoretical physics at the University of Berlin in 1889. He was to hold the Berlin chair for 38 years until he retired in 1927.

While in Berlin Planck did his most brilliant work and delivered outstanding lectures. He studied thermodynamics in particular examining the distribution of energy according to wavelength. By combining the formulas of Wien and Rayleigh, Planck announced in 1900 a formula now known as Planck's radiation formula. In a letter written a year later Planck described proposing the formula saying ... *the whole procedure was an act of despair because a theoretical interpretation had to be found at any price, no matter how high that might be.*

Within two months Planck made a complete theoretical deduction of his formula renouncing classical physics and introducing the quanta of energy. At first the theory met resistance but due to the successful work of Niels Bohr in 1913, calculating positions of spectral lines using the theory, it became generally accepted. Planck himself[26] explains how despite having invented quantum theory he did not understand it himself at first: *I tried immediately to weld the elementary quantum of action somehow in the framework of classical theory. But in the face of all such attempts this constant showed itself to be obdurate ... My futile attempts to put the elementary quantum of action into the classical theory continued for a number of years and they cost me a great deal of effort.*

Planck received the Nobel Prize for Physics in 1918. He took little part in the further development of quantum theory, this being left to Paul Dirac and others. Planck took on administrative duties such as Secretary of the Mathematics and Natural Science Section of the Prussian Academy of Sciences, a post he held from 1912 until 1943. He had been elected to the Academy in 1894.

Planck was president of the Kaiser Wilhelm Gesellschaft, the main German research organization, from 1930 until 1937. He remained in Germany

---

[26] M. Planck, *Scientific Autobiography, and Other Papers*, (1949).

during World War II through what must have been times of the deepest difficulty since his son Erwin was executed for plotting to assassinate Hitler. Heilbron[27] describes the impact of wars on Planck and his family: *He would remember, even in his old age, the sight of Prussian and Austrian troops marching into his native town when he was six years old. Throughout his life, war would cause him deep personal sorrow. He lost his eldest son during World War I. In World War II, his house in Berlin was burned down in an air raid. In 1945 his other son was executed when declared guilty of complicity in a plot to kill Hitler.*

After World War II, he again became president of the Kaiser Wilhelm Gesellschaft in 1945-1946 for the second time defending German science through another period of exceptional difficulty.

Figure 10.17: Andrei Nikolaevich Kolmogorov (1903-1987).

Kolmogorov   Andrei Kolmogorov was born out of wedlock, and his father took no part in his upbringing. His father Nikolai Kataev, the son of a priest, was an agriculturist who was exiled. He returned after the Revolution to head a Department in the Agricultural Ministry but died in fighting in 1919. Kolmogorov's mother also, tragically, took no part in his upbringing since she died in childbirth at Kolmogorov's birth. His

---

[27] J.L. Heilbron, *The Dilemmas of an Upright Man: Max Planck as Spokesman for German Science (Berkeley, 1986).*

10.6. MARKOV PROCESS-BASED MODELS

mother's sister, Vera Yakovlena, brought Kolmogorov up and he always had the deepest affection for her.

In fact it was chance that had Kolmogorov born in Tambov since the family had no connections with that place. Kolmogorov's mother had been on a journey from the Crimea back to her home in Tunoshna near Yaroslavl and it was in the home of his maternal grandfather in Tunoshna that Kolmogorov spent his youth. Kolmogorov's name came from his grandfather, Yakov Stepanovich Kolmogorov, and not from his own father. Yakov Stepanovich was from the nobility, a difficult status to have in Russia at this time, and there are certainly stories told that an illegal printing press was operated from his house.

After Kolmogorov left school he worked for a while as a conductor on the railway. In his spare time he wrote a treatise on Newton's laws of mechanics. Then, in 1920, Kolmogorov entered Moscow State University but at this stage he was far from committed to mathematics. He studied a number of subjects, for example in addition to mathematics he studied metallurgy and Russian history. Nor should it be thought that Russian history was merely a topic to fill out his course, indeed he wrote a serious scientific thesis on the owning of property in Novgorod in the 15th and 16th centuries.

In mathematics Kolmogorov was influenced at an early stage by a number of outstanding mathematicians. P.S. Aleksandrov was beginning his research at Moscow around the time Kolmogorov began his undergraduate career. Luzin and Egorov were running their impressive research group at this time which the students called "Luzitania." It included M. Ya Suslin and P.S. Urysohn, in addition to Aleksandrov. However, the person who made the deepest impression on Kolmogorov at this time was Stepanov, who lectured to him on trigonometric series.

It is remarkable that Kolmogorov, although only an undergraduate, began research and produced results of international importance at this stage. He had finished writing a paper on operations on sets by the spring of 1922 which was a major generalization of results obtained by Suslin. By June of 1922, he had constructed a summable function which diverged almost everywhere. This was wholly unexpected by the experts and Kolmogorov's name began to be known around the world. Gnedenko[28] notes that: *Almost simultaneously [Kolmogorov] exhibited his interest in a number of other areas of classical analysis: in problems of differentiation and integration, in measures of sets, for example. In every one of his papers, dealing with*

---

[28] B.V. Gnedenko, "Andreii Nikolaevic Kolmogorov" (on the occation of his seventieth birthday) (Russian), *Uspehi Mat. Nauk* 28 5(173) (1973), 5-15.

B.V. Gnedenko, "Andreii Nikolaevic Kolmogorov" (on the occation of his seventieth birthday), *Russian Math. Surveys* 28 (5) (1973), 5-17.

such a variety of topics, he introduced an element of originality, a breadth of approach, and a depth of thought.

Kolmogorov graduated from Moscow State University in 1925 and began research under Luzin's supervision in that year. He published eight papers in 1925, all written while he was still an undergraduate. Another milestone occurred in 1925, namely Kolmogorov's first paper on probability appeared. This was published jointly with Khinchine and contains the "three series" theorem as well as results on inequalities of partial sums of random variables, which would become the basis for martingale inequalities and the stochastic calculus.

In 1929 Kolmogorov completed his doctorate. By this time he had 18 publications and Kendall writes: *These included his versions of the strong law of large numbers and the law of the iterated logarithm, some generalizations of the operations of differentiation and integration, and a contribution to intuitional logic. His papers ... on this last topic are regarded with awe by specialists in the field.* The Russian language edition of Kolmogorov's collected works contains a retrospective commentary on these papers which [Kolmogorov] evidently regarded as marking an important development in his philosophical outlook.

An important event for Kolmogorov was his friendship with Aleksandrov which began in the summer of 1929 when they spent three weeks together. On a trip starting from Yaroslavl, they went by boat down the Volga then across the Caucasus mountains to Lake Sevan in Armenia. There Aleksandrov worked on the topology book which he coauthored with Hopf, while Kolmogorov worked on Markov processes with continuous states and continuous time. Kolmogorov's results from his work by the lake were published in 1931 and mark the beginning of diffusion theory. In the summer of 1931 Kolmogorov and Aleksandrov made another long trip. They visited Berlin, Göttingen, Munich, and Paris where Kolmogorov spent many hours in deep discussions with Paul Lévy. After this they spent a month at the seaside with Fréchet.

Kolmogorov was appointed a professor at Moscow University in 1931. His monograph on probability theory *Grundbegriffe der Wahrscheinlichkeitsrechnung* published in 1933 built up probability theory in a rigorous way from fundamental axioms in a way comparable with Euclid's treatment of geometry. One success of this approach is that it provides a rigorous definition of conditional expectation. As noted by Kendall: *The year 1931 can be regarded as the beginning of the second creative stage in Kolmogorov's life. Broad general concepts advanced by him in various branches of mathematics are characteristic of this stage.*

After mentioning the highly significant paper, *Analytic Methods in Probability Theory*, which Kolmogorov published in 1938 laying the founda-

10.6. MARKOV PROCESS-BASED MODELS 585

tions of the theory of Markov random processes, they continue to describe: *... his ideas in set-theoretic topology, approximation theory, the theory of turbulent flow, functional analysis, the foundations of geometry, and the history and methodology of mathematics. [His contributions to] each of these branches ... [is] a single whole, where a serious advance in one field leads to a substantial enrichment of the others.*

Aleksandrov and Kolmogorov bought a house in Komarovka, a small village outside Moscow, in 1935. Many famous mathematicians visited Komarovka: Hadamard, Fréchet, Banach, Hopf, Kuratowski, and others. Gnedenko and other graduate students went on: *... mathematical outings [which] ended in Komarovka, where Kolmogorov and Aleksandrov treated the whole company to dinner. Tired and full of mathematical ideas, happy from the consciousness that we had found out something which one cannot find in books, we would return in the evening to Moscow.*

Around this time Malcev and Gelfand and others were graduate students of Kolmogorov along with Gnedenko who describes what it was like being supervised by Kolmogorov: *The time of their graduate studies remains for all of Kolmogorov's students an unforgettable period in their lives, full of high scientific and cultural striving, outbursts of scientific progress and a dedication of all one's powers to the solutions of the problems of science. It is impossible to forget the wonderful walks on Sundays to which [Kolmogorov] invited all his own students (graduates and undergraduates), as well as the students of other supervisors. These outings in the environs of Bolshevo, Klyazma, and other places about 30-35 kilometers away, were full of discussions about the current problems of mathematics (and its applications), as well as discussions about the questions of the progress of culture, especially painting, architecture and literature.*

In 1938-1939 a number of leading mathematicians from the Moscow University joined the Steklov Mathematical Institute of the Academy of Sciences while retaining their positions at the University. Among them were Aleksandrov, Gelfand, Kolmogorov, Petrovsky, and Khinchine. The Department of Probability and Statistics was set up at the Institute and Kolmogorov was appointed as Head of Department.

Kolmogorov later extended his work to study the motion of the planets and the turbulent flow of air from a jet engine. In 1941 he published two papers on turbulence that are of fundamental importance. In 1954 he developed his work on dynamical systems in relation to planetary motion. He thus demonstrated the vital role of probability theory in physics.

We must mention just a few of the numerous other major contributions which Kolmogorov made in a whole range of different areas of mathematics. In topology Kolmogorov introduced the notion of cohomology groups at much the same time, and independently of, Alexander. In 1934 Kolmogorov investigated chains, cochains, homology, and cohomology of a finite cell complex. In further papers, published in 1936, Kolmogorov defined cohomology groups for an arbitrary locally compact topological space. Another contribution of the highest significance in this area was his definition of the cohomology ring, which he announced at the International Topology Conference in Moscow in 1935. At this conference both Kolmogorov and Alexander lectured on their independent work on cohomology.

In 1953 and 1954 two papers by Kolmogorov, each of four pages in length, appeared. These are on the theory of dynamical systems with applications to Hamiltonian dynamics. These papers mark the beginning of KAM-theory, which is named after Kolmogorov, Arnold, and Moser. Kolmogorov addressed the International Congress of Mathematicians in Amsterdam in 1954 on this topic with his important talk, *General Theory of Dynamical Systems and Classical Mechanics*.

N.H. Bingham notes Kolmogorov's major part in setting up the theory to answer the probability part of Hilbert's Sixth Problem "to treat ... by means of axioms those physical sciences in which mathematics plays an important part; in the first rank are the theory of probability and mechanics" in his 1933 monograph *Grundbegriffe der Wahrscheinlichkeitsrechnung*. Bingham also notes: ... *Paul Lévy writes poignantly of his realization, immediately on seeing the Grundbegriffe, of the opportunity which he himself had neglected to take. A rather different perspective is supplied by the eloquent writings of Mark Kac on the struggles that Polish mathematicians of the calibre Steinhaus and himself had in the 1930s, even armed with the Grundbegriffe, to understand the (apparently perspicuous) notion of stochastic independence.*

If Kolmogorov made a major contribution to Hilbert's Sixth Problem, he completely solved Hilbert's Thirteenth Problem in 1957 when he showed that Hilbert was wrong in asking for a proof that there exist continuous functions of three variables that could not be represented by continuous functions of two variables.

Kolmogorov took a special interest in a project to provide special education for gifted children: *To this school he devoted a major proportion of his time over many years, planning syllabuses, writing textbooks, spending a large number of teaching hours with the children themselves, introducing them to literature and music, joining in their recreations and taking them on hikes, excursions, and expeditions. ... [Kolmogorov] sought to ensure for these children a broad and natural development of personality, and it did not worry him if the children in his school did not become mathematicians. Whatever profession they ultimately followed, he would be content if their outlook remained broad and their curiosity unstifled. Indeed it must have been wonderful to belong to this extended family of [Kolmogorov].*

Such an outstanding scientist as Kolmogorov naturally received a whole host of honors from many different countries. In 1939 he was elected to the Academy of Sciences of the USSR. He received one of the first State Prizes to be awarded in 1941, the Lenin Prize in 1965, the Order of Lenin on six separate occasions, and the Lobachevsky Prize in 1987. He was also elected to many other academies and societies including the Romanian Academy of Sciences (1956), the Royal Statistical Society of London (1956), the Loopoldina Academy of Germany (1959), the American Academy of Arts and Sciences (1959), the London Mathematical Society (1959), the American Philosophical Society (1961), the Indian Statistical Institute (1962), the Netherlands Academy of Sciences (1963), the Royal Society of London (1964), the National Academy of the United States (1967), the French Academy of Sciences (1968).

In addition to the prizes mentioned above, Kolmogorov was awarded the Bolzano International Prize in 1963. Many universities awarded him an honorary degree including Paris, Stockholm, and Warsaw.

Kolmogorov had many interests outside mathematics, in particular he was interested in the form and structure of the poetry of the Russian author Pushkin.

10.7 Concluding Summary

In this chapter we introduced deterministic and stochastic approaches to derive the approximate solutions to nonlinear oscillators. Linearization and perturbation methods are derived, and this class of methods convert nonlinear equations into a sequence of linear equations.

Additionally, the Markov process is introduced, and methods related to the stochastic analysis of nonlinear equations driven by Markov processes. In particular, the Chapman-Kolmogorov equation and the related Fokker-Planck equation are derived for special equations.

This chapter demonstrates how important it is to understand linear systems. A significant solution method of nonlinear analysis, the perturbation method, depends on the ability to solve linear equations. It will be observed in the next chapter that, in the same vein, nonstationary solutions are based in many instances on stationary solutions.

10.8 Problems

Section 10.1: Examples of Nonlinear Vibration

1. Figure 10.1 compares $\sin\theta$ versus θ, showing the effect of the nonlinearity as θ increases. At what value of θ is the error 5%, 10%, and

15%. The percent error can be defined as

$$\% \text{ error} = \frac{\sin\theta - \theta}{\theta} \times 100.$$

2. Figure 10.3 compares a hard and a soft spring. Which is likely to be a better choice for a design? Discuss with a specific design in mind.

Section 10.3: Statistical Equivalent Linearization

3. Derive Equations 10.7 and 10.8.

Section 10.4: Perturbation or Expansion Methods

4. Derive Equation 10.23.

5. Solve for Equations 10.27 and 10.28.

6. Suppose Γ is a random variable in Equation 10.41. How should we proceed to estimate the statistics of $\theta(\varepsilon, \tau)$?

7. Explain in physical terms what Equation 10.45 implies about the oscillator $\chi(T)$.

8. What kind of random process does Equation 10.48 represent? Discuss in physical terms.

9. Derive Equations 10.49-10.51.

Section 10.5: The van der Pol Equation

10. Consider van der Pol Equation 10.55. Suppose α is a random variable. How do the statistics of α affect the trajectories shown in Figure 10.11 and the time traces in Figure 10.12.

Section 10.6: Markov Process-Based Models

11. Give examples of physical processes or mechanical behavior that are suitably modeled as Markov processes. Explain the suitability.

12. Discuss what a transition probability density function is in physical terms.

13. *(i)* Can a Markov process be stationary?

    (ii) Is the steady-state assumption for the Fokker-Planck transition density the same as a stationarity assumption?

10.8. PROBLEMS

14. In Example 10.9, work from the three first-order equations to derive the dynamical equations.

15. In Example 10.10, derive the original dynamical equations beginning with the first-order equations.

Chapter 11

Nonstationary Models

Nonstationary random processes are very complex. The statistics of these processes vary with time and/or space. To be able to model such processes requires a tremendous amount of data, much of which is usually not easy to create.

However, many applications cannot be realistically modeled as stationary processes, and therefore efforts are required to capture some key aspects of the nonstationarity in the mathematical model. In dynamic systems, unless the behavior is steady state, the statistics vary with time as well. Examples include explosive loading, earthquake forces, and gust aerodynamic forces. Thus, though there may be scant information available, efforts to model nonstationary processes are important, even if such models are approximate.

Evolutionary spectral analysis involves the same physical ideas as those used in conventional, stationary spectral analysis, that is, the characterization of the energy distributions across frequencies for the random process. The difference is that these distributions can vary with time and/or space. Often the simplest approach to modeling nonstationary processes is to somehow base those models on extensions of stationary models, in the same way that nonlinear models were formulated in terms of linear models in the last chapter.

This chapter is an introduction to nonstationary modeling, with relatively simple applications provided as examples. Motivating examples are provided next in important and interesting disciplines: jet noise spectra, seismological data, and ground acceleration. With this introduction, the reader can proceed to the more specialized texts that focus on nonstationary models and data analysis.

11.1 Some Applications

Jet Noise Spectra

Priestly[1] cites an application discussed by Hammond where jet engine noise spectra are analyzed for single and multidegree-of-freedom systems driven by nonstationary random excitation. As the velocity $V(t)$ of the jet varies, so do the spectral patterns of the sound pressure field. These time varying spectral properties are nonstationary.

A reference jet velocity $V_0 = 2000$ ft/s is taken and the changing spectral patterns are analyzed as the velocity varies between 1700 ft/s and 2000 ft/s, with the purpose of estimating the evolutionary spectra over the frequency range 100-1000 Hz. The evolutionary spectra were then compared with the stationary spectra obtained when the jet velocity was held fixed, and this was done for fixed velocities in increments of 50 ft/s between 1700 ft/s and 2000 ft/s.

By matching the resulting seven stationary spectra with the corresponding evolutionary spectra, a relation was found relating the stationary spectrum $S_{V_0 V_0}(\omega)$ to the evolutionary spectrum $S_{VV}(t, \omega)$,

$$S_{VV}(t,\omega) = |A(t,\omega)|^2 S_{V_0 V_0}(\omega), \qquad (11.1)$$

where

$$A(t,\omega) = \left\{1 - \frac{v(t)}{V_0}\right\}^4 10^{x_1(t)} \left\{\frac{720}{900 - x_3(t)}\right\} 10^{x_2(t)}$$

$$x_1(t) = \frac{v(t)}{4000}$$

$$x_2(t) = -\frac{(f - f_c)v(t)}{5300 \times 10^3}$$

$$x_3(t) = \frac{(V_0 - v(t))^2}{22222.2},$$

where $v(t) = V_0 - V(t)$, $f = \omega/2\pi$ Hz, and $f_c = 365$ Hz. f_c depends on the position of the point at which sound pressure measurements are made, relative to the jet exhaust. Equation 11.1 is reminiscent of fundamental Equation 6.43: $S_{XX}(\omega) = |H(\omega)|^2 S_{FF}(\omega)$.

The idea presented in this example, that the evolutionary spectrum is related to a reference stationary spectrum, is widely used as a nonstationary modeling approach.

---

[1] *Nonlinear and Nonstationary Time Series Analysis*, M.B. Priestly, Academic Press, 1988.

11.1. SOME APPLICATIONS

Seismological Data

Priestly also cites the work of Dargahi-Noubary and Laycock on the problem of analyzing seismological data. They observe that many seismological series exhibit clearly nonstationary behavior, and suggest that such series may be more realistically modelled as modulated processes of the form,

$$X(t) = c(t)Y(t),$$

where $c(t)$ is a deterministic function and $Y(t)$ is a stationary process. Examples of $c(t)$ that have been used to fit earthquake and explosion data are

$$\begin{aligned} c(t) &= a_1 \exp(a_2 t) + a_3 \exp(a_4 t) \\ c(t) &= (a + bt) \exp(ct) \\ c(t) &= t^\alpha a^{-\beta t}. \end{aligned}$$

These functions are fit to specific application data using the constant parameters.

Ground Acceleration

Yang[2] discusses various models for the response of a dam-reservoir system to vertical and horizontal ground motion. This problem is formulated in different ways, as deterministic, stationary, and nonstationary models. The horizontal ground acceleration is modeled as a zero-mean, nonstationary process. The nonstationary spectral density of the response shear force is given by

$$S_{FF}(t,\omega) = \left| \int_0^t A(t-\theta) \exp(-i\omega\theta) g_F(\theta) d\theta \right|^2 S_{AA}(\omega),$$

where $S_{AA}(\omega)$ is a reference stationary spectrum, $A(t-\theta)$ is a deterministic modulating function, and $g_F(\theta)$ is the impulse response function for the shear force $F(t)$. It is assumed that the modulating function is a unit step function, and the stationary spectrum is white noise, $S_{AA}(\omega) = S_0$. Then

$$S_{FF}(t,\omega) = \left| \int_0^t \exp(-i\omega\theta) g_F(\theta) d\theta \right|^2 S_0,$$

where

$$g_F(\theta) = \frac{16c}{\pi^2 g H} F_0 \sum_{n=1}^\infty \frac{J_0(k_n ct)}{(2n-1)^2},$$

---

[2] *Random Vibration of Structures*, C.Y. Yang, Wiley-Interscience, 1986.

J_0 = Bessel function of the first kind of order zero, $F_0 = \omega H^2/2$ is the hydrostatic shear force, $k_n = \pi(2n-1)/2H$, n is the wave number of the vibrating water in the reservoir, c is the wave speed in water, and H is the water depth.

Figure 11.1: Friedrich Wilhelm Bessel (1784-1846).

Bessel   Friedrich Wilhelm Bessel, whose father was a civil servant in Minden, attended the Gymnasium there for four years, but he did not appear to be very talented, finding Latin difficult. The fact that he later became proficient in Latin, teaching himself the language, probably suggests that the Gymnasium failed to inspire Bessel. In January 1799, at the age of 14, he left school to become an apprentice to the commercial firm of Kulenkamp in Bremen. The firm was involved in the import-export business.

At first Bessel received no salary from the firm but, as his accounting skills became appreciated by the firm, he received a small salary. Interest in the countries his firm dealt with led Bessel to spend his evenings studying geography, Spanish and English. His interests turned towards navigation and he considered the problem of finding the position of a ship at sea. This in turn led him to study astronomy and mathematics and he began to make observations to determine longitude.

In 1804 Bessel wrote a paper on Halley's comet, calculating the orbit using data from observations made by Harriot in 1607. He sent his results to Heinrich Olbers, the leading comet expert of his time, who recognized at once the quality of Bessel's work, and Olbers gave Bessel the task of making further observations to carry his work further. The resulting paper, at the level required for a doctoral dissertation, was published on Olbers'

11.1. SOME APPLICATIONS

recommendation. From that time on Bessel concentrated on astronomy, celestial mechanics and mathematics.

Olbers suggested to Bessel, who was still an apprentice to the import-export firm, that he should become a professional astronomer. In 1806 he accepted the post of assistant at the Lilienthal Observatory, a private observatory near Bremen. It was only after some considerable thought that Bessel left the affluence that was guaranteed in his commercial job choosing instead the near poverty of the Observatory post. However the Lilienthal Observatory gave him valuable experience observing planets, in particular Saturn, its rings and satellites. He also observed comets and continued his study of celestial mechanics. In 1807 he began to work on reducing James Bradley's observations (Bradley was English Astronomer Royal from 1742 to 1762) of the positions of 3222 stars made around 1750 at Greenwich.

Bessel's brilliant work was quickly recognized and both Leipzig and Greifswald offered him posts. However he declined both. In 1809, at the age of 26, Bessel was appointed director of Frederick William III of Prussia's new Königsberg Observatory and professor of astronomy. It was not possible for Bessel to receive a professorship without first being granted the title of doctor. A doctorate was awarded by the University of Göttingen on the recommendation of Gauss, who had met Bessel in Bremen in 1807 and recognized his talents.

Although the Observatory at Königsberg was still under construction, Bessel took up his new post on May 10, 1810. He continued to work on Bradley's observations while work continued on the observatory from 1810 to 1813. Bessel's work had now become known internationally and he was honored with the award of the Lalande Prize from the Institut de France for his tables of refraction based on Bradley's observations. Also during this period, in 1812, he was elected to the Berlin Academy.

The Königsberg Observatory was completed in 1813 and Bessel began observing there. Fricke writes: *Bessel remained in Königsberg for the rest of his life, pursuing his research and teaching without interruption, although he often complained about the limited possibilities for observations because of the unfavorable climate. He declined the directorship of the Berlin Observatory, fearing greater administrative and social responsibilities ..... It was in Königsberg that Bessel undertook his monumental task of determining the positions and proper motions of over 50000 stars which led to the discovery in 1838 of the parallax of 61 Cygni. However his life did not run very smoothly although he made a happy marriage in 1812: ... they had two sons and three daughters. [The marriage] was clouded by sickness and by the early death of both sons.... From 1840 on, Bessel's health deteriorated. His last long trip, in 1842, was to England, where he participated in the Congress of the British Association in Manchester. His meeting with important English scientists, including Herschel, impressed him deeply and stimulated him to finish and publish, despite his weakened health, a series of works. After two years he died of cancer....*

Let us examine Bessel's work in more detail. He used Bradley's data to give a reference system for the positions of stars and planets and also to determine the positions of stars. He had to deduce errors in Bradley's instruments and errors caused by refraction. He had to reduce the positions to one fixed date and eliminate the effects of the Earth's motion, the precession of the Earth and other effects.

Bessel was one of the first astronomers to realize that, before a positional observation could be fully relied upon, one must have quantitative knowledge of every possible error that might enter into the finished result. He came to use Bradley's and Maskelyne's eighteenth-century Greenwich observations because these two astronomers were the first to provide exhaustive analyses of their own instrumental errors, along with temperature and pressure of the atmosphere through which the measurements had been made. By eliminating all sources of error – optical, mechanical and meteorological – Bessel was able to obtain astronomical results of astonishing delicacy from which a great deal of new data could be extracted.

Bessel's work in determining the constants of precession, nutation, and aberration won him further honors, such as a prize from the Berlin Academy in 1815. Bessel published Bradley's stellar positions in 1818 in a work which gives the proper motion of stars. In 1825 he was honored by election as a Fellow of the Royal Society.

In 1830 Bessel published the mean and apparent positions of 38 stars over the 100 year period 1750-1850. These 38 stars were the 36 "fundamental stars" of Maskelyne together with two further polar stars. From periodic variations in the proper motions of Sirius and Procyon, two of Maskelyne's 36 fundamental stars, Bessel deduced that they had companion stars in orbit which had not been observed. He announced that Sirius had a companion in 1841 thus being the first to predict the existence of "dark stars." Ten years later the orbit of the companion was computed and it was observed in 1862.

Bessel used parallax to determine the distance to 61 Cygni announcing his result in 1838. Clearly to succeed it was important to choose a star that was close to the Sun. His method of selecting a star was based on his own data for he chose the star which had the greatest proper motion of all the stars he had studied, correctly deducing that this would mean that the star was nearby. Since 61 Cygni is a relatively dim star it was a bold choice based on his correct understanding of the cause of the proper motions. Bessel, using a Fraunhofer heliometer to make the measurements, announced his value of 0.314" which, given the diameter of the Earth's

11.1. SOME APPLICATIONS

orbit, gave a distance of about 10 light years. The correct value of the parallax of 61 Cygni is 0.292".

John Herschel, when he learned of Bessel's achievement, wrote to him describing it as: ... *the greatest and most glorious triumph which practical astronomy has ever witnessed.* Olbers, told of Bessel's achievement on his 80th birthday, said it was a gift that ... *put our ideas about the universe for the first time on a sound basis.*

The Royal Astronomical Society awarded him their gold medal to mark this achievement. Bessel also worked out a method of mathematical analysis involving what is now known as the Bessel function. He introduced this in 1817 in his study of a problem of Kepler of determining the motion of three bodies moving under mutual gravitation. This mathematical achievement is described as follows: *Bessel was also an outstanding mathematician whose name became generally known through a special class of functions that have become an indispensable tool in applied mathematics, physics and engineering. The interest in these functions ... arose in the treatment of the problem of the perturbation in the planetary system.*

Bessel functions appear as coefficients in the series expansion of the indirect perturbation of a planet, that is, the motion caused by the motion of the Sun caused by the perturbing body. In 1824 he developed Bessel functions more fully in a study of planetary perturbations and published a treatise on them in Berlin. It was not the first time that special cases of the functions had appeared, Jacob Bernoulli, Daniel Bernoulli, Euler and Lagrange having studied special cases of them earlier. In fact, it was probably Lagrange's work on elliptical orbits that first suggested to Bessel to work on the Bessel functions.

This remarkable man who left formal education at the age of 14 made contributions beyond astronomy and mathematics. His contributions to geodesy include a correction in 1826 to the seconds pendulum, the length of which is precisely calculated so that it requires exactly one second for a swing. During 1831-1832 he directed geodetical measurements of meridian arcs in East Prussia, and in 1841 he deduced a value of $1/299$ for the ellipticity of the Earth, the amount of elliptical distortion by which the Earth's shape departs from a perfect sphere.

Bessel also had a very significant impact on university teaching despite the fact that he never had a university education. Klein describes how the name of Bessel, together with the names of Jacobi and Franz Neumann, is intimately linked to the reform of teaching at universities, first in Germany and then throughout the world.

11.2 Envelope Function Model

Suppose the nonstationary random process $X(t)$ is defined by the product of a deterministic function of time $A(t)$ and a zero-mean stationary process $F(t)$,
$$X(t) = A(t)F(t).$$
The reason for defining the nonstationary process in this way is that it is easier to work with the equivalent product of a deterministic function and a stationary process. Selecting an appropriate modulating function $A(t)$ and a suitable stationary process $F(t)$, the product of which is the required nonstationary function $X(t)$, is a difficult procedure. It is necessary to show, usually experimentally, that the product $A(t)F(t)$ is equivalent to the nonstationary process. We saw in the jet noise example above that the stationary process is first defined, and then much data is required to derive the modulating function by which it is multiplied.

This modeling approach implies that a nonstationary random process can be separated into a product of deterministic and stationary random processes. This is not always the case or possible.

For a linear system with input $X(t)$, the output $Y(t)$ is given by
$$Y(t) = \int_{-\infty}^{\infty} g(t-\tau)X(\tau)d\tau.$$

The time-dependent autocorrelation function and the power spectrum of the response can be derived as follows,
$$\begin{aligned} R_{YY}(t,\tau) &= E\{Y(t)Y(t+\tau)\} \\ &= \int_0^t \int_0^{t+\tau} g(t-u)g(t+\tau-v)E\{X(u)X(v)\}du\,dv, \end{aligned}$$
where $E\{X(u)X(v)\} = R_{XX}(v-u) = A(u)A(v)R_{FF}(v-u)$, and
$$\begin{aligned} R_{FF}(v-u) &= \int_{-\infty}^{\infty} S_{FF}(\omega)e^{i\omega(v-u)}d\omega \\ &= \frac{1}{A(u)A(v)} \int_{-\infty}^{\infty} S_{XX}(\omega)e^{i\omega(v-u)}d\omega. \end{aligned}$$

Then,
$$\begin{aligned} R_{YY}(t,\tau) = \int_0^t \int_0^{t+\tau} & A(u)A(v)g(t-u)g(t+\tau-v) \\ & \cdot \left[\int_{-\infty}^{\infty} S_{FF}(\omega)e^{i\omega(v-u)}d\omega\right]du\,dv. \end{aligned} \quad (11.2)$$

11.2. ENVELOPE FUNCTION MODEL

We also know that the time-dependent autocorrelation function is given by definition,
$$R_{YY}(t,\tau) = \int_{-\infty}^{\infty} S_{YY}(t,\omega) e^{i\omega\tau} d\omega. \tag{11.3}$$

Setting Equations 11.2 and 11.3 equal to each other leads to the relation,

$$\int_{-\infty}^{\infty} S_{YY}(t,\omega) e^{i\omega\tau} d\omega = \int_0^t \int_0^{t+\tau} A(u)A(v)g(t-u)\,g(t+\tau-v)$$
$$\cdot \left[\int_{-\infty}^{\infty} S_{FF}(\omega) e^{i\omega(v-u)} d\omega\right] du\,dv$$

$$S_{YY}(t,\omega) e^{i\omega\tau} = S_{FF}(\omega) \int_0^t \int_0^{t+\tau} A(u)A(v)g(t-u)$$
$$\cdot g(t+\tau-v) e^{i\omega(v-u)} du\,dv$$

$$S_{YY}(t,\omega) = S_{FF}(\omega) e^{-i\omega\tau} \int_0^t \int_0^{t+\tau} A(u)A(v)g(t-u)$$
$$\cdot g(t+\tau-v) e^{i\omega(v-u)} du\,dv. \tag{11.4}$$

Since the response spectral density $S_{YY}(t,\omega)$ is independent of time lag τ, it may be viewed as arbitrary and can be set equal to zero: $\tau = 0$. $R_{FF}(\tau)$ is an even function and $S_{XX}(\omega)$ must be real and even. Thus,

$$S_{YY}(t,\omega) = S_{FF}(\omega) \int_0^t \int_0^t A(u)A(v)g(t-u)\,g(t-v)\cos[\omega(v-u)] du\,dv. \tag{11.5}$$

This result is a relation between the stationary spectrum $S_{FF}(\omega)$ of the forcing and the evolutionary spectrum $S_{YY}(t,\omega)$ of the response, where $A(t)$ must be specified (using data) for a particular problem.

11.2.1 Transient Response

The transient response of a simple harmonic oscillator to stationary random input having an arbitrary spectrum is studied here.[3] This system is subjected to a stationary random force that is applied at time $t = 0$. Therefore, the structure experiences a suddenly applied force and responds with initial nonstationarity.

The oscillator is governed by the equation,
$$\ddot{X} + 2\zeta\omega_n \dot{X} + \omega_n^2 X = F(t),$$

---

[3] This section is based on "Transient Response of a Dynamic System under Random Excitation," T.K. Caughey, H.J. Stumpf, *Journal of Applied Mechanics*, pp. 563-566, December 1961.

where the random force per unit mass $F(t)$ has the following properties: *(i)* stationary, *(ii)* Gaussian, *(iii)* zero mean, *(iv)* power spectrum $S_{FF}^o(\omega)$.

The impulse response of the system is

$$g(t) = \begin{cases} (1/\omega_d)\exp(-\zeta\omega_n t)\sin\omega_d t, & t \geq 0 \\ 0, & t < 0, \end{cases} \quad (11.6)$$

where $\omega_d = \omega_n\sqrt{1-\zeta^2}$, and only the case where $\zeta < 1$ is considered. The initial conditions are taken to be $X(0) = a$ and $\dot{X}(0) = b$. The force is assumed to be mean-square continuous, and the response is given in general,

$$\begin{aligned} X(t) &= a\exp(-\zeta\omega_n t)\left[\cos\omega_d t + \frac{\zeta\omega_n}{\omega_d}\sin\omega_d t\right] \\ &+ \frac{b}{\omega_d}\exp(-\zeta\omega_n t)\sin\omega_d t + \int_0^t g(t-\tau)F(\tau)\,d\tau. \end{aligned} \quad (11.7)$$

This solution is used to determine the statistics of response $X(t)$.

Mean Value

For a linear system, if the input $F(t)$ is Gaussian, then so is the output $X(t)$. The Gaussian process requires only the mean value and variance for its characterization. Therefore, once $E\{X\}$ and $Var\{X\}$ are calculated, the probabilistic description of $X(t)$ is fully known. From Equation 11.7,

$$\begin{aligned} E\{X(t)\} &= a\exp(-\zeta\omega_n t)\left[\cos\omega_d t + \frac{\zeta\omega_n}{\omega_d}\sin\omega_d t\right] \\ &+ \frac{b}{\omega_d}\exp(-\zeta\omega_n t)\sin\omega_d t + \int_0^t g(t-\tau)E\{F(\tau)\}\,d\tau. \end{aligned} \quad (11.8)$$

This is a time-dependent mean value, a property of a nonstationary process. The time dependence arises from the first two transient terms. As time increases, $E\{X(t)\}$ approaches a limit, with the result that the mean value is asymptotically stationary. How fast stationarity is approached is a function of how rapidly the exponentials decay.

In this problem $E\{F(\tau)\} = 0$. Therefore,

$$\lim_{t\to\infty} E\{X(t)\} = 0.$$

Variance

The variance is given by the difference,

$$\sigma_X^2(t) = E\{X^2(t)\} - [E\{X(t)\}]^2. \quad (11.9)$$

11.2. ENVELOPE FUNCTION MODEL

Substituting Equations 11.7 and 11.8 into Equation 11.9, we find

$$\sigma_X^2(t) = \int_0^t \int_0^t g(t-\tau) g(t-\lambda) E\{F(\tau) F(\lambda)\} d\tau\, d\lambda, \qquad (11.10)$$

where

$$E\{F(\tau) F(\lambda)\} = R_{FF}(\tau, \lambda) = R_{FF}(\tau - \lambda),$$

and where the last equality is due to the assumption that $F(t)$ is a stationary process. The power spectrum of the force is related to the correlation function by the Fourier cosine transform,

$$R_{FF}(\tau - \lambda) = \int_0^\infty S_{FF}^o(\omega) \cos\omega(\tau - \lambda) d\omega.$$

Substitute this expression into Equation 11.10 to find

$$\sigma_X^2(t) = \int_0^t \int_0^t \int_0^\infty S_{FF}^o(\omega) \cos\omega(\tau - \lambda) g(t-\tau) g(t-\lambda) d\omega\, d\tau\, d\lambda.$$

Since the integrals are uniformly convergent, the integration order may be changed. Using Equation 11.6,

$$\sigma_X^2(t) = \int_0^\infty \frac{1}{\omega_d^2} S_{FF}^o(\omega) \left[\int_0^t \int_0^t \exp(-\zeta\omega_n[2t - \tau - \lambda]) \sin\omega_d(t-\tau) \right.$$
$$\left. \cdot \sin\omega_d(t-\lambda) \cos\omega(\tau - \lambda)\, d\tau\, d\lambda \right] d\omega.$$

After some tedious algebra the double integral can be evaluated, giving

$$\sigma_X^2(t)$$
$$= \int_0^\infty \frac{1}{(\omega_n^2 - \omega^2)^2 + (2\zeta\omega_n\omega)^2} S_{FF}^o(\omega)$$
$$\cdot \left[1 + \exp(-2\zeta\omega_n t) \left\{ 1 + \frac{2\omega_n}{\omega_d} \zeta \sin\omega_d t \cos\omega_d t \right. \right.$$
$$- \exp(\zeta\omega_n t) \left(2\cos\omega_d t + \frac{2\omega_n}{\omega_d} \zeta \sin\omega_d t \right) \cos\omega t$$
$$\left. \left. - \exp(\zeta\omega_n t) \frac{2\omega}{\omega_d} \sin\omega_d t \sin\omega t + \frac{(\zeta\omega_n)^2 - \omega_d^2 + \omega^2}{\omega_d^2} \sin^2\omega_d t \right\} \right] d\omega.$$
$$(11.11)$$

Equation 11.11 exhibits the following properties;

1. As $t \to 0$, $\sigma_X^2(t) \to 0$, as expected.

2. As $t \to \infty$,

$$\sigma_X^2(t) \to \int_0^\infty \frac{1}{(\omega_n^2 - \omega^2)^2 + (2\zeta\omega_n\omega)^2} S_{FF}^o(\omega)\, d\omega.$$

This result arises from the analysis of the equivalent stationary problem.

3. If $F(t)$ is a white noise process with $S_{FF}^o(\omega) = 2D/\pi$, Equation 11.11, evaluated by contour integration, becomes

$$\sigma_X^2(t) = \frac{D}{2\zeta\omega_n^3}\left[1 - \frac{\exp(-2\zeta\omega_n t)}{\omega_d^2}\left\{\omega_d^2 + \zeta\omega_n\omega_d \sin 2\omega_d t \right.\right.$$
$$\left.\left. + \frac{(2\zeta\omega_n)^2}{2}\sin^2\omega_d t\right\}\right], \qquad (11.12)$$

which is the known result for the white noise forcing case.

Evaluation of the Variance

Equation 11.11 can be evaluated in a number of ways, depending on the accuracy desired and the form of the function $S_{FF}^o(\omega)$. If the spectrum is available numerically, then the integral can be evaluated numerically as well. If the spectrum is a smooth function of ω, and damping constant ζ is small, then $(\omega_n^2 - \omega^2)^2 + (2\zeta\omega_n\omega)^2$ is sharply peaked around $\omega = \omega_n$ and the main contribution to the integral is around the region $\omega = \omega_n$. Then $S_{FF}^o(\omega_n)$ can be taken outside the integral and approximated by

$$\sigma_X^2(t) \simeq \frac{\pi S_{FF}^o(\omega_n)}{4\zeta\omega_n^3}\left[1 - \frac{\exp(-2\zeta\omega_n t)}{\omega_d^2}\right.$$
$$\left. \cdot \left\{\omega_d^2 + \frac{(2\zeta\omega_n)^2}{2}\sin^2\omega_d t + \zeta\omega_n\omega_d\sin 2\omega_d t\right\}\right]. \qquad (11.13)$$

For the purpose of plotting results, let $\theta = \omega_n t$. Equation 11.13 can be

11.2. ENVELOPE FUNCTION MODEL

written as

$$\frac{2\sigma_X^2(t)}{\pi} \frac{\omega_n^3}{S_{FF}^o(\omega_n)}$$
$$\simeq \frac{1}{2\zeta}\left[1 - \exp(-2\zeta\theta)\right.$$
$$\left.\cdot\left\{1 + \frac{2\zeta^2}{1-\zeta^2}\sin^2\sqrt{1-\zeta^2}\theta + \frac{\zeta}{\sqrt{1-\zeta^2}}\sin 2\sqrt{1-\zeta^2}\theta\right\}\right].$$

Figure 11.2 depicts plots of $2\sigma_X^2(t)\omega_n^3/\pi S_{FF}^o(\omega_n)$ as functions of $\theta = \omega_n t$ for increasing values of the damping constant ζ. The larger the value of ζ, the smaller the magnitude of the evolution of the variance with time. All the damped response statistics have asymptotic values.

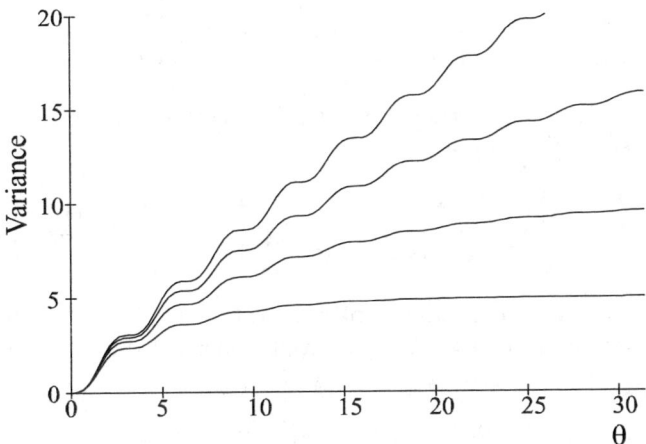

Figure 11.2: $\frac{2\sigma_X^2(t)}{\pi}\frac{\omega_n^3}{S_{FF}^o(\omega_n)}$ as a function of θ for $\zeta = 0.1, 0.05, 0.025,$ and 0.005. The lowest variance curve is for $\zeta = 0.1$ and the highest variance curve for $\zeta = 0.005$.

If damping can be ignored, then $\lim_{\zeta \to 0}$ of the right-hand side results in

$$\sigma_X^2(t) \simeq \frac{\pi S_{FF}^o(\omega_n)}{4\omega_n^3}\left[2\omega_n t - \sin 2\omega_n t\right]. \tag{11.14}$$

In this undamped case, the response variance grows without bounds due to the secular term $2\omega_n t$. Since $F(t)$ is a force per unit mass, a units balance for Equation 11.14 is

$$\sigma_X^2(t) [=] [\text{m}^2]$$

$$S_{FF}^o(\omega_n) [=] \left[\frac{1}{\text{m}}\frac{\text{kg-m}}{\text{s}^2}\right]\left[\frac{1}{\text{m}}\frac{\text{kg-m}}{\text{s}^2}\right][\text{s}]$$

$$\omega_n^3 [=] \frac{1}{[\text{s}^3]},$$

which, of course, satisfies the equation.

Density Function and its Uses

The output probability density function is a Gaussian for this linear oscillator. The probability that $x < X \leq x + dx$ is approximately

$$f_X(x)\,dx = \frac{1}{\sqrt{2\pi}\sigma_X(t)}\exp\left\{-\frac{(x - E\{X(t)\})^2}{2\sigma_X^2(t)}\right\}dx.$$

With the density function, it is possible to calculate the probability that $X(t)$ exceeds a given value $k\sigma$,

$$\Pr(X(t) > k\sigma) = \int_{k\sigma}^{\infty}\frac{1}{\sqrt{2\pi}\sigma_X(t)}\exp\left\{-\frac{(x - E\{X(t)\})^2}{2\sigma_X^2(t)}\right\}dx.$$

For earthquake engineering applications, Equation 11.13 can be used to estimate the variance of the relative displacement of the roof with respect to the base of the structure,

$$\sigma_X^2(t) \simeq \frac{\pi S_{GG}(\omega_n)}{4\zeta\omega_n^3}\left[1 - \frac{\exp(-2\zeta\omega_n t)}{\omega_d^2}\right.$$

$$\left.\cdot\left\{\omega_d^2 + \frac{(2\zeta\omega_n)^2}{2}\sin^2\omega_d t + \zeta\omega_n\omega_d\sin 2\omega_d t\right\}\right],$$

where the Kanai earthquake spectral density is a reasonable choice,

$$S_{GG}(\omega) = B\frac{1 + 4h_g^2(\omega/\nu_g)^2}{\left(1 - (\omega/\nu_g)^2\right)^2 + 4h_g^2(\omega/\nu_g)^2},$$

and B = spectral density at bedrock, ν_g, h_g are parameters depending on local geology. Example parameter values are $h_g = 0.5$, $\nu_g = 4\pi$ and $B = 1$ $(\text{ft}^2/\text{s}^4)(\text{s}/\text{rad})$.

11.2.2 MS Nonstationary Response

Here we examine the mean-square response of a SDOF system to amplitude modulated random noise.[4] Begin with the single-degree-of-freedom system,

$$\ddot{Y} + 2\zeta\omega_n\dot{Y} + \omega_n^2 Y = \frac{1}{m}F(t),$$

where $F(t) = e(t)N(t)$; $e(t)$ is a well-defined deterministic envelope function and $N(t)$ is a Gaussian broadband stationary process with zero mean value.

This excitation is a nonstationary process created by the product of sample functions from a stationary process and $e(t)$. Here, instead of formulating the solution in terms of system impulse response function $g(t)$, the mean-square solution is developed in terms of the system frequency response function and the generalized spectral density function.

Assume that the system is initially at rest. In this instance, since $E\{F(t)\} = 0$, then $E\{Y(t)\} = 0$, and we need only consider the mean-square response, $E\{Y^2(t)\}$. This is derived for the cases where

- $e(t)$ is the unit step $u(t)$

- $e(t)$ is a rectangular step, and where

- the correlation function of $N(t)$ is delta correlated, and

- the correlation function of $N(t)$ is exponentially correlated.

Response Formulation

The response autocorrelation is given by

$$R_{YY}(t_1, t_2) = E\{Y(t_1)Y(t_2)\}. \qquad (11.15)$$

In the Fourier transform domain,

$$Y(\omega) = H(\omega)F(\omega),$$

where the system response function is

$$H(\omega) = \frac{1}{m}\frac{1}{\omega_n^2 - \omega^2 + i2\zeta\omega\omega_n}.$$

---

[4] This section is based on the paper, "Mean-Square Response of Simple Mechanical Systems to Nonstationary Random Excitation," R.L. Barnoski, J.R. Maurer, *Journal of Applied Mechanics*, pp. 221-227, June 1969.

Some additional studies are in the paper, "Mean-Square Response of a Second-Order System to Nonstationary, Random Excitation," L.L. Bucciarelli, C. Kuo, *Journal of Applied Mechanics*, pp. 612-616, Sept. 1970.

Then, the response $Y(t)$ is

$$Y(t) = \frac{1}{2\pi} \int_{-\infty}^{\infty} H(\omega) F(\omega) \exp(i\omega t) d\omega. \tag{11.16}$$

Upon substitution of Equation 11.16 into 11.15, we find

$$R_{YY}(t_1, t_2) = \int_{-\infty}^{\infty} \int_{-\infty}^{\infty} S_{YY}(\omega_1, \omega_2) \exp(-i[\omega_1 t_1 - \omega_2 t_2]) d\omega_1 d\omega_2, \tag{11.17}$$

where

$$S_{YY}(\omega_1, \omega_2) = H^*(\omega_1) H(\omega_2) S_{FF}(\omega_1, \omega_2) \tag{11.18}$$
$$S_{FF}(\omega_1, \omega_2) = \frac{1}{(2\pi)^2} E\{F^*(\omega_1) F(\omega_2)\}.$$

The mean-square response is given by setting $t_1 = t_2 = t$,

$$E\{Y^2(t)\} = R_{YY}(t,t)$$
$$R_{YY}(t,t) = \int_{-\infty}^{\infty} \int_{-\infty}^{\infty} H^*(\omega_1) H(\omega_2) S_{FF}(\omega_1, \omega_2)$$
$$\cdot \exp(-i[\omega_1 - \omega_2]t) d\omega_1 d\omega_2.$$

From the Wiener-Khinchine relations,

$$S_{FF}(\omega_1, \omega_2) = \frac{1}{(2\pi)^2} \int_{-\infty}^{\infty} \int_{-\infty}^{\infty} R_{FF}(t_1, t_2) \exp(i[\omega_1 t_1 - \omega_2 t_2]) dt_1 dt_2,$$

where

$$R_{FF}(t_1, t_2) = e(t_1) e(t_2) R_{NN}(\tau),$$

and $N(t)$ is a stationary process, thus,

$$R_{NN}(\tau) = \int_{-\infty}^{\infty} S_{NN}(\omega) \exp(i\omega\tau) d\omega,$$

where $\tau = t_2 - t_1$. Then,

$$R_{FF}(t_1, t_2) = e(t_1) e(t_2) \int_{-\infty}^{\infty} S_{NN}(\omega) \exp(i\omega[t_2 - t_1]) d\omega,$$

11.3. NONSTATIONARY GENERALIZATIONS

and

$$\begin{aligned}
S_{FF}(\omega_1,\omega_2) &= \frac{1}{(2\pi)^2}\int_{-\infty}^{\infty}\int_{-\infty}^{\infty} e(t_1)e(t_2) \\
&\quad \cdot \left[\int_{-\infty}^{\infty} S_{NN}(\omega)\exp(i\omega[t_2-t_1])d\omega\right] \\
&\quad \cdot \exp(i[\omega_1 t_1 - \omega_2 t_2])dt_1 dt_2 \\
&= \int_{-\infty}^{\infty} S_{NN}(\omega)\cdot\left[\frac{1}{2\pi}\int_{-\infty}^{\infty} e(t_1)\exp\left[-i(\omega-\omega_1)t_1\right]dt_1 \right. \\
&\quad \left. \cdot \frac{1}{2\pi}\int_{-\infty}^{\infty} e(t_2)\exp\left[-i(-\omega+\omega_2)t_2\right]dt_2\right]d\omega.
\end{aligned}$$

(11.19)

Equation 11.18 can now be evaluated given $e(t)$ and $S_{NN}(\omega)$,

$$S_{YY}(\omega_1,\omega_2) = H^*(\omega_1)H(\omega_2)\int_{-\infty}^{\infty} S_{NN}(\omega)\mathcal{E}(\omega-\omega_1)\mathcal{E}(-\omega+\omega_2)d\omega$$

$$\mathcal{E}(\omega-\omega_1) = \frac{1}{2\pi}\int_{-\infty}^{\infty} e(t_1)\exp\left[-i(\omega-\omega_1)t_1\right]dt_1$$

$$\mathcal{E}(-\omega+\omega_2) = \frac{1}{2\pi}\int_{-\infty}^{\infty} e(t_2)\exp\left[-i(-\omega+\omega_2)t_2\right]dt_2.$$

$\mathcal{E}(\omega)$ is the Fourier transform of $e(t)$.

11.3 Nonstationary Generalizations

There are several ways by which functions of time can be represented. A single harmonic function of time is given by

$$x(t) = a\cos(\omega t - \theta),$$

where the parameters a, ω and θ represent amplitude, frequency, and phase angle, respectively. It is frequently necessary to represent a more complex function of time using sums of harmonic functions,

$$x(t) = \sum_{n=1}^{N} a_n \cos(\omega_n t - \theta_n), \qquad (11.20)$$

or in the limit using the integral,

$$x(t) = \int_{-\infty}^{\infty} a(\omega)\cos(\omega t - \theta(\omega))d\omega.$$

We also know that it can be useful to represent harmonic functions by the real or imaginary part of a complex exponential,

$$x(t) = \int_{-\infty}^{\infty} a(\omega) e^{i(\omega t - \theta(\omega))} d\omega. \tag{11.21}$$

Finally, another model replaces $a(\omega)d\omega$ by a complex valued function $dZ(\omega)$, that also incorporates the phase $\theta(\omega)$,

$$x(t) = \int_{-\infty}^{\infty} e^{i\omega t} dZ(\omega). \tag{11.22}$$

This type of integral is discussed in more detail subsequently.

11.3.1 Discrete Model

Suppose in Equation 11.20 the phase angles are random variables with the uniform probability density function,

$$f(\theta_n) = \begin{cases} 1/2\pi, & 0 \leq \theta_n \leq 2\pi \\ 0, & \text{otherwise.} \end{cases}$$

Then, $x(t)$ becomes a random process $X(t)$ with expected value,

$$E\{X(t)\} = \sum_{n=1}^{N} a_n E\{\cos(\omega_n t - \theta_n)\} = 0,$$

since the mathematical expectations of $\cos \theta_n$ and $\sin \theta_n$ are 0, that is,

$$\begin{aligned}
E\{\cos(\omega_n t - \theta_n)\} &= E\{\cos \omega_n t \cos \theta_n + \sin \omega_n t \sin \theta_n\} \\
&= \cos \omega_n t E\{\cos \theta_n\} + \sin \omega_n t E\{\sin \theta_n\} \\
&= \cos \omega_n t \int_0^{2\pi} \cos \theta_n \left(\frac{1}{2\pi}\right) d\theta_n \\
&\quad + \sin \omega_n t \int_0^{2\pi} \cos \theta_n \left(\frac{1}{2\pi}\right) d\theta_n \\
&= 0.
\end{aligned}$$

The ensemble mean-square value is given by

$$\begin{aligned}
E\{X^2(t)\} &= E\left\{\sum_{n=1}^{N} a_n \cos(\omega_n t - \theta_n) \sum_{m=1}^{N} a_n \cos(\omega_m t - \theta_m)\right\} \\
&= \sum_{n=1}^{N} \frac{a_n^2}{2}.
\end{aligned}$$

11.3. NONSTATIONARY GENERALIZATIONS

$X(t)$ is therefore stationary.

Suppose in Equation 11.20 $X(t)$ is nonstationary and modeled as

$$X(t) = \sum_{n=1}^{N} A(t, \omega_n) a_n \cos(\omega_n t - \theta_n),$$

where $A(t, \omega_n)$ is a deterministic function of time and frequency. By modifying the model in this way we obtain a discrete model for a nonstationary random process. The moments for this process and time-dependent spectral density can be derived in terms of the stationary process spectral density. The ensemble mean is

$$E\{X(t)\} = \sum_{n=1}^{N} A(t, \omega_n) a_n E\{\cos(\omega_n t - \theta_n)\} = 0.$$

The ensemble mean-square value is given by

$$\begin{aligned} E\{X^2(t)\} &= E\left\{\sum_{n=1}^{N} A(t, \omega_n) a_n \cos(\omega_n t - \theta_n) \right. \\ &\quad \left. \cdot \sum_{m=1}^{N} A(t, \omega_m) a_m \cos(\omega_m t - \theta_m)\right\} \\ &= \sum_{n=1}^{N} \frac{1}{2} A^2(t, \omega_n) a_n^2. \end{aligned}$$

The mean-square value also equals the area under the power spectral density function,

$$E\{X^2(t)\} = \sum_{n=1}^{N} 2S(t, \omega_n) \Delta\omega,$$

where $\Delta\omega = \omega_{n+1} - \omega_n$, and the factor 2 is due to the one-sided nature of the summation. The power spectrum is symmetric about the origin and therefore one may consider one side and multiply it by 2 when calculating the mean-square energy. Comparing and equating the last two equations leads to the relation,

$$\frac{1}{2} A^2(t, \omega_n) a_n^2 = 2S(t, \omega_n) \Delta\omega. \tag{11.23}$$

Thus, the nonstationary discrete random process becomes

$$X(t) = \sum_{n=1}^{N} \sqrt{4S(t, \omega_n) \Delta\omega} \cos(\omega_n t - \theta_n).$$

Then, $a_n = \sqrt{4S(\omega_n)\Delta\omega}$, where $S(\omega)$ is the spectral density of the process $X(t)$ without the time varying coefficient $A(t,\omega)$. From Equation 11.23

$$S(t,\omega_n) = A^2(t,\omega_n)S(\omega_n).$$

Note that the stationary process is recovered if $A(t,\omega_n) = 1$.

11.3.2 Complex-Valued Stochastic Processes

A complex-valued stochastic process, $Z(t) = X(t) + iY(t)$, where $i = \sqrt{-1}$, can be regarded either as a process the state space of which is a subset of the complex plane or, equivalently, as a two-component vector-valued process. The two components are the real part $X(t)$ and the imaginary part $Y(t)$ of $Z(t)$. Its mean function is complex-valued,

$$\begin{aligned} E\{Z(t)\} &= E\{X(t)\} + iE\{Y(t)\} \\ \mu_Z(t) &= \mu_X(t) + i\mu_Y(t). \end{aligned}$$

The variance is, however, real:

$$\begin{aligned} \sigma_Z^2(t) &= E\left\{|Z(t) - \mu_Z(t)|^2\right\} \\ &= E\left\{|X(t) + iY(t) - \mu_X(t) - i\mu_Y(t)|^2\right\} \\ &= E\left\{|X(t) - \mu_X(t) + i[Y(t) - \mu_Y(t)]|^2\right\} \\ &= E\{[X(t) - \mu_X(t)]^2 + [Y(t) - \mu_Y(t)]^2\} \\ &= \sigma_X^2(t) + \sigma_Y^2(t). \end{aligned}$$

The modulus of a complex function $|Z(t)|^2$ is given by the product of the function and its complex conjugate $Z(t)Z^*(t)$. The correlation function is defined by

$$R_{ZZ}(t_1, t_2) = E\{Z(t_1)Z^*(t_2)\},$$

a complex-valued function that exhibits Hermitian symmetry,

$$R_{ZZ}(t_1, t_2) = R_{ZZ}^*(t_2, t_1).$$

11.3.3 Continuous Model

Begin with Equation 11.21, insert the deterministic factor $A(t,\omega)$ to obtain the expression for a continuous nonstationary random process,

$$X(t) = \int_{-\infty}^{\infty} A(t,\omega)a(\omega)e^{i(\omega t - \theta(\omega))}d\omega.$$

11.3. NONSTATIONARY GENERALIZATIONS

The ensemble mean is defined as

$$E\{X(t)\} = \int_{-\infty}^{\infty} A(t,\omega)a(\omega)E\{e^{i(\omega t - \theta(\omega))}\}d\omega = 0.$$

The ensemble autocorrelation function $R(t,\tau)$ is now time dependent and given by

$$\begin{aligned} R(t,\tau) &= E\{X^*(t)X(t+\tau)\} \\ &= E\left\{\int_{-\infty}^{\infty} A(t,\omega_1)a(\omega_1)e^{i(-\omega_1 t + \theta(\omega_1))}d\omega_1 \right. \\ &\quad \left. \cdot \int_{-\infty}^{\infty} A(t,\omega_2)a(\omega_2)e^{i(\omega_2[t+\tau] - \theta(\omega_2))}d\omega_2\right\} \\ &= \int_{-\infty}^{\infty}\int_{-\infty}^{\infty} A(t,\omega_1)a(\omega_1)A(t,\omega_2)a(\omega_2) \\ &\quad \cdot E\{\exp i[\theta(\omega_1) - \theta(\omega_2)]\}\exp i[-\omega_1 t + \omega_2 t + \omega_2 \tau]d\omega_1 d\omega_2 \\ &= \int_{-\infty}^{\infty} [A(t,\omega)a(\omega)d\omega]^2 \exp(i\omega\tau), \quad (11.24) \end{aligned}$$

since, for the integration with respect to ω_2, the expectation is given by

$$E\{\exp i[\theta(\omega_1) - \theta(\omega_2)]\} = \begin{cases} 0 & \omega_2 \neq \omega_1 \\ 1 & \omega_2 = \omega_1. \end{cases}$$

Comparing Equation 11.24 with

$$R(t,\tau) = \int_{-\infty}^{\infty} S(t,\omega)e^{i\omega\tau}d\omega$$

yields the relation,

$$A(t,\omega)a(\omega)d\omega = \sqrt{S(t,\omega)d\omega}.$$

The resulting nonstationary random process is

$$X(t) = \int_{-\infty}^{\infty} \sqrt{S(t,\omega)d\omega}\exp\left[i(\omega t - \theta(\omega))\right].$$

As before, for the stationary case, $A(t,\omega) = 1$, and $a(\omega)d\omega = \sqrt{S(\omega)d\omega}$, and

$$S(t,\omega) = A^2(t,\omega)S(\omega),$$

which relates the stationary and nonstationary spectral densities.

11.4 Priestley's Model

From Equation 11.22, a stationary random process $X(t)$ can be represented by the following complex integral,

$$X(t) = \int_{-\infty}^{\infty} e^{i\omega t} dZ(\omega), \qquad (11.25)$$

where $dZ(\omega)$ is defined as a zero-mean random orthogonal complex process such that

$$E\{dZ(\omega)\} = 0$$
$$E\{dZ^*(\omega) dZ(\omega_1)\} = 0, \quad \text{for } \omega \neq \omega_1.$$

11.4.1 The Stieltjes Integral: An Aside

A bit of discussion is warranted regarding this functional form for a random process. Our discussion is based on Priestley.[5] When a spectral analysis of a stationary process is performed, that is, decompose the random process into its frequency components, the difficulty in using a Fourier integral representation is that the integral does not in general exist. This can be circumvented by considering a finite interval and then performing a limit.

A more general Fourier expansion is available, known as the *Fourier-Stieltjes integral*, defined by Equation 11.25. This equation is called the spectral representation of the process $X(t)$, and any stationary process can be represented as (the limit of) the sum of sine and cosine functions with random coefficients $dZ(\omega)$, or, with random amplitudes $|dZ(\omega)|$ and random phases $\arg\{dZ(\omega)\}$. This representation is valid for processes with continuous and discontinuous spectra.

This type of integral is understandable as an extension of the Riemann definition of an integral. For example, the integral of the form,

$$\int_a^b g(t)X(t)dt,$$

where $g(t)$ is a deterministic function and $X(t)$ is a stochastic process, is the *limit in the mean* of the sequence,

$$\sum_{i=1}^{n} g(t_i)X(t_i)(t_i - t_{i-1}),$$

---

[5] *Spectral Analysis and Time Series*, M.B. Priestly, Academic Press, London 1981, Sec. 4.11.

11.4. PRIESTLEY'S MODEL

as $\max(t_i - t_{i-1}) \to 0$. The classical, deterministic Riemann-Stieltjes integral for two deterministic functions $g(t)$ and $F(t)$,

$$R = \int_a^b g(t) dF(t),$$

is defined as the limiting value of the discrete summation,

$$\sum_{i=1}^n g(t_i) \{F(t_i) - F(t_{i-1})\}$$

as $\max(t_i - t_{i-1}) \to 0$. Note that if $f(t) = F'(t)$ exists for all $t \in (a, b)$, then $dF(t) = f(t)dt$, and R may be written in the familiar form $\int_a^b g(t)f(t)dt$. But for functions $F(t)$ that are not differentiable, the Riemann-Stieltjes integral is defined in a meaningful way.

If $F(t)$ is a stochastic process, then a similar expression is available for R except that the limit of the sequence is in the mean square sense. The essential theorem follows without proof.

Theorem (Spectral representation of continuous parameter stationary process). *Let $X(t)$, $-\infty < t < \infty$, be a wide sense stationary, zero mean stochastically continuous process with the power spectrum,*

$$S_{XX}(\omega), \quad -\infty < \omega < \infty.$$

Then there exists a complex-valued finite-variance process $Z(\omega)$ on the frequency domain $-\infty < \omega < \infty$ such that, for all t, $X(t)$ may be written in the form,

$$X(t) = \int_{-\infty}^{\infty} e^{i\omega t} dZ(\omega). \tag{11.26}$$

$Z(\omega)$ is called the spectral process of $X(t)$, and has the following properties:

$$E\{dZ(\omega)\} = 0 \quad \text{for all } \omega$$
$$E\{|dZ(\omega)|^2\} = dS_{XX}(\omega) \quad \text{for all } \omega.$$

$Z(\omega)$ is also orthogonal, that is, it has uncorrelated increments, and for any two distinct frequencies, ω and ω', $\omega \neq \omega'$,

$$Cov\{dZ(\omega), dZ(\omega')\} = E\{dZ^*(\omega) dZ(\omega')\} = 0.$$

For our purposes $dS_{XX}(\omega) = S_{XX}(\omega) d\omega$.

11.4.2 Priestley's Model

The autocorrelation function for stationary process $X(t)$ is defined as

$$\begin{aligned}
R_{XX}(\tau) &= \operatorname{Re} E\{X^*(t) X(t+\tau)\} \\
&= \operatorname{Re}\left[E\left\{\int_{-\infty}^{\infty}\int_{-\infty}^{\infty} e^{-i\omega t} dZ^*(\omega) e^{i\omega_1(t+\tau)} dZ(\omega_1)\right\}\right] \\
&= \operatorname{Re}\left[\int_{-\infty}^{\infty}\int_{-\infty}^{\infty} e^{-i\omega t} e^{i\omega_1(t+\tau)} E\{dZ^*(\omega) dZ(\omega_1)\}\right] \\
&= \operatorname{Re}\int_{-\infty}^{\infty} e^{i\omega\tau} E\{|dZ(\omega)|^2\},
\end{aligned} \quad (11.27)$$

since the only nonzero value is for $\omega_1 = \omega$, and where Re is defined as the real part of the argument. Following some of our earlier discussions, we let

$$E\{|dZ(\omega)|^2\} \equiv S_{XX}(\omega) d\omega. \quad (11.28)$$

Then Equation 11.27 becomes

$$R_{XX}(\tau) = \int_{-\infty}^{\infty} e^{i\omega\tau} S_{XX}(\omega) d\omega,$$

where the notation Re has been removed since $X(t)$ is restricted to real functions of t and thus $dX(\omega)$ and $S_{XX}(\omega)$ are even functions of ω.

Extending the above for a nonstationary process requires the introduction of the deterministic modulation function $A(t,\omega)$ in Equation 11.25,

$$X(t) = \int_{-\infty}^{\infty} A(t,\omega) e^{i\omega t} dZ(\omega). \quad (11.29)$$

Then, following the above procedure, we find

$$R_{XX}(t,\tau) = \operatorname{Re}\int_{-\infty}^{\infty} |A(t,\omega)|^2 e^{i\omega\tau} E\{|dZ(\omega)|^2\}. \quad (11.30)$$

Define

$$|A(t,\omega)|^2 E\{|dZ(\omega)|^2\} \equiv S_{XX}(t,\omega) d\omega, \quad (11.31)$$

and assuming real processes with even functions $S_{XX}(t,\omega)$ of ω, obtain

$$R_{XX}(t,\tau) = \int_{-\infty}^{\infty} e^{i\omega\tau} S_{XX}(t,\omega) d\omega.$$

Compare Equations 11.28 and 11.31 to find that

$$S_{XX}(t,\omega) = |A(t,\omega)|^2 S_{XX}(\omega). \quad (11.32)$$

This is another relation between stationary and nonstationary spectral densities in terms of an assumed modulation function. Apply this approach next to a simple harmonic oscillator.

11.5 SDoF Oscillator Response

11.5.1 Stationary Case

Let $F(t)$, $-\infty < t < \infty$, be a wide sense, stationary, mean square continuous process with power spectrum $S_{FF}(\omega)$, $-\infty < t < \infty$. Consider the convolution response,

$$Y(t) = \int_{-\infty}^{\infty} g(t-\tau)F(\tau)d\tau, \quad -\infty < t < \infty.$$

$g(t)$ is the impulse response function. Now replace $F(\tau)$ by its spectral representation, Equation 11.26,

$$F(\tau) = \int_{-\infty}^{\infty} e^{i\omega\tau} dZ_F(\omega), \quad -\infty < t < \infty,$$

where $dZ_F(\omega)$ is the spectral process of $F(t)$. We obtain

$$\begin{aligned} Y(t) &= \int_{-\infty}^{\infty} g(t-\tau) \int_{-\infty}^{\infty} e^{i\omega\tau} dZ_F(\omega)\, d\tau \\ &= \int_{-\infty}^{\infty} e^{i\omega t}\left[\int_{-\infty}^{\infty} g(t-\tau)e^{-i\omega(t-\tau)} d\tau\right] dZ_F(\omega), \quad (11.33)\end{aligned}$$

since $\exp(i\omega\tau) = \exp(i\omega t)\exp(-i\omega(t-\tau))$. The integral in the brackets can be rewritten in the following way,

$$\int_{-\infty}^{\infty} g(t-\tau)e^{-i\omega(t-\tau)} d\tau = \int_{-\infty}^{\infty} g(s)e^{-i\omega s} ds = H(\omega).$$

Equation 11.33 can now be written as

$$Y(t) = \int_{-\infty}^{\infty} e^{i\omega t} H(\omega) dZ_F(\omega).$$

Compare this with the spectral representation of the output process,

$$Y(t) = \int_{-\infty}^{\infty} e^{i\omega t} dZ_Y(\omega), \quad -\infty < t < \infty,$$

and we conclude the following relation holds between the input and output spectral processes for a linear system,

$$dZ_Y(\omega) = H(\omega) dZ_F(\omega), \quad -\infty < t < \infty.$$

This is the same relationship as the one between the Fourier transforms of the output and the input for linear systems. From this relation, along with the spectral properties,

$$E\{|dZ_F(\omega)|^2\} = dS_{FF}(\omega)$$
$$E\{|dZ_Y(\omega)|^2\} = dS_{YY}(\omega),$$

the following relation holds,

$$dS_{YY}(\omega) = |H(\omega)|^2 dS_{FF}(\omega), \quad -\infty < t < \infty.$$

This result is used next for the nonstationary problem.

11.5.2 Nonstationary Case

We follow the ideas of the last section as well as those of Section 11.4. Consider first a generic oscillator with impulse response function $g(t)$ and frequency response function $H(\omega)$. Let the input force be given by the zero mean nonstationary random process $F(t)$, as given by Equation 11.29. Assume that the response $Y(t)$ is also zero mean, but nonstationary, and given by

$$Y(t) = \int_{-\infty}^{\infty} B(t,\omega) e^{i\omega t} dZ_Y(\omega), \tag{11.34}$$

where $B(t,\omega)$ is the respective modulation function, with correlations and spectral densities defined by

$$S_{YY}(t,\omega) = |B(t,\omega)|^2 S_{YY}(\omega) \tag{11.35}$$
$$R_{YY}(t,\tau) = \int_{-\infty}^{\infty} e^{i\omega\tau} S_{YY}(t,\omega) d\omega. \tag{11.36}$$

For a linear system, the input-output relation is

$$Y(t) = \int_{-\infty}^{\infty} g(t-\tau) F(\tau) d\tau.$$

Substituting for $F(t)$ from Equation 11.29 results in

$$Y(t) = \int_0^t g(t-\tau) \left[\int_{-\infty}^{\infty} A(\tau,\omega) e^{i\omega\tau} dZ_F(\omega) \right] d\tau.$$

Let $\theta = t - \tau$ and interchange the order of integration to find

$$Y(t) = \int_{-\infty}^{\infty} e^{i\omega t} \left[\int_0^t A(t-\theta,\omega) g(\theta) e^{-i\omega\theta} d\theta \right] dZ_F(\omega). \tag{11.37}$$

11.5. SDOF OSCILLATOR RESPONSE

From the equivalent stationary problem, the following relations exist,
$$dZ_Y(\omega) = H(\omega) dZ_F(\omega)$$
$$S_{YY}(\omega) = |H(\omega)|^2 S_{FF}(\omega). \quad (11.38)$$

Equation 11.37 becomes
$$Y(t) = \frac{1}{H(\omega)} \int_{-\infty}^{\infty} e^{i\omega t} \left[\int_0^t A(t-\theta,\omega) g(\theta) e^{-i\omega\theta} d\theta \right] dZ_Y(\omega).$$

Comparing this equation with Equation 11.34 yields the expression for $B(t,\omega)$,
$$B(t,\omega) = \frac{1}{H(\omega)} \int_0^t A(t-\theta,\omega) g(\theta) e^{-i\omega\theta} d\theta. \quad (11.39)$$

Now substitute Equations 11.38 and 11.39 into Equation 11.35 to find
$$S_{YY}(t,\omega) = \left| \frac{1}{H(\omega)} \int_0^t A(t-\theta,\omega) g(\theta) e^{-i\omega\theta} d\theta \right|^2 |H(\omega)|^2 S_{FF}(\omega)$$
$$= \left| \int_0^t A(t-\theta,\omega) g(\theta) e^{-i\omega\theta} d\theta \right|^2 S_{FF}(\omega). \quad (11.40)$$

When $A(t,\omega) = 1$, and $t \to \infty$, Equation 11.40 reduces to Equation 11.38. From Equation 11.36, we can obtain the output correlation function
$$R_{YY}(t,\tau) = \int_{-\infty}^{\infty} \left| \int_0^t A(t-\theta,\omega) g(\theta) e^{-i\omega\theta} d\theta \right|^2 S_{FF}(\omega) e^{i\omega\tau} d\omega, \quad (11.41)$$

where by definition,
$$S_{FF}(\omega) = \frac{1}{2\pi} \int_{-\infty}^{\infty} R_{FF}(\tau) e^{-i\omega\tau} d\tau.$$

Again, when $A(t,\omega) = 1$, and $t \to \infty$, Equation 11.41 reduces to the stationary case.

Equation 11.40 is equivalent to Equation 11.5. In the general relation,
$$S_{YY}(t,\omega) = \left| \int_0^t A(t-\theta,\omega) g(\theta) e^{-i\omega\theta} d\theta \right|^2 S_{XX}(\omega),$$

make the following substitutions, $A(t-\theta,\omega) = A(t-\tau)$ and $S_{XX}(\omega) = S_{FF}(\omega)$. Replace the magnitude squared factor by the product of the respective complex conjugates,
$$S_{YY}(t,\omega) = S_{XX}(\omega) \int_0^t A(t-\theta_1) g(\theta_1) e^{+i\omega\theta_1} d\theta_1$$
$$\cdot \int_0^t A(t-\theta_2) g(\theta_2) e^{-i\omega\theta_2} d\theta_2.$$

Then, let $t - \theta_1 = v$ and $t - \theta_2 = u$,

$$S_{YY}(t,\omega) = S_{XX}(\omega) \int_0^t A(v) g(t-v) e^{+i\omega(t-v)} dv$$
$$\cdot \int_0^t A(u) g(t-u) e^{-i\omega(t-u)} du$$
$$= S_{XX}(\omega) \int_0^t \int_0^t A(u) A(v) g(t-u) g(t-v) e^{-i\omega(v-u)} du\, dv,$$

which is identical to Equation 11.4 with $\tau = 0$.

11.5.3 Undamped Oscillator

Consider the undamped linear oscillator with impulse response function,

$$g(t) = \begin{cases} (1/\omega_n) \sin \omega_n t, & t \geq 0 \\ 0, & t < 0. \end{cases}$$

Assume the excitation $F(t)$ to be a zero mean and nonstationary random process where $A(t,\omega) = u(t)$, the unit step function, and $S_{FF}(\omega) = S_0$. Then, from Equation 11.32,

$$S_{FF}(t,\omega) = u(t) S_0.$$

From Equation 11.40,

$$S_{YY}(t,\omega) = \left| \int_0^t u(t-\tau) e^{-i\omega\tau} \sin \omega_n \tau\, d\tau \right|^2 \frac{S_0}{\omega_n^2}.$$

Since $t \geq \tau$ for the integration range, $u(t - \tau) = 1$ and the expression becomes

$$S_{YY}(t,\omega) = \left| \int_0^t e^{-i\omega\tau} \sin \omega_n \tau\, d\tau \right|^2 \frac{S_0}{\omega_n^2}. \qquad (11.42)$$

11.5. SDOF OSCILLATOR RESPONSE

The integral is easily evaluated as follows,

$$\int_0^t e^{-i\omega\tau}\sin\omega_n\tau\, d\tau = \frac{1}{2i}\int_0^t e^{-i\omega\tau}\left(e^{i\omega_n\tau} - e^{-i\omega_n\tau}\right)d\tau$$

$$= \frac{1}{2i}\int_0^t \left(e^{i(\omega_n-\omega)\tau} - e^{-i(\omega_n+\omega)\tau}\right)d\tau$$

$$= \frac{1}{2}\left(-\frac{1}{(\omega_n-\omega)}e^{i(\omega_n-\omega)t} - \frac{1}{(\omega_n+\omega)}e^{-i(\omega_n+\omega)t} + 2\omega_n\right)$$

$$= \frac{1}{2}\frac{1}{(\omega_n^2-\omega^2)}\left\{-(\omega_n+\omega)e^{i(\omega_n-\omega)t} - (\omega_n-\omega)e^{-i(\omega_n+\omega)t} + 2\omega_n\right\}$$

$$= \frac{1}{(\omega_n^2-\omega^2)}\{\omega_n - \omega\sin t\omega\sin t\omega_n - \omega_n\cos t\omega\cos t\omega_n$$
$$+ i(\omega_n\cos\omega_n t\sin\omega t - \omega\cos\omega t\sin\omega_n t)\}$$

Evaluating its magnitude requires rewriting this expression in terms of its real and imaginary parts. The magnitude of a complex number equals the square root of the sum of the squares of the real and imaginary parts. Performing this set of operations and substituting into Equation 11.42 yields

$$S_{YY}(t,\omega) = \frac{S_0}{(-\omega^2+\omega_n^2)^2}\left(1 + \cos^2\omega_n t + \frac{\omega^2}{\omega_n^2}\sin^2\omega_n t\right.$$
$$\left. - 2\cos\omega t\cos\omega_n t - 2\frac{\omega}{\omega_n}\sin\omega t\sin\omega_n t\right). \quad (11.43)$$

Using the spectral density, the mean-square response is derived,

$$\sigma_Y^2(t) = R_{YY}(t,\tau=0) = \int_{-\infty}^{\infty} S_{YY}(t,\omega)\, d\omega$$

$$= \frac{\pi S_0}{2\omega_n^3}\left(2\omega_n t - \sin 2\omega_n t\right), \quad t \geq 0,$$

and is plotted in Figure 11.3.

11.5.4 Underdamped Oscillator

Suppose now that the system is an underdamped oscillator. Then,

$$g(t) = \begin{cases} (1/\omega_d)\exp(-\zeta\omega_n t)\sin\omega_d t, & t \geq 0 \\ 0, & t < 0. \end{cases}$$

Assume the excitation $F(t)$ to be a zero-mean and nonstationary random process, where $A(t,\omega) = u(t)$, the unit step function, and $S_{FF}(\omega) = S_0$. Then, from Equation 11.32,

$$S_{FF}(t,\omega) = u(t)S_0.$$

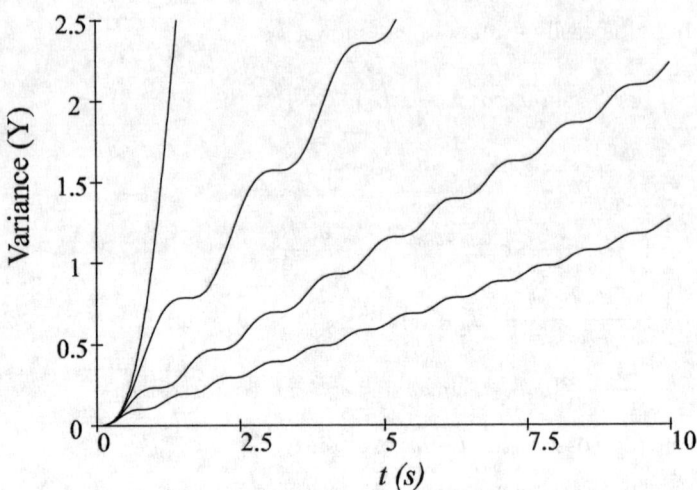

Figure 11.3: $2\sigma_Y^2/(\pi S_0)$ versus time for $\omega_n = 1, 2, 3, 4$.

From Equation 11.40,

$$S_{YY}(t,\omega) = \left|\int_0^t u(t-\tau)e^{-i\omega\tau}\exp(-\zeta\omega_n\tau)\sin\omega_d\tau\,d\tau\right|^2 \frac{S_0}{\omega_d^2}.$$

Since $t \geq \tau$ for the integration range, $u(t-\tau) = 1$ and the expression becomes

$$S_{YY}(t,\omega) = \left|\int_0^t e^{-i\omega\tau}\exp(-\zeta\omega_n\tau)\sin\omega_d\tau\,d\tau\right|^2 \frac{S_0}{\omega_d^2}$$

$$= \left|\int_0^t \exp(-i\omega-\zeta\omega_n)\tau\sin\omega_d\tau\,d\tau\right|^2 \frac{S_0}{\omega_d^2}.$$

Performing the integral, and taking the magnitude squared, we find

$$S_{YY}(t,\omega) = \frac{S_0}{\omega_d}|I(t,\omega)|^2$$
$$|I(t,\omega)|^2 = \omega_d^2|H(\omega)|^2\left\{1+\exp\left(-2\zeta\omega_n t\right)\left[1+a(t)+\omega^2 b(t)\right]\right.$$
$$\left.-\exp\left(-\zeta\omega_n t\right)\left[c(t)\cos\omega t+\omega d(t)\sin\omega t\right]\right\},$$

11.5. SDOF OSCILLATOR RESPONSE

where

$$a(t) = \left(\frac{\zeta^2 \omega_n^2 - \omega_d^2}{\omega_d^2}\right) \sin^2 \omega_d t + \frac{\zeta \omega_n}{\omega_d} \sin 2\omega_d t$$

$$b(t) = \frac{\sin^2 \omega_n t}{\omega_d^2}$$

$$c(t) = \frac{2\zeta \omega_n}{\omega_d} \sin \omega_d t + 2 \cos \omega_d t$$

$$d(t) = \frac{2 \sin \omega_d t}{\omega_d}.$$

For small damping,

$$a(t) \simeq -\sin^2 \omega_n t \qquad b(t) \simeq \sin^2 \omega_n t / \omega_n^2$$
$$c(t) \simeq 2\cos \omega_n t \qquad d(t) \simeq 2 \sin \omega_n t / \omega_n,$$

and

$$S_{YY}(t,\omega) \simeq \frac{S_0}{(\omega_n^2 - \omega^2)^2} \left\{ 1 + \exp(-2\zeta \omega_n t) \left[\cos^2 \omega_n t + \frac{\omega^2}{\omega_n^2} \sin^2 \omega_n t \right] \right.$$
$$\left. - \exp(-\zeta \omega_n t) \left[2 \cos \omega_n t \cos \omega t + \frac{2\omega}{\omega_n} \sin \omega_n t \sin \omega t \right] \right\}.$$

Where $\zeta = 0$, Equation 11.43 is recovered. Using the spectral density, the mean square response is given by the relation,

$$\sigma_Y^2(t) = R_{YY}(t, \tau = 0) = \int_{-\infty}^{\infty} S_{YY}(t, \omega) \, d\omega$$
$$= \frac{\pi S_0}{2\zeta \omega_n^3} \left\{ 1 - \exp(-2\zeta \omega_n t) \left[1 + 2 \left(\frac{\zeta \omega_n}{\omega_d}\right)^2 \sin^2 \omega_d t + \frac{\zeta \omega_n}{\omega_d} \sin 2\omega_d t \right] \right\},$$

which is the same as Equation 11.12 if S_0 is replaced by $2D/\pi$. Equation 11.12 was derived via time domain; here we used a frequency domain approach. For plotting purposes, the last equation can be written as

$$\frac{2\sigma_Y^2}{\pi S_0} = \frac{1}{2\zeta \omega_n^3} \left\{ 1 - \exp(-2\zeta \omega_n t) \left[1 + 2 \left(\frac{\zeta}{\sqrt{1-\zeta^2}}\right)^2 \sin^2 \omega_n \sqrt{1-\zeta^2} t \right. \right.$$
$$\left. \left. + \frac{\zeta}{\sqrt{1-\zeta^2}} \sin 2\omega_n \sqrt{1-\zeta^2} t \right] \right\}.$$

See Figure 11.4, plotted for $\zeta = 0.05$ and 0.15.

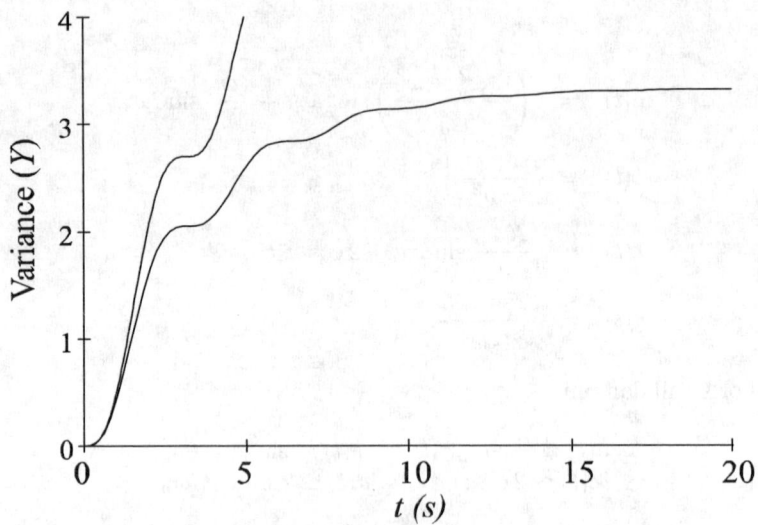

Figure 11.4: $2\sigma_Y^2/\pi S_0$ versus time for $\zeta = 0.05, 0.15$.

11.6 Multi DoF Oscillator Response

Just as it is straightforward to extend deterministic SDoF vibration to multi DoF using matrix theory, we can extend similarly to the response of general linear systems subjected to nonstationary random excitation.[6]

11.6.1 Input Characterization

Begin with the matrix equation of motion of a linear multi DoF system,

$$\mathbf{m\ddot{X} + c\dot{X} + kX = pF}(t), \tag{11.44}$$

where vectors and matrices are shown in bold for simplicity. Matrix **p** is an $n \times m$ load distribution matrix. The property matrices are $n \times n$. Vector $\mathbf{F}(t)$ is a random $m \times 1$ vector process. While any of the other methods outlined above can be used here, we use Priestley's definition of an evolutionary, zero mean vector process $\mathbf{F}(t)$ as a Fourier-Stieltjes integral,

$$\mathbf{F}(t) = \int_{-\infty}^{\infty} \mathbf{A_F}(t,\omega) \exp(i\omega t) \mathbf{dZ}(\omega), \tag{11.45}$$

---

[6]This section is based on the paper, "An Explicit Closed-Form Solution for Linear Systems Subjected to Nonstationary Random Excitation," J.P. Conte, B-F. Peng, *Probabilistic Engineering Mechanics* **11** (1996) 37-50.

11.6. MULTI DOF OSCILLATOR RESPONSE

where
$$\mathbf{A_F}(t,\omega) = \begin{bmatrix} \mathbf{a}_{F_1}(t,\omega) \\ \mathbf{a}_{F_2}(t,\omega) \\ \vdots \\ \mathbf{a}_{F_m}(t,\omega) \end{bmatrix}_{(m \times k)}$$

is a matrix of frequency-time modulating functions of $\mathbf{F}(t)$, $\mathbf{a}_{F_j}(t,\omega)$ is a $k \times 1$ row vector of frequency-time modulating functions of $F_j(t)$, and $\mathbf{dZ}(\omega)$ is a $k \times 1$ row orthogonal-increment vector process having the properties,

$$E\{\mathbf{dZ}(\omega)\} = \mathbf{0}$$

and

$$E\left\{\mathbf{dZ}^*(\omega_1)\mathbf{dZ}^\mathbf{T}(\omega_2)\right\} = \mathbf{S}(\omega_1)\delta(\omega_1 - \omega_2)d\omega_1 d\omega_2, \quad (11.46)$$

in which $\mathbf{S}(\omega)$ is a $k \times k$ Hermitian matrix, and $\delta(\cdot)$ is the Dirac delta function. If $\mathbf{A_F}(t,\omega) = \mathbf{A_F}$, a constant matrix, then $\mathbf{F}(t)$ is a stationary vector process.

The autocorrelation function of $\mathbf{F}(t)$ is defined by

$$\begin{aligned}
\mathbf{R_{FF}}(t,\tau) &= E\{\mathbf{F}^*(t)\mathbf{F}^\mathbf{T}(t+\tau)\} \\
&= E\left\{\int_{-\infty}^{\infty} \mathbf{A_F^*}(t,\omega_1)\exp(-i\omega_1 t)\mathbf{dZ}^*(\omega_1) \right. \\
&\quad \left. \cdot \int_{-\infty}^{\infty} \mathbf{A_F^T}(t+\tau,\omega_2)\exp(i\omega_2(t+\tau))\mathbf{dZ}^\mathbf{T}(\omega_2)\right\} \\
&= \int_{-\infty}^{\infty}\int_{-\infty}^{\infty} \mathbf{A_F^*}(t,\omega_1)\mathbf{A_F^T}(t+\tau,\omega_2)\exp(-i\omega_1 t)\exp(i\omega_2(t+\tau)) \\
&\quad \cdot E\left\{\mathbf{dZ}^*(\omega_1)\mathbf{dZ}^\mathbf{T}(\omega_2)\right\}.
\end{aligned}$$

Using Equation 11.46, we obtain

$$\mathbf{R_{FF}}(t,\tau) = \int_{-\infty}^{\infty} \mathbf{A_F^*}(t,\omega)\mathbf{S}(\omega)\mathbf{A_F^T}(t+\tau,\omega)\exp(i\omega\tau)d\omega.$$

For $\tau = 0$,

$$\mathbf{R_{FF}}(t,0) = \int_{-\infty}^{\infty} \mathbf{A_F^*}(t,\omega)\mathbf{S}(\omega)\mathbf{A_F^T}(t,\omega)d\omega.$$

This defines the $m \times m$ evolutionary power spectral density matrix of $\mathbf{F}(t)$ as

$$\mathbf{S_{FF}}(t,\omega) = \mathbf{A_F^*}(t,\omega)\mathbf{S}(\omega)\mathbf{A_F^T}(t,\omega).$$

$\mathbf{S}(\omega)$ can be seen to be the corresponding stationary power spectral density matrix.

11.6.2 Response Characterization

Suppose we go through the procedures of expanding the matrix governing Equation 11.44 in orthogonal modes, and assume that the damping is of a proportional nature,
$$\mathbf{X}(t) = \mathbf{U}\mathbf{Z}(t),$$
where $\mathbf{U}$ is the modal matrix,
$$\mathbf{U} = [\mathbf{u}_1 \; \mathbf{u}_2 \; \ldots \mathbf{u}_n],$$
and $\mathbf{Z}(t)$ is the vector of modal responses,
$$\mathbf{Z}(t) = \{Z_1(t), Z_2(t), \ldots, Z_n(t)\}^T,$$
governed by
$$\ddot{Z}_i + 2\zeta_i \omega_i \dot{Z}_i + \omega_i^2 Z_i = \mathbf{\Gamma}_i \mathbf{F}(t), \quad i = 1, 2, \ldots, n,$$
where $\mathbf{\Gamma}_i = U_i^T \mathbf{p} / (U_i^T \mathbf{m} U_i)$ is a length m row vector of modal participations factors.

This results in decoupled differential equations that govern the normalized modal response vector $\boldsymbol{\nu}_i(t)$,
$$\ddot{\boldsymbol{\nu}}_i + 2\zeta_i \omega_i \dot{\boldsymbol{\nu}}_i + \omega_i^2 \boldsymbol{\nu}_i = F_i(t), \quad i = 1, 2, \ldots, n, \qquad (11.47)$$
in which $\boldsymbol{\nu}_i(t) = \{\nu_{i1}(t), \nu_{i2}(t), \ldots, \nu_{im}(t)\}^T$ and $\nu_{ij}(t)$ can be interpreted as the response of a single degree of freedom oscillator of unit mass, natural frequency ω_i and damping ratio ζ_i to the forcing function $F_j(t)$. We see that $Z_i(t) = \mathbf{\Gamma}_i \boldsymbol{\nu}_i(t)$.

The solution to Equations 11.47 is the Duhamel integral,
$$\boldsymbol{\nu}_i(t) = \int_0^t g_i(t-\tau) F_i(\tau) d\tau, \quad i = 1, 2, \ldots, n, \qquad (11.48)$$
where the impulse response function is given by
$$g_i(t) = \frac{1}{\omega_d} \exp(-\zeta_i \omega_i t) \sin \omega_d t, \quad t \geq 0.$$

Substitute Equation 11.45 into Equation 11.48, and change the order of integration,
$$\boldsymbol{\nu}_i(t) = \int_{-\infty}^{\infty} \left[\int_0^t g_i(t-\tau) \mathbf{A_F}(t, \omega) \exp(-i\omega(t-\tau)) d\tau \right] \exp(i\omega t) \, \mathbf{dZ}(\omega).$$

Let
$$\mathbf{M}_i(\omega, t) = \int_0^t g_i(t-\tau) \mathbf{A_F}(t, \omega) \exp(-i\omega(t-\tau)) d\tau,$$

an $m \times k$ matrix of frequency-time modulating functions of $\boldsymbol{\nu}_i(t)$. It is useful to generalize this definition to the pth time derivative of the ith modal response,

$$\frac{\partial^p}{\partial t^p}\boldsymbol{\nu}_i(t) \equiv \boldsymbol{\nu}_i^{(p)}(t) = \frac{\partial^p}{\partial t^p}\left[\int_0^t g_i(t-\tau)F_i(\tau)d\tau\right]$$
$$= \int_{-\infty}^{\infty} \tilde{\mathbf{M}}_i^{(p)}(t,\omega)\exp(i\omega t)\,\mathbf{dZ}(\omega),$$

in which

$$\tilde{\mathbf{M}}_i^{(p)}(t,\omega) = \exp(-i\omega t)\frac{\partial^p}{\partial t^p}\left[\mathbf{M}_i(t,\omega)\exp(i\omega t)\right].$$

Using these relations, the $m \times m$ matrix of evolutionary cross power spectral density functions between derivatives of order p and q of the ith and jth normalized modal responses is

$$\mathbf{S}_{\boldsymbol{\nu}_i^{(p)}\boldsymbol{\nu}_j^{(q)}}(t,\omega) = \left[\tilde{\mathbf{M}}_i^{(p)}(t,\omega)\right]^*\mathbf{S}(\omega)\left[\tilde{\mathbf{M}}_i^{(q)}(t,\omega)\right]^T.$$

Then,

$$S_{\mathbf{Z}_i^{(p)}\mathbf{Z}_j^{(q)}}(t,\omega) = \boldsymbol{\Gamma}_i \mathbf{S}_{\boldsymbol{\nu}_i^{(p)}\boldsymbol{\nu}_j^{(q)}}(t,\omega)\boldsymbol{\Gamma}_j^T.$$

In terms of physical response,

$$R_{\mathbf{X}_i^{(p)}\mathbf{X}_j^{(q)}}(t,\tau) = \sum_{k=1}^{n}\sum_{l=1}^{n}U_{ik}U_{jl}\boldsymbol{\Gamma}_k\mathbf{R}_{\boldsymbol{\nu}_k^{(p)}\boldsymbol{\nu}_l^{(q)}}(t,\tau)\boldsymbol{\Gamma}_l^T$$
$$S_{\mathbf{X}_i^{(p)}\mathbf{X}_j^{(q)}}(t,\omega) = \sum_{k=1}^{n}\sum_{l=1}^{n}U_{ik}U_{jl}\boldsymbol{\Gamma}_k\mathbf{S}_{\boldsymbol{\nu}_k^{(p)}\boldsymbol{\nu}_l^{(q)}}(t,\omega)\boldsymbol{\Gamma}_l^T.$$

Similar expressions can be derived for the accelerations. These are useful for ground motion problems.

11.7 Nonstationary and Nonlinear Oscillator

Adding to the complexity of nonstationarity is the possibility that the system behaves nonlinearly.[7] In this section, we utilize the technique of equivalent linearization, introduced in Section 10.3. The general single-degree-of-freedom equation is

$$\ddot{X} + \beta\dot{X} + \omega_n^2 X + \varepsilon h\left(X,\dot{X},t\right) = F(t), \quad (11.49)$$

---

[7]This section is based on the paper, "Mean Square Response of a Duffing Oscillator to a Modulated White Noise Excitation by the Generalized Method of Equivalent Linearization," G. Ahmadi, *J. Sound and Vibration* (1980) **71**(1) 9-15.

where $F(t)$ is taken to be a nonstationary Gaussian force. The nonlinear differential equation is to be approximated by an equivalent linear equation,

$$\ddot{X} + \beta_e \dot{X} + \omega_e^2 X = F(t), \tag{11.50}$$

where β_e and ω_e are the equivalent damping and frequency. The error Δ introduced by replacing Equation 11.49 with Equation 11.50 is given by

$$\Delta = (\beta - \beta_e)\dot{X} + (\omega_n^2 - \omega_e^2)X + \varepsilon h\left(X, \dot{X}, t\right).$$

The mean-square error $E\{\Delta^2\}$ must be minimized, resulting in equations used to solve for β_e and ω_e,

$$-\frac{1}{2}\frac{\partial E\{\Delta^2\}}{\partial \omega_e^2} = (\beta - \beta_e)E\{X\dot{X}\} + (\omega_n^2 - \omega_e^2)E\{X^2\} + \varepsilon E\{Xh\}$$
$$= 0 \tag{11.51}$$

$$-\frac{1}{2}\frac{\partial E\{\Delta^2\}}{\partial \beta_e} = (\beta - \beta_e)E\{\dot{X}^2\} + (\omega_n^2 - \omega_e^2)E\{X\dot{X}\} + \varepsilon E\{\dot{X}h\}$$
$$= 0. \tag{11.52}$$

Solving Equations 11.51 and 11.52, the equivalent parameters are derived,

$$\omega_e^2 = \omega_n^2 + \varepsilon \frac{E\{Xh\}E\{\dot{X}^2\} - E\{\dot{X}h\}E\{X\dot{X}\}}{E\{X^2\}E\{\dot{X}^2\} - \left(E\{X\dot{X}\}\right)^2} \tag{11.53}$$

$$\beta_e = \beta + \varepsilon \frac{E\{\dot{X}h\}E\{X^2\} - E\{Xh\}E\{X\dot{X}\}}{E\{X^2\}E\{\dot{X}^2\} - \left(E\{X\dot{X}\}\right)^2}, \tag{11.54}$$

which is valid except when $E\{X^2\}E\{\dot{X}^2\} - \left(E\{X\dot{X}\}\right)^2 = 0$. Since $X(t)$ is in general a nonstationary process, $E\{X\dot{X}\} \neq 0$ and β_e and ω_e are time dependent.

At this point the system is assumed to be lightly damped and weakly nonlinear. Otherwise numerical approaches such as the Monte Carlo are better suited to such problems. Based on these assumptions, $\beta_e(t)$ and $\omega_e(t)$ become slowly varying functions of time. Also approximate the impulse response function by $g(t, \tau) = g(t - \tau)$. The approximate solution to Equation 11.50 is

$$X(t) \simeq \int_0^t g(t - \tau)F(\tau)d\tau, \tag{11.55}$$

where the impulse response function is given by

$$g(t) = \begin{cases} (1/\overline{\omega}_e)\exp(-\beta t/2)\sin\overline{\omega}_e t, & t \geq 0 \\ 0 & t < 0, \end{cases}$$

11.7. NONSTATIONARY AND NONLINEAR OSCILLATOR

and $\overline{\omega}_e^2 = \omega_e^2 - \beta_e^2/4$. The initial conditions are

$$X(0) = 0 \quad g(0) = 0$$
$$\dot{X}(0) = 0 \quad \dot{g}(0^+) = 1,$$

where

$$\dot{X}(t) \simeq \int_0^t \dot{g}(t-\tau)F(\tau)d\tau. \tag{11.56}$$

Equations 11.55 and 11.56 can be used to generate some of various moments needed to evaluate the equivalent damping and frequency expressions. These are

$$E\{X^2\} = \int_0^t d\tau_1 \int_0^t d\tau_2 g(t-\tau_1)g(t-\tau_2)E\{F(\tau_1)F(\tau_2)\}$$

$$E\{\dot{X}^2\} = \int_0^t d\tau_1 \int_0^t d\tau_2 \dot{g}(t-\tau_1)\dot{g}(t-\tau_2)E\{F(\tau_1)F(\tau_2)\}$$

$$E\{X\dot{X}\} = \int_0^t d\tau_1 \int_0^t d\tau_2 g(t-\tau_1)\dot{g}(t-\tau_2)E\{F(\tau_1)F(\tau_2)\}.$$

In addition, expressions for $E\{Xh\}$ and $E\{\dot{X}h\}$ are needed, and can be evaluated once h is defined for a specific problem.

This general theory is now applied to a Duffing oscillator.

11.7.1 The Nonstationary and Nonlinear Duffing

Consider the Duffing equation,

$$\ddot{X} + \beta\dot{X} + \omega_n^2 X + \varepsilon\omega_n^2 X^3 = F(t),$$

where $F(t) = e(t)n(t)$ is taken to be shaped white noise, with $e(t)$ being a slowly varying envelope function, $n(t)$ a white noise process with autocorrelation

$$E\{n(t_1)n(t_2)\} = 2\pi S_0 \delta(t_1 - t_2),$$

and S_0 is a constant power spectrum. The moments are given by

$$E\{X^2\} = 2\pi S_0 \int_0^t g^2(t-\tau)e^2(\tau)\,d\tau$$

$$E\{\dot{X}^2\} = 2\pi S_0 \int_0^t \dot{g}^2(t-\tau)e^2(\tau)\,d\tau$$

$$E\{X\dot{X}\} = 2\pi S_0 \int_0^t g(t-\tau)\dot{g}(t-\tau)e^2(\tau)\,d\tau.$$

Furthermore, knowing $h\left(X,\dot{X},t\right) = \omega_n^2 X^3$, the following cross moments are found,

$$\begin{aligned}E\{Xh\} &= \omega_n^2 E\{X^4\} &&= 3\omega_n^2 E^2\{X^2\}\\ E\{\dot{X}h\} &= \omega_n^2 E\{\dot{X}X^3\} &&= 3\omega_n^2 E\{\dot{X}X\}E\{X^2\},\end{aligned}$$

where it has been assumed that X and $\dot{X}$ are jointly Gaussian. Substitute these moments into Equations 11.53 and 11.54 to find

$$\begin{aligned}\omega_e^2 &= \omega_n^2\left(1+3\varepsilon E\{X^2\}\right)\\ \beta_e &= \beta,\end{aligned}$$

where $E\{X^2\}$ can be calculated once the envelope function $e(t)$ is defined or assumed.

Suppose

$$e(t) = \begin{cases} 1, & 0 < t \leq t_0 \\ 0, & \text{otherwise}, \end{cases}$$

and $\omega_e^2(0) = \omega_n^2$. Then, without providing the details here, the following expressions are derived,

$$\omega_e^2(t) = \omega_n^2\sqrt{1+\frac{6\varepsilon\alpha(t)}{\omega_n^2}}, \quad 0 < t \leq t_0,$$

$$E\{X^2\} = \left(\frac{1}{3\varepsilon}\right)\left[\sqrt{1+\frac{6\varepsilon\alpha(t)}{\omega_n^2}}-1\right], \quad 0 < t \leq t_0,$$

where

$$\alpha(t) = \left(\frac{\pi S_0}{\beta}\right)\left[1-\exp(-\beta t)\right].$$

Suppose that $S_0 = 1$, $\omega_n^2 = 4$ rad/s, $\varepsilon = 0.05$, $\zeta = 0.1$, $\beta = 2\zeta\omega_n = 0.4$, then

$$\alpha(t) = \left(\frac{\pi}{0.4}\right)\left[1-\exp(-0.4t)\right]$$

$$\omega_e^2(t) = 4\sqrt{1+\frac{6(0.05)\alpha(t)}{4}}$$

and

$$\begin{aligned}E\{X^2\} &= \left(\frac{1}{3(0.05)}\right)\left[\sqrt{1+\frac{6(0.05)\alpha(t)}{4}}-1\right]\\ &= \left(\frac{1}{3(0.05)}\right)\left[\sqrt{1+\frac{6(0.05)(\pi/0.4)\left[1-\exp(-0.4t)\right]}{4}}-1\right]\\ &= 6.67\left[\sqrt{1+0.59\left[1-\exp(-0.4t)\right]}-1\right].\end{aligned}$$

11.8. CONCLUDING SUMMARY

See Figure 11.5 for a plot of the evolution of the mean square value $E\{X^2(t)\}$.

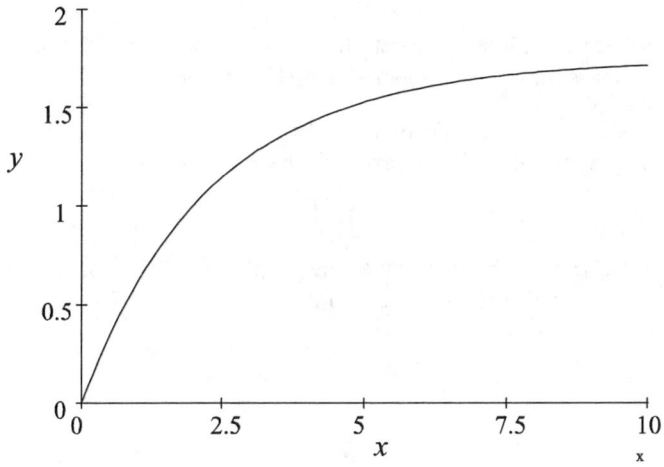

Figure 11.5: $E\{X^2\}$ for the case $S_0 = 1$, $\omega_n^2 = 4$ rad/s, $\varepsilon = 0.05$, $\zeta = 0.1$, $\beta = 2\zeta\omega_n = 0.4$.

In the limit for large t,

$$\lim_{t\to\infty} E\{X^2\} = \lim_{t\to\infty} 6.67(\sqrt{1 + 0.59(1 - \exp(-0.4t))} - 1)$$
$$= 1.7405.$$

Thus, transients in the mean-square response can be approximated for a relatively complicated system.

11.8 Concluding Summary

This chapter introduces some elementary methods for including nonstationary effects in a stochastic model. Various approaches of the evolutionary type are introduced and tied into one framework. It is noted that nonstationary solutions can be built upon stationary solutions.

11.9 Problems

1. Verify Equation 11.30.

2. What does Figure 11.2 demonstrate about the significance of ζ to the nonstationary response? Discuss with respect to Equation 11.14.

3. For Equation 11.43, assume $S_0 = 1$, $\omega_n = 0.5$, and plot as a function of ω.

4. Proceed through the derivation of the equations in Section 11.6.1 explicitly assuming a two-degree-of-freedom system.

5. Proceed through the derivation of the equations in Section 11.6.2 explicitly assuming a two-degree-of-freedom system.

6. Derive Equations 11.53 and 11.54.

7. Perform the necessary calculations to plot $E\{X^2(t)\}$ as in Figure 11.5 for the cases $\zeta = 0.15$ and $\zeta = 0.2$.

Chapter 12
The Monte Carlo Method

12.1 Introduction

The Monte Carlo method[1] is a general framework for probabilistic computational methods that are based on the repetitive sampling of distributions of random variables. The term Monte Carlo was first introduced by Metropolis and Ulam[2] during World War II. The method was named after the resort city on the Riviera where gambling is a major industry. Monte Carlo methods were used in the development of the atomic bomb, where they were applied to problems involving random neutron diffusion in fissile material.

The first documented use of the Monte Carlo method is Buffon's needle problem in 1777.[3] Suppose that a needle of length L is thrown on a board with vertical lines with uniform spacing $a > L$ as shown in Figure 12.1. What is the probability of the needle intersecting one of these lines?

In order to answer this question, let X be the random variable representing the distance between the midpoint of the needle and the nearest line, and Θ be the random angle between the needle and the line. X ranges from zero to $a/2$, and Θ ranges from zero to $\pi/2$. Physically, X and Θ are independent and distributed uniformly. That is,

$$f_{X\Theta}(x, \theta) = f_X(x) f_\Theta(\theta)$$

---

[1] See J.M. Hammersley and D.C. Handscomb, *Monte Carlo Methods*, John Wiley and Sons, 1964. It is one of the most readable introductions to Monte Carlo methods.

[2] N. Metropolis and S. Ulam, "The Monte Carlo Method," *Journal of the American Statistical Association* 44, No. 247 (1949): 335-41.

[3] M. Kalos and P. Whitlock, *Monte Carlo Methods*, Wiley-Interscience, 1986. See p. 4.

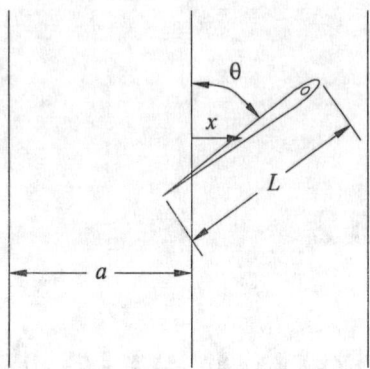

Figure 12.1: Buffon's needle problem. π can be estimated experimentally by this method.

and

$$f_X(x) = \frac{2}{a}, \quad 0 < x < \frac{a}{2}$$
$$f_\Theta(\theta) = \frac{2}{\pi}, \quad 0 < \theta < \frac{\pi}{2}.$$

The needle will cross one of these lines if $X < \frac{L}{2} \sin \Theta$. The probability that this happens is

$$p_o = \Pr\left(X < \frac{L}{2} \sin \Theta\right)$$
$$= \int_0^{\pi/2} \int_0^{\frac{L}{2}\sin\theta} f_{X\Theta}(x,\theta)\, dx\, d\theta = \frac{2L}{a\pi}.$$

The Monte Carlo method can be used to simulate any process influenced by random factors. However, it can also be used in nonprobabilistic problems such as this one. Laplace later suggested that this experiment can be used to estimate the value of π using the relation,

$$\pi = \frac{2L}{ap_o}. \tag{12.1}$$

The value of π does not have any stochastic content. However, we can still estimate π by throwing a needle on the board. This experiment can be repeated many times to produce the mean and variance of the estimate for π. The value of π obtained in this way has artificial stochastic properties. It should be noted that the needle must be thrown randomly without any

12.1. INTRODUCTION

preference given to any special location on the table. That is, the location of the needle is *uniformly* distributed.

Figure 12.2: Georges Buffon (1707-1788).

Buffon   At the age of 20 Georges Louis Leclerc Comte de Buffon discovered the binomial theorem. He corresponded with Cramer on mechanics, geometry, probability, number theory, and the differential and integral calculus. His first work, *Sur le Jeu de Franc-Carreau*, introduced differential and integral calculus into probability theory. He next wrote *Théorie de la Terre* and became the most important natural historian of his day having great influence across a wide scientific field. He is remembered most in mathematics for a probability experiment which he carried out calculating p by throwing sticks over his shoulder onto a tiled floor and counting the number of times the sticks fell across the lines between the tiles. This experiment caused much discussion among mathematicians which helped toward an understanding of probability.

Random Number Generation

To perform a simulation of the Buffon's needle problem to calculate the value of π, it is necessary to generate uniform random numbers for X and Θ and to test whether $x < \frac{L}{2}\sin\theta$ for each pair. Then calculate the probability p_o and estimate π using Equation 12.1. The value of π will be estimated numerically in Example 12.3.

It is essential to be able to generate the uniform random numbers. We can certainly generate them by mechanical means such as a roulette wheel. However, random numbers generated using physical devices often have statistical properties slightly different from the ideal, and the simulation process may be too slow, requiring a large amount of data for the statistics to converge.

Therefore, random numbers are often generated systematically from already developed programs. Such *random numbers* are called *pseudo-random* because they are attained via deterministic equations.

Consider first generating uniform random numbers. The task of generating satisfactory uniform pseudo-random numbers was not initially a trivial matter. The first algorithm, called the *midsquare method* for obtaining pseudo-random numbers, was proposed by Neyman in 1951.[4] However, this method turned out to be unsatisfactory because the random numbers did not have the desired statistics of a uniform distribution. The *linear congruential method* proposed by Lehmer in 1951[5] proved to be the best method for generating uniform random numbers. We will examine this method in Section 12.2.1.

In some cases, preference should be given to more probable values. That is, not all distributions are uniform. For example, if a magnet is placed underneath the table, Buffon's needle will tend to gather near the magnet. In that case, the probability density functions, $f_X(x)$ and $f_\Theta(\theta)$, will no longer be uniform. They should be modified to accommodate this change. Now, the question is how the random numbers for X and Θ should be selected so that they are consistent with the probability density functions. Random numbers with nonuniform distributions can be generated from uniform random numbers. We will learn how this can be done in Section 12.2.2.

Once we estimate the value of the simulated variable, for example π in Buffon's needle problem, it is necessary to know how precise is this value. The estimate is useless if we do not know its *quality*. Generally, the error becomes smaller (the variance of the estimate gets smaller) with an increasing number of experiments.

However, a large number of experiments can be computationally time-consuming and yield only an 'estimate' of the true solution. Therefore, the Monte Carlo method is viewed as a last resort. Cheap computational power and theoretical advances have made it possible for the Monte Carlo method to be recognized as a full-fledged numerical method for analyzing complex probabilistic problems.[6] Recent theoretical advances include error (variance) estimation using the central limit theorem. Error estimation

---

[4] See J. Von Neyman, "Various Techniques used in Connection with Random Digits," *National Bureau of Standards, Applied Mathematics Series*, (1951), **12**, 36-38.

The midsquare method can be also found in books by I.M. Sobol, *The Monte Carlo Method*, The University of Chicago Press, pp. 18-23, and I. Elishakoff, *Probabilistic Theory of Structures*, Dover Publications, pp. 436-438.

[5] See D.H. Lehmer, "Mathematical Methods in Large-Scale Computing Units," *Ann. Comp. Lab, Harvard University*, (1951), **26**, pp. 141-146.

[6] An advanced text on this subject is by R.Y. Rubinstein, *Simulation and the Monte Carlo Method*, John Wiley and Sons, 1981.

12.2. RANDOM-NUMBER GENERATION

enables us to find the number of simulation cycles needed for a desired level of error. This will be discussed in Section 12.4.

In summary, sets of random values for the simulated variable are required in the Monte Carlo process. Each generated number produces a corresponding result, until a family of results exists for the set of random input variables. Finally, all the results are averaged to obtain a mean value and higher moments, from which confidence bounds can be obtained. The flow chart in Figure 12.3 illustrates the procedure.

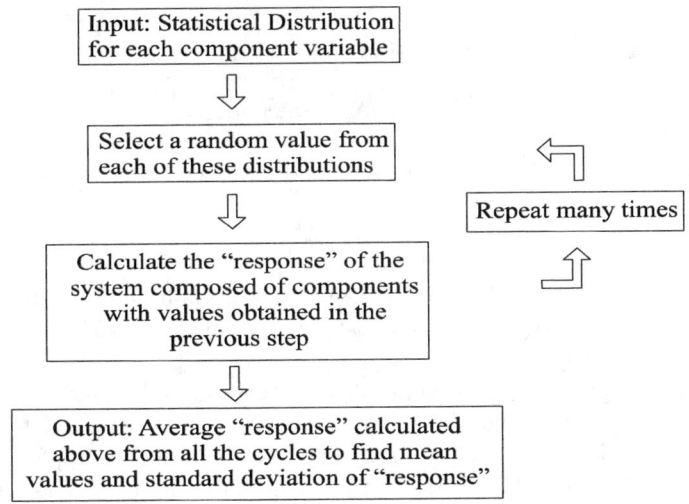

Figure 12.3: Flow chart of the Monte Carlo simulation procedure.

In Section 12.5, some practical examples are presented to illustrate how the Monte Carlo method is used to evaluate integrals and for solving nonlinear differential equations.

12.2 Random-Number Generation

12.2.1 Standard Uniform Random Numbers

In this section, a procedure is outlined on how *pseudo-random numbers* can be generated systematically with a standard uniform distribution. The most frequently used present-day method for generating standard uniform pseudo-random numbers is Lehmer's *linear congruential generator*. This generator is based on recursive calculations of the residues of modulus m of a linear transformation. Such a recursive relation may be expressed as

$$X_{i+1} = (aX_i + c) \,(\mathrm{mod}\, m),$$

where a, c, and m are nonnegative integers and mod is the *modulus*, defined such that X_{i+1} is the remainder when $aX_i + c$ is divided by m. The process is repeated to obtain the next value X_{i+2}. If k_i is the quotient of the division $(aX_i + c)/m$, that is,

$$k_i = \mathrm{Int}\left(\frac{aX_i + c}{m}\right),$$

then the corresponding residue of modulus m is

$$X_{i+1} = aX_i + c - mk_i. \tag{12.2}$$

Normalizing the values obtained from Equation 12.2 by the modulus m, we obtain

$$u_{i+1} = \frac{X_{i+1}}{m},$$

which constitute a set of random values on the unit interval $(0, 1)$ with the standard uniform probability distribution (mean = 0.5 and standard deviation = 0.289).[7]

Such a sequence of pseudo-random numbers is cyclic, and will repeat itself in at most m steps. To ensure randomness, the period of the cycle should be as long as possible, and therefore, in practical applications a large value of m should be assigned in the generation of u_i. The selection of values for m, a, and c is the most important step for creating a generator of this sort. Table 12.1 lists some choices for these constants that yield a large period. These values have been tested statistically and shown to give satisfactory results. The choice of which sets of constants to select from Table 12.1, as well as the size of the randomizing shuffle, is essentially arbitrary. However, the larger the shuffling array, the less likely it is that sequential correlation will occur. Ultimately this choice must be balanced with computational time and required storage space for the array.

---

[7]The probability density function of a standard uniform distribution is

$$f_X(x) = 1,\ 0 < x < 1.$$

The mean and the standard deviations are

$$\mu_X = \int_0^1 \hat{x} f_X(\hat{x})\, d\hat{x} = 0.5 \text{ and } \sigma_X = \sqrt{\int_0^1 (\hat{x} - \mu_X)^2 f_X(\hat{x})\, d\hat{x}} = 0.289,$$

where $\hat{x}$ is a dummy variable.

12.2. RANDOM-NUMBER GENERATION

Table 12.1: Random Number Generator

m	a	c
7875	421	1663
11979	859	2531
21870	1291	4621
81000	421	17117
86436	1093	18257
117128	1277	24749
121500	4081	25673
134456	8121	28411
243000	4561	51349
259200	7141	54773

Example 12.1 Standard Uniform Random Numbers

For $m = 7875$, $a = 421$, $c = 1663$, and $X_0 = 1000$, the next three pseudo-random numbers are $X_1 = 5288$, $X_2 = 7131$, $X_3 = 3409$. The corresponding normalized pseudo-random numbers are $u_0 = 0.1270$, $u_1 = 0.6715$, $u_2 = 0.9055$, $u_3 = 0.4329$.

✳

12.2.2 Generation of Nonuniform Random Variates

The next question that needs addressing is how to generate random numbers according to a particular probability density function. This generation of random variates can be accomplished systematically from the uniform distribution on the interval $(0, 1)$, determined in the preceding section. This is done through one of several methods, such as the *inverse transform* method, *composition* method, and *acceptance-rejection* method. These methods will be studied in this section.

Inverse Transform Method

Suppose that X is a random variable with the standard uniform distribution and Y is a random variable with the probability density function $f_Y(y)$ on (a, b) and cumulative distribution function $F_Y(y)$. Then what is the functional relationship between Y and X? Once we know the relationship, we can obtain realization y that corresponds to each uniform random number x.

Previously, in Chapter 4, we were interested in the probability density function $f_Y(y)$ when $f_X(x)$ and the functional relationship between X and

Y, were known: $Y = g(X)$. Here, we are looking for the functional relationship $Y = g(X)$ when both $f_Y(y)$ and $f_X(x)$ are known. Note that there are many functional relationships that correspond to a given set of $f_Y(y)$ and $f_X(x)$. Assume that Y is a monotonically increasing function of X so that we can write Equation 4.1 as

$$f_Y(y)\, dy = f_X(x)\, dx. \tag{12.3}$$

Integrating Equation 12.3, the cumulative distribution functions must equal each other:

$$F_Y(y) = F_X(x), \quad \text{if} \quad \frac{dy(x)}{dx} > 0, \tag{12.4}$$

where the standard uniform density of X is

$$\begin{aligned} f_X(x) &= 1, & 0 \le x < 1 \\ F_X(x) &= \int_0^x 1\, d\hat{x} = x, & 0 \le x < 1. \end{aligned} \tag{12.5}$$

Equation 12.4 can then be written as

$$F_Y(y) = x, \quad \text{if} \quad \frac{dy(x)}{dx} > 0. \tag{12.6}$$

To find y, $F_Y(y)$ in Equation 12.6 is inverted,

$$y = F_Y^{-1}(x). \tag{12.7}$$

The inverse transform can be described graphically as follows. Suppose that the probability density function is as shown in Figure 12.4(a). The corresponding cumulative distribution function is plotted in Figure 12.4(b). Uniformly distributed random numbers $x_1, \cdots, x_n$ between 0 and 1 are plotted on the vertical axis. The inversion process in Equation 12.7 constitutes finding the corresponding $y_1, \cdots, y_n$ for each x_i. Since F_Y and f_y are related by

$$f_Y(y) = \frac{dF_Y(y)}{dy},$$

$F_Y(y)$ has an inflection point (concavity changes from up to down) where $f_Y(y)$ reaches its maximum. Therefore, more points are sampled where $f_Y(y)$ is maximum.

Naturally, in order to apply the inverse transform method, the distribution function must exist in a form for which the corresponding inverse transform can be found analytically.

12.2. RANDOM-NUMBER GENERATION

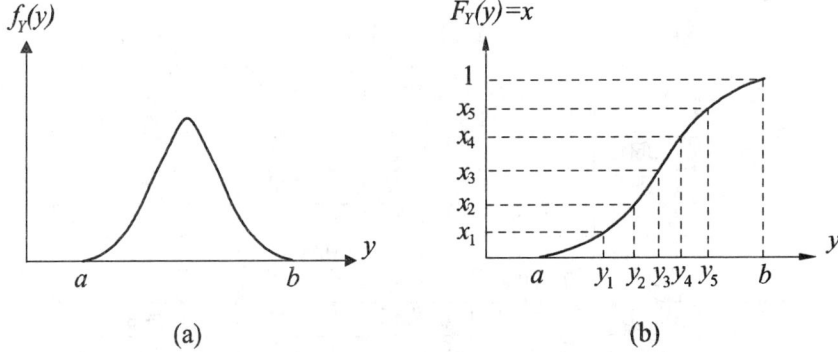

Figure 12.4: Schematic for the inverse transform method.

Example 12.2 Nonstandard Uniform Distribution

Consider a nonstandard uniform distribution on an arbitrary interval (a, b). The cumulative distribution function is

$$F_Y(y) = \begin{cases} 0, & y < a \\ (y-a)/(b-a), & a \leq y \leq b \\ 1, & y > b, \end{cases}$$

and the inverse function is found using Equation 12.5, setting $F_Y(y) = x$, and then

$$y = F_Y^{-1}(x) = a + (b-a)x, \qquad (12.8)$$

where x is uniformly distributed on $(0, 1)$.
✻

Example 12.3 Buffon's Needle Problem

Buffon's needle problem is revisited here. The distance between the midpoint of the needle and the nearest line, X, is uniformly distributed on $(0, a/2)$, and the smaller of the angles between the needle and the line, Θ, is uniformly distributed on $(0, \pi/2)$. We demonstrate how the value of π can be obtained. For numerical purposes, use $a = 1$ and $L = 0.5$.

Recall that the probability density functions are given by

$$\begin{aligned} f_X(x) &= \frac{2}{a}, & 0 < x < \frac{a}{2} \\ f_\Theta(\theta) &= \frac{2}{\pi}, & 0 < \theta < \frac{\pi}{2}. \end{aligned}$$

Table 12.2: Sample Results for Buffon's Needle Problem

Trial	x_1	$x = \frac{a}{2}x_1$	x_2	$\theta = \frac{\pi}{2}x_2$	Cross? $(x < \frac{L}{2}\sin\theta?)$
1	0.9501	0.4751	0.4447	0.6985	no
2	0.4860	0.2430	0.9218	1.4480	yes
3	0.4565	0.2282	0.4057	0.6373	no
4	0.2311	0.1156	0.6154	0.9667	yes
5	0.8913	0.4456	0.7382	1.1596	no
6	0.0185	0.0093	0.9355	1.4694	yes
7	0.6068	0.3034	0.7919	1.2440	no
8	0.7621	0.3810	0.1763	0.2769	no
9	0.8214	0.4107	0.9169	1.4403	no
10	0.6854	0.3427	0.0136	0.0214	no

First, generate two sets of standard uniform numbers, x_1 and x_2, using the *rand* command in MATLAB. Assuming that x and θ are increasing functions of x_1 and x_2, respectively,

$$x = \frac{a}{2}x_1 \quad \text{and} \quad \theta = \frac{\pi}{2}x_2.$$

Note that Equation 12.8 is used.

Table 12.2 shows sample results for ten trials. The needle crosses three times out of nine, and therefore, the probability that the needle will cross is estimated as 0.3. The estimate for π in Equation 12.1 ($\pi = 2L/p_o a$) is 3.333. This estimate will improve with an increasing number of trials. When 2500 samples are used, the estimate for π is 3.1211. Note that this value will vary slightly depending on the particular 2500 random numbers used.

Example 12.4 The Weibull Distribution

Consider the Weibull probability density function, used in reliability theory to predict life span,

$$f_T(t) = \alpha\beta t^{\beta-1} \exp\left[-\alpha t^\beta\right], \qquad 0 < t < \infty,$$

where $\alpha, \beta > 0$. Obtain the inverse of the Weibull probability density function.

12.2. RANDOM-NUMBER GENERATION

Solution The cumulative distribution is given by

$$F_T(t) = \int_0^t \alpha\beta \hat{t}^{\beta-1} \exp\left[-\alpha\hat{t}^\beta\right] d\hat{t}$$
$$= 1 - \exp\left[-\alpha t^\beta\right].$$

Setting the cumulative distribution equal to x, we obtain

$$t = \left[-\frac{1}{\alpha}\ln(1-x)\right]^{1/\beta}.$$

If x is uniform on $(0,1)$, then so is $(1-x)$. Therefore,

$$t = \left[-\frac{1}{\alpha}\ln x\right]^{1/\beta}, \qquad 0 < x < 1.$$

⊛

Example 12.5 Cauchy Density Function

Obtain the inverse of the Cauchy density function,

$$f_Y(y) = \frac{1}{\pi(1+y^2)}, \qquad -\infty < y < \infty.$$

Solution The cumulative distribution is given by

$$F_Y(y) = \int_0^y \frac{1}{\pi(1+\hat{y}^2)} d\hat{y}$$
$$= \frac{1}{\pi}\tan^{-1}(y).$$

Then,

$$y = \tan \pi x.$$

⊛

Example 12.6 Exponential Density I

Obtain the inverse of the probability density function,

$$f_Y(y) = \frac{1}{\lambda}e^{-y/\lambda}, \qquad 0 < y < \infty,$$

where λ is a constant.

Solution   The exponential density can be thought of as a special case of the Weibull density with $\alpha = 1/\lambda$ and $\beta = 1$. To find the expression for y, solve for $F_Y(y)$, set this expression equal to x, and then solve for y. The random variable y is given by

$$y = -\lambda \ln x, \qquad 0 < x < 1.$$

✳

Example 12.7 Exponential Density II

Obtain the inverse of the probability density function,

$$f_Y(y) = y \exp\left(-\frac{y^2}{2}\right), \qquad 0 < y < \infty.$$

Solution   The corresponding cumulative distribution is given by

$$F_Y(y) = \int_0^y \hat{y} \exp\left(-\frac{\hat{y}^2}{2}\right) d\hat{y} = 1 - \exp\left(-\frac{y^2}{2}\right), \qquad 0 < y < \infty.$$

Equating the cumulative distribution functions (Equation 12.6), we obtain

$$y = +\sqrt{-2\ln(1-x)}, \qquad 0 < x < 1.$$

Note that only the positive solution is taken since y is defined on $0 < y < \infty$. This expression can also be written as

$$y = \sqrt{-2\ln x}, \qquad 0 < x < 1,$$

since $1 - x$ is uniformly distributed if x is uniformly distributed. This result will be useful when generating the random variables of a normal density.

✳

The inverse transform method requires the analytical expression for the inverse of the cumulative density function. However, this may be difficult to obtain for some cases. For example, other techniques must be employed to generate the random variates in a normal density,

$$f_Y(y) = \frac{1}{\sigma\sqrt{2\pi}} \exp\left[-\frac{(y-\mu)^2}{2\sigma^2}\right], \qquad -\infty < y < \infty.$$

This probability density function $f_Y(y)$ is denoted by $R_N(\mu, \sigma)$, where μ is the mean value and σ is the standard deviation. There is no analytical expression for the inverse of the cumulative distribution function,

$$F_Y(y) = \frac{1}{\sigma\sqrt{2\pi}} \int_{-\infty}^y \exp\left[-\frac{(\hat{y}-\mu)^2}{2\sigma^2}\right] d\hat{y}, \qquad -\infty < y < \infty.$$

12.2. RANDOM-NUMBER GENERATION

It turns out to be easier to sample two independent normal random variables simultaneously using the *Box-Muller* method, as shown in the following example.

Example 12.8 Box-Muller Method for Standard Normal Density

Let Y_1 and Y_2 be independent standard normal random variables. The joint probability density function is given by

$$f_{Y_1 Y_2}(y_1, y_2) = \frac{1}{2\pi} \exp\left[-\frac{1}{2}(y_1^2 + y_2^2)\right], \quad -\infty < y_1 < \infty, \; -\infty < y_2 < \infty.$$

The cumulative distribution is given by

$$F_{Y_1 Y_2}(y_1, y_2) = \frac{1}{2\pi} \int_{-\infty}^{y_1} \int_{-\infty}^{y_2} \exp\left[-\frac{1}{2}(\hat{y}_1^2 + \hat{y}_2^2)\right] d\hat{y}_2 \, d\hat{y}_1,$$

for $-\infty < y_1 < \infty$ and $-\infty < y_2 < \infty$. The random variables Y_1 and Y_2 are related to R and Θ by the coordinate transformation,

$$Y_1 = R \cos \Theta, \quad Y_2 = R \sin \Theta, \tag{12.9}$$

or

$$Y_1^2 + Y_2^2 = R^2$$
$$\frac{Y_2}{Y_1} = \tan \Theta.$$

Using the probability equality,

$$\Pr\left((r < R \leq r + dr) \cap (\theta < \Theta \leq \theta + d\theta)\right)$$
$$= \Pr\left((y_1 < Y_1 \leq y_1 + dy_1) \cap (y_2 < Y_2 \leq y_2 + dy_2)\right),$$

or

$$F_{R\Theta}(r, \theta) = F_{Y_1 Y_2}(y_1, y_2),$$

we can write

$$F_{R\Theta}(r, \theta) = \frac{1}{2\pi} \int_{-\infty}^{y_1} \int_{-\infty}^{y_2} \exp\left[-\frac{1}{2}(\hat{y}_1^2 + \hat{y}_2^2)\right] d\hat{y}_2 \, d\hat{y}_1$$
$$= \frac{1}{2\pi} \int_{0}^{\theta} \int_{0}^{r} \exp\left[-\frac{1}{2}\hat{r}^2\right] J \, d\hat{r} \, d\hat{\theta},$$
$$0 < r < \infty, \quad 0 < \theta < 2\pi,$$

where $d\hat{y}_2 \, d\hat{y}_1$ is replaced by $J \, d\hat{r} \, d\hat{\theta}$ and J is the Jacobian, defined by

$$J = \left| \begin{array}{cc} \partial y_1 / \partial r & \partial y_1 / \partial \theta \\ \partial y_2 / \partial r & \partial y_2 / \partial \theta \end{array} \right|.$$

In this case, $J = r$.

The probability density function in terms of r and θ is given by

$$f_{R\Theta}(r,\theta) = \frac{1}{2\pi} r \exp\left[-\frac{1}{2}r^2\right], \qquad 0 < r < \infty, \ 0 < \theta < 2\pi.$$

Since R and Θ are independent, the probability density function $f_{R\Theta}(r,\theta)$ can be thought of as product of the density function of each variable,

$$f_R(r) = r \exp\left[-\frac{1}{2}r^2\right], \qquad 0 < r < \infty$$

$$f_\Theta(\theta) = \frac{1}{2\pi}, \qquad 0 < \theta < 2\pi.$$

The random variable R can be sampled using a uniform distribution of x_1 on $(0,1)$, as shown in Example 12.7,

$$r = \sqrt{-2\ln x_1},$$

and the random variable Θ can be sampled using a uniform distribution of x_2 on $(0,1)$, as shown in Example 12.2,

$$\theta = 2\pi x_2.$$

Using the transformation relations in Equation 12.9, we can write

$$y_1 = \sqrt{-2\ln x_1}\cos 2\pi x_2 \qquad (12.10)$$

$$y_2 = \sqrt{-2\ln x_1}\sin 2\pi x_2, \qquad (12.11)$$

where x_1 and x_2 are random numbers uniformly distributed on $(0,1)$.

⊛

Similarly, independent Gaussian random numbers with mean values μ_1 and μ_2 and standard deviations σ_1 and σ_2 can be generated using the relations,

$$y_1 = \sigma_1\sqrt{-2\ln x_1}\cos 2\pi x_2 + \mu_1 \qquad (12.12)$$

$$y_2 = \sigma_2\sqrt{-2\ln x_1}\sin 2\pi x_2 + \mu_2. \qquad (12.13)$$

Note that either Equation 12.12 or 12.13 can be used if only one set of random numbers is needed.

Another common density, the lognormal, can be derived from the normal density. The random variable y is said to have a lognormal density if $\ln y$ is normal with mean value μ and standard deviation σ. The probability density function is given by

$$f_Y(y) = \frac{1}{\sqrt{2\pi}\sigma y}\exp\left[-\frac{1}{2}\left(\frac{\ln y - \mu}{\sigma}\right)^2\right], \qquad y > 0.$$

12.2. RANDOM-NUMBER GENERATION

Two independent lognormal variables can be generated using the relations,

$$y_1 = \exp\left[\sigma_1\sqrt{-2\ln x_1}\cos 2\pi x_2 + \mu_1\right]$$
$$y_2 = \exp\left[\sigma_2\sqrt{-2\ln x_2}\cos 2\pi x_2 + \mu_2\right],$$

where μ_i and σ_i are the mean values and standard deviations of $\ln y_i$, respectively, and x_1 and x_2 are realizations of random variables with a uniform density on $(0,1)$.

Example 12.9 Generation of Normal Random Numbers

Here, we verify that the random numbers generated by Equation 12.12 are normally distributed with mean μ and standard deviation σ. Use $\mu = 5$ and $\sigma = 1$ to generate 4900 uniform random numbers x_1 and x_2.

The solid line in Figure 12.5 shows the normal distribution plotted using the exact expression,

$$f_Y(y) = \frac{1}{\sqrt{2\pi}\sigma}\exp\left[-\frac{1}{2}\left(\frac{y-\mu}{\sigma}\right)^2\right].$$

The histogram in the same figure shows the frequency of occurrence of the random numbers y_1 in Equation 12.12. The frequency of occurrence, n, is normalized by the product of the total number of random numbers N (4900 in this case) and the length of the interval $h = 0.5$ ($\sigma/2$ in this case).

For example, 974 out of 4900 random numbers appear in the range $5 < y < 5.5$. The normalized frequency is $974/(4900 \times 0.5) = 0.398$.

The frequency is normalized so that two plots can be compared more easily. Comparing the two plots, we see that they are of the same shape. Therefore, the random numbers are generated according to the desired probability density function. Note that the lognormal density can be produced using the same procedure.

Numerical Inverse Transformation

In most cases, it is not possible to obtain an analytical expression for the inverse. Therefore, the random numbers need to be generated numerically.

First, find the variates y of $f_Y(y)$ that correspond to N equally spaced numbers x on $(0,1)$. Mathematically, we are looking for y_i that satisfy

$$\int_{-\infty}^{y_i} f_Y(\tilde{y})\,d\tilde{y} = x_i \text{ or } F_Y(y_i) = x_i, \qquad \text{for } i = 0, 1, \cdots, N, \qquad (12.14)$$

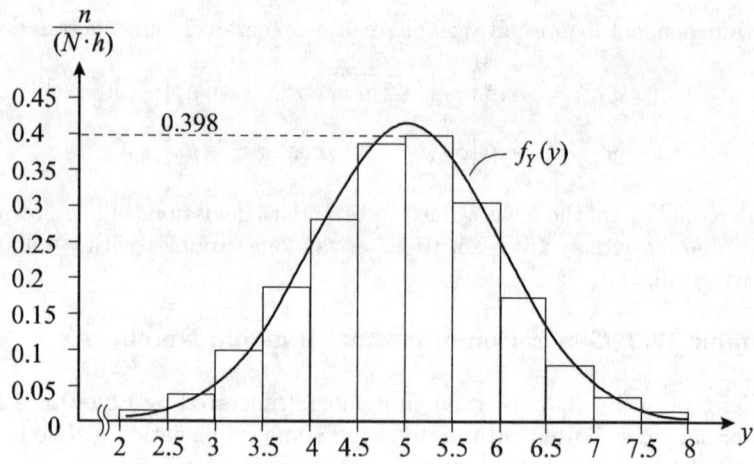

Figure 12.5: Normal distribution with $\mu = 5$ and $\sigma = 1$ using the exact expression and the Monte Carlo method.

where
$$x_i = \frac{i}{N}, \qquad \text{for } i = 0, 1, \cdots, N. \tag{12.15}$$

In order to find a y corresponding to an x that is not one of the sampled points, first find the interval that encloses x. Suppose x is between x_n and x_{n+1}, where $0 \leq n < N$. The random variable y is then linearly interpolated such that

$$\begin{aligned} y &= y_n + (y_{n+1} - y_n) \frac{x - x_n}{x_{n+1} - x_n} \\ &= y_n + (y_{n+1} - y_n)(Nx - n), \end{aligned} \tag{12.16}$$

using Equation 12.15. This procedure is demonstrated in the next example problem.

Example 12.10 Numerical Inverse Transformation of the Gamma Density

Generate random numbers according to the gamma density using the numerical inverse transformation method.[8]

---

[8] The gamma probability density function is often used in reliability theory. See Section 9.2.6 for details.

12.2. RANDOM-NUMBER GENERATION

Solution The random variable Y follows the gamma distribution when its probability density function is given by

$$f_Y(y) = \frac{\beta}{\Gamma(\alpha)} (\beta y)^{\alpha-1} e^{-y/\beta}, \qquad y > 0,$$

where α and β are positive constants, and $\Gamma(\alpha)$ is the gamma function, defined by

$$\Gamma(\alpha) = \int_0^\infty x^{\alpha-1} e^{-x} dx. \tag{12.17}$$

The gamma density is often used in reliability theory to predict the time to failure.[9] When α is an integer, the gamma function is given by

$$\Gamma(\alpha) = (\alpha-1)!, \qquad \alpha = 1, 2, 3, \cdots.$$

This can be shown by integrating Equation 12.17 by parts $n-1$ times. The cumulative distribution is given by[10]

$$F_Y(y) = 1 - \sum_{k=0}^{\alpha-1} \frac{e^{-\beta y}}{k!} (\beta y)^k, \qquad y > 0 \text{ and } \alpha = 1, 2, 3, \cdots.$$

Note that the gamma distribution becomes the exponential distribution when $\alpha = 1$.

The object is to find the random numbers according to the gamma density. Consider the case when $\alpha = 2$ and $\beta = 1$. The probability density and the cumulative distribution are given by

$$\begin{aligned} f_Y(y) &= y e^{-y}, & y > 0 \\ F_Y(y) &= 1 - e^{-y} - y e^{-y}, & y > 0. \end{aligned}$$

There is no analytical expression for the inverse cumulative distribution. Therefore, random numbers are derived numerically.

First choose twenty-one equally spaced numbers between 0 and 1 as in Equation 12.15. That is, $N = 20$. Then find y_i that correspond to each x_i for $i = 0, \cdots, 20$ using Equation 12.14. In this case,

$$1 - e^{-y_i} - y_i e^{-y_i} = x_i.$$

---

[9]For a short summary on the reliability theory, see Chapter 11, pp. 225-243, P.I. Meyer, *Introductory Probability and Statistical Applications*, Addison-Wesley Publishing Company, 1970. This chapter includes the normal, exponential, gamma, and Weibull failure laws.

[10]The cumulative distribution can be found easily by writing

$$F_Y(y) = 1 - \int_y^\infty f_Y(\hat{y}) d\hat{y}.$$

Table 12.3: y_i that Correspond to x_i

i	x_i	y_i	i	x_i	y_i
1	0.05	0.355	11	0.55	1.844
2	0.1	0.532	12	0.6	2.022
3	0.15	0.683	13	0.65	2.219
4	0.2	0.824	14	0.7	2.439
5	0.25	0.961	15	0.75	2.693
6	0.3	1.100	16	0.8	2.994
7	0.35	1.235	17	0.85	3.372
8	0.4	1.376	18	0.9	3.890
9	0.45	1.523	19	0.95	4.744
10	0.5	1.678	20	1.0	∞

Table 12.4: Gamma-Distributed Random Numbers with $\alpha = 2$, $\beta = 1$

x	0.4447	0.9218	0.4057	0.6154	0.7382
y	1.507	4.262	1.393	2.083	2.633
x	0.9355	0.7919	0.1763	0.9169	0.0136
y	4.496	2.906	0.767	4.179	0.0966

Each y_i is solved numerically using a root-finding program.[11] The corresponding y_i are given in Table 12.3.

The next step is to select uniform random numbers x on $(0, 1)$. Use the second set of uniform random numbers of Example 12.3 in Table 12.2. The random numbers y are obtained by linear interpolation Equation 12.16. For example, $x = 0.4447$ is between x_8 and x_9 so that $n = 8$. The corresponding y is then

$$y = y_8 + \frac{y_9 - y_8}{0.05} (0.4447 - 0.4).$$

This is tabulated in the first column of Table 12.4.

⊛

12.2.3 Composition Method

Sometimes, the probability density function to be sampled is a weighted sum of other density functions whose inverse cumulative distributions are known analytically. For example, consider the probability density function

---

[11] A root-finding algorithm such as the bisection method will be sufficient. Commercial programs such as MAPLE, MATHCAD, and MATLAB can also find the roots.

12.2. RANDOM-NUMBER GENERATION

given by
$$f_Y(y) = \frac{1}{4} + 3y^3, \qquad 0 < y < 1. \qquad (12.18)$$

The cumulative distribution is given by
$$F_Y(y) = \frac{1}{4}y + \frac{3}{4}y^4.$$

The inverse of this cumulative distribution is not trivial to obtain. However, the inverse of the each term, $y/4$ and $3y^4/4$, is easy to find. Equation 12.18 can be thought of as
$$f_Y(y) = \frac{1}{4}f_I(y) + \frac{3}{4}f_{II}(y),$$
where
$$f_I(y) = 1, \quad f_{II}(y) = 4y^3, \qquad 0 < y < 1,$$
with the known cumulative distributions,
$$F_I(y) = y, \quad F_{II}(y) = y^4, \qquad 0 < y < 1.$$

Then, the cumulative distribution of $f_Y(y)$ can be written as
$$F_Y(y) = \int_0^y \left[\frac{1}{4}f_I(\hat{y}) + \frac{3}{4}f_{II}(\hat{y})\right] d\hat{y}.$$

Note that the coefficients of f_I and f_{II} must add up to 1. *The cumulative distribution, $F_Y(y)$, can be thought of as a combination of $F_I(y)$ with 25% weight and $F_{II}(y)$ with 75% weight.*

This procedure involves generating two independent uniform random numbers, x_1 and x_2, on $(0,1)$. The uniform random number x_1 is used to determine which distribution should be used. For instance, $F_I(y)$ is used if $0 < x_1 < 1/4$, and $F_{II}(y)$ is used if $1/4 < x_1 < 1$.[12] Once we determine which distribution should be used, x_2 is used with this distribution (either $F_I(y)$ or $F_{II}(y)$) to find the corresponding y. Mathematically, we can write

$$y = F_I^{-1}(x_2) = x_2, \qquad \text{if } 0 \le x_1 < \frac{1}{4}$$
$$y = F_{II}^{-1}(x_2) = \sqrt[4]{x_2}, \qquad \text{if } \frac{1}{4} \le x_1 < \frac{3}{4}.$$

In more general terms, if the probability density function is given by
$$f_Y(y) = \sum_{i=1}^N \beta_i g_i(y), \qquad a < y < b,$$

---

[12] Because x_1 is a uniform random variable, the specific limits are unimportant as long as $F_I(y)$ is used 25% of the time and $F_{II}(y)$ is used 75% of the time.

and satisfies

$$\sum_{i=1}^{N} \beta_i = 1,$$

$$G_i(y) = \int_a^y g_i(\hat{y})\,d\hat{y} \text{ and } G_i(b) = 1,$$
$$\text{for } i = 1,\cdots,N,$$

then

$$y = G_1^{-1}(x_2) \qquad 0 \le x_1 < \beta_1$$
$$y = G_2^{-1}(x_2), \qquad \beta_1 \le x_1 < \beta_1 + \beta_2$$
$$\vdots$$
$$y = G_N^{-1}(x_2), \qquad \beta_1 + \cdots + \beta_{N-1} \le x_1 < 1.$$

Example 12.11 Composition Method

Find the random numbers governed by probability density function,

$$f_Y(y) = y^2 + \frac{\pi}{3}\sin\frac{\pi}{2}y, \qquad 0 < y < 1. \tag{12.19}$$

Solution Equation 12.19 can be decomposed into

$$f_Y(y) = \frac{1}{3}\cdot 3y^2 + \frac{2}{3}\cdot\frac{\pi}{2}\sin\frac{\pi}{2}y,$$

so that

$$\beta_1 = \frac{1}{3}, \qquad \beta_2 = \frac{2}{3}$$
$$g_1(y) = 3y^2, \qquad g_2(y) = \frac{\pi}{2}\sin\frac{\pi}{2}y, \qquad 0 < y < 1.$$

The cumulative distributions of g_1 and g_2 are given by

$$G_1(y) = y^3, \qquad G_2(y) = 1 - \cos\frac{\pi}{2}y, \qquad 0 < y < 1.$$

The random variate y is given by

$$y = \sqrt[3]{x_2}, \qquad 0 \le x_1 < \frac{1}{3}$$
$$y = \frac{2}{\pi}\cos^{-1}(1 - x_2), \qquad \frac{1}{3} \le x_1 < 1,$$

where x_1 and x_2 are independent uniform random numbers on $(0,1)$. Table 12.5 shows the sample results when 9 uniform random numbers x_1 and x_2 are used.

⊛

12.2. RANDOM-NUMBER GENERATION

Table 12.5: Sample Results for the Composite Method

x_1	x_2	y
0.9501	0.4447	0.6252
0.4860	0.9218	0.8506
0.4565	0.4057	0.8665
0.2311	0.6154	0.9502
0.8913	0.7382	0.8314
0.0185	0.9355	0.3838
0.6068	0.7919	0.5949
0.7621	0.1763	0.9780
0.8214	0.9169	0.9470

12.2.4 Von Neumann's Rejection-Acceptance Method

The *rejection-acceptance* method is a general technique that generates random numbers with an arbitrary probability density function. A random number is proposed, and is accepted or rejected depending on the probability density at that selected random number. The method can be very slow since not all the proposed random numbers are accepted. This method can be thought of as a generalized composition method.

Suppose y is a random variable with an arbitrary probability density function $f_Y(y)$ on (a,b). Select a uniform random number y_1 on (a,b). This can be accomplished by generating standard uniform random numbers x_1 on $(0,1)$ and transforming to y_1 using Equation 12.8,

$$y_1 = a + (b-a)x_1.$$

We are more likely to accept this number, y_1, if $f_Y(y_1)$ is large and less likely if $f_Y(y_1)$ is small. In order to determine whether y_1 should be accepted or rejected, we generate another standard uniform random number, x_2. If x_2 is less than $f_Y(y_1)/\max(f_Y)$, we accept y_1. If $f_Y(y_1)/\max(f_Y)$ is 0.3, we have a 30% of chance of accepting y_1. $\max(f_Y)$ is the largest ordinate of $f_Y(y)$.

Note that the probability density function $f_Y(y)$ does not need to be normalized because only the ratio $f_Y(y_1)/\max(f_Y)$ is used. Let us demonstrate the rejection-acceptance method with an example.

Example 12.12 Rejection-Acceptance Method

Let Y be a random variable with the probability density function,

$$f_Y(y) = 3.661 e^{-10y^2}, \qquad 0 < y < 0.5.$$

Find the corresponding random numbers.

Solution Figure 12.6 is a graph of the probability density function. The random numbers y that correspond to $f_Y(y)$ are generated using the rejection-acceptance method. Note that there is no analytical expression for the cumulative distribution. Therefore, it is not possible to invert the cumulative distribution analytically. However, it is possible to do so numerically following the procedure of Section 12.2.2, but it may not be trivial to obtain y_i in Equation 12.14 even numerically.

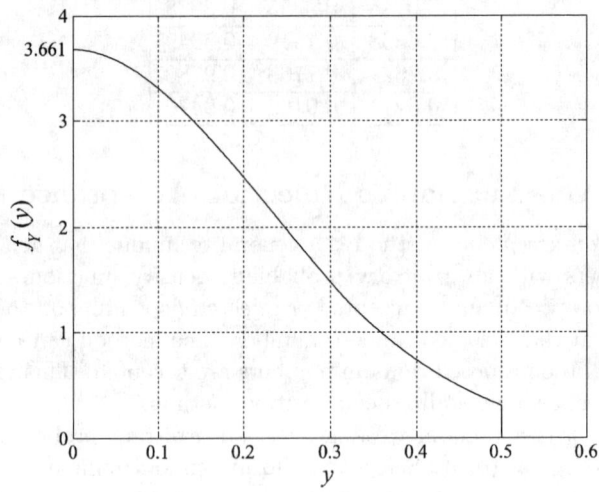

Figure 12.6: $f_Y(y) = 3.661 e^{-10y^2}$.

In order to use the rejection-acceptance method, two sets of uniform random numbers, x_1 and x_2, on $(0,1)$, are first generated. The uniform random numbers generated in Table 12.5 are used. The first set of random numbers x_1 is used to find uniform random numbers y_1 on $(0, 0.5)$ using the relation,

$$y_1 = 0.5 x_1.$$

The second set of standard uniform random numbers x_2 is then used to test whether $f_Y(y_1) / \max(f_Y)$ is greater than x_2. If $f_Y(y_1) / \max(f_Y)$ is greater than x_2, then y_1 is accepted. The results are shown in Table 12.6. The accepted values of random numbers y are

$$0.2282, \ 0.1156, \ 0.0093, \ 0.3810.$$

In practice, many more than nine trials should be used in order to obtain random numbers y with satisfactory statistical properties. Note that only

12.2. RANDOM-NUMBER GENERATION

Table 12.6: Sample Results for the Rejection-Acceptance Method

Trial	x_1	$y_1 = 0.5x_1$	x_2	$f_Y(y_1)/\text{Max}(f_Y)$	y_1 accepted?
1	0.9501	0.4751	0.4447	0.1047	no
2	0.4860	0.2430	0.9218	0.5541	no
3	0.4565	**0.2282**	0.4057	0.5940	yes
4	0.2311	**0.1156**	0.6154	0.8750	yes
5	0.8913	0.4456	0.7382	0.1372	no
6	0.0185	**0.0093**	0.9355	0.9991	yes
7	0.6068	0.3034	0.7919	0.3983	no
8	0.7621	**0.3810**	0.1763	0.2341	yes
9	0.8214	0.4107	0.9169	0.1851	no

four random numbers y were accepted out of nine, demonstrating the slow nature of this method.

von Neumann   John von Neumann was born János von Neumann. He was called Jancsi as a child, a diminutive form of János, then later he was called Johnny in the United States. His father, Max Neumann, was a top banker and he was brought up in a extended family, living in Budapest where as a child he learned languages from the German and French governesses that were employed. Although the family were Jewish, Max Neumann did not observe the strict practices of that religion and the household seemed to mix Jewish and Christian traditions.

It is also worth explaining how Max Neumann's son acquired the "von" to become János von Neumann. In 1913 Max Neumann purchased a title but did not change his name. His son, however, used the German form von Neumann where the "von" indicated the title.

As a child von Neumann showed he had an incredible memory. Poundstone writes: *At the age of six, he was able to exchange jokes with his father in classical Greek. The Neumann family sometimes entertained guests with demonstrations of Johnny's ability to memorize phone books. A guest would select a page and column of the phone book at random. Young Johnny read the column over a few times, then handed the book back to the guest. He could answer any question put to him (who has number such and such?) or recite names, addresses, and numbers in order.*

In 1911 von Neumann entered the Lutheran Gymnasium. The school had a strong academic tradition that seemed to count for more than the religious affiliation both in the Neumann's eyes and in those of the school. His mathematics teacher quickly recognized von Neumann's genius and special

CHAPTER 12. THE MONTE CARLO METHOD

Figure 12.7: John von Neumann (1903-1957).

tuition was put on for him. The school had another outstanding mathematician one year ahead of von Neumann, namely Eugene Wigner.

World War I had relatively little effect on von Neumann's education but, after the war ended, Béla Kun controlled Hungary for five months in 1919 with a Communist government. The Neumann family fled to Austria as the affluent came under attack. However, after a month, they returned to face the problems of Budapest. When Kun's government failed, the fact that it had been largely composed of Jews meant that Jewish people were blamed. Such situations are devoid of logic and the fact that the Neumann's were opposed to Kun's government did not save them from persecution.

In 1921 von Neumann completed his education at the Lutheran Gymnasium. His first mathematics paper, written jointly with Fekete, the assistant at the University of Budapest who had been tutoring him, was published in 1922. However Max Neumann did not want his son to take up a subject that would not bring him wealth. Max Neumann asked Theodore von Kármán to speak to his son and persuade him to follow a career in business. Perhaps von Kármán was the wrong person to ask to undertake such a task but in the end all agreed on the compromise subject of chemistry for von Neumann's university studies.

Hungary was not an easy country for those of Jewish descent for many reasons and there was a strict limit on the number of Jewish students who could enter the University of Budapest. Of course, even with a strict quota, von Neumann's record easily won him a place to study mathematics in 1921, but he did not attend lectures. Instead he also entered the University of

12.2. RANDOM-NUMBER GENERATION 655

Berlin in 1921 to study chemistry.

Von Neumann studied chemistry at the University of Berlin until 1923 when he went to Zurich. He achieved outstanding results in the mathematics examinations at the University of Budapest despite not attending any courses. Von Neumann received his diploma in chemical engineering from the Technische Hochschule in Zürich in 1926. While in Zurich he continued his interest in mathematics, despite studying chemistry, and interacted with Weyl and Pólya who were both at Zurich. He even took over one of Weyl's courses when he was absent from Zurich for a time. Pólya said: *Johnny was the only student I was ever afraid of. If in the course of a lecture I stated an unsolved problem, the chances were he'd come to me as soon as the lecture was over, with the complete solution in a few scribbles on a slip of paper.*

Von Neumann received his doctorate in mathematics from the University of Budapest, also in 1926, with a thesis on set theory. He published a definition of ordinal numbers when he was 20; the definition is the one used today.

Von Neumann lectured at Berlin from 1926 to 1929 and at Hamburg from 1929 to 1930. However he also held a Rockefeller Fellowship to enable him to undertake postdoctoral studies at the University of Göttingen. He studied under Hilbert at Göttingen during 1926-1927. By this time von Neumann had achieved celebrity status: By his mid-twenties, von Neumann's fame had spread worldwide in the mathematical community. At academic conferences, he would find himself pointed out as a young genius.

Veblen invited von Neumann to Princeton to lecture on quantum theory in 1929. Replying to Veblen that he would come after attending to some personal matters, von Neumann went to Budapest where he married his fiancée Marietta Kovesi before setting out for the United States. In 1930 von Neumann became a visiting lecturer at Princeton University, being appointed professor there in 1931.

Between 1930 and 1933 von Neumann taught at Princeton but this was not one of his strong points. His fluid line of thought was difficult for those less gifted to follow. He was notorious for dashing out equations on a small portion of the available blackboard and erasing expressions before students could copy them.

In contrast, however, he had an ability to explain complicated ideas in physics. For a man to whom complicated mathematics presented no difficulty, he could explain his conclusions to the uninitiated with amazing lucidity. After a talk with him one always came away with a feeling that the problem was really simple and transparent.

He became one of the original six mathematics professors (J.W. Alexander, A. Einstein, M. Morse, O. Veblen, J. von Neumann and H. Weyl) in

1933 at the newly founded Institute for Advanced Study in Princeton, a position he kept for the remainder of his life.

During the first years that he was in the United States, von Neumann continued to return to Europe during the summers. Until 1933 he still held academic posts in Germany but resigned these when the Nazis came to power. Unlike many others, von Neumann was not a political refugee but rather he went to the United States mainly because he thought that the prospect of academic positions there was better than in Germany.

In 1933 von Neumann became coeditor of the *Annals of Mathematics* and, two years later, he became coeditor of *Composition Mathematica*. He held both these editorships until his death.

Von Neumann and Marietta had a daughter Marina in 1936 but their marriage ended in divorce in 1937. The following year he married Klára Dán, also from Budapest, whom he met on one of his European visits. After marrying, they sailed to the United States and made their home in Princeton. There von Neumann lived a rather unusual lifestyle for a top mathematician. He had always enjoyed parties. Parties and nightlife held a special appeal for von Neumann. While teaching in Germany, von Neumann had been a denizen of the Cabaret-era Berlin nightlife circuit.

Now married to Klára the parties continued. The parties at the von Neumann's house were frequent, and famous, and long.

Ulam summarizes von Neumann's work. He writes: *In his youthful work, he was concerned not only with mathematical logic and the axiomatics of set theory, but, simultaneously, with the substance of set theory itself, obtaining interesting results in measure theory and the theory of real variables. It was in this period also that he began his classical work on quantum theory, the mathematical foundation of the theory of measurement in quantum theory and the new statistical mechanics.*

His text, *Mathematische Grundlagen der Quantenmechanik* (1932), built a solid framework for the new quantum mechanics. Van Hove writes: *Quantum mechanics was very fortunate indeed to attract, in the very first years after its discovery in 1925, the interest of a mathematical genius of von Neumann's stature. As a result, the mathematical framework of the theory was developed and the formal aspects of its entirely novel rules of interpretation were analyzed by one single man in two years (1927-1929).*

Self-adjoint algebras of bounded linear operators on a Hilbert space, closed in the weak operator topology, were introduced in 1929 by von Neumann in a paper in *Mathematische Annalen* . Kadison explains: *His interest in ergodic theory, group representations, and quantum mechanics contributed significantly to von Neumann's realization that a theory of operator algebras was the next important stage in the development of this area of mathematics.*

12.2. RANDOM-NUMBER GENERATION

Such operator algebras were called "rings of operators" by von Neumann and later they were called W*-algebras by some other mathematicians. J. Dixmier, in 1957, called them "von Neumann algebras" in his monograph, *Algebras of Operators in Hilbert Space,* (von Neumann algebras). In the second half of the 1930s and the early 1940s von Neumann, working with his collaborator F.J. Murray, laid the foundations for the study of von Neumann algebras in a fundamental series of papers.

However, von Neumann is known for a wide variety of different scientific studies. Ulam explains how he was led towards game theory: *Von Neumann's awareness of results obtained by other mathematicians and the inherent possibilities which they offer is astonishing. Early in his work, a paper by Borel on the minimax property led him to develop ... ideas which culminated later in one of his most original creations, the theory of games.*

In game theory von Neumann proved the minimax theorem. He gradually expanded his work in game theory, and with coauthor Oskar Morgenstern, he wrote the classic text *Theory of Games and Economic Behavior* (1944). Ulam continues: *An idea of Koopman on the possibilities of treating problems of classical mechanics by means of operators on a function space stimulated him to give the first mathematically rigorous proof of an ergodic theorem. Haar's construction of measure in groups provided the inspiration for his wonderful partial solution of Hilbert's fifth problem, in which he proved the possibility of introducing analytical parameters in compact groups.*

In 1938 the American Mathematical Society awarded the Bôcher Prize to John von Neumann for his memoir, *Almost Periodic Functions and Groups*. This was published in two parts in the *Transactions of the American Mathematical Society*, the first part in 1934 and the second part in the following year.

Around this time von Neumann turned to applied mathematics. In the middle 1930s, Johnny was fascinated by the problem of hydrodynamic turbulence. It was then that he became aware of the mysteries underlying the subject of nonlinear partial differential equations. His work, from the beginnings of the Second World War, concerns a study of the equations of hydrodynamics and the theory of shocks. The phenomena described by these nonlinear equations are baffling analytically and defy even qualitative insight by present methods. Numerical work seemed to him the most promising way to obtain a feeling for the behavior of such systems. This impelled him to study new possibilities of computation on electronic machines.

Von Neumann was one of the pioneers of computer science, making significant contributions to the development of logical design. Shannon writes: *Von Neumann spent a considerable part of the last few years of his* life working in [automata theory]. It represented for him a synthesis of his early interest in logic and proof theory and his later work, during World War II and after, on large scale electronic computers. Involving a mixture of pure and applied mathematics as well as other sciences, automata theory was an ideal field for von Neumann's wide-ranging intellect. He brought to it many new insights and opened up at least two new directions of research.

He advanced the theory of cellular automata, advocated the adoption of the bit as a measurement of computer memory, and solved problems in obtaining reliable answers from unreliable computer components.

During and after World War II, von Neumann served as a consultant to the armed forces. His valuable contributions included a proposal of the implosion method for bringing nuclear fuel to explosion and his participation in the development of the hydrogen bomb. From 1940 he was a member of the Scientific Advisory Committee at the Ballistic Research Laboratories at the Aberdeen Proving Ground in Maryland. He was a member of the Navy Bureau of Ordnance from 1941 to 1955, and a consultant to the Los Alamos Scientific Laboratory from 1943 to 1955. From 1950 to 1955 he was a member of the Armed Forces Special Weapons Project in Washington, D.C. In 1955 President Eisenhower appointed him to the Atomic Energy Commission, and in 1956 he received its Enrico Fermi Award, knowing that he was incurably ill with cancer.

Eugene Wigner wrote of von Neumann's death: *When von Neumann realized he was incurably ill, his logic forced him to realize that he would cease to exist, and hence cease to have thoughts .... It was heartbreaking to watch the frustration of his mind, when all hope was gone, in its struggle with the fate which appeared to him unavoidable but unacceptable .... his mind, the amulet on which he had always been able to rely, was becoming less dependable. Then came complete psychological breakdown; panic, screams of uncontrollable terror every night.*

His friend Edward Teller said, "I think that von Neumann suffered more when his mind would no longer function, than I have ever seen any human being suffer."

Von Neumann's sense of invulnerability, or simply the desire to live, was struggling with unalterable facts. He seemed to have a great fear of death until the last. No achievements and no amount of influence could save him now, as they always had in the past. Johnny von Neumann, who knew how to live so fully, did not know how to die.

It would be almost impossible to give even an idea of the range of honors that were given to von Neumann. He was Colloquium Lecturer of the American Mathematical Society in 1937 and received the its Bôcher Prize as mentioned above. He held the Gibbs Lectureship of the American Mathematical Society in 1947 and was President of the Society in 1951-

1953.

He was elected to many academies including the Academia Nacional de Ciencias Exactas (Lima, Peru), Academia Nazionale dei Lincei (Rome, Italy), American Academy of Arts and Sciences (U.S.), American Philosophical Society (U.S.), Instituto Lombardo di Scienze e Lettere (Milan, Italy), National Academy of Sciences (U.S.) and Royal Netherlands Academy of Sciences and Letters (Amsterdam, The Netherlands).

Von Neumann received two Presidential Awards, the Medal for Merit in 1947 and the Medal for Freedom in 1956. Also in 1956 he received the Albert Einstein Commemorative Award and the Enrico Fermi Award mentioned above. Peierls writes: *He was the antithesis of the "long-haired" mathematics don. Always well groomed, he had as lively views on international politics and practical affairs as on mathematical problems.*

12.3 Random Numbers for Joint Probability Densities

So far, we have discussed the generalization of random numbers for models with one random variable. What happens if the outcome is controlled by more than one variable? How would those variables be sampled?

Let Y_1 and Y_2 be the random variables of a joint probability density function $f_{Y_1 Y_2}(y_1, y_2)$. The problem has been briefly dealt with when normal random numbers were obtained in Example 12.8. y_1 and y_2 were 'transformed' into r and θ, variables that are "independent" of each other, so that the joint probability density function can be written as a product of probability density functions of each variable,

$$f_{Y_1 Y_2}(y_1, y_2) \longrightarrow f_R(r) f_\Theta(\theta),$$

with known cumulative distribution functions, $F_R(r)$ and $F_\Theta(\theta)$. Then, r and θ are generated independently.

The most important step of this Box-Muller method is to write the joint probability density function as a product of probability density functions with known inverse cumulative distribution functions. In the following section, we consider a general case when the random variables of the joint probability function are not independent.

12.3.1 Inverse Transform Method

Generally, a joint probability density function can be written as[13]

$$f_{Y_1 Y_2}(y_1 y_2) = f(y_1) f(y_2|y_1).$$

The random variables Y_1 and $Y_2|Y_1$ are independent of each other. Note that Y_1 and Y_2 do not have to be independent of each other. The cumulative distributions are set equal to two independent standard uniform random numbers, x_1 and x_2,

$$F(y_1) = x_1$$
$$F(y_2|y_1) = x_2.$$

Then, the corresponding y_1 and y_2 can be generated.

An example is the joint probability density function,

$$f_{Y_1 Y_2}(y_1, y_2) = y_1 + y_2, \qquad 0 < y_1, y_2 < 1.$$

The marginal probability density function $f_{Y_1}(y_1)$ and the conditional probability density function $f_{Y_2|Y_1}(y_2|y_1)$ are given by

$$f_{Y_1}(y_1) = \int_0^1 f_{Y_1 Y_2}(y_1, y_2)\, dy_2 = y_1 + \frac{1}{2}, \qquad 0 < y_1 < 1$$

$$f_{Y_2|Y_1}(y_2|y_1) = \frac{f_{Y_1 Y_2}(y_1, y_2)}{f_{Y_1}(y_1)} = \frac{2(y_1 + y_2)}{2y_1 + 1}, \qquad 0 < y_1, y_2 < 1.$$

The cumulative distributions are given by

$$F_{Y_1}(y_1) = \int_0^{y_1} f_{Y_1}(\tilde{y}_1)\, d\tilde{y}_1 = \frac{1}{2}y_1^2 + \frac{1}{2}y_1, \qquad 0 < y_1 < 1$$

$$F_{Y_2|Y_1}(y_2|y_1) = \int_0^{y_2} f_{Y_2|Y_1}(\tilde{y}_2|y_1)\, d\tilde{y}_2$$

$$= \frac{2 y_1 y_2 + y_2^2}{2y_1 + 1}, \qquad 0 < y_1, y_2 < 1.$$

Equate the cumulative distribution functions to the independent random numbers x_1 and x_2 with the standard uniform distribution,

$$\frac{1}{2}y_1^2 + \frac{1}{2}y_1 = x_1$$

$$\frac{2 y_1 y_2 + y_2^2}{2y_1 + 1} = x_2.$$

---

[13] The reader is encouraged to review the concepts related to the conditional probability density in Section 2.9.

12.3. JOINT RANDOM NUMBERS

This system of equations can be solved for the random numbers y_1 and y_2, where y_1 must be obtained first in order to obtain y_2. This method is efficient when the inverse cumulative distributions are known analytically.

We next extend this method to problems with a joint probability density function of more than two random variables. The joint probability density $f(y_1, \cdots, y_n)$ can be written as

$$f(y_1, \cdots, y_n) = f(y_1) f(y_2|y_1) f(y_3|y_2, y_1) \cdots f(y_n|y_{n-1}, \cdots, y_1).$$

The random variables are obtained from the set of equations,

$$\begin{aligned} F(y_1) &= x_1 \\ F(y_2|y_1) &= x_2 \\ &\vdots \\ F(y_n|y_{n-1}, \cdots, y_1) &= x_n, \end{aligned} \qquad (12.20)$$

where $x_1, \cdots, x_n$ are independent random numbers with the standard uniform distribution. Equations 12.20 are solved for $y_1, \cdots, y_n$.

12.3.2 Linear Transform Method

Let $Y_1, Y_2, \cdots, Y_n$ be random variables with known probability distributions. One way to generate the random numbers[14] is to use the method described in the preceding section. This procedure can be time consuming because it involves obtaining each conditional probability density function, inverting it, and solving for each random variable. If there are no analytical expressions for the inverse distributions, this method could not be used.

The linear transform method enables us to find the random numbers $y_1, \cdots, y_n$ of known distributions from *independent* random numbers of the same respective probability density functions.

For argument's sake, let $Y_1, \cdots, Y_n$ be random variables of a normal distribution such that

$$\begin{aligned} E\{Y_i\} &= \mu_i, \\ Cov(Y_i Y_j) &= \sigma_{ij}^2, \qquad \text{for } i, j = 1, \cdots, n, \end{aligned} \qquad (12.21)$$

where μ_i and σ_{ij} are known. Let $Z_1, \cdots, Z_n$ be random variables of an independent standard normal distribution with

$$\begin{aligned} E\{Z_i\} &= 0, \\ Cov(Z_i Z_j) &= \delta_{ij}, \qquad \text{for } i, j = 1, \cdots, n. \end{aligned} \qquad (12.22)$$

---

[14] This section is based on Section 11.5 of I. Elishakoff, *Probabilistic Theory of Structures*, Dover Publications, 1999.

The n sets of independent random numbers with the standard normal distribution can be generated from the uniform random numbers as shown in Example 12.8.

Assume that $Y_1, \cdots, Y_n$ can be obtained by the linear transformation of $Z_1, \cdots, Z_n$,

$$\{Y\} = [C]\{Z\} + \{\mu\}, \qquad (12.23)$$

where

$$\begin{aligned} \{Y\} &= [Y_1, Y_2, \cdots, Y_{n-1}, Y_n]^T \\ \{Z\} &= [Z_1, Z_2, \cdots, Z_{n-1}, Z_n]^T \\ \{\mu\} &= [\mu_1, \mu_2, \cdots, \mu_{n-1}, \mu_n]^T . \end{aligned}$$

This problem is now recast to finding the transformation matrix $[C]$ such that Equation 12.21 is satisfied. Equations 12.21 and 12.22 can be written in matrix form as

$$\begin{aligned} E\{\{Y\}\} &= \{\mu\} & (12.24) \\ Cov\{\{Y\}\{Y\}^T\} &= [\sigma^2], & (12.25) \end{aligned}$$

and

$$\begin{aligned} E\{\{Z\}\} &= \{0\} \\ Cov\{\{Z\}\{Z\}^T\} &= [I], \end{aligned}$$

where

$$Cov\{\{Y\}\{Y\}^T\} = E\{(\{Y\} - \mu_Y)(\{Y\} - \mu_Y)^T\}.$$

Replacing $\{Y\}$ with Equation 12.23, Equation 12.24 is satisfied automatically, and Equation 12.25 is reduced to

$$[C][C]^T = [\sigma^2]. \qquad (12.26)$$

The matrix $[C]$ is obtained by equating element by element in Equation 12.26. Due to symmetry, Equation 12.26 produces $(n^2 + n)/2$ independent equations. Therefore matrix $[C]$ should have only $(n^2 + n)/2$ unknown elements (not n^2). We use the lower-triangular coefficient matrix for $[C]$,

$$[C] = \begin{bmatrix} c_{11} & 0 & \cdots & 0 \\ c_{21} & c_{22} & & \vdots \\ \vdots & \vdots & & 0 \\ c_{n1} & c_{n2} & \cdots & c_{nn} \end{bmatrix}.$$

12.3. JOINT RANDOM NUMBERS

In this way, $[C]$ has the right number of unknowns, and the zero elements make the computation easier. The first few equations are written below:

$$\begin{aligned} c_{11}^2 &= \sigma_{11}^2 \\ c_{11}c_{21} &= \sigma_{21}^2 \\ c_{21}^2 + c_{22}^2 &= \sigma_{22}^2 \\ &\vdots \end{aligned}$$
(12.27)

Let us demonstrate the procedure with an example.

Example 12.13 Linear Transform Method

Suppose that a cantilever beam with length $L = 1$ m shown in Figure 12.8 is loaded with forces Y_2 and Y_1 at $x = L/2$ and L, respectively. Assuming that the point loads are normally distributed, generate the random point loads with mean vector and covariance matrix given by

$$\{\mu\} = \begin{bmatrix} 15 \\ 20 \end{bmatrix}, \quad [\sigma^2] = \begin{bmatrix} 2 & 0.5 \\ 0.5 & 1 \end{bmatrix}.$$

Find the reaction moment at $x = 0$. Note that the off-diagonal terms in the covariance matrix $[\sigma^2]$ are not zeros. That is, the point loads Y_1 and Y_2 are not statistically independent. The value of one affects the value of the other.

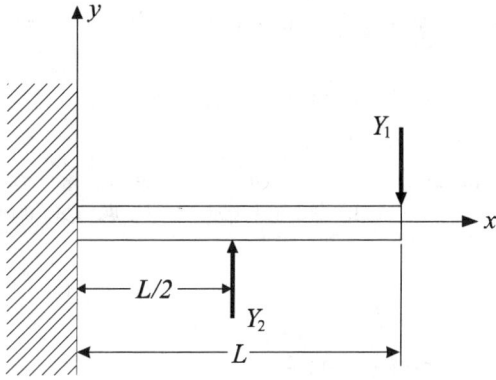

Figure 12.8: Cantilever beam loaded at $x = L/2$ and $x = L$ by correlated forces Y_1 and Y_2.

Solution In order to obtain random samples of the point loads with means and covariances given above, the transform matrix $[C]$ is obtained using Equations 12.27. The individual elements are

$$
\begin{aligned}
c_{11}^2 &= 2 \rightarrow c_{11} = \sqrt{2} \\
c_{11}c_{21} &= 0.5 \rightarrow c_{21} = 0.5/\sqrt{2} = 0.354 \\
c_{21}^2 + c_{22}^2 &= 1 \rightarrow c_{22} = \sqrt{1 - \left(0.5/\sqrt{2}\right)^2} = 0.935,
\end{aligned}
$$

and

$$[C] = \begin{bmatrix} 2 & 0 \\ 0.354 & 0.935 \end{bmatrix}.$$

Then, two sets of independent uniform random numbers, x_1 and x_2, are generated. We use the uniform random numbers generated in Table 12.5. Using these uniform random numbers, generate two sets of independent standard normal random numbers z_1 and z_2 using Equations 12.10 and 12.11. For each pair of normal random numbers, a pair of random numbers y_1 and y_2 is obtained using the transformation Equation 12.23.

From statics, the reaction moment at the clamped end is

$$M = -Y_1 L + Y_2 \frac{L}{2}. \tag{12.28}$$

The moment is calculated for each pair of point loads. The numerical results are shown in Table 12.7.

Note that the moment obtained from each pair of point loads has its own statistical properties such as mean and variance. The sample mean for the moment is

$$\bar{M} = \sum_{i=1}^{N} \frac{M\left(y_1^i, y_2^i\right)}{N} = -5.6302 \text{ Nm},$$

where N is the number of the samples and the superscript i refers to the ith row in the table. The arithmetic mean or the sample mean is called the "estimator."

⊛

12.4 Error Estimates

In the previous example, we found the value of the estimator $\bar{M}$ for the moment. However, the estimator itself is not a sufficient result. It is necessary to know the *quality* of this estimator. To do so, find the variance of the estimator. The quality of the estimator improves with decreasing variance. In this section, the upper bounds for the error are found using the Chebyshev inequality and the central limit theorem.

12.4. ERROR ESTIMATES

Table 12.7: Sample Results for Example 12.13

x_1	x_2	z_1	z_2	y_1	y_2	M
0.9501	0.4447	−0.3008	0.1089	14.3985	20.0704	−4.3633
0.4860	0.9218	1.0592	−0.5667	12.4384	18.7136	−3.0816
0.4565	0.4057	−1.0389	0.6993	15.5207	19.0753	−5.9830
0.2311	0.6154	−1.2808	−1.1354	17.1185	19.5702	−7.3334
0.8913	0.7382	−0.0355	−0.4784	14.9290	19.5210	−5.1685
0.0185	0.9355	2.5958	−1.1142	15.6588	20.6953	−5.3112
0.6068	0.7919	0.2603	−0.9650	12.9221	20.5639	−2.6402
0.7621	0.1763	0.3294	0.6594	20.1916	19.2192	−10.5820
0.8214	0.9169	0.5437	−0.3128	16.0874	19.7576	−6.2086

Note that M in Example 12.13 is a random variable defined by Equation 12.28 with a mean value and a variance. The mean value and the variance of the random variable (M in this case) are likely to be known. In Example 12.13, the mean value is given by

$$E\{M\} = E\left\{-Y_1 L + Y_2 \frac{L}{2}\right\} = -L E\{Y_1\} + \frac{L}{2} E\{Y_2\} = -5,$$

and the variance is given by

$$\begin{aligned} Var\{M\} &= Var\left\{-Y_1 L + Y_2 \frac{L}{2}\right\} \\ &= L^2 Var\{Y_1\} - L^2 Cov\{Y_1 Y_2\} + \frac{L^2}{4} Var\{Y_2\} = 1.75. \end{aligned}$$

Now, let us estimate M with $\bar{M}$. Estimator $\bar{M}$ is also a random variable with its own mean and variance. The mean of this estimator is given by[15]

$$\begin{aligned} E\{\bar{M}\} &= E\left\{\frac{1}{N} \sum_{i=1}^{N} M\left(y_1^i, y_2^i\right)\right\} \\ &= \frac{1}{N} \sum_{i=1}^{N} E\{M(y_1, y_2)\}. \end{aligned}$$

Therefore, the expected value of $\bar{M}$ is same as the expected value of M,

$$E\{\bar{M}\} = E\{M\}. \quad (12.29)$$

---

[15] The expressions for the mean and variance of the estimator are given more generally by M.H. Kalos and P.A. Whitlock, *Monte Carlo Methods*, John Wiley and Sons, 1986. See Section 2.6.

Similarly, the variance of $\bar{M}$ is given by

$$Var\left\{\bar{M}\right\} = Var\left\{\frac{1}{N}\sum_{i=1}^{N} M\left(y_1^i, y_2^i\right)\right\}$$

$$= \frac{1}{N^2}\sum_{i=1}^{N} Var\left\{M\left(y_1, y_2\right)\right\},$$

so that

$$Var\left\{\bar{M}\right\} = \frac{1}{N}Var\left\{M\right\}. \qquad (12.30)$$

$Var\left\{\bar{M}\right\}$ decreases with number of samples N, and the quality of the estimate becomes better. The numerical values of $E\left\{\bar{M}\right\}$ and $Var\left\{\bar{M}\right\}$ for Example 12.13 are -5.000 and 0.1944, respectively.

Although the distribution of M is known,[16] it is not known how $\bar{M}$ is distributed. If the probability density function of $\bar{M}$ is known, then the probability that $\bar{M}$ lies within some threshold is

$$\Pr\left(\left|\bar{M} - E\left\{\bar{M}\right\}\right| < \delta\right) = \int_{E\{\bar{M}\}-\delta}^{E\{\bar{M}\}+\delta} f_{\bar{M}}\left(\bar{m}\right) d\bar{m}.$$

Although such probabilities cannot be obtained precisely, the upper limit bound can be given using the *Chebyshev inequality*,[17]

$$\Pr\left(\left|\bar{M} - E\left\{\bar{M}\right\}\right| > \delta\right) \leq \frac{\sqrt{Var\left\{\bar{M}\right\}}}{\delta^2}.$$

Let $\delta = k\sqrt{Var\left\{\bar{M}\right\}}$, and then

$$\Pr\left(\bar{M} - E\left\{\bar{M}\right\} > k\sqrt{Var\left\{\bar{M}\right\}}\right) \leq \frac{1}{k^2},$$

or

$$\Pr\left(\frac{\left|\bar{M} - E\left\{\bar{M}\right\}\right|}{\sqrt{Var\left\{\bar{M}\right\}}} > k\right) \leq \frac{1}{k^2}. \qquad (12.31)$$

Equation 12.31 is a statement that the probability that the estimator is outside the kth standard deviation from its mean is less than $1/k^2$. Then

---

[16] Random variable M is normally distributed since Y_1 and Y_2 are normally distributed.

[17] For the proof of the Chebyshev inequality, the reader is referred to P.L. Meyer, *Introductory Probability and Statistical Applications*, Addison-Wesley Publishing Company, 1970, p. 142.

12.4. ERROR ESTIMATES

the probability that the estimator is within k standard deviations is

$$\Pr\left(\frac{|\bar{M} - E\{\bar{M}\}|}{\sqrt{Var\{\bar{M}\}}} < k\right) \geq 1 - \frac{1}{k^2}.$$

The probability that the estimator is within 2 standard deviations, or $k = 2$, is 3/4. Replacing $E\{\bar{M}\}$ with $E\{M\}$ and $\sqrt{Var\{\bar{M}\}}$ with $\sqrt{Var\{M\}/N}$ by making use of Equations 12.29 and 12.30, we can also write

$$\Pr\left(\frac{|\bar{M} - E\{M\}|}{\sqrt{Var\{M\}}} < \frac{k}{\sqrt{N}}\right) \geq 1 - \frac{1}{k^2}. \qquad (12.32)$$

Example 12.14 Chebyshev Inequality

This example is a continuation of Example 12.13. Here, find the number of samples required such that probability that the estimator is within one tenth of a standard deviation of M, $\sqrt{Var\{M\}}/10$, is at least 0.9544, or

$$\Pr\left(|\bar{M} - E\{M\}| < \frac{\sqrt{Var\{M\}}}{10}\right) \geq 0.9544. \qquad (12.33)$$

Solution   Equation 12.32 can be written as

$$\Pr\left(|\bar{M} - E\{M\}| < \frac{k}{\sqrt{N}}\sqrt{Var\{M\}}\right) \geq 1 - \frac{1}{k^2}.$$

Compare this equation with Equation 12.33 to find

$$\frac{k}{\sqrt{N}} = \frac{1}{10}$$

$$1 - \frac{1}{k^2} = 0.9544,$$

from which

$$k = 4.6529$$
$$N = 2164.9.$$

Since N is not a whole number, $N = 2165$ is the minimum number of samples required to achieve the specified probability.

❋

A better way to provide the upper limit for the probability that the estimator is outside a certain threshold about its mean, comes from the central limit theorem.[18] The central limit theorem states that if $\bar{M}$ is the summation of a large number of random variables with identical or similar distributions, then $\bar{M}$ is normally distributed. That is, the probability density function for $\bar{M}$ is

$$f_{\bar{M}}(\bar{m}) = \frac{1}{\sqrt{2\pi \, Var\{\bar{M}\}}} \exp\left[-\frac{1}{2}\frac{(\bar{m} - E\{\bar{M}\})^2}{Var\{\bar{M}\}}\right], \quad -\infty < \bar{m} < \infty.$$

This can be transformed into a standard normal distribution using the transformation,

$$x_n = \frac{\bar{m} - E\{\bar{M}\}}{\sqrt{Var\{\bar{M}\}}},$$

so that

$$f_{X_n}(x_n) = \frac{1}{\sqrt{2\pi}} \exp\left[-\frac{1}{2}x_n^2\right], \quad -\infty < x_n < \infty.$$

If $\bar{M}$ is distributed normally, the probability that $\bar{M}$ is within one standard deviation of M, $\sqrt{Var\{M\}/N}$, is 0.683, within two standard deviations is 0.954, and within three standard deviations is 0.997.

Example 12.15 The Central Limit Theorem

Now answer the same question as in Example 12.14: How many samples are needed such that the probability that the estimator is within $\sqrt{Var\{M\}}/10$ is greater than 0.9544?

Solution Again, look for N that satisfies

$$\Pr\left(|\bar{M} - E\{M\}| < \frac{\sqrt{Var\{M\}}}{10}\right) > 0.9544.$$

This equation can be rewritten as

$$\Pr\left(\frac{|\bar{M} - E\{M\}|}{\sqrt{Var\{\bar{M}\}}} < \frac{\sqrt{N}}{10}\right) > 0.9544,$$

or

$$\Pr\left(|X_n| < \frac{\sqrt{N}}{10}\right) > 0.9544, \qquad (12.34)$$

---

[18] The proof will be omitted here. The reader is encouraged to read the proof given by I.M. Sobol, *Monte Carlo Method*, University of Chicago Press, 1974. See p. 15.

12.4. ERROR ESTIMATES

where $Var\{M\} = Var\{\bar{M}\}/N$. The cumulative distribution function of the standard normal distribution is tabulated by $\Phi(x_n)$. That is,

$$\Pr(X_n < x_n) = \Phi(x_n)$$
$$\Pr(|X_n| < x_n) = 2\Phi(x_n) - 1.$$

Therefore,

$$\Pr\left(|X_n| < \sqrt{N}/10\right) = 2\Phi\left(\sqrt{N}/10\right) - 1.$$

Equation 12.34 can then be written as

$$2\Phi\left(\sqrt{N}/10\right) - 1 \geq 0.9544.$$

From the standard normal distribution table, $\sqrt{N}/10 \geq 2$. Therefore, N has to be at least 400.

According to the central limit theorem, we need at least 400 samples. This number is much smaller than the number of 2165 obtained by the Chebyshev inequality. Having the extra information that the estimator has the normal distribution for large N, we are able to give a tighter upper limit for the required sample size.

⊛

This can be generalized. Let G be the estimator defined by

$$G = \frac{1}{N}\sum_{i=1}^{N} g\left(y_1^i, \cdots, y_n^i\right), \tag{12.35}$$

where $y_1, \cdots, y_n$ are random variables. The mean value and variance of this estimator are given by

$$E\{G\} = E\{g\} \quad \text{and} \quad Var\{G\} = \frac{1}{N}Var\{g\}.$$

When the sample size is small, use the Chebyshev inequality,

$$\Pr\left(\frac{|G - E\{g\}|}{\sqrt{Var\{g\}}} < \frac{k}{\sqrt{N}}\right) \geq 1 - \frac{1}{k^2}.$$

When the sample size is large, use the central limit theorem,

$$\Pr\left(\frac{|G - E\{g\}|}{\sqrt{Var\{g\}}} < \frac{k}{\sqrt{N}}\right) = 2\Phi(k) - 1, \tag{12.36}$$

where $\Phi(k)$ can be found from the standard normal distribution table.

12.5 Applications

12.5.1 Evaluation of Finite-Dimensional Integrals

The Monte Carlo method is used to evaluate integrals that have the form,

$$I = \int_D F(y_1, \cdots, y_n) \, dy_1 \cdots dy_n.$$

First rewrite the integrand as a product[19] of $g(y_1, \cdots, y_n)$ and $h(y_1, \cdots, y_n)$,

$$I = \int_D [g(y_1, \cdots, y_n) \, h(y_1, \cdots, y_n)] \, dy_1 \cdots dy_n,$$

where D stands for the domain of integration, $h(y_1, \cdots, y_n) > 0$ on D, and

$$\int_D h(y_1, \cdots, y_n) \, dy_1 \cdots dy_n = 1. \tag{12.37}$$

The value of the integral can be thought of as the expected value of $g(y_1, \cdots, y_n)$ with the probability density function $h(y_1, \cdots, y_n)$, that is,

$$I = E\{g(y_1, \cdots, y_n)\}.$$

Recall that the expected value of g is same as the expected value of its estimator G, $E\{g\} = E\{G\}$, where G is defined in Equation 12.35. For large N, apply the central limit theorem so that Equation 12.36 can be applied. For a fixed number k, say $k = 3$,

$$\Pr\left(\frac{|G - E\{g\}|}{\sqrt{Var\{g\}}} < \frac{3}{\sqrt{N}}\right) = 0.997.$$

The equation also implies that as $N \to \infty$, $E\{g\} \to G$. That is, for large N,

$$\begin{aligned} I &\simeq G \\ &= \frac{1}{N} \sum_{i=1}^{N} g(y_1^i, \cdots, y_n^i), \end{aligned} \tag{12.38}$$

where $y_1^i, \cdots, y_n^i$ are chosen according to the probability density function $h(y_1, \cdots, y_n)$.

---

[19] This can always be done. Once we choose $h(y_1, ..., y_n)$, then

$$g(y_1, ..., y_n) = \frac{F(y_1, ..., y_n)}{h(y_1, ..., y_n)}.$$

12.5. APPLICATIONS

In most cases the probability density $h(y_1, \cdots, y_n)$ will not be given. Instead, it is up to us to choose a function $h(y_1, \cdots, y_n)$ that satisfies Equation 12.37. Any function, as long as it satisfies Equation 12.37, will be sufficient. However, $h(y_1, \cdots, y_n)$ can be chosen such that the variance of the estimator is minimum, therefore achieving the minimum error. It can be found that the variance will be minimum if $h(y_1, \cdots, y_n)$ is chosen such that it is proportional to the integrand[20] $F(y_1, \cdots, y_n)$,

$$h(y_1, \cdots, y_n) \propto |F(y_1, \cdots, y_n)|.$$

In practice, when dealing with one-dimensional problems, quadrature formulas such as the trapezoidal rule and Simpson's rule are preferred over Monte Carlo integration since they are more efficient and precise. However, when multivalued integrals are considered, Monte Carlo integration may be preferred since it does not become much more complicated with increasing dimension.

Example 12.16 Monte Carlo Integration

Compute the integral,

$$I = \int_0^2 y \exp(y^2)\, dy,$$

using the following choices of $h(y)$:

(a) $h_1(y) = 1/2$
(b) $h_2(y) = 0.0243 y e^{2y}$.

These functions are plotted in Figure 12.9. $F(y) = y \exp(y^2)$.

Solution   We expect the estimate obtained using $h_2(y)$ to be better than that obtained using $h_1(y)$ since the shape of $h_2(y)$ is closer to $F(y)$. The corresponding $g(y)$ are obtained using the fact that

$$h_1(y) g_1(y) = h_2(y) g_2(y) = y e^{y^2},$$

where

$$g_1(y) = 2y e^{y^2}$$
$$g_2(y) = 41.15 e^{y^2 - 2y}.$$

The exact value for this integral is 26.80.

---

[20] The proof is shown by M.H. Kalos and P.A. Whitlock, *Monte Carlo Methods*, John Wiley and Sons, 1986, pp. 92-94, and I.M. Sobol, Section 9.6.

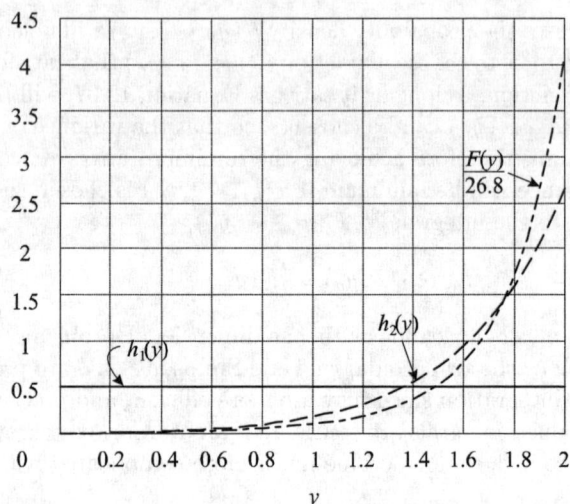

Figure 12.9: Arbitrary probability density functions, $h_1(y)$, $h_2(y)$, and $F(y)/26.8$.

First generate the uniform random numbers x on $(0,1)$, then random numbers y_1 and y_2 are generated according to the probability density functions $h_1(y)$ and $h_2(y)$. These are tabulated in Table 12.8. The cumulative distribution functions are given by

$$H_1(y) = \frac{1}{2}y$$
$$H_2(y) = 0.0243\left(\frac{1}{2}ye^{2y} - \frac{1}{4}e^{2y} + \frac{1}{4}\right).$$

The integral is approximated using Equation 12.38 or

$$I \simeq \frac{1}{N}\sum_{i=1}^{N} g(y).$$

The approximation for the integral in cases *(a)* and *(b)* are 36.34 and 29.16, respectively. As we expected, case *(b)* gives a better approximation to the true value of 26.80.

✳

Table 12.8: Sample Results for Example 12.16

x	$y_1 = H_1^{-1}(x)$	$g(y_1)$	$y_2 = H_2^{-1}(x)$	$g(y_2)$
0.9501	1.900	140.6	1.980	39.57
0.4860	0.9720	5.000	1.733	25.92
0.4565	0.9130	4.203	1.711	25.09
0.2311	0.4622	1.145	1.473	18.94
0.8913	1.783	85.53	1.956	37.78
0.0185	0.03700	0.07410	0.7351	16.24
0.6068	1.214	10.59	1.814	29.36
0.7621	1.524	31.12	1.898	33.90
0.8214	1.643	48.83	1.926	35.67

12.5.2 Generating a Time History for a Stationary Random Process Defined by a Power Spectral Density

When computing the vibratory response of a machine or component in a random environment, it is sometimes necessary to work in the time domain rather than in the frequency domain. This is especially true when the differential equation governing structural behavior is *nonlinear*. Such systems are solved numerically.

Consider a system subjected to a random force. The random force is assumed to be a stationary Gaussian process with a spectral density function $S_{FF}(\omega)$. In Chapter 6, we learned how to deal with such problems in the frequency domain if the system is linear. That is, for a given autocorrelation function, spectral density function, mean value, and variance of the input force, we were able to find those of the output response. In Chapter 10, approximate analytical techniques were developed for the study of nonlinear equations.

Here, we develop a technique to find the output response in time by simulating the input force in the time domain. This is done by utilizing the information stored in a power spectral density as follows.

It was shown in Section 5.8.1 that if $F(t)$ is stationary Gaussian, its sample time history can be represented by

$$F(t) = \sum_{n=1}^{N} \sqrt{\frac{2}{N}} \sigma_{FF} \cos(\bar{\omega}_n t - \varphi_n),$$

674   CHAPTER 12. THE MONTE CARLO METHOD

where the random phase angle φ is chosen uniformly on $[0, 2\pi)$[21] and

$$\bar{\omega}_n = \frac{\omega_n + \omega_{n-1}}{2}.$$

The frequencies ω_n are chosen such that the area between ω_{n-1} and ω_n are equal for all n. Shinozuka[22] proposed that ω_n are random frequencies distributed according to the density function $f(\omega) \equiv S_{FF}^o(\omega)/\sigma_{FF}^2$, where S_{FF}^o is the one-sided power spectrum of $F(t)$. Note that the total area under $S_{FF}^o(\omega)$ is σ_{FF}^2. Therefore, $S_{FF}^o(\omega)/\sigma_{FF}^2$ can act as the probability density function.

Example 12.17 Simulation of Ocean Wave Elevation Using Shinozuka's Method

For the ocean wave elevation considered in Section 5.8.1 the one-sided spectral density is given by

$$S_{\eta\eta}^o(\omega) = \frac{A}{\omega^5} \exp\left(-\frac{B}{\omega^4}\right) \text{ m}^2\text{s/rad, for } \omega > 0,$$

where

$$A = 0.7795 \text{ m}^4 (\text{rad/s})^4$$
$$B = 0.0175 \ (\text{rad/s})^4$$

for the wind velocity at 19.5 m above the still water level, taken to be 25 m/s. The wave elevation at $x = 0$ (zero horizontal location) is then given by

$$\eta(t) = \sum_{n=1}^{N} \sqrt{\frac{2}{N}} \sigma \cos(\omega_n t - \varphi_n),$$

where it was previously found that $\sigma^2 = A/4B$. ω_n are chosen according to the probability density $f(\omega) = S_{\eta\eta}^o(\omega)/\sigma^2$,

$$f(\omega) = \frac{4B}{\omega^5} \exp\left(-\frac{B}{\omega^4}\right), \qquad \omega \geq 0.$$

The cumulative distribution function is

$$F(\omega) = \exp\left(-\frac{B}{\omega^4}\right), \qquad \omega \geq 0.$$

---

[21] 2π is excluded since 2π rad is the same as 0 rad.
[22] M. Shinozuka, "Monte Carlo Solution of Structure Dynamics," *Computers and Structures*, **2**, 855-874, 1972.

12.5. APPLICATIONS

Table 12.9: Random Frequencies and Phase Angles for Ocean Wave Height

x_1	$\varphi = 2\pi x_1$	x_2	$\omega = F^{-1}(x_2) = (-B/\ln(x_2))^{0.25}$
0.9501	5.970	0.4447	0.3834
0.4860	3.054	0.9218	0.6809
0.4565	2.868	0.4057	0.3732
0.2311	1.452	0.6154	0.4357
0.8913	5.600	0.7382	0.4900
0.0185	0.1162	0.9355	0.7158
0.6068	3.813	0.7919	0.5233
0.7621	4.788	0.1763	0.3169
0.8214	5.161	0.9169	0.6702

The random frequencies and the random phase angles are generated from standard uniform random numbers x_1 and x_2 and tabulated in Table 12.9.

Figure 12.10 shows the time history of a sample wave derived using the above procedure for wind velocity $V_{19.5} = 25$ m/s and $N = 10$ at $x = 0$.

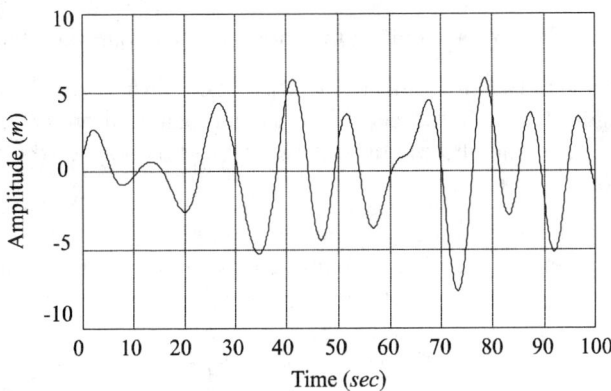

Figure 12.10: Sample wave profile derived by Shinozuka's method applied to the Pierson-Moskowitz wave height spectrum.

12.6 Concluding Summary

This chapter introduced the fundamental ideas underlying the numerical methods grouped by the term Monte Carlo methods. Such methods are widely used in both probabilistic and deterministic problems. The methods are based on the generation of numbers according to specific probability density functions. These are introduced and discussed. In addition, as a numerical method, it is necessary to establish error estimates. All these are introduced briefly. With this background the reader can move to more advanced studies.

12.7 Problems

Section 12.2: Random Number Generation

1. Generate uniform random numbers, using available software such as MATLAB, either using the linear congruential generator or the program's built-in random number generator. Plot the average value and the standard deviation and compare them with the theoretical values for the uniform distribution.

2. Show that random numbers that are distributed according to the lognormal density function with mean value $\exp(\mu)$, and standard deviation $\exp(\sigma)$, can be obtained by using

$$y = \exp\left\{\sigma\sqrt{-2\ln x_1}\cos 2\pi x_2 + \mu\right\},$$

where x_1 and x_2 are uniform random numbers between 0 and 1.

3. Verify that the random numbers produced in the previous problem are indeed distributed according to the lognormal density. Plot the mean value and the standard deviation as functions of the number of random numbers used.

4. Show that random numbers that are distributed according to the lognormal density function with mean value $\exp(\mu)$, and standard deviation $\exp(\sigma)$, can be obtained by using

$$y = \exp\left\{\sigma\sqrt{-2\ln x_1}\cos 2\pi x_2 + \mu\right\},$$

where x_1 and x_2 are uniform random numbers between 0 and 1.

5. Verify that the random numbers produced in the previous problem are indeed distributed according to the lognormal density. Plot the mean value and the standard deviation as functions of the number of random numbers used.

12.7. PROBLEMS

6. Use the inverse numerical transformation method to generate normal random numbers with $\mu = 0$ and $\sigma = 1$. Use ten equally spaced intervals between $y = -3\sigma$ and $y = 3\sigma$.

7. Use the composition method to generate random numbers distributed according to

$$f_Y(y) = \frac{1}{4}\sin y + \frac{9}{32\pi^3}y^2, \qquad 0 < y < 2\pi.$$

8. Use Von Neumann's rejection-acceptance method to generate normally distributed random numbers. How does the method compare to the Box-Muller method?

Section 12.3: Random Numbers for Joint Probability Densities

9. Use the inverse transform method to generate random numbers for the joint probability density function,

$$f_{Y_1 Y_2}(y_1, y_2) = y_1 y_2, \qquad 0 < y_1, y_2 < 1.$$

10. Use the linear transform method to generate random numbers y_1 and y_2 for the joint probability density function given in the previous problem.

Chapter 13

Fluid-Induced Vibration

In modeling offshore structures, one needs to account for the forces exerted by the surrounding fluid. The vibrational characteristics of a structure can be significantly altered when it is surrounded by water. For example, damping by the fluid lowers the natural frequency of vibration.

The purpose of this chapter is to show how the fluid forces on an offshore structure due to current and random waves are modeled. In Section 13.1, we discuss how the fluid forces in an ocean can be described. In Section 13.2, we find how the fluid force on an offshore structure can be estimated. In Section 13.3, some example problems are presented to show how the first two sections can be applied. Readers who are interested in more detailed studies should refer to texts by Chakrabarti,[1] Faltinsen,[2] Kinsman,[3] Sarpkaya,[4] and Wilson.[5] This chapter is meant to be self-contained and therefore repeats small parts of earlier chapters for the ease of the reader.

13.1 Ocean Currents and Waves

When one considers the dynamics of an offshore structure, one must also consider the forces due to the surrounding fluid. The two important sources of fluid motion are ocean waves and ocean currents.

---

[1] *Hydrodynamics of Offshore Structures*, S. Chakrabarti, Computational Mechanics Publications, 1987.

[2] *Sea Loads on Ships and Offshore Structures*, O. Faltinsen, Cambridge University Press, 1993.

[3] *Wind Waves*, B. Kinsman, Prentice-Hall, 1965, now available in a Dover edition.

[4] *Mechanics of Wave Forces on Offshore Structures*, T. Sarpkaya, M. Isaacson, Van Nostrand Reinhold, 1981.

[5] *Dynamics of Offshore Structures*, J. Wilson, John Wiley & Sons, Second Edition, 2003.

Most steady large currents are generated by the drag of the wind passing over the surface of the water, with the currents confined to a region near the ocean surface. Tidal currents are generated by the gravitational attraction of the Sun and the Moon, and they are most significant near coasts. The ultimate source of the ocean circulation is the uneven radiative heating of the Earth by the Sun.

Isaacson[6] suggested an empirical formula for the current velocity in the horizontal direction as a function of depth,

$$U_c(x) = (U_{tide}(d) + U_{circulation}(d)) \left(\frac{x}{d}\right)^{1/7} + U_{drift}(d) \left(\frac{x - d + d_o}{d_o}\right),$$

where U_{drift} is the wind-induced drift current, U_{tide} is the tidal current, $U_{circulation}$ is the low-frequency long-term circulation, x is the vertical distance measured from the ocean bottom, d is the depth of the water, and d_o is the smaller of the depth of the thermocline or 50 m. The value of U_{tide} is obtained from tide tables, and U_{drift} is about 3% of the 10 minute mean wind velocity at 10 m above sea level.

It should be noted that these currents evolve slowly compared to the time scales of engineering interests. Therefore, they can be treated as quasi-steady phenomena. Waves, on the other hand, cannot be treated as steady phenomena. The underlying physics that govern wave dynamics are too complex, and therefore waves must be modeled stochastically. In the subsequent section, we discuss the concept of the spectral density, the available ocean wave spectral densities, a method to obtain the spectral density from wave time-histories, methods to obtain a sample time history from a spectral density, short-term and long-term statistics, and a method to obtain fluid velocities and accelerations from wave elevation using linear wave theory.

13.1.1 Spectral Density

Here, only surface gravity waves are considered. A regular wave is examined first in order to become familiar with the terms that are used to describe a wave. The wave surface elevation is denoted as $\eta(x,t)$ and can be written as $\eta(x,t) = A\cos(kx - \omega t)$, where k is the wave number and ω is the angular frequency. Figure 13.1(a) shows the surface elevation at two time instances ($t = 0$ and $t = \tau$) and (b) the surface elevation at a fixed location ($x = 0$). A is the amplitude, H is the wave height or the distance between the maximum and minimum wave elevation or twice the amplitude, and T is the period, given by $T = 2\pi/\omega$.

---

[6]"Wave and Current Forces on Fixed Offshore Structures," M. Isaacson, *Canadian Journal of Civil Engineering*, 15:937-947, 1988.

13.1. OCEAN CURRENTS AND WAVES

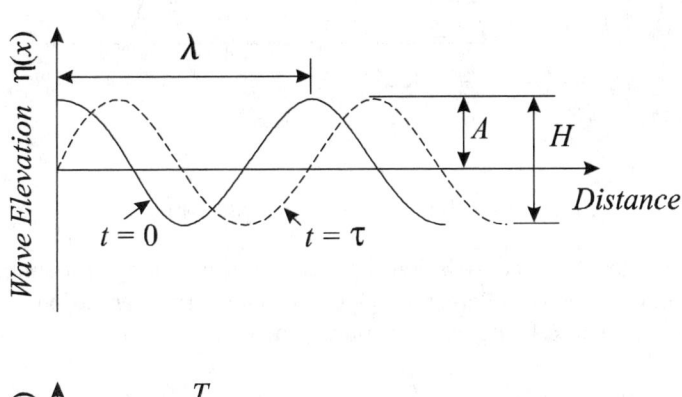

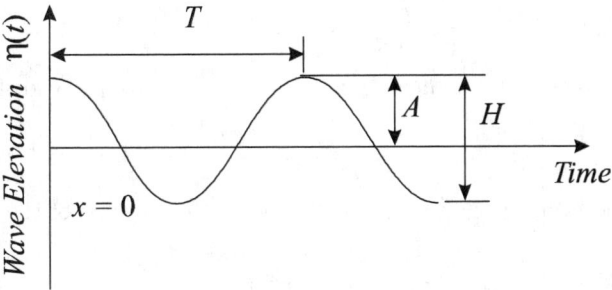

Figure 13.1: Defining parameters of a regular wave.

In practice, waves are not regular. Figure 13.2 shows a schematic time history of an irregular wave surface elevation. The wave height and frequency are not easily defined. Therefore, the wave height spectral density is utilized for a statistical description of the wave elevation.

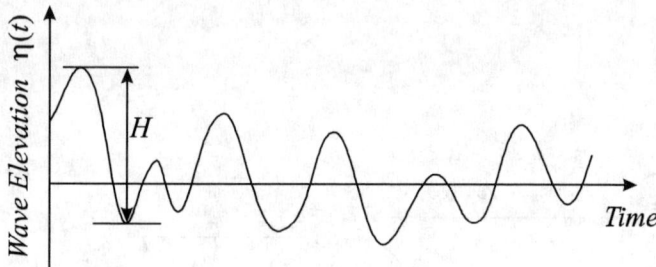

Figure 13.2: Time history of a random wave.

The random surface elevation $\eta(t)$ can be thought of as a summation of regular waves with different frequencies. The surface elevation $\eta(t)$ can therefore be related to its Fourier transform $X(\omega)$ by

$$\eta(t) = \frac{1}{2\pi} \int_{-\infty}^{\infty} X(\omega) \exp(-i\omega t) \, d\omega.$$

Suppose that the energy of the system is proportional to $\eta^2(t)$ so that the energy becomes

$$\mathcal{E} = \frac{1}{2} C \eta^2(t),$$

where C is the proportionality constant. The expected value of the energy is then given by

$$E\{\mathcal{E}\} = \frac{1}{2} C E\{\eta^2(t)\},$$

where $E\{\eta^2(t)\}$ is the mean-square value of $\eta(t)$. If $\eta(t)$ is an ergodic process, the mean-square value of $\eta(t)$ can be approximated by the time average over a long period of time,

$$E\{\eta^2(t)\} = \lim_{T_s \to \infty} \frac{1}{T_s} \int_{-T_s/2}^{T_s/2} \eta^2(t) \, dt$$

$$= \lim_{T_s \to \infty} \frac{1}{T_s} \frac{1}{2\pi} \int_{-\infty}^{\infty} |X(\omega)|^2 \, d\omega,$$

where Parseval's theorem has been utilized,

$$\int_{-\infty}^{\infty} \eta^2(t) \, dt = \frac{1}{2\pi} \int_{-\infty}^{\infty} |X(\omega)|^2 \, d\omega,$$

13.1. OCEAN CURRENTS AND WAVES

where

$$|X(\omega)|^2 = X(\omega) X^*(\omega)$$

$$X(\omega) = \int_{-\infty}^{\infty} \eta(t) \exp(-i\omega t)\, dt$$

$$X^*(\omega) = \int_{-\infty}^{\infty} \eta(t) \exp(i\omega t)\, dt.$$

The power spectral density (or simply the spectrum) is defined as

$$S_{\eta\eta}(\omega) \equiv \frac{1}{2\pi T_s} |X(\omega)|^2, \qquad (13.1)$$

so that $E\{\eta^2(t)\}$ is given by

$$E\{\eta^2(t)\} = \int_{-\infty}^{\infty} S_{\eta\eta}(\omega)\, d\omega. \qquad (13.2)$$

For a zero-mean process, $E\{\eta^2(t)\}$ equals the variance σ_η^2. The spectral density has units of $\eta^2 t$, or m²s.

$S_{\eta\eta}(\omega)$ is related to the autocorrelation function, $R_{\eta\eta}(\tau)$, by the Wiener-Khinchine relations,[7,8]

$$S_{\eta\eta}(\omega) = \frac{1}{2\pi} \int_{-\infty}^{\infty} R_{\eta\eta}(\tau) \exp(-i\omega\tau)\, d\tau \qquad (13.3)$$

$$R_{\eta\eta}(\tau) = \int_{-\infty}^{\infty} S_{\eta\eta}(\omega) \exp(i\omega\tau)\, d\omega. \qquad (13.4)$$

It should be noted that in some textbooks, the factor $1/2\pi$ appears in the second equation instead of the first.

There are a few properties of the spectral density with which the reader should become familiar. The first property is that the spectral density function of a real-valued stationary process is both real and symmetric. That is, $S_{\eta\eta}(\omega) = S_{\eta\eta}(-\omega)$ (Equation 13.1). Secondly, the area under the spectral density is equal to $E\{\eta^2(t)\}$ (Equation 13.2) and also equal to $R_{\eta\eta}(0) = \sigma_\eta^2 - \mu_\eta^2$, where σ_η^2 is the variance and μ_η is the mean of $\eta(t)$. In most cases, we only consider a zero mean process so that the area under the spectral density equals σ_η^2. If the process does not have a zero mean, the mean value can be subtracted from the process so that it now has a zero mean value.

---

[7] "Generalized Harmonic Analysis," N. Wiener, *Acta Math.*, 55:117-258, 1930.

[8] "Korrelations Theorie der Stationaren Stochastischen Prozesse," A. Khinchine, *Math. Ann.*, 109:604-615, 1934.

For ocean applications, a one-sided spectrum in terms of cycles per second or *Hertz* (Hz) is often used. The one-sided spectrum is given a superscript *o*, and can be obtained from the two-sided spectrum by the relation,

$$S^o_{\eta\eta}(\omega) = 2S_{\eta\eta}(\omega), \quad \omega \geq 0.$$

The two-sided spectrum in terms of ω can be transformed to the spectrum in terms of f (where $\omega = 2\pi f$) by the relation,

$$S_{\eta\eta}(f) = 2\pi S_{\eta\eta}(\omega).$$

Then, the two-sided spectrum in terms of ω can be transformed to the one-sided spectrum in units of Hz by the relation,

$$S^o_{\eta\eta}(f) = 4\pi S_{\eta\eta}(\omega), \quad \text{for } f, \omega > 0.$$

It should be noted that the spectral density that we have defined in this text is the *amplitude half-spectrum*. For an amplitude half-spectrum $S(\omega)$, the amplitude, height, and height double spectra are related by

$$S^A(\omega) = 2S(\omega)$$
$$S^H(\omega) = 8S(\omega)$$
$$S^{2H}(\omega) = 16S(\omega).$$

13.1.2 Ocean Wave Spectral Densities

In this section, spectral density models for a random sea are introduced. An excellent review of existing spectral density models is given in Chapter 4 of Chakrabarti.

The ocean wave spectrum models are semiempirical formulas. That is, they are derived mathematically, but the formulation requires one or more experimentally determined parameters. The accuracy of the spectrum depends significantly on the choice of these parameters.

In formulating spectral densities, the parameters that influence the spectrum are *fetch limitations, decaying versus developing seas, water depth, current,* and *swell.* The fetch is the distance over which a wind blows in a wave-generating phase. Fetch limitation refers to the limitation on the distance due to some physical boundaries so that full wave development is prohibited. In a developing sea, wave development has not yet reached its stationary state under a stationary wind. In contrast, the wind has blown a sufficient time in a fully developed sea, and the sea has reached its stationary state. In a decaying sea, the wind has dropped off from its stationary value. Swell is the wave motion caused by a distant storm that persists even after the storm has died down or moved away.

13.1. OCEAN CURRENTS AND WAVES

The Pierson-Moskowitz (P-M) spectrum[9] is the most extensively used spectrum for representing a fully developed sea. It is a one-parameter model in which the sea severity can be specified in terms of the wind velocity. The P-M spectrum is given by

$$S_{\eta\eta}^o(\omega) = \frac{8.1 \times 10^{-3} g^2}{\omega^5} \exp\left(-0.74 \left(\frac{g}{U_{w,19.5}}\right)^4 \omega^{-4}\right),$$

where g is the gravitational constant, and $U_{w,19.5}$ is the wind speed at a height of 19.5 m above the still water. The Pierson-Moskowitz spectrum is also called the wind-speed spectrum because it requires wind data. It can be written in terms of the modal frequency ω_m as well,

$$S_{\eta\eta}^o(\omega) = \frac{8.1 \times 10^{-3} g^2}{\omega^5} \exp\left(-1.25 \left(\frac{\omega_m}{\omega}\right)^4\right).$$

Note that the modal frequency is the frequency at which the spectrum is the maximum.

In some cases, it may be more convenient to express the spectrum in terms of significant wave height rather than the wind speed or modal frequency. For a narrow-band Gaussian process,[10] the significant wave height is related to the standard deviation by $H_s = 4\sigma_\eta$. The standard deviation is the square root of the area under the spectral density, $\int_0^\infty S_{\eta\eta}^o(\omega)\,d\omega = \sigma_\eta^2$. Then, the spectrum can be written as

$$S_{\eta\eta}^o(\omega) = \frac{8.1 \times 10^{-3} g^2}{\omega^5} \exp\left(-\frac{0.0324 g^2}{H_s^2} \omega^{-4}\right),$$

and the peak frequency and the significant wave height are related by

$$\omega_m = 0.4\sqrt{g/H_s}.$$

The Pierson-Moskowitz spectrum is applicable for deep water, unidirectional seas, fully developed and local-wind-generated with unlimited fetch, and was developed for the North Atlantic. The effect of swell is not accounted for in this spectrum. It is found that even though it is derived for the North Atlantic, the spectrum is valid for other locations. However, the limitation that the sea is fully developed may be too restrictive because it cannot model the effect of waves generated at a distance. Therefore, in such instances a two-parameter spectrum, such as the Bretschneider spectrum, is used to model a nonfully developed sea as well as a fully developed sea.

---

[9] "A Proposed Spectral Form for Fully Developed Wind Seas Based on the Similarity Theory of S.A. Kitaigorodskii," W. Pierson, L. Moskowitz, *Journal of Geophysical Research*, 69(24):5181-5203, 1964.

[10] See Section 13.1.5 for details.

The Bretschneider spectrum[11,12] is a two-parameter spectrum in which both the sea severity and the state of development can be specified. The Bretschneider spectrum is given by

$$S_{\eta\eta}^o(\omega) = 0.169 \frac{\omega_s^4}{\omega^5} H_s^2 \exp\left(-0.675 \left(\frac{\omega_s}{\omega}\right)^4\right),$$

where $\omega_s = 2\pi/T_s$ and T_s is the significant period. The sea severity can be specified by H_s and the state of development can be specified by ω_s. It can be shown that for $\omega_s = 1.167\omega_m$ (equivalently $\omega_s = 1.46/\sqrt{H_s}$) the Bretschneider spectrum and the Pierson-Moskowitz spectrum are equivalent. Figure 13.3 plots the Bretschneider spectra for $H_s = 4$ m and various values of ω_s. For $\omega_s = 0.731$ rad/s, the Pierson-Moskowitz and the Bretschneider spectra are identical. It should be noted that the developing sea will have a slightly higher modal frequency than the fully developed sea and can be described by ω_s greater than $1.461/\sqrt{H_s}$.

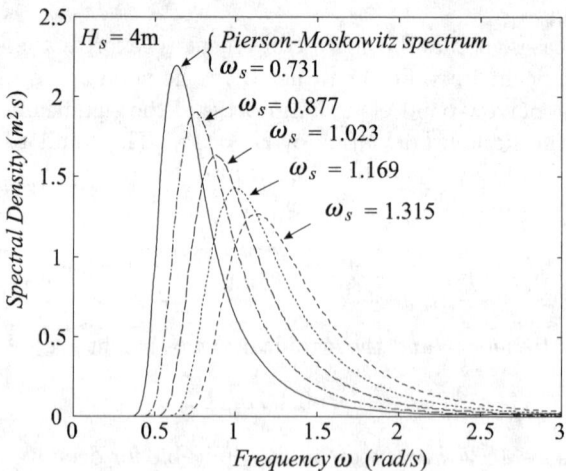

Figure 13.3: Bretschneider spectra for various values of ω_s.

Other two-parameter spectral densities that are often used are the ISSC (International Ship Structures Congress) and the ITTC (International Towing Tank Conference) spectra. The ISSC spectrum is written in terms of the

---

[11] *Wave Variability and Wave Spectra for Wind-Generated Gravity Waves*, C. Bretschneider, Technical Memorandum No. 118, Beach Erosion Board, U.S. Army Corps of Engineers, Washington, D.C., 1959.

[12] "Wave Forcasting," in *Handbook of Ocean and Underwater Engineering*, John Meyers, Editor, McGraw-Hill, 1969.

13.1. OCEAN CURRENTS AND WAVES

Table 13.1: The parameters for the generalized two-parameter spectrum, $S_{\eta\eta}^o(\omega) = \frac{A}{4} H_s^2 \tilde{\omega}^4 / \omega^5 \exp\left(-A(\omega/\tilde{\omega})^{-4}\right)$

Model	A	$\tilde{\omega}$
Bretschneider	0.675	ω_s
ITTC	0.318	ω_z
ISSC	0.4427	$\bar{\omega}$

significant wave height and the mean frequency, where the mean frequency is given by

$$\bar{\omega} = \sqrt{\frac{\int_0^\infty \omega S(\omega)\,d\omega}{\int_0^\infty S(\omega)}} = 1.30\,\omega_m.$$

The ISSC spectrum is given by

$$S_{\eta\eta}^o(\omega) = 0.111 \frac{\bar{\omega}^4}{\omega^5} H_s^2 \exp\left(-0.444\left(\frac{\bar{\omega}}{\omega}\right)^4\right).$$

The ITTC spectrum is based on the significant wave height and the zero crossing frequency and is given by

$$S_{\eta\eta}^o(\omega) = 0.0795 \frac{\omega_z^4}{\omega^5} H_s^2 \exp\left(-0.318\left(\frac{\omega_z}{\omega}\right)^4\right),$$

where the zero crossing frequency ω_z is given by

$$\omega_z = \sqrt{\frac{\int_0^\infty \omega^2 S(\omega)\,d\omega}{\int_0^\infty S(\omega)}} = 1.41\,\omega_m.$$

The Bretschneider, ITTC, and ISSC spectra are called two-parameter spectra. A generalized equation for such two-parameter spectra is

$$S_{\eta\eta}^o(\omega) = \frac{A}{4} H_s^2 \frac{\tilde{\omega}^4}{\omega^5} \exp\left(-A\left(\frac{\tilde{\omega}}{\omega}\right)^4\right),$$

with A and $\tilde{\omega}$ given in Table 13.1 for the examples discussed here.

The spectra that we have discussed so far do not permit spectra with two peaks to represent local or distant storms, nor do they allow specification of the sharpness of the peaks. The Ochi-Hubble spectrum[13] is a six-parameter

---

[13] "Six Parameter Wave Spectra," M. Ochi, E. Hubble, *Proc. Fifteenth ASCE Coastal Engineering Conference*, Honolulu, 301-328, 1976.

spectrum that provides such modeling flexibility. It has the form,

$$S_{\eta\eta}^o(\omega) = \frac{1}{4}\sum_{i=1}^{2} \frac{\left(\frac{4\lambda_i+1}{4}\omega_{m_i}^4\right)^{\lambda_i}}{\Gamma(\lambda_i)} \frac{H_{si}^2}{\omega^{4\lambda_i+1}} \exp\left(-\left(\frac{4\lambda_i+1}{4}\right)\left(\frac{\omega_{mi}}{\omega}\right)^4\right),$$

where $\Gamma(\lambda_i)$ is the gamma function, H_{s1}, ω_{m1}, and λ_1 are the significant wave height, modal frequency, and shape factor for the lower frequency components, respectively, and H_{s2}, ω_{m2}, and λ_2 are those for the higher frequency components. Assuming that the entire spectrum is that of a narrow band, the equivalent significant wave height is given by

$$H_s = \sqrt{H_{s1}^2 + H_{s2}^2}.$$

For $\lambda_1 = 1$ and $\lambda_2 = 0$, the spectrum reduces to the Pierson-Moskowitz spectrum. With the assumption that the entire spectrum is narrow band, the value of λ_1 is much larger than λ_2. The Ochi-Hubble spectrum represents unidirectional seas with unlimited fetch. The sea severity and the state of development can be specified by H_{si} and ω_{mi}, respectively. In addition, λ_i can be selected appropriately to control the frequency width of the spectrum. For example, a small λ_i (wider frequency range) describes a developing sea and a large λ_i (narrower frequency range) describes a swell condition. Figure 13.4 shows the Ochi-Hubble spectrum with $\lambda_1 = 2.72$, $\omega_{m1} = 0.626$ rad/s, $H_{s1} = 3.35$ m, $\lambda_2 = 2.72$, $\omega_{m2} = 1.25$ rad/s, and $H_{s2} = 2.19$ m.

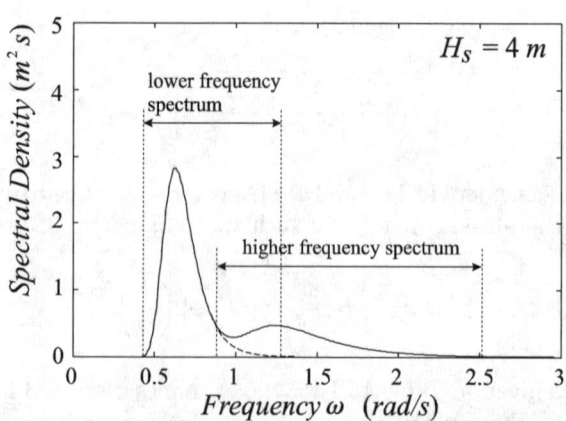

Figure 13.4: Ochi-Hubble spectrum.

Finally, another spectrum that is commonly used is the JONSWAP

13.1. OCEAN CURRENTS AND WAVES

(Joint North Sea Wave Project) spectrum developed by Hasselmann et al.[14] It is a fetch-limited spectrum because the growth over only a limited fetch is taken into account. In addition, the attenuation in shallow water is taken into account. The JONSWAP spectrum is

$$S^o_{\eta\eta}(\omega) = \frac{\alpha g^2}{\omega^5} \exp\left(-1.25 \left(\frac{\omega_m}{\omega}\right)^4\right) \gamma^{\exp\left(-(\omega-\omega_m)^2/2\tau^2\omega_m^2\right)},$$

where γ = peakedness parameter, and τ = shape parameter. The peakedness parameter γ is the ratio of the maximum spectral energy to the maximum spectral energy of the corresponding Pierson-Moskowitz spectrum. That is, when $\gamma = 7$, the peak spectral energy is 7 times that of the Pierson-Moskowitz spectrum. The parameters are valued as follows:

$$\gamma = \begin{cases} 7.0 \text{ for very peaked data} \\ 3.3 \text{ for mean of selected JONSWAP data} \\ 1.0 \text{ for P-M spectrum} \end{cases}$$

$$\tau = \begin{cases} 0.07 \text{ for } \omega \leq \omega_m \\ 0.09 \text{ for } \omega > \omega_m \end{cases}$$

$\alpha = 0.076 \left(\bar{X}\right)^{-0.22}$ or 0.0081 if fetch independent

$\bar{X} = gX/U_w^2$

X = fetch length (nautical miles)

U_w = wind speed (knots)

$\omega_m = 2\pi \cdot 3.5 \cdot (g/U_w)\bar{X}^{-0.33}$.

Figure 13.5 depicts the JONSWAP spectrum for $\alpha = 0.0081$ and $\omega_m = 0.626$ rad/s, for three-peakedness parameters.

13.1.3 Approximation of Spectral Density from Time Series

From the time history of the wave elevation, the spectral density function can be obtained by two methods. The first method is to use the autocorrelation function $R_{\eta\eta}(\tau) = E\{\eta(t)\eta(t+\tau)\}$, which is related to the spectral density function $S_{\eta\eta}(\omega)$ by the Wiener-Khinchine relations (Equations 13.3 and 13.4).

---

[14] *Measurement of Wind-Wave Growth and Swell Decay During the Joint North Sea Wave Project (JONSWAP)*, K. Hasselmann, T.P. Barnett, E. Bouws, H. Carlson, D.E. Cartwright, K. Enke, J.A. Ewing, H. Gienapp, D.E Hasselmann, P. Kruseman, A. Meerburg, P. Muller, D.J. Olbers, K. Richter, W. Sell, and H. Walden, Deutschen Hydrographischen Zeitschrift, Technical Report 13A, 1973.

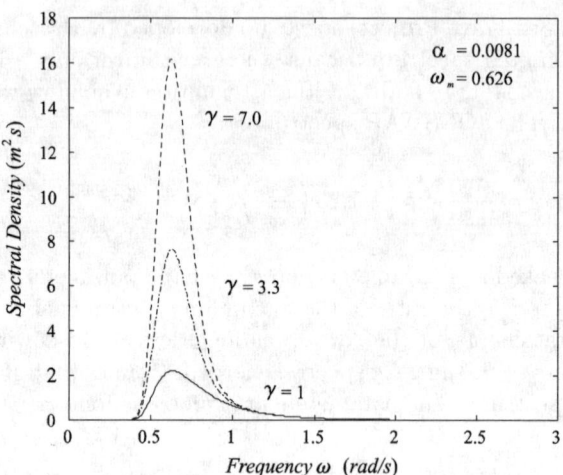

Figure 13.5: JONSWAP spectrum for $\gamma = 1.0$, 3.3, and 7.0.

Assuming that the process is ergodic, the autocorrelation function for a given time history of length T_s can be approximated by

$$\hat{R}_{\eta\eta}(\tau) = \lim_{T_s \to \infty} \frac{1}{T_s - \tau} \int_0^{T_s - \tau} \eta(t)\eta(t+\tau)\,dt, \quad \text{for } 0 < \tau < T_s,$$

where the notation $\hat{}$ is used to emphasize that the variable is an approximation based on a sample time history of length T_s. Then, the spectral density is obtained by taking the Fourier cosine transform of $\hat{R}_{\eta\eta}(\tau)$,

$$\hat{S}_{\eta\eta}(\omega) = \frac{1}{\pi} \int_0^{T_s} \hat{R}_{\eta\eta}(\tau) \cos\omega\tau\,d\tau. \tag{13.5}$$

The second method for obtaining the spectral density function is to use the relationship between the spectral density and the Fourier transform of the time series. They are related by

$$\hat{S}_{\eta\eta}(\omega) = \lim_{T_s \to \infty} \frac{1}{2\pi T_s} \left| \hat{X}(\omega) \hat{X}^*(\omega) \right|, \tag{13.6}$$

where $\hat{X}(\omega)$ is given by

$$\hat{X}(\omega) = \int_0^{T_s} \eta(t) \exp(-i\omega t)\,dt,$$

13.1. OCEAN CURRENTS AND WAVES

and $\hat{X}^*(\omega)$ is the complex conjugate, given by

$$\hat{X}^*(\omega) = \int_0^{T_s} \eta(t) \exp(i\omega t)\, dt.$$

In order to obtain the Fourier transforms of the time series, the Discrete Fourier Transform (DFT) or the Fast Fourier Transform (FFT) procedure can be used. For detailed descriptions of how this is done, refer to Appendix 1 in Tucker.[15] Spectral analysis is almost always carried out via FFTs because it is easier to use and faster than the formal method via correlation function.

It should be noted that the length of the sample time history only needs to be long enough so that the limits converge. Taking a longer sample will not improve the accuracy of the estimate. Instead, one should take many samples or break one long sample into many parts. For n samples, the spectral densities are obtained for each sample time history using either Equation 13.5 or 13.6, and they are averaged to give the estimate.

The determination of the spectral density from wave records depends on the details of the procedure, such as the length of the record, sampling interval, degree and type of filtering and smoothing, and time discretization.

13.1.4 Generation of Time Series from a Spectral Density

In a nonlinear analysis, the structural response is found by a numerical integration in time. Therefore, one needs to convert the wave elevation spectrum into an equivalent time history. The wave elevation can be represented as a sum of many sinusoidal functions with different angular frequencies and random phase angles. That is, write $\eta(t)$ as

$$\eta(t) = \sum_{i=1}^{N} \cos(\omega_i t - \varphi_i)\sqrt{2S_{\eta\eta}(\omega_i)\Delta\omega_i}, \qquad (13.7)$$

where φ_i is a uniformly distributed random number between 0 and 2π, ω_i are discrete sampling frequencies, $\Delta\omega_i = \omega_i - \omega_{i-1}$, and N is the number of partitions. Recall that the area under the spectrum is equal to the variance, σ_η^2. The incremental area under the spectrum, $S_{\eta\eta}(\omega_i)\Delta\omega_i$, can be denoted as σ_i^2 such that the sum of all the incremental areas equals the variance of the wave elevation or $\sigma_\eta^2 = \sum_{i=1}^{N} \sigma_i^2$. The time history can then be written as

$$\eta(t) = \sum_{i=1}^{N} \cos(\omega_i t - \varphi_i)\sqrt{2}\sigma_i.$$

---

[15] *Waves in Ocean Engineering: Measurement, Analysis, Interpretation*, M. Tucker, Ellis Horwood Limited, England, 1991.

The sampling frequencies, ω_i, can be chosen at equal intervals such that $\omega_i = i\omega_1$. However, the time history will then have the lowest frequency of ω_1 and will have a period of $T = 2\pi/\omega_1$. In order to avoid this unwanted periodicity, Borgman[16] suggested that the frequencies be chosen so that the area under the spectrum curve for each interval is equal or $\sigma_i^2 = \sigma^2 = \sigma_\eta^2/N$. The time history is written as

$$\eta(t) = \sqrt{\frac{2}{N}}\sigma_\eta \sum_{i=1}^{N} \cos(\bar{\omega}_i t - \varphi_i), \qquad (13.8)$$

where $\bar{\omega}_i = (\omega_i + \omega_{i-1})/2$. The discrete frequencies ω_i are chosen such that the area between the interval $0 < \omega < \omega_i$ is equal to i/N of the total area under the curve between the interval $0 < \omega < \omega_N$ or

$$\int_0^{\omega_i} S_{\eta\eta}(\omega)\,d\omega = \frac{i}{N}\int_0^{\omega_N} S_{\eta\eta}(\omega)\,d\omega \quad \text{for } i = 1, ..., N,$$

where it is assumed that the area under the spectrum beyond ω_N is negligible. If $\eta(t)$ is a narrow-band Gaussian process, the standard deviation can be replaced by $\sigma_\eta = H_s/4$, and the time history can be written as

$$\eta(t) = \frac{H_s}{4}\sqrt{\frac{2}{N}}\sum_{i=1}^{N}\cos(\bar{\omega}_i t - \varphi_i).$$

Shinozuka[17] proposed that the sampling frequencies $\bar{\omega}_i$ in Equation 13.8 should be randomly chosen according to the density function $f(\omega) \equiv S_{\eta\eta}^o(\omega)/\sigma_\eta^2$. This is equivalent to performing an integration using the Monte Carlo method. The random frequencies ω distributed according to $f(\omega)$ can be obtained from uniformly distributed random numbers x by $\omega = F^{-1}(x)$, where $F(\omega)$ is the cumulative distribution of $f(\omega)$.

The random frequencies obtained in this way are used in Equation 13.8 to generate a sample time series. It should be noted that many sample time histories should be obtained and averaged to synthesize a time history for use in numerical simulations.

13.1.5 Short-Term Statistics

In discussing wave statistics, we often use the term, significant wave, to describe an irregular sea surface. The significant wave is not a physical wave

---

[16] "Ocean Wave Simulation for Engineering Design," L. Borgman, *Journal of the Waterways and Harbors Division, ASCE*, 95:557-583, 1969.

[17] "Monte Carlo Solution of Structural Dynamics," M. Shinozuka, *Computers and Structures*, 2:855-874, 1972.

13.1. OCEAN CURRENTS AND WAVES

that can be seen, but a statistical description of random waves. The concept of significant wave height was first introduced by Sverdrup and Munk[18] as the average height of the highest one-third of all waves. Usually ships cooperate in programs to find sea statistics by reporting a rough estimate of the storm severity in terms of an observed wave height. It is found that this observed wave height is consistently very close to the significant wave height.

Two assumptions that are made in describing short-term wave statistics are stationarity and ergodicity. These assumptions are valid only for "short" time intervals, on the order of a couple of hours or the duration of a storm, but not for weeks or years. The wave elevation is assumed to be weakly stationary so that its autocorrelation is a function of time lag only. As a result, the mean and the variance are constant, and the spectral density is invariant with time. Therefore, the significant wave height and the significant wave period are constant when considering short term statistics. In this case, the individual wave height and wave period are the stochastic variables.

Consider a sample time history of a zero mean random process, as shown in Figure 13.6. The questions that we ask are: how often is a certain level (for example, Z in the figure) exceeded, and how are the maxima distributed? Equivalently, we can ask when can we expect to see that a certain level is exceeded for the first time, and what are the values of the peaks of a random process. The first question is important for determining when a structure may fail due to a one-time excessive load, and the second question is important for establishing when a structure may fail due to cyclic loads.

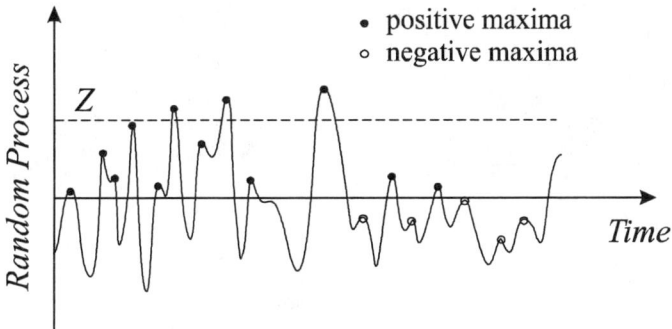

Figure 13.6: A sample time history with maxima highlighted.

---

[18] *Wind, Sea, and Swell: Theory of Relations for Forecasting*, H. Sverdrup, W. Munk, Technical Report 601, U.S. Navy Hydrographic Office, 1947.

It is found that the rate at which a random process $X(t)$ crosses the random variable Z with a positive slope can be calculated using the relation,

$$\nu_{z+} = \int_0^\infty v f_{X\dot{X}}(z, v) \, dv,$$

where $f_{X\dot{X}}(x, \dot{x})$ is the joint probability density function of X and $\dot{X}$. The expected time of the first up-crossing is then the inverse of the crossing rate, or

$$E\{T\} = 1/\nu_{z+}.$$

The probability density function of the maxima, A, can be calculated from

$$f_A(a) = \frac{\int_{-\infty}^0 -w f_{X\dot{X}\ddot{X}}(a, 0, w) \, dw}{\int_{-\infty}^0 w f_{\dot{X}\ddot{X}}(0, w) \, dw},$$

where $f_{X\dot{X}\ddot{X}}(x, \dot{x}, \ddot{x})$ is the joint probability density function of X, $\dot{X}$, and $\ddot{X}$.

If $X(t)$ is a Gaussian process, the joint probability density functions are

$$f_{X\dot{X}}(x, \dot{x}) = \frac{1}{2\pi \sigma_X \sigma_{\dot{X}}} \exp\left[-\frac{1}{2}\left(\frac{x}{\sigma_X}\right)^2 - \frac{1}{2}\left(\frac{\dot{x}}{\sigma_{\dot{X}}}\right)^2\right],$$
$$-\infty < x < \infty, \ -\infty < \dot{x} < \infty,$$

and

$$f_{X\dot{X}\ddot{X}}(x, \dot{x}, \ddot{x})$$
$$= \frac{1}{(2\pi)^{3/2} |M|^{1/2}} \exp\left(-\frac{1}{2}(\{x\} - \{\mu_X\})^T [M]^{-1} (\{x\} - \{\mu_X\})\right),$$

where

$$[M] = \begin{bmatrix} \sigma_X^2 & 0 & \sigma_{\ddot{X}}^2 \\ 0 & \sigma_{\dot{X}}^2 & 0 \\ \sigma_{\ddot{X}}^2 & 0 & \sigma_{\ddot{X}}^2 \end{bmatrix}$$

$$\{x\} - \{\mu_X\} = \begin{bmatrix} x - \mu_X \\ \dot{x} - \mu_{\dot{X}} \\ \ddot{x} - \mu_{\ddot{X}} \end{bmatrix}.$$

13.1. OCEAN CURRENTS AND WAVES

Then, for a stationary Gaussian process, the up-crossing rate is given by

$$\nu_z^+ = \int_0^\infty f_{X\dot{X}}(Z, \dot{x}) \dot{x}\, d\dot{x}.$$

$$= \frac{1}{2\pi\sigma_X\sigma_{\dot{X}}} \exp\left[-\frac{1}{2}\left(\frac{Z}{\sigma_X}\right)^2\right] \int_0^\infty \exp\left[-\frac{1}{2}\left(\frac{\dot{x}}{\sigma_{\dot{X}}}\right)^2\right] \dot{x}\, d\dot{x}$$

$$= \frac{\sigma_{\dot{X}}}{2\pi\sigma_X} \exp\left[-\frac{1}{2}\left(\frac{Z}{\sigma_X}\right)^2\right],$$

and the probability density function of maxima is given by the Rice density function,[19]

$$f_A(a) = \frac{\sqrt{1-\alpha^2}}{\sqrt{2\pi}\sigma_\eta} \exp\left(\frac{-a^2}{2\sigma_\eta^2(1-\alpha^2)}\right)$$
$$+ a\frac{\alpha}{\sigma_\eta^2} \Phi\left(\frac{a\alpha}{\sigma_\eta\sqrt{(\alpha^2-1)}}\right) \exp\left(\frac{-a^2}{2\sigma_\eta^2}\right),$$

where $\Phi(x)$ is the cumulative distribution function of the standard normal random variable,

$$\Phi(x) = \frac{1}{\sqrt{2\pi}} \int_{-\infty}^x \exp\left(-z^2/2\right) dz,$$

and α is the irregularity factor, equivalent to the ratio of the number of zero up-crossings (number of times that $\eta(t)$ crosses zero with a positive slope) to the number of peaks. α ranges from 0 to 1, and it is also equal to

$$\alpha = \frac{\sigma_{\dot{\eta}}^2}{\sigma_\eta \sigma_{\ddot{\eta}}}.$$

If $X(t)$ is a broad-band process, $\alpha = 0$ and the Rice distribution is reduced to the Gaussian probability density function,

$$f_A(a) = \frac{1}{\sqrt{2\pi}\sigma_\eta} \exp\left(\frac{-a^2}{2\sigma_\eta^2}\right) \quad \text{for } -\infty < a < \infty.$$

If $X(t)$ is a narrow-band process, it is guaranteed that it will have a peak whenever $\eta(t)$ crosses its mean. In this case, the irregularity factor is close to unity, and the Rice distribution is reduced to the Rayleigh probability density function given by

$$f_A(a) = \frac{a}{\sigma_\eta^2} \exp\left(-\frac{1}{2}\frac{a^2}{\sigma_\eta^2}\right) \quad \text{for } 0 < a < \infty.$$

---

[19] *Mathematical Analysis of Random Noise*, S. Rice, Dover Publications, 1954.

696   CHAPTER 13. FLUID-INDUCED VIBRATION

In other words, the amplitudes of a *narrow-band stationary Gaussian process* are distributed according to the Rayleigh distribution.

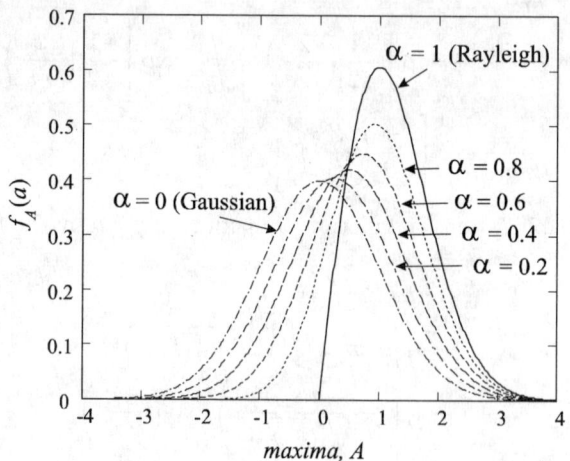

Figure 13.7: Rice distribution for maxima.

Figure 13.7 shows the Rice distribution for various values of α. Note that the Rice distribution includes both positive and negative maxima except when $\alpha = 1$, in which case all the maxima are positive. The positive maxima are the local maxima that occur above the mean of $X(t)$, and the negative maxima are the local maxima that occur below the mean, as shown in Figure 13.6. In some cases, the negative maxima may not have physical meaning. In those cases, we can use the truncated Rice distribution, where only the positive portion of $f_A(a)$ is used. $f_A(a)$ is normalized by the area under the probability density for positive maxima,[20,21]

$$f_A^{trunc}(a) = \frac{f_A(a)}{\int_0^\infty f_A(a)\,da}, \quad a \geq 0.$$

The truncated Rice distribution is shown in Figure 13.8.

If $X(t)$ is the wave elevation, its maxima, A, are the amplitudes of the

---

[20] "On the Statistical Distribution of the Height of Sea Waves," M. Longuet-Higgins, *Journal of Marine Research*, 11(3):245-266, 1952.

[21] "On Prediction of Extreme Values," M. Ochi, *Journal of Ship Research*, 17:29-37, 1973.

13.1. OCEAN CURRENTS AND WAVES

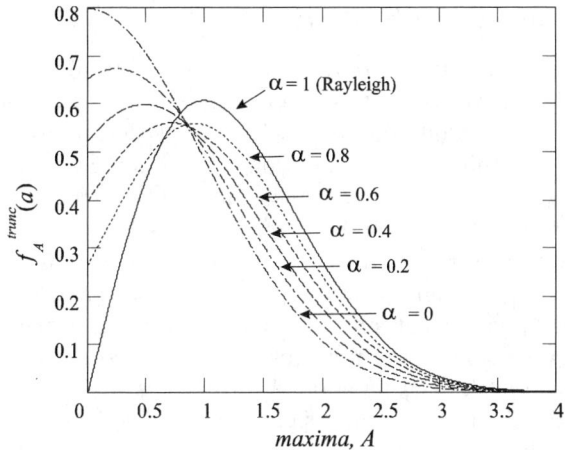

Figure 13.8: Truncated Rice distributions for various α.

wave elevation. The wave height, $H = 2A$, is then distributed according to

$$\begin{aligned} f_H(h) &= f_A(H/2)\frac{dA}{dH} \\ &= \frac{h}{4\sigma_\eta^2}\exp\left(-\frac{1}{2}\frac{h^2}{4\sigma_\eta^2}\right) \quad \text{for } 0 < h < \infty. \end{aligned}$$

For any given wave, the probability that the height is less than h (the cumulative distribution) is

$$F_H(h) = 1 - \exp\left(-\frac{1}{2}\frac{h^2}{4\sigma_\eta^2}\right) \quad \text{for } 0 < h < \infty.$$

If $\eta(t)$ is a stationary narrow-band process so that the peaks are distributed according to the Rayleigh distribution, we find that the root-mean-square wave height, $\sqrt{E\{H^2\}}$, is given by

$$\sqrt{E\{H^2\}} = \int_0^\infty h^2 f_H(h)\, dh = 2\sqrt{2}\sigma_\eta.$$

In addition, it can be shown that the average and the significant wave heights are given by

$$H_o \equiv E\{H\} = \sqrt{2\pi}\sigma_\eta$$
$$H_s \equiv E\{H_{1/3}\} - 4\sigma_\eta,$$

where $E\{H_{1/3}\}$ is the expectation of the highest one-third of the waves.

13.1.6 Long-Term Statistics

Because offshore structures are designed for long life spans, we must also consider long-term wave statistics. Previously, when we considered the short term statistics, the significant wave height and spectrum were assumed to be invariant with time. This assumption is valid only over time periods on the order of days at most. For longer time periods, the significant wave height has its own statistics and is a random variable.

When one uses short-term statistics to describe long-term events, improbable events seem unjustifiably probable. For example, consider the probability that the wave height, distributed according to the Rayleigh distribution as discussed in the previous section, exceeds a certain extreme value. Assume that the mean period of this wave is 10 s and the probability that the height of any given wave is greater than 300 ft is 10^{-10}. The value is small and the occurrence of a 300 ft wave seems improbable. However, the probability that the height will exceed 300 ft at least once in ten years (3×10^8 s) is given by

$$1 - \left(1 - 10^{-10}\right)^{3 \times 10^8 / 10} = 0.997.$$

Thus, the statistical description states that it is almost certain that the wave height will exceed 300 ft at least once in ten years. This prediction is incorrect because short-term statistics are used to make a long-term prediction. Waves of this magnitude do not arrive with this probability.

In order to compute the probability that a wave height will exceed a certain extreme value, we must have statistics for these extreme events. The actual maximum amplitude in a sequence of random amplitudes is a random variable itself. It has a probability distribution with mean value, standard deviation, and other statistical properties. In fact, the distributions of these maximum values are called the Extreme Value Distributions (EVD). Gumbel[22] derived three methods of extrapolation known as three asymptotes. They are the Gumbel, Fretchét, and Weibull distributions. We will discuss the Gumbel and Weibull distributions in the next section. For the moment, consider the concept of the N-year storm.

In long term statistics, we often speak of an N-year storm. This means that for any given year, the probability that an N-year storm will occur is

$$p = \frac{1}{N}.$$

It follows that the probability that m storms will occur in n years is given by

$$\Pr\left(m \ N\text{-year storms in } n \text{ years}\right) = {_nP_m} \left(\frac{1}{N}\right)^m \left(1 - \frac{1}{N}\right)^{n-m},$$

---

[22] *Statistics of Extremes*, E. Gumbel, Columbia University Press, 1958.

13.1. OCEAN CURRENTS AND WAVES

where $_nP_m$ is the permutation, given by

$$_nP_m = \frac{n!}{(n-m)!}.$$

The probability of at least one N-year storm in n years is

$$\Pr\left(\text{at least one } N\text{-year storm in } n \text{ years}\right) = 1 - \left(1 - \frac{1}{N}\right)^n.$$

For large N, the probability can be approximated by $1 - \exp\{-n/N\}$. It should be noted that the probability that exactly one N-year storm will occur in N years is not one but

$$\Pr\left(\text{one } N\text{-year storm in } N \text{ years}\right) = \left(1 - \frac{1}{N}\right)^{N-1}.$$

As $N \to \infty$, we find that

$$\Pr\left(\text{one } N\text{-year storm in } N \text{ years}\right) = 1/e \simeq 0.3679.$$

The probability that at least one N-year storm will occur in N years is

$$\Pr\left(\text{at least one } N\text{-year storm in } N \text{ years}\right) = 1 - \left(1 - \frac{1}{N}\right)^N.$$

As $N \to \infty$, we find that

$$\Pr\left(\text{at least one } N\text{-year storm in } N \text{ years}\right) = 1 - 1/e \simeq 0.6321.$$

Weibull Distribution

The Weibull distribution is found to fit probabilities of extremes quite satisfactorily. For long-term statistics, the significant wave height follows the Weibull distribution closely. The probability density and the cumulative distribution are given by

$$f(h) = \frac{m}{\beta}\left(\frac{h-\gamma}{\beta}\right)^{m-1} \exp\left(-\left(\frac{h-\gamma}{\beta}\right)^m\right)$$
$$F(h) = 1 - \exp\left(-\left(\frac{h-\gamma}{\beta}\right)^m\right) \quad \text{for } \gamma < h, \tag{13.9}$$

where m is called the shape parameter. Manipulating the cumulative distribution, we can write

$$\ln\left(-\ln\{1 - F(h)\}\right) = m\{\ln(h-\gamma) - \ln\beta\},$$

where the left-hand side is known from data. Let $y = \ln(-\ln\{1 - F(h)\})$ and $x = \ln(h - \gamma)$, then y is a straight line with slope m and y-intercept $-m \ln \beta$,

$$y = mx - m \ln \beta.$$

Suppose we have significant wave height data over a long period of time, and our goal is to find the Weibull parameters, γ, β, and m that best fit the distribution of the significant wave heights. These parameters can be determined by the least-squares method or using Weibull paper. Using the latter method, we first guess γ so that the discrete points (x,y) or $[\ln(-\ln\{1 - F(h)\}), \ln(h - \gamma)]$ form a straight line. The slope of this line is m, and the value of y when the line intersects the y axis is $-m \ln \beta$. This method will be illustrated in Section 13.3.3.

Gumbel and Lognormal Distributions

The Gumbel density and distribution are given by

$$\begin{aligned} f(h) &= \alpha \exp[-\alpha(h - \beta)] \exp\{-\exp[-\alpha(h - \beta)]\} \\ F(h) &= \exp\{-\exp[-\alpha(h - \beta)]\} \quad \text{for } -\infty < h < \infty. \end{aligned} \quad (13.10)$$

When $\ln[-\ln F(h)]$ is plotted against h, the result is a line with slope $-\alpha$ and y-intercept $\alpha\beta$.

Another distribution that may be used is the lognormal distribution given by[23]

$$\begin{aligned} f(h) &= \frac{1}{\sqrt{2\pi}\sigma h} \exp\left\{\frac{-(\ln h - \mu)^2}{2\sigma^2}\right\} \\ F(h) &= \Phi\left(\frac{\ln h - \mu}{\sigma}\right) \quad \text{for } 0 \leq h. \end{aligned}$$

13.1.7 Wave Velocities via Linear Wave Theory

The wave velocities that correspond to the wave elevation given in Equation 13.7 can be obtained by using linear wave theory. Linear wave theory, also called Airy wave theory, sinusoidal wave theory, and small-amplitude wave theory, is the simplest wave theory. It is also the most important wave theory because it forms the basis for the probabilistic spectral description of waves.

Linear wave theory assumes that the wave height is small compared to the wavelength and water depth. In addition, fluid particles are assumed

---

[23] "Statistical Distribution Patterns of Ocean Waves and of Wave Induced Ship Stresses and Motions with Engineering Applications," N. Jasper, *Transactions of the Society of Naval Architects and Marine Engineers*, 64:375-432, 1954.

13.1. OCEAN CURRENTS AND WAVES

to follow a circular orbit. The reader should refer to LeMehaute[24] and Kinsman for detailed descriptions.

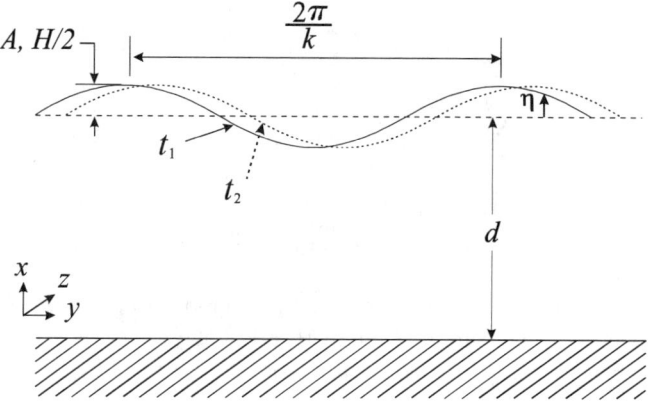

Figure 13.9: A schematic of a simple sinusoidal wave shown at two different times.

In linear wave theory, the surface elevation is given by

$$\eta(y,t) = A\cos(\omega t - ky),$$

which is a plane wave traveling to the right in Figure 13.9. Linear wave theory relates this sinusoidal surface elevation to the wave velocities given by

$$w_y(x,y,t) = A\omega \frac{\cosh kx}{\sinh kd} \cos(\omega t - ky)$$

$$w_x(x,y,t) = A\omega \frac{\sinh kx}{\sinh kd} \sin(\omega t - ky),$$

where k, ω, and A are wave number, angular frequency, and amplitude of a surface wave. The velocities vary with time, horizontal coordinate y, and depth x measured from the ocean floor. The wave velocities are sinusoidal in y and t, but exponentially decrease with the distance from the surface. The frequency ω is related to the wave number k by the dispersion relation,

$$\omega^2 = gk \tanh kd,$$

where d is the water depth. For deep water, $\tanh kd$ approaches unity and the frequency-squared is given by

$$\lim_{d\to\infty} \omega^2 = gk.$$

---

[24] *Introduction to Hydrodynamics and Water Waves*, B. LeMehaute, Springer-Verlag, New York, 1976.

For the surface elevation defined in Equation 13.7 in terms of a spectral density, the linear wave profile and wave velocities are given by

$$\eta(y,t) = \sum_{i=1}^{N} \cos(\omega_i t - k_i y - \varphi_i)\sqrt{2S_{\eta\eta}(\omega_i)\Delta\omega_i}$$

$$w_y(x,y,t) = \sum_{i=1}^{N} \omega_i \frac{\cosh k_i x}{\sinh k_i d} \cos(\omega_i t - k_i y - \varphi_i)\sqrt{2S_{\eta\eta}(\omega_i)\Delta\omega_i}$$

$$w_x(x,y,t) = \sum_{i=1}^{N} \omega_i \frac{\sinh k_i x}{\sinh k_i d} \sin(\omega_i t - k_i y - \varphi_i)\sqrt{2S_{\eta\eta}(\omega_i)\Delta\omega_i}.$$

The wave accelerations can be obtained by differentiating the wave velocities with respect to time. Sample time histories of the wave velocity and acceleration can be obtained by using either Borgman's or Shinozuka's method.

13.2 Fluid Forces in General

Following is a list of the major types of forces that a fluid can exert on a body.

1. *Drag Force* – The drag force is due to the pressure difference between the downstream and upstream flow region. It can be thought of as the force required to hold a body stationary in a fluid of constant velocity. The drag force is proportional to the square of the velocity of the fluid relative to the structure.

2. *Inertia Force* – The inertia force is the force exerted by the fluid while it accelerates and decelerates as it passes the structure. It is also the force required to hold a rigid structure in a uniformly accelerating flow, and it is proportional to the fluid acceleration. The concept of the inertia force in an inviscid flow was first formulated by Lamb.[25]

3. *Added Mass* – As the body accelerates or decelerates in a stationary fluid, the body carries a certain amount of the surrounding fluid along with it. This entrained fluid is called the *added, apparent* or *virtual mass*. In order to accelerate the body, additional force is required to accelerate or decelerate the added mass.

4. *Diffraction Force* – The diffraction force is due to the scattering of an incident wave on the surface of the structure. It is important when the body is large compared to the wavelength of the incident wave.

---

[25] *Hydrodynamics*, H. Lamb, Cambridge University Press, New York, Sixth Edition, 1945. Now available in a Dover edition.

13.2. FLUID FORCES IN GENERAL

5. *Froude-Kryloff Force* – The Froude-Kryloff force is the pressure force on the structure due to the incident wave assuming that the structure does not exist and does not interfere with the incident wave.

6. *Lift Force* – The lift force is due to nonsymmetrical separation of the fluid or due to vortices that are shed in a nonsymmetrical way. The component of the force perpendicular to the flow direction is the lift force.

7. *Wave Slamming Force* – The wave slamming force is due to a single occasional wave with a particularly high amplitude and energy, and it may be important at the free surface. Sarpkaya and Isaacson reviewed the research on slamming of water against circular cylinders. Miller[26,27] found that the peak wave slamming force on a rigidly held horizontal circular cylinder is proportional to the square of the horizontal water particle velocity.

13.2.1 Wave Force Regime

Various types of forces are caused by waves and currents. In some cases, one type of force may be dominant. Hogben[28] reviewed the literature on fluid forces in various regimes.

The load regime of importance is shown schematically, for the case of a vertical cylinder, in Figure 13.10 in terms of H/D and $\pi D/\lambda$, where H is the wave height, D is the cylinder diameter, and λ is the wavelength. When linear wave theory is used, H/D is related to the *Keulegan-Carpenter* number K by

$$K = \pi H/D.$$

The Keulegan-Carpenter number gives a measure of the importance of drag force relative to the inertia force. The term $\pi D/\lambda$ is called the diffraction parameter, and it determines the importance of the diffraction effect. As H/D increases, the drag force becomes more important and the inertia force becomes less important. As $\pi D/\lambda$ increases, the diffraction force becomes more important.

---

[26] *Wave Slamming Loads on Horizontal Circular Elements of Offshore Structures*, B. Miller, Technical Report RINA Paper No. 5, Journal of the Royal Institute of Naval Architecture, 1977.

[27] *Wave Slamming on Offshore Structures*, B. Miller, Technical Report No. NMI-R81, National Maritime Institute, 1980.

[28] "Wave Loads on Structures," N. Hogben, *Marine Science Communications,* Vol. 4, No. 2, 1978, pp. 89-119,

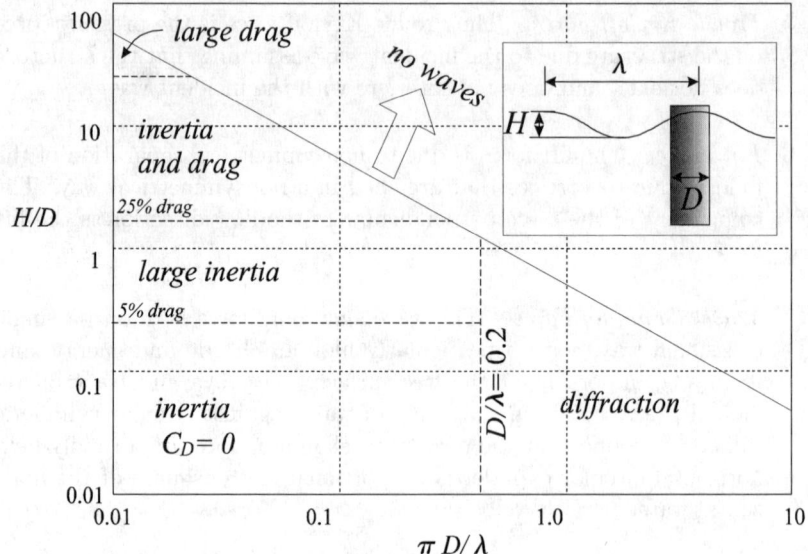

Figure 13.10: Load regimes near surface.

Using linear wave theory, the maximum drag force to the maximum inertia force can be written as

$$\frac{f_{drag}}{f_{inertia}} = \frac{1}{2\pi}\frac{H}{D} = \frac{K}{2\pi^2}.$$

Using this relation, we find that the drag force is 5% of the inertial force when $H/D = 0.314$. The Morison equation may be used for $D/\lambda < 0.2$ and $f_{drag}/f_{inertia} > 0.1$, or so. It should be noted that Figure 13.10 is valid only near the surface. The drag force is predominant for a cylinder that extends from the bottom to the near surface, so that the Morison equation may be used.

As an example, consider a fixed jacket platform with legs of diameter 10 m and bracings of diameter 0.8 m. For a 10-year storm with $\lambda = 100$ m and $H = 8$ m, $H/D = 0.8$ and $D/\lambda = 0.1$. Similarly, for the bracings, $H/D = 10$ and $D/\lambda = 0.08$. Figure 13.10 shows that the inertia force is dominant for the legs, and both inertia and the drag forces are important for the bracings.

13.2.2 Wave Forces on Small Structures – Morison Equation

The added mass, M_A, can be written as

$$M_A = C_A M_{disp},$$

where C_A, is called the *added-mass coefficient*, and M_{disp} is the mass of the fluid displaced by the structure. For a cylinder with a diameter D and height h, the displaced fluid mass is $\pi D^2 h/4$.

It should be noted that the added mass is a tensor quantity. That is, we can speak of the added mass force in the x_i direction due to the acceleration of the body in the x_j direction, denoted as M_{ij}^A. M_{ij}^A; these are symmetric and so the added mass force in the x_i direction due to the acceleration in the x_j direction is equal to the added mass force in the x_j direction due to the acceleration in the x_i direction. The off-diagonal terms are not zero if the cross section is not symmetric.

Similarly, the inertia force can be written as

$$F_M = C_M M_{disp} \dot{w}, \qquad (13.11)$$

where the proportionality constant, C_M, is called the *inertia coefficient*.

The added mass and the inertia effects are often neglected for a body vibrating in air since the displaced air mass is negligible.

The drag force is proportional to the square of the fluid velocity w, the density of the fluid ρ, and the area of the body projected onto the plane perpendicular to the flow direction A_f,

$$F_D = \frac{1}{2} C_D \rho A_f w \left| w \right|,$$

where C_D is the *drag coefficient*. The absolute value sign is used to ensure that the drag force always acts in the direction of the flow. For a cylinder with a diameter D and height h, the projected area $A_f = Dh$.

For a body with non-zero velocity, the drag force is given by

$$F_D = \frac{1}{2} C_D \rho A_f \left(w - v \right) \left| w - v \right|, \qquad (13.12)$$

where $w - v$ is the velocity of the fluid relative to the body.

Morison[29] and his colleagues combined the inertia and drag terms (Equations 13.11 and 13.12) so that the fluid force on a body is given by

$$f = \frac{1}{2} C_D \rho A_f w \left| w \right| + C_M M_{disp} \dot{w}.$$

---

[29] "The Force Exerted by Surface Waves on Piles," J. Morison, M. O'Brien, J. Johnson, and S. Schaaf, *Petroleum Transactions AIME*, 189:149-157, 1950.

For a cylinder, the fluid force per unit length can be written as

$$f = C_D \rho \frac{D}{2} w |w| + C_M \rho \pi \frac{D^2}{4} \dot{w}.$$

For a cylinder moving with velocity v, the Morison force is given by

$$f = C_D \rho \frac{D}{2} (w - v) |w - v| + C_M \rho \pi \frac{D^2}{4} \dot{w}.$$

Inclined Cylinder

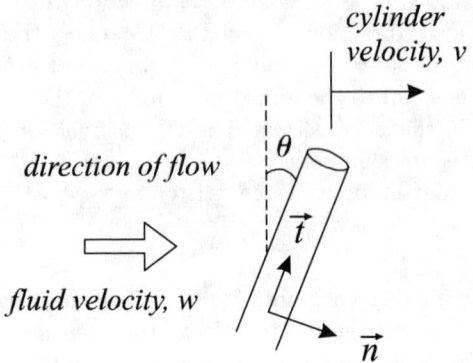

Figure 13.11: Inclined cylinder.

Consider the inclined cylinder shown in Figure 13.11. The direction of the flow makes an angle of θ with the cylinder. Often, only the fluid force in the normal direction is considered. The normal component is given by

$$f^n = C_D \rho \frac{D}{2} (w^n - v^n) |w^n - v^n| + C_M \rho \pi \frac{D^2}{4} \dot{w}^n, \tag{13.13}$$

where the superscript is used for the normal component. The term, $w^n - v^n$, is the normal component of the relative velocity of the fluid with respect to the structure. Suppose that fluid is flowing to the right, and the cylinder is also moving to the right as shown in Figure 13.11. The normal components of the fluid and cylinder velocities are

$$w^n = |w| \cos \theta, \quad v^n = |v| \cos \theta.$$

In three dimensions, it may be difficult to picture what the normal component should be. In that case, the normal component is determined using the formula

$$(w^n - v^n) \vec{n} = \vec{t} \times (\vec{w} - \vec{v}) \times \vec{t}, \tag{13.14}$$

13.2. FLUID FORCES IN GENERAL

where $\vec{t}$ is the unit vector tangent to the cylinder, $\times$ is the cross product, and $\vec{n}$ is the unit vector normal to the cylinder. Note that the normal direction depends on the direction of the flow as well as the inclination of the cylinder.

In some cases, the tangential drag force may be included, and it can be written as

$$f^t = C_T \rho \frac{D}{2} \left(w^t - v^t\right) \left|w^t - v^t\right|, \qquad (13.15)$$

where C_T is the tangential drag coefficient. Note that C_T is usually a very small number.

The normal component of the fluid force is more dominant than the tangential component. It may seem strange that the fluid force does not act in the direction of the fluid motion. Instead, the force is predominantly in the normal direction defined by Equation 13.14. In Section 13.3.1, we will demonstrate what this means by looking at a towing cable.

Determination of Fluid Coefficients

The drag, inertia- and added-mass coefficients must be obtained by experiment. However, for a long cylinder, C_M approaches its theoretical limiting value (uniformly accelerated inviscid flow) of 2 and C_A approaches unity [Wilson, Lamb]. In reality, the inertia and drag coefficients are functions of at least three parameters [Wilson],

$$C_M = C_M(\text{Re}, K, \text{ cylinder roughness})$$
$$C_D = C_D(\text{Re}, K, \text{ cylinder roughness}),$$

where Re is the Reynolds number, and K is the Keulegan-Carpenter number,

$$\text{Re} \equiv \frac{\rho_f U D}{\mu}$$
$$K \equiv \frac{UT}{D},$$

where ρ_f is the density of the fluid, U is the free stream velocity, D is the diameter of the structure, μ is the dynamic or absolute viscosity, and T is the wave period.

Sarpkaya looked at the variation of these hydrodynamic coefficients extensively and obtained the plots reproduced in Figures 13.12, 13.13, and 13.14. Figure 13.12 shows the inertia and drag coefficients for a smooth cylinder as a function of K for various values of Reynolds number and the reduced frequency β, defined by $\beta = \text{Re}/K$. From this figure, we find that for low Re and β, the inertial coefficient decreases and the drag coefficient

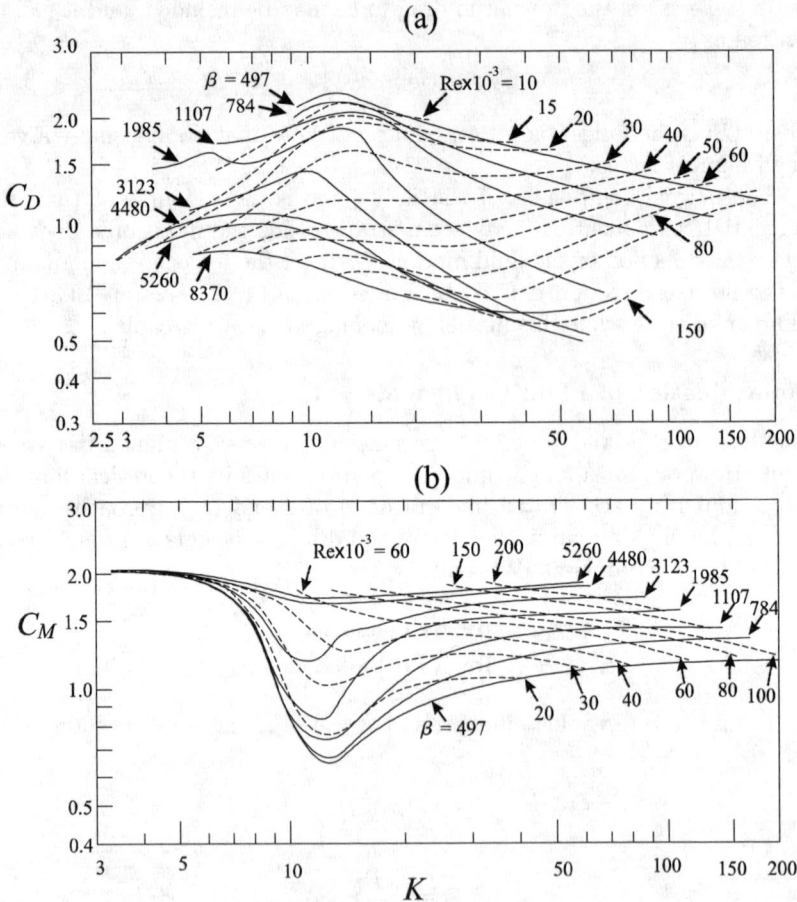

Figure 13.12: Drag and inertia coefficients as functions of K for various values of Re and β. ["In-Line and Transverse Forces on Cylinders in Oscillating Flow at High Reynolds Numbers," T. Sarpkaya, *Proceedings of the Eighth Offshore Technology Conference*, Houston, OTC 2533:95-108, 1976.]

13.2. FLUID FORCES IN GENERAL

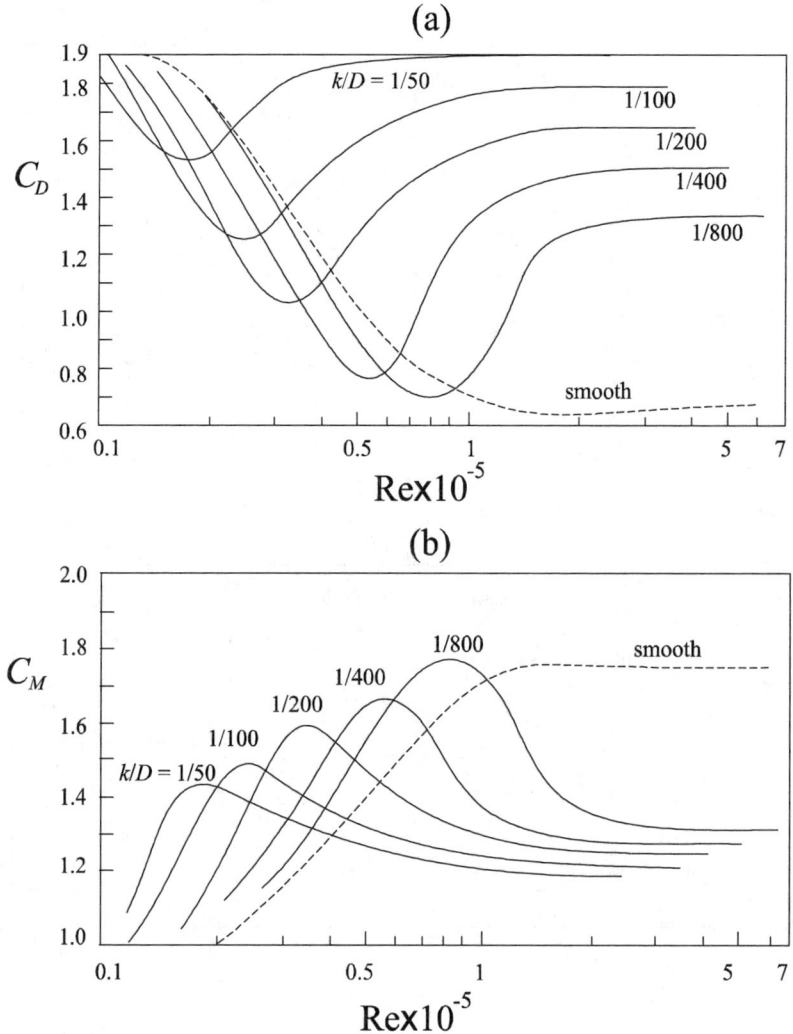

Figure 13.13: Drag and inertia coefficients for a rough cylinder as functions of Re for various values of cylinder roughness (as measured by k/D) for $K = 20$. ["Wave Forces on Rough-Walled Cylinders at High Reynolds Numbers," T. Sarpkaya, N. Collins, S. Evans, *Proceedings of the Ninth Offshore Technology Conference*, Houston, OTC 2901:175-184, 1977.]

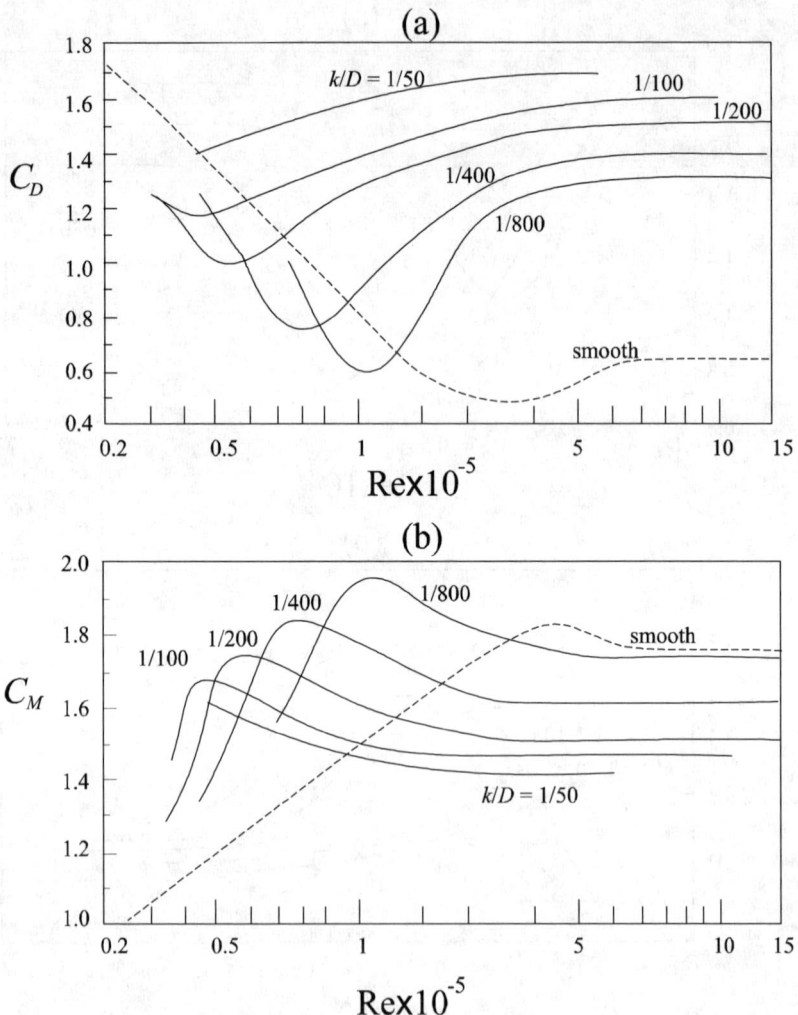

Figure 13.14: Drag and inertia coefficients for a rough cylinder as functions of Re for various values of cylinder roughness (as measured by k/D) for $K = 60$. ["Wave Forces on Rough-Walled Cylinders at High Reynolds Numbers," T. Sarpkaya, N. Collins, S. Evans, *Proceedings of the Ninth Offshore Technology Conference*, Houston, OTC 2901:175-184, 1977.]

13.2. FLUID FORCES IN GENERAL

increases at about $10 < K < 15$. It is found that the drop and the increase in these coefficients are due to shedding vortices, which also exert forces perpendicular to the structure and the flow.

Figures 13.13 and 13.14 show the inertia and drag coefficients for a rough cylinder, whose roughness is measured by k/D. k is a roughness parameter. Figure 13.13(a) shows a drop in the drag coefficient for Reynolds number between 10^4 and 10^5, a behavior called the "drag crisis." For a larger Reynolds number, the drag coefficient stays constant. As the surface becomes rougher, the drop occurs at lower Reynolds number, and the drag coefficients for the larger Reynolds number increases.

Figures 13.12, 13.13, and 13.14 can be used to obtain proper values of the drag and inertia coefficients for a fluid with known Reynolds number, Keulegan-Carpenter number, and cylinder roughness.

13.2.3 Vortex-Induced Vibration

When the flow passes around a fixed cylinder, for a very low Reynolds number ($0 < \text{Re} < 4$), the flow separates and reunites smoothly. When the Reynolds number is between 4 and 40, eddies are formed and are attached to the downstream side of cylinder. They are stable, and there is no oscillation in the flow. For a flow with a Reynolds number greater than about 40, the fluid near the cylinder starts to oscillate due to shedding vortices. These shedding vortices exert an oscillatory force on the cylinder in the direction perpendicular to both the flow and the structure. The frequency of oscillation is related to a nondimensionalized parameter, the Strouhal number, defined by

$$St \equiv \frac{f_v D}{U},$$

where f_v is the frequency of oscillation, U is the steady velocity of the flow, and D is the diameter of the cylinder. For circular cylinders, the Strouhal number stays roughly at 0.22 for laminar flow $\left(10^3 < \text{Re} < 2 \times 10^5\right)$ and 0.3 for turbulent flow.[30]

The lift force due to these shedding vortices can be written as

$$f_L(t) = \frac{1}{2}\rho A_f C_L U^2 \cos 2\pi f_v t,$$

where C_L is the lift coefficient, which is also a function of Re, K, and the surface roughness. Experimental data on lift coefficients show considerable scatter with typical values ranging from 0.25 to 1. For smooth cylinders, the lift coefficient approaches about 0.25 as Re and K increase. It should be noted that the vortex shedding forces are not generally correlated on

---

[30] *Dynamics of Offshore Structures*, M. Patel, Butterworths and Co., 1989.

the entire cylinder length. That is, the phase of the vortex shedding forces varies over the length. The correlation length, the length over which vortex shedding is synchronized, for a stationary cylinder is about three to seven diameters for laminar flow. If sectional forces are randomly phased, the net effect will be small. The total force on a cylinder of length L will be only a fraction of Lf_L. This fraction is called the *joint acceptance* and depends on the ratio of the correlation length to the total length.

When the flow passes by a cylinder that is free to vibrate, the shedding frequency is also controlled by the movement of the cylinder. When the shedding frequency is close to the first natural frequency of the cylinder (± 25 to 30% of the natural frequency [Sarpkaya and Isaacson]), the cylinder takes control of the vortex shedding frequency. The vortices will shed at the fundamental structural natural frequency instead of at the frequency determined by the Strouhal number. This is called lock-in or synchronization, which is a result of nonlinear interaction between the oscillation of the body and the action of the fluid. Figure 13.15 shows the shedding frequency as a function of flow velocity in the presence of a structure. f_1 and f_2 are the first two natural frequencies of the structure.

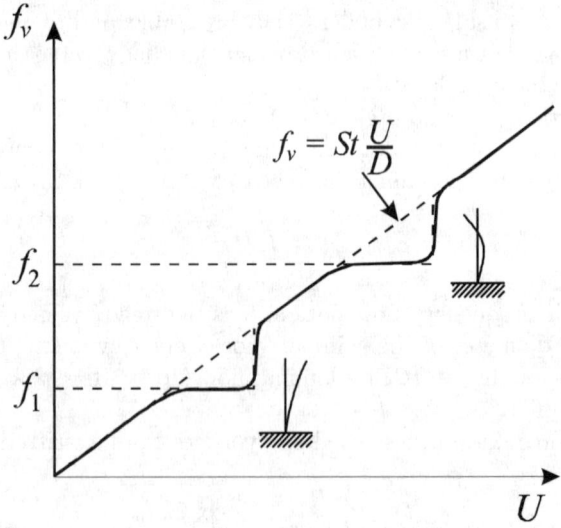

Figure 13.15: An example of fluid elastic synchronization, with lock-in at the fundamental two natural frequencies of the beam.

The amplitude of the structural response and the range of the fluid velocity over which the lock-in phenomenon persists are functions of a reduced damping parameter, which is the ratio of the damping force to the exciting force.[31,32] If the reduced damping parameter is small, the lock-in can persist over a greater range of flow velocity.

The existing models for vortex induced oscillation for a rigid cylinder include single degree of freedom models and coupled models. The single degree of freedom models assume that the effect of vortex shedding is an external forcing function, which is not affected by the motion of the body. The coupled models assume that the equations that govern the motion of the structure and the lift coefficients are coupled so that the fluid and the structure affect each other.[33,34]

13.3 Examples

Four examples are given in this section. The first example illustrates the roles of the normal and the tangential components of the drag force on the static configuration of a towing cable. The second example shows how the equation of motion of an articulated tower can be formulated in the presence of surrounding fluid. The third example shows how to choose a single significant wave height to represent a certain condition from significant wave height data over a long period of time. The last example shows how to reconstruct a time series from a given spectrum.

13.3.1 Static Configuration of a Towing Cable

For the purposes of ocean surveillance, oceanographic or geographic measurements, or ocean exploration, marine cables with instrument packages or Remotely Operated Vehicles are often towed behind ships or submarines. For example, the goal of the VENTS program by the National Oceanic and Atmospheric Administration (NOAA) is to conduct research on the impact and consequences of submarine volcanoes and hydrothermal venting on the global ocean. In attempts to locate and map the distributions of hydrothermal plumes in the Mid-Ocean Ridge system, an instrument package called a CTD (Conductivity, Temperature and Depth Sensors) is towed behind a ship.

---

[31] "Prediction of Lock-In Vibration on Flexible Cylinders in Sheared Flow," J. Vandiver, *Proceedings of the 1985 Offshore Technology Conference*, Paper No. 5006, Houston, 1985.

[32] "Dimensionless Parameters Important to the Prediction of Vortex-Induced Vibration of Long, Flexible Cylinders in Ocean Currents," J. Vandiver, *Journal of Fluids and Structures*, 7(5):423-455, 1993.

[33] *A Study of Vortex-Induced Vibration*, K. Billah, Ph.D. Dissertation, Princeton University, 1989.

[34] "An Overview of Modeling and Experiments of Vortex-Induced Vibration of Circular Cylinders," R.D. Gabbai, H. Benaroya, *Journal of Sound and Vibration*, 282 (2005) 575-616.

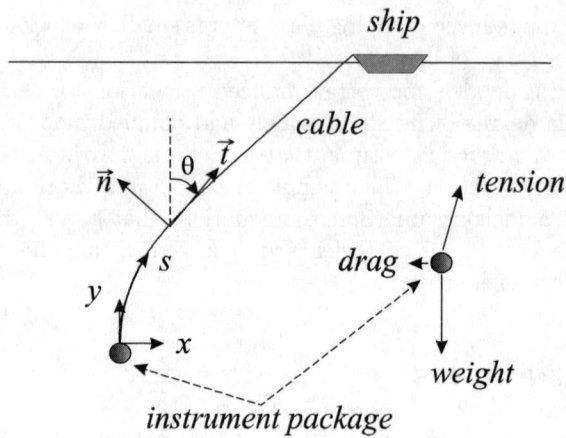

Figure 13.16: Towed system in equilibrium and the forces acting on the towed free body.

Consider a cable and a body towed behind a ship at a constant velocity with no current, as shown in Figure 13.16. What kind of shape will the cable take? What will be the distance between the ship and the towed body?

We immediately recognize that this is equivalent to having a stationary ship with a steady current in the opposite direction. The equation of motion is given by

$$\sum \vec{F} = m\vec{a}(s,t) = \vec{0}$$
$$= \frac{\partial}{\partial s}\left(T\vec{t}\right) + f^n\vec{n} + f^t\vec{t} + mg\vec{k},$$

where m is the mass of the cable per unit length, $\vec{a}(s,t)$ is the acceleration of the cable, s is the coordinate along the cable, T is the tension that is a function of s, $\left(\vec{t},\vec{n},\vec{b}\right)$, is the set of unit vectors of the curvilinear coordinate system, $\vec{k}$ is the unit vector downward in the direction of gravity, g is the gravitational acceleration, f^n is the normal drag force, and f^t is the tangential drag force. The added mass and the inertial terms are zero because the fluid acceleration and the cable acceleration are zero. The normal and tangential drag forces are given by Equations 13.13 and 13.15. In this example, they are given by

$$f^n = C_D\rho\frac{D}{2}U^2\cos^2\theta$$
$$f^t = -C_T\rho\frac{D}{2}U^2\sin^2\theta.$$

13.3. EXAMPLES

The corresponding scalar equations are given by

$$\frac{dT}{ds} - C_T \rho \frac{D}{2} U^2 \sin^2 \theta - mg \cos \theta = 0$$
$$-T\frac{d\theta}{ds} + C_D \rho \frac{D}{2} U^2 \cos^2 \theta - mg \sin \theta = 0, \qquad (13.16)$$

where θ is the angle that the tangential vector makes with the vertical, measured positive clockwise. Note that we have used the following relations,

$$\partial \vec{t}/\partial s = -(\partial \theta/\partial s)\,\vec{n}$$
$$\vec{k} = -\cos\theta \vec{t} - \sin\theta \vec{n}.$$

Equation 13.16 shows that the tangential components of the external forces act to increase the tension, while the normal components cause the towline to bend. Since the normal component of the drag force is much larger than the tangential component, most of the fluid force is used to *turn* the cable.

From the force diagram (in Figure 13.16), the angle that the cable makes with the vertical where it is connected to the towed body is given by

$$|T(s)\cos\theta(s)|_{s=0} = W$$
$$|T(s)\sin\theta(s)|_{s=0} = Drag.$$

Once we know the weight and the drag force on the towed body, the tension and the angle at $s = 0$ can be found. If the drag is negligible compared to the weight, the cable must be near vertical and the tension must be equal to the weight of the towed body at $s = 0$,

$$T(0) \simeq W \quad \text{and} \quad \theta(0) \simeq 0.$$

For now, assume that this is the case. Then, with these initial conditions, the system of ordinary differential equations (Equation 13.16) can be solved numerically for $T(s)$ and $\theta(s)$.[35] The Cartesian coordinates, x and y, are related to θ by

$$\frac{dx}{ds} = \sin\theta \quad \text{and} \quad \frac{dy}{ds} = \cos\theta,$$

and can also be obtained by integrating numerically.

---

[35] Even very simple finite difference equations will work. For example, a set of equations,

$$T_{i+1} = T_i + (mg\cos\theta_i - C_T \rho \frac{D}{2} U^2 \sin^2 \theta_i)\Delta s$$
$$\theta_{i+1} = \theta_i - \left(mg\sin\theta_i + C_D \rho \frac{D}{2} U^2 \cos^2 \theta_i\right) \Delta s/T_i,$$

where $T_i = T(i\Delta s)$, is used here, and it works very well for $\Delta s = 0.05$.

716 CHAPTER 13. FLUID-INDUCED VIBRATION

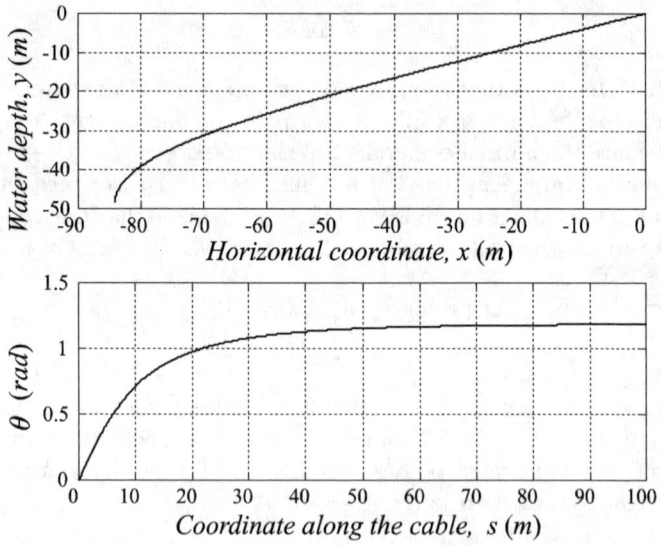

Figure 13.17: The equilibrium configuration of a towed cable and the angle that the cable makes with the vertical when $mg = 1.5$ N/m, $C_D \rho \frac{D}{2} U^2 = 10$ N/m, $C_T \rho \frac{D}{2} U^2 = 0.1$ N/m, and $W = 100$ N.

13.3. EXAMPLES

Figure 13.17 plots the cable configuration for $mg = 1.5 \text{ N/m}$, $C_D \rho \frac{D}{2} U^2 = 10 \text{ N/m}$, $C_T \rho \frac{D}{2} U^2 = 0.1 \text{ N/m}$, $W = 100 \text{ N}$, and cable length 100 m. Care is taken so that the ship is located at $x = 0$ and $y = 0$.

It is interesting to note that θ approaches a critical value, and the shape gradually becomes linear toward the ship. Mathematically, $d\theta/ds$ becomes zero. This is when the drag force is completely balanced by the normal component of the cable weight. The angle at which this occurs, θ_{cr}, can be obtained from the second governing equation, and

$$mg \sin \theta_{cr} = -f^n$$
$$\frac{\sin \theta_{cr}}{\cos^2 \theta_{cr}} = C_D \rho \frac{D}{2} U^2 \frac{1}{mg}.$$

In our case, $\theta_{cr} = 1.184$ rad, and this value agrees with Figure 13.17.

13.3.2 Fluid Forces on an Articulated Tower

Offshore structures are used in the oil industry as exploratory, production, oil storage, and oil landing facilities. They are designed to be self-supporting and sufficiently stable for offshore activities such as drilling and production. An articulated tower, as seen in Figure 13.18, is an example of an offshore platform that consists of a base, shaft, universal joint that connects the base and the shaft, ballast chamber, buoyancy chamber, and deck. The ballast chambers provide the extra weight so that the tower's bottom stays on the ocean floor, and the buoyancy chamber adds to the buoyancy so that the tower does not fall.

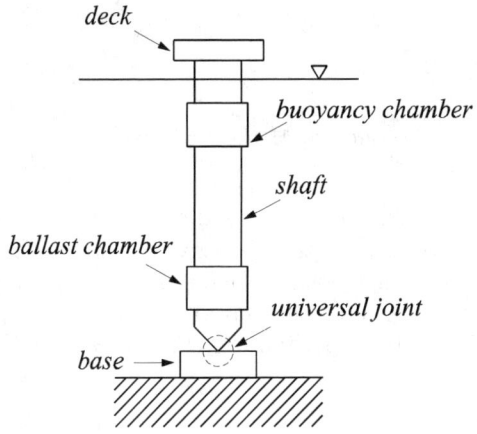

Figure 13.18: Schematic of an articulated tower.

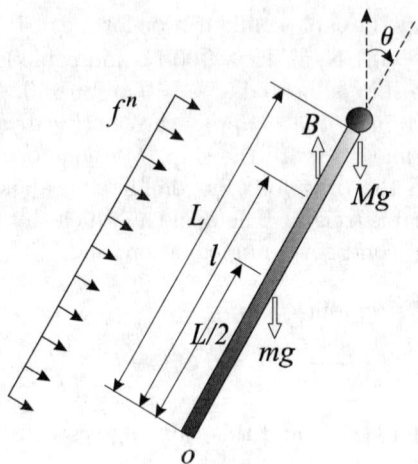

Figure 13.19: Free body diagram.

An articulated tower can effectively be modeled as a rigid inverted pendulum, where the deck is modeled as a point mass, the shaft as a uniform rigid bar, and the buoyancy chamber by a point buoyancy.

In two dimensions, the motion of the tower can be described with a single degree of freedom.[36,37] The equation of motion in terms of the tower's deflection angle is obtained by summing the moment about the point o in Figure 13.19, and is given by

$$I\frac{d^2\theta}{dt^2} = \sum M_o$$
$$= mg\frac{L}{2}\sin\theta + MgL\sin\theta - Bl\sin\theta + \int_0^L f^n x\, dx,$$

where I is the mass moment of inertia about the point o, given by the expression $I = mL^2/3 + ML^2$, m is the mass of the shaft, g is the gravitational acceleration, L is the length of the shaft, M is the point mass at the top, B is the buoyancy provided by the buoyancy chamber, l is its moment arm, f^n is the normal fluid force per unit length, and x is the coordinate along the shaft from o.

---

[36] "Motion Analysis of Articulated Tower," S. Chakrabarti, D. Cotter, *J. of the Waterway, Port, Coastal, and Ocean Division, ASCE*, 105:281-292, 1979.

[37] *Dynamic Response of an Offshore Articulated Tower*, P. Bar-Avi, Ph.D. Dissertation, Rutgers University, 1996.

13.3. EXAMPLES

The fluid force per unit length in the normal direction is given by

$$f^n = C_D \rho \frac{D}{2} (w^n - v^n) |w^n - v^n| + C_M \rho \pi \frac{D^2}{4} \dot{w}^n - C_A \rho \pi \frac{D^2}{4} a^n,$$

where the last term is the force in the normal direction due to the added mass. v^n and a^n are the velocity and the acceleration of the body in the normal direction and are given by

$$v^n = x \frac{d\theta}{dt} \quad \text{and} \quad a^n = x \frac{d^2\theta}{dt^2}.$$

If the surrounding fluid is stationary, the normal velocity and the acceleration of the fluid (w and $\dot{w}$) are each zero. Then, the moment due to the fluid force is given by

$$\int_0^L f^n x \, dx = \int_0^L \left(-C_D \rho \frac{D}{2} x^2 \left(\frac{d\theta}{dt}\right)^2 \text{sign}\left(\frac{d\theta}{dt}\right) \right.$$
$$\left. + C_A \rho \pi \frac{D^2}{4} x \frac{d^2\theta}{dt^2} \right) x \, dx$$
$$= \left(-C_D \rho \frac{D}{2} \frac{L^4}{4} \left(\frac{d\theta}{dt}\right)^2 \text{sign}\left(\frac{d\theta}{dt}\right) + C_A \rho \pi \frac{D^2}{4} \frac{L^3}{3} \frac{d^2\theta}{dt^2} \right),$$

and the equation of motion is given by

$$\left(m \frac{L^2}{3} + ML^2 + C_A \rho \pi \frac{D^2}{4} \frac{L^3}{3} \right) \frac{d^2\theta}{dt^2} = \left(mg\frac{L}{2} + MgL - Bl_b \right) \sin \theta$$
$$- C_D \rho \frac{D}{2} \frac{L^4}{4} \left(\frac{d\theta}{dt}\right)^2 \text{sign}\left(\frac{d\theta}{dt}\right).$$

Note that the normal fluid drag force adds directly to the restoring moment in the case of a rigid bar. The equation of motion can be solved numerically given the initial conditions, $\theta(0)$ and $d\theta(0)/dt$.

The equation of motion can be simplified if it is assumed that the angle of rotation θ is small. More specifically, assume that θ^2 is negligible when compared to 1, then[38] $\sin \theta \simeq \theta$. The equation of motion can then be simplified to

$$\left(m \frac{L^2}{3} + ML^2 + C_A \rho \pi \frac{D^2}{4} \frac{L^3}{3} \right) \frac{d^2\theta}{dt^2} + \left(Bl_b - mg\frac{L}{2} - MgL \right) \theta$$
$$+ C_D \rho \frac{D}{2} \frac{L^4}{4} \left(\frac{d\theta}{dt}\right)^2 \text{sign}\left(\frac{d\theta}{dt}\right) = 0,$$

---

[38] This is called the small-angle assumption.

720   CHAPTER 13. FLUID-INDUCED VIBRATION

Table 13.2: Number of Occurrences of Various Sea States

Significant Wave Height, h m	Number of Occurrences	Sum
<1	2,367	2,367
1-2	46,353	48,720
2-3	3,4285	83,005
3-4	1,3181	96,186
4-5	3,813	99,999
5-6	716	100,715
6-7	145	100,860
7-8	32	100,892
8-9	8	100,900
9-10	2	100,902
Total	100,902	

which resembles the equation for a linear oscillator with a nonlinear damping term. Note that the system becomes unstable when the stiffness term (the coefficient of θ) becomes negative. This occurs when the buoyancy is not sufficient, that is,

$$B < \frac{1}{l_b}\left(mg\frac{L}{2} + MgL\right).$$

Design procedures for such a structure requires strength as well as stability studies. A number of design iterations are required before the structural "sizing" is determined.

13.3.3 Weibull and Gumbel Wave Height Distributions

The National Buoy Data Center (NBDC) run by NOAA collects ocean data such as wind, current, wave, pressure, and temperature in various locations and the records are made public. Suppose we are to design the articulated tower of Section 13.3.2 in one of these locations where the data is available. The first task is to characterize the environment. Using all the information that is collected is inefficient and impractical. Instead, we are interested in choosing a single number that can represent typical and extreme situations such as 10-yr and 50-year storms. A consideration of only random waves requires a determination of the significant wave heights representing 10-year and 50-year storms.

From NBDC data for a buoy outside Monterrey Bay, the number of occurrences of significant wave heights are constructed in Table 13.2. The measurements were taken every hour for about twelve years. We first construct the corresponding Weibull distribution using the method described

13.3. EXAMPLES

in Section 13.1.6. We first guess γ so that a plot of

$$\ln\left(-\ln\left\{1 - F(h)\right\}\right) \quad \text{vs.} \quad \ln(h - \gamma)$$

forms a straight line. Figure 13.20 shows that the pair yields nearly a straight line when $\gamma \simeq 0.84$. The slope and the y-intercept of this line are 1.6 and -0.78, respectively. The Weibull parameters are then $m = 1.6$ and $\beta = 1.6$.

Similarly, we can find the corresponding Gumbel probability density function by plotting pairs of $(h, \ln(-\ln\{F(h)\}))$ to form a line. For the data shown in Table 13.2, the line has a slope of -1.52 and y-intercept of 2.84 so that $\alpha = -1.52$ and $\beta = 1.87$.

Figure 13.21 shows the Weibull probability density and the cumulative distribution (Equation 13.9) in solid lines, the Gumbel probability density and the cumulative distribution in dotted lines (Equation 13.10), and the discrete probability density and the cumulative distribution derived from Table 13.2 in symbols.

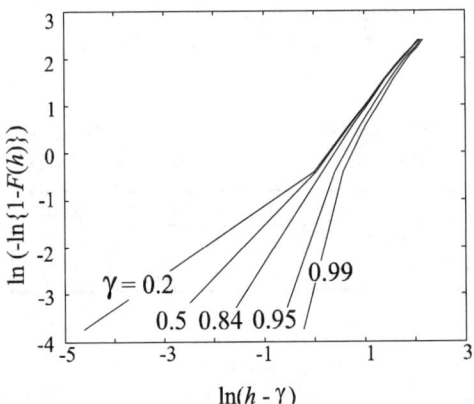

Figure 13.20: Plots of $(\ln(h - \gamma), \ln[-\ln\{1 - F(h)\}])$ for various values of γ.

The next step is to find a significant wave height that can represent an N-year storm, h_N. The probability that there will not be an N-year storm in any given year is $1 - 1/N$, and is equivalent to the probability that the significant wave height will not exceed h_N in the same year. The probability that $h < h_N$ in a single measurement is $F(h_N)$, and the probability that $h < h_N$ in every measurement taken in a year is $F(h_N)^{24 \times 365}$. Then,

$$1 - \frac{1}{N} = F(h_N)^{24 \times 365}.$$

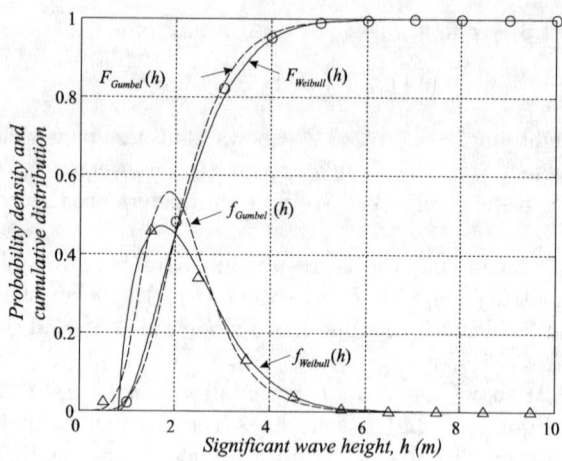

Figure 13.21: Weibull approximations to the probability density and cumulative distribution of significant wave heights measured in the outer Monterey Bay area. The symbols are the values given in Table 13.2.

Table 13.3: Comparison of representative significant wave heights for long term predictions from Gumbel and Weibull distributions

	5-year	10-year	50-year
Weibull	7.84 m	8.15 m	8.79 m
Gumbel	8.83 m	9.33 m	10.4 m

Table 13.3 shows significant wave heights that represent 5-year, 10-year and 50-year storms obtained using the Weibull and Gumbel distributions.

The Gumbel probability distribution results in higher significant wave height estimates. For this particular set of data, the Weibull distribution seems to fit the data better (Figure 13.21), and the Weibull distribution is the most often used distribution in the offshore industry.

13.3.4 Reconstructing Time Series for a Given Significant Wave Height

Previously, we found significant wave heights that could represent 5-year, 10-year, and 25-year storms for a given site. Recall that the significant wave height can entirely characterize the Pierson-Moskowitz spectra. Once the spectral density is determined, a sample time history of the wave profile, $\eta(t)$, can be determined using either Borgman's or Shinozuka's method

13.4. AVAILABLE NUMERICAL CODES

(Section 13.1.4). Here, Shinozuka's method is used to generate the random wave elevations.

First find the random frequencies distributed according to $S_{\eta\eta}^o(\omega)/\sigma_\eta^2$. The Pierson-Moskowitz spectrum in terms of the significant wave height is given by

$$S_{\eta\eta}^o(\omega) = 0.7795\omega^{-5} \exp\left(-\frac{3.118}{H_s^2}\omega^{-4}\right).$$

The variance is given by

$$\sigma_\eta^2 = \int_0^\infty S_{\eta\eta}^o(\omega)\,d\omega = \frac{H_s^2}{16}.$$

The probability density and the cumulative distribution functions are given by

$$f(\omega) = \frac{1}{\sigma_\eta^2} S_{\eta\eta}^o(\omega) = \frac{12.472}{H_s^2}\omega^{-5} \exp\left(-\frac{3.118}{H_s^2}\omega^{-4}\right)$$

$$F(\omega) = 1 - \exp\left(-\frac{3.118}{H_s^2}\omega^{-4}\right).$$

The inverse of the cumulative distribution function is given by

$$F^{-1}(x) = \left(-\frac{H_s^2}{3.118}\ln(1-x)\right)^{-1/4}.$$

The random frequencies distributed according to $f(\omega)$ can be obtained from uniformly distributed random numbers X between 0 and 1. Table 13.4 shows uniform random numbers[39] between 0 and 1 and the random frequencies distributed according to $f(\omega)$. The significant wave height of 7.84 m is used.

One hundred random frequencies are obtained in this way, and the wave elevation is obtained using Equation 13.8. The random phases φ_i are obtained by multiplying uniform random numbers (different from the ones used to generate the random frequencies) by 2π.

Figure 13.22 shows the surface elevation as a function of time, the corresponding wave velocities at the water surface (Section 13.1.7) as functions of time, and the wave velocities at $t = 0$ as functions of the water depth. Note that the wave velocities decay with depth.

13.4 Available Numerical Codes

Many numerical codes are available for modeling the dynamics of slender structures such as risers, tether, umbilicals, and mooring lines. The first

---
[39] The uniform random numbers can be generated by the MATLAB *rand* function.

Table 13.4: Generation of Random Frequencies Distributed According to $f(\omega)$ from Uniform Random Numbers

Uniform Random Numbers, $0 < X < 1$	Random Frequencies ω Distributed According to $f(\omega)$
0.950	$(-19.713 \ln(1 - 0.950))^{-1/4} = 0.360$
0.231	$(-19.713 \ln(1 - 0.231))^{-1/4} = 0.662$
0.606	$(-19.713 \ln(1 - 0.606))^{-1/4} = 0.483$
$\vdots$	$\vdots$

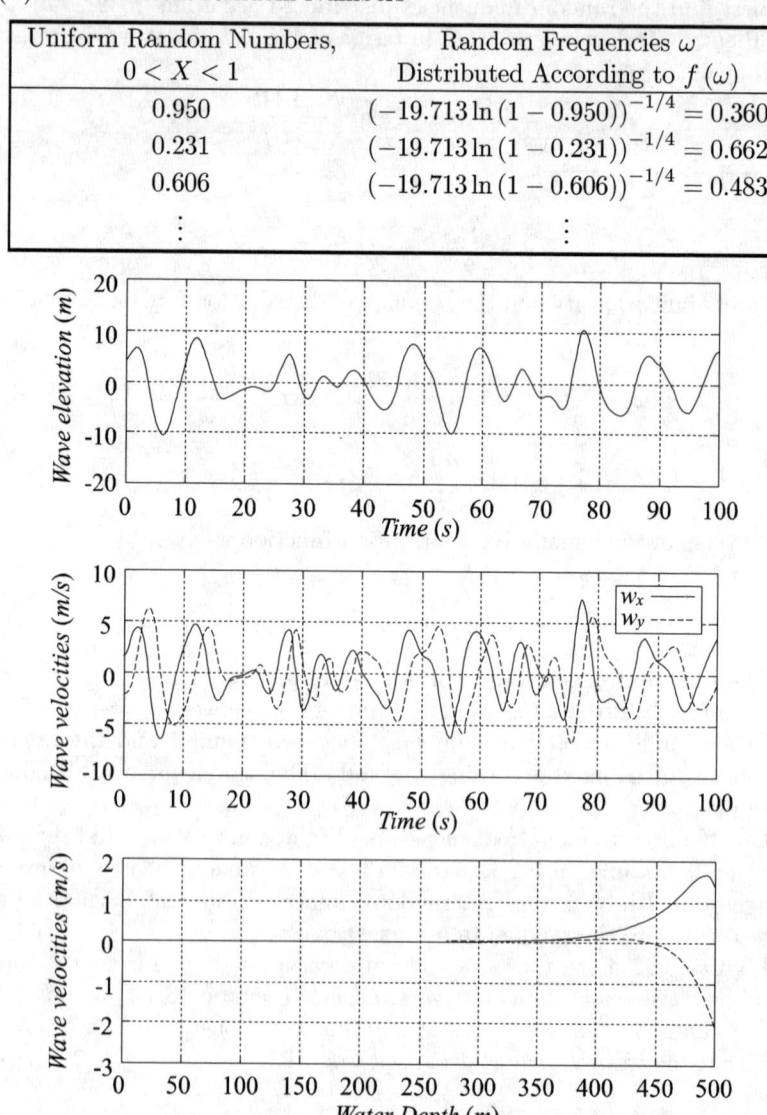

Figure 13.22: Wave elevation and the corresponding wave velocities at the water surface as functions of time, and the wave velocities at $t = 0$ as functions of the water depth.

13.4. AVAILABLE NUMERICAL CODES

example in this section was solved by a numerical code, WHOI Cable, developed at Woods Hole Oceanographic institution. WHOI Cable is a time domain program that can be used for analyzing the dynamics of towed and moored cable systems in both two and three dimensions. It takes into account bending and torsion as well as extension.

Comparative studies investigating flexible risers were carried out by ISSC Committee V7 from computer programs developed by eleven different institutions in the period between 1988 and 1991, and the results were reported by Larsen.[40] More recently, Brown and Mavrakos[41] conducted a comparative study on the dynamic analysis of suspended wire and chain mooring lines and reported results from fifteen different numerical codes. The participants included engineering consultancies, and academic and research institutions involved in marine technology. Some of the time domain programs that were included in the comparative study are MODEX by Chalmers University of Technology, FLEXAN-C by Institute Français du Petrole, DYWFLX95 by MARIN, R.FLEX by MARINTEK, CABLEDYN by National Technical University of Athens, DMOOR by Noble Denton Consultancy Services Ltd, V.ORCAFLEX by Orcina Ltd Consulting Engineers, ANFLEX by Petrobras SA, TDMOOR-DYN by University College London, FLEXRISER by Zentech International. Some of these programs are available to academic institutions and government labs at no cost.

---

[40] "Flexible Riser Analysis – Comparison of Results from Computer Programs," C. Larsen, *Marine Structures*, 5:103-119, 1992.

[41] "Comparative Study on Mooring Line Dynamic Loading," D. Brown, S. Mavrakos, *Marine Structures*, 12:131-151, 1999.

Index

assembly line, 63
autocorrelation function, 198, 232, 262, 265
   derivatives of, 209
autocovariance, 200
autonomous equation, 520
   nonautonomous, 520, 544
axioms of probability, 29, 50

Bayes' theorem, 38
   loaded dice, 40
Bayes, Thomas, 43
beam
   axial vibration, 406
   axially loaded beam on elastic foundation, 444
   random transverse vibration, 436
   transverse vibration on an elastic foundation, 442
   transverse vibration to traveling force, 446, 448
   transverse vibration with axial force, 439
   transversely vibrating beam, 427
beam under static random load, 162
Bessel, Friedrich W., 594
binomial distribution, 90
Borgman's method, 267, 690
   ocean wave elevation simulation, 269
Box-Muller method, 642, 659

Bretschneider spectrum, 684
broad-band random process, 254
Buffon's needle, 631, 639
Buffon, Georges, 633

cable failure, 76, 86
central limit theorem, 71, 261, 665, 667
Chapman-Kolmogorov, 560, 563, 565, 566
characteristic function, 100
Chebyshev inequality, 665, 666
coefficient of variation, 60
colored noise, 258
   response to, 318
composition method, 648
conditional probability, 32
convolution integral, 299, 342
correlation coeff, 110, 111, 204
   reliability, 116
correlation time, 204
covariance, 110
covariance matrix, 501
critical damping, 291
cross-correlation function, 203, 353
cross-spectra, 328
cross-spectral density, 233, 353
crossing rate
   calculation of, 469

De Morgan's rules, 22
De Morgan, Augustus, 24
detection of fracture cracks, 61
Dirac, P.A.M., 219

discrete random variable, 90
Duffing equation, 532
Duffing oscillator
    random forcing, 552, 570
Duhamel integral, 300
Duhamel, Jean-Marie, 300

earthquake forces, 41
earthquake spectrum, 242
    Kanai-Tajimi spectrum, 243
    nonstationary, 593
eigenvalue, 343
    random, 389
    random two-degree-of-freedom model, 393
eigenvector, 343
ensemble averaging, 197
    second-order, 198
envelope function
    probability density function, 484
    Rice's envelope function, 505
equivalent linearization, 521
equivalent nonlinear, 530
ergodicity, 205
error function, 503
event, 15
    basic, 16
    certain, 50
    complementary, 16
    disjoint or mutually exclusive, 18
    impossible, 16, 50
    intersection, 18
    union, 18
    universal, 16
extreme value distribution, 696

failure
    exponential failure law, 462
    failure probability, 69, 70
    fatigue failure, 456
    fatigue life prediction, 3, 493
    first excursion failure, 455, 458
    first passage failure, 455
    life distribution function, 496
failure law
    gamma, 488
    Weibull, 491
first passage failure, 457
first passage, Gaussian, 485
flow-induced vibration of circular cylinder, 523
fluid forces, 5, 700
    on an articulated tower, 715
Fokker-Planck equation, 560, 565, 569
    coupled linear-Duffing oscillator, 576
    Duffing oscillator excited by colored noise, 575
    transition density, 562
forced vibration
    damping, 289, 292
Fourier series, 210
    Fourier coefficients, 211
    representation of a stationary random process, 261
Fourier transform, 213, 304, 306, 341, 354
    convolution theorem, 225
    of time derivative, 224
    truncated Fourier transform, 231
Fourier, Joseph, 213
fracture, 493
free body diagram, 278
free vibration
    damping, 291
    no damping, 288
frequency response function, 304, 342
frequency spectrum, 230

Galerkin, Boris G., 527
gamma function, 507

INDEX

Gauss, Carl F., 79
Gaussian random process, 263
Gaussian table, 73, 75, 87
generating a time history, 689
    Borgman's method, 672
    Shinozuka's method, 673
Gumbel probability density, 719
Gumbel probability distribution, 698, 718

histogram, 27, 51, 63

impulse response function, 297, 342
inclusion principle, 382
inverse vibration, 378
    two-degree-of-freedom, 383
irregularity factor, 504
ISSC spectrum, 684

jet noise spectra, 592
joint density function, 102
joint distribution function, 102, 112
JONSWAP spectrum, 687
jump phenomenon, 543

Khinchine, Aleksandr Y., 249
kinetic energy distribution, 145
Kolmogorov, Andrei N., 582

Leibniz, Gottfried, 137
likelihood function, 38
Lindstedt-Poincaré method, 534
linear ocean wave theory, 698
liquid column damper
    vibration control, 528
liquid sloshing in container, 41
lognormal distribution, 698
long-term ocean wave statistics, 695

magnification factor, 294
marginal density function, 107
marginal probability mass function, 103
Markov process, 560

initial conditions, 567
Markov, Andrei A., 565
mathematical expectation
    continuous variable, 57
    discrete variable, 58
    mean, mode and median, 59
mean and variance of a general function, 168
mean value, 57
mean-square value, 203
memoryless system, 463
Miner's rule, 4, 494, 505, 508
modal analysis, 343
    advantages, 349
    random vibration, mdof, 359
modal coordinates, 348
modal matrix, 345
moment-generating function, 99
Monte Carlo methods
    error estimates, 665
    for a static beam, 181
    generation of normal random numbers, 183
    generation of uniform random numbers, 179
    random numbers, 633
    static problems, 179
Morison equation, 6, 703
    fluid force coefficients, 705
multi-dof vibration, 339
multiple random forces, 356

narrow-band random process, 254, 693
    envelope function, 473
    Rice's envelope function, 474
natural frequency, 279
Newton's second law of motion, 278
Newton, Isaac, 280
nondestructive testing, 61
nonlinear oscillator, 519
    Duffing equation, 520

equivalent nonlinear, 530
  statistical equivalent linearization, 521
nonlinear pendulum, 517, 532
  forced response, 544
nonlinear response of plate, 525
nonlinearities
  examples, 516
nonstationary random process, 598
  Duffing oscillator, 627
  envelope function, 598
  Fourier-Stieltjes integral, 612
  generalized models, 607
  mdof response, 622
  nonlinear oscillator, 625
  nonstationary oscillator, 616
  Priestley's model, 614
  sdof response, 605

ocean currents, 677
ocean wave statistics
  long-term, 695
  short-term, 690
ocean waves, 677
Ochi-Hubble spectrum, 685
offshore structural dynamics, 6, 530
orthogonality of modes, 345, 413

peak distribution
  Gaussian process, 501
  stationary process, 497
periodic structures, 373
  coupling stiffness ratio, 376
  localization, 378
  stiffness imperfection ratio, 378
  weakly-coupled structure, 376
perturbation methods, 532
  generating solution, 533
  stochastic parametric excitation, 549
phase angle, 289
phase diagram, 295

Pierson-Moskowitz spectrum, 238, 239, 269, 531, 683, 721
Planck, Max, 580
Poincaré, Jules H., 537
Poisson distribution, 92, 93
Poisson, Siméon-Denis, 93
probability, 26
probability density function, 52
  derivation of, 230
  exponential, 54, 69
  Gaussian or normal, 71
    time-dependent, 200
  lognormal, 84, 698
  normalization property, 53
  Poisson, 458
  quadratic, 68
  Rayleigh, 87
  standard normal, 72
    table of, 73
  uniform, 66
probability distribution function, 50
probability mass function, 63
projectile motion, 175
proportional damping, 347

quasiharmonic oscillator, 539

random numbers, 633
  for joint probability densities, 659
  linear congruential generator, 635
  pseudo-random, 634
  pseudo-random numbers, 635
random process, 28, 53, 195
  cumulative distribution, 196
  first-order density, 196
  second-order density, 196
  second-order distribution, 196
random variable, 28, 49
random vibration fundamental result, 314

INDEX

Rayleigh, John W. Strutt, 88
realization, 15, 50
rejection-acceptance method, 650
reliability, 456
resonance, 295
response spectrum due to two random loads, 328
Reynolds number, 709
Rice distribution, 504
rocket, 277
rocket ship traffic volume, 77

S-N curve, 494, 495, 509
safety, 16
sample point, 15
sample space, 15
satellite transport, 277
second joint moment, 110
separation of variables, 411
set, 15
set theory, 15
short-term statistics of ocean waves, 690
simply supported beam, 114
single-degree-of-freedom, 278
spectral density, 230, 232, 307
    derivatives of, 259
    for oscillator response, 312
    one-sided, 230, 252
standard deviation, 60
standard normal density, 72
stationarity, 201
    derivatives, 208
    derivatives of a stationary process, 502
    ergodicity, 205
    forcing, 307
    stationary autocorrelation, 203
    stationary density, 202
    strictly stationary, 201
    weakly stationary, 202
statistical independence, 33, 106, 551

statistics, 275
statistics of a tension element, 170
statistics of an area, 167
stochastic parametric excitation, 549
string
    deterministic forced vibration, 425
    DNA, 404
    eigenvalues, eigenvectors, 414
    free vibration, 421
    governing equation, 405
    orthogonality, 413
    random vibration, 432
    wave equation, 405
Strouhal number, 709
Sturm-Liouville eigenvalue problem, 411
superposition principle, 300

table of significant wave heights, 718
table of uniform random numbers, 180
Taylor series, 385
    approximation of expectation, 165
    four random variables with correlation, 172
    mean and variance of a general function, 168
Taylor, Brook, 157
theorem of total probability, 35, 61
transfer function, 304, 320
transforming random variables
    general functions, 149
    harmonic relations, 131
    inverse relations, 128
    lognormal density, 127
    one variable, 125
    parabolic relations, 129
    sums of random variables, 128

two or more variables, 132

uncorrelated forces, 330
up-crossing rate, 466, 692
    Gaussian process, 470

van der Pol equation, 554
    entrainment, 558
    forced response, 558
    isoclines, 556
    limit cycle, 556
    random forcing, 574
van der Pol, Balthazar, 560
variance, 57, 60
    mean square value, 60
Venn diagram, 17
Venn, John, 20
viscous damping factor, 279
von Neumann, John, 653
vortex-induced vibration, 709

Weibull distribution, 506, 640, 697, 718
white noise, 257
    approximation, 319
    Gaussian, 568
    oscillator response to, 314
Wiener, Norbert, 243
Wiener-Khinchine relations, 231, 318, 681
wind forces, 8
wind spectrum, 241
wind speed, 105